S0-ARA-868

OXFORD Reference Collection

The **Oxford Reference Collection** uses sustainable print-on-demand technology to make the acclaimed backlist of the Oxford Reference programme perennially available in hardback format.

Oxford Dictionary of

Scientific Quotations

edited by **W.F. Bynum** and **Roy Porter**

Assistant Editors
Sharon Messenger
Caroline Overy

W.F. Bynum is Professor Emeritus of the history of medicine at the Wellcome Trust Centre for the History of Medicine at University College London. He has edited many books, including (with Roy Porter), the *Companion Encyclopedia of the History of Medicine*, 2 vols. (1993). He is the author of *Science and the Practice of Medicine in the Nineteenth Century* (1994).

Roy Porter was until his retirement Professor of the Social History of Medicine at the Wellcome Trust Centre for the History of Medicine at University College London. His books include *The Greatest Benefit to Mankind: A Medical History of Humanity* (1997), winner of the Los Angeles Times Book Award, *Enlightenment: Britain and the Creation of the Modern World* (2000), winner of a Wolfson Literary Award for History, *Madness: A Brief History* (2002), and *Flesh and the Age of Reason* (2003). He died in March 2002.

Great Clarendon Street, Oxford, OX2 6DP,
United Kingdom

Oxford University Press is a department of the University of Oxford.
It furthers the University's objective of excellence in research, scholarship,
and education by publishing worldwide. Oxford is a registered trade mark of
Oxford University Press in the UK and in certain other countries

First published in paperback 2006
Reissued in Oxford Reference Collection 2016

Published in the United States of America by Oxford University Press
198 Madison Avenue, New York, NY 10016, United States of America

British Library Cataloguing in Publication Data
Data available

Library of Congress Cataloging in Publication Data
Data available

ISBN 978-0-19-880485-7

In memory of

Roy Porter,

1946-2002

Editorial Board

Contents

Introduction

> There is no form of prose more difficult to understand and more tedious to read than the average scientific paper.
>
> **Francis Crick** *The Astonishing Hypothesis: The Scientific Search for the Soul* (1995), xiii

Science and its attendant technologies are essential to modern life. Virtually everything we do has been shaped and facilitated by our scientific world. We participate in it when we switch on the television, read the newspapers, or check out in the supermarket. Universities have faculties of both arts and sciences, often equal in their staffing levels and student intake. Nevertheless, the 'two cultures' phenomenon persists. It was most famously identified a generation back by C. P. Snow, although he was hardly the first to notice the divide between the culture of science and the culture of everyday literacy. Fortunately for us, there is a lot of scientific writing, *pace* Crick's quotation at the head of this page, that is neither average nor tedious.

This volume was conceived almost fifteen years ago, by Judith May (now Lady May), when she was with the Science Department of Oxford University Press. She had noticed that most dictionaries of quotations give short shrift to science, a reflection of the cultural chasm that Snow had lamented. She came to me, to discuss her plan to give science equal voice in a quotations dictionary. Far from merely suggesting appropriate editors, I was intrigued by the project and wanted to do the job myself, assuming that I could persuade my friend and colleague Roy Porter to join me. It was in the heyday of what was sometimes called the Bynum and Porter Industry, at the then Wellcome Institute for the History of Medicine. As was his wont, Roy agreed, and so the project was born.

We agreed a structure with OUP, recruited advisory editors, and began collecting material. Our terms of reference remained pretty fixed throughout the long years of this volume's development. We wanted the *Oxford Dictionary of Scientific Quotations* (its abbreviation ODSQ has a fine Latin flavour) to reflect the richness of the history of science: scientists reflecting on their craft, and reporting key moments of insights they had had on nature and nature's laws. We wanted, too, a generous representation of other voices—poets, novelists, critics, theologians—who had things of substance to say about science and its ambitions. Although we did not exclude humour (there are some funny things here), we did not want a science joke book, or simply a series of anecdotes. There are a few anecdotes here, but this genre has in the meantime been covered very well by the recent volume edited by Walter Gratzer (*Eurekas and Euphorias: the Oxford Book of Scientific Anecdotes*, OUP, 2002). What Judith May had in mind, with which Roy and I entirely concurred, was a volume that looked rather like *The Oxford Dictionary of Quotations* (then in its 3rd edition, published in 1979), in both length and feel. This reflected our common feeling that there really are two broad areas of human knowledge and activity, namely, the arts and the sciences.

Judith's willingness to entrust this enterprise to two historians of science and medicine guaranteed that the final product would have a historical orientation. Modern science is, I hope, adequately covered, but this volume's centre of gravity lies within science's traditions. Aristotle, Kepler, Newton, Faraday, Darwin, and Einstein receive fuller coverage than many of the giants of contemporary science. But Gerald Holton's felicitous invocation of Newton's famous 'standing on the shoulders of giants' quotation can be found here, as well as a couple of reflections on the fact that the majority of scientists who have ever lived are still alive.

One of the pleasures of compiling this book has been the cooperation we have received from people whose quotations we wished to use. The late Stephen Jay Gould wrote a typically charming letter, in response to a request for a couple of references to quotations of his we wished to use. He thanked us for including him but confessed that he could not remember where he had written what we had quoted (we subsequently found references). He suggested that we attribute the quotation to 'that gentleman who must be the brother of the famous Mr. Anon—that is, Mr. No Ref'. The late Francis Crick, Richard Dawkins, and E. O. Wilson, among others, helped us with their entries, and Noam Chomsky quietly supplied us with a missing reference. The bald list of acknowledgements does scant justice to the aid that has been freely and generously offered by so many people.

Most of that help has come within the last few years, when the project's end seemed always in sight, yet ever receding. Roy and I began thinking about what makes a good scientific quotation and collecting of material all those years ago. We made lots of mistakes, among others failing to realise that an unreferenced quotation (as Sharon Messenger will attest) can be worse than no quotation at all, since we were determined to provide authoritative sources whenever possible or sensible.

We began as all scholars do, with the known. In 1990, that was the existing volumes of scientific quotations, as well as the trawl of general volumes, looking for quotations by scientists or about science. When we started, Alan MacKay's two useful volumes, plus Isaac Asimov's larger volume, were available. Asimov's turned out to be a red herring, since it contains many famous and memorable scientific quotations, but they are all unreferenced and, as we discovered, often not reliably transcribed. More recently, we have discovered that the Web is a useful source of scientific quotations, but these are mostly, like those in the Asimov volume, unreferenced and often inaccurate. There are still a few unreferenced quotations in this volume, but in most instances, there is at least a Web verification that the quotation is attributed to the individual in question. Since we began, the series of volumes of thematic scientific quotations edited by Carl C. Gaither and Alma E. Cavazos-Gaither have appeared. Quotations dictionaries in philosophy and the social sciences also exist. We had decided to exclude clinical medicine, since Maurice B. Strauss's *Familiar Medical Quotations* provided a model for medicine. Since then, Edward J. Huth and Jock Murray's *Medicine in Quotations* and Peter McDonald's *Oxford Dictionary of Medical Quotations* (OUP, 2003) have reinforced the medical scene.

Despite these volumes and the Web, we have been determined that ODSQ would be distinctive, and researched largely *de novo*. For this, we looked to our editorial board. During the project's long gestation, death and inactivity have led to changes, but this *apologia pro vita sua* would be inadequate were I not to highlight the loyal contributions of Bill Brock, Ivor Grattan-Guinness, David Oldroyd, and William Outhwaite from the very beginning. Nathan Reingold was energetic before he fell ill. Among later advisors, Hasok Chang, John Henry, Desmond King-Hele, and Robert Sharples deserve special mention.

After four years of initial activity, this volume was put on the back burner, for a variety of reasons. Parkinson's Third Law, quoted here, p. 479, provides a framework, without specifying the causes. It was resurrected in 1999, when Sharon Messenger became the full-time assistant editor. Sharon infused new energy into the project, and many people know of the ODSQ from her charming, literate, but probing communications. Three years ago, Caroline Overy, my long time research assistant, was also commandeered to the project. These two have acquired a certain hallowed status at the British Library, for the range of volumes they require and the speed with which they go through them. The necessity to read a lot of books in the search for the right quotation from the right person has been much aided by the fact that Helen Bynum is an avid reader.

The result, after all these years, is a volume which I believe justifies the vision of Judith May, and the patience of Oxford University Press. Elizabeth Knowles and Susan Ratcliffe have enthusiastically guided us to a conclusion. This is a large book, but one which I believe will set the standard for quotations dictionaries in that other culture. Careful readers will notice recurrent themes in the history of science, but they will also become aware of the richness of science's traditions.

This volume carries Roy Porter's name as one of the co-editors. His shockingly unexpected death delayed further its completion, but he had been active on it to the very end. When he took early retirement from the Wellcome Trust Centre and moved to St. Leonards-on-Sea, he hardly lost contact with the ODSQ, or with London. On his regular commutes up, he would spend time searching for appropriate quotations. One of his final chores was to mine the printed volumes of Nobel Prize speeches, and one of his fellow commuters expressed her puzzlement at the eclecticism of his interests. One week he was reading the speeches in physics, the next in chemistry or medicine or physiology. What, exactly, was he about? Editing this volume has been an eclectic education for all of us involved, and we dedicate it to his memory.

W. F. BYNUM

Wellcome Trust Centre for the History of Medicine
University College London

Acknowledgements

A very large number of people have helped with the production of this volume, and this acknowledgement is bald tribute to the knowledge, skill and generosity of the many individuals listed below. Editing this book has been wonderfully revealing of the best side of academic and scholarly life.

Without input from the editorial board, the volume would have been much the poorer. They are appropriately listed elsewhere, and each will know how much we owe to them. We are particularly grateful that they were willing to pick up the threads after the book had spent several years on the back burner. Professor Abdus Salam died shortly after the book was begun, but he and his colleagues had begun to contribute to it. We hope that they will all find the finished product worth their individual and collective efforts.

In addition to our Board members, many people have helped to supply quotations and references. The following alphabetical list does scant justice to the variety of help we have received, often on several occasions and with much expenditure of time and sharing of expertise. We are grateful to the following:

David Allen, Cristina Alvarez Millán, Hugo Adam Bedau, Janet Browne, Stephen G. Brush, John Carson, Noam Chomsky, Frances Collin, Stefan Collini, Hal Cook, Michael Cooper, Pietro Corsi, the late Francis Crick, M. P. Crosland, Ivan Crozier, Albert Carozzi, Marguerite Carozzi, Peter Day, Richard Dawkins, Maria Pia Donato, Gregg Easterbrook, John Forrester, John Kenneth Galbraith, Peter Gay, Tony Gould, the late Stephen Jay Gould, Mario di Gregorio, Roger Guillieum, Michael Hagner, Lesley Hall, Peter Harman, Rhodri Hayward, John Heilbron, John Henderson, Richard Holmes, Rod Home, Michael Hoskin, Michael Hunter, Karl Hufbauer, Anthony Hyman, Stephen Jacyna, Frank James, Martin Jay, Gwyn Jones, Hywel Jones, Steve Jones, Harmke Kamminga, Martin Kemp, Linda Lear, Joshua Lederberg, Joan Leopold, Vivienne Lo, Eileen Magnello, Kan-wen Ma, Diana Manuel, Joan Mason, Robert May, Ernst Mayr, David Miller, Michael Neve, Marianne Offereins, Cornelius O'Boyle, Pilar Pérez Cañizares, Robert Perrin, Massimo Piattelli-Palmarini, David Picken, Felicity Pors, Karen Reich, Tracey Rihill, Erik Rüdinger, Charles Rowland, Martin Rudwick, Nicolaas Rupke, J. P. Schaer, Sonu Shamdasani, Ruth Sime, David Singmaster, Peter Skelton, Charles Smith, Roger Smith, Philip Sloan, the late William Smeaton, Thomas Söderqvist, Tilli Tansey, Sam Ullman, Mel Usselman, Ezio Vaccari, John Waller, Michael Walton, Andrew Wear, Adam Wilkinson, Lise Wilkinson, Paul Wood, Richard Yeo, John Ziman.

Despite the computer age, this is a book about the printed page. Our offices are located five minutes' walk from the Wellcome Trust Library in one direction and the British Library in the other. Staff in the Rare Books and the Music and Humanities 2 Reading rooms of the British Library have been uniformly helpful. Among those at the Wellcome Trust Library for the History and Understanding of Medicine, we should like to thank Cath Bergin, Ed Bishop, Wendy Fish, Alice Ford-Smith, Sue

Gold, Phoebe Harkins, Mike Jackson, Simon Jones, Douglas Knock, Damian Nicolau, Jette Nielsen, Danny Rees, and Jim Williamson. We have also used the libraries of the University of London (Senate House), University College London, Imperial College, and the University of Oxford (Bodleian). We have also had occasion to use the interlibrary loan service of the Wellcome Trust Library, when even the resources of these great libraries did not throw up what we were looking for.

The transcription of a passage from the printed page to our database is called 'data capture'. Over the years, several individuals have been involved in this process, and we would like to thank Carol Bowen, Sally Bragg, Jacqui Carter, Anna Crozier, Debra Scallan, and Mohsina Somji. When the volume began, computers and computer programs were not what they have more recently become. For guiding us through the morass of databases, backups and the peculiarities of the computer, Michael Clark (Wellcome Trust), Gwyn Griffiths (Wellcome Trust Centre), and Ken Moore (Oxford University Press) deserve special mention. Thanks are also due to Bill Trumble for reading and commenting on the draft text, to Carolyn Garwes for indexing and proofreading, and to Penny Trumble for proofreading. We have also been able to call on the help of John Driver (Wellcome Trust), and Vera Man (Wellcome Trust Centre). The Centre's administrator, Alan Shiel, has invariably been helpful. We are also grateful for the financial support of the Governors of the Wellcome Trust.

The long gestation of this book has already been described, and over the years editors at Oxford University Press have been supportive. Special thanks, then, to Elizabeth Knowles, Susan Ratcliffe, Judith May, Martin Baum, and Lisa Begley.

How to Use the Dictionary

The sequence of entries is by alphabetical order of author. Author names are followed by dates of birth and death (where known) and brief descriptions; where appropriate, cross-references are then given to quotations relating to that author elsewhere in the text (*See* ***Hobbes, Thomas*** 464:9). Within each author entry, quotations are in chronological order, and in general spelling and capitalization follow the original. Quotations by others about an author appear at the end of the entry for that author, preceded by a note: ON **NEWTON.**

Contextual information regarded as essential to a full appreciation of the quotation precedes the text in an italicized note; information seen as providing useful amplification follows in an italicized note. Each quotation is accompanied by a full bibliographical note of the source from which it is taken. Cross-references to specific quotations are used to direct the reader to another related item. In each case a reference is given to the author's name, followed by the page number and then the unique quotation number on that page ('See **Pope, Alexander** 473:17').

Index

The most significant words from each quotation appear in the keyword index, allowing individual quotations to be traced. Both the keywords and the context lines following each keyword, including those in foreign languages, are in strict alphabetical order. Singular and plural nouns (with their possessive forms) are grouped separately. In the index, spelling of keywords has been standardized.

References are to the author's name (usually in abbreviated form, as EINS for Albert Einstein) followed by the page number and the number of the unique quotation on the page. Thus EINS 39:18 means quotation number 18 on page 39, in the entry for Albert Einstein.

Peter Abelard 1079–1142

French theologian and philosopher

1 *Dubitando quippe ad inquisitionem venimus; inquirendo veritatem percipimus.*
By doubting we come to enquiry, and through enquiry we perceive truth.

Sic et Non [*c.*1120], Preface, in *Sic et Non: A Critical Edition*, B. Boyer and R. McKeon (eds.) (1970), 103, lines 338–339, quoted in M. T. Clanchy, *Abelard, A Medieval Life* (1997), 107.

Franz Karl Achard 1753–1821

German chemist and experimental physicist

2 Everyone now agrees that a physics lacking all connection with mathematics . . . would only be an historical amusement, fitter for entertaining the idle than for occupying the mind of a philosopher.

Quoted in J. L. Heilbron, *Electricity in the 17th and 18th centuries: A Study of Early Modern Physics* (1979), 74.

Russell Lincoln Ackoff 1919–

American systems sciences educator

3 Common sense . . . has the very curious property of being more correct retrospectively than prospectively. It seems to me that one of the principal criteria to be applied to successful science is that its results are almost always obvious retrospectively; unfortunately, they seldom are prospectively. Common sense provides a kind of ultimate validation *after* science has completed its work; it seldom anticipates what science is going to discover.

Quoted in A. De Reuck, M. Goldsmith and J. Knight (eds.), *Decision Making in National Science Policy* (1968), 96.

John Emerich Edward Dalberg Acton (Lord Acton) 1834–1902

British historian and moralist

4 It is they [men of science] who hold the secret of the mysterious property of the mind by which error ministers to truth, and truth slowly but irrevocably prevails. Theirs is the logic of discovery, the demonstration of the advance of knowledge and the development of ideas, which as the earthly wants and passions of men remain almost unchanged, are the charter of progress, and the vital spark in history.

Inaugural Lecture on the Study of History, Cambridge, 11 June 1895. *Lectures on Modern History* (1906), 21.

Henry Brooks Adams 1838–1918

American historian and man of letters

5 Man has mounted science, and is now run away with. I firmly believe that before many centuries more, science will be the master of men. The engines he will have invented will be beyond his strength to control. Someday science may have the existence of mankind in its power, and the human race commit suicide, by blowing up the world. Not only shall we be able to cruise in space, but I'll be hanged if I see any reason why some future generation shouldn't walk off like a beetle with the world on its back, or give it another rotary motion so that every zone should receive in turn its due portion of heat and light.

Letter to Charles Francis Adams Jr., London, 11 April 1862. In J. C. Levenson, E. Samuels, C. Vandersee and V. Hopkins Winner (eds.), *The Letters of Henry Adams: 1858–1868* (1982), Vol. 1, 290.

6 [Adams] supposed that, except musicians, every one thought Beethoven a bore, as every one except mathematicians thought mathematics a bore.

The Education of Henry Brooks Adams: An Autobiography (1919), 80.

7 *After viewing the Palace of Electricity at the 1900 Trocadero Exposition in Paris:*
All the steam in the world could not, like the Virgin, build Chartres.

The Education of Henry Brooks Adams: An Autobiography (1919), 388.

Roger Adams 1889–1971

American organic chemist

8 Many thanks for the sending me the book *Biology of the Striped Skunk* . . .

Frankly, I doubt whether I shall read it or not, unless I happen to have some intimate contact with a skunk which may induce me to learn more about him.

Undated letter to a member of the Natural History Survey. In D. S. Tarbell and A. Tarbell, *Roger Adams, Scientist and Statesman* (1981), 192.

Joseph Addison 1672–1719

British essayist and poet

1 The Spacious Firmament on high,
With all the blue Etherial Sky,
And spangled Heav'ns, a Shining Frame,
Their great Original proclaim:
Th'unwearied Sun, from day to day,
Does his Creator's Pow'r display,
And publishes to every Land
The Work of an Almighty Hand.
Soon as the Evening Shades prevail,
The Moon takes up the wondrous Tale,
And nightly to the listning Earth
Repeats the Story of her Birth:
Whilst all the Stars that round her burn,
And all the Planets, in their turn,
Confirm the Tidings as they rowl,
And spread the Truth from Pole to Pole.
What though, in solemn Silence, all
Move round the dark terrestrial Ball?
What tho' nor real Voice nor Sound
Amid their radiant Orbs be found?
In Reason's Ear they all rejoice,
And utter forth a glorious Voice,
For ever singing, as they shine,
'The Hand that made us is Divine'.

The Spectator, no. 465, Saturday 23 August 1712. In D. F. Bond (ed.) *The Spectator* (1965), Vol. 4, 144–5.

Homer Burton Adkins 1892–1949

American organic chemist

2 Basic research is like shooting an arrow into the air and, where it lands, painting a target.

Quoted in A. Mackay (ed.), *A Dictionary of Scientific Quotations* (1991), 2.

Theodor Wiesengrund Adorno 1903–69

German philosopher

3 In psycho-analysis nothing is true except the exaggerations.

Minima Moralia: Reflections from Damaged Life (1974), 28.

Edgar Douglas Adrian (Lord Adrian) 1889–1977

British physiologist

4 We come back then to our records of nervous messages with a reasonable assurance that they do tell us what the message is like. It is a succession of brief waves of surface breakdown, each allowing a momentary leakage of ions from the nerve fibre. The waves can be set up so that they follow one another in rapid or in slow succession, and this is the only form of gradation of which the message is capable. Essentially the same kind of activity is found in all sorts of nerve fibres from all sorts of animals and there is no evidence to suggest that any other kind of nervous transmission is possible. In fact we may conclude that the electrical method can tell us how the nerve fibre carries out its function as the conducting unit of the nervous system, and that it does so by reactions of a fairly simple type.

The Mechanism of Nervous Action (1932), 21.

5 Unless social sciences can be as creative as natural science, our new tools are not likely to be of much use to us.

Proceedings of the 3rd Congress of Psychiatry, Montreal, 1961, 42.

Franz Ulrich Theodosius Aepinus 1724–1802

German mathematician and physicist

6 I think that considerable progress can be made in the analysis of the operations of nature by the scholar who reduces rather complicated phenomena to their proximate causes and primitive forces, even though the causes of those causes have not yet been detected.

R.W. Home (ed.), *Aepinus's Essay on the Theory of Electricity and Magnetism* (1979), 240.

1 The belief that all things are created solely for the utility of man has stained with many errors that most noble part of physics which deals with the ends of things.

R.W. Home (ed.), *Aepinus's Essay on the Theory of Electricity and Magnetism* (1979), 399.

Jean Louis Rodolphe Agassiz

1807–73

Swiss-born American geologist and natural historian

2 The epoch of intense cold which preceded the present creation has been only a temporary oscillation of the earth's temperature, more important than the century-long phases of cooling undergone by the Alpine valleys. It was associated with the disappearance of the animals of the diluvial epoch of the geologists, as still demonstrated by the Siberian mammoths; it preceded the uplifting of the Alps and the appearance of the present-day living organisms, as demonstrated by the moraines and the existence of fishes in our lakes. Consequently, there is complete separation between the present creation and the preceding ones, and if living species are sometimes almost identical to those buried inside the earth, we nevertheless cannot assume that the former are direct descendants of the latter or, in other words, that they represent identical species.

Discours de Neuchâtel (1837), trans. A.V. Carozzi, *Studies on Glaciers Preceded by the Discourse of Neuchâtel* (1967), lviii.

3 The earth was covered by a huge ice sheet which buried the Siberian mammoths, and reached just as far south as did the phenomenon of erratic boulders. This ice sheet filled all the irregularities of the surface of Europe before the uplift of the Alps, the Baltic Sea, all the lakes of Northern Germany and Switzerland. It extended beyond the shorelines of the Mediterranean and of the Atlantic Ocean, and even covered completely North America and Asiatic Russia. When the Alps were uplifted, the ice sheet was pushed upwards like the other rocks, and the debris, broken loose from all the cracks generated by the uplift, fell over its surface and, without becoming rounded (since they underwent no friction), moved down the slope of the ice sheet.

Études sur Les Glaciers (1840), trans. A. V. Carozzi, *Studies on Glaciers Preceded by the Discourse of Neuchâtel* (1967), 166.

4 The surface of the earth is not simply a stage on which the thousands of present and past inhabitants played their parts in turn. There are much more intimate relations between the earth and the living organisms which populated it, and it may even be demonstrated that the earth was developed because of them.

Études sur Les Glaciers (1840), trans. A. V. Carozzi, *Studies on Glaciers Preceded by the Discourse of Neuchâtel* (1967), 175.

5 While a glacier is moving, it rubs and wears down the bottom on which it moves, scrapes its surface (now smooth), triturates the broken-off material that is found between the ice and the rock, pulverizes or reduces it to a clayey paste, rounds angular blocks that resist its pressure, and polishes those having a larger surface. At the surface of the glacier, other processes occur. Fragments of rocks that are broken-off from the neighbouring walls and fall on the ice, remain there or can be transported to the sides; they advance in this way on the top of the glacier, without moving or rubbing against each other . . . and arrive at the extremity of the glacier with their angles, sharp edges, and their uneven surfaces intact.

La théorie des glaciers et ses progrès les plus récents. Bibl. universelle de Genève, (3), Vol. 41, p. 127. Trans. Karin Verrecchia.

1 In-depth studies have an influence on general ideas, whereas theories, in turn, in order to maintain themselves, push their spectators to search for new evidence. The mind's activity that is maintained by the debates about these works, is probably the source of the greatest joys given to man to experience on Earth.

La théorie des glaciers et ses progrès les plus récents. Bibl. universelle de Genève, (3), Vol. 41, p. 139. Trans. Karin Verrecchia.

2 Embryology furnishes, also, the best measure of true affinities existing between animals.

Essay on Classification (1857). *Contributions to the Natural History of the United States of America* (1857), Vol. 1, 85.

3 *Branches* or *types* are characterized by the plan of their structure,

Classes, by the manner in which that plan is executed, as far as ways and means are concerned,

Orders, by the degrees of complication of that structure,

Families, by their form, as far as determined by structure,

Genera, by the details of the execution in special parts, and

Species, by the relations of individuals to one another and to the world in which they live, as well as by the proportions of their parts, their ornamentation, etc.

Essay on Classification (1857). *Contributions to the Natural History of the United States of America* (1857), Vol. 1, 170.

4 Philosophers and theologians have yet to learn that a physical fact is as sacred as a moral principle. Our own nature demands from us this double allegiance.

Contributions to the Natural History of the United States of America (1857).

5 The resources of the Deity cannot be so meagre, that, in order to create a human being endowed with reason, he must change a monkey into a man.

Methods of Study in Natural History (1863), Preface, iv.

6 I have devoted my whole life to the study of Nature, and yet a single sentence may express all that I have done. I have shown that there is a correspondence between the succession of Fishes in geological times and the different stages of their growth in the egg,—this is all. It chanced to be a result that was found to apply to other groups and has led to other conclusions of a like nature.

Methods of Study in Natural History (1863), 23.

7 It must be for truth's sake, and not for the sake of its usefulness to humanity, that the scientific man studies Nature. The application of science to the useful arts requires other abilities, other qualities, other tools than his; and therefore I say that the man of science who follows his studies into their practical application is false to his calling. The practical man stands ever ready to take up the work where the scientific man leaves it, and adapt it to the material wants and uses of daily life.

Methods of Study in Natural History (1863), 24.

8 America, so far as her physical history is concerned, has been falsely denominated the *New World*. Hers was the first dry land lifted out of the waters, hers the first shore washed by the ocean that enveloped all the earth beside; and while Europe was represented only by islands rising here and there above the sea, America already stretched an unbroken line of land from Nova Scotia to the Far West.

Geological Sketches (1866), 1.

9 The world is the geologist's great puzzle-box; he stands before it like the child to whom the separate pieces of his puzzle remain a mystery till he detects their relation and sees where they fit, and then his fragments grow at once into a connected picture beneath his hand.

Geological Sketches (1866), 11.

10 When chemists have brought their knowledge out of their special laboratories into the laboratory of the world, where chemical combinations are and have been through all time going on in such vast proportions,—when physicists study the laws of moisture, of clouds and storms, in past periods as well as in the present,—when, in short, geologists and zoologists are chemists and physicists, and *vice versa*,—then we shall learn more of the changes the world has undergone than is possible now that they are separately studied.

Geological Sketches (1866), 73.

1 The long summer was over. For ages a tropical climate had prevailed over a great part of the earth, and animals whose home is now beneath the Equator roamed over the world from the far South to the very borders of the Arctics . . . But their reign was over. A sudden intense winter, that was also to last for ages, fell upon our globe.

Geological Sketches (1866), 208.

2 One naturally asks, what was the use of this great engine set at work ages ago to grind, furrow, and knead over, as it were, the surface of the earth? We have our answer in the fertile soil which spreads over the temperate regions of the globe. The glacier was God's great plough.

Geological Sketches (1875), 99.

3 The office of science is not to record possibilities; but to ascertain what nature does . . . As far as Darwinism deals with mere arguments of possibilities or even probabilities, without a basis of fact, it departs from the true scientific method and injures science, as most of the devotees of the new ism have already done.

'Professor Agassiz on the Darwinian Theory . . . Interesting Facsimile Letter from the Great Naturalist', *Scientific American*, 1874, 30, 85.

4 Absorbed in the special investigation, I paid no heed to the edifice which was meanwhile unconsciously building itself up. Having however completed the comparison of the fossil species in Paris, I wanted, for the sake of an easy revision of the same, to make a list according to their succession in geological formations, with a view of determining the characteristics more exactly and bringing them by their enumeration into bolder relief. What was my joy and surprise to find that the simplest enumeration of the fossil fishes according to their geological succession was also a complete statement of the natural relations of the families among themselves; that one might therefore read the genetic development of the whole class in the history of creation, the representation of the genera and species in the several families being therein determined; in one word, that the genetic succession of the fishes corresponds perfectly with their zoological classification, and with just that classification proposed by me.

Quoted in Elizabeth Cary Agassiz (ed.), *Louis Agassiz: His Life and Correspondence* (1885), Vol. 1, 203–4.

5 I cannot afford to waste my time making money.

A reply to an offer of a lecture tour.
Attributed.

6 Every great scientific truth goes through three states: first, people say it conflicts with the Bible; next, they say it has been discovered before; lastly, they say they always believed it.

Attributed.

7 ON AGASSIZ The beauty of his better self lives on
In minds he touched with fire, in many an eye
He trained to Truth's exact severity;
He was a teacher: why be grieved for him
Whose living word still stimulates the air?

James Russell Lowell, 'Ode on the Death of Agassiz' (1888). In *The Poetical Works of James Russell Lowell* (1978), 381.

Derek Victor Ager 1923–92

British geologist

8 *Palaeontologists cannot live by uniformitarianism alone.* This may be termed the 'Phenomenon of the Fallibility of the Fossil Record'.

The Nature of the Stratigraphical Record, 3rd edn. (1993), 40.

1 *Sedimentation in the past has often been very rapid indeed and very spasmodic.* This may be called the 'Phenomenon of the Catastrophic Nature of the Stratigraphical Record'.

The Nature of the Stratigraphical Record, 3rd edn. (1993), 70.

2 *Let us make an arbitrary decision* (by a show of hands if necessary) *to define the base of every stratigraphical unit in a selected section.* This may be called the 'Principle of the Golden Spike'. Then stratigraphical nomenclature can be forgotten and we can get on with the real work of stratigraphy, which is correlation and interpretation.

The Nature of the Stratigraphical Record, 3rd edn. (1993), 110.

3 Though the theories of plate tectonics now provide us with a *modus operandi*, they still seem to me to be a periodic phenomenon. Nothing is world-wide, but everything is episodic. In other words, the history of any one part of the earth, like the life of a soldier, consists of long periods of boredom and short periods of terror.

The Nature of the Stratigraphical Record, 3rd edn. (1993), 141.

Georgius Agricola see Bauer, Georg

Heinrich Cornelius Agrippa

(Agrippa von Nettesheim) 1486–1535

German philosopher and alchemist

4 Every Alchymist is a Physician or a Sope-boyler.

The Vanity of the Arts and Sciences (1530), in 1676 edn, 313.

Mark Akenside 1721–70

British poet and physician

5 SCIENCE! thou fair effusive ray
From the great source of mental Day,
Free, generous, and refin'd!
Descend with all thy treasures fraught,
Illumine each bewilder'd thought,
And bless my labour'g mind.

'Hymn to Science' (1739). In Robin Dix (ed.), *The Poetical Works of Mark Akenside* (1996), 406.

Saint Albertus Magnus

C.1200–1280

German theologian and philosopher

6 Now it must be asked if we can comprehend why comets signify the death of magnates and coming wars, for writers of philosophy say so. The reason is not apparent, since vapor no more rises in a land where a pauper lives than where a rich man resides, whether he be king or someone else. Furthermore, it is evident that a comet has a natural cause not dependent on anything else; so it seems that it has no relation to someone's death or to war. For if it be said that it does relate to war or someone's death, either it does so as a cause or effect or sign.

De Cometis (*On Comets*) [before 1280], trans. Lynn Thorndike, from ed. Borgnet, IV, 499–508, quoted in Lynn Thorndike (ed.), *Latin Treatises on Comets between 1238 and 1368 A.D.* (1950), 75.

7 To say that there is a soul in stones simply in order to account for their production is unsatisfactory: for their production is not like the reproduction of living plants, and of animals which have senses. For all these we see reproducing their own species from their own seeds; and a stone does not do this at all. We never see stones reproduced from stones; . . . because a stone seems to have no reproductive power at all.

De Mineralibus (*On Minerals*) [c.1261/63], Book I, tract I, chapter 4, trans. D. Wyckoff (1967), 20.

8 It seems wonderful to everyone that sometimes stones are found that have figures of animals inside and outside. For outside they have an outline, and when they are broken open, the shapes of the internal organs are found inside.

And Avicenna says that the cause of this is that animals, just as they are, are sometimes changed into stones, and especially [salty] stones. For he says that just as the Earth and Water are material for stones, so animals, too, are material for stones. And in places where a petrifying force is exhaling, they change into their elements and are attacked by the properties of the qualities [hot, cold, moist, dry] which are present in those places, and in the elements in the bodies of such animals are changed into the dominant element, namely Earth mixed with Water; and then the mineralizing power converts [the mixture] into stone, and the parts of the body retain their shape, inside and outside, just as they were before. There are also stones of this sort that are [salty] and frequently not hard; for it must be a strong power which thus transmutes the bodies of animals, and it slightly burns the Earth in the moisture, so it produces a taste of salt.

De Mineralibus (*On Minerals*) [c.1261/63], Book I, tract 2, chapter 8, trans. D. Wyckoff (1967), 52–53.

1 The beaver is an animal which has feet like those of a goose for swimming and front teeth like a dog, since it frequently walks on land. It is called the *castor* from 'castration,' but not because it castrates itself as Isodore says, but because it is especially sought for castration purposes. As has been ascertained frequently in our regions, it is false that when it is bothered by a hunter, it castrates itself with its teeth and hurls its musk [*castoreum*] away and that if one has been castrated on another occasion by a hunter, it raises itself up and shows that it lacks its musk.

De Animalibus (*On Animals*) [1258/62], Book XXII, tract 2, chapter 1 (22), trans. K. F. Kitchell Jr. and I. M. Resnick (1999), Vol. 2, 1467.

2 Evidence of this [transformation of animals into fossils] is that parts of aquatic animals and perhaps of naval gear are found in rock in hollows on mountains, which water no doubt deposited there enveloped in sticky mud, and which were prevented by coldness and dryness of the stone from petrifying completely. Very striking evidence of this kind is found in the stones of Paris, in which one very often meets round shells the shape of the moon.

De Causis Proprietatum Elementorum (*On the Causes of the Properties of the Elements*) [before 1280], Book II, tract 3, chapter 5, quoted in A. C. Crombie, *Augustine to Galileo* (1959), Vol. 1, 126.

Al-Biruni 973–c.1050

Arabic natural philosopher

3 But if you have seen the soil of India with your own eyes and meditate on its nature—if you consider the rounded stones found in the earth however deeply you dig, stones that are huge near the mountains and where the rivers have a violent current; stones that are of smaller size at greater distance from the mountains, and where the streams flow more slowly; stones that appear pulverised in the shape of sand where the streams begin to stagnate near their mouths and near the sea—if you consider all this, you could scarcely help thinking that India has once been a sea which by degrees has been filled up by the alluvium of the streams.

Alberuni's India, trans. E. C. Sachau (1888), Vol. 1, 198.

4 For it is the same whether you take it that the Earth is in motion or the Sky. For, in both the cases, it does not affect the Astronomical Science. It is just for the Physicist to see if it is possible to refute it.

In Syed Hasan Barani, 'Al-Biruni's Scientific Achievements', *Indo-Iranica*, 1952, **5** (4), 47.

Alcmaeon of Crotonia c.535 BC

Greek natural philosopher

5 ON ALCMAEON Alcmaeon was the first to define the difference between man and animals, saying that man differs from the latter in the fact that he alone has the power of understanding.

Theophrastus, *On Sense Perceptions*, section 25. In Edwin Clarke and C. D. O'Malley, *The Human Brain and Spinal Cord* (1968), 3.

Paul Alderson 1926–

1 I have yet to see any problem, however complicated, which, when you looked at it in the right way, did not become still more complicated.

New Scientist, 25 September 1969, 638.

Alfonso X 1221–84

Spanish King and astronomer

2 If the Lord Almighty had consulted me before embarking upon his creation, I should have recommended something simpler.

A response attributed to Alfonso X, King of Leon and Castile, on having the Ptolemaic system of astronomy explained to him.

In John Esten Keller, *Alfonso X, El Sabio* (1967), Preface.

Al Iraqi c.1211–1289

Iranian poet

3 This prime matter which is proper for the form of the Elixir is taken from a single tree which grows in the lands of the West. It has two branches, which are too high for whoso seeks to eat the fruit thereof to reach them without labour and trouble; and two other branches, but the fruit of these is drier and more tanned than that of the two preceding. The blossom of one of the two is red [corresponding to gold], and the blossom of the second is between white and black [corresponding to silver]. Then there are two other branches weaker and softer than the four preceding, and the blossom of one of them is black [referring to iron] and the other between white and yellow [probably tin]. And this tree grows on the surface of the ocean [the *material prima* from which all metals are formed] as plants grow on the surface of the earth. This is the tree of which whosoever eats, man and jinn obey him; it is also the tree of which Adam (peace be upon him!) was forbidden to eat, and when he ate thereof he was transformed from his angelic form to human form. And this tree may be changed into every animal shape.

'Cultivation of Gold', trans. E. J. Holmyard (1923), 23. Quoted and annotated in Seyyed Hossein Nasr, *Science and Civilization in Islam* (1968), 279.

Ismail ibn al-Razzaz al-Jazari fl. c.1206

Arabic mechanical engineer

4 The book of knowledge of ingenious mechanical devices.

The Book of Knowledge of Ingenious Mechanical Devices (c.1204 or 1206).

Al-Jili 1365–c.1424

Arab mystic

5 The world is comparable to ice, and the Truth to water, the origin of this ice. The name 'ice' is only lent to this coagulation; it is the name of water which is restored to it, according to its essential reality.

Universal Man. In Seyyed Hossein Nasr, *Science and Civilisation in Islam* (1968), 341.

Al Khayyami c.1048–1131

Iranian natural philosopher and poet

6 I was unable to devote myself to the learning of this *al-jabr* [algebra] and the continued concentration upon it, because of obstacles in the vagaries of Time which hindered me; for we have been deprived of all the people of knowledge save for a group, small in number, with many troubles, whose concern in life is to snatch the opportunity, when Time is asleep, to devote themselves meanwhile to the investigation and perfection of a science; for the majority of people who imitate philosophers confuse the true with the false, and they do nothing but deceive and pretend knowledge, and they do not use what they know of the sciences except for base and material purposes; and if they see a certain person seeking for the right and preferring the truth, doing his best to refute the false and untrue and leaving aside hypocrisy and deceit, they make a fool of him and mock him.

A. P. Youschkevitch and B. A. Rosenfeld, 'Al-Khayyami', in C. C. Gillispie (ed.), *Dictionary of Scientific Biography* (1973), Vol. 7, 324.

Claude J. Allègre 1937–

French geologist

1 The formation of planets is like a gigantic snowball fight. The balls bounce off, break apart, or stick together, but in the end they are rolled up into one enormous ball, a planet-ball that has gathered up all the snowflakes in the surrounding area.

From Stone to Star: A View of Modern Geology, trans. Deborah Kurmes van Dam (1992), 110.

David Elliston Allen 1932–

British botanist and historian of science

2 Selborne is the secret, private parish inside each one of us.

The Naturalist in Britain: A Social History (1976), 51.

Gordon Allen see Dobzhansky, Theodosius and Allen, Gordon

Gordon Willard Allport 1897–1967

American psychologist

3 The dog [in Pavlov's experiments] does not continue to salivate whenever it hears a bell unless sometimes at least an edible offering accompanies the bell. But there are innumerable instances in human life where a single association, *never* reinforced, results in the establishment of a life-long dynamic system. An experience associated only once with a bereavement, an accident, or a battle, may become the center of a permanent phobia or complex, not in the least dependent on a recurrence of the original shock.

Personality: A Psychological Interpretation (1938), 199.

4 Each person is an idiom unto himself, an apparent violation of the syntax of the species.

Becoming: Basic Considerations for a Psychology of Personality (1955), 19.

Luis W. Alvarez 1911–88

American physicist
See ***Warden, Ian*** *607:3*

5 The bomb took forty-five seconds to drop thirty thousand feet to its detonation point, our three parachute gauges drifting down above. For half that time we were diving away in a two-g turn. Before we leveled off and flew directly away, we saw the calibration pulses that indicated our equipment was working well. Suddenly a bright flash lit the compartment, the light from the explosion reflecting off the clouds in front of us and back through the tunnel. The pressure pulse registered its N-shaped wave on our screen, and then a second wave recorded the reflection of the pulse from the ground. A few moments later two sharp shocks slammed the plane.

Alvarez: Adventures of a Physicist (1987), 7.

6 The world of mathematics and theoretical physics is hierarchical. That was my first exposure to it. There's a limit beyond which one cannot progress. The differences between the limiting abilities of those on successively higher steps of the pyramid are enormous. I have not seen described anywhere the shock a talented man experiences when he finds, late in his academic life, that there are others enormously more talented than he. I have personally seen more tears shed by grown men and women over this discovery than I would have believed possible. Most of those men and women shift to fields where they can compete on more equal terms. The few who choose not to face reality have a difficult time.

Alvarez: Adventures of a Physicist (1987), 20.

1 Dirac politely refused Robert's [Robert Oppenheimer] two proffered books: reading books, the Cambridge theoretician announced gravely, 'interfered with thought'.

Alvarez: Adventures of a Physicist (1987), 87.

2 The last few centuries have seen the world freed from several scourges—slavery, for example; death by torture for heretics; and, most recently, smallpox. I am optimistic enough to believe that the next scourge to disappear will be large-scale warfare—killed by the existence and nonuse of nuclear weapons.

Alvarez: Adventures of a Physicist (1987), 152.

3 My observations of the young physicists who seem to be most like me and the friends I describe in this book tell me that they feel as we would if we had been chained to those same oars. Our young counterparts aren't going into nuclear or particle physics (they tell me it's too unattractive); they are going into condensed-matter physics, low-temperature physics, or astrophysics, where important work can still be done in teams smaller than ten and where everyone can feel that he has made an important contribution to the success of the experiment that every other member of the collaboration is aware of. Most of us do physics because it's fun and because we gain a certain respect in the eyes of those who know what we've done. Both of those rewards seem to me to be missing in the huge collaborations that now infest the world of particle physics.

Alvarez: Adventures of a Physicist (1987), 198.

4 In my considered opinion the peer review system, in which proposals rather than proposers are reviewed, is the greatest disaster visited upon the scientific community in this century. No group of peers would have approved my building the 72-inch bubble chamber. Even Ernest Lawrence told me he thought I was making a big mistake. He supported me because he knew my track record was good. I believe that U.S. science could recover from the stultifying effects of decades of misguided peer reviewing if we returned to the tried-and-true method of evaluating experimenters rather than experimental proposals. Many people will say that my ideas are elitist, and I certainly agree. The alternative is the egalitarianism that we now practice and I've seen nearly kill basic science in the USSR and in the People's Republic of China.

Alvarez: Adventures of a Physicist (1987), 200–1.

5 Most of us who become experimental physicists do so for two reasons; we love the tools of physics because to us they have intrinsic beauty, and we dream of finding new secrets of nature as important and as exciting as those uncovered by our scientific heroes. But we walk a narrow path with pitfalls on either side. If we spend all our time developing equipment, we risk the appellation of 'plumber', and if we merely use the tools developed by others, we risk the censure of our peers for being parasitic.

'Recent Developments in Particle Physics', Nobel Lecture, December 11th, 1968. In *Nobel Lectures: Physics 1963–1970* (1972), 241.

6 There is no democracy in physics. We can't say that some second-rate guy has as much right to opinion as Fermi.

Quoted in Daniel S. Greenberg, *The Politics of American Science* (1969), 72.

André-Marie Ampère 1775–1836

French mathematician, chemist and physicist

7 ON AMPÈRE Ampere was a mathematician of various resources & I think might rather be called excentric [sic] than original. He was as it were always mounted upon a hobby horse of a monstrous character pushing the most remote & distant analogies. This hobby horse was sometimes like that of a child ['s] made of heavy wood, at other times it resembled those [?] shapes [?] used in the theatre [?] & at other times it was like a hypogrif in a pantomime de imagie. He had a sort of faith in animal magnetism & has published some refined & ingenious memoirs to prove the identity of electricity & magnetism but even in these views he is rather as I said before excentric than original. He has always appeared to me to possess a very discursive imagination & but little accuracy of observation or acuteness of research.

J. Z. Fullmer, 'Davy's Sketches of his Contemporaries', *Chymia*, 1967, 12, 135–6.

Anaxagoras c.500–428 BC

Greek natural philosopher

1 The Greeks are wrong to recognize coming into being and perishing; for nothing comes into being nor perishes, but is rather compounded or dissolved from things that are. So they would be right to call coming into being composition and perishing dissolution.

Simplicius, *Commentary on Aristotle's Physics*, 163, 20–4. In G. S. Kirk, J. E. Raven and M. Schofield (eds.), *The Presocratic Philosophers: A Critical History with a Selection of Texts* (1983), p. 358.

2 Neither is there a smallest part of what is small, but there is always a smaller (for it is impossible that what is should cease to be). Likewise there is always something larger than what is large.

Simplicius, *Commentary on Aristotle's Physics*, 164, 17–9. In G. S. Kirk, J. E. Raven and M. Schofield (eds.), *The Presocratic Philosophers: A Critical History with a Selection of Texts* (1983), p. 360.

3 All other things have a portion of everything, but Mind is infinite and self-ruled, and is mixed with nothing but is all alone by itself.

Simplicius, *Commentary on Aristotle's Physics*, 164, 24–5. In G. S. Kirk, J. E. Raven and M. Schofield (eds.), *The Presocratic Philosophers: A Critical History with a Selection of Texts* (1983), p. 363.

4 And since the portions of the great and the small are equal in number, so too all things would be in everything. Nor is it possible that they should exist apart, but all things have a portion of everything.

Simplicius, *Commentary on Aristotle's Physics*, 164, 26–8. In G. S. Kirk, J. E. Raven and M. Schofield (eds.), *The Presocratic Philosophers: A Critical History with a Selection of Texts* (1983), p. 365–6.

5 Appearances are a glimpse of the obscure.

Anaxagoras, fr. 21a. Trans. R. W. Sharples.

6 ON **ANAXAGORAS** He (Anaxagoras) is said to have been twenty years old at the time of Xerxes' crossing, and to have lived to seventy-two. Apollodorus says in his *Chronicles* that he was born in the seventieth Olympiad (500–497 B.C.) and died in the first year of the eighty-eighth (428/7). He began to be a philosopher at Athens in the archonship of Callias (456/5), at the age of twenty, as Demetrius Phalereus tells us in his *Register of Archons*, and they say he spent thirty years there . . . There are different accounts given of his trial. Sotion, in his *Succession of Philosophers*, says that he was prosecuted by Cleon for impiety, because he maintained that the sun was a red hot mass of metal, and after that Pericles, his pupil, had made a speech in his defence, he was fined five talents and exiled. Satyrus in his *Lives*, on the other hand, says that the charge was brought by Thucydides in his political campaign against Pericles; and he adds that the charge was not only for the impiety but for Medism as well; and he was condemned to death in his absence . . . Finally he withdrew to Lampsacus, and there died. It is said that when the rulers of the city asked him what privilege he wished to be granted, he replied that the children should be given a holiday every year in the month in which he died. The custom is preserved to the present day.

When he died the Lampsacenes buried him with full honours.

Diogenes Laertius 2.7. In G. S. Kirk, J. E. Raven and M. Schofield (eds.), *The Presocratic Philosophers: A Critical History with a Selection of Texts* (1983), p. 353.

1 ON **ANAXAGORAS** Anaxagoras of Clazomenae, son of Hegesiboulos, held that the first principles of things were the homoeomeries. For it seemed to him quite impossible that anything should come into being from the non-existent or be dissolved into it. Anyhow we take in nourishment which is simple and homogeneous, such as bread or water, and by this are nourished hair, veins, arteries, flesh, sinews, bones and all the other parts of the body. Which being so, we must agree that everything that exists is in the nourishment we take in, and that everything derives its growth from things that exist. There must be in that nourishment some parts that are productive of blood, some of sinews, some of bones, and so on—parts which reason alone can apprehend. For there is no need to refer the fact that bread and water produce all these things to sense-perception; rather, there are in bread and water parts which only reason can apprehend.

Aetius 1.3.5. In G. S. Kirk, J. E. Raven and M. Schofield (eds), *The Presocratic Philosophers: A Critical History with a Selection of Texts* (1983), p. 375.

Anaximander c.610–546/7 BC

Greek philosopher and astronomer

2 ON **ANAXIMANDER** Anaximander the Milesian, a disciple of Thales, first dared to draw the inhabited world on a tablet; after him Hecataeus the Milesian, a much-travelled man, made the map more accurate, so that it became a source of wonder.

Agathemerus 1.1. In G. S. Kirk, J. E. Raven and M. Schofield (eds.), *The Presocratic Philosophers: A Critical History with a Selection of Texts* (1983), p. 104.

3 ON **ANAXIMANDER** Anaximander son of Praxiades, of Miletus: he said that the principle and element is the Indefinite, not distinguishing air or water or anything else . . . he was the first to discover a *gnomon*, and he set one up on the Sundials (?) in Sparta, according to Favorinus in his *Universal History*, to mark solstices and equinoxes; and he also constructed hour indicators. He was the first to draw an outline of earth and sea, but also constructed a [celestial] globe. Of his opinions he made a summary exposition, which I suppose Apollodorus the Athenian also encountered. Apollodorus says in his *Chronicles* that Anaximander was sixty-four years old in the year of the fifty-eighth Olympiad [547/6 B.C.], and that he died shortly afterwards (having been near his prime approximately during the time of Polycrates, tyrant of Samos).

Diogenes Laertius 2.1–2. In G. S. Kirk, J. E. Raven and M.Schofield (eds), *The Presocratic Philosophers: A Critical History with a Selection of Texts* (1983), p. 100.

Anaximenes of Miletus fl. c.546/545 BC

Greek philosopher of nature

4 ON **ANAXIMENES** Anaximenes . . . said that infinite air was the principle, from which the things that are becoming, and that are, and that shall be, and gods and things divine, all come into being, and the rest from its products. The form of air is of this kind: whenever it is most equable it is invisible to sight, but is revealed by the cold and the hot and the damp and by movement. It is always in motion; for things that change do not change unless there be movement. Through becoming denser or finer it has different appearances; for when it is dissolved into what is finer it becomes fire, while winds, again, are air that is becoming condensed, and cloud is produced from air by felting. When it is condensed still more, water is produced; with a further degree of condensation earth is produced, and when condensed as far as possible, stones. The result is that the most influential components of the generation are opposites, hot and cold.

Hippolytus, *Refutation*, 1.7.1. In G. S. Kirk, J. E. Raven and M. Schofield (eds.), *The Presocratic Philosophers: A Critical History with a Selection of Texts* (1983), p. 145.

1 The earth is flat, being borne upon air, and similarly the sun, moon and the other heavenly bodies, which are all fiery, ride upon the air through their flatness.

Hippolytus, *Refutation* 1.7.4. In G. S. Kirk, J. E. Raven and M. Schofield (eds.), *The Presocratic Philosophers: A Critical History with a Selection of Texts* (1983), p. 154.

2 ON **ANAXIMENES** Anaximenes . . . also says that the underlying nature is one and infinite . . . but not undefined as Anaximander said but definite, for he identifies it as air; and it differs in its substantial nature by rarity and density. Being made finer it becomes fire; being made thicker it becomes wind, then cloud, then (when thickened still more) water, then earth, then stones; and the rest come into being from these.

Simplicius, *Commentary on Aristotle's Physics*, 24, 26–31, quoting Theophrastus on Anaximenes. In G. S. Kirk, J. E. Raven and M. Schofield (eds.), *The Presocratic Philosophers: A Critical History with a Selection of Texts* (1983), p. 145.

3 ON **ANAXIMENES** Anaximenes and Anaxagoras and Democritus say that its [the earth's] flatness is responsible for it staying still: for it does not cut the air beneath but covers it like a lid, which flat bodies evidently do: for they are hard to move even for the winds, on account of their resistance.

Aristotle, *On the Heavens*, 294b, 13. In G. S. Kirk, J. E. Raven and M. Schofield (eds.), *The Presocratic Philosophers: A Critical History with a Selection of Texts* (1983), p. 153.

4 ON **ANAXIMENES** Anaximenes son of Eurystratus, of Miletus, was a pupil of Anaximander; some say he was also a pupil of Parmenides. He said that the material principle was air and the infinite; and that the stars move, not under the earth, but round it. He used simple and economical Ionic speech. He was active, according to what Apollodorus says, around the time of the capture of Sardis, and died in the 63rd Olympiad.

Diogenes Laertius 2.3. In G. S. Kirk, J. E. Raven and M. Schofield (eds), *The Presocratic Philosophers: A Critical History with a Selection of Texts* (1983), p. 143.

5 ON **ANAXIMENES** Anaximenes . . . declared that air is the principle of existing things; for from it all things come-to-be and into it they are again dissolved. As our soul, he says, being air holds us together and controls us, so does wind [or *breath*] and air enclose the whole world.

Aetius, 1, 3. 4. In G. S. Kirk, J. E. Raven and M. Schofield (eds.), *The Presocratic Philosophers: A Critical History with a Selection of Texts* (1983), p. 158–9.

Herbert George Andrewartha

1907–92

Australian zoologist and entomologist

6 Model-making, the imaginative and logical steps which precede the experiment, may be judged the most valuable part of scientific method because skill and insight in these matters are rare. Without them we do not know what experiment to do. But it is the experiment which provides the raw material for scientific theory. Scientific theory cannot be built directly from the conclusions of conceptual models.

Introduction to the Study of Animal Population (1961), 181.

Frederick William Andrewes

1859–1932

British bacteriologist and pathologist

7 It may very properly be asked whether the attempt to define distinct species, of a more or less permanent nature, such as we are accustomed to deal with amongst the higher plants and animals, is not altogether illusory amongst such lowly organised forms of life as the bacteria. No biologist nowadays believes in the absolute fixity of species . . . but there are two circumstances which here render the problem of specificity even more difficult of solution. The bacteriologist is deprived of the test of mutual fertility or sterility, so valuable in determining specific limits amongst organisms in which sexual reproduction prevails. Further, the extreme rapidity with which generation succeeds generation amongst bacteria offers to the forces of variation and natural selection a field for their operation wholly unparalleled amongst higher forms of life.

'The Evolution of the Streptococci', *The Lancet*, 1906, **2**, 1415–6.

Roy Chapman Andrews
1884–1960

American explorer, zoologist and museum director

1 Palaeontology is the Aladdin's lamp of the most deserted and lifeless regions of the earth; it touches the rocks and there spring forth in orderly succession the monarchs of the past and the ancient river streams and savannahs wherein they flourished. The rocks usually hide their story in the most difficult and inaccessible places.

On the Trail of Ancient Man (1926), x.

2 Today there remain but a few small areas on the world's map unmarked by explorers' trails. Human courage and endurance have conquered the Poles; the secrets of the tropical jungles have been revealed. The highest mountains of the earth have heard the voice of man. But this does not mean that the youth of the future has no new worlds to vanquish. It means only that the explorer must change his methods.

On the Trail of Ancient Man (1926), 5.

Thomas Andrews 1813–85

British chemist

3 We may indeed live yet to see, or at least we may feel some confidence that those who come after us will see, such bodies as oxygen and hydrogen in the liquid, perhaps even in the solid state.

The Scientific Papers of the late Thomas Andrews (1889), lx.

Anonymous

4 *Post coitum omne animal triste.*

After coition every animal is sad.

Post-classical saying.

5 *Post hoc, ergo propter hoc.*

After this, therefore because of this.

Latin Proverb.

6 Alas! That partial Science should approve
The sly rectangle's too licentious love!
From *three* bright Nymphs the wily wizard burns;—
Three bright-ey'd Nymphs requite his flame by turns.
Strange force of magic skill! Combined of yore.

'The Loves of the Triangles. A Mathematical and Philosophical Poem', in *The Anti-Jacobean or Weekly Examiner*, Monday 16 April 1798, 182. [Written by George Canning, Hookham Frere, and George Ellis].

7 *Tierchemie ist Schmierchemie.*

Animal chemistry is messy chemistry.

Quoted without source in Joseph S. Fruton, *Proteins, Enzymes, Genes: The Interplay of Chemistry and Biology* (1999), 57.

8 An astronomer is a guy who stands around looking at heavenly bodies.

Quoted in M. Goran, *A Treasury of Science Jokes* (1986), 29.

9 A bacteriologist is a man whose conversation always starts with the germ of an idea.

Quoted in M. Goran, *A Treasury of Science Jokes* (1986), 37.

10 Better Things for Better Living Through Chemistry.

Advertising campaign slogan for the DuPont Company from 1935.

11 Botany is the science in which plants are known by their aliases.

Quoted in M. Goran, *A Treasury of Science Jokes* (1986), 49.

12 The central dogma, enunciated by Crick in 1958 and the keystone of molecular biology ever since, is likely to prove a considerable over-simplification. That is the heretical but inescapable conclusion stemming from experiments done in the past few months in two laboratories in the United States.

'News and Views', *Nature*, 1970, 226, 1198.

1 Experiment adds to knowledge,
Credulity leads to error.

Arabic Proverb.

2 Fiction tends to become 'fact' simply by serial passage via the printed page.

Saying.

3 Garbage in, garbage out.

Saying.

4 A geologist is a fault-finder.

Quoted in M. Goran, *A Treasury of Science Jokes* (1986), 73.

5 Half of the secret of resistance to disease is cleanliness; the other half is dirtiness.

Saying.

6 Here are the opinions on which my facts are based.

Saying.

7 If a research project is not worth doing at all, it is not worth doing properly.

Saying.

8 *Gnothi seauton.*

Know thyself.

From The Temple of Apollo at Delphi; Pausanias 10.24.1; Juvenal 11.27.

9 Laws of Thermodynamics

1) You cannot win, you can only break even.

2) You can only break even at absolute zero.

3) You cannot reach absolute zero.

Folklore amongst physicists.

10 Like the statistician who was drowned in a lake of average depth six inches.

Saying.

11 Man occasionally stumbles on the truth, but then just picks himself up and hurries on regardless.

Saying.

12 The most powerful antigen in human biology is a new idea.

Saying.

13 Nature is by nature perverse.

Saying.

14 A physicist learns more and more about less and less, until he knows everything about nothing; whereas a philosopher learns less and less about more and more, until he knows nothing about everything.

Saying.

15 Sex is the best form of fusion at room temperature.

Saying.

16 *Nihil est in intellectu quod non prius fuerit in sensu.*

There is nothing in the mind that has not previously been in the senses.

Saying.

17 There is one thing stronger than all the armies in the world; and that is an idea whose time has come.

The Nation, 15 April 1943.

18 *Hic locus est ubi mors gaudet succurere vitae.*

This place is where death rejoices to come to the aid of life.

In the anatomical dissection theatre of the University of Bologna.

19 To-day, science has withdrawn into realms that are hardly understanded of the people. Biology means very largely histology, the study of the cell by difficult and elaborate microscopical processes. Chemistry has passed from the mixing of simple substances with ascertained reactions, to an experimentation of these processes under varying conditions of temperature, pressure, and electrification—all requiring complicated apparatus and the most delicate measurement and manipulation. Similarly, physics has outgrown the old formulas of gravity, magnetism, and pressure; has discarded the molecule and atom for the ion, and may in its recent generalizations be followed only by an expert in the higher, not to say the transcendental, mathematics.

'Exit the Amateur Scientist.' Editorial, *The Nation*, 23 August 1906, 83, 160.

1 *Magna opera Domini exquisita in omnes voluntates eius.*

The works of the Lord are great; sought out of all those that have pleasure therein.

Over the entrance to the Cavendish Laboratory, Cambridge.

St. Thomas Aquinas c.1225–74

Italian theologian and philosopher

2 Practical sciences proceed by building up; theoretical sciences by resolving into components.

Sententia libri Ethicorum (*Commentary on the Nicomachean Ethics*) [1271], Book 1, chapter 3, number 4, trans. C. I. Litzinger (1993).

3 For it is necessary in every practical science to proceed in a composite (i.e. deductive) manner. On the contrary in speculative science, it is necessary to proceed in an analytical manner by breaking down the complex into elementary principles.

Sententia libri Ethicorum (*Commentary on the Nicomachean Ethics*) [1271], Book 1, lecture 3, section 35, trans. C. I. Litzinger (1993), 12.

4 It must be understood that prime matter, and form as well, is neither generated nor corrupted, because every generation is from something to something. Now that from which generation proceeds is matter, and that to which it proceeds is form. So that, if matter or form were generated, there would be a matter for matter and a form for form, endlessly. Whence, there is generation only of the composite, properly speaking.

De Principiis Naturae (*On the Principles of Nature*) [before 1256], chapter 2, section 12, trans. J. Bobik, *Aquinas on Nature and Form and the Elements: A Translation and Interpretation of the de Principiis Naturae and the De Mixtione Elementorum of St. Thomas Aquinas* (1998), 29.

5 Reason may be employed in two ways to establish a point: first for the purpose of furnishing sufficient proof of some principle, as in natural science, where sufficient proof can be brought to show that the movement of the heavens is always of uniform velocity. Reason is employed in another way, not as furnishing a sufficient proof of a principle, but as confirming an already established principle, by showing the congruity of its results, as in astrology the theory of eccentrics and epicycles is considered as established because thereby the sensible appearances of the heavenly movements can be explained; not, however, as if this reason were sufficient, since some other theory might explain them.

Summa Theologica [1266–1273], Part 1, question 32, article 2 (reply to objection 2), trans. Fathers of the English Dominican Province (i.e. L. Shapcote), revised D. J. Sullivan (1952), Vol. 1, 177.

6 If there were some solitary or feral man, the passions of the soul would be sufficient for him; by them he would be conformed to things in order that he might have knowledge of them. But because man is naturally political and social, there is need for one man to make his conceptions known to others, which is done with speech. So significant speech was needed if men were to live together. Which is why those of different tongues do not easily live together.

Sententia super libri Perihermeneias (*Commentary on Aristotle's On Interpretation*) [1270–1271], Book I, lesson 2, number 2, trans. R. McInerny, quoted in R. McInerny (ed.) *Thomas Aquinas, Selected Writings* (1998), 460.

Michael A. Arbib 1940–

British-born neuroscientist

7 In the beginning was the word
WORD
WORE
GORE
GONE
GENE

and by the mutations came the gene.

Appendix: notes on the Second Symposium. In C. H. Waddington (ed.), *Towards a Theoretical Biology: An IUBS Symposium* (1969), Vol. 2, 323.

John Arbuthnot 1667–1735

British physician and essayist

1 The Reader may here observe the Force of Numbers, which can be successfully applied, even to those things, which one would imagine are subject to no Rules. There are very few things which we know, which are not capable of being reduc'd to a Mathematical Reasoning, and when they cannot, it's a sign our Knowledge of them is very small and confus'd; and where a mathematical reasoning can be had, it's as great folly to make use of any other, as to grope for a thing in the dark when you have a Candle standing by you.

Of the Laws of Chance, or, a Method of the Hazards of Game (1692), Preface.

2 Among innumerable footsteps of divine providence to be found in the works of nature, there is a very remarkable one to be observed in the exact balance that is maintained, between the numbers of men and women; for by this means is provided, that the species never may fail, nor perish, since every male may have its female, and of proportionable age. This equality of males and females is not the effect of chance but divine providence, working for a good end.

'An Argument for Divine Providence, taken from the Constant Regularity observ'd in the Births of both Sexes', *Philosophical Transactions of the Royal Society*, 1710–12, 27, 186.

Archilochus c.680–640 BC

Greek lyric poet

3 The Fox knows many things—the hedgehog one big one.

E. Diehl (ed.), *Anthologia Lyrica Graeca* (1925), 241, no.103.

Archimedes c.287–212 BC

Italian-Greek mathematician and inventor

4 Give me a place to stand, and I will move the earth.

F. Hultsch (ed.) *Pappus Alexandrinus: Collectio* (1876–8), Vol. 3, book 8, section 10, ix.

5 Having been the discoverer of many splendid things, he is said to have asked his friends and relations that, after his death, they should place on his tomb a cylinder enclosing a sphere, writing on it the proportion of the containing solid to that which is contained.

Plutarch, *Life of Marcellus*, 17.12. Trans. R. W. Sharples.

6 ON **ARCHIMEDES** Hieron asked Archimedes to discover, without damaging it, whether a certain crown or wreath was made of pure gold, or if the goldsmith had fraudulently alloyed it with some baser metal. While Archimedes was turning the problem over in his mind, he chanced to be in the bath house. There, as he was sitting in the bath, he noticed that the amount of water that was flowing over the top of it was equal in volume to that part of his body that was immersed. He saw at once a way of solving the problem. He did not delay, but in his joy leaped out of the bath. Rushing naked through the streets towards his home, he cried out in a loud voice that he had found what he sought. For, as he ran, he repeatedly shouted in Greek; '*Eureka! Eureka!* I've found it! I've found it!'

Vitrivius Pollio, *De Architectura*, ix, prologue, section 10.

Robert Ardrey 1908–80

American author and popular science writer

7 The miracle of man is not how far he has sunk but how magnificently he has risen. We are known among the stars by our poems, not our corpses. No creature who began as a mathematical improbability, who was selected through millions of years of unprecedented environmental hardship and change for ruggedness, ruthlessness, cunning, and adaptability, and who in the short ten thousand years of what we may call civilization has achieved such wonders as we find about us, may be regarded as a creature without promise.

African Genesis: A Personal Investigation into the Animal Origins and Nature of Man (1961), 348.

Giovanni Arduino (Arduini) 1714–95

Italian geologist

1 So far as I have been able to observe thus far, the series of strata which compose the earth's visible crust, seem to me to be divided into four general or successive, orders, without taking into consideration the sea. These four orders can be thought of as being four enormous strata . . . which, wherever they are found, are seen to be placed one above the other, in a consistently uniform manner.

'Lettere Seconda . . . sopra varie sue Osservazioni fatti in diverse parti del Territorio di Vicenza, ad altrove, appartenenti alla Teoria terrestre, ed alla Mineralogia', *Nuova Raccolta di Opuscoli Scientificie Filologici*, 1760, 6, 158, trans. Ezio Vaccari.

2 I shall conclude, for the time being, by saying that until Philosophers make observations (especially of mountains) that are longer, more attentive, orderly, and interconnected, and while they fail to recognize the two great agents, fire and water, in their distinct affects, they will not be able to understand the causes of the great natural variety in the disposition, structure, and other matter that can be observed in the terrestrial globe in a manner that truly corresponds to the facts and to the phenomena of Nature.

'Aleune Osservazioni Orittologiche fatte nei Monti del Vicentino', *Giornale d'Italia*, 1769, 5, 411, trans. Ezio Vaccari.

3 With the sole guidance of our practical knowledge of those physical agents which we see actually used in the continuous workings of nature, and of our knowledge of the respective effects induced by the same workings, we can with reasonable basis surmise what the forces were which acted even in the remotest times.

Quoted in Francesco Rodolico, 'Arduino'. In Charles Coulston Gillispie (ed.), *Dictionary of Scientific Biography* (1970), Vol. 1, 234.

Émile Argand 1879–1940

Swiss geologist

4 The volumes, the surfaces, the lines—in one word, the structures that build a tectonic construction—do not represent the whole picture: there is also the movement that animated and still animates these bodies because the history continues and we live under no particular privileged conditions at any given time in this great process.

Tectonics of Asia (1924), 2, trans. Albert V. and Marguerite Carozzi.

5 The universe flows, carrying with it milky ways and worlds, Gondwanas and Eurasias, inconsistent visions and clumsy systems. But the good conceptual models, these *serena templa* of intelligence on which several masters have worked, never disappear entirely. They are the great legacy of the past. They linger under more and more harmonious forms and actually never cease to grow. They bring solace by the great art that is inseparable from them. Their permanence relies on the immortal poetry of truth, of the truth that is given to us in minute amounts, foretelling an order whose majesty dominates time.

Tectonics of Asia (1924), 164, trans. Albert V. and Marguerite Carozzi.

6 How many times did the sun shine, how many times did the wind howl over the desolate tundras, over the bleak immensity of the Siberian taigas, over the brown deserts where the Earth's salt shines, over the high peaks capped with silver, over the shivering jungles, over the undulating forests of the tropics! Day after day, through infinite time, the scenery has changed in imperceptible features. Let us smile at the illusion of eternity that appears in these things, and while so many temporary aspects fade away, let us listen to the ancient hymn, the spectacular song of the seas, that has saluted so many chains rising to the light.

Tectonics of Asia (1924), 165, trans. Albert V. and Marguerite Carozzi.

Aristotle 384–22 BC

Greek natural philosopher
See ***Dante Alighieri*** *151:1*

1 The same ideas, one must believe, recur in men's minds not once or twice but again and again.

On the Heavens, 270b, 19–20. In Jonathan Barnes (ed.), *The Complete Works of Aristotle* (1984), Vol. 1, 451.

2 As to the position of the earth, then, this is the view which some advance, and the views advanced concerning its *rest or motion* are similar. For here too there is no general agreement. All who deny that the earth lies at the centre think that it revolves about the centre, and not the earth only but, as we said before, the counter-earth as well. Some of them even consider it possible that there are several bodies so moving, which are invisible to us owing to the interposition of the earth. This, they say, accounts for the fact that eclipses of the moon are more frequent than eclipses of the sun; for in addition to the earth each of these moving bodies can obstruct it.

On the Heavens, 293b, 15–25. In Jonathan Barnes (ed.), *The Complete Works of Aristotle* (1984), Vol. 1, 483.

3 For any two portions of fire, small or great, will exhibit the same ratio of solid to void; but the upward movement of the greater is quicker than that of the less, just as the downward movement of a mass of gold or lead, or of any other body endowed with weight, is quicker in proportion to its size.

On the Heavens, 309b, 11–5. In Jonathan Barnes (ed.), *The Complete Works of Aristotle* (1984), Vol. 1, 505.

4 We, on the other hand, must take for granted that the things that exist by nature are, either all or some of them, in motion.

Physics, 185a, 12–3. In Jonathan Barnes (ed.), *The Complete Works of Aristotle* (1984), Vol. 1, 316.

5 Now, the causes being four, it is the business of the student of nature to know about them all, and if he refers his problems back to all of them, he will assign the 'why' in the way proper to his science—the matter, the form, the mover, that for the sake of which.

Physics, 198a, 22–4. In Jonathan Barnes (ed.), *The Complete Works of Aristotle* (1984), Vol. 1, 338.

6 It is clear, then, that though there may be countless instances of the perishing of unmoved movers, and though many things that move themselves perish and are succeeded by others that come into being, and though one thing that is unmoved moves one thing while another moves another, nevertheless there is something that comprehends them all, and that as something apart from each one of them, and this it is that is the cause of the fact that some things are and others are not and of the continuous process of change; and this causes the motion of the other movers, while they are the causes of the motion of other things. Motion, then, being eternal, the first mover, if there is but one, will be eternal also; if there are more than one, there will be a plurality of such eternal movers.

Physics, 258b, 32–259a, 8. In Jonathan Barnes (ed.), *The Complete Works of Aristotle* (1984), Vol. 1, 432.

7 But the whole vital process of the earth takes place so gradually and in periods of time which are so immense compared with the length of our life, that these changes are not observed, and before their course can be recorded from beginning to end whole nations perish and are destroyed.

Meteorology, 351b, 8–13. In Jonathan Barnes (ed.), *The Complete Works of Aristotle* (1984), Vol. 1, 573.

1 So it is clear, since there will be no end to time and the world is eternal, that neither the Tanais nor the Nile has always been flowing, but that the region whence they flow was once dry; for their action has an end, but time does not. And this will be equally true of all other rivers. But if rivers come into existence and perish and the same parts of the earth were not always moist, the sea must needs change correspondingly. And if the sea is always advancing in one place and receding in another it is clear that the same parts of the whole earth are not always either sea or land, but that all this changes in the course of time.

Meteorology, 353a, 14–24. In Jonathan Barnes (ed.), *The Complete Works of Aristotle* (1984), Vol. 1, 575.

2 We maintain that there are two exhalations, one vaporous the other smoky, and these correspond to two kinds of bodies that originate in the earth, things quarried and things mined. The heat of the dry exhalation is the cause of all things quarried. Such are the kinds of stones that cannot be melted, and realgar, and ochre, and ruddle, and sulphur, and the other things of that kind, most things quarried being either coloured lye or, like cinnabar, a stone compounded of it. The vaporous exhalation is the cause of all things mined—things which are either fusible or malleable such as iron, copper, gold.

Meteorology, 378a, 19–28. In Jonathan Barnes (ed.), *The Complete Works of Aristotle* (1984), Vol. 1, 607.

3 Nature proceeds little by little from things lifeless to animal life in such a way that it is impossible to determine the exact line of demarcation, nor on which side thereof an intermediate form should lie. Thus, next after lifeless things comes the plant, and of plants one will differ from another as to its amount of apparent vitality; and, in a word, the whole genus of plants, whilst it is devoid of life as compared with an animal, is endowed with life as compared with other corporeal entities. Indeed, as we just remarked, there is observed in plants a continuous scale of ascent towards the animal. So, in the sea, there are certain objects concerning which one would be at a loss to determine whether they be animal or vegetable. For instance, certain of these objects are fairly rooted, and in several cases perish if detached.

History of Animals, 588b, 4–14. In Jonathan Barnes (ed.) *The Complete Works of Aristotle* (1984), Vol. 1, 922.

4 If any person thinks the examination of the rest of the animal kingdom an unworthy task, he must hold in like disesteem the study of man. For no one can look at the elements of the human frame—blood, flesh, bones, vessels, and the like—without much repugnance. Moreover, when any one of the parts or structures, be it which it may, is under discussion, it must not be supposed that it is its material composition to which attention is being directed or which is the object of the discussion, but rather the total form. Similarly, the true object of architecture is not bricks, mortar or timber, but the house; and so the principal object of natural philosophy is not the material elements, but their composition, and the totality of the substance, independently of which they have no existence.

Parts of Animals, 645a, 26–36. In Jonathan Barnes (ed.), *The Complete Works of Aristotle* (1984), Vol. 1, 1004.

5 Plants, again, inasmuch as they are without locomotion, present no great variety in their heterogeneous parts. For, when the functions are but few, few also are the organs required to effect them. . . . Animals, however, that not only live but perceive, present a great multiformity of parts, and this diversity is greater in some animals than in others, being most varied in those to whose share has fallen not mere life but life of high degree. Now such an animal is man.

Parts of Animals, 655b, 37–656a, 7. In Jonathan Barnes (ed.), *The Complete Works of Aristotle* (1984), Vol. 1, 1021–2.

1 But nature flies from the infinite; for the infinite is imperfect, and nature always seeks an end.

Generation of Animals, 715b, 114–6. In Jonathan Barnes (ed.), *The Complete Works of Aristotle* (1984), Vol. 1, 1112.

2 If there is any kind of animal which is female and has no male separate from it, it is possible that this may generate a young one from itself. No instance of this worthy of any credit has been observed up to the present at any rate, but one case in the class of fishes makes us hesitate. No male of the so-called erythrinus has ever yet been seen, but females, and specimens full of roe, have been seen. Of this, however, we have as yet no proof worthy of credit.

Generation of Animals, 741a, 32–8. In Jonathan Barnes (ed.), *The Complete Works of Aristotle* (1984), Vol. 1, 1150.

3 For nature by the same cause, provided it remain in the same condition, always produces the same effect, so that either coming-to-be or passing-away will always result.

On Generation and Corruption, 336a, 27–9. In Jonathan Barnes (ed.), *The Complete Works of Aristotle* (1984), Vol. 1, 551.

4 At present we must confine ourselves to saying that soul is the source of these phenomena and is characterized by them, viz. by the powers of self-nutrition, sensation, thinking, and movement.

On the Soul, 413b, 11–3. In Jonathan Barnes (ed.), *The Complete Works of Aristotle* (1984), Vol. 1, 658.

5 Since we think we understand when we know the explanation, and there are four types of explanation (one, what it is to be a thing; one, that if certain things hold it is necessary that this does; another, what initiated the change; and fourth, the aim), all these are proved through the middle term.

Posterior Analytics, 94a, 20–4. In Jonathan Barnes (ed.), *The Complete Works of Aristotle* (1984), Vol. 1, 155.

6 A likely impossibility is always preferable to an unconvincing possibility.

Poetics, 1460a, 26–7. In Jonathan Barnes (ed.), *The Complete Works of Aristotle* (1984), Vol. 2, 2337.

7 It is evidently equally foolish to accept probable reasoning from a mathematician and to demand from a rhetorician demonstrative proofs.

Nicomachean Ethics, 1094b, 25–7. In Jonathan Barnes (ed.), *The Complete Works of Aristotle* (1984), Vol. 2, 1730.

8 Everything that depends on the action of nature is by nature as good as it can be.

Nicomachean Ethics, 1099b, 21–22. In Jonathan Barnes (ed.), *The Complete Works of Aristotle* (1984), Vol. 2, 1738.

9 For the things we have to learn before we can do, we learn by doing.

Nicomachean Ethics, 1103a, 32–3. In Jonathan Barnes (ed.), *The Complete Works of Aristotle* (1984), Vol. 2, 1743.

10 All men by nature desire to know.

Metaphysics, 980a, 21. In Jonathan Barnes (ed.), *The Complete Works of Aristotle* (1984), Vol. 2, 1552.

11 At first he who invented any art that went beyond the common perceptions of man was naturally admired by men, not only because there was something useful in the inventions, but because he was thought wise and superior to the rest. But as more arts were invented, and some were directed to the necessities of life, others to its recreation, the inventors of the latter were always regarded as wiser than the inventors of the former, because their branches of knowledge did not aim at utility.

Metaphysics, 981b, 13–20. In Jonathan Barnes (ed.), *The Complete Works of Aristotle* (1984), Vol. 2, 1553.

1 For it is owing to their wonder that men now both begin and at first began to philosophize; they wondered originally at the obvious difficulties, then advanced little by little and stated difficulties about the greater matters, e.g. about the phenomena of the moon and those of the sun and the stars, and about the genesis of the universe. And a man who is puzzled and wonders thinks himself ignorant (whence even the lover of myth is in a sense a lover of wisdom, for myth is composed of wonders); therefore since they philosophized in order to escape from ignorance, evidently they were pursuing science in order to know, and not for any utilitarian end. And this is confirmed by the facts; for it was when almost all the necessities of life and the things that make for comfort and recreation were present, that such knowledge began to be sought. Evidently then we do not seek it for the sake of any advantage; but as the man is free, we say, who exists for himself and not for another, so we pursue this as the only free science, for it alone exists for itself.

Metaphysics, 982b, 12–27. In Jonathan Barnes (ed.), *The Complete Works of Aristotle* (1984), Vol. 2, 1554.

2 The investigation of the truth is in one way hard, in another easy. An indication of this is found in the fact that no one is able to attain the truth adequately, while, on the other hand, no one fails entirely, but every one says something true about the nature of things, and while individually they contribute little or nothing to the truth, by the union of all a considerable amount is amassed. Therefore, since the truth seems to be like the proverbial door, which no one can fail to hit, in this way it is easy, but the fact that we can have a whole truth and not the particular part we aim at shows the difficulty of it.

Perhaps, as difficulties are of two kinds, the cause of the present difficulty is not in the facts but in us.

Metaphysics, 993a, 30–993b, 9. In Jonathan Barnes (ed.), *The Complete Works of Aristotle* (1984), Vol. 2, 1569–70.

3 The same thing may have all the kinds of causes, e.g. the moving cause of a house is the art or the builder, the final cause is the function it fulfils, the matter is earth and stones, and the form is the definitory formula.

Metaphysics, 996b, 5–8. In Jonathan Barnes (ed.), *The Complete Works of Aristotle* (1984), Vol. 2, 1574.

4 The saying of Protagoras is like the views we have mentioned; he said that man is the measure of all things, meaning simply that that which seems to each man assuredly is. If this is so, it follows that the same thing both is and is not, and is bad and good, and that the contents of all other opposite statements are true, because often a particular thing appears beautiful to some and ugly to others, and that which appears to each man is the measure.

Metaphysics, 1062b, 12–19. In Jonathan Barnes (ed.), *The Complete Works of Aristotle* (1984), Vol. 2, 1678.

Henry Edward Armstrong

1848–1937

British chemist

5 After all, we scientific workers . . . like women, are the victims of fashion: at one time we wear dissociated ions, at another electrons; and we are always loth to don rational clothing; some fixed belief we must have manufactured for us: we are high or low church, of this or that degree of nonconformity, according to the school in which we are brought up—but the agnostic is always rare of us and of late years the critic has been taboo.

'The Thirst of Salted Water or the Ions Overboard', *Science Progress*, 1909, 3, 643.

1 When the chemist makes gloves, he usually cannot help making them in pairs for both hands.

'The Origin of Life: A Chemist's Fantasy', *Science Progress*, 1912, 7, 318.

2 *His motion to the meeting of the Council of the Chemical Society:*

That henceforth the absurd game of chemical noughts and crosses be tabu within the Society's precincts and that, following the practice of the Press in ending a correspondence, it be an instruction to the officers to give notice 'That no further contributions to the mysteries of Polarity will be received, considered or printed by the Society'. His challenge was not accepted.

From the personal and other items column of *Chemistry and Industry*, 1925, **44**, 1050.

3 I notice that, in the lecture . . . which Prof. Lowry gave recently, in Paris . . . he brought forward certain freak formulae for tartaric acid, in which hydrogen figures as bigamist . . . I may say, he but follows the loose example set by certain Uesanians, especially one G. N. Lewis, a Californian thermodynamiter, who has chosen to disregard the fundamental canons of chemistry—for no obvious reason other than that of indulging in premature speculation upon electrons as the cause of valency . . .

'Bigamist Hydrogen. A Protest', *Nature*, 1926, 117, 553.

4 The fact is the physical chemists never use their eyes and are most lamentably lacking in chemical culture. It is essential to cast out from our midst, root and branch, this physical element and return to our laboratories.

'Ionomania in Extremis', *Chemistry and Industry*, 1936, **14**, 917.

5 Hypotheses like professors, when they are seen not to work any longer in the laboratory, should disappear.

Sir Harold Hartley, 'Henry Armstrong', in *Studies in the History of Chemistry* (1971), 199.

John Armstrong 1709–79

British physician and poet

6 The blood, the fountain whence the spirits flow,
The generous stream that waters every part,
And motion, vigour, and warm life conveys
To every Particle that moves or lives;
This vital fluid, thro' unnumber'd tubes
Pour'd by the heart, and to the heart again
Refunded; scourg'd forever round and round;
Enrag'd with heat and toil, at last forgets
Its balmy nature; virulent and thin
It grows; and now, but that a thousand gates
Are open to its flight, it would destroy
The parts it cherish'd and repair'd before.

The Art of Preserving Health (1744), book 2, l. 12–23, p.15–16.

Neil A. Armstrong 1930–

American astronaut

7 We came in peace for all mankind.

Plaque left on the moon, 20 July 1969.

8 That's one small step for a man, one giant leap for mankind.

Said as he stepped onto the Moon at 10.56 (G.M.T) on July 21st 1969. However, this is often misquoted as 'That's one small step for man, one giant leap for mankind.' This was due to static during Armstrong's transmission to earth. The 'a' was left out of his statement, ruining the contrast he had made between one man 'a man' and all mankind 'man'.

Quoted in *Nature*, 1974, **250**, 451. Information on the frequent misuse of this quote is given in Paul F. Boller and John George (eds.), *They Never Said it: A Book of Fake Quotes, Misquotes and Misleading Attributions* (1989), 4–5.

Friedrich Arnold 1803–90

German anatomist

1 We have in the spinal cord the antetype (*Vorbild*) and the foundation for the entire structure of the brain.

'Handbuch der Anatomie des Menschen mit besonderer Rücksicht auf Physiologie und praktische Medicin' (1851), Vol. 2, 682. Trans. Edwin Clarke and L. S. Jacyna, *Nineteenth Century Origins of Neuroscientific Concepts* (1987), 52.

Matthew Arnold 1822–88

British poet and social critic

2 The interpretations of science do not give us this intimate sense of objects as the interpretations of poetry give it; they appeal to a limited faculty, and not to the whole man. It is not Linnaeus or Cavendish or Cuvier who gives us the true sense of animals, or water, or plants, who seizes their secret for us, who makes us participate in their life; it is Shakspeare [sic] . . . Wordsworth . . . Keats . . . Chateaubriand . . . Senancour.

'Maurice de Guérin' *Essays in Criticism* (1865), in R.H. Super (ed.) *The Complete Prose Works of Matthew Arnold: Lectures and Essays in Criticism* (1962), Vol. 3, 13.

3 The highest reach of science is, one may say, an inventive power, a faculty of divination, akin to the highest power exercised in poetry; therefore, a nation whose spirit is characterised by energy may well be eminent in science; and we have Newton. Shakspeare[sic] and Newton: in the intellectual sphere there can be no higher names. And what that energy, which is the life of genius, above everything demands and insists upon, is freedom; entire independence of all authority, prescription and routine, the fullest room to expand as it will.

'The Literary Influence of Academies' *Essays in Criticism* (1865), in R.H. Super (ed.) *The Complete Prose Works of Matthew Arnold: Lectures and Essays in Criticism* (1962), Vol. 3, 238.

4 The bent of our time is towards science, towards knowing things as they are . . .

On the Study of Celtic Literature (1867), in R.H. Super (ed.) *The Complete Prose Works of Matthew Arnold: Lectures and Essays in Criticism* (1962), Vol. 3, 298.

5 Nor bring, to see me cease to live,
Some doctor full of phrase and fame,
To shake his sapient head, and give
The ill he cannot cure a name.

'A Wish' (1867). In Kenneth Allot (ed.), *Matthew Arnold: A Selection* (1954), 194.

6 But the idea of science and systematic knowledge is wanting to our whole instruction alike, and not only to that of our business class . . . In nothing do England and the Continent at the present moment more strikingly differ than in the prominence which is now given to the idea of science there, and the neglect in which this idea still lies here; a neglect so great that we hardly even know the use of the word science in its strict sense, and only employ it in a secondary and incorrect sense.

Schools and Universities on the Continent (1868), 278–9.

7 The study of letters is the study of the operation of human force, of human freedom and activity; the study of nature is the study of the operation of non-human forces, of human limitation and passivity. The contemplation of human force and activity tends naturally to heighten our own force and activity; the contemplation of human limits and passivity tends rather to check it. Therefore the men who have had the humanistic training have played, and yet play, so prominent a part in human affairs, in spite of their prodigious ignorance of the universe.

Schools and Universities on the Continent (1868), in R. H. Super (ed.) *The Complete Prose Works of Matthew Arnold: Schools and Universities on the Continent* (1964), Vol. 4, 292.

8 Our society distributes itself into Barbarians, Philistines, and Populace; and America is just ourselves, with the

Barbarians quite left out, and the Populace nearly.

Culture and Anarchy: An Essay in Political and Social Criticism (1869), Preface, xxx.

1 The love of *science*, and the energy and honesty in the pursuit of science, in the best of the Aryan races do seem to correspond in a remarkable way to the love of *conduct*, and the energy and honesty in the pursuit of *conduct*, in the best of the Semitic.

Literature and Dogma: An Essay Towards a Better Apprehension of the Bible (1873), 386.

2 The 'hairy quadruped furnished with a tail and pointed ears, probably arboreal in his habits,' this good fellow carried hidden in his nature, apparently, something destined to develop into a necessity for humane letters.

'Literature and Science', delivered as a lecture during Arnold's tour of the United States in 1883 and published in *Discourses in America* (1885). Taken from M. H. Abrams (ed.), *The Norton Anthology of English Literature* (1993), Vol. 2, 1441.

Svante August Arrhenius

1859–1927

Swedish chemist and physicist

3 I was led to the conclusion that at the most extreme dilutions all salts would consist of simple conducting molecules. But the conducting molecules are, according to the hypothesis of Clausius and Williamson, dissociated; hence at extreme dilutions all salt molecules are completely disassociated. The degree of dissociation can be simply found on this assumption by taking the ratio of the molecular conductivity of the solution in question to the molecular conductivity at the most extreme dilution.

Letter to Van't Hoff, 13 April 1887. In J. R. Partington, *A History of Chemistry* (1961), Vol. 4, 678.

4 The title affixed to it is 'The Chemical Theory of Electrolytes,' but it is a bigger thing than this: it really is an attempt at *an electrolytic theory of chemistry*.

On Svante Arrhenius' Theorie Chemique des Electrolytes, *abstract and report by Oliver Lodge*.

56th Report of The British Association for the Advancement of Science, 1886, 362.

5 At first sight nothing seems more obvious than that everything has a beginning and an end, and that everything can be subdivided into smaller parts. Nevertheless, for entirely speculative reasons the philosophers of Antiquity, especially the Stoics, concluded this concept to be quite unnecessary. The prodigious development of physics has now reached the same conclusion as those philosophers, Empedocles and Democritus in particular, who lived around 500 B.C. and for whom even ancient man had a lively admiration.

'Development of the Theory of Electrolytic Dissociation', Nobel Lecture, 11 December 1903. In *Nobel Lectures: Chemistry 1901–1921* (1966), 45.

6 Humanity stands . . . before a great problem of finding new raw materials and new sources of energy that shall never become exhausted. In the meantime we must not waste what we have, but must leave as much as possible for coming generations.

Chemistry in Modern Life (1925), trans. Clifford Shattuck-Leonard, vii.

Isaac Asimov 1920–92

Russian-born American writer and biochemist

7 1—A robot may not injure a human being, or, through inaction, allow a human being to come to harm.
2—A robot must obey the orders given it by human beings except where such orders would conflict with the first law.
3—A robot must protect its own existence as long as such protection does not conflict with the First or Second Law.

'The Three Laws of Robotics', in *I, Robot* (1950), Frontispiece.

8 A number of years ago, when I was a freshly-appointed instructor, I met, for the first time, a certain eminent historian of science. At the time I could only regard him with tolerant condescension.

I was sorry of the man who, it seemed to me, was forced to hover about the edges of science. He was compelled to shiver endlessly in the outskirts, getting only feeble warmth from the distant sun of science-in-progress; while I, just beginning my research, was bathed in the heady liquid heat up at the very center of the glow.

In a lifetime of being wrong at many a point, I was never more wrong. It was I, not he, who was wandering in the periphery. It was he, not I, who lived in the blaze.

I had fallen victim to the fallacy of the 'growing edge'; the belief that only the very frontier of scientific advance counted; that everything that had been left behind by that advance was faded and dead.

But is that true? Because a tree in spring buds and comes greenly into leaf, are those leaves therefore the tree? If the newborn twigs and their leaves were all that existed, they would form a vague halo of green suspended in mid-air, but surely that is not the tree. The leaves, by themselves, are no more than trivial fluttering decoration. It is the trunk and limbs that give the tree its grandeur and the leaves themselves their meaning.

There is not a discovery in science, however revolutionary, however sparkling with insight, that does not arise out of what went before. 'If I have seen further than other men,' said Isaac Newton, 'it is because I have stood on the shoulders of giants.'

Adding A Dimension: Seventeen Essays on the History of Science (1964), Introduction.

1 The young specialist in English Lit, having quoted me, went on to lecture me severely on the fact that in *every* century people have thought they understood the Universe at last, and in *every* century they were proved to be wrong. It follows that the one thing we can say about our modern 'knowledge' is that it is *wrong*.

The young man then quoted with approval what Socrates had said on learning that the Delphic oracle had proclaimed him the wisest man in Greece. 'If I am the wisest man,' said Socrates, 'it is because I alone know that I know nothing.' The implication was that I was very foolish because I was under the impression I knew a great deal.

Alas, none of this was new to me. (There is very little that is new to me; I wish my correspondents would realize this.) This particular theme was addressed to me a quarter of a century ago by John Campbell, who specialized in irritating me. He also told me that all theories are proven wrong in time.

My answer to him was, 'John, when people thought the Earth was flat, they were wrong. When people thought the Earth was spherical, they were wrong. But if *you* think that thinking the Earth is spherical is *just as wrong* as thinking the Earth is flat, then your view is wronger than both of them put together.'

The Relativity of Wrong (1989), 214.

William Thomas Astbury

1898–1961

British x-ray crystallographer and molecular biologist

2 It [molecular biology] is concerned particularly with the *forms* of biological molecules and with the evolution, exploitation and ramification of these forms in the ascent to higher and higher levels of organization. Molecular biology is predominantly three-dimensional and structural—which does not mean, however, that it is merely a refinement of morphology. It must at the same time inquire into genesis and function.

Quoted in Gunther S. Stent, 'That was the Molecular Biology that was', *Science*, 1968, 160, 390.

3 We are at the dawn of a new era, the era of 'molecular biology' as I like to call it, and there is an urgency about the need for more intensive application of physics and chemistry, and specially of structure analysis, that is still not sufficiently appreciated.

'On the Structure of Biological Fibres and the Problem of Muscle', *Proceedings of the Royal Society of London*, 1947, 134, 326.

William Thomas Astbury 1898–1961 and Florence O. Bell

British x-ray crystallographers

See ***Bell, Florence O.***

1 Knowing what we know from X-ray and related studies of the fibrous proteins, how they are built from long polypeptide chains with linear patterns drawn to a grand scale, how these chains can contract and take up different configurations by intramolecular folding, how the chain-groups are penetrated by, and their sidechains react with, smaller co-operating molecules, and finally how they can combine so readily with nucleic acid molecules and still maintain the fibrous configuration, it is but natural to assume, as a first working hypothesis at least, that they form the long scroll on which is written the pattern of life. No other molecules satisfy so many requirements.

'Some Recent Developments in the X-Ray Study of Proteins and Related Structures', *Cold Spring Harbor Symposia on Quantitative Biology*, 1938, 6, 114.

Francis William Aston 1877–1945

British chemist and physicist

2 Should the research worker of the future discover some means of releasing this [atomic] energy in a form which could be employed, the human race will have at its command powers beyond the dream of scientific fiction, but the remotest possibility must always be considered that the energy once liberated will be completely uncontrollable and by its intense violence detonate all neighbouring substances. In this event, the whole of the hydrogen on earth might be transformed at once and the success of the experiment published at large to the universe as a new star.

'Mass Spectra and Isotopes', Nobel Lecture, 12 December 1922. In *Nobel Lectures, Chemistry, 1922–1941* (1966), 20.

Peter William Atkins 1940–

British theoretical chemist

3 A great deal of the universe does not need any explanation. Elephants, for instance. Once molecules have learnt to compete and to create other molecules in their own image, elephants, and things resembling elephants, will in due course be found roaming around the countryside . . . Some of the things resembling elephants will be men.

The Creation (1981), 3.

Wallace W. Atwood 1872–1949

American geologist and geographer

4 There is ample evidence of suspicion, due to ignorance, directing the thoughts of American people when they consider the actions of peoples in other lands.

'Research and Educational Work in Geography', *The Journal of Geography*, 1928, 27, 269.

John Aubrey 1626–97

English antiquarian

See ***Boyle, Robert*** *83:3;* ***Harvey, William*** *267:5, 267:6, 267:7;* ***Hobbes, Thomas*** *287:2, 287:3;* ***Hooke, Robert*** *297:1, 297:2*

Wystan Hugh Auden 1907–73

British-born American poet

5 And make us as Newton was, who in his garden watching
The apple falling towards England, became aware
Between himself and her of an eternal tie.

'Prologue' in *Look Stranger!* (1936), 11.

6 To the man-in-the-street who, I'm sorry to say,

Is a keen observer of life,
The word *intellectual* suggests right away
A man who's untrue to his wife.

'Note on intellectuals', in *Collected Shorter Poems 1927–1957* (1966), 190.

1 How happy the lot of the mathematician! He is judged solely by his peers, and the standard is so high that no colleague or rival can ever win a reputation he does not deserve.

The Dyer's Hand and Other Essays (1965), Prologue, 'Writing', 15.

2 The subject matter of the scientist is a crowd of natural events at all times; he presupposes that this crowd is not real but apparent, and seeks to discover the true place of events in the system of nature. The subject matter of the poet is a crowd of historical occasions of feeling recollected from the past; he presupposes that this crowd is real but should not be, and seeks to transform it into a community. Both science and art are primarily spiritual activities, whatever practical applications may be derived from their results. Disorder, lack of meaning, are spiritual not physical discomforts, order and sense spiritual not physical satisfactions.

The Dyer's Hand and Other Essays (1965), 66.

3 The true men of action in our time, those who transform the world, are not the politicians and statesmen, but the scientists. Unfortunately poetry cannot celebrate them because their deeds are concerned with things, not persons, and are, therefore, speechless. When I find myself in the company of scientists, I feel like a shabby curate who has strayed by mistake into a drawing room full of dukes.

The Dyer's Hand and Other Essays (1965), 81.

4 Of course, Behaviourism 'works'. So does torture. Give me a no-nonsense, down-to-earth behaviourist, a few drugs, and simple electrical appliances, and in six months I will have him reciting the Athanasian Creed in public.

A Certain World: A Commonplace Book (1971), 33.

Aurelius Augustinus Augustine (Saint Augustine of Hippo) 354–430

Early Christian theologian

5 *Nisi credideritis, non intelligitis.*

Unless you believe, you will not understand.

De Libero Arbitrio (On Free Choice of the Will) [386], Book I, chapter 2, section 4 (Augustine quoting from Isaiah 7:9).

6 Wherever it was, I did not come to know it through the bodily senses; the only things we know through the bodily senses are material objects, which we have found are not truly and simply one. Moreover, if we do not perceive *one* by the bodily sense, then we do not perceive *any number* by that sense, at least of those numbers that we grasp by understanding.

De Libero Arbitrio (On Free Choice of the Will) [386], trans. T. Williams (1993), 45.

7 The believer has the whole world of wealth (Prov. 17: 6 LXX) and 'possesses all things as if he had nothing' (2 Cor. 6: 10) by virtue of his attachment to you whom all things serve; yet he may know nothing about the circuits of the Great Bear. It is stupid to doubt that he is better than the person who measures the heaven and counts the stars and weighs the elements, but neglects you who have disposed everything 'by measure and number and weight' (Wisd. 11: 21).

Confessions [c.397], Book V, chapter 4 (7), trans. H. Chadwick (1991), 76.

8 Yes indeed: the human mind, so blind and languid, shamefully and dishonourably wishes to hide, and yet does not wish anything to be concealed from itself. But it is repaid on the principle that while the human mind lies open to the truth, truth remains hidden from it. Yet even thus, in its miserable condition, it prefers to find joy in true rather than false things. It will be happy if it comes to find joy only in that truth by which all things are true—without any distraction interfering.

Confessions [c.397], Book X, chapter 23 (34), trans. H. Chadwick (1991), 200.

1 To this I may add another form of temptation, manifold in its dangers . . . There exists in the soul . . . a cupidity which does not take delight in the carnal pleasure but in perceptions acquired through the flesh. It is a vain inquisitiveness dignified with the title of knowledge and science. As this is rooted in the appetite for knowing, and as among the senses the eyes play a leading role in acquiring knowledge, the divine word calls it 'the lust of the eyes' (I John, 2: 16) . . . To satisfy this diseased craving . . . people study the operations of nature, which lie beyond our grasp when there is no advantage in knowing and the investigators simply desire knowledge for its own sake. This motive is again at work if, using a perverted science for the same end, people try to achieve things by magical arts.

Confessions [c.397], Book X, chapter 35 (54–55), trans. H. Chadwick (1991), 210–212.

2 I do not study to understand the transit of the stars. My soul has never sought for responses from ghosts. I detest all sacrilegious rites.

Confessions [c.397], Book X, chapter 35 (56), trans. H. Chadwick (1991), 212.

3 The truth is rather in what God reveals than in what groping men surmise.

De Genesi ad Literam (On The Literal Interpretation of Genesis) [401/415], Book II, chapter 9, section 21, trans. J. H. Taylor (1982), Vol. 1, 59.

4 Hence, a devout Christian must avoid astrologers and all impious soothsayers, especially when they tell the truth, for fear of leading his soul into error by consorting with demons and entangling himself with the bonds of such association.

De Genesi ad Litteram (On The Literal Interpretation of Genesis) [401/415], Book II, chapter 17, section 37, trans. J. H. Taylor (1982), Vol. 1, 72–3.

5 But how is it that they [astrologers] have never been able to explain why, in the life of twins, in their actions, in their experiences, their professions, their accomplishments, their positions—in all the other circumstances of human life, and even in death itself, there is often found such a diversity that in those respects many strangers show more resemblance to them than they show to one another, even though the smallest possible interval separated their births and though they were conceived at the same moment, by a single act of intercourse.

De Civitate Dei (The City of God) [413–426], Book V, chapter 1, trans. H. Bettenson (1972), 180–181.

6 Thus there can be no doubt that the world was not created *in* time but *with* time. An event in time happens after one time and before another, after the past and before the future. But at the time of creation there could have been no past, because there was nothing created to provide the change and movement which is the condition of time.

De Civitate Dei (The City of God) [413–426], Book XI, chapter 6, trans. H. Bettenson (1972), 436.

7 Now let us look at man's happiness which is signified by the name of paradise. Men usually find delightful rest in meadows, and the light rises for our corporeal senses from the east and then climbs the heaven which is a body higher and more excellent than our body. Hence, these words also express figuratively the spiritual delights the happy life contains, and the garden is planted in the east. We should understand that our spiritual joys are signified by every tree beautiful to the gaze of the intelligence and good for the food which is not corrupted and by which the happy souls are fed. For the Lord also says, 'Work for the food which is not corrupted.' Such is every idea which is the food of the soul. He set the light of wisdom to the east in Eden, that is, immortal and intelligible delights. For this word is said to signify delights, or pleasure, or a feast if it is translated from Hebrew to Latin. It is set down in this way without translation so that it seems to signify a particular place and to make the expression more figurative. We take every tree that the earth produced as every spiritual joy; for such joys rise above the earth and are not caught and overwhelmed by the tangles of earthly desires. The tree of Life planted in the middle of paradise signifies the wisdom by which the soul should understand that it is ordered in a certain middle range of things.

'The Meaning of the Paradise of Delights', in *Saint Augustine on Genesis: Two Books on Genesis against the Manichees and on the Literal Interpretation of Genesis: An Unfinished Book*, trans. R. J. Teske (1991), 107–8.

Antoninus Marcus Aurelius

121–180

Roman emperor and philosopher

1 Either an ordered Universe or a medley heaped together mechanically but still an order; or can order subsist in you and disorder in the Whole! And that, too, when all things are so distinguished and yet intermingled and sympathetic.

A. S. L. Farquharson (ed.), *The Meditations of the Emperor Marcus Antoninus Aurelius* (1944), Vol. 1, Book IV, 63.

2 What does not benefit the hive is no benefit to the bee.

A. S. L. Farquharson (ed.), *The Meditations of the Emperor Marcus Antoninus Aurelius* (1944), Vol. 1, Book VI, 119.

Oswald T. Avery 1877–1955, Colin MacLeod 1909–72 and Maclyn McCarty 1911–2005

Canadian-born American biochemist and biologist; Canadian physician; American physician and bacteriologist

3 Biologists have long attempted by chemical means to induce in higher organisms predictable and specific changes which thereafter could be transmitted in series as hereditary characters. Among microorganisms the most striking example of inheritable and specific alterations in cell structure and function that can be experimentally induced and are reproducible under well defined and adequately controlled conditions is the transformation of specific types of Pneumococcus.

'Studies on the Chemical Nature of the Substance Inducing Transformation of Pneumococcal Types', *Journal of Experimental Medicine*, 1944, 79, 137.

4 The inducing substance, on the basis of its chemical and physical properties, appears to be a highly polymerized and viscous form of sodium desoxyribonucleate. On the other hand, the Type III capsular substance, the synthesis of which is evoked by this transforming agent, consists chiefly of a non-nitrogenous polysaccharide constituted of glucose-glucuronic acid units linked in glycosidic union. The presence of the newly formed capsule containing this type-specific polysaccharide confers on the transformed cells all the distinguishing characteristics of Pneumococcus Type III. Thus, it is evident that the inducing substance and the substance produced in turn are chemically distinct and biologically specific in their action and that both are requisite in determining the type of specificity of the cell of which they form a part. The experimental data presented in this paper strongly suggest that nucleic acids, at least those of the desoxyribose type, possess different specificities as evidenced by the selective action of the transforming principle.

'Studies in the Chemical Nature of the Substance Inducing Transformation of Pneumococcal Types', *Journal of Experimental Medicine* 1944, 79, 152.

1 If the results of the present study on the chemical nature of the transforming principle are confirmed, then nucleic acids must be regarded as possessing biological specificity the chemical basis of which is as yet undetermined.

'Studies in the Chemical Nature of the Substance Inducing Transformation of Pneumococcal Types', *Journal of Experimental Medicine* 1944, 79, 155.

Avicenna 980–1037

Persian physician and philosopher

2 I devoted myself to studying the texts—the original and commentaries—in the natural sciences and metaphysics, and the gates of knowledge began opening for me. Next I sought to know medicine, and so read the books written on it. Medicine is not one of the difficult sciences, and therefore, I excelled in it in a very short time, to the point that distinguished physicians began to read the science of medicine under me. I cared for the sick and there opened to me some of the doors of medical treatment that are indescribable and can be learned only from practice. In addition I devoted myself to jurisprudence and used to engage in legal disputations, at that time being sixteen years old.

W. E. Gohlman, *The Life of Ibn Sina: A Critical Edition and Annotated Translation* (1974), 25–7.

3 At night I would return home, set out a lamp before me, and devote myself to reading and writing. Whenever sleep overcame me or I became conscious of weakening, I would turn aside to drink a cup of wine, so that my strength would return to me. Then I would return to reading. And whenever sleep seized me I would see those very problems in my dream; and many questions became clear to me in my sleep. I continued in this until all of the sciences were deeply rooted within me and I understood them as is humanly possible. Everything which I knew at the time is just as I know it now; I have not added anything to it to this day. Thus I mastered the logical, natural, and mathematical sciences, and I had now reached the science.

W. E. Gohlman, *The Life of Ibn Sina: A Critical Edition and Annotated Translation* (1974), 29–31.

4 Pure earth does not petrify, because the predominance of dryness over [i.e. in] the earth endows it not with coherence but rather with crumbliness. In general, stone is formed in two ways only (a) through the hardening of clay, and (b) by the congelation [of waters].

Congelatione et Conglutinatione Lapidium (1021–23), trans. E. J. Holmyard and D. C. Mandeville (1927), 18.

5 It is in the nature of water . . . to become transformed into earth through a predominating earthy virtue; . . . it is in the nature of earth to become transformed into water through a predominating aqueous virtue.

Congelatione et Conglutinatione Lapidum (1021–23), trans. E. J. Holmyard and D. C. Mandeville (1927), 20.

6 Mountains have been formed by one [or other] of the causes of the formation of stone, most probably from agglutinative clay which slowly dried and petrified during ages of which we have no record. It seems likely that this habitable world was in former days uninhabitable and, indeed, submerged beneath the ocean. Then, becoming exposed little by little, it petrified in the course of ages.

Congelatione et Conglutinatione Lapidium (1021–1023), trans. E. J. Holmyard and D. C. Mandeville (1927), 28.

7 Medicine is the science by which we learn the various states of the human body in health and when not in health, and the means by which health is likely to be lost and, when lost, is likely to be restored back to health. In other words, it is the art whereby health is conserved and the art whereby it is restored after being lost. While some divide medicine into a theoretical and a practical [applied] science, others may assume that it is only theoretical because they see it as a pure science. But, in truth, every science has both a theoretical and a practical side.

'The Definition of Medicine', in *The Canon of Medicine*, adapted by L. Bakhtiar (1999), 9.

1 The theory of medicine, therefore, presents what is useful in thought, but does not indicate how it is to be applied in practice—the mode of operation of these principles. The theory, when mastered, gives us a certain kind of knowledge. Thus we say, for example, there are three forms of fevers and nine constitutions. The practice of medicine is not the work which the physician carries out, but is that branch of medical knowledge which, when acquired, enables one to form an opinion upon which to base the proper plan of treatment.

'The Definition of Medicine', in *The Canon of Medicine*, adapted by L. Bakhtiar (1999), 10.

2 Medicine deals with the states of health and disease in the human body. It is a truism of philosophy that a complete knowledge of a thing can only be obtained by elucidating its causes and antecedents, provided, of course, such causes exist. In medicine it is, therefore, necessary that causes of both health and disease should be determined.

'Concerning the Subject-Matter of Medicine', in *The Canon of Medicine*, adapted by L. Bakhtiar (1999), 11.

3 But the fact is that when wine is taken in moderation, it gives rise to a large amount of breath, whose character is balanced, and whose luminosity is strong and brilliant. Hence wine disposes greatly to gladness, and the person is subject to quite trivial exciting agents. The breath now takes up the impression of agents belonging to the present time more easily than it does those which relate to the future; it responds to agents conducive to delight rather than those conducive to a sense of beauty.

'The External Causes of Delight and Sadness', in *The Canon of Medicine*, adapted by L. Bakhtiar (1999), 149–50.

4 Disease is an abnormal state of the body which primarily and independently produces a disturbance in the normal functions of the body. It may be an abnormality of temperament or form (structure). Symptom is a manifestation of some abnormal state in the body. It may be harmful as a colic pain or harmless as the flushing of cheeks in peripneumonia.

'A Discussion of the Cause of Disease and Symptoms', in *The Canon of Medicine*, adapted by L. Bakhtiar (1999), 171.

5 Pain is a sensation produced by something contrary to the course of nature and this sensation is set up by one of two circumstances: either a very sudden change of the temperament (or the bad effect of a contrary temperament) or a solution of continuity.

'A General Discussion of the Causes of Pain', in *The Canon of Medicine*, adapted by L. Bakhtiar (1999), 246.

6 Signs and symptoms indicate the present, past and future states of the three states of the body (health, illness, neutrality). According to Galen, knowledge of the present state is of advantage only to the patient as it helps him to follow the proper course of management. Knowledge of the past state is useful only to the physician inasmuch as its disclosure by him to the patient brings him a greater respect for his professional advice. Knowledge of the future state is useful to both. It gives an opportunity to the patient to be forewarned to adopt necessary preventative measures and it enhances the reputation of the physician by correctly forecasting the future developments.

'The Signs and Symptoms (Diagnosis): General Remarks,' in *The Canon of Medicine*, adapted by L. Bakhtiar (1999), 259.

1 When you do not know the nature of the malady, leave it to nature; do not strive to hasten matters. For either nature will bring about the cure or it will itself reveal clearly what the malady really is.

'General Therapeutics', in *The Canon of Medicine*, adapted by L. Bakhtiar (1999), 468.

Amedeo Avogadro 1776–1856

Italian physicist and chemist

2 It must . . . be admitted that very simple relations . . . exist between the volumes of gaseous substances and the numbers of simple or compound molecules which form them. The first hypothesis to present itself in this connection, and apparently even the only admissible one, is the supposition that the number of integral molecules in any gases is always the same for equal volumes, or always proportional to the volumes. Indeed, if we were to suppose that the number of molecules contained in a given volume were different for different gases, it would scarcely be possible to conceive that the law regulating the distance of molecules could give in all cases relations so simple as those which the facts just detailed compel us to acknowledge between the volume and the number of molecules.

'Essay on a Manner of Determining the Relative Masses of the Elementary Molecules of Bodies, and the Proportions in which they enter into these Compounds', *Journal de Physique*, 1811, 73, 58–76. In *Foundations of the Molecular Theory: Alembic Club Reprints, Number 4* (1923), 28–9.

3 We suppose . . . that the constituent molecules of any simple gas whatever (i.e., the molecules which are at such a distance from each other that they cannot exercise their mutual action) are not formed of a solitary elementary molecule, but are made up of a certain number of these molecules united by attraction to form a single one.

'Essay on a Manner of Determining the Relative Masses of the Elementary Molecules of Bodies, and the Proportions in which they enter into these Compounds', *Journal de Physique*, 1811, 73, 58–76. In *Foundations of the Molecular Theory: Alembic Club Reprints, Number 4* (1923), 31.

Alfred Jules Ayer 1910–89

British philosopher

4 The traditional disputes of philosophers are, for the most part, as unwarranted as they are unfruitful. The surest way to end them is to establish beyond question what should be the purpose and method of a philosophical enquiry. And this is by no means so difficult a task as the history of philosophy would lead one to suppose. For if there are any questions which science leaves it to philosophy to answer, a straightforward process of elimination must lead to their discovery.

Language, Truth and Logic (1960), 33.

5 It is time, therefore, to abandon the superstition that natural science cannot be regarded as logically respectable until philosophers have solved the problem of induction. The problem of induction is, roughly speaking, the problem of finding a way to prove that certain empirical generalizations which are derived from past experience will hold good also in the future.

Language, Truth and Logic (1960), 49.

6 The principles of logic and mathematics are true universally simply because we never allow them to be anything else. And the reason for this is that we cannot abandon them without contradicting ourselves, without sinning against the rules which govern the use of language, and so making our utterances self- stultifying. In other words, the truths of logic and mathematics are analytic propositions or tautologies.

Language, Truth and Logic (1960), 77.

Charles Babbage 1792–1871

British mathematician

7 That a country, eminently distinguished for its mechanical and manufacturing ingenuity, should be indifferent to the progress of inquiries which form the highest departments of that knowledge on whose more elementary truths its wealth and rank depend, is a fact which is well deserving the attention of those who shall inquire into the causes that influence the progress of nations.

Reflections on the Decline of Science in England and on Some of its Causes (1830), 1.

1 A young man passes from our public schools to the universities, ignorant almost of the elements of every branch of useful knowledge.

Reflections on the Decline of Science in England and on Some of its Causes (1830), 3.

2 For one person who is blessed with the power of invention, many will always be found who have the capacity of applying principles.

Reflections on the Decline of Science in England and on Some of its Causes (1830), 18.

3 He will also find that the high and independent spirit, which usually dwells in the breast of those who are deeply versed in scientific pursuits, is ill adapted for administrative appointments; and that even if successful, he must hear many things he disapproves, and raise no voice against them.

Reflections on the Decline of Science in England and on Some of its Causes, (1830), 38.

4 The errors which arise from the absence of facts are far more numerous and more durable than those which result from unsound reasoning respecting true data.

On the Economy of Machinery and Manufactures (1832), 119.

5 Precedents are treated by powerful minds as fetters with which to bind down the weak, as reasons with which to mistify the moderately informed, and as reeds which they themselves fearlessly break through whenever new combinations and difficult emergencies demand their highest efforts.

A Word to the Wise (1833), 3–6. Quoted in Anthony Hyman (ed.), *Science and Reform: Selected Works of Charles Babbage* (1989), 202.

6 The gradual advance of Geology, during the last twenty years, to the dignity of a science, has arisen from the laborious and extensive collection of facts, and from the enlightened spirit in which the inductions founded on those facts have been deduced and discussed. To those who are unacquainted with this science, or indeed to any person not deeply versed in the history of this and kindred subjects, it is impossible to convey a just impression of the nature of that evidence by which a multitude of its conclusions are supported:—evidence in many cases so irresistible, that the records of the past ages, to which it refers, are traced in language more imperishable than that of the historian of any human transactions; the relics of those beings, entombed in the strata which myriads of centuries have heaped upon their graves, giving a present evidence of their past existence, with which no human testimony can compete.

The Ninth Bridgewater Treatise (1838), 47–8.

7 'Every moment dies a man,/ Every moment one is born':

I need hardly point out to you that this calculation would tend to keep the sum total of the world's population in a state of perpetual equipoise whereas it is a well-known fact that the said sum total is constantly on the increase. I would therefore take the liberty of suggesting that in the next edition of your excellent poem the erroneous calculation to which I refer should be corrected as follows:

'Every moment dies a man / And one and a sixteenth is born.' I may add that the exact figures are 1.167, but something must, of course, be conceded to the laws of metre.

Unpublished letter to Tennyson in response to his *Vision of Sin* (1842). Quoted in Philip and Emily Morrison, *Charles Babbage and his Calculating Engines: Selected Writings by Charles Babbage and Others* (1961), xxiii.

8 Science in England is not a profession: its cultivators are scarcely recognised even as a class. Our language itself contains no *single* term by which their occupation can be expressed. We borrow a foreign word [*Savant*] from another country whose high ambition it is to advance science, and whose deeper policy, in accord with more generous feelings, gives to the intellectual labourer reward and honour, in return for services which crown the nation with imperishable renown, and ultimately enrich the human race.

The Exposition of 1851: Or the Views of Industry, Science and Government of England (1851), 171.

1 Remember that accumulated knowledge, like accumulated capital, increases at compound interest: but it differs from the accumulation of capital in this; that the increase of knowledge produces a more rapid rate of progress, whilst the accumulation of capital leads to a lower rate of interest. Capital thus checks its own accumulation: knowledge thus accelerates its own advance. Each generation, therefore, to deserve comparison with its predecessor, is bound to add much more largely to the common stock than that which it immediately succeeds.

The Exposition of 1851: Or the Views of Industry, Science and Government of England (1851), 192–3.

2 I have no desire to write my own biography, as long as I have strength and means to do better work.

Passages From the Life of a Philosopher (1864), vii.

3 What is there in a name? It is merely an empty basket, until you put something into it.

Passages From the Life of a Philosopher (1864), 1.

4 The whole of the developments and operations of analysis are now capable of being executed by machinery . . . As soon as an Analytical Engine exists, it will necessarily guide the future course of science.

Passages From the Life of a Philosopher (1864), 136–7.

5 Whenever a man can get hold of numbers, they are invaluable: if correct, they assist in informing his own mind, but they are still more useful in deluding the minds of others. Numbers are the masters of the weak, but the slaves of the strong.

Passages From the Life of a Philosopher (1864), 410.

6 I look upon it [mechanical notation] as one of the most important additions I have made to human knowledge. It has placed the construction of machinery in the rank of a demonstrative science. The day will arrive when no school of mechanical drawing will be thought complete without teaching it.

Passages From the Life of a Philosopher (1864), 452.

Gaston Bachelard 1884–1962

French philosopher and historian of science

7 Any work of science, no matter what its point of departure, cannot become fully convincing until it crosses the boundary between the theoretical and the experimental: *Experimentation must give way to argument, and argument must have recourse to experimentation.*

The New Scientific Spirit (1934), trans. A. Goldhammer (1984), 3–4.

Francis Bacon 1561–1626

English essayist and philosopher of science

See ***Hobbes, Thomas*** *287:2*

8 *Scientia potestas est.*

For also knowledge itself is power.

'Meditationes Sacrae' (1597), in James Spedding, Robert Ellis and Douglas Heath (eds.), *The Works of Francis Bacon* (1887–1901), Vol. 7, 253.

9 Never any knowledge was delivered in the same order it was invented.

'Of the Interpretation of Nature' (*c.*1603) in James Spedding, Robert Ellis and Douglas

Heath (eds.), *The Works of Francis Bacon* (1887–1901), Vol. 3, 248.

1 And yet surely to alchemy this right is due, that it may be compared to the husbandman whereof Æsop makes the fable, that when he died he told his sons that he had left unto them gold buried under the ground in his vineyard: and they digged over the ground, gold they found none, but by reason of their stirring and digging the mould about the roots of their vines, they had a great vintage the year following: so assuredly the search and stir to make gold hath brought to light a great number of good and fruitful inventions and experiments, as well for the disclosing of nature as for the use of man's life.

The Advancement of Learning (1605) in James Spedding, Robert Ellis and Douglas Heath (eds.), *The Works of Francis Bacon* (1887–1901), Vol. 3, 289.

2 If a man will begin with certainties, he shall end in doubts; but if he will be content to begin with doubts, he shall end in certainties.

The Advancement of Learning (1605) in James Spedding, Robert Ellis and Douglas Heath (eds.), *The Works of Francis Bacon* (1887–1901), Vol. 3, 293.

3 They are ill discoverers that think there is no land, when they can see nothing but sea.

The Advancement of Learning (1605) in James Spedding, Robert Ellis and Douglas Heath (eds.), *The Works of Francis Bacon* (1887–1901), Vol. 3, 355.

4 We come therefore now to that knowledge whereunto the ancient oracle directeth us, which is *the knowledge of ourselves*; which deserveth the more accurate handling, by how much it toucheth us more nearly. This knowledge, as it is the end and term of natural philosophy in the intention of man, so notwithstanding it is but a portion of natural philosophy in the continent of nature. And generally let this be a rule, that all partitions of knowledges be accepted rather for lines and veins, than for sections and separations; and that the continuance and entireness of knowledge be preserved. For the contrary hereof hath made particular sciences to become barren, shallow, and erroneous; while they have not been nourished and maintained from the common fountain. So we see Cicero the orator complained of Socrates and his school, that he was the first that separated philosophy and rhetoric; whereupon rhetoric became an empty and verbal art. So we may see that the opinion of Copernicus touching the rotation of the earth, which astronomy itself cannot correct because it is not repugnant to any of the phenomena, yet natural philosophy may correct. So we see also that the science of medicine, if it be destituted and forsaken by natural philosophy, it is not much better than an empirical practice. With this reservation therefore we proceed to Human Philosophy or Humanity, which hath two parts: the one considereth man segregate, or distributively; the other congregate, or in society. So as Human Philosophy is either Simple and Particular, or Conjugate and Civil. Humanity Particular consisteth of the same parts whereof man consisteth; that is, of knowledges that respect the Body, and of knowledges that respect the Mind. But before we distribute so far, it is good to constitute. For I do take the consideration in general and at large of Human Nature to be fit to be emancipate and made a knowledge by itself; not so much in regard of those delightful and elegant discourses which have been made of the dignity of man, of his miseries, of his state and life, and the like *adjuncts of his common and undivided nature*; but chiefly in regard of the knowledge concerning the *sympathies and concordances between the mind and body*, which, being mixed, cannot be properly assigned to the sciences of either.

The Advancement of Learning (1605) in James Spedding, Robert Ellis and Douglas Heath (eds.), *The Works of Francis Bacon* (1887–1901), Vol. 3, 366–7.

1 Medicine is a science which hath been (as we have said) more professed than laboured, and yet more laboured than advanced: the labour having been, in my judgment, rather in circle than in progression. For I find much iteration, but small addition. It considereth *causes of diseases*, with the *occasions or impulsions* ; the *diseases themselves*, with the *accidents* ; and the *cures*, with the *preservation*.

The Advancement of Learning (1605) in James Spedding, Robert Ellis and Douglas Heath (eds.), *The Works of Francis Bacon* (1887–1901), Vol. 3, 373.

2 Men are rather beholden . . . generally to chance or anything else, than to logic, for the invention of arts and sciences.

The Advancement of Learning (1605) in James Spedding, Robert Ellis and Douglas Heath (eds.), *The Works of Francis Bacon* (1887–1901), Vol. 3, 386.

3 Nevertheless if any skillful Servant of Nature shall bring force to bear on matter, and shall vex it and drive it to extremities as if with the purpose of reducing it to nothing, then will matter (since annihilation or true destruction is not possible except by the omnipotence of God) finding itself in these straits, turn and transform itself into strange shapes, passing from one change to another till it has gone through the whole circle and finished the period.

De Sapientia Veterum (1609) XIII 'Proteus; or matter' in James Spedding, Robert Ellis and Douglas Heath (eds.), *The Works of Francis Bacon* (1887–1901), Vol. 6, 726.

4 Books must follow sciences, and not sciences books.

A Proposition Touching the Compiling and Amendment of the Laws of England (written 1616).

5 Man, being the servant and interpreter of Nature, can do and understand so much and so much only as he has observed in fact or thought of the course of nature; beyond this he neither knows anything nor can do anything.

The New Organon (1620) in James Spedding, Robert Ellis and Douglas Heath (eds.), *The Works of Francis Bacon* (1887–1901), Vol. 4, 47.

6 The Syllogism consists of propositions, propositions consist of words, words are symbols of notions. Therefore if the notions themselves (which is the root of the matter) are confused and over-hastily abstracted from the facts, there can be no firmness in the superstructure. Our only hope therefore lies in a true induction.

The New Organon (1620) in James Spedding, Robert Ellis and Douglas Heath (eds.), *The Works of Francis Bacon* (1887–1901), Vol. 4, 49.

7 It is idle to expect any great advancement in science from the superinducing and engrafting of new things upon old. We must begin anew from the very foundations, unless we would revolve for ever in a circle with mean and contemptible progress.

The New Organon (1620) in James Spedding, Robert Ellis and Douglas Heath (eds.), *The Works of Francis Bacon* (1887–1901), Vol. 4, 52.

8 There are four classes of Idols which beset men's minds. To these for distinction's sake I have assigned names,—calling the first class *Idols of the Tribe*; the second, *Idols of the Cave*; the third, *Idols of the Market Place*; the fourth, *Idols of the Theatre* . . .

The Idols of the Tribe have their foundation in human nature itself, and in the tribe or race of men. For it is a false assertion that the sense of man is the measure of things. On the contrary, all perceptions as well of the sense as of the mind are according to the measure of the individual and not according to the measure of the universe. And the human understanding is like a false mirror, which, receiving rays irregularly, distorts and discolours the nature of things by mingling its own nature with it.

The Idols of the Cave are the idols of the individual man. For every one (besides the errors common to human nature in general) has a cave or den of his own, which refracts and discolours the light of nature; owing either to his own proper and peculiar nature; or to his education and conversation with others; or to the reading of books, and the authority of those whom he esteems and admires; or to the differences of impressions, accordingly as they take place in a mind preoccupied and predisposed or in a mind indifferent and settled; or the like.

There are also Idols formed by the intercourse and association of men with each other, which I call Idols of the Market-place, on account of the commerce and consort of men there. For it is by discourse that men associate; and words are imposed according to the apprehension of the vulgar, and therefore the ill and unfit choice of words wonderfully obstructs the understanding. Nor do the definitions or explanations where with in some things learned men are wont to guard and defend themselves, by any means set the matter right. But words plainly force and overrule the understanding, and throw all into confusion, and lead men away into numberless empty controversies and idle fancies.

Lastly, there are Idols which have immigrated into men's minds from the various dogmas of philosophies, and also from wrong laws of demonstration. These I call Idols of the Theatre; because in my judgment all the received systems are but so many stage-plays, representing worlds of their own creation after an unreal and scenic fashion.

The New Organon (1620) in James Spedding, Robert Ellis and Douglas Heath (eds.), *The Works of Francis Bacon* (1887–1901), Vol. 4, 53–55.

1 Those who have handled sciences have been either men of experiment or men of dogmas. The men of experiment are like the ant; they only collect and use; the reasoners resemble spiders, who make cobwebs out of their own substance. But the bee takes a middle course; it gathers its material from the flowers of the garden and of the field, but transforms and digests it by a power of its own. Not unlike this is the true business of philosophy; for it neither relies solely or chiefly on the powers of the mind, nor does it take the matter which it gathers from natural history and mechanical experiments and lay it up in the memory whole, as it finds it; but lays it up in the understanding altered and digested.

The New Organon (1620) in James Spedding, Robert Ellis and Douglas Heath (eds.), *The Works of Francis Bacon* (1887–1901), Vol. 4, 92–3.

2 The understanding must not . . . be allowed to jump and fly from particulars to axioms remote and of almost the highest generality (such as the first principles, as they are called, of arts and things), and taking stand upon them as truths that cannot be shaken, proceed to prove and frame the middle axioms by reference to them; which has been the practice hitherto, the understanding being not only carried that way by a natural impulse, but also by the use of syllogistic demonstration trained and inured to it. But then, and then only, may we hope well of the sciences when in a just scale of ascent, and by successive steps not interrupted or broken, we rise from particulars to lesser axioms; and then to middle axioms, one above the other; and last of all to the most general. For the lowest axioms differ but slightly from bare experience, while the highest and most general (which we now have) are notional and abstract and without solidity. But the middle are the true and solid and living axioms, on which depend the affairs and fortunes of men; and above them again, last of all, those which are indeed the most general; such, I mean, as are not abstract, but of which those intermediate axioms are really limitations.

The New Organon (1620) in James Spedding, Robert Ellis and Douglas Heath (eds.), *The*

Works of Francis Bacon (1887–1901), Vol. 4, 97.

1 It is well to observe the force and virtue and consequence of discoveries, and these are to be seen nowhere more conspicuously than in those three which were unknown to the ancients, and of which the origins, although recent, are obscure and inglorious; namely, printing, gunpowder, and the magnet. For these three have changed the whole face and state of things throughout the world; the first in literature, the second in warfare, the third in navigation; whence have followed innumerable changes, insomuch that no empire, no sect, no star seems to have exerted greater power and influence in human affairs than these mechanical discoveries. . . . But if a man endeavour to establish and extend the power and dominion of the human race itself over the universe, his ambition (if ambition it can be called) is without doubt both a more wholesome thing and a more noble than the other two. Now the empire of man over things depends wholly on the arts and sciences. For we cannot command nature except by obeying her.

The New Organon (1620) in James Spedding, Robert Ellis and Douglas Heath (eds.), *The Works of Francis Bacon* (1887–1901), Vol. 4, 114.

2 And yet since truth will sooner come out of error than from confusion.

The New Organon (1620) in James Spedding, Robert Ellis and Douglas Heath (eds.), *The Works of Francis Bacon* (1887–1901), Vol. 4, 149.

3 Heat is a motion; expansive, restrained, and acting in its strife upon the smaller particles of bodies. But the expansion is thus modified; while it expands all ways, it has at the same time an inclination upward. And the struggle in the particles is modified also; it is not sluggish, but hurried and with violence.

The New Organon (1620) in James Spedding, Robert Ellis and Douglas Heath (eds.), *The Works of Francis Bacon* (1887–1901), Vol. 4, 154–5.

4 Men fear death as children fear to go in the dark; and as that natural fear in children is increased with tales, so is the other.

'Of Death' (1625) in James Spedding, Robert Ellis and Douglas Heath (eds.), *The Works of Francis Bacon* (1887–1901), Vol. 6, 379.

5 Above all things, good policy is to be used that the treasure and moneys in a state be not gathered into few hands. For otherwise a state may have a great stock, and yet starve. And money is like muck, not good except it be spread.

'Of Seditions and Troubles' (1625) in James Spedding, Robert Ellis and Douglas Heath (eds.), *The Works of Francis Bacon* (1887–1901), Vol. 6, 410.

6 I had rather believe all the Fables in the *Legend*, and the *Talmud*, and the *Alcoran*, then that this universall Frame, is without a Minde. And therefore, God never wrought Miracle, to convince *Atheisme*, because his Ordinary Works Convince it. It is true, that a little Philosophy inclineth Mans Minde to *Atheisme*; But depth in Philosophy, bringeth Mens Mindes about to *Religion*.

'Of Atheisme' (1625) in James Spedding, Robert Ellis and Douglas Heath (eds.), *The Works of Francis Bacon* (1887–1901), Vol. 6, 413.

7 Time is the greatest innovator.

'Of Innovations' (1625) in James Spedding, Robert Ellis and Douglas Heath (eds.), *The Works of Francis Bacon* (1887–1901), Vol. 6, 433.

8 Nature is often hidden, sometimes overcome, seldom extinguished.

'Of Nature in Men' (1625) in James Spedding, Robert Ellis and Douglas Heath (eds.), *The Works of Francis Bacon* (1887–1901), Vol. 6, 469.

9 Some books are to be tasted, others to be swallowed, and some few to be chewed and digested; that is, some books are to be read only in parts; other to be read, but not curiously; and some few to be read wholly, and with diligence and attention. Some books also may be read by deputy, and extracts made of them by others; but that would be only in the less important arguments, and the meaner sort of books; else distilled books are like common distilled waters, flashy things. Reading maketh a full man; conference a ready man; and writing an exact man. And therefore, if a man write little, he had need have a great memory; if he confer little, he had need have a present wit: and if he read little, he had need have much cunning, to seem to know that he doth not. Histories make men wise; poets witty; the mathematics subtile; natural philosophy deep; moral grave; logic and rhetoric able to contend. *Abeunt studia in mores*. [The studies pass into the manners.]

'Of Studies' (1625) in James Spedding, Robert Ellis and Douglas Heath (eds.), *The Works of Francis Bacon* (1887–1901), Vol. 6, 498.

1 I would by all means have men beware, lest Æsop's pretty fable of the fly that sate [sic] on the pole of a chariot at the Olympic races and said, 'What a dust do I raise,' be verified in them. For so it is that some small observation, and that disturbed sometimes by the instrument, sometimes by the eye, sometimes by the calculation, and which may be owing to some real change in the heaven, raises new heavens and new spheres and circles.

'Of Vain Glory' (1625) in James Spedding, Robert Ellis and Douglas Heath (eds.), *The Works of Francis Bacon* (1887–1901), Vol. 6, 503.

2 Aristippus said; *That those that studied particular sciences, and neglected philosophy, were like Penelope's wooers, that made love to the waiting women.*

Apophthegms: New and Old (1625), no. 189 in James Spedding, Robert Ellis and Douglas Heath (eds.), *The Works of Francis Bacon* (1887–1901), Vol. 7, 151.

3 The End of our Foundation is the knowledge of Causes; and secret motions of things; and the enlarging of the bounds of Human Empire, to the effecting of all things possible.

'New Atlantis' (1626) in James Spedding, Robert Ellis and Douglas Heath (eds.), *The Works of Francis Bacon* (1887–1901), Vol. 3, 156.

4 The world hath been much abused by the opinion of making gold; the work itself I judge to be possible; but the means (hitherto propounded) to effect it are, in the practice, full of error and imposture.

Sylva Sylvarum (1627), Century IV 'Experiment Solitary Touching the Making of Gold' in James Spedding, Robert Ellis and Douglas Heath (eds.), *The Works of Francis Bacon* (1887–1901), Vol. 2, 448.

Roger Bacon c.1219–92

English natural philosopher

5 *Omnes scientiae sunt connexae et fovent auxiliis sicut partes ejusdem totius, quarum quaelibet opus suum peragit non propter se sed pro aliis.*

All sciences are connected; they lend each other material aid as parts of one great whole, each doing its own work, not for itself alone, but for the other parts; as the eye guides the body and the foot sustains it and leads it from place to place.

Opus Tertium [1266–1268], chapter 4, Latin text quoted in J. B. Bury, *The Idea of Progress* (1920), 355 (footnote to page 25). In J. S. Brewer (ed.), *Fr. Rogeri Bacon Opera . . . inedita* (1859), 18.

6 There are four great sciences, without which the other sciences cannot be known nor a knowledge of things secured . . . Of these sciences the gate and key is mathematics . . . He who is ignorant of this [mathematics] cannot know the other sciences nor the affairs of this world.

Opus Majus [1266–1268], Part IV, distinction 1, chapter 1, trans. R. B. Burke, *The Opus Majus of Roger Bacon* (1928), Vol. 1, 116.

7 As regards authority I so proceed. Boetius says in the second prologue to his Arithmetic, 'If an inquirer lacks the four parts of mathematics, he has very little ability to discover truth.' And again, 'Without this theory no one can have a correct insight into truth.' And he says also, 'I warn the man who spurns these paths of knowledge that he cannot philosophize correctly.' And

Again, 'It is clear that whosoever passes these by, has lost the knowledge of all learning.'

Opus Majus [1266–1268], Part IV, distinction 1, chapter 2, trans. R. B. Burke, *The Opus Majus of Roger Bacon* (1928), Vol. 1, 117.

1 But concerning vision alone is a separate science formed among philosophers, namely, optics, and not concerning any other sense . . . It is possible that some other science may be more useful, but no other science has so much sweetness and beauty of utility. Therefore it is the flower of the whole of philosophy and through it, and not without it, can the other sciences be known.

Opus Majus [1266–1268], Part V, distinction 1, chapter 1, trans. R. B. Burke, *The Opus Majus of Roger Bacon* (1928), Vol. 2, 420.

2 But we must here state that we should not see anything if there were a vacuum. But this would not be due to some nature hindering species, and resisting it, but because of the lack of a nature suitable for the multiplication of species; for species is a natural thing, and therefore needs a natural medium; but in a vacuum nature does not exist.

Opus Majus [1266–1268], Part V, distinction 9, chapter 2, trans. R. B. Burke, *The Opus Majus of Roger Bacon* (1928), Vol. 2, 485.

3 For there are two modes of acquiring knowledge, namely, by reasoning and experience. Reasoning draws a conclusion and makes us grant the conclusion, but does not make the conclusion certain, nor does it remove doubt so that the mind may rest on the intuition of truth, unless the mind discovers it by the path of experience; since many have the arguments relating to what can be known, but because they lack experience they neglect the arguments, and neither avoid what is harmful nor follow what is good. For if a man who has never seen fire should prove by adequate reasoning that fire burns and injures things and destroys them, his mind would not be satisfied thereby, nor would he avoid fire, until he placed his hand or some combustible substance in the fire, so that he might prove by experience that which reasoning taught. But when he has had actual experience of combustion his mind is made certain and rests in the full light of truth. Therefore reasoning does not suffice, but experience does.

Opus Majus [1266–1268], Part VI, chapter 1, trans. R. B. Burke, *The Opus Majus of Roger Bacon* (1928), Vol. 2, 583.

4 But there is another alchemy, operative and practical, which teaches how to make the noble metals and colours and many other things better and more abundantly by art than they are made in nature. And science of this kind is greater than all those preceding because it produces greater utilities. For not only can it yield wealth and very many other things for the public welfare, but it also teaches how to discover such things as are capable of prolonging human life for much longer periods than can be accomplished by nature . . . Therefore this science has special utilities of that nature, while nevertheless it confirms theoretical alchemy through its works.

Opus Tertium [1266–1268], chapter 12, quoted in A. C. Crombie, *Augustine to Galileo* (1959), Vol. 1, 69.

5 *Sed tamen salis petrae. VI. Part V. NOV. CORVLI. ET V. sulphuris, et sic facies toniitrum et coruscationem: sic facies artificium.*

But, however, of saltpetre take six parts, five of young willow (charcoal), and five of sulphur, and so you will make thunder and lightning, and so you will turn the trick.

Bacon's recipe for Gunpowder, partly expressed as an anagram in the original Latin.

Roger Bacon's Letter Concerning the Marvelous Power of Art and of Nature and Concerning the Nullity of Magic, trans. T. L. Davis (1922), 48.

Karl Ernst von Baer 1792–1876

German-Russian embryologist and zoologist

6 Let us only imagine that birds had studied their own development and that it was they in turn who investigated the structure of the adult mammal and of man. Wouldn't their physiological textbooks teach the following? 'Those four and two-legged animals bear many resemblances to embryos, for their cranial bones are separated, and they have no beak, just as we do in the first five or six days of incubation; their extremities are all very much alike, as ours are for about the same period; there is not a single true feather on their body, rather only thin feather-shafts, so that we, as fledgelings in the nest, are more advanced than they shall ever be . . . And these mammals that cannot find their own food for such a long time after their birth, that can never rise freely from the earth, want to consider themselves more highly organized than we?'

Über Entwicklungsgeschichte der Thiere: Beobachtung und Reflexion (1828), 203. Trans. Stephen Jay Gould, *Ontogeny and Phylogeny* (1977), 54.

1 *That the general characters of the big group to which the embryo belongs appear in development earlier than the special characters.* In agreement with this is the fact that the vesicular form is the most general form of all; for what is common in a greater degree to all animals than the opposition of an internal and an external surface?

The less general structural relations are formed after the more general, and so on until the most special appear.

The embryo of any given form, instead of passing through the state of other definite forms, on the contrary separates itself from them.

Fundamentally the embryo of a higher animal form never resembles the adult of another animal form, but only its embryo.

Über Entwicklungsgeschichte der Thiere: Beobachtung und Reflexion (1828), 224. Trans. E. S. Russell, *Form and Function: A Contribution to the History of Animal Morphology* (1916), 125–6.

2 Vertebrate development consists in the formation, in the median plane, of four leaflets two of which are above the axis and two below. During this evolution the germ subdivides in layers, and this has the effect of dividing the primordial tubes into secondary masses. The latter, included in the other masses, are the fundamental organs with the faculty of forming all the other organs.

Über Entwicklungsgeschichte der Thiere: Beobachtung und Reflexion (1828). Quoted in François Jacob, *The Logic of Life* (1993), 121–122.

Alexander Bain 1818–1903

British philosopher and psychologist

3 The arguments for the two substances [mind and body] have, we believe, entirely lost their validity; they are no longer compatible with ascertained science and clear thinking. The one substance, with two sets of properties, two sides, the physical and the mental—*a double-faced unity*—would appear to comply with all the exigencies of the case.

Mind and Body: The Theories of their Relation (1872), 196.

4 The most essential nature of a sentient being, which is to move *to* pleasure and *from* pain.

On the Study of Character, Including An Estimate in Phrenology (1861), 202.

Kenneth T. Bainbridge 1904–96

American nuclear physicist

5 Now we're all sons-of-bitches.

After the test explosion at Alamogordo.
Quoted in Lansing Lamont, *Day of Trinity* (1966), 242.

Antoine Jérome Balard 1802–76

French chemist

6 ON BALARD Balard did not discover bromine, rather bromine discovered Balard.

J. Liebig on Balard.
Quoted in R. Oesper, *The Human Side of Scientists* (1975), 124.

Arthur James Balfour 1848–1930

British statesman and philosopher

1 We sound the future, and learn that after a period, long compared with the divisions of time open to our investigation, the energies of our system will decay, the glory of the sun will be dimmed and the earth, tideless and inert, will no longer tolerate the race which has for a moment disturbed its solitude. Man will go down into the pit, and all his thoughts will perish.

The Foundations of Belief: Being Notes Introductory to the Study of Theology (1895), 30–1.

2 Science preceded the theory of science, and is independent of it. Science preceded naturalism, and will survive it.

The Foundations of Belief: Being Notes Introductory to the Study of Theology (1895), 134.

3 But science is the great instrument of social change, all the greater because its object is not change but knowledge, and its silent appropriation of this dominant function, amid the din of political and religious strife, is the most vital of all the revolutions which have marked the development of modern civilisation.

Decadence: Henry Sidgwick Memorial Lecture (1908), 55–6.

Francis Maitland Balfour 1851–82

British embryologist

4 The embryological record is almost always abbreviated in accordance with the tendency of nature (to be explained on the principle of survival of the fittest) to attain her needs by the easiest means.

A Treatise on Comparative Embryology (1880), Vol. 1, 3–4.

Eric Glendinning Ball 1904–79

British-born American biochemist

5 The energy liberated when substrates undergo air oxidation is not liberated in one large burst, as was once thought, but is released in stepwise fashion. At least six separate steps seem to be involved. The process is not unlike that of locks in a canal. As each lock is passed in the ascent from a lower to a higher level a certain amount of energy is expended. Similarly, the total energy resulting from the oxidation of foodstuffs is released in small units or parcels, step by step. The amount of free energy released at each step is proportional to the difference in potential of the systems comprising the several steps.

'Oxidative Mechanisms in Animal Tissues', *A Symposium on Respiratory Enzymes* (1942), 22.

Honoré de Balzac 1799–1850

French novelist

6 A man cannot marry before he has studied anatomy and has dissected at the least one woman.

The Physiology of Marriage (1826), trans. Sharon Marcus (1997), Aphorism XXVIII, 63.

7 Have you ever plunged into the immensity of space and time by reading the geological treatises of Cuvier? Borne away on the wings of his genius, have you hovered over the illimitable abyss of the past as if a magician's hand were holding you aloft? As one penetrates from seam to seam, from stratum to stratum and discovers, under the quarries of Montmartre or in the schists of the Urals, those animals whose fossilized remains belong to antediluvian civilizations, the mind is startled to catch a vista of the milliards of years and the millions of peoples which the feeble memory of man and an indestructible divine tradition have forgotten and whose ashes heaped on the surface of our globe, form the two feet of earth which furnish us with bread and flowers. Is not Cuvier the greatest poet of our century? Certainly Lord Byron has expressed in words some aspects of spiritual turmoil; but our immortal natural historian has reconstructed worlds from bleached bones.

La Peau de Chagrin (1831), trans. Herbert J. Hunt, *The Wild Ass's Skin* (1977), 40–1.

Wilder Dwight Bancroft

1867–1953

American chemist

1 There is one experiment which I always like to try, because it proves something whichever way it goes. A solution of iodine in water is shaken with bone-black, filtered and tested with starch paste. If the colorless solution does not turn the starch blue, the experiment shows how completely charcoal extracts iodine from aqueous solution. If the starch turns blue, the experiment shows that the solution, though apparently colorless, still contains iodine which can be detected by means of a sensitive starch test.

Applied Colloid Chemistry (1921), 111.

2 We can distinguish three groups of scientific men. In the first and very small group we have the men who discover fundamental relations. Among these are van't Hoff, Arrhenius and Nernst. In the second group we have the men who do not make the great discovery but who see the importance and bearing of it, and who preach the gospel to the heathen. Ostwald stands absolutely at the head of this group. The last group contains the rest of us, the men who have to have things explained to us.

'Ostwald', *Journal of Chemical Education*, 1933, 10, 612.

Frederick Grant Banting

1891–1941

Canadian physician

3 Diabetus [sic]. Ligate pancreatic ducts of dog. Keep dogs alive till acini degenerate leaving Islets. Try to isolate the internal secretion of these to try to relieve glycosurea [sic].

Frederick Banting's notebook, entry for 31 October 1920. Notebook in Toronto Academy of Medicine. Quoted in Michael Bliss, *The Discovery of Insulin* (1982), 50.

4 Insulin is not a cure for diabetes; it is a treatment. It enables the diabetic to burn sufficient carbohydrates, so that proteins and fats may be added to the diet in sufficient quantities to provide energy for the economic burdens of life.

'Diabetes and Insulin', Nobel Lecture, 15 September 1925. In *Nobel Lectures: Physiology or Medicine, 1922–1941* (1965), 68.

Frederick Grant Banting

1891–1941 and

Charles Herbert Best 1899–1978

Canadian physician and Canadian physiologist

5 Intravenous injections of extract from dog's pancreas, removed from seven to ten weeks after ligation of the ducts, invariably exercises a reducing influence upon the percentage sugar of the blood and the amount of sugar excreted in the urine . . . the extent and duration of the reduction varies directly with the amount of extract injected.

'The Internal Secretion of the Pancreas', *Journal of Laboratory and Clinical Medicine*, 1922, 7, 251–266.

Lincoln Kinnear Barnett 1909–

American writer

6 The aims of pure basic science, unlike those of applied science, are neither fast-flowing nor pragmatic. The quick harvest of applied science is the useable process, the medicine, the machine. The shy fruit of pure science is understanding.

Life, 9 January 1950.

Joseph Barrell 1869–1919

American geologist

7 Nature vibrates with rhythms, climatic and diastrophic, those finding stratigraphic expression ranging in period from the rapid oscillation of surface waters, recorded in ripple-mark, to those long-deferred stirrings of the deep imprisoned titans which have divided earth history into periods and eras. The flight of time is measured by the weaving of composite rhythms—day and night, calm and storm, summer and winter, birth and death—such as these are sensed in the brief life of man. But the career of the earth recedes into a remoteness against which these lesser cycles are as unavailing for the measurement of that abyss of time as would be for human history the beating of an insect's wing. We must seek out, then, the nature of those longer rhythms whose very existence was unknown until man by the light of science sought to understand the earth. The larger of these must be measured in terms of the smaller, and the smaller must be measured in terms of years.

'Rhythm and the Measurement of Geologic Time', *Bulletin of the Geological Society of America*, 1917, **28**, 746.

James Matthew Barrie
1860–1937

British dramatist and novelist

1 I am aware that those hateful persons called Original Researchers now maintain that Raleigh was not the man; but to them I turn a deaf ear.

On who offered his coat for Queen Elizabeth I.

My Lady Nicotine (1890), 107.

Roland Barthes 1915–80

French philosopher and literary critic

2 In the hierarchy of the major poetic substances, it [plastic] figures as a disgraced material, lost between the effusiveness of rubber and the flat hardness of metal.

'Plastic', in *Mythologies* (1957), trans. Annette Lavers (1972), 98.

Paul Doughty Bartlett 1907–97

American chemist

3 Among nonclassical ions the ratio of conceptual difficulty to molecular weight reaches a maximum with the cyclopropylcarbinyl-cyclobutyl system.

Nonclassical Ions (1965), 272.

Derek Harold Richard Barton
1918–98

British chemist

4 I do not believe that the present flowering of science is due in the least to a real appreciation of the beauty and intellectual discipline of the subject. It is due simply to the fact that power, wealth and prestige can only be obtained by the correct application of science.

'Some Reflections on the Present Status of Organic Chemistry', in *Science and Human Progress: Addresses at the Celebrations of the 50th Anniversary of the Mellon Institute* (1963), 90.

Jacques Barzun 1907–

French-born American writer and historian

5 The philosophical implication of race-thinking is that by offering us the mystery of heredity as an explanation, it diverts our attention from the social and intellectual factors that make up personality.

Race (1937), 282.

6 It is not clear to anyone, least of all the practitioners, how science and technology in their headlong course do or should influence ethics and law, education and government, art and social philosophy, religion and the life of the affections. Yet science is an all-pervasive energy, for it is at once a mode of thought, a source of strong emotion, and a faith as fanatical as any in history.

Science: The Glorious Entertainment (1964), 3.

7 Out of man's mind in free play comes the creation Science. It renews itself, like the generations, thanks to an activity which is the best game of *homo ludens*: science is in the strictest and best sense a glorious entertainment.

Science: The Glorious Entertainment (1964), 110.

Agostino Maria Bassi 1773–1856

Italian life scientist

1 All infections, of whatever type, with no exceptions, are products of parasitic beings; that is, by living organisms that enter in other living organisms, in which they find nourishment, that is, food that suits them, here they hatch, grow and reproduce themselves.

Quoted in English in Paolo Mazzarello, *The Hidden Structure: A Scientific Biography of Camillo Golgi* (1999), trans. and ed. Henry A. Buchtel and Aldo Badiani, 19.

Henry Walter Bates 1825–92

British naturalist

2 I was obliged, at last, to come to the conclusion that the contemplation of nature alone is not sufficient to fill the human heart and mind.

The Naturalist on the River Amazons (1863), Vol. 2, Ch. 3, 186.

William Bateson 1861–1926

British geneticist and biologist

3 Variation, whatever may be its cause, and however it may be limited, is the essential phenomenon of Evolution. Variation, in fact, *is* Evolution. The readiest way, then, of solving the problem of Evolution is to study the facts of Variation.

Materials for the Study of Variation Treated with Especial Regard to Discontinuity in the Origin of Species (1894), 6.

4 The Study of Adaptation ceases to help us at the exact point at which help is most needed. We are seeking for the cause of the differences between species and species, and it is precisely on the utility of Specific Differences that the students of Adaptation are silent. For, as Darwin and many others have often pointed out, the characters which visibly differentiate species are not as a rule capital facts in the constitution of vital organs, but more often they are just those features which seem to us useless and trivial, such as the patterns of scales, the details of sculpture on chitin or shells, differences in numbers of hairs or spines, differences between the sexual prehensile organs, and so forth. These differences are often complex and are strikingly constant, but their utility is in almost every case problematical.

Materials for the Study of Variation Treated with Especial Regard to Discontinuity in the Origin of Species (1894), 11.

5 [The data of variation] attract men of two classes, in tastes and temperament distinct, each having little sympathy or even acquaintance with the work of the other. Those of the one class have felt the attraction of the problem. It is the challenge of Nature that calls them to work. But disgusted with the superficiality of 'naturalists' they sit down in the laboratory to the solution of the problem, hoping that the closer they look the more truly will they see. For the living things out of doors, they care little. Such work to them is all vague. With the other class it is the living thing that attracts, not the problem. To them the methods of the first school are frigid and narrow. Ignorant of the skill and of the accurate, final knowledge that the other school has bit by bit achieved, achievements that are the real glory of the method, the 'naturalists' hear only those theoretical conclusions which the laboratories from time to time ask them to accept. With senses quickened by the range and fresh air of their own work they feel keenly how crude and inadequate are these poor generalities, and for what a small and conventional world they are devised. Disappointed with the results they condemn the methods of the others, knowing nothing of their real strength. So it happens that for them the study of the problems of life and of Species becomes associated with crudity and meanness of scope. Beginning as naturalists they end as collectors, despairing of the problem, turning for relief to the tangible business of classification, accounting themselves happy if they can keep their species apart, caring little how they became so, and rarely telling us how they may be brought together. Thus each class misses that which in the other is good.

Materials for the Study of Variation Treated with Especial Regard to Discontinuity in the Origin of Species (1894), 574–5.

1 This *Aa* is the hybrid, or 'mule' form, or as I have elsewhere called it, the *heterozygote*, as distinguished from *AA* or *aa* the *homozygotes*.

Mendel's Principles of Heredity: A Defence (1902), 23.

2 Treasure your exceptions! . . . Keep them always uncovered and in sight. Exceptions are like the rough brickwork of a growing building which tells that there is more to come and shows where the next construction is to be.

The Methods and Scope of Genetics: An Inaugural Lecture Delivered 23 October 1908 (1908), 22.

3 If the Quick Fund were used for the foundation of a Professorship relating to Heredity and Variation the best title would, I think, be 'The Quick Professorship of the study of Heredity'. No single word in common use gives this meaning. Such a word is badly wanted, and if it were desirable to coin one, 'GENETICS' might do.

Letter to Adam Sedgwick, 18 April 1905. In Beatrice Bateson (ed.), *William Bateson F.R.S. Naturalist. His Essays & Addresses Together with an Account of his Life* (1928) 93.

4 To [Walter Frank Raphael] Weldon I owe the chief awakening of my life. It was through him that I first learnt that there was work in the world which I could do. Failure and uselessness had been my accepted destiny before. Such a debt is perhaps the greatest that one man can feel towards another; nor have I have been backward in owning it. But his is the personal, private obligation of my soul.

Letter to Beatrice Bateson (*c.* April 1906). In Beatrice Bateson (ed.), *William Bateson, F.R.S., Naturalist: His Essays & Addresses Together with a Short Account of his Life* (1928), 103.

5 I would trust Shakespeare, but I would not trust a committee of Shakespeares.

Quoted in J. K. Brierley, *Biology and the Social Crisis* (1967), 75.

Georg Bauer (Georgius Agricola) 1494–1555

German mining technologist

6 Mineral substances vary greatly in color, transparency, luster, brilliance, odor, taste, and other properties which are shown by their strength and weakness, shape, and form. They do not have the variety of origins that we find not only in living matter but also in original matter. Moreover they have not been classified like the latter on the basis of the place where they pass their life since mineral substances lack life and with rare exceptions are found only within the earth. They do not have the differences in characters and actions which nature has given to living things alone. Great differences are not the essential features of minerals as they are of living and original matter.

De Natura Fossilium (1546), trans. M. C. and J. A. Bandy (1955), 1.

7 There are many arts and sciences of which a miner should not be ignorant. First there is Philosophy, that he may discern the origin, cause, and nature of subterranean things; for then he will be able to dig out the veins easily and advantageously, and to obtain more abundant results from his mining. Secondly there is Medicine, that he may be able to look after his diggers and other workman . . . Thirdly follows astronomy, that he may know the divisions of the heavens and from them judge the directions of the veins. Fourthly, there is the science of Surveying that he may be able to estimate how deep a shaft should be sunk . . . Fifthly, his knowledge of Arithmetical Science should be such that he may calculate the cost to be incurred in the machinery and the working of the mine. Sixthly, his learning must comprise Architecture, that he himself may construct the various machines and timber work required underground . . . Next, he must have knowledge of Drawing, that he can draw plans of his machinery. Lastly, there is the Law, especially that dealing with metals, that he may claim his own rights, that he may undertake the duty of giving others his opinion on legal matters, that he may not take another man's property and so make trouble for himself, and that he may fulfil his obligations to others according to the law.

De Re Metallica (1556), trans. H. C. and L. H. Hoover (1950), 3–4.

1 Albertus [Magnus] . . . debased the doctrine of Aristotle with the itch of the chemists flowing with the bloody flux of quicksilver and the stench of sulphur.

De Orta et Causis Subterraneorum Lib. V (1546), 46, trans. John Howes.

William Maddock Bayliss 1860–1924

British physiologist

2 But, as Bacon has well pointed out, truth is more likely to come out of error, if this is clear and definite, than out of confusion, and my experience teaches me that it is better to hold a well-understood and intelligible opinion, even if it should turn out to be wrong, than to be content with a muddle-headed mixture of conflicting views, sometimes miscalled impartiality, and often no better than no opinion at all.

Principles of General Physiology (1915), x.

3 But at the same time, there must never be the least hesitation in giving up a position the moment it is shown to be untenable. It is not going too far to say that the greatness of a scientific investigator does not rest on the fact of his having never made a mistake, but rather on his readiness to admit that he has done so, whenever the contrary evidence is cogent enough.

Principles of General Physiology (1915), x-xi.

4 ON BAYLISS A true anecdote which illustrates his unworldly nature is of the instruction he received in 1922 to appear at Buckingham Palace to receive the accolade of the Order of Knighthood; he replied that as the date coincided with that of a meeting of the Physiological Society, he would be unable to attend.

Charles Lovatt Evans, *Reminiscences of Bayliss and Starling* (1964), 3.

George Wells Beadle 1903–89

American geneticist

5 We spend long hours discussing the curious situation that the two great bodies of biological knowledge, genetics and embryology, which were obviously intimately interrelated in development, had never been brought together in any revealing way. An obvious difficulty was that the most favorable organisms for genetics, *Drosophila* as a prime example, were not well suited for embryological study, and the classical objects of embryological study, sea urchins and frogs as examples, were not easily investigated genetically. What might we do about it? There were two obvious approaches: one to learn more about the genetics of an embryologically favourable organism, the other to better understand the development of *Drosophila*. We resolved to gamble up to a year of our lives on the latter approach, this in Ephrussi's laboratory in Paris which was admirably equipped for tissue culture, tissue or organ transplantation, and related techniques.

'Recollections', *Annual Review of Biochemistry*, 1974, 43, 6.

6 Since many cases are known in which the specificities of antigens and enzymes appear to bear a direct relation to gene specificities, it seems reasonable to suppose that the gene's primary and possibly sole function is in directing the final configurations of protein molecules.

Assuming that each specific protein of the organism has its unique configuration copied from that of a gene, it follows that every enzyme whose specificity depends on a protein should be subject to modification or inactivation through gene mutation. This would, of course, mean that the reaction normally catalyzed by the enzyme in question would either have its rate or products modified or be blocked entirely.

Such a view does not mean that genes directly 'make' proteins. Regardless of precisely how proteins are synthesized, and from what component parts, these parts must themselves be synthesized by reactions which are enzymatically catalyzed and which in turn depend on the functioning of many genes. Thus in the synthesis of a single protein molecule, probably at least several hundred different genes contribute. But the final molecule corresponds to only one of them and this is the gene we visualize as being in primary control.

'Genetics and Metabolism in Neurospora', *Physiological Reviews*, 1945, 25, 660.

1 In a sense, genetics grew up as an orphan. In the beginning botanists and zoologists were often indifferent and sometimes hostile toward it. 'Genetics deals only with superficial characters', it was often said. Biochemists likewise paid it little heed in its early days. They, especially medical biochemists, knew of Garrod's inborn errors of metabolism and no doubt appreciated them in the biochemical sense and as diseases; but the biological world was inadequately prepared to appreciate fully the significance of his investigations and his thinking. Geneticists, it should be said, tended to be preoccupied mainly with the mechanisms by which genetic material is transmitted from one generation to the next.

'Genes and chemical reactions in Neurospora', Nobel Lecture, 11 December 1958. In *Nobel Lectures: Physiology or Medicine 1942–1962* (1964), 598.

2 It is, I believe, justifiable to make the generalization that anything an organic chemist can synthesize can be made without him. All he does is increase the probability that given reactions will 'go.' So it is quite reasonable to assume that given sufficient time and proper conditions, nucleotides, amino acids, proteins, and nucleic acids will arise by reactions that, though less probable, are as inevitable as those by which the organic chemist fulfills his predictions. So why not self-duplicating virus-like systems capable of further evolution?

The Place of Genetics in Modern Biology (1959), 18.

3 Knowing what we now know about living systems—how they replicate and how they mutate—we are beginning to know how to control their evolutionary futures. To a considerable extent we now do that with the plants we cultivate and the animals we domesticate. This is, in fact, a standard application of genetics today. We could even go further, for there is no reason why we cannot in the same way direct our own evolutionary futures. I wish to emphasize, however—and emphatically—that *whether* we should do this and, if so, *how*, are not questions science alone can answer. They are for society as a whole to think about. Scientists can say what the consequences might be, but they are not justified in going further except as responsible members of society.

The Place of Genetics in Modern Biology (1959), 20.

4 Beadle believed that genetics were inseparable from chemistry—more precisely, biochemistry. They were, he said, 'two doors leading to the same room'.

In Warren Weaver, *Science and Imagination* (1967), xii. Quoted in Thomas Hager, *Force of Nature: The Life of Linus Pauling* (1995), 276.

Tim Beardsley 1957–

British-born writer living in U.S.

5 And Lady Luck remains tantalizingly in the shadows, pulling some of evolution's strings—but nobody knows how many.

'Weird Wonders', *Scientific American*, June 1992, 14.

William Beaumont 1785–1853

American army surgeon

1 I submit a body of facts which cannot be invalidated. My opinions may be doubted, denied, or approved, according as they conflict or agree with the opinions of each individual who may read them; but their worth will be best determined by the foundation on which they rest—the incontrovertible facts.

Experiments and Observations on the Gastric Juice, and the Physiology of Digestion (1833), Preface.

Simone de Beauvoir 1908–86

French novelist and feminist

2 It is among the psychoanalysts in particular that man is defined as a human being and woman as a female—whenever she behaves as a human being she is said to imitate the male.

The Second Sex (1949). Trans. H. M. Parshley (1953), 83.

3 To be a woman, if not a defect, is at least a peculiarity.

The Second Sex (1949). Trans. H. M. Parshley (1953).

Giambatista Beccaria 1716–81

Italian physicist

4 For you teach very clearly by your behaviour how slowly and how meagerly our senses proceed in the investigation of ever inexhaustible nature.

Elettricismo artificiale (1772), vii-viii, trans. in Antonio Pace, *Franklin and Italy* (1958), 58.

5 Moral certainty is never more than probability.

On Crimes and Punishments (1764), Chapter 14.

Johann Joachim Becher 1635–82

German chemist

6 The chemists are a strange class of mortals, impelled by an almost insane impulse to seek their pleasures amid smoke and vapour, soot and flame, poisons and poverty; yet among all these evils I seem to live so sweetly that may I die if I were to change places with the Persian king.

Physica subterranea (1667). Quoted in R. Oesper, *The Human Side of Scientists* (1973), 11.

Johann Beckmann 1739–1811

German economist and technologist

7 Technology.

Coinage of the word in his textbook of 1777.

Anleitung zur Technologie (1777).

Martinus Willem Beijerinck 1851–1931

Dutch botanist and virologist

8 In its most primitive form, life is, therefore, no longer bound to the cell, the cell which possesses structure and which can be compared to a complex wheel-work, such as a watch which ceases to exist if it is stamped down in a mortar. No, in its primitive form life is like fire, like a flame borne by the living substance;—like a flame which appears in endless diversity and yet has specificity within it;—which can adopt the form of the organic world, of the lank grass-leaf and of the stem of the tree.

Address given at the 1913 meeting of the Koninklijke Akademie van Wetenschappen in Amsterdam. Trans. in G. Van Iterson, Jr, L. E. Den Dooren De Jong and A. J. Kluyver, *Martinus Willem Beijerinck: His Life and Work* (1940), 120.

Georg von Békésy 1899–1972

Hungarian-born American physicist and psychologist

9 One way of dealing with errors is to have friends who are willing to spend the time necessary to carry out a critical examination of the experimental design beforehand and the results after the experiments have been completed. An even better way is to have an enemy. An enemy is willing to devote a vast amount of time and brain power to ferreting out errors both large and small, and this without any compensation. The trouble is that really capable enemies are scarce; most of them are only ordinary. Another trouble with enemies is that they sometimes develop into friends and lose a great deal of their zeal. It was in this way the writer lost his three best enemies. Everyone, not just scientists, needs a good few enemies.

Quoted in George A. Olah, *A Life of Magic Chemistry* (2001), 146.

Vissarion Grigorievich Belinskii 1811–48

Russian critic and essayist

1 In science one must search for ideas. If there are no ideas, there is no science. A knowledge of facts is only valuable in so far as facts conceal ideas: facts without ideas are just the sweepings of the brain and the memory.

Collected Works (1948), Vol.2, 348.

Alexander Graham Bell 1847–1922

British-born American inventor

2 'Watson, . . . if I can get a mechanism which will make a current of electricity vary in its intensity, as the air varies in density when a sound is passing through it, I can telegraph any sound, even the sound of speech.'

Quoted in Thomas A. Watson, *Exploring Life: The Autobiography of Thomas A. Watson* (1926), 62.

3 I have read somewhere that the resistance offered by a wire . . . is affected by the *tension of the wire.* If this is so, a *continuous current of electricity* passed through a vibrating wire should meet with a varying resistance, and hence a pulsatory action should be induced in the current . . . [corresponding] in *amplitude*, as well as in rate of movement, to the vibrations of the string . . . [Thus] the *timbre* of a sound [a quality essential to intelligible speech] could be transmitted . . . [and] the strength of the current can be increased *ad libitum* without destroying the *relative intensities of the vibrations.*

Letter to Gardiner Greene Hubbard, 4 May 1875. Quoted in Robert V. Bruce, *Bell: Alexander Graham Bell and the Conquest of Solitude* (1973), 144–5.

4 'Grand telegraphic discovery today . . . Transmitted *vocal sounds* for the first time . . . With some further modification I hope we may be enabled to distinguish . . . the 'timbre' of the sound. Should this be so, conversation *viva voce* by telegraph will be a *fait accompli.*'

Letter to Sarah Fuller, 1 July 1875. Quoted in Robert V. Bruce, *Bell: Alexander Graham Bell and the Conquest of Solitude* (1973), 149.

5 I then shouted into M [the mouthpiece] the following sentence: 'Mr. Watson—Come here—I want to see you.' To my delight he came and declared that he had heard and understood what I said. I asked him to repeat the words. He answered 'You said—'Mr. Watson—come here—I want to see you.'' We then changed places and I listened at S [the reed receiver] while Mr. Watson read a few passages from a book into the mouth piece M. It was certainly the case that articulate sounds proceeded from S. The effect was loud but indistinct and muffled. If I had read beforehand the passage given by Mr. Watson I should have recognized every word. As it was I could not make out the sense—but an occasional word here and there was quite distinct. I made out 'to' and 'out' and 'further'; and finally the sentence 'Mr. Bell do you understand what I say? Do—you—un—der—stand—what—I—say' came quite clearly and intelligibly. No sound was audible when the armature S was removed.

Notebook, 'Experiments made by A. Graham Bell, vol. 1'. Entry for 10 March 1876.

Quoted in Robert V. Bruce, *Bell: Alexander Graham Bell and the Conquest of Solitude* (1973), 181.

Charles Bell 1774–1842

British anatomist and surgeon

1 On laying bare the roots of the spinal nerves, I found that I could cut across the posterior fasciculus of nerves, which took its origin from the posterior portion of the spinal marrow without convulsing the muscles of the back; but that on touching the anterior fasciculus with the point of a knife, the muscles of the back were immediately convulsed.

Idea of a New Anatomy of the Brain (1811), 22.

2 I took this view of the subject. The *medulla spinalis* has a central division, and also a distinction into anterior and posterior fasciculi, corresponding with the anterior and posterior portions of the brain. Further we can trace down the crura of the *cerebrum* into the anterior fasciculus of the spinal marrow, and the crura of the *cerebellum* into the posterior fasciculus. I thought that here I might have an opportunity of touching the *cerebellum*, as it were, through the posterior portion of the spinal marrow, and the *cerebrum* by the anterior portion. To this end I made experiments which, though they were not conclusive, encouraged me in the view I had taken. I found that injury done to the anterior portion of the spinal marrow, convulsed the animal more certainly than injury done to the posterior portion; but I found it difficult to make the experiment without injuring both portions.

Idea of a New Anatomy of the Brain (1811), 21–22.

3 The cerebrum I consider as the grand organ by which the mind is united to the body. Into it all the nerves from the external organs of the senses enter; and from it all the nerves which are agents of the will pass out.

Idea of a New Anatomy of the Brain (1811), 27.

4 In France, where an attempt has been made to deprive me of the originality of these discoveries, experiments without number and without mercy have been made on living animals; not under the direction of anatomical knowledge, or the guidance of just induction, but conducted with cruelty and indifference, in hope to catch at some of the accidental facts of a system, which, is evident, the experimenters did not fully comprehend.

An Exposition of the Natural System of the Nerves of the Human Body (1824), 2–3.

5 Man has two conditions of existence in the body. Hardly two creatures can be less alike than an infant and a man. The whole fetal state is a preparation for birth . . . The human brain, in its earlier stage, resembles that of a fish: as it is developed, it resembles more the cerebral mass of a reptile; in its increase, it is like that of a bird, and slowly, and only after birth, does it assume the proper form and consistence of the human encephalon.

Attributed.

Eric Temple Bell 1883–1960

British-American mathematician

6 These estimates may well be enhanced by one from F. Klein (1849–1925), the leading German mathematician of the last quarter of the nineteenth century. 'Mathematics in general is fundamentally the science of self-evident things.' . . . If mathematics is indeed the science of self-evident things, mathematicians are a phenomenally stupid lot to waste the tons of good paper they do in proving the fact. Mathematics is abstract and it is hard, and any assertion that it is simple is true only in a severely technical sense—that of the modern postulational method which, as a matter of fact, was exploited by Euclid. The assumptions from which mathematics starts are simple; the rest is not.

Mathematics: Queen and Servant of Science (1952), 19–20.

Florence O. Bell

British x-ray crystallographer

1 Possibly the most pregnant recent development in molecular biology is the realization that the beginnings of life are closely associated with the interactions of proteins and nucleic acids.

'X-ray and Related Studies of the Structure of the Proteins and Nucleic Acids', Leeds PhD Thesis (1939), quoted in Robert Olby *The Path to the Double Helix: The Discovery of DNA* (1994), 70.

Thomas Bell 1792–1880

British dental surgeon and naturalist

2 The year which has passed . . . has not been unproductive in contributions of interest and value, in those sciences to which we are professedly more particularly addicted, as well as in every other walk of scientific research. It has not, indeed, been marked by any of those striking discoveries which at once revolutionize, so as to speak, the department of science on which they bear.

Summary of the year in which the Darwin-Wallace communication was read.

'Presidential Address', *Proceedings of the Linnean Society*, 24 May (1859), viii.

Hilaire Belloc 1870–1953

French-born British historian and essayist

3 The Microbe is so very small
You cannot make him out at all.
But many sanguine people hope
To see him down a microscope.
His jointed tongue that lies beneath
A hundred curious rows of teeth;
His seven tufted tails with lots
Of lovely pink and purple spots
On each of which a pattern stands,
Composed of forty separate bands;
His eyebrows of a tender green;
All these have never yet been seen—
But Scientists, who ought to know,
Assure us they must be so . . .
Oh! let us never, never doubt
What nobody is sure about!

More Beasts for Worse Children (1897), 47–8.

4 Physicians of the Utmost Fame
Were called at once; but when they came
They answered, as they took their Fees,
'There is no Cure for this Disease.'

'Henry King', in *Cautionary Tales for Children* (first published 1907), (1908 edition), 18–19.

5 Statistics are the triumph of the quantitative method, and the quantitative method is the victory of sterility and death.

'On Statistics'. In *On the Silence of the Sea and Other Essays* (1941), 199–200.

Walter Benjamin 1892–1940

German critic and philosopher

6 With whom [do] the adherents of historicism actually empathize[?] The answer is inevitable: with the victor. And all rulers are the heirs of those who conquered before them. Hence, empathy with the victor invariably benefits the rulers. Historical materialists know what that means. Whoever has emerged victorious participates to this day in the triumphal procession in which the present rulers step over those who are lying prostrate. According to traditional practice, the spoils are carried along in the procession. They are called cultural treasures, and a historical materialist views them with cautious detachment. For without exception the cultural treasures he surveys have an origin which he cannot contemplate without horror. They owe their existence not only to the efforts of the great minds and talents who have created them, but also to the anonymous toil of their contemporaries. There is no document of civilization which is not at the same time a document of barbarism.

'Theses on the Philosophy of History' (completed 1940, first published 1950). In *Illuminations*, ed. Hannah Arendt and trans. Harry Zohn (1970), 258.

Jeremy Bentham 1748–1832

British utilitarian philosopher

1 It is the greatest happiness of the greatest number that is the measure of right and wrong.

A Fragment on Government: An Introduction to the Principles of Morals and Legislation (1776), Preface, ii.

2 Nature has placed mankind under the governance of two sovereign masters, *pain and pleasure*. It is for them alone to point out what we ought to do, as well as to determine what we shall do. On the one hand the standard of right and wrong, on the other the chain of causes and effects, are fastened to their throne. They govern us in all we do, in all we say, in all we think: every effort we can make to throw off our subjection, will serve but to demonstrate and confirm it. In words a man may pretend to abjure their empire: but in reality he will remain subject to it all the while.

An Introduction to the Principles of Morals and Legislation (1789), i.

3 By the principle of utility is meant that principle which approves or disapproves of every action whatsoever, according to the tendency which it appears to have to augment or diminish the happiness of the party whose interest is in question: or, what is that same thing in other words, to promote or to oppose that happiness.

An Introduction to the Principles of Morals and Legislation (1789), ii.

4 The French have already discovered that the blackness of the skin is no reason why a human being should be abandoned without redress to the caprice of the tormentor. It may come one day to be recognized, that the number of the legs, the villosity of the skin, or the termination of the *os sacrum*, are reasons equally insufficient for abandoning a sensitive being to the same fate? What else is it that should trace the insuperable line? Is it the faculty of reason, or, perhaps, the faculty of discourse? But a full-grown horse, or dog, is beyond comparison a more rational, as well as a more conversible animal, than an infant of a day, or a week, or even a month, old. But suppose the case were otherwise, what would it avail? the question is not, Can they *reason?* nor, Can they *talk?* But, Can they *suffer?*

An Introduction to the Principles of Morals and Legislation (1789), cccviii-ix.

5 The utility of all these arts and sciences,—I speak both of those of amusement and curiosity,—the value which they possess, is exactly in proportion to the pleasure they yield. Every other species of pre-eminence which may be attempted to be established among them is altogether fanciful. Prejudice apart, the game of push-pin is of equal value with the arts and sciences of music and poetry. If the game of push-pin furnish more pleasure, it is more valuable than either. Everybody can play at push-pin: poetry and music are relished only by a few.

The Rationale of Reward (1825), 206.

Edmund Clerihew Bentley see

Comte, Isidore 131:7; **Davy, Humphry** 168:3; **Dewar, James** 177:2; **Jeans, James Hopwood** 323:7; **Wren, Christopher** 635:4

Richard Bentley 1662–1742

British clergyman

6 The Atoms or Particles, which now constitute Heaven and Earth, being once separate and diffused in the Mundane Space, like the supposed *Chaos*, could never *without a God by their Mechanical affections* have convened into this present Frame of Things or any other like it.

A Confutation of Atheism from the Origin and Frame of the World (1693), Part II, 7.

Seymour Benzer 1921–

American molecular biologist

7 The genes are the atoms of heredity.

'Genetic Fine Structure', Lecture delivered 15

September 1960. Quoted in *The Harvey Lectures*, Series 56 (1961), 1.

Peter L. Berger 1929– and **Thomas Luckmann** 1927–

American sociologists

1 The sociology of knowledge is concerned with the analysis of the social construction of reality.

The Social Construction of Reality (1966), 15.

Torbern Olof Bergman 1735–84

Swedish chemist and naturalist

2 A scientist strives to understand the work of Nature. But with our insufficient talents as scientists, we do not hit upon the truth all at once. We must content ourselves with tracking it down, enveloped in considerable darkness, which leads us to make new mistakes and errors. By diligent examination, we may at length little by little peel off the thickest layers, but we seldom get the core quite free, so that finally we have to be satisfied with a little incomplete knowledge.

Lecture to the Royal Swedish Academy of Science, 23 May 1764. Quoted in J. A. Schufle 'Torbern Bergman, Earth Scientist', *Chymia*, 1967, **12**, 78.

3 The mineral kingdom consists of the fossil substances found in the earth. These are either entirely destitute of organic structure, or, having once possessed it, possess it no longer: such are the petrefactions.

Outlines of Mineralogy (1783), trans. W. Withering, 5.

4 Fossils are of four kinds, viz. *saline*, *earthy*, *inflammable* and *metallic*; hence arise four classes.

Outlines of Mineralogy (1783), trans. W. Withering, 12.

5 Finally, *I aim at giving denominations to things, as agreeable to truth as possible.* I am not ignorant that words, like money, possess an ideal value, and that great danger of confusion may be apprehended from a change of names; in the mean time it cannot be denied that chemistry, like the other sciences, was formerly filled with improper names. In different branches of knowledge, we see those matters long since reformed: why then should chemistry, which examines the real nature of things, still adopt vague names, which suggest false ideas, and favour strongly of ignorance and imposition? Besides, there is little doubt but that many corrections may be made without any inconvenience.

Physical and Chemical Essays (1784), Vol. 1, xxxvii.

Henri Louis Bergson 1859–1941

French philosopher

6 It is with our entire past . . . that we desire, will and act . . . from this survival of the past it follows that consciousness cannot go through the same state twice. The circumstances may still be the same, but they will act no longer on the same person . . . that is why our duration is irreversible.

Creative Evolution (1911), trans. Arthur Mitchell, 6.

7 Mankind lies groaning, half-crushed beneath the weight of its own progress. Men do not sufficiently realize that their future is in their own hands. Theirs is the task of determining first of all whether they want to go on living or not. Theirs the responsibility, then, for deciding if they want merely to live, or intend to make just the extra effort required for fulfilling, even on their refractory planet, the essential function of the universe, which is a machine for the making of gods.

The Two Sources of Morality and Religion, trans. by R. Ashley Audra and Cloudesley Brereton (1935), 275.

8 It is of man's essence to create materially and morally, to fabricate things and to fabricate himself. Homo faber is the definition I propose . . . Homo faber, Homo sapiens, I pay my respects to both, for they tend to merge.

The Creative Mind (1946), 84–5.

George Berkeley 1685–1753

British philosopher

See **Boswell, James** 78:4

1 It is indeed an Opinion strangely prevailing amongst Men, that Houses, Mountains, Rivers, and in a word all sensible Objects have an Existence Natural or Real, distinct from their being perceived by the Understanding. But with how great an Assurance and Acquiescence soever this Principle may be entertained in the World; yet whoever shall find in his Heart to call it in Question, may, if I mistake not, perceive it to involve a manifest Contradiction. For what are the forementioned Objects but the things we perceive by Sense, and what do we perceive besides our own Ideas or Sensations; and is it not plainly repugnant that any one of these or any Combination of them should exist unperceived?

A Treatise Concerning the Principles of Human Knowledge [first published 1710], (1734), 38.

2 There are certain general Laws that run through the whole Chain of natural Effects: these are learned by the Observation and Study of Nature, and are by Men applied as well to the framing artificial things for the Use and Ornament of Life, as to the explaining the various *Phænomena*: Which Explication consists only in shewing the Conformity any particular Phænomenon hath to the general Laws of Nature, or, which is the same thing, in discovering the *Uniformity* there is in the production of natural Effects; as will be evident to whoever shall attend to the several Instances, wherin Philosophers pretend to account for Appearances.

A Treatise Concerning the Principles of Human Knowledge [first published 1710], (1734), 87–8.

3 Colour, Figure, Motion, Extension and the like, considered only so many *Sensations* in the Mind, are perfectly known, there being nothing in them which is not perceived. But if they are looked on as notes or Images, referred to *Things* or *Archetypes* existing without the Mind, then are we involved all in *Scepticism*.

A Treatise Concerning the Principles of Human Knowledge [first published 1710], (1734), 109.

4 After what has been premised, I think we may lay down the following Conclusions. First, It is plain Philosophers amuse themselves in vain, when they inquire for any natural efficient Cause, distinct from a *Mind* or *Spirit*. Secondly, Considering the whole Creation is the Workmanship of a *wise and good Agent*, it should seem to become Philosophers, to employ their Thoughts (contrary to what some hold) about the final Causes of Things: And I must confess, I see no reason, why pointing out the various Ends, to which natural Things are adapted and for which they were originally with unspeakable Wisdom contrived, should not be thought one good way of accounting for them, and altogether worthy a Philosopher.

A Treatise Concerning the Principles of Human Knowledge [first published 1710], (1734), 126–7.

John Desmond Bernal 1901–71

Irish physicist and x-ray crystallographer

5 Men will not be content to manufacture life: they will want to improve on it.

The World, The Flesh and The Devil: An Enquiry into the Future of the Three Enemies of the Rational Soul (1929), 56.

6 The events of the past few years have led to a critical examination of the function of science in society. It used to be believed that the results of scientific investigation would lead to continuous progressive improvements in conditions of life; but first the War and then the economic crisis have shown that science can be used as easily for destructive and wasteful purposes, and voices have been raised demanding the cessation of scientific research as the only means of preserving a tolerable civilization. Scientists themselves, faced with these criticisms, have been forced to consider, effectively for the first time, how the work they are doing is connected around them. This book is an attempt to analyse this connection; to investigate how far scientists, individually and collectively, are responsible for this state of affairs, and to suggest what possible steps could be taken which would lead to a fruitful and not to a destructive utilization of science.

The Social Function of Science (1939), xiii.

1 The very bulk of scientific publications is itself delusive. It is of very unequal value; a large proportion of it, possibly as much as three-quarters, does not deserve to be published at all, and is only published for economic considerations which have nothing to do with the real interests of science.

The Social Function of Science (1939), 118.

2 In England, more than in any other country, science is felt rather than thought . . . A defect of the English is their almost complete lack of systematic thinking. Science to them consists of a number of successful raids into the unknown.

The Social Function of Science (1939), 197.

3 But if capitalism had built up science as a productive force, the very character of the new mode of production was serving to make capitalism itself unnecessary.

Marx and Science (1952), 39.

4 Published papers may omit important steps and the memory of men of science, even the greatest, is sadly fallible.

Science and Industry in the Nineteenth Century (1953), 199.

5 The greater the man, the more he is soaked in the atmosphere of his time; only thus can he get a wide enough grasp of it to be able to change substantially the pattern of knowledge and action.

Science in History (1954), 22.

6 We should admit in theory what is already very largely a case in practice, that the main currency of scientific information is the secondary sources in the forms of abstracts, reports, tables, &c., and that the primary sources are only for detailed reference by very few people. It is possible that the fate of most scientific papers will be not to be read by anyone who uses them, but with luck they will furnish an item, a number, some facts or data to such reports which may, but usually will not, lead to the original paper being consulted. This is very sad but it is the inevitable consequence of the growth of science. The number of papers that can be consulted is absolutely limited, no more time can be spent in looking up papers, by and large, than in the past. As the number of papers increase the chance of any one paper being looked at is correspondingly diminished. This of course is only an average, some papers may be looked at by thousands of people and may become a regular and fixed part of science but most will perish unseen.

'The Supply of Information to the Scientist: Some Problems of the Present Day', *The Journal of Documentation*, 1957, 13, 195.

7 All that glisters may not be gold, but at least it contains free electrons. [But consider the Golden Scarab Beetle which has a metallic lustre without metal.]

Lecture at Birkbeck College, University of London, 1960.

8 We academic scientists move within a certain sphere, we can go on being useless up to a point, in the confidence that sooner or later some use will be found for our studies. The mathematician, of course, prides himself on being totally useless, but usually turns out to be the most useful of the lot. He finds the solution but he is not interested in what the problem is: sooner or later, someone will find the problem to which his solution is the answer.

'Concluding Remarks', *Proceedings of the Royal Society of London*, Series A, *A Discussion of New Materials*, 1964, **282**, 152–3.

1 The beauty of life is, therefore, geometrical beauty of a type that Plato would have much appreciated.

The Origin of Life (1967), xiii.

2 The question of the origin of life is essentially speculative. We have to construct, by straightforward thinking on the basis of very few factual observations, a plausible and self-consistent picture of a process which must have occurred before any of the forms which are known to us in the fossil record could have existed.

The Origin of Life (1967), 2.

3 In fact, we will have to give up taking things for granted, even the apparently simple things. We have to learn to understand nature and not merely to observe it and endure what it imposes on us. Stupidity, from being an amiable individual defect, has become a social crime.

The Origin of Life (1967), 163.

4 In my own field, x-ray crystallography, we used to work out the structure of minerals by various dodges which we never bothered to write down, we just used them. Then Linus Pauling came along to the laboratory, saw what we were doing and wrote out what we now call Pauling's Rules. We had all been using Pauling's Rules for about three or four years before Pauling told us what the rules were.

The Extension of Man (1972), 116.

Bernard of Chartres d. c.1130

French theologian and philosopher

5 We are like dwarfs sitting on the shoulders of giants. Our glance can thus take in more things and reach farther than theirs. It is not because our sight is sharper nor our height greater than theirs; it is that we are carried and elevated by the high stature of the giants.

Attributed to Bernard of Chartres in John of Salisbury, *Metalogicon* [1159], Book III, chapter 4, quoted in E. Jeaneau, 'Bernard of Chartres', in C. C. Gillispie, *Dictionary of Scientific Biography* (1971), Vol. 3, 19.

Claude Bernard 1813–78

French physiologist

6 The first entirely vital action, so termed because it is not effected outside the influence of life, consists in the creation of the glycogenic material in the living hepatic tissue. The second entirely chemical action, which can be effected outside the influence of life, consists in the transformation of the glycogenic material into sugar by means of a ferment.

Sur le Méchanisme de la Fonction du Sucre dans le Foie (1857), 583. Translated in Joseph S. Fruton, *Proteins, Enzymes, Genes: The Interplay of Chemistry and Biology* (1999), 340.

7 In the philosophic sense, observation shows and experiment teaches.

An Introduction to the Study of Experimental Medicine (1865), trans. Henry Copley Green (1957), 5.

8 Effects vary with the conditions which bring them to pass, but laws do not vary. Physiological and pathological states are ruled by the same forces; they differ only because of the special conditions under which the vital laws manifest themselves.

An Introduction to the Study of Experimental Medicine (1865), trans. Henry Copley Green (1957), 10.

9 Speaking concretely, when we say 'making experiments or making observations,' we mean that we devote ourselves to investigation and to research, that we make attempts and trials in order to gain facts from which the mind, through reasoning, may draw knowledge or instruction.

Speaking in the abstract, when we say 'relying on observation and gaining experience,' we mean that observation is the mind's support in reasoning, and experience the mind's support in deciding, or still better, the fruit of exact reasoning applied to the interpretation of facts. It follows from this that we can gain experience without making experiments, solely by reasoning appropriately about well-established facts, just as we can make experiments and observations without gaining experience, if we limit ourselves to noting facts.

Observation, then, is what shows facts; experiment is what teaches about facts and gives experience in relation to anything.

An Introduction to the Study of Experimental Medicine (1865), trans. Henry Copley Green (1957), 11.

1 But while I accept specialization in the practice, I reject it utterly in the theory of science.

An Introduction to the Study of Experimental Medicine (1865), trans. Henry Copley Green (1957), 25.

2 We see, then, that the elements of the scientific method are interrelated. Facts are necessary materials; but their working up by experimental reasoning, i.e., by theory, is what establishes and really builds up science. Ideas, given form by facts, embody science. A scientific hypothesis is merely a scientific idea, preconceived or previsioned. A theory is merely a scientific idea controlled by experiment. Reasoning merely gives a form to our ideas, so that everything, first and last, leads back to an idea. The idea is what establishes, as we shall see, the starting point or the *primum movens* of all scientific reasoning, and it is also the goal in the mind's aspiration toward the unknown.

An Introduction to the Study of Experimental Medicine (1865), trans. Henry Copley Green (1957), 26.

3 The great experimental principle, then, is doubt, that philosophic doubt which leaves to the mind its freedom and initiative, and from which the virtues most valuable to investigators in physiology and medicine are derived.

An Introduction to the Study of Experimental Medicine (1865), trans. Henry Copley Green (1957), 37.

4 We must never make experiments to confirm our ideas, but simply to control them.

An Introduction to the Study of Experimental Medicine (1865), trans. Henry Copley Green (1957), 38.

5 A contemporary poet has characterized this sense of the personality of art and of the impersonality of science in these words,—'Art is myself; science is ourselves.'

Victor Hugo in William Shakespeare, *1864.*

An Introduction to the Study of Experimental Medicine (1865), trans. Henry Copley Green (1957), 43.

6 Now, a living organism is nothing but a wonderful machine endowed with the most marvellous properties and set going by means of the most complex and delicate mechanism.

An Introduction to the Study of Experimental Medicine (1865), trans. Henry Copley Green (1957), 63.

7 First causes are outside the realm of science.

An Introduction to the Study of Experimental Medicine (1865), trans. Henry Copley Green (1957), 66.

8 If I had to define life in a single phrase, I should clearly express my thought of throwing into relief one characteristic which, in my opinion, sharply differentiates biological science. I should say: life is creation.

An Introduction to the Study of Experimental Medicine (1865), trans. Henry Copley Green (1957), 93.

9 By destroying the biological character of phenomena, the use of *averages* in physiology and medicine usually gives only apparent accuracy to the results. From our point of view, we may distinguish between several kinds of averages: physical averages, chemical averages and physiological and pathological averages. If, for instance, we observe the number of pulsations and the degree of blood pressure by means of the oscillations of a manometer throughout one day, and if we take the average of all our figures to get the true or average blood pressure and to learn the true or average number of pulsations, we shall simply have wrong numbers. In fact, the pulse decreases in number and intensity when we are fasting and increases during digestion or under different influences of movement and rest; all the biological characteristics of the phenomenon disappear in the average. Chemical averages are also often used. If we collect a man's urine during twenty-four hours and mix all this urine to analyze the average, we get an analysis of a urine which simply does not exist; for urine, when fasting, is different from urine during digestion. A startling instance of this kind was invented by a physiologist who took urine from a railroad station urinal where people of all nations passed, and who believed he could thus present an analysis of *average* European urine! Aside from physical and chemical, there are physiological averages, or what we might call average descriptions of phenomena, which are even more false. Let me assume that a physician collects a great many individual observations of a disease and that he makes an average description of symptoms observed in the individual cases; he will thus have a description that will never be matched in nature. So in physiology, we must never make average descriptions of experiments, because the true relations of phenomena disappear in the average; when dealing with complex and variable experiments, we must study their various circumstances, and then present our most perfect experiment as a type, which, however, still stands for true facts. In the cases just considered, averages must therefore be rejected, because they confuse, while aiming to unify, and distort while aiming to simplify. Averages are applicable only to reducing very slightly varying numerical data about clearly defined and *absolutely simple* cases.

An Introduction to the Study of Experimental Medicine (1865), trans. Henry Copley Green (1957), 134–5.

1 A physician's subject of study is necessarily the patient, and his first field for observation is the hospital. But if clinical observation teaches him to know the form and course of diseases, it cannot suffice to make him understand their nature; to this end he must penetrate into the body to find which of the internal parts are injured in their functions. That is why dissection of cadavers and microscopic study of diseases were soon added to clinical observation. But to-day these various methods no longer suffice; we must push investigation further and, in analyzing the elementary phenomena of organic bodies, must compare normal with abnormal states. We showed elsewhere how incapable is anatomy alone to take account of vital phenenoma, and we saw that we must add study of all physico-chemical conditions which contribute necessary elements to normal or pathological manifestations of life. This simple suggestion already makes us feel that the laboratory of a physiologist-physician must be the most complicated of all laboratories, because he has to experiment with phenomena of life which are the most complex of all natural phenomena.

An Introduction to the Study of Experimental Medicine (1865), trans. Henry Copley Green (1957), 140–1.

2 In a word, I consider hospitals only as the entrance to scientific medicine; they are the first field of observation which a physician enters; but the true sanctuary of medical science is a laboratory; only there can he seek explanations of life in the normal and pathological states by means of experimental analysis.

An Introduction to the Study of Experimental

Medicine (1865), trans. Henry Copley Green (1957), 146.

1 In every enterprise . . . the mind is always reasoning, and, even when we seem to act without a motive, an instinctive logic still directs the mind. Only we are not aware of it, because we begin by reasoning before we know or say that we are reasoning, just as we begin by speaking before we observe that we are speaking, and just as we begin by seeing and hearing before we know what we see or what we hear.

An Introduction to the Study of Experimental Medicine (1865), trans. Henry Copley Green (1957), 158–9.

2 *Tout est poison, rien n'est poison, tout est une question de dose.*

Everything is poisonous, nothing is poisonous, it is all a matter of dose.

Pathologie expérimentale (1872), 72.

3 Descriptive anatomy is to physiology what geography is to history, and just as it is not enough to know the typography of a country to understand its history, so also it is not enough to know the anatomy of organs to understand their functions.

Lectures on the Phenomena of Life Common to Animals and Plants (1878), trans. Hebbel E. Hoff, Roger Guillemin and Lucienne Guillemin (1974), 7.

4 Constant, or free, life is the third form of life; it belongs to the most highly organized animals. In it, life is not suspended in any circumstance, it unrolls along a constant course, apparently indifferent to the variations in the cosmic environment, or to the changes in the material conditions that surround the animal. Organs, apparatus, and tissues function in an apparently uniform manner, without their activity undergoing those considerable variations exhibited by animals with an oscillating life. This because in reality the *internal environment* that envelops the organs, the tissues, and the elements of the tissues does not change; the variations in the atmosphere stop there, so that it is true to say that *physical conditions of the environment* are constant in the higher animals; it is enveloped in an invariable medium, which acts as an atmosphere of its own in the constantly changing cosmic environment. It is an organism that has placed itself in a hot-house. Thus the perpetual changes in the cosmic environment do not touch it; it is not chained to them, it is free and independent.

Lectures on the Phenomena of Life Common to Animals and Plants (1878), trans. Hebbel E. Hoff, Roger Guillemin and Lucienne Guillemin (1974), 83.

5 *The constancy of the internal environment is the condition for free and independent life:* the mechanism that makes it possible is that which assured the maintenance, with the *internal environment*, of all the conditions necessary for the life of the elements.

Lectures on the Phenomena of Life Common to Animals and Plants (1878), trans. Hebbel E. Hoff, Roger Guillemin and Lucienne Guillemin (1974), 84.

6 Put off your imagination, as you put off your overcoat, when you enter the laboratory. But put it on again, as you put on your overcoat, when you leave.

Attributed.

7 The experimenter who does not know what he is looking for will never understand what he finds.

Attributed.

Daniel Bernoulli 1700–82

Swiss mathematician

8 The determination of the value of an item must not be based on its price, but rather on the utility it yields. The price of the item is dependent only on the thing itself and is equal for everyone; the utility, however, is dependent on the particular circumstances of the person making the estimate. Thus there is no doubt that a gain of one thousand ducats is more significant to a pauper than to a rich man though both gain the same amount.

Exposition of a New Theory on the Measurement of Risk (1738), 24.

Jakob Bernoulli 1654–1705

Swiss mathematician and astronomer

1 It often happens that men, even of the best understandings and greatest circumspection, are guilty of that fault in reasoning which the writers on logick call *the insufficient, or imperfect enumeration of parts, or cases*: insomuch that I will venture to assert, that this is the chief, and almost the only, source of the vast number of erroneous opinions, and those too very often in matters of great importance, which we are apt to form on all the subjects we reflect upon, whether they relate to the knowledge of nature, or the merits and motives of human actions. It must therefore be acknowledged, that the art which affords a cure to this weakness, or defect, of our understandings, and teaches us to enumerate all the possible ways in which a given number of things may be mixed and combined together, that we may be certain that we have not omitted any one arrangement of them that can lead to the object of our inquiry, deserves to be considered as most eminently useful and worthy of our highest esteem and attention. And this is the business of *the art, or doctrine of combinations* . . . It proceeds indeed upon mathematical principles in calculating the number of the combinations of the things proposed: but by the conclusions that are obtained by it, the sagacity of the natural philosopher, the exactness of the historian, the skill and judgement of the physician, and the prudence and foresight of the politician, may be assisted; because the business of all these important professions is but *to form reasonable conjectures* concerning the several objects which engage their attention, and all wise conjectures are the results of a just and careful examination of the several different effects that may possibly arise from the causes that are capable of producing them.

Ars conjectandi (1713). In F. Maseres, *The Doctrine of Permutations and Combinations* (1795), 36.

Norman John Berrill 1903–96

British-born American biologist

2 Life can be thought of as water kept at the right temperature in the right atmosphere in the right light for a long enough period of time.

You and The Universe (1958), 145.

Pierre Eugéne Marcellin Berthelot 1827–1907

French chemist

3 We all teach . . . the chemistry of Lavoisier and Gay-Lussac.

Comment made in 1877. Cited in Maurice Crosland, *Gay-Lussac, Scientist and Bourgeois* (1972), 248.

4 Within a hundred years of physical and chemical science, men will know what the atom is. It is my belief when science reaches this stage, God will come down to earth with His big ring of keys and will say to humanity, 'Gentlemen, it is closing time.'

Quoted in R. Oesper, *The Human Side of Scientists* (1975), 17.

5 I do not want chemistry to degenerate into a religion; I do not want the chemist to believe in the existence of atoms as the Christian believes in the existence of Christ in the communion wafer.

Said to Alfred Nacquet in a debate on chemical education. Quoted in 'Berthelot', *Berichte*, 1908, **41**, 4855.

Jöns Jacob Berzelius 1779–1848

Swedish chemist

6 As mineralogy constitutes a part of chemistry, it is clear that this arrangement [of minerals] must derive its principles from chemistry. The most perfect mode of arrangement would certainly be to allow bodies to follow each other according to the order of their electro-chemical properties, from the most electro-negative, oxygen, to the most electro-positive, potassium; and to place every compound body according to its most electro-positive ingredient.

An Attempt to Establish a Pure Scientific System of Mineralogy (1814), trans. J. Black, 48.

1 Chemical signs ought to be letters, for the greater facility of writing, and not to disfigure a printed book . . . I shall take therefore for the chemical sign, the *initial letter of the Latin name of each elementary substance:* but as several have the same initial letter, I shall distinguish them in the following manner:— 1. In the class which I shall call *metalloids*, I shall employ the initial letter only, even when this letter is common to the metalloid and to some metal. 2. In the class of metals, I shall distinguish those that have the same initials with another metal, or a metalloid, by writing the first two letters of the word. 3. If the first two letters be common to two metals, I shall, in that case, add to the initial letter the first consonant which they have not in common: for example, S = sulphur, Si = silicium, St = stibium (antimony), Sn = stannum (tin), C = carbonicum, Co = colbaltum (colbalt), Cu = cuprum (copper), O = oxygen, Os = osmium, &c.

'Essay on the Cause of Chemical Proportions, and on some circumstances relating to them: together with a short and easy method of expressing them', *Annals of Philosophy*, 1814, 3, 51–2.

2 The experiments made on the mutual electrical relations of bodies have taught us that they can be divided into two classes: *electropositive* and *electronegative*. The simple bodies which belong to the first class, as well as their oxides, always take up positive electricity when they meet simple bodies or oxides belonging to the second class; and the oxides of the first class always behave with the oxides of the other like salifiable bases with acids.

Essai sur le théorie des proportions chimiques (1819). Translated in Henry M. Leicester and Herbert S. Klickstein, *A Source Book in Chemistry 1400–1900* (1952), 260.

3 In arranging the bodies in order of their electrical nature, there is formed an electro-chemical system which, in my opinion, is more fit than any other to give an idea of chemistry.

Essai sur le théorie des proportions chimiques (1819). Translated in Henry M. Leicester and Herbert S. Klickstein, *A Source Book in Chemistry 1400–1900* (1952), 260.

4 Since it is necessary for specific ideas to have definite and consequently as far as possible selected terms, I have proposed to call substances of similar composition and dissimilar properties *isomeric*, from the Greek ἰσομερης.

Jahrebericht, 1832. Translated in Henry M. Leicester and Herbert S. Klickstein, *A Source Book in Chemistry 1400–1900* (1952), 265.

5 This new force, which was unknown until now, is common to organic and inorganic nature. I do not believe that this is a force entirely independent of the electrochemical affinities of matter; I believe, on the contrary, that it is only a new manifestation, but since we cannot see their connection and mutual dependence, it will be easier to designate it by a separate name. I will call this force *catalytic force*. Similarly, I will call the decomposition of bodies by this force *catalysis*, as one designates the decomposition of bodies by chemical affinity analysis.

'Some Ideas on a New Force which Acts in Organic Compounds', *Annales chimie physiques*, 1836, 61,146. Translated in Henry M. Leicester and Herbert S. Klickstein, *A Source Book in Chemistry 1400–1900* (1952), 267.

6 A tidy laboratory means a lazy chemist.

Berzelius to Nils Sefström, 8th July 1812. In C. G. Bernard, 'Berzelius as a European Traveller', in E. M. Melhardo and T. Frängsmyr (eds.), *Enlightenment Science in the Romantic Era* (1992), 225.

Henry Beston 1888–1968

American author

7 We need another and a wiser and perhaps a more mystical concept of animals. Remote from universal nature, and living by complicated artifice, man in civilization surveys the creature through the glass of his knowledge and sees thereby a feather magnified and the whole image in distortion. We patronize them for their incompleteness, for their tragic fate of having taken form so far below ourselves. And therein we err, and greatly err. For the animal shall not be measured by man. In a world older and more complete than ours they move finished and complete, gifted with extensions of the senses we have lost or never attained, living by voices we shall never hear. They are not brethren, they are not underlings; they are other nations, caught with ourselves in the net of life and time, fellow prisoners of the splendour and travail of the earth.

The Outermost House (1928), 25.

Hans Albrecht Bethe 1906–2005

German-born American physicist

1 Finally I got to carbon, and as you all know, in the case of carbon the reaction works out beautifully. One goes through six reactions, and at the end one comes back to carbon. In the process one has made four hydrogen atoms into one of helium. The theory, of course, was not made on the railway train from Washington to Ithaca . . . It didn't take very long, it took about six weeks, but not even the Trans-Siberian railroad [has] taken that long for its journey.

'Pleasure from Physics', From *A Life of Physics: Evening Lectures at the International Centre for Theoretical Physics, Trieste, Italy. A Special Supplement of the IAEA Bulletin* (1968), 14.

William Ian Beardmore Beveridge 1908–

Australian microbiologist

2 The rôle of hypothesis in research can be discussed more effectively if we consider first some examples of discoveries which originated from hypotheses. One of the best illustrations of such a discovery is provided by the story of Christopher Columbus' voyage; it has many of the features of a classic discovery in science. (*a*) He was obsessed with an idea—that since the world is round he could reach the Orient by sailing West, (*b*) the idea was by no means original, but evidently he had obtained some additional evidence from a sailor blown off his course who claimed to have reached land in the west and returned, (*c*) he met great difficulties in getting someone to provide the money to enable him to test his idea as well as in the actual carrying out of the experimental voyage, (*d*) when finally he succeeded he did not find the expected new route, but instead found a whole new world, (*e*) despite all evidence to the contrary he clung to the bitter end to his hypothesis and believed that he had found the route to the Orient, (*f*) he got little credit or reward during his lifetime and neither he nor others realised the full implications of his discovery, (*g*) since his time evidence has been brought forward showing that he was by no means the first European to reach America.

The Art of Scientific Investigation (1950), 41.

3 No one believes an hypothesis except its originator but everyone believes an experiment except the experimenter. Most people are ready to believe something based on experiment but the experimenter knows the many little things that could have gone wrong in the experiment. For this reason the discoverer of a new fact seldom feels quite so confident of it as others do. On the other hand other people are usually critical of an hypothesis, whereas the originator identifies himself with it and is liable to become devoted to it.

The Art of Scientific Investigation (1950), 47.

4 More discoveries have arisen from intense observation of very limited material than from statistics applied to large groups. The value of the latter lies mainly in testing hypotheses arising from the former. While observing one should cultivate a speculative, contemplative attitude of mind and search for clues to be followed up. Training in observation follows the same principles as training in any activity. At first one must do things consciously and laboriously, but with practice the activities gradually become automatic and unconscious and a habit is established. Effective scientific observation also requires a good background, for only by being familiar with the usual can we notice something as being unusual or unexplained.

The Art of Scientific Investigation (1950), 101.

The Bible

1 In the beginning God created the heaven and earth. And the earth was waste and void; and darkness was upon the face of the deep: and the spirit of God moved upon the face of the waters.

Genesis 1:1 in *The Holy Bible Containing the Old and New Testaments Translated Out of the Original Tongues. Printed for the Universities of Oxford and Cambridge* (1895), 1.

2 And there was evening and there was morning, one day.

Genesis 1:5 in *The Holy Bible Containing the Old and New Testaments Translated Out of the Original Tongues. Printed for the Universities of Oxford and Cambridge* (1895), 1.

3 And God saw that it was good.

Genesis 1:11 in *The Holy Bible Containing the Old and New Testaments Translated Out of the Original Tongues. Printed for the Universities of Oxford and Cambridge* (1895), 1.

4 And God made the two great lights; the greater light to rule the day, and the lesser light to rule the night: *he made* the stars also.

Genesis 1:16 in *The Holy Bible Containing the Old and New Testaments Translated Out of the Original Tongues. Printed for the Universities of Oxford and Cambridge* (1895), 2.

5 And God said, Let us make man in our image, after our likeness: and let them have dominion over the fish of the sea, and over the fowl of the air, and over the cattle, and over all the earth, and over every creeping thing that creepeth upon the earth.

Genesis 1:26 in *The Holy Bible Containing the Old and New Testaments Translated Out of the Original Tongues. Printed for the Universities of Oxford and Cambridge* (1895), 3.

6 Male and female created he.

Genesis 1:27 in *The Holy Bible Containing the Old and New Testaments Translated Out of the Original Tongues. Printed for the Universities of Oxford and Cambridge* (1895), 3.

7 God said unto them, Be fruitful, and multiply, and replenish the earth, and subdue it.

Genesis 1:28 in *The Holy Bible Containing the Old and New Testaments Translated Out of the Original Tongues. Printed for the Universities of Oxford and Cambridge* (1895), 3.

8 While the earth remaineth, seedtime and harvest, and cold and heat, and summer and winter, and day and night shall not cease.

Genesis 8:22 in *The Holy Bible Containing the Old and New Testaments Translated Out of the Original Tongues. Printed for the Universities of Oxford and Cambridge* (1895), 16.

9 Thou shalt not suffer a sorceress to live.

Exodus 22:18 in *The Holy Bible Containing the Old and New Testaments Translated Out of the Original Tongues. Printed for the Universities of Oxford and Cambridge* (1895),173.

10 What hath God wrought!

Numbers 23:23 in *The Holy Bible Containing the Old and New Testaments Translated Out of the Original Tongues. Printed for the Universities of Oxford and Cambridge* (1895), 366. See **Morse** 448:4

11 To every thing there is a season, and a time to every purpose under heaven: a time to be born, and a time to die; a time to plant and a time to pluck up that which is planted; a time to kill, and a time to heal; a time to break down, and a time to build up; a time to weep, and a time to laugh; a time to mourn, and a time to dance; a time to cast away stones, and a time to gather stones together; a time to embrace, and a time to refrain from embracing; a time to seek, and a time to lose; a time to keep, and a time to cast away; a time to rend, and a time to sew; a time to keep silence, and a time to speak; a time to love, and a time for hate; a time for war and a time for peace.

Ecclesiastes 3:1 in *The Holy Bible Containing the Old and New Testaments Translated Out of the Original Tongues. Printed for the Universities of Oxford and Cambridge* (1895), 1466–7.

1 And further, my son, be admonished: of making many books there is no end; and much study is a weariness of flesh.

Ecclesiastes 12:12 in *The Holy Bible Containing the Old and New Testaments Translated Out of the Original Tongues. Printed for the Universities of Oxford and Cambridge* (1895), 1480.

Marie-François-Xavier Bichat 1771–1802

French physiologist and surgeon

2 Open up a few corpses: you will dissipate at once the darkness that observation alone could not dissipate.

Anatomie générale appliquée à la physiologie à la médecine (1801), avant-propos, xic.

3 Life consists in the sum of the functions, by which death is resisted.

Physiological Researches on Life and Death (1815), trans. F. Gold, 21.

4 Thus it might be said, that the vegetable is only the sketch, nor rather the ground-work of the animal; that for the formation of the latter, it has only been requisite to clothe the former with an apparatus of external organs, by which it might be connected with external objects.

From hence it follows, that the functions of the animal are of two very different classes. By the one (which is composed of an habitual succession of assimilation and excretion) it lives within itself, transforms into its proper substance the particles of other bodies, and afterwards rejects them when they are become heterogeneous to its nature. By the other, it lives externally, is the inhabitant of the world, and not as the vegetable of a spot only; it feels, it perceives, it reflects on its sensations, it moves according to their influence, and frequently is enabled to communicate by its voice its desires, and its fears, its pleasures, and its pains.

The aggregate of the functions of the first order, I shall name the organic life, because all organized beings, whether animal or vegetable, enjoy it more or less, because organic texture is the sole condition necessary to its existence. The sum of the functions of the second class, because it is exclusively the property of the animal, I shall denominate the animal life.

Physiological Researches on Life and Death (1815), trans. F. Gold, 22–3.

John Shaw Billings 1838–1913

American surgeon and librarian

5 Statistics are somewhat like old medical journals, or like revolvers in newly opened mining districts. Most men rarely use them, and find it troublesome to preserve them so as to have them easy of access; but when they do want them, they want them badly.

'On Vital and Medical Statistics', *The Medical Record*, 1889, 36, 589.

Christian Albert Theodor Billroth 1829–94

Austrian surgeon

6 It is a most gratifying sign of the rapid progress of our time that our best text-books become antiquated so quickly.

The Medical Sciences in the German Universities (1924), 49.

Alfred Binet 1857–1911

French psychologist

7 Heredity, to our understanding is not capable of giving to this illness (paraphilia) its characteristic form . . . Heredity invents nothing, creates nothing anew; it has no imagination.

Études de psychologie expérimentale: Le fétichisme dans l'amour (1888), 42.

1 Mere numbers cannot bring out . . . the intimate essence of the experiment. This conviction comes naturally when one watches a subject at work. . . . What things can happen! What reflections, what remarks, what feelings, or, on the other hand, what blind automatism, what absence of ideas! . . . The experimenter judges what may be going on in [the subject's] mind, and certainly feels difficulty in expressing all the oscillations of a thought in a simple, brutal number, which can have only a deceptive precision. How, in fact, could it sum up what would need several pages of description!

La suggestibilité (1900), 119–20.

2 I wish that one would be persuaded that psychological experiments, especially those on the complex functions, are not improved [by large studies]; the statistical method gives only mediocre results; some recent examples demonstrate that. The American authors, who love to do things big, often publish experiments that have been conducted on hundreds and thousands of people; they instinctively obey the prejudice that the persuasiveness of a work is proportional to the number of observations. This is only an illusion.

L'Etude expérimentale de l'intelligence (1903), 299.

3 Comprehension, inventiveness, direction, and criticism: intelligence is contained in these four words.

Les idées modernes sur les enfants (1909), 118.

4 Some recent philosophers seem to have given their moral approval to these deplorable verdicts that affirm that the intelligence of an individual is a fixed quantity, a quantity that cannot be augmented. We must protest and react against this brutal pessimism; we will try to demonstrate that it is founded on nothing.

Les idées modernes sur les enfants (1909), 141.

Alfred Binet 1857–1911 and Théodore Simon 1873–1961

French psychologists

5 Our purpose is to be able to measure the intellectual capacity of a child who is brought to us in order to know whether he is normal or retarded. . . . We do not attempt to establish or prepare a prognosis and we leave unanswered the question of whether this retardation is curable, or even improveable. We shall limit ourselves to ascertaining the truth in regard to his present mental state.

'New Methods for the Diagnosis of the Intellectual Level of Subnormals' (1905), in *The Development of Intelligence in Children*, trans. Elizabeth Kite (1916), 37.

Joseph Black 1728–99

British physicist and chemist

See ***Brougham, Henry Peter*** *93:4, 93:5;* ***Buckle, Henry Thomas*** *99:6*

6 The opinion I formed from attentive observation of the facts and phenomena, is as follows. When ice, for example, or any other solid substance, is changing into a fluid by heat, I am of opinion that it receives a much greater quantity of heat than that what is perceptible in it immediately after by the thermometer. A great quantity of heat enters into it, on this occasion, without making it apparently warmer, when tried by that instrument. This heat, however, must be thrown into it, in order to give it the form of a fluid; and I affirm, that this great addition of heat is the principal, and most immediate cause of the fluidity induced. And, on the other hand, when we deprive such a body of its fluidity again, by a diminution of its heat, a very great quantity of heat comes out of it, while it is assuming a solid form, the loss of which heat is not to be perceived by the common manner of using the thermometer. The apparent heat of the body, as measured by that instrument, is not diminished, or not in proportion to the loss of heat which the body actually gives out on this occasion; and it appears from a number of facts, that the state of solidity cannot be induced without the abstraction of this great quantity of heat. And this confirms the opinion, that this quantity of heat, absorbed, and, as it were, concealed in the composition of fluids, is the most necessary and immediate cause of their fluidity.

Lectures on the Elements of Chemistry, delivered in the University of Edinburgh (1803), Vol. 1, 116–7.

1 What I have related is sufficient for establishing the main principle, namely, that the heat which disappears in the conversion of water into vapour, is not lost, but is retained in vapour, and indicated by its expansive form, although it does not affect the thermometer. This heat emerges again from this vapour when it becomes water, and recovers its former quality of affecting the thermometer; in short, it appears again as the cause of heat and expansion.

Lectures on the Elements of Chemistry, delivered in the University of Edinburgh (1803), Vol. 1, 173.

Patrick Maynard Stuart Blackett (Lord Blackett) 1897–1974

British physicist

2 A first-rate laboratory is one in which mediocre scientists can produce outstanding work.

Quoted by M. G. K. Menon in his commemoration lecture on H. J. Bhabba, Royal Institution 1967.

Richard Blackmore 1654–1729

British physician and writer

3 *Copernicus*, who rightly did condemn
This eldest systeme, form'd a wiser scheme;
In which he leaves the Sun at Rest, and rolls
The Orb Terrestial on its proper Poles;
Which makes the Night and Day by this Career,
And by its slow and crooked Course the Year.
The famous *Dane*, who oft the Modern guides,
To Earth and Sun their Provinces divides:
The Earth's Rotation makes the Night and Day,
The Sun revolving through th'Eccliptic Way
Effects the various seasons of the Year,
Which in their Turn for happy Ends appear.
This Scheme or that, which pleases best, embrace,
Still we the Fountain of their Motion trace.
Kepler asserts these Wonders may be done
By the Magnetic Vertue of the Sun,
Which he, to gain his End, thinks fit to place
Full in the Center of that mighty Space,
Which does the Spheres, where Planets roll, include,
And leaves him with Attractive Force endu'd.
The Sun, thus seated, by Mechanic Laws,
The Earth, and every distant Planet draws;
By which Attraction all the Planets found
Within his reach, are turn'd in *Ether* round.

Creation: A Philosophical Poem in Seven Books (1712), book 2, l. 430–53, p.78–9.

Richard Eliot Blackwelder 1880–1969

American zoologist and entomologist

4 But as a nation we have not yet come to have a proper respect for the forest and to regard it as an indispensable part of our resources—one which is easily destroyed but difficult to replace; one which confers great benefits while it endures, but whose disappearance is accompanied by a train of evil consequences not readily foreseen and positively irreparable.

'A Country that has Used up its Trees', *The Outlook*, 1906, **82**, 700.

William Blake 1757–1827
British poet, painter and visionary

1 Energy is Eternal Delight.
'The Marriage of Heaven and Hell' (1790). In W. H. Stevenson (ed.), *The Poems of William Blake* (1971), 106.

2 Mock on, mock on, Voltaire, Rousseau!
Mock on, mock on: 'Tis all in vain!
You throw the sand against the wind,
And the wind blows it back again.
And every sand becomes a gem
Reflected in the beams divine;
Blown back they blind the mocking eye,
But still in Israel's paths they shine.
The atoms of Democritus
And Newton's particles of light
Are sands upon the Red Sea shore,
Where Israel's tents do shine so bright.
Notebook Drafts (*c*.1804). In W. H. Stevenson (ed.), *The Poems of William Blake* (1971), 481.

3 To see a world in a grain of sand
And a heaven in a wild flower,
Hold infinity in the palm of your hand
And eternity in an hour.
'The Pickering Manuscript'—Auguries of Innocence (*c*.1805). In W. H. Stevenson (ed.), *The Poems of William Blake* (1971), 585.

4 I turn my eyes to the schools & universities of Europe
And there behold the loom of Locke whose woof rages dire,
Washed by the water-wheels of Newton. Black the cloth
In heavy wreaths folds over every nation; cruel works
Of many wheels I view, wheel without wheel, with cogs tyrannic
Moving by compulsion each other: not as those in Eden, which
Wheel within wheel in freedom revolve, in harmony & peace.
'Jerusalem, The Emanation of the Giant Albion' (1804–20), First Chapter, Pl.15, lines 14–20. In W. H. Stevenson (ed.), *The Poems of William Blake* (1971), 654–55.

5 Pray God us keep
From Single vision & Newton's sleep!
Letter to Thomas Butt, 22 November 1802. Quoted in Geoffrey Keynes (ed.), *The Letters of William Blake* (1956), 79.

Gilbert Blane 1749–1834
British physician

6 And it has been sarcastically said, that there is a wide difference between a good physician and a bad one, but a small difference between a good physician and no physician at all; by which it is meant to insinuate, that the mischievous officiousness of art does commonly more than counterbalance any benefit derivable from it.
Elements of Medical Logic (1819), 8–9.

Mathilde Blind 1841–96
British poet

7 War rages on the teeming earth;
The hot and sanguinary fight
Begins with each new creature's birth:
A dreadful war where might is right;
Where still the strongest slay and win,
Where weakness is the only sin.
The Ascent of Man (1889), 13.

Baroness Karen Blixen (Isak Dinesen) 1885–1962
Danish writer

8 What is man, when you come to think upon him, but a minutely set, ingenious machine for turning, with infinite artfulness, the red wine of Shiraz into urine?
'The Dreamers', *Seven Gothic Tales* (1943), 332.

Leonard Bloomfield 1887–1949
American linguist

9 If it were possible to transfer the methods of physical or of biological science directly to the study of man, the transfer would long ago have been made . . . We have failed not for lack of hypotheses which equate man with the rest of the universe, but for lack of a hypothesis (short of animism) which provides for the peculiar divergence of man . . . Let me now state my belief that the peculiar factor in man which forbids our explaining his actions upon the ordinary plane of biology is a highly specialized and unstable biological complex, and that this factor is none other than language.

Linguistics as a Science (1930), 555.

Franz Boas 1858–1942

German-born American anthropologist

1 Anthropology has reached that point of development where the careful investigation of facts shakes our firm belief in the far-reaching theories that have been built up. The complexity of each phenomenon dawns on our minds, and makes us desirous of proceeding more cautiously. Heretofore we have seen the features common to all human thought. Now we begin to see their differences. We recognize that these are no less important than their similarities, and the value of detailed studies becomes apparent. Our aim has not changed, but our method must change. We are still searching for the laws that govern the growth of human culture, of human thought; but we recognize the fact that before we seek for what is common to all culture, we must analyze each culture by careful and exact methods, as the geologist analyzes the succession and order of deposits, as the biologist examines the forms of living matter. We see that the growth of human culture manifests itself in the growth of each special culture. Thus we have come to understand that before we can build up the theory of the growth of all human culture, we must know the growth of cultures that we find here and there among the most primitive tribes of the Arctic, of the deserts of Australia, and of the impenetrable forests of South America; and the progress of the civilization of antiquity and of our own times. We must, so far as we can, reconstruct the actual history of mankind, before we can hope to discover the laws underlying that history.

The Jesup North Pacific Expedition: Memoir of the American Museum of Natural History (1898), Vol. 1, 4.

2 It is clear, from these considerations, that the three methods of classifying mankind—that according to physical characters, according to language, and according to culture—all reflect the historical development of races from different standpoints; and that the results of the three classifications are not comparable, because the historical facts do not affect the three classes of phenomena equally. A consideration of all these classes of facts is needed when we endeavour to reconstruct the early history of the races of mankind.

'Summary of the Work of the Committee in British Columbia', *Report of the Sixty-Eighth Meeting of the British Association for the Advancement of Science*, 1899, 670.

3 Its [the anthropological method] power to make us understand the roots from which our civilization has sprung, that it impresses us with the relative value of all forms of culture, and thus serves as a check to an exaggerated valuation of the standpoint of our own period, which we are only too liable to consider the ultimate goal of human evolution, thus depriving ourselves of the benefits to be gained from the teachings of other cultures and hindering an objective criticism of our own work.

'The History of Anthropology', *Science*, 1904, 20, 524.

4 There is no fundamental difference in the ways of thinking of primitive and civilized man. A close connection between race and personality has never been established.

The Mind of Primitive Man (1938), preface, v.

Hermann Boerhaave 1668–1738

Dutch physician, botanist and chemist

5 Generation by male and female is a law common to animals and plants.

Preface to Louis Ferdinand Compte de Marsilli's *Histoire Physique de la Mer* (1725), ix.

6 For chemistry is no science form'd *à priori*; 'tis no production of the human mind, framed by reasoning and deduction: it took its rise from a number of experiments casually made, without any expectation of what follow'd; and was only reduced into an art or system, by collecting and comparing the effects of such unpremeditated experiments, and observing the uniform tendency thereof. So far, then, as a number of experimenters agree to establish any undoubted truth; so far they may be consider'd as constituting the theory of chemistry.

From 'The Author's Preface', in *A New Method of Chemistry* (1727), vi.

1 If this fire determined by the sun, be received on the blackest known bodies, its heat will be long retain'd therein; and hence such bodies are the soonest and the strongest heated by the flame fire, as also the quickest dried, after having been moisten'd with water; and it may be added, that they also burn by much the readiest: all which points are confirm'd by daily observations. Let a piece of cloth be hung in the air, open to the sun, one part of it dyed black, another part of a white colour, others of scarlet, and diverse other colours; the black part will always be found to heat the most, and the quickest of all; and the others will each be found to heat more slowly, by how much they reflect the rays more strongly to the eye; thus the white will warm the slowest of them all, and next to that the red, and so of the rest in proportion, as their colour is brighter or weaker.

A New Method of Chemistry, 2nd edition (1741), 262.

2 It were indeed to be wish'd that our art had been less ingenious, in contriving means destructive to mankind; we mean those instruments of war, which were unknown to the ancients, and have made such havoc among the moderns. But as men have always been bent on seeking each other's destruction by continual wars; and as force, when brought against us, can only be repelled by force; the chief support of war, must, after money, be now sought in chemistry.

A New Method of Chemistry, 3rd edition (1753), Vol. 1, trans. P. Shaw, 189–90.

Anicius Manlius Severinus Boethius c.480–c.524/5

Roman scholar and philosopher

3 Compare the length of a moment with the period of ten thousand years; the first, however minuscule, does exist as a fraction of a second. But that number of years, or any multiple of it that you may name, cannot even be compared with a limitless extent of time, the reason being that comparisons can be drawn between finite things, but not between finite and infinite.

The Consolation of Philosophy [before 524], Book II, trans. P. G. Walsh (1999), 36.

4 If one were to define chance as the outcome of a random movement which interlocks with no causes, I should maintain that it does not exist at all, that it is a wholly empty term denoting nothing substantial.

The Consolation of Philosophy [before 524], Book V, trans. P. G. Walsh (1999), 97.

Niels Henrik David Bohr 1885–1962

Danish physicist

See **Eddington, Arthur Stanley** 194:6

5 The present state of atomic theory is characterized by the fact that we not only believe the existence of atoms to be proved beyond a doubt, but also we even believe that we have an intimate knowledge of the constituents of the individual atoms.

'The structure of the atom', Nobel Lecture, 11 December 1922. In *Nobel Lectures: Physics 1922–1941* (1998), 5.

6 When searching for harmony in life one must never forget that in the drama of existence we are ourselves both actors and spectators.

Niels Bohr, 'Discussion with Einstein on Epistemological Problems in Atomic Physics', in P. A. Schilpp (ed.), *Albert Einstein: Philosopher-Scientist* (1949), 236.

1 The old saying of the two kinds of truth. To the one kind belongs statements so simple and clear that the opposite assertion obviously could not be defended. The other kind, the so-called 'deep truths', are statements in which the opposite also contains deep truth.

Niels Bohr, 'Discussion with Einstein on Epistemological Problems in Atomic Physics', in P. A. Schilpp (ed.), *Albert Einstein: Philosopher-Scientist* (1949), 240.

2 How wonderful that we have met with a paradox. Now we have some hope of making progress.

Quoted in R. Moore, *Niels Bohr, the Man and the Scientist* (1967), 140.

3 What is that we human beings ultimately depend on? We depend on our words. We are suspended in language. Our task is to communicate experience and ideas to others.

Quoted in Aage Petersen, 'The Philosophy of Niels Bohr', *Bulletin of the Atomic Scientists*, 1963, 19, 10.

4 It is wrong to think that the task of physics is to find out how nature *is*. Physics concerns what we can say about Nature.

Quoted in Aage Petersen, 'The Philosophy of Niels Bohr', *Bulletin of the Atomic Scientists*, 1963, 19, 12.

5 The existence of life must be considered as an elementary fact that can not be explained, but must be taken as a starting point in biology, in a similar way as the quantum of action, which appears as an irrational element from the point of view of classical mechanical physics, taken together with the existence of elementary particles, forms the foundation of atomic physics. The asserted impossibility of a physical or chemical explanation of the function peculiar to life would in this sense be analogous to the insufficiency of the mechanical analysis for the understanding of the stability of atoms.

'Light and Life', *Nature*, 1933, 131, 458.

6 Predictions can be very difficult—especially about the future.

Quoted in H. Rosovsky, *The University: An Owners Manual* (1991), 147. It is said that Bohr used to quote this saying to illustrate the differences between Danish and Swedish humour. Bohr always attributed the saying to Robert Storm Petersen (1882–1949), a well-known Danish artist and writer. However, the saying did NOT originate from Petersen. It may have been said in the Danish Parliament between 1935 and 1939 [Information supplied courtesy of Professor Erik Rüdinger, Niels Bohr Archive, Copenhagen].

7 If we couldn't laugh at ourselves, that would be the end of everything.

Comment made to Professor Erik Rüdinger, 1962. Quotation supplied and translated by Professor Erik Rüdinger, Niels Bohr Archive.

8 When it comes to atoms, language can be used only as in poetry.

Quoted in K. C. Cole, 'On Imagining the Unseeable', *Discover*, 1982, 3, 70.

9 You must come to Copenhagen to work with us. We like people who can actually perform thought experiments!

Said to Otto Frisch. Quoted in Otto R. Frisch, *What Little I Remember* (1979), 76.

10 But, but, but . . . if anybody says he can think about quantum theory *without* getting giddy it merely shows that he hasn't understood the first thing about it!

Quoted in Otto R. Frisch, *What Little I Remember* (1979), 95.

11 ON BOHR When Bohr is about everything is somehow different. Even the dullest gets a fit of brilliancy.

Isidor I. Rabi in Daniel J. Kevles, *The Physicists* (1978), 201.

12 ON BOHR *Contraria sunt complementa.*

Opposites are complementary.

Motif on Niels Bohr's coat of arms.

Louis Bolk 1866–1930

Dutch anatomist

13 If I wished to express the basic principle of my ideas in a somewhat strongly worded sentence, I would say that man, in his bodily development, is a primate fetus that has become sexually mature [*einen zur Geschlechsreife gelangten Primatenfetus*].

Das Problem der Menschwerdung (1926), 8. Trans. in Stephen Jay Gould, *Ontogeny and Phylogeny* (1977), 361.

Ludwig Boltzmann 1844–1906

Austrian physicist

1 We must make the following remark: a proof, that after a certain time t_1 the spheres must necessarily be mixed uniformly, whatever may be the initial distribution of states, cannot be given. This is in fact a consequence of probability theory, for any non-uniform distribution of states, no matter how improbable it may be, is still not absolutely impossible. Indeed it is clear that any individual uniform distribution, which might arise after a certain time from some particular initial state, is just as improbable as an individual non-uniform distribution; just as in the game of Lotto, any individual set of five numbers is as improbable as the set 1, 2, 3, 4, 5. It is only because there are many more uniform distributions than non-uniform ones that the distribution of states will become uniform in the course of time. One therefore cannot prove that, whatever may be the positions and velocities of the spheres at the beginning, the distributions must become uniform after a long time; rather one can only prove that infinitely many more initial states will lead to a uniform one after a definite length of time than to a non-uniform one. Loschmidt's theorem tells us only about initial states which actually lead to a very non-uniform distribution of states after a certain time t_1; but it does not prove that there are not infinitely many more initial conditions that will lead to a uniform distribution after the same time. On the contrary, it follows from the theorem itself that, since there are infinitely many more uniform distributions, the number of states which lead to uniform distributions after a certain time t_1 is much greater than the number that leads to non-uniform ones, and the latter are the ones that must be chosen, according to Loschmidt, in order to obtain a non-uniform distribution at t_1.

'On the Relation of a General Mechanical Theorem to the Second Law of Thermodynamics' (1877), in Stephen G. Brush (ed.), *Selected Readings in Physics* (1966), Vol. 2, Irreversible Processes, 191–2.

2 Since a given system can never of its own accord go over into another equally probable state but into a more probable one, it is likewise impossible to construct a system of bodies that after traversing various states returns periodically to its original state, that is a perpetual motion machine.

'The Second Law of Thermodynamics', *Populäre Schriften*, Essay 3. Address to a Formal meeting of the Imperial Academy of Science, 29 May 1886. In Brian McGuinness (ed.), *Ludwig Boltzmann: Theoretical Physics and Philosophical Problems, Selected Writings* (1974), 30.

3 A closer look at the course followed by developing theory reveals for a start that it is by no means as continuous as one might expect, but full of breaks and at least apparently not along the shortest logical path. Certain methods often afforded the most handsome results only the other day, and many might well have thought that the development of science to infinity would consist in no more than their constant application. Instead, on the contrary, they suddenly reveal themselves as exhausted and the attempt is made to find other quite disparate methods. In that event there may develop a struggle between the followers of the old methods and those of the newer ones. The former's point of view will be termed by their opponents as out-dated and outworn, while its holders in turn belittle the innovators as corrupters of true classical science.

'On the Development of the Methods of Theoretical Physics in Recent Times', *Populäre Schriften*, Essay 14. Address to the Meeting of Natural Scientists at Munich, 22 September 1899. In Brian McGuinness (ed.), *Ludwig Boltzmann: Theoretical Physics and*

Philosophical Problems, Selected Writings (1974), 79.

1 The most ordinary things are to philosophy a source of insoluble puzzles. In order to explain our perceptions it constructs the concept of matter and then finds matter quite useless either for itself having or for causing perceptions in a mind. With infinite ingenuity it constructs a concept of space or time and then finds it absolutely impossible that there be objects in this space or that processes occur during this time . . . The source of this kind of logic lies in excessive confidence in the so-called laws of thought.

'On Statistical Mechanics' (1904), in *Theoretical Physics and Philosophical Problems* (1974), 164–5.

2 ON **BOLTZMANN** $S = k \log \Omega$

Carved above his name on his tombstone in the Zentralfriedhof in Vienna.

Image in Stephen Brush, *The Kind of Motion we Call Heat: A History of the Kinetic Theory of Gases in the 19th Century* (1976), 609.

Marie Bonaparte 1882–1962

French psychoanalyst

3 On the one hand, then, in the reproductive functions proper—menstruation, defloration, pregnancy and parturition—woman is biologically doomed to suffer. Nature seems to have no hesitation in administering to her strong doses of pain, and she can do nothing but submit passively to the regimen prescribed. On the other hand, as regards sexual attraction, which is necessary for the act of impregnation, and as regards the erotic pleasures experienced during the act itself, the woman may be on an equal footing with the man.

'Passivity, Masochism and Femininity', *Journal of Psychoanalysis*, 1935, 16, 327.

Napoléon Bonaparte 1769–1821

French emperor

See ***Laplace, Pierre-Simon*** *366:1*

4 *To Laplace, on receiving a copy of the* Mécanique Céleste*:*

The first six months which I can spare will be employed in reading it.

Correspondance de Napoléon Ier, 27 vendémiaire an VIII [19 October 1799] no. 4384 (1861), Vol. 6, 1. Trans. Charles Coulston Gillispie, *Pierre-Simon Laplace 1749–1827: A Life in Exact Science* (1997), 176.

5 The advancement and perfection of mathematics are intimately connected with the prosperity of the State.

Quoted in Q. Mushtaq and A. L. Tan, *Mathematics: The Islamic Legacy* (1993), 11.

Hermann Bondi 1919–2005

Austrian-born British mathematician and cosmologist

6 An observer situated in a nebula and moving with the nebula will observe the same properties of the universe as any other similarly situated observer at any time.

'Review of Cosmology', *Monthly Notices of the Royal Astronomical Society*, 1948, **108**, 107.

7 Sometimes I am a little unkind to all my many friends in education . . . by saying that from the time it learns to talk every child makes a dreadful nuisance of itself by asking 'Why?'. To stop this nuisance society has invented a marvellous system called education which, for the majority of people, brings to an end their desire to ask that question. The few failures of this system are known as scientists.

'The Making of a Scientist', *Journal of the Royal Society of Arts*, June 1983, 403.

8 If you walk along the street you will encounter a number of scientific problems. Of these, about 80 per cent are insoluble, while 19½ per cent are trivial. There is then perhaps half a per cent where skill, persistence, courage, creativity and originality can make a difference. It is always the task of the academic to swim in that half a per cent, asking the questions through which some progress can be made.

'The Making of a Scientist', *Journal of the Royal Society of Arts*, June 1983, 406.

Charles Bonnet 1720–93

Swiss naturalist

1 Between the lowest and the highest degree of spiritual and corporal perfection, there is an almost infinite number of intermediate degrees. The succession of degrees comprises the *Universal Chain*. It unites all beings, ties together all worlds, embraces all the spheres. One SINGLE BEING is outside this chain, and this is HE who made it.

Contemplation de la nature (1764), Vol. 1, 27. Trans. Stephen Jay Gould, *Ontogeny and Phylogeny* (1977), 23.

Thomas George Bonney 1833–1923

British geologist

2 Perfect concordance among reformers is not to be expected; and men who are honestly struggling towards the light cannot hope to attain at one bound to the complete truth. There is always a danger lest the fascination of a new discovery should lead us too far. Men of science, being human, are apt, like lovers, to exaggerate the perfections and be a little blind to the faults of the object of their choice.

'The Anniversary Address of the President', *Quarterly Journal of the Geological Society of London*, 1885, 41, 55.

3 [Microscopic] evidence cannot be presented *ad populum*. What is seen with the microscope depends not only upon the instrument and the rock-section, but also upon the brain behind the eye of the observer. Each of us looks at a section with the accumulated experience of his past study. Hence the veteran cannot make the novice see with his eyes; so that what carries conviction to the one may make no appeal to the other. This fact does not always seem to be sufficiently recognized by geologists at large.

'The Anniversary Address of the President', *Quarterly Journal of the Geological Society of London*, 1885, 41, 59.

4 Pressure, no doubt, has always been a most important factor in the metamorphism of rocks; but there is, I think, at present some danger in over-estimating this, and representing a partial statement of truth as the whole truth. Geology, like many human beings, suffered from convulsions in its infancy; now, in its later years, I apprehend an attack of pressure on the brain.

'The Foundation-Stones of the Earth's Crust', *Nature*, 1888, 39, 93.

William Bowen Bonnor 1920–

British physicist

5 It is the business of science to offer rational explanations for all the events in the real world, and any scientist who calls on God to explain something is falling down on his job. This applies as much to the start of the expansion as to any other event. If the explanation is not forthcoming at once, the scientist must suspend judgment: but if he is worth his salt he will always maintain that a rational explanation will eventually be found. This is the one piece of dogmatism that a scientist can allow himself—and without it science would be in danger of giving way to superstition every time that a problem defied solution for a few years.

The Mystery of the Expanding Universe (1964), 122.

George Boole 1815–64

British mathematician

6 A distinguished writer [Siméon Denis Poisson] has thus stated the fundamental definitions of the science:

'The probability of an event is the reason we have to believe that it has taken place, or that it will take place.'

'The measure of the probability of an event is the ratio of the number of cases favourable to that event, to the total number of cases favourable or contrary, and all equally possible' (equally like to happen).

From these definitions it follows that the word *probability*, in its mathematical acceptation, has reference to the state of our knowledge of the circumstances under which an event may happen or fail. With the degree of information which we possess concerning the circumstances of an event, the reason we have to think that it will occur, or, to use a single term, our *expectation* of it, will vary. Probability is expectation founded upon partial knowledge. A perfect acquaintance with *all* the circumstances affecting the occurrence of an event would change expectation into certainty, and leave neither room nor demand for a theory of probabilities.

An Investigation of the Laws of Thought (1854), 258.

Émile Borel 1871–1956

French mathematician

1 It may seem rash indeed to draw conclusions valid for the whole universe from what we can see from the small corner to which we are confined. Who knows that the whole visible universe is not like a drop of water at the surface of the earth? Inhabitants of that drop of water, as small relative to it as we are relative to the Milky Way, could not possibly imagine that beside the drop of water there might be a piece of iron or a living tissue, in which the properties of matter are entirely different.

Space and Time (1926), 227.

Jorge Luis Borges 1899–1986

Argentine writer

2 Consider the eighth category, which deals with stones. Wilkins divides them into the following classifications: ordinary (flint, gravel, slate); intermediate (marble, amber, coral); precious (pearl, opal); transparent (amethyst, sapphire); and insoluble (coal, clay, and arsenic). The ninth category is almost as alarming as the eighth. It reveals that metals can be imperfect (vermilion, quicksilver); artificial (bronze, brass); recremental (filings, rust); and natural (gold, tin, copper). The whale appears in the sixteenth category: it is a viviparous, oblong fish. These ambiguities, redundances, and deficiencies recall those attributed by Dr. Franz Kuhn to a certain Chinese encyclopedia entitled *Celestial Emporium of Benevolent Knowledge*. On those remote pages it is written that animals are divided into (a) those that belong to the Emperor, (b) embalmed ones, (c) those that are trained, (d) suckling pigs, (e) mermaids, (f) fabulous ones, (g) stray dogs, (h) those that are included in this classification, (i) those that tremble as if they were mad, (j) innumerable ones, (k) those drawn with a very fine camel's hair brush, (l) others, (m) those that have just broken a flower vase, (n) those that resemble flies from a distance.

Other Inquisitions 1937–1952 (1964), trans. Ruth L. C. Simms, 103.

Edwin Garrigues Boring 1886–1968

American psychologist

3 Scientific truth, like puristic truth, must come about by controversy. Personally this view is abhorrent to me. It seems to mean that scientific truth must transcend the individual, that the best hope of science lies in its greatest minds being often brilliantly and determinedly wrong, but in opposition, with some third, eclectically minded, middle-of-the-road nonentity seizing the prize while the great fight for it, running off with it, and sticking it into a textbook for sophomores written from no point of view and in defense of nothing whatsoever. I hate this view, for it is not dramatic and it is not fair; and yet I believe that it is the verdict of the history of science.

'The Psychology of Controversy', (1929). In *History, Psychology and Science: Selected Papers* (1963), 68.

Max Born 1882–1970

German physicist

1 It is odd to think that there is a word for something which, strictly speaking, does not exist, namely, 'rest'. We distinguish between living and dead matter; between moving bodies and bodies at rest. This is a primitive point of view. What seems dead, a stone or the proverbial 'door-nail', say, is actually forever in motion. We have merely become accustomed to judge by outward appearances; by the deceptive impressions we get through our senses.

The Restless Universe (1935), 1.

2 The ultimate origin of the difficulty lies in the fact (or philosophical principle) that we are compelled to use the words of common language when we wish to describe a phenomenon, not by logical or mathematical analysis, but by a picture appealing to the imagination. Common language has grown by everyday experience and can never surpass these limits. Classical physics has restricted itself to the use of concepts of this kind; by analysing visible motions it has developed two ways of representing them by elementary processes; moving particles and waves. There is no other way of giving a pictorial description of motions—we have to apply it even in the region of atomic processes, where classical physics breaks down.

Atomic Physics (1957), 97.

3 I have tried to read philosophers of all ages and have found many illuminating ideas but no steady progress toward deeper knowledge and understanding. Science, however, gives me the feeling of steady progress: I am convinced that theoretical physics is actual philosophy. It has revolutionized fundamental concepts, e.g., about space and time (relativity), about causality (quantum theory), and about substance and matter (atomistics), and it has taught us new methods of thinking (complementarity) which are applicable far beyond physics.

My Life & My Views (1968), 48.

4 To present a scientific subject in an attractive and stimulating manner is an artistic task, similar to that of a novelist or even a dramatic writer. The same holds for writing textbooks.

My Life & My Views (1968), 48.

5 During my span of life science has become a matter of public concern and the *l'art pour l'art* standpoint of my youth is now obsolete. Science has become an integral and most important part of our civilization, and scientific work means contributing to its development. Science in our technical age has social, economic, and political functions, and however remote one's own work is from technical application it is a link in the chain of actions and decisions which determine the fate of the human race. I realized this aspect of science in its full impact only after Hiroshima.

My Life & My Views (1968), 49.

6 All attempts to adapt our ethical code to our situation in the technological age have failed.

My Life & My Views (1968), 52.

7 But in practical affairs, particularly in politics, men are needed who combine human experience and interest in human relations with a knowledge of science and technology. Moreover, they must be men of action and not contemplation. I have the impression that no method of education can produce people with all the qualities required. I am haunted by the idea that this break in human civilization, caused by the discovery of the scientific method, may be irreparable.

My Life & My Views (1968), 57–8.

Jean Baptiste Georges Marie Bory de Saint-Vincent 1778–1846

French biologist

8 Nature becomes fertile only by virtue of laws that oblige matter to organize itself into one of a number of necessarily very simple primitive forms. Because of their very simplicity, these are capable of constituting the basis for increasingly complex bodies, by the addition of organs calculated according to identical laws of possibility.

'Matière', *Dictionnaire Classique d'Histoire Naturelle* (1822–31), Vol. 10, 277, trans. J. Mandelbaum. Quoted in Pietro Corsi, *The Age of Lamarck* (1988), 225.

Roger Joseph Boscovich 1711–87

Croatian natural philosopher

1 Prejudice for regularity and simplicity is a source of error that has only too often infected philosophy.

'De litteraria expeditione per pontificiam ditionem', Accademia della scienze, Bologna, *Commentarii*, 1757, 4, 353, 361. Trans. J. L. Heilbron, *Weighing Imponderables and Other Quantitative Science around 1800* (1993), 227.

2 It will be found that everything depends on the composition of the forces with which the particles of matter act upon one another; and from these forces, as a matter of fact, all phenomena of Nature take their origin.

Philosophiae Naturalis Theoria (1758), sec. 1.5

3 Were it not for gravity one man might hurl another by a puff of his breath into the depths of space, beyond recall for all eternity.

Philosophiae Naturalis Theoria (1758), par. 552.

James Boswell 1740–95

British biographer and diarist

4 After we came out of the church, we stood talking for some time together of Bishop Berkeley's ingenious sophistry to prove the non-existence of matter, and that every thing in the universe is merely ideal. I observed, that though we are satisfied his doctrine is not true, it is impossible to refute it. I never shall forget the alacrity with which Johnson answered, striking his foot with mighty force against a large stone, till he rebounded from it, 'I refute it *thus*.'

Entry for Saturday 6th August 1763. In George Birkbeck-Hill (ed.), *Boswell's Life of Johnson* (1934–50), Vol. 1, 471.

Daniel B. Botkin 1937–

American ecologist

5 Wherever we seek to find constancy we discover change. Having looked at the old woodlands in Hutcheson Forest, at Isle Royale, and in the wilderness of the boundary waters, in the land of the moose and the wolf, and having uncovered the histories hidden within the trees and within the muds, we find that nature undisturbed is not constant in form, structure, or proportion, but changes at every scale of time and space. The old idea of a static landscape, like a single musical chord sounded forever, must be abandoned, for such a landscape never existed except in our imagination. Nature undisturbed by human influence seems more like a symphony whose harmonies arise from variation and change over many scales of time and space, changing with individual births and deaths, local disruptions and recoveries, larger scale responses to climate from one glacial age to another, and to the slower alterations of soils, and yet larger variations between glacial ages.

Discordant Harmonies (1990), 62.

Henri Bouasse 1866–1953

French physicist

6 [The French Academy of Sciences is] the receptacle of a crowd of mediocrities and ignoramuses whose places have been made as college professors, herb collectors, village veterinarians and assistant engineers of bridges and roads.

Said at a meeting at the University of Toulouse, 1 Feb 1911. Quoted in M. J. Nye, *Science in the Provinces* (1986), 136.

Kenneth E. Boulding 1910–93

American economist

7 The proposition that the meek (that is the adaptable and serviceable), inherit the earth is not merely a wishful sentiment of religion, but an iron law of evolution.

The Organizational Revolution (1953), 252.

1 DNA was the first three-dimensional Xerox machine.

'Energy and the Environment,' A paper given at Laramie College of Commerce and Industry, University of Wyoming, January 1976. In Richard P. Beilock (ed.) *Beasts, Ballads and Bouldingisms: A Selection of Writings by Kenneth E. Boulding* (1976), 160.

Nicolas-Antoine Boullanger

1722–59

French geologist

2 Nothing is sudden in nature: whereas the slightest storms are forecasted several days in advance, the destruction of the world must have been announced several years beforehand by heat waves, by winds, by meteorites, in short, by an infinity of phenomena.

L'Antiquité dévoilée par ses usages ou Examen critique des principales Opinions, Cérémonies & Institutions religieuses & politiques des différens Peuples de la Terre (1766), 373–4.

Matthew Boulton 1728–1809

British engineeer

3 I sell here, Sir, what all the world desires to have—POWER.

About the improved steam engine invented by James Watt and brought into production at Boulton's manufactory.

Entry for Friday 22 March 1776. In George Birkbeck-Hill (ed.), *Boswell's Life of Johnson* (1934–50), Vol. 2, 459.

Theodor Boveri 1862–1915

German biologist

4 At fertilization, these two 'haploid' nuclei are added together to make a 'diploid' nucleus that now contains 2a, 2b and so on; and, by the splitting of each chromosome and the regulated karyokinetic separation of the daughter chromosomes, this double series is inherited by both of the primary blastomeres. In the resulting resting nuclei the individual chromosomes are apparently destroyed. But we have the strongest of indications that, in the stroma of the resting nucleus, every one of the chromosomes that enters the nucleus survives as a well-defined region; and as the cell prepares for its next division this region again gives rise to the same chromosome (*Theory of the Individuality of the Chromosomes*). In this way the two sets of chromosomes brought together at fertilization are inherited by all the cells of the new individual. It is only in the germinal cells that the so called reduction division converts the double series into a single one. Out of the diploid state, the haploid is once again generated.

Arch. Zellforsch, 1909, 3, 181, trans. Henry Harris, *The Birth of the Cell* (1999), 171–2.

5 For it is not cell nuclei, not even individual chromosomes, but certain parts of certain chromosomes from certain cells that must be isolated and collected in enormous quantities for analysis; that would be the precondition for placing the chemist in such a position as would allow him to analyse [the hereditary material] more minutely than [can] the morphologists . . . For the morphology of the nucleus has reference at the very least to the gearing of the clock, but at best the chemistry of the nucleus refers only to the metal from which the gears are formed.

Ergebnisse über die Konstitution der chromatischen Substanz des Zellkerns (1904), 123. Translated in Robert Olby, *The Path to the Double Helix: The Discovery of DNA* (1994), xx.

Henry Pickering Bowditch

1840–1911

American physiologist

6 An induction shock results in a contraction or fails to do so according to its strength; if it does so at all, it produces in the muscle at that time the maximal contraction that can result from stimuli of any strength.

'Über die Eigentümlichkeiten der Reizbarkeit welche die Muskelfasern des Herzen zeigen', *Ber. sächs. Akad. Wiss., Math.-nat Klasse*, 1871, 23, 652–689. Trans. Edwin Clarke and C. D. O'Malley, *The Human Brain and Spinal Cord* (1968), 218.

Nathaniel Bowditch 1773–1838

American mathematician and astronomer

1 Whenever I meet in Laplace with the words 'Thus it plainly appears', I am sure that hours and perhaps days, of hard study will alone enable me to discover *how* it plainly appears.

Mécanique céleste (1829–39), *Celestial mechanics* (1966).

George E. P. Box 1919–

American engineer

2 To find out what happens to a system when you interfere with it you have to interfere with it (not just passively observe it).

Use and Abuse of Regression (1966), 629.

Robert Boyle 1627–91

British natural philosopher and chemist

See **Burnet, Gilbert** 106:5; **Locke, John** 392:2

3 And when with excellent Microscopes I discern in otherwise invisible Objects the inimitable Subtlety of Nature's Curious Workmanship; And when, in a word, by the help of Anatomicall Knives, and the light of Chymicall Furnaces, I study the Book of Nature, and consult the Glosses of *Aristotle*, *Epicurus*, *Paracelsus*, *Harvey*, *Helmont*, and other learn'd Expositors of that instructive Volumne; I find my self oftentimes reduc'd to exclaim with the Psalmist, *How manifold are thy works, O Lord? In wisdom hast thou made them all.*

Some Motives and Incentives to the Love of God (1659), 56–7.

4 That there is a Spring, or Elastical power in the Air we live in. By which ἐλατηρ [elater] or Spring of the Air, that which I mean is this: That our Air either consists of, or at least abounds with, parts of such a nature, that in case they be bent or compress'd by the weight of the incumbent part of the Atmosphere, or by any other Body, they do endeavour, as much as in them lies, to free themselves from that pressure, by bearing against the contiguous Bodies that keep them bent.

New Experiments Physico-Mechanical Touching the Spring of the Air (1660), 22.

5 'Tis evident, that as common Air when reduc'd to half its wonted extent, obtained near about twice as forcible a Spring as it had before; so this thus-comprest Air being further thrust into half this narrow room, obtained thereby a Spring about as strong again as that it last had, and consequently four times as strong as that of the common Air. And there is no cause to doubt, that if we had been here furnisht with a greater quantity of Quicksilver and a very long Tube, we might by a further compression of the included Air have made it counter-balance the pressure of a far taller and heavier Cylinder of *Mercury*. For no man perhaps yet knows how near to an infinite compression the Air may be capable of, if the compressing force be competently increast.

A Defense of the Doctrine Touching the Spring and Weight of the Air (1662), 62.

6 And, to prevent mistakes, I must advertize you, that I now mean by elements, as those chymists that speak plainest do by their principles, certain primitive or simple, or perfectly unmingled bodies; which not being made of any other bodies, or of one another, are the ingredients of which all those called perfectly mixt bodies are immediately compounded, and into which they are ultimately resolved: now whether there be any such body to be constantly met with in all, and each, of those that are said to be elemented bodies, is the thing I now question.

The Sceptical Chemist (1661), 187.

7 If the omniscient author of nature knew that the study of his works tends to make men disbelieve his Being or Attributes, he would not have given them so many invitations to study and contemplate Nature.

'Some considerations touching the usefulness of experimental philosophy' (1663). Quoted in Peter Gay, *The Enlightenment* (1977), 140.

1 And let me adde, that he that throughly understands the nature of Ferments and Fermentations, shall probably be much better able than he that ignores them, to give a fair account of divers *phenomena* of severall diseases (as well Feavers and others) which will perhaps be never throughly understood, without an insight into the doctrine of Fermentation.

'Offering some Particulars relating to the Pathologicall Part of Physick', in *Of the Usefulnesse of Naturall Philosophy. The Second Part* (1663), 43.

2 I will not now discuss the Controversie betwixt some of the Modern Atomists, and the *Cartesians*; the former of whom think, that betwixt the Earth and the Stars, and betwixt these themselves there are vast Tracts of Space that are empty, save where the beams of Light do pass through them; and the later of whom tell us, that the Intervals betwixt the Stars and Planets (among which the Earth may perhaps be reckon'd) are perfectly fill'd, but by a Matter far subtiler than our Air, which some call Celestial, and others *Æther*. I shall not, I say, engage in this controversie, but thus much seems evident, That if there be such a Celestial Matter, it must make up far the Greatest part of the Universe known to us. For the Interstellar part of the world (if I may so stile it) bears so very great a proportion to the Globes, and their Atmospheres too, (if other Stars have any as well as the Earth,) that it is almost incomparably Greater in respect of them, than all our Atmosphere is in respect of the Clouds, not to make the comparison between the Sea and the Fishes that swim in it.

A Continuation of New Experiments Physico-Mechanical, Touching the Spring and Weight of the Air, and their Effects (1669), 127.

3 The Requisites of a good Hypothesis are:

That it be Intelligible.

That it neither Assume nor Suppose anything Impossible, unintelligible, or demonstrably False.

That it be consistent with itself.

That it be fit and sufficient to Explicate the Phaenomena, especially the chief.

That it be, at least, consistent, with the rest of the Phaenomena it particularly relates to, and do not contradict any other known Phaenomena of nature, or manifest Physical Truth.

The Qualities and Conditions of an Excellent Hypothesis are:

That it be not Precarious, but have sufficient Grounds in the nature of the Thing itself or at least be well recommended by some Auxiliary Proofs.

That it be the Simplest of all the good ones we are able to frame, at least containing nothing that is superfluous or Impertinent.

That it be the only Hypothesis that can Explicate the Phaenomena; or at least, that do's Explicate them so well.

That it enable a skilful Naturalist to foretell future Phaenomena by the Congruity or Incongruity to it; and especially the event of such Experim'ts as are aptly devis'd to examine it, as Things that ought, or ought not, to be consequent to it.

Boyle Papers, 37. Quoted in Barbara Kaplan (ed.), *Divulging of Useful Truths in Physick: The Medical Agenda of Robert Boyle* (1993), 50.

4 The inspired and expired air may be sometimes very useful, by condensing and cooling the blood that passeth through the lungs; I hold that the depuration of the blood in that passage, is not only one of the ordinary, but one of the principal uses of respiration.

New Experiments . . . Touching the Spring of Air. In *Works*, Vol. 1, 113. Quoted in Barbara Kaplan (ed.), *Divulging of Useful Truths in*

Physick: The Medical Agenda of Robert Boyle (1993), 85.

1 If the juices of the body were more chymically examined, especially by a naturalist, that knows the ways of making fixed bodies volatile, and volatile fixed, and knows the power of the open air in promoting the former of those operations; it is not improbable, that both many things relating to the nature of the humours, and to the ways of sweetening, actuating, and otherwise altering them, may be detected, and the importance of such discoveries may be discerned.

Quoted in Barbara Kaplan (ed.) *Divulging of Useful Truths in Physick: The Medical Agenda of Robert Boyle* (1993), 92.

2 I look upon a good physician, not so properly as a servant to nature, as one, that is a counsellor and friendly assistant, who, in his patient's body, furthers those motions and other things, that he judges conducive to the welfare and recovery of it; but as to those, that he perceives likely to be hurtful, either by increasing the disease, or otherwise endangering the patient, he thinks it is his part to oppose or hinder, though nature do manifestly enough seem to endeavour the exercising or carrying on those hurtful motions.

Quoted in Barbara Kaplan (ed.) *Divulging of Useful Truths in Physick: The Medical Agenda of Robert Boyle* (1993), 125.

3 The main thing that induces me to question the safeness of the vulgar *methodus medendi* in many cases is the consideration of the nature of those Helps they usually employ, and some of which are honoured with the title of *Generous Remedies*. These helps are Bleeding, Vomiting, Purging, Sweating, and Spitting, of which I briefly observe in General, that they are sure to weaken or discompose when they are imployed, but do not certainly cure afterwards.

RSMS 199, Folio 177v. Michael Hunter identifies as passages of a suppressed work, *Considerations and Doubts Touching the Vulgar Method of Physick*. Quoted in Barbara Kaplan (ed.), *Divulging of Useful Truths in Physick: The Medical Agenda of Robert Boyle* (1993), 138.

4 I consider then, that generally speaking, to render a reason of an effect or Phaenomenon, is to deduce it from something else in Nature more known than it self, and that consequently there may be divers kinds of Degrees of Explication of the same thing. For although such Explications be the most satisfactory to the Understanding, wherein 'tis shewn how the effect is produc'd by the more primitive and Catholick Affection of Matter, namely bulk, shape and motion, yet are not these Explications to be despis'd, wherein particular effects are deduc'd from the more obvious and familiar Qualities or States of Bodies, . . . For in the search after Natural Causes, every new measure of Discovery does both instruct and gratifie the Understanding.

Physiological Essays (1669), 20.

5 But the World being once fram'd, and the course of Nature establish'd, the Naturalist, (except in some few cases, where God, or Incorporeal Agents interpose), has recourse to the first Cause but for its general and ordinary Support and Influence, whereby it preserves Matter and Motion from Annihilation or Desition; and in explicating *particular phenomena*, considers onely the *Size*, *Shape*, *Motion*, (or *want of it*) *Texture*, and the resulting Qualities and Attributes of the small particles of Matter.

The Origine of Formes and Qualities (1666), 194.

6 The veneration, wherewith Men are imbued for what they call *Nature*, has been a discouraging impediment to the Empire of Man over the inferior Creatures of God. For many have not only look'd upon it, as an *impossible* thing to compass, but as something *impious* to attempt.

A Free Enquiry into the Vulgarly Receiv'd Notion of Nature Made in an Essay, Address'd to a Friend (1686), 18–9.

7 The subsequent course of nature, teaches, that God, indeed, gave motion to matter; but that, in the beginning, he so guided the various motion of the parts of it, as to contrive them into the world he design'd they should compose; and establish'd those rules of motion, and that order amongst things corporeal, which we call the laws of nature. Thus, the universe being once fram'd by God, and the laws of motion settled, and all upheld by his perpetual concourse, and general providence; the same philosophy teaches, that the phenomena of the world, are physically produced by the mechanical properties of the parts of matter; and, that they operate upon one another according to mechanical laws. 'Tis of this kind of corpuscular philosophy, that I speak.

'The Excellence and Grounds of the Mechanical Philosophy', in P. Shaw (ed.), *The Philosophical Works of Robert Boyle* (1725), Vol. 1, 187.

1 Acid Salts have the Power of Destroying the Blewness of the Infusion of our Wood [lignum nephreticum], and those Liquors indiscriminatly that abound with Sulphurous Salts, (under which I comprehend the Urinous and Volatile Salts of Animal Substances, and the Alcalisate or fixed Salts that are made by Incineration) have the virtue of Restoring it.

Experiments and Considerations Touching Colours (1664), 212.

2 Divers of Hermetic Books have such involv'd Obscuritys that they may justly be compar'd to Riddles written in Cyphers. For after a Man has surmounted the difficulty of decyphering the Words & Terms, he finds a new & greater difficulty to discover ye meaning of the seemingly plain Expression.

Fragment in Boyle papers. Cited by Lawrence Principe, 'Boyle's Alchemical Pursuits', in M. Hunter (ed.), *Robert Boyle Reconsidered* (1994), 95.

3 ON **BOYLE** He [Robert Boyle] is very tall (about six foot high) and straight, very temperate, and vertuouse, and frugall: a batcheler; keepes a Coach; sojournes with his sister, the Lady Ranulagh. His greatest delight is Chymistrey. He haz at his sister's a noble laboratory, and severall servants (Prentices to him) to look to it. He is charitable to ingeniose men that are in want, and foreigne Chymists have had large proofe of his bountie, for he will not spare for cost to get any rare Secret.

John Aubrey, *Brief Lives* (1680), edited by Oliver Lawson Dick (1949), 37.

C. Loring Brace 1930–

American anthropologist

4 We live in a cultural milieu . . . The idea that culture is our ecological niche is still applicable. The impact and force of natural selection on the human physique are conditioned by the dimensions of culture.

Interview with Pat Shipman, 21 January 1991. Quoted in Erik Trinkaus and Pat Shipman, *The Neanderthals: Changing the Image of Mankind* (1993), 334.

Francis Herbert Bradley 1846–1924

British philosopher

5 Metaphysics is the finding of bad reasons for what we believe upon instinct, but to find these reasons is no less an instinct.

Appearance and Reality: A Metaphysical Essay (1893), preface, xiv.

William Henry Bragg 1862–1942

British physicist

6 No known theory can be distorted so as to provide even an approximate explanation [of wave-particle duality]. There must be some fact of which we are entirely ignorant and whose discovery may revolutionize our views of the relations between waves and ether and matter. For the present we have to work on both theories. On Mondays, Wednesdays, and Fridays we use the wave theory; on Tuesdays, Thursdays, and Saturdays we think in streams of flying energy quanta or corpuscles.

'Electrons and Ether Waves', The Robert Boyle Lecture 1921, *Scientific Monthly*, 1922, 14, 158. See **Bragg, William Lawrence** 84:5.

1 But in its [the corpuscular theory of radiation] relation to the wave theory there is one extraordinary and, at present, insoluble problem. It is not known how the energy of the electron in the X-ray bulb is transferred by a wave motion to an electron in the photographic plate or in any other substance on which the X-rays fall. It is as if one dropped a plank into the sea from the height of 100 ft. and found that the spreading ripple was able, after travelling 1000 miles and becoming infinitesimal in comparison with its original amount, to act upon a wooden ship in such a way that a plank of that ship flew out of its place to a height of 100 ft. How does the energy get from one place to the other?

'Aether Waves and Electrons' (Summary of the Robert Boyle Lecture), *Nature*, 1921, 107, 374.

2 Light brings us the news of the Universe.

The Universe of Light (1933), 1.

3 The dividing line between the wave or particle nature of matter and radiation is the moment 'Now'. As this moment steadily advances through time it coagulates a wavy future into a particle past.

Attributed.

4 ON BRAGG [Professor Bragg asserts that] In sodium chloride there appear to be no molecules represented by NaCl. The equality in number of sodium and chlorine atoms is arrived at by a chess-board pattern of these atoms; it is a result of geometry and not of a pairing-off of the atoms.

In Henry E. Armstrong, 'Poor Common Salt!', *Nature*, 1927, **120**, 478.

William Lawrence Bragg

1890–1971

British physicist

5 God runs electromagnetics on Monday, Wednesday, and Friday by the wave theory, and the devil runs it by quantum theory on Tuesday, Thursday, and Saturday.

In Daniel J. Kevles, *The Physicists* (1978), 159. See **Bragg, William Henry** 83:6.

Tycho Brahe 1546–1601

Danish astronomer

6 In the evening, after sunset . . . I noticed that a new and unusual star, surpassing all the others in brilliancy, was shining almost directly above my head; and since I had, almost from boyhood, known all the stars of the heavens perfectly . . . it was quite evident to me that there had never before been any star in that place in the sky, even the smallest, to say nothing of a star so conspicuously bright as this . . . But when I observed that others, too, on having the place pointed out to them, could see that there really was a star there, I had no further doubts. A miracle indeed, either the greatest of all that have occurred in the whole range of nature since the beginning of the world, or one certainly that is to be classed with those attested by the Holy Oracles.

De Stella Nova (On the New Star) (1573). Quoted in H. Shapley and A. E. Howarth (eds.), *Source Book in Astronomy* (1929), 13.

7 Because the region of the Celestial World is of so great and such incredible magnitude as aforesaid, and since in what has gone before it was at least generally demonstrated that this comet continued within the limits of the space of the Aether, it seems that the complete explanation of the whole matter is not given unless we are also informed within narrower limits in what part of the widest Aether, and next to which orbs of the Planets [the comet] traces its path, and by what course it accomplishes this.

De Mundi Aetherei Recentioribus Phaenomenis (On Recent Phenomena in the Aetherial World) (1588). Quoted in M. Boas Hall, *The Scientific Renaissance 1450–1630* (1962), 115.

1 The body of the Earth, large, sluggish and inapt for motion, is not to be disturbed by movement (especially three movements), any more than the Aetherial Lights [stars] are to be shifted, so that such ideas are opposed both to physical principles and to the authority of the Holy Writ which many times confirms the stability of the Earth (as we shall discuss more fully elsewhere).

De Mundi Aetherei Recentioribus Phaenomenis (On Recent Phenomena in the Aetherial World) (1588). Quoted in M. Boas Hall, *The Scientific Renaissance 1450–1630* (1962), 115.

2 That the machine of Heaven is not a hard and impervious body full of various real spheres, as up to now has been believed by most people. It will be proved that it extends everywhere, most fluid and simple, and nowhere presents obstacles as was formerly held, the circuits of the Planets being wholly free and without the labour and whirling round of any real spheres at all, being divinely governed under a given law.

De Mundi Aetherei Recentioribus Phaenomenis (On Recent Phenomena in the Aetherial World) (1588). Quoted in M. Boas Hall, *The Scientific Renaissance 1450–1630* (1962), 117.

3 For those [observations] that I made in Leipzig in my youth and up to my 21st year, I usually call childish and of doubtful value. Those that I took later until my 28th year [i.e., until 1574] I call juvenile and fairly serviceable. The third group, however, which I made at Uraniborg during approximately the last 21 years with the greatest care and with very accurate instruments at a more mature age, until I was fifty years of age, those I call the observations of my manhood, completely valid and absolutely certain, and this is my opinion of them.

H. Raeder, E. and B. Strömgren (eds. and trans.), *Tycho Brahe's Description of his Instruments and Scientific Work: as given in Astronomiae Instauratae Mechanica, Wandesburgi 1598* (1946), 110.

4 There really are not any spheres in the heavens . . . Those which have been devised by the experts to save the appearances exist only in the imagination, for the purpose of enabling the mind to conceive the motion which the heavenly bodies trace in their course and, by the aid of geometry, to determine the motion numerically through the use of arithmetic.

J. L. E. Dreyer (ed.), *Opera Omnia* (1913–29), Vol. 4, 222. Trans. Edward Rosen, 'Nicholas Copernicus', in Charles C. Gillispie (ed.), *Dictionary of Scientific Biography* (1971), Vol. 3, 409.

Georges Braque 1882–1963

French cubist painter

5 *L'Art est fait pour troubler, la Science rassure.*

Art is meant to disturb, science reassures.

Georges Braque Illustrated Notebooks: 1917–1955, trans. S. Appelbaum (1971), 10.

Ernest Braun 1925–

British physicist

6 The history of semiconductor physics is not one of grand heroic theories, but one of painstaking intelligent labor. Not strokes of genius producing lofty edifices, but great ingenuity and endless undulation of hope and despair. Not sweeping generalizations, but careful judgment of the border between perseverance and obstinacy. Thus the history of solid-state physics in general, and of semiconductors in particular, is not so much about great men and women and their glorious deeds, as about the unsung heroes of thousands of clever ideas and skillful experiments—reflection of an age of organization rather than of individuality.

'Selected Topics from the History of Semiconductor Physics and Its Applications', in Lillian Hoddeson et al. (eds.), *Out of the Crystal Maze* (1992), 474.

Wernher von Braun 1912–77

German-born American rocket scientist

1 Basic research is what I am doing when I don't know what I am doing.

Attributed.

2 It [space travel] will free man from his remaining chains, the chains of gravity which still tie him to this planet. It will open to him the gates of heaven.

Attributed.

Scipione Breislak 1750–1826

Italian geologist

3 In geology we cannot dispense with conjectures: [but] because we are condemned to dream let us ensure that our dreams are like those of sane men—e.g. that they have their foundations in truth—and are not like the dreams of the sick, formed by strange combinations of phantasms, contrary to nature and therefore incredible.

Introduzione alla Geologia, Part 1 (1811), trans. Ezio Vaccari, 81.

Sydney Brenner 1927–

South African-born British molecular biologist

4 It is now widely realized that nearly all the 'classical' problems of molecular biology have either been solved or will be solved in the next decade. The entry of large numbers of American and other biochemists into the field will ensure that all the chemical details of replication and transcription will be elucidated. Because of this, I have long felt that the future of molecular biology lies in the extension of research to other fields of biology, notably development and the nervous system.

Letter to Max Perutz, 5 June 1963. Quoted in William B. Wood (ed.), *The Nematode Caenorhabditis Elegans* (1988), x-xi.

5 As was predicted at the beginning of the Human Genome Project, getting the sequence will be the easy part as only technical issues are involved. The hard part will be finding out what it means, because this poses intellectual problems of how to understand the participation of the genes in the functions of living cells.

Loose Ends from Current Biology (1997), 71.

6 I learnt very quickly that the only reason that would be accepted for not attending a committee meeting was that one already had a previous commitment to attend a meeting of another organization on the same day. I therefore invented a society, the Orion Society, a highly secret and very exclusive society that spawned a multitude of committees, sub-committees, working parties, evaluation groups and so on that, regrettably, had a prior claim on my attention. Soon people wanted to know more about this club and some even decided that they would like to join it. However, it was always made clear to them that applications were never entertained and that if they were deemed to qualify for membership they would be discreetly approached at the appropriate time.

Loose Ends from Current Biology (1997), 14.

7 Have you tried neuroxing papers? It's a very easy and cheap process. You hold the page in front of your eyes and you let it go through there into the brain. It's much better than xeroxing.

Quoted in L. Wolpert and A. Richards (eds.), *A Passion for Science* (1988), 104.

8 Progress in science depends on new techniques, new discoveries and new ideas, probably in that order.

Attributed.

David Brewster 1781–1868

British physicist

9 And why does England thus persecute the votaries of her science? Why does she depress them to the level of her hewers of wood and her drawers of water? Is it because science flatters no courtier, mingles in no political strife? . . . Can we behold unmoved the science of England, the vital principle of her arts, struggling for existence, the meek and unarmed victim of political strife?

Reviewing Babbage's Book, Reflections on the Decline of Science in England *(1830).*

Quarterly Review, 1830, **43**, 323–4.

1 There is no profession so incompatible with original enquiry as is a Scotch Professorship, where one's income depends on the numbers of pupils. Is there one Professor in Edinburgh pursuing science with zeal? Are they not all occupied as showmen whose principal object is to attract pupils and make money?

Brewster to J. D. Forbes, 11 February 1830 (St. Andrew's University Library). Quoted in William Cochran, 'Sir David Brewster: An Outline Biography', in J. R. R. Christie (ed.), *Martyr of Science: Sir David Brewster, 1781–1868* (1984), 13.

2 Prophetic of infidel times, and indicating the unsoundness of our general education, 'The Vestiges of the Natural History of Creation', has started into public favour with a fair chance of poisoning the fountains of science, and sapping the foundations of religion.

Review of Chambers' Book, 'Vestiges of the Natural History of Creation', *The North British Review*, 1845, 3, 471.

3 Thus identified with astronomy, in proclaiming truths supposed to be hostile to Scripture, Geology has been denounced as the enemy of religion. The twin sisters of terrestrial and celestial physics have thus been joint-heirs of intolerance and persecution—unresisting victims in the crusade which ignorance and fanaticism are ever waging against science. When great truths are driven to make an appeal to reason, knowledge becomes criminal, and philosophers martyrs. Truth, however, like all moral powers, can neither be checked nor extinguished. When compressed, it but reacts the more. It crushes where it cannot expand—it burns where it is not allowed to shine. Human when originally divulged, it becomes divine when finally established. At first, the breath of a rage—at last it is the edict of a god. Endowed with such vital energy, astronomical truth has cut its way through the thick darkness of superstitious times, and, cheered by its conquests, Geology will find the same open path when it has triumphed over the less formidable obstacles of a civilized age.

More Worlds than One: The Creed of the Philosopher and the Hope of the Christian (1854), 42.

Percy Williams Bridgman

1882–1961

American physicist and philosopher of science

4 The results have exhibited one striking feature which has been frequently emphasized, namely that at high pressures all twelve liquids become more nearly like each other. This suggests that it might be useful in developing a theory of liquids to arbitrarily construct a 'perfect liquid' and to discuss its properties. Certainly the conception of a 'perfect gas' has been of great service in the kinetic theory of gases; and the reason is that all actual gases approximate closely to the 'perfect gas.' In the same way, at high pressures all liquids approximate to one and the same thing, which may be called by analogy the 'perfect liquid.' It seems to offer at least a promising line of attack to discuss the properties of this 'perfect liquid,' and then to invent the simplest possible mechanism to explain them.

'Thermodynamic Properties of Twelve Liquids Between 20° and 80° and up to 1200 KGM. Per Sq. Cm.', *Memoirs of the American Academy of Arts and Sciences*, 1913, **49**, 113.

5 The first business of a man of science is to proclaim the truth as he finds it, and let the world adjust itself as best it can to the new knowledge.

Letter to R. M. Hunter, 23 October 1919. In Maila L. Walter, *Science and Cultural Crisis: An Intellectual Biography of Percy Williams Bridgman* (1990), 32.

1 Not only are there meaningless questions, but many of the problems with which the human intellect has tortured itself turn out to be only 'pseudo problems,' because they can be formulated only in terms of questions which are meaningless. Many of the traditional problems of philosophy, of religion, or of ethics, are of this character. Consider, for example, the problem of the freedom of the will. You maintain that you are free to take either the right- or the left-hand fork in the road. I defy you to set up a single objective criterion by which you can prove after you have made the turn that you might have made the other. The problem has no meaning in the sphere of objective activity; it only relates to my personal subjective feelings while making the decision.

The Nature of Physical Theory (1936), 12.

2 Every new theory as it arises believes in the flush of youth that it has the long sought goal; it sees no limits to its applicability, and believes that at last it is the fortunate theory to achieve the 'right' answer. This was true of electron theory—perhaps some readers will remember a book called *The Electrical Theory of the Universe* by de Tunzelman. It is true of general relativity theory with its belief that we can formulate a mathematical scheme that will extrapolate to all past and future time and the unfathomed depths of space. It has been true of wave mechanics, with its first enthusiastic claim a brief ten years ago that no problem had successfully resisted its attack provided the attack was properly made, and now the disillusionment of age when confronted by the problems of the proton and the neutron. When will we learn that logic, mathematics, physical theory, are all only inventions for formulating in compact and manageable form what we already know, like all inventions do not achieve complete success in accomplishing what they were designed to do, much less complete success in fields beyond the scope of the original design, and that our only justification for hoping to penetrate at all into the unknown with these inventions is our past experience that sometimes we have been fortunate enough to be able to push on a short distance by acquired momentum?

The Nature of Physical Theory (1936), 136.

3 The operational approach demands that we make our reports and do our thinking in the freshest terms of which we are capable, in which we strip off the sophistications of millenia of culture and report as directly as we can on what happens.

'Rejoinders and Second Thoughts'. In a Symposium on Operationism, *Psychological Review*, 1945, 52, 283.

4 The result is that a generation of physicists is growing up who have never exercised any particular degree of individual initiative, who have had no opportunity to experience its satisfactions or its possibilities, and who regard cooperative work in large teams as the normal thing. It is a natural corollary for them to feel that the objectives of these large teams must be something of large social significance.

'Science and Freedom: Reflections of a Physicist', *Isis*, 1947, 37, 130.

5 It is profitable nevertheless to permit ourselves to talk about 'meaningless' terms in the narrow sense if the preconditions to which all profitable operations are subject are so intuitive and so universally accepted as to form an almost unconscious part of the background of the public using the term. Physicists of the present day do constitute a homogenous public of this character; it is in the air that certain sorts of operation are valueless for achieving certain sorts of result. If one wants to know how many planets there

are one counts them but does not ask a philosopher what is the perfect number.
Reflections of a Physicist (1950), 6.

1 We have here no esoteric theory of the ultimate nature of concepts, nor a philosophical championing of the primacy of the 'operation'. We have merely a pragmatic matter, namely that we have observed after much experience that if we want to do certain kinds of things with our concepts, our concepts had better be constructed in certain ways. In fact one can see that the situation here is no different from what we always find when we push our analysis to the limit; operations are not ultimately sharp or irreducible any more than any other sort of creature. We always run into a haze eventually, and all our concepts are describable only in spiralling approximation.
Reflections of a Physicist (1950), 9.

2 My point of view is that science is essentially private, whereas the almost universal counter point of view, explicitly stated in many of the articles in the Encyclopaedia, is that it *must* be public.
Reflections of a Physicist (1950), 44.

3 The process that I want to call scientific is a process that involves the continual apprehension of meaning, the constant appraisal of significance accompanied by a running act of checking to be sure that I am doing what I want to do, and of judging correctness or incorrectness. This checking and judging and accepting, that together constitute understanding, are done by me and can be done for me by no one else. They are as private as my toothache, and without them science is dead.
Reflections of a Physicist (1950), 50.

4 The feeling of understanding is as private as the feeling of pain. The act of understanding is at the heart of all scientific activity; without it any ostensibly scientific activity is as sterile as that of a high school student substituting numbers into a formula. For this reason, science, when I push the analysis back as far as I can, must be private.
Reflections of a Physicist (1950), 72.

5 The attitude which the man in the street unconsciously adopts towards science is capricious and varied. At one moment he scorns the scientist for a highbrow, at another anathematizes him for blasphemously undermining his religion; but at the mention of a name like Edison he falls into a coma of veneration. When he stops to think, he does recognize, however, that the whole atmosphere of the world in which he lives is tinged by science, as is shown most immediately and strikingly by our modern conveniences and material resources. A little deeper thinking shows him that the influence of science goes much farther and colors the entire mental outlook of modern civilised man on the world about him.
Reflections of a Physicist (1950), 81.

6 On careful examination the physicist finds that in the sense in which he uses language no meaning at all can be attached to a physical concept which cannot ultimately be described in terms of some sort of measurement. A body has position only in so far as its position can be measured; if a position cannot in principle be measured, the concept of position applied to the body is meaningless, or in other words, a position of the body does not exist. Hence if both the position and velocity of electron cannot in principle be measured, the electron cannot have the same position and velocity; position and velocity as expressions of properties which an electron can simultaneously have are meaningless.
Reflections of a Physicist (1950), 90.

7 The man in the street will, therefore, twist the statement that the scientist has come to the end of meaning into the statement that the scientist has penetrated as far as he can with the tools at his command, and that there is something beyond the ken of the scientist. This imagined beyond, which the scientist has proved he cannot penetrate, will become the playground of the imagination of every mystic and dreamer. The existence of such a domain will be made the basis of an orgy of rationalizing. It will be made the substance of the soul; the spirits of the dead will populate it; God will lurk in its shadows; the principle of vital processes will have its seat here; and it will be the medium of telepathic communication. One group will find in the failure of the physical law of cause and effect the solution of the age-long problem of the freedom of the will; and on the other hand the atheist will find the justification of his contention that chance rules the universe.

Reflections of a Physicist (1950), 102–3.

1 I believe it to be of particular importance that the scientist have an articulate and adequate social philosophy, even more important than the average man should have a philosophy. For there are certain aspects of the relation between science and society that the scientist can appreciate better than anyone else, and if he does not insist on this significance no one else will, with the result that the relation of science to society will become warped, to the detriment of everybody.

Reflections of a Physicist (1950), 287.

2 In general, we mean by any concept nothing more than a set of operations; *the concept is synonymous with the corresponding set of operations.*

The Logic of Modern Physics (1960), 5.

3 If a specific question has meaning, it must be possible to find operations by which an answer may be given to it . . . I believe that many of the questions asked about social and philosophical subjects will be found to be meaningless when examined from the point of view of operations.

The Logic of Modern Physics (1960), 28.

4 By far the most important consequence of the conceptual revolution brought about in physics by relativity and quantum theory lies not in such details as that meter sticks shorten when they move or that simultaneous position and momentum have no meaning, but in the insight that we had not been using our minds properly and that it is important to find out how to do so.

'Quo Vadis'. In Gerald Holton (ed.), *Science and the Modern Mind* (1971), 84.

Jean-Anthelme Brillat-Savarin

1755–1826

French lawyer and gourmet

5 Tell me what you eat: I will tell you what you are.

The Philosopher in the Kitchen (1825), Aphorism iv.

6 The discovery of a new dish does more for the happiness of mankind than the discovery of a star.

The Philosopher in the Kitchen (1825), Aphorism ix.

Charlie Dunbar Broad 1887–1971

British philosopher

7 I shall no doubt be blamed by certain scientists, and, I am afraid, by some philosophers, for having taken serious account of the alleged facts which are investigated by Psychical Researchers. I am wholly impenitent about this. The scientists in question seem to me to confuse the Author of Nature with the Editor of *Nature*; or at any rate to suppose that there can be no productions of the former which would not be accepted for publication by the latter. And I see no reason to believe this.

The Mind and its Place in Nature (1925), viii.

8 It is worth remembering (though there is nothing that we can do about it) that the world as it really is may easily be a far nastier place than it would be if scientific materialism were the whole truth and nothing but the truth about it.

Lectures on Psychical Research, Incorporating the Perrot Lectures Given in Cambridge University in 1959 and 1960 (1962), 430.

Pierre Paul Broca 1824–80

French surgeon and anthropologist

1 Ethnologists regard man as the primitive element of tribes, races, and peoples. The anthropologist looks at him as a member of the fauna of the globe, belonging to a zoölogical classification, and subject to the same laws as the rest of the animal kingdom. To study him from the last point of view only would be to lose sight of some of his most interesting and practical relations; but to be confined to the ethnologist's views is to set aside the scientific rule which requires us to proceed from the simple to the compound, from the known to the unknown, from the material and organic fact to the functional phenomenon.

'Paul Broca and the French School of Anthropology'. Lecture delivered in the National Museum, Washington, D.C., 15 April 1882, by Dr. Robert Fletcher. In *The Saturday Lectures* (1882), 118.

2 As for me . . . I would much rather be a perfected ape than a degraded Adam. Yes, if it is shown to me that my humble ancestors were quadrupedal animals, arboreal herbivores, brothers or cousins of those who were also the ancestors of monkeys and apes, far from blushing in shame for my species because of its genealogy and parentage, I will be proud of all that evolution has accomplished, of the continuous improvement which takes us up to the highest order, of the successive triumphs that have made us superior to all of the other species . . . the splendid work of progress.

I will conclude in saying: the fixity of species is almost impossible, it contradicts the mode of succession and of the distribution of species in the sequence of extant and extinct creatures. It is therefore extremely likely that species are variable and are subject to evolution.

But the causes, the mechanisms of this evolution are still unknown.

'Discussion sur la Machoire Humaine de la Naulette (Belgique)', *Bulletin de la Société d'Anthropologie de Paris*, 2nd Series, 1 (1866), 595. Trans. Erik Trinkaus and Pat Shipman, *The Neanderthals: Changing the Image of Mankind* (1993), 103–4.

Giovanni Battista Brocchi 1772–1826

Italian geologist

3 The science of fossil shells is the first step towards the study of the earth.

Conchiologia Fossile Subappennina (1814), Vol. 1, trans. Ezio Vaccari, 13.

Bertram Neville Brockhouse 1918–2003

Canadian physicist

4 For strictly scientific or technological purposes all this is irrelevant. On a pragmatic view, as on a religious view, theory and concepts are held in faith. On the pragmatic view the only thing that matters is that the theory is efficacious, that it 'works' and that the necessary preliminaries and side issues do not cost too much in time and effort. Beyond that, theory and concepts go to constitute a language in which the scientistic matters at issue can be formulated and discussed.

'Slow Neutron Spectroscopy and the Grand Atlas of the Physical World', Nobel Lecture, 8 December 1994. In *Nobel Lectures: Physics 1991–1995* (1997), 111.

Peter Bellinger Brodie 1815–97

British clergyman and geologist

5 Fifty years ago Geology was in its infancy; there were but few who cultivated it as a Science . . . If an unfortunate lover of nature was seen hammering in a stone quarry, he was generally supposed to be slightly demented.

'Geology Considered with Reference to its Utility and Practical Effects', *The Geologist*, 1858, 1, 6.

Louis Victor de Broglie
1892–1987
French physicist

1 Many scientists have tried to make determinism and complementarity the basis of conclusions that seem to me weak and dangerous; for instance, they have used Heisenberg's uncertainty principle to bolster up human free will, though his principle, which applies exclusively to the behavior of electrons and is the direct result of microphysical measurement techniques, has nothing to do with human freedom of choice. It is far safer and wiser that the physicist remain on the solid ground of theoretical physics itself and eschew the shifting sands of philosophic extrapolations.
New Perspectives in Physics (1962), viii.

2 Thus with every advance in our scientific knowledge new elements come up, often forcing us to recast our entire picture of physical reality. No doubt, theorists would much prefer to perfect and amend their theories rather than be obliged to scrap them continually. But this obligation is the condition and price of all scientific progress.
New Perspectives in Physics (1962), 31.

3 There is no reason why the history and philosophy of science should not be taught in such a way as to bring home to all pupils the grandeur of science and the scope of its discoveries.
New Perspectives in Physics (1962), 195.

4 Science itself, no matter whether it is the search for truth or merely the need to gain control over the external world, to alleviate suffering, or to prolong life, is ultimately a matter of feeling, or rather, of desire—the desire to know or the desire to realize.
New Perspectives in Physics (1962), 196.

5 Vulnerable, like all men, to the temptations of arrogance, of which intellectual pride is the worst, he [the scientist] must nevertheless remain sincere and modest, if only because his studies constantly bring home to him that, compared with the gigantic aims of science, his own contribution, no matter how important, is only a drop in the ocean of truth.
New Perspectives in Physics (1962), 215.

Jacob Bronowski 1908–74
Polish-born British mathematician and science writer

6 Man masters nature not by force but by understanding. That is why science has succeeded where magic failed: because it has looked for no spell to cast on nature.
Science and Human Values (1961), 20.

7 The problem of values arises only when men try to fit together their need to be social animals with their need to be free men. There is no problem, and there are no values, until men want to do both. If an anarchist wants only freedom, whatever the cost, he will prefer the jungle of man at war with man. And if a tyrant wants only social order, he will create the totalitarian state.
Science and Human Values (1961), 63.

8 Science has nothing to be ashamed of, even in the ruins of Nagasaki.
Science and Human Values (1961), 83.

9 Among the multitude of animals which scamper, fly, burrow and swim around us, man is the only one who is not locked into his environment. His imagination, his reason, his emotional subtlety and toughness, make it possible for him not to accept the environment, but to change it. And that series of inventions, by which man from age to age has remade his environment, is a different kind of evolution—not biological, but cultural evolution. I call that brilliant sequence of cultural peaks *The Ascent of Man*. I use the word *ascent* with a precise meaning. Man is distinguished from other animals by his imaginative gifts. He makes plans, inventions, new discoveries, by putting different talents together; and his discoveries become more subtle and penetrating, as he learns to combine his talents in more complex and intimate ways. So the great discoveries of different ages and different cultures, in technique, in science, in the arts, express in their progression a richer and more intricate conjunction of human faculties, an ascending trellis of his gifts.

The Ascent of Man (1973), 19–20.

1 We have to understand that the world can only be grasped by action, not by contemplation. The hand is more important than the eye . . . The hand is the cutting edge of the mind.

The Ascent of Man (1973), 115–6.

2 The most powerful drive in the ascent of man is his pleasure in his own skill. He loves to do what he does well and, having done it well, he loves to do it better. You see it in his science. You see it in the magnificence with which he carves and builds, the loving care, the gaiety, the effrontery. The monuments are supposed to commemorate kings and religions, heroes, dogmas, but in the end the man they commemorate is the builder.

The Ascent of Man (1973), 116.

3 [John] Dalton was a man of regular habits. For fifty-seven years he walked out of Manchester every day; he measured the rainfall, the temperature—a singularly monotonous enterprise in this climate. Of all that mass of data, nothing whatever came. But of the one searching, almost childlike question about the weights that enter the construction of these simple molecules—out of that came modern atomic theory. That is the essence of science: ask an impertinent question, and you are on the way to the pertinent answer.

The Ascent of Man (1973), 153.

Henry Peter Brougham

(Lord Brougham) 1778–1868

British statesman and science promoter

4 In one department of his [Joseph Black's] lecture he exceeded any I have ever known, the neatness and unvarying success with which all the manipulations of his experiments were performed. His correct eye and steady hand contributed to the one; his admirable precautions, foreseeing and providing for every emergency, secured the other. I have seen him pour boiling water or boiling acid from a vessel that had no spout into a tube, holding it at such a distance as made the stream's diameter small, and so vertical that not a drop was spilt. While he poured he would mention this adaptation of the height to the diameter as a necessary condition of success. I have seen him mix two substances in a receiver into which a gas, as chlorine, had been introduced, the effect of the combustion being perhaps to produce a compound inflammable in its nascent state, and the mixture being effected by drawing some string or wire working through the receiver's sides in an air-tight socket. The long table on which the different processes had been carried on was as clean at the end of the lecture as it had been before the apparatus was planted upon it. Not a drop of liquid, not a grain of dust remained.

Lives of Men of Letters and Science, who flourished in the time of George III (1845), 346–7.

5 Nothing could be more admirable than the manner in which for forty years he [Joseph Black] performed this useful and dignified office. His style of lecturing was as nearly perfect as can well be conceived; for it had all the simplicity which is so entirely suited to scientific discourse, while it partook largely of the elegance which characterized all he said or did . . . I have heard the greatest understandings of the age giving forth their efforts in its most eloquent tongues—have heard the commanding periods of Pitt's majestic oratory—the vehemence of Fox's burning declamation—have followed the close-compacted chain of Grant's pure reasoning—been carried away by the mingled fancy, epigram, and argumentation of Plunket; but I should without hesitation prefer, for mere intellectual gratification (though aware how much of it is derived from association), to be once more allowed the privilege which I in those days enjoyed of being present while the first philosopher of his age was the historian of his own discoveries, and be an eye-witness of those experiments by which he had formerly made them, once more performed with his own hands.

'Philosophers of the Time of George III'. In *The Works of Henry, Lord Brougham, F.R.S.* (1855), Vol. 1, 19–21.

Alexander Crum Brown

1838–1922

British chemist and physiologist

1 Unless the chemist learns the language of mathematics, he will become a provincial and the higher branches of chemical work, that require reason as well as skill, will gradually pass out of his hands.

Quoted in *Journal of the Chemical Society*, 1929, 6, 254.

Herbert Charles Brown 1912–

British-born American chemist

2 Why did I decide to undertake my doctorate research in the exotic field of boron hydrides? As it happened, my girl friend, Sarah Baylen, soon to become my wife, presented me with a graduation gift, Alfred Stock's book, *The Hydrides of Boron and Silicon.* I read this book and became interested in the subject. How did it happen that she selected this particular book? This was the time of the Depression. None of us had much money. It appears she selected as her gift the most economical chemistry book ($2.06) available in the University of Chicago bookstore. Such are the developments that can shape a career.

'From Little Acorns Through to Tall Oaks—From Boranes Through Organoboranes', Nobel Lecture, 8 December 1979. In *Nobel Lectures—Chemistry, 1971–1980* (1993), 341.

3 TO MY WIFE—who made the writing of my previous book a pleasure and writing of the present one a necessity.

Boranes in Organic Chemistry (1972), dedication.

4 Professor Brown: 'Since this slide was made,' he opined, 'My students have re-examined the errant points and I am happy to report that all fall close to the [straight] line.' Questioner: 'Professor Brown, I am delighted that the points which fell off the line proved, on reinvestigation, to be in compliance. I wonder, however, if you have had your students reinvestigate all these points that previously fell on the line to find out how many no longer do so?'

Quoted in D. A. Davenport, 'The Invective Effect', *Chemtech*, September 1987, 530.

5 As usual, the author in his thorough, unobjective fashion has marshalled up all the good, indifferent and bad arguments . . . I offer the following detailed comments . . . though I realize that many of them will arouse him to a vigorous, if not violent rebuttal. In order to preserve the pH of Dr. Brown's digestive system I would not require a rebuttal as a condition of publication . . .

With heartiest greetings of the season to you and yours!Jack Roberts

PS The above comments should (help) to reduce your winter heating bill!

Jack Roberts' referee's report on Brown's paper with Rachel Kornblum on the role of steric strain in carbonium ion reactions.

Quoted in D. A. Davenport, 'The Invective Effect', *Chemtech*, September 1987, 530.

Robert Brown 1773–1858

Scottish botanist

6 These motions were such as to satisfy me, after frequently repeated observation, that they arose neither from currents in the fluid, nor from its gradual evaporation, but belonged to the particle itself.

Summary of Brownian motion.
'A Brief Account of Microscopical Observations made in the Middle of June, July, and August, 1827, on the Particles Contained in the Pollen of Plants', *Philosophical Magazine*, 1828, NS 4, 162–3.

Thomas Browne 1605–82

British natural philosopher and physician

1 For who can speake of eternitie without a solœcisme, or thinke thereof without an extasie? Time we may comprehend, 'tis but five dayes elder then our selves.
Religio Medici (1642), Part 1, Section 11. In L. C. Martin (ed.), *Thomas Browne: Religio Medici and Other Works* (1964), 11.

2 Thus there are two bookes from whence I collect my Divinity; besides that written one of God, another of his servant Nature, that universall and publik Manuscript, that lies expans'd unto the eyes of all; those that never saw him in the one, have discovered him in the other.
Religio Medici (1642), Part 1, Section 16. In L. C. Martin (ed.), *Thomas Browne: Religio Medici and Other Works* (1964), 15.

3 For God is like a skilfull Geometrician.
Religio Medici (1642), Part 1, Section 16. In L. C. Martin (ed.), *Thomas Browne: Religio Medici and Other Works* (1964), 16.

4 In briefe, all things are artificiall, for nature is the Art of God.
Religio Medici (1642), Part 1, Section 16. In L. C. Martin (ed.), *Thomas Browne: Religio Medici and Other Works* (1964), 16.

5 Thus is Man that great and true Amphibium, whose nature is disposed to live . . . in divided and distinguished worlds.
Religio Medici (1642), Part 1, Section 34. In L. C. Martin (ed.), *Thomas Browne: Religio Medici and Other Works* (1964), 33.

6 To call ourselves a Microcosme, or little world, I thought it onely a pleasant trope of Rhetorick, till my neare judgement and second thoughts told me there was a reall truth therein: for first wee are a rude masse, and in the ranke of creatures, which only are, and have a dull kinde of being not yet priviledged with life, or preferred to sense or reason; next we live the life of plants, the life of animals, the life of men, and at last the life of spirits, running on in one mysterious nature those five kinds of existence, which comprehend the creatures not onely of world, but of the Universe.
Religio Medici (1642), Part 1, Section 34. In L. C. Martin (ed.), *Thomas Browne: Religio Medici and Other Works* (1964), 33.

7 Men that looke no further than their outsides thinke health an appertinance unto life, and quarrell with their constitutions for being sick; but I that have examined the parts of man, and know upon what tender filaments that Fabrick hangs, doe wonder what we are not alwayes so; and considering the thousand dores that lead to death doe thanke my God that we can die but once.
Religio Medici (1642), Part 1, Section 44. In L. C. Martin (ed.), *Thomas Browne: Religio Medici and Other Works* (1964), 42.

8 It is common wonder of all men, how among so many millions of faces, there should be none alike.
Religio Medici (1642), Part 2, Section 2. In L. C. Martin (ed.), *Thomas Browne: Religio Medici and Other Works* (1964), 57.

9 There are infirmities, not onely of body, but of soule, and fortunes, which doe require the mercifull hand of our abilities. I cannot contemn a man for ignorance but behold him with as much pity as I doe *Lazarus*.
Religio Medici (1642), Part 2, Section 3. In L. C. Martin (ed.), *Thomas Browne: Religio Medici and Other Works* (1964), 58.

10 This triviall and vulgar way of coition; It is the foolishest act a wise man commits in all his life, nor is there any thing that will deject his coold imagination, when hee shall consider what an odde and unworthy piece of folly hee hath committed.
Religio Medici (1642), Part 2, Section 9. In L. C. Martin (ed.), *Thomas Browne: Religio Medici and Other Works* (1964), 67.

1 I boast nothing, but plainely say, we all labour against our owne cure, for death is the cure of all diseases.

Religio Medici (1642), Part 1, Section 9. In L. C. Martin (ed.), *Thomas Browne: Religio Medici and Other Works* (1964), 68.

2 There is surely a peece of Divinity in us, something that was before the Elements, and owes no homage unto the Sun.

Religio Medici (1642), Part 2, Section 11. In L. C. Martin (ed.), *Thomas Browne: Religio Medici and Other Works* (1964), 70.

3 But man is a Noble Animal, splendid in ashes, and pompous in the grave.

Hydriotaphia (Urn Burial, 1658), Chapter 5. In L. C. Martin (ed.), *Thomas Browne: Religio Medici and Other Works* (1964), 123.

Elizabeth Barrett Browning
1806–61

British poet

4 Men of science, osteologists
And surgeons, beat some poets, in respect
For nature,—count nought common or unclean,
Spend raptures upon perfect specimens
Of indurated veins, distorted joints,
Or beautiful new cases of curved spine;
While we, we are shocked at nature's falling off,
We dare to shrink back from her warts and blains.

'Aurora Leigh' (1856), Sixth Book, lines 171–8. In John Robert Glorey Bolton and Julia Bolton Holloway (eds.), *Aurora Leigh and Other Poems* (1995), 178.

Robert Browning 1812–89

British poet

5 Thou at first prompting of what I call God,
And fools call Nature, didst hear, comprehend,
Accept the obligation laid on thee.

'The Ring and the Book', Book X. In James F. Loucks (ed.), *Robert Browning's Poetry* (1979), 385.

Ernst Wilhelm von Brücke
1819–92

German physiologist

6 Teleology is a lady without whom no biologist can live. Yet he is ashamed to show himself with her in public.

Quoted in H.A. Krebs, 'Excursion into the Borderland of Biochemistry and Philosophy', *Bulletin of the Johns Hopkins Hospital*, 1954, 95, 45.

Giordano Bruno 1548–1600

Italian philosopher and hermeticist

7 There is a single general space, a single vast immensity which we may freely call void: in it are unnumerable globes like this on which we live and grow, this space we declare to be infinite, since neither reason, convenience, sense-perception nor nature assign to it a limit.

Quoted in Joseph Silk, *The Big Bang* (1997), 89.

8 To a body of infinite size there can be ascribed neither center nor boundary . . . Just as we regard ourselves as at the center of that universally equidistant circle, which is the great horizon and the limit of our own encircling ethereal region, so doubtless the inhabitants of the moon believe themselves to be at the center (of a great horizon) that embraces this earth, the sun, and the stars, and is the boundary of the radii of their own horizon. Thus the earth no more than any other world is at the center; moreover no points constitute determined celestial poles for our earth, just as she herself is not a definite and determined pole to any other point of the ether, or of the world-space; and the same is true for all other bodies. From various points of view these may all be regarded either as centers, or as points on the circumference, as poles, or zeniths and so forth. Thus the earth is not in the center of the universe; it is central only to our own surrounding space.

Irving Louis Horowitz, *The Renaissance Philosophy of Giordano Bruno* (1952), 60.

1 Unless you make yourself equal to God, you cannot understand God: for the like is not intelligible save to the like. Make yourself grow to a greatness beyond measure, by a bound free yourself from the body; raise yourself above all time, become Eternity; then you will understand God. Believe that nothing is impossible for you, think yourself immortal and capable of understanding all, all arts, all sciences, the nature of every living being. Mount higher than the highest height; descend lower than the lowest depth. Draw into yourself all sensations of everything created, fire and water, dry and moist, imagining that you are everywhere, on earth, in the sea, in the sky, that you are not yet born, in the maternal womb, adolescent, old, dead, beyond death. If you embrace in your thought all things at once, times, places, substances, qualities, quantities, you may understand God.

Quoted in F. A. Yates, *Giordano Bruno and the Hermetic Tradition* (1964), 198.

Christian Leopold von Buch 1774–1853

German geologist

2 Geognosy urgently needs the instruction of zoology.

Abhandlungen der königlichen Akademie der Wissenschaften (1830), 135.

Scott Milross Buchanan 1895–1968

American philosopher

3 Science is an allegory that asserts that the relations between the parts of reality are similar to the relations between terms of discourse.

Poetry and Mathematics (1929), 96–7.

Eduard Buchner 1860–1917

German chemist

4 The initiation of the fermentation process does not require so complicated an apparatus as is represented by the living cell. The agent responsible for the fermenting action of the press juice is rather to be regarded as a dissolved substance, doubtless a protein; this will be denoted *zymase*.

'Gährung ohne Hefezellen', *Berichte der deutschen chemischen Gesellschaft*, 1897, 30, 119–20. Trans. in Joseph S. Fruton, *Proteins, Enzymes, Genes: The Interplay of Chemistry and Biology* (1999), 117.

5 This [discovery of a cell-free yeast extract] will make him famous, even though he has no talent for chemistry.

Baeyer to Willstätter. Quoted in R. Willstätter, *From My Life* (1965), trans. J. S. Fruton.

Friedrich Karl Christian Ludwig Büchner 1824–99

German physician and philosopher

6 What we still designate as chance, merely depends on a concatenation of circumstances, the internal connection and final causes of which we have as yet been unable to unravel.

Force and Matter (1884), 226.

7 *Ohne Phosphor, Kein Gedanke.*

Without phosphorus there would be no thoughts.

Attributed.

William Buckland 1784–1856

British geologist and palaeontologist

8 The field of the Geologist's inquiry is the Globe itself, . . . [and] it is his study to decipher the monuments of the mighty revolutions and convulsions it has suffered.

Vindiciae Geologicae (1820), 5.

9 The human mind has a natural tendency to explore what has passed in distant ages in scenes with which it is familiar: hence the taste for National and Local Antiquities. Geology gratifies a larger taste of this kind; it inquires into what may appropriately be termed the Antiquities of the Globe itself, and collects and deciphers what may be considered as the monuments and medals of its remoter eras.

Vindiciae Geologicae (1820), 6.

1 Geology holds the keys of one of the kingdoms of nature; and it cannot be said that a science which extends our Knowledge, and by consequence our Power, over a third part of nature, holds a low place among intellectual employments.

Vindiciae Geologicae (1820), 7.

2 It is demonstrable from Geology that there was a period when no organic beings had existence: these organic beings must therefore have had a beginning subsequently to this period; and where is that beginning to be found, but in the will and *fiat* of an intelligent and all-wise Creator?

Vindiciae Geologicae (1820), 21.

3 With respect to those points, on which the declaration of Scripture is positive and decisive, as, for instance, in asserting the low antiquity of the human race; the evidence of all facts that have yet been established in Geology coincides with the records of Sacred History and Profane Tradition to confirm the conclusion that *the existence of mankind* can on no account be supposed to have taken its beginning before that time which is assigned to it in the Mosaic writings.

Vindiciae Geologicae (1820), 23.

4 A . . . hypothesis may be suggested, which supposes the word 'beginning' as applied by Moses in the first of the Book of Genesis, to express an undefined period of time which was antecedent to the last great change that affected the surface of the earth, and to the creation of its present animal and vegetable inhabitants; during which period a long series of operations and revolutions may have been going on, which, as they are wholly unconnected with the history of the human race, are passed over in silence by the sacred historian, whose only concern with them was largely to state, that the matter of the universe is not eternal and self-existent but was originally created by the power of the Almighty.

Vindiciae Geologicae (1820), 31–2.

5 The days of the Mosaic creation are not to be strictly construed as implying the same length of time which is at present occupied by a single revolution of our globe, but PERIODS of a much longer extent.

Vindiciae Geologicae (1820), 32.

6 Thus far I have produced a various and, in my judgement, incontrovertible body of facts, to show that the whole earth has been subjected to a recent and universal inundation.

Reliquire Diluvianae (1824), 224.

7 Geology has shared the fate of other infant sciences, in being for a while considered hostile to revealed religion; so like them, when fully understood, it will be found a potent and consistent auxiliary to it, exalting our conviction of the Power, and Wisdom, and Goodness of the Creator.

Geology and Mineralogy Considered with Reference to Natural Theology (1836), Vol. 1, 9.

8 The successive series of stratified formations are piled on one another, almost like courses of masonry.

Geology and Mineralogy, Considered with Reference to Natural Theology (1836), Vol. 1, 37.

9 At the voice of comparative anatomy, every bone, and fragment of a bone, resumed its place. I cannot find words to express the pleasure I have in seeing, as I discovered one character, how all the consequences, which I predicted from it, were successively confirmed; the feet were found in accordance with the characters announced by the teeth; the teeth in harmony with those indicated beforehand by the feet; the bones of the legs and thighs, and every connecting portion of the extremities, were found set together precisely as I had arranged them, before my conjectures were verified by the discovery of the parts entire: in short, each species was, as it were, reconstructed from a single one of its component elements.

Geology and Mineralogy (1836), Vol. 1, 83–4.

1 No conclusion is more fully established, than the important fact of the total absence of any vestiges of the human species throughout the entire series of geological formations.

Geology and Mineralogy, Considered with Reference to Natural Theology (1836), Vol. 1, 103.

2 To the mind which looks not to general results in the economy of Nature, the earth may seem to present a scene of perpetual warfare, and incessant carnage: but the more enlarged view, while it regards individuals in their conjoint relations to the general benefit of their own species, and that of other species with which they are associated in the great family of Nature, resolves each apparent case of individual evil, into an example of subserviency to universal good.

Geology and Mineralogy, Considered with Reference to Natural Theology (1836), Vol. 1, 131–2.

3 Thus the great drama of universal life is perpetually sustained; and though the individual actors undergo continual change, the same parts are ever filled by another and another generation; renewing the face of the earth, and the bosom of the deep, with endless successions of life and happiness.

Geology and Mineralogy, Considered with Reference to Natural Theology (1836), Vol. 1, 134.

4 Shall it any longer be said that a science [geology], which unfolds such abundant evidence of the Being and Attributes of God, can reasonably be viewed in any other light than as the efficient Auxiliary and Handmaid of Religion?

Geology and Mineralogy, Considered with Reference to Natural Theology (1836), Vol. 1, 593.

Henry Thomas Buckle 1821–62

British historian

5 Suicide is merely the product of the general condition of society, and that the individual felon only carries into effect what is a necessary consequence of preceding circumstances. In a given state of society, a certain number of persons must put an end to their own life. This is the general law; and the special question as to who shall commit the crime depends of course upon special laws; which, however, in their total action, must obey the large social law to which they are all subordinate. And the power of the larger law is so irresistible, that neither the love of life nor the fear of another world can avail any thing towards even checking its operation.

History of Civilization in England (1857), Vol. 1, 25–6.

6 Unconscious, perhaps, of the remote tendency of his own labours, he [Joseph Black] undermined that doctrine of material heat, which he seemed to support. For, by his advocacy of latent heat, he taught that its movements constantly battle, not only some of our senses, but all of them; and that, while our feelings make us believe that heat is lost, our intellect makes us believe that it is not lost. Here, we have apparent destructability, and real indestructibility. To assert that a body received heat without its temperature rising, was to make the understanding correct the touch, and defy its dictates. It was a bold and beautiful paradox, which required courage as well as insight to broach, and the reception of which marks an epoch in the human mind, because it was an immense step towards idealizing matter into force.

History of Civilization in England (1861), Vol. 2, 494.

George-Louis Leclerc, Comte de Buffon 1707–88

French naturalist and philosopher

1 It appears that all that can be, is. The Creator's hand does not appear to have been opened in order to give existence to a certain determinate number of species, but it seems that it has thrown out all at once a world of relative and non-relative creatures, an infinity of harmonic and contrary combinations and a perpetuity of destructions and replacements. What idea of power is not given us by this spectacle! What feeling of respect for its Author is not inspired in us by this view of the universe!

'Premier Discours: De la Manière d'Étudier et de Traiter l'Histoire naturelle', *Histoire Naturelle, Générale et Particulière, Avec la Description du Cabinet du Roi* (1749), Vol. 1, 11. Trans. Phillip R. Sloan.

2 There are several kinds of truths, and it is customary to place in the first order mathematical truths, which are, however, only truths of definition. These definitions rest upon simple, but abstract, suppositions, and all truths in this category are only constructed, but abstract, consequences of these definitions . . . Physical truths, to the contrary, are in no way arbitrary, and do not depend on us.

'Premier Discours: De la Manière d'Étudier et de Traiter l'Histoire naturelle', *Histoire Naturelle, Générale et Particulière, Avec la Description du Cabinet du Roi* (1749), Vol. 1, 53–4. Trans. Phillip R. Sloan.

3 One can descend by imperceptible degree from the most perfect creature to the most shapeless matter, from the best-organised animal to the roughest mineral.

'De La Maniére d'étudier et de Traiter l'Histoire Naturelle'. In *Oeuvres Complètes* (1774–79), Vol. 1, 17.

4 In general, the more one augments the number of divisions of the productions of nature, the more one approaches the truth, since in nature only individuals exist, while genera, orders, and classes only exist in our imagination.

Histoire Naturelle (1749), trans. by John Lyon, ' The 'Initial Discourse' to Buffon's *Histoire Naturelle:* The First Complete English Translation', *Journal of the History of Biology*, 9(1), 1976, 164.

5 We can only penetrate the rind of the earth.

'Second Discours: Histoire & Théorie de la Terre', *Histoire Naturelle, Générale et Particulière, Avec la Description du Cabinet du Roi* (1749),Vol. 1, 70; *Natural History, General and Particular* (1785), Vol. 1, trans. W. Smellie, 6.

6 I am convinced, by repeated observation, that marbles, lime-stones, chalks, marls, clays, sand, and almost all terrestrial substances, wherever situated, are full of shells and other spoils of the ocean.

'Second Discours: Histoire & Théorie de la Terre', *Histoire Naturelle, Générale et Particulière, Avec la Description du Cabinet du Roi* (1749), Vol. 1, 76–77; *Natural History, General and Particular* (1785), Vol. 1, trans. W. Smellie, 13.

7 Let us suppose, that the Old and New worlds were formerly but one continent, and that, by a violent earthquake, the ancient Atalantis [sic] of Plato was sunk . . . The sea would necessarily rush in from all quarters, and form what is now called the Atlantic ocean.

'Second Discours: Histoire et Théorie de la Terre', *Histoire Naturelle, Générale et Particulière, Avec la Description du Cabinet du Roi* (1749), Vol. 1, 96; *Natural History, General and Particular* (1785), Vol. 1, trans. W. Smellie, 31.

8 As historians, we refuse to allow ourselves these vain speculations which turn on possibilities that, in order to be reduced to actuality, suppose an overturning of the Universe, in which our globe, like a speck of abandoned matter, escapes our vision and is no longer an object worthy of our regard. In order to fix our vision, it is necessary to take it such as it is, to observe well all parts of it, and by indications infer from the present to the past.

'Second Discours: Histoire et Théorie de la

Terre', *Histoire Naturelle, Générale et Particulière, Avec la Description du Cabinet du Roi* (1749), Vol. 1, 98–9. Trans. Phillip R. Sloan.

1 We may conclude, that the flux and reflux of the ocean have produced all the mountains, valleys, and other inequalities on the surface of the earth; that currents of the sea have scooped out the valleys, elevated the hills, and bestowed on them their corresponding directions; that that same waters of the ocean, by transporting and depositing earth, &c., have given rise to the parallel strata; that the waters from the heavens gradually destroy the effects of the sea, by continually diminishing the height of the mountains, filling up the valleys, and choking the mouths of rivers; and, by reducing every thing to its former level, they will, in time, restore the earth to the sea, which, by its natural operations, will again create new continents, interspersed with mountains and valleys, every way similar to those we inhabit.

'Second Discours: Histoire et Théorie de la Terre', *Histoire Naturelle, Générale et Particulière, Avec la Description du Cabinet du Roi* (1749), Vol. 1, 124; *Natural History, General and Particular* (1785), Vol. 1, trans. W. Smellie, 57–8.

2 When I read an Italian letter [*Saggio* by Voltaire] on changes which had occurred on the surface of the earth, published in Paris this year (1746), I believed that these facts were reported by La Loubère. Indeed, they correspond perfectly with the author's ideas. Petrified fish are according to him merely rare fish thrown away by Roman cooks because they were spoiled; and with respect to shells, he said that they were from the sea of the Levant and brought back by pilgrims from Syria at the time of the crusades. These shells are found today petrified in France, in Italy and in other Christian states. Why did he not add that monkeys transported shells on top of high mountains and to every place where humans cannot live? It would not have harmed his story but made his explanation even more plausible.

'Preuves de la Théorie de la Terre', *Histoire Naturelle, Générale et Particulière, Avec la Description du Cabinet du Roi* (1749), Vol. 1, 281. Trans. Albert V. and Marguerite Carozzi. See **Voltaire** 600:3.

3 In Ireland, there are the same fossils, the same shells and the same sea bodies, as appear in America, and some of them are found in no other part of Europe.

'Preuves de la Théorie de la Terre', *Histoire Naturelle, Générale et Particulière, Avec la Description du Cabinet du Roi* (1749), Vol. 1, 606; *Natural History: Theory of the Earth* (1749), Vol. 1, trans. W. Smellie (1785), 507.

4 Nature progresses by unknown gradations and consequently does not submit to our absolute division when passing by imperceptible nuances, from one species to another and often from one genus to another. Inevitably there are a great number of equivocal species and in-between specimens that one does not know where to place and which throw our general systems into turmoil.

Jean Piveteau (ed.), *Oeuvres Philosophiques de Buffon* (1965), 10. Trans. in Paul Farber, 'Buffon and the Concept of Species', in *Journal of the History of Biology*, 1972, 5, 260.

5 Although the works of the Creator may be in themselves all equally perfect, the animal is, as I see it, the most complete work of nature, and man is her masterpiece.

'Histoire des Animaux', *Histoire Naturelle, Générale et Particulière, avec la Description du Cabinet du Roi* (1749), Vol. 2, 2. Quoted in Jacques Roger, *The Life Sciences in Eighteenth-Century French Thought*, ed. Keith R. Benson and trans. Robert Ellrich (1997), 437.

6 The greatest marvel is not in the individual. It is in the succession, in the renewal and in the duration of the species that Nature would seem quite inconceivable. This power of producing its likeness that resides in animals and plants, this form of unity, always subsisting and appearing eternal, this procreative virtue which is perpetually expressed without ever being destroyed, is for us a mystery which, it seems, we will never be able to fathom.

'Histoire des Animaux', *Histoire Naturelle, Générale et Particulière, avec la Description du Cabinet du Roi* (1749), Vol. 2, 3. Trans. Phillip R. Sloan.

1 *Rassemblons des faits pour nous donner des idées.*

Let us gather facts in order to get ourselves thinking.

'Histoire des Animaux', *Histoire Naturelle, Générale et Particulière, avec la Description du Cabinet du Roi* (1749), Vol. 2, 18. Quoted in Jacques Roger, *The Life Sciences in Eighteenth-Century French Thought*, ed. Keith R. Benson and trans. Robert Ellrich (1997), 440.

2 Let us investigate more closely this property common to animal and plant, this power of producing its likeness, this chain of successive existences of individuals, which constitutes the real existence of the species.

'De la Reproduction en Générale et particuliére', *Histoire Naturelle, Générale et Particulière, Avec la Description du Cabinet du Roi* (1749), Vol. 2, 18. Trans. Phillip R. Sloan.

3 For the little that one has reflected on the origin of our knowledge, it is easy to perceive that we can acquire it only by means of comparison. That which is absolutely incomparable is wholly incomprehensible. God is the only example that we could give here. He cannot be comprehended, because he cannot be compared. But all which is susceptible of comparison, everything that we can perceive by different aspects, all that we can consider relatively, can always be judged according to our knowledge.

'Histoire naturelle de l'Homme', *Histoire Naturelle, Générale et Particulière, avec la Description du Cabinet du Roi* (1749), Vol. 2, 431. Trans. Phillip R. Sloan.

4 To be and to think are one and the same for us.

'De la Nature de l'Homme', *Histoire Naturelle, Générale et Particulière, avec la Description du Cabinet du Roi* (1749), Vol. 2, 432. Quoted in Jacques Roger, *The Life Sciences in Eighteenth-Century French Thought*, ed. Keith R. Benson and trans. Robert Ellrich (1997), 434.

5 There is in Nature a general prototype in each species on which each individual is modeled, but which seems, in realizing itself, to alter itself or perfect itself according to circumstances. So that, relative to certain qualities, this is an extraordinary appearing variation in the succession of these individuals, and at the same time a constancy which appears wonderful in the entire species. The first animal, the first horse, for example, has been the external model and the interieur mold on which all horses which have been born, all those which now exist, and all those which will be born have been formed.

'Le Cheval', *Histoire Naturelle, Générale et Particulière, avec la Description du Cabinet du Roi* (1753), Vol. 4, 215–6. Trans. Phillip R. Sloan.

6 The great workman of nature is time.

'Les Animaux Sauvages', *Histoire Naturelle, Générale et Particulière, avec la Description du Cabinet du Roi* (1756), Vol. 6, 60. Quoted in Jacques Roger, *The Life Sciences in Eighteenth-Century French Thought*, ed. Keith R. Benson and trans. Robert Ellrich (1997), 468.

7 Only those works which are well-written will pass to posterity: the amount of knowledge, the uniqueness of the facts, even the novelty of the discoveries are no guarantees of immortality . . . These things are exterior to a man but style is the man himself.

'Discours prononcé dans l'Académie française, Le Samedi 25 Aout 1753', *Histoire Naturelle, Générale et Particulière, Avec la Description du Cabinet du Roi* (1753), Vol. 7, xvi-xvii.

8 Just as, in civil History, one consults title-deeds, one studies coins, one deciphers ancient inscriptions, in order to determine the epochs of human revolutions and to fix the dates of moral [i.e. human] events; so, in Natural History, one must excavate the archives of the world, recover ancient monuments from the depths of the earth, collect their remains, and assemble in one body of proofs all the evidence of physical changes that enable us to reach back to the different ages of Nature. This, then, is the order of the times indicated by facts and monuments: these are six epochs in the succession of the first ages of Nature; six spaces of duration, the limits of which although indeterminate are not less real; for these epochs are not like those of civil History . . . that we can count and measure exactly; nevertheless we can compare them with each other and estimate their relative duration.

'Des Époques de la Nature', *Histoire Naturelle, Générale et Particulière contenant les Époques de la Nature* (1778), Supplément Vol. 9, 1–2, 41. Trans. Martin J. Rudwick.

1 Nature is the system of laws established by the Creator for the existence of things and for the succession of creatures. Nature is not a thing, because this thing would be everything. Nature is not a creature, because this creature would be God. But one can consider it as an immense vital power, which encompasses all, which animates all, and which, subordinated to the power of the first Being, has begun to act only by his order, and still acts only by his concourse or consent . . . Time, space and matter are its means, the universe its object, motion and life its goal.

'De la Nature: Première Vue', *Histoire naturelle, générale et particulière, avec la description du cabinet du roi* (1764), Vol. 12, iii-iv. Trans. Phillip R. Sloan.

2 An Individual, whatever species it might be, is nothing in the Universe. A hundred, a thousand individuals are still nothing. The species are the only creatures of Nature, perpetual creatures, as old and as permanent as it. In order to judge it better, we no longer consider the species as a collection or as a series of similar individuals, but as a whole independent of number, independent of time, a whole always living, always the same, a whole which has been counted as one in the works of creation, and which, as a consequence, makes only a unity in Nature.

'De la Nature: Seconde Vue', *Histoire naturelle, générale et particulière, avec la description du cabinet du roi* (1765), Vol. 13, i. Trans. Phillip R. Sloan.

3 Far from becoming discouraged, the philosopher should applaud nature, even when she appears miserly of herself or overly mysterious, and should feel pleased that as he lifts one part of her veil, she allows him to glimpse an immense number of other objects, all worthy of investigation. For what we already know should allow us to judge of what we will be able to know; the human mind has no frontiers, it extends proportionately as the universe displays itself; man, then, can and must attempt all, and he needs only time in order to know all. By multiplying his observations, he could even see and foresee all phenomena, all of nature's occurrences, with as much truth and certainty as if he were deducing them directly from causes. And what more excusable or even more noble enthusiasm could there be than that of believing man capable of recognizing all the powers, and discovering through his investigations all the secrets, of nature!

'Des Mulets', *Oeuvres Philosophiques*, ed. Jean Piveteau (1954), 414. Quoted in Jacques Roger, *The Life Sciences in Eighteenth-Century French Thought*, ed. Keith R. Benson and trans. Robert Ellrich (1997), 458.

4 Nature, displayed in its full extent, presents us with an immense tableau, in which all the order of beings are each represented by a chain which sustains a continuous series of objects, so close and so similar that their difference would be difficult to define. This chain is not a simple thread which is only extended in length, it is a large web or rather a network, which, from interval to interval, casts branches to the side in order to unite with the networks of another order.

'Les Oiseaux Qui Ne Peuvent Voler', *Histoire Naturelle des Oiseaux* (1770), Vol. 1, 394. Trans. Phillip R. Sloan.

1 The sublime can only be found in the great subjects. Poetry, history and philosophy all have the same object, and a very great object—Man and Nature. Philosophy describes and depicts Nature. Poetry paints and embellishes it. It also paints men, it aggrandizes them, it exaggerates them, it creates heroes and gods. History only depicts man, and paints him such as he is.

'Discours Prononcé a L'Académie française par M. De Buffon. Le Jour de sa Réception 25 Aout 1753'. Supplément a T.iv (1753), *Histoire naturelle, générale et particulière, avec la description du cabinet du roi* (1777), 11. Trans. Phillip R. Sloan.

2 All the work of the crystallographers serves only to demonstrate that there is only variety everywhere where they suppose uniformity . . . that in nature there is nothing absolute, nothing perfectly regular.

Histoire Naturelle des Minéraux (1783–88), Vol. 3, 433.

Edward Crisp Bullard 1907–80

British geophysicist

3 The transition from sea-floor spreading to plate tectonics is largely a change of emphasis. Sea-floor spreading is a view about the method of production of new oceans floor on the ridge axis. The magnetic lineations give the history of this production back into the late Mesozoic and illuminate the history of the new aseismic parts of the ocean floor. This naturally directed attention to the relation of the sea-floor to the continents. There are two approaches: in the first, one looks back in time to earlier arrangements of the continents; in the second, one considers the current problem of the disposal of the rapidly growing sea floor.

'The Emergence of Plate Tectonics: A Personal View', *Annual Review of Earth and Planetary Sciences*, 1975, 3, 20.

4 Most of the scientists in their twenties and thirties who went in 1939 to work on wartime problems were profoundly affected by their experience. The belief that Rutherford's boys were the best boys, that we could do anything that was do-able and could master any subject in a few days was of enormous value.

'The Effect of World War II on the Development of Knowledge in the Physical Sciences', *Proceedings of the Royal Society of London*, 1975, Series A, 342, 531.

5 I suspect that the most important effect of World War II on physical science lay in the change in the attitude of people to science. The politicians and the public were convinced that science was useful and were in no position to argue about the details. A professor of physics might be more sinister than he was in the 1930s, but he was no longer an old fool with a beard in a comic-strip. The scientists or at any rate the physicists, had changed their attitude. They not only believed in the interest of science for themselves, they had acquired also a belief that the tax-payer should and would pay for it and would, in some unspecified length of run, benefit by it.

'The Effect of World War II on the Development of Knowledge in the Physical Sciences', *Proceedings of the Royal Society of London*, 1975, Series A, 342, 532.

Theodore Holmes Bullock 1915–

American neuroscientist

6 Nature has provided two great gifts: life and then the diversity of living things, jellyfish and humans, worms and crocodiles. I don't undervalue the investigation of commonalities but can't avoid the conclusion that diversity has been relatively neglected, especially as concerns the brain.

'Theodore H. Bullock', in Larry R. Squire (ed.), *The History of Science in Autobiography* (1996), Vol. 1, 144.

Edward George Bulwer-Lytton (Earl Lytton) 1803–73

British novelist and politician

7 In science, read, by preference, the newest works; in literature, the oldest.

Caxtoniana: A Series of Essays on Life, Literature, and Manners (1863), Vol. 1, 169.

1 In science, address the few; in literature, the many. In science, the few must dictate opinion to the many; in literature, the many, sooner or later, force their judgement on the few. But the few and the many are not necessarily the few and the many of the passing time: for discoverers in science have not un-often, in their own day, had the few against them; and writers the most permanently popular not unfrequently found, in their own day, a frigid reception from the many. By the few, I mean those who must ever remain the few, from whose dieta we, the multitude, take fame upon trust; by the many, I mean those who constitute the multitude in the long-run. We take the fame of a Harvey or a Newton upon trust, from the verdict of the few in successive generations; but the few could never persuade us to take poets and novelists on trust. We, the many, judge for ourselves of Shakespeare and Cervantes.

Caxtoniana: A Series of Essays on Life, Literature, and Manners (1863), Vol. 2, 329–30.

Robert Wilhelm Eberhard Bunsen 1811–99

German chemist

2 *Ein Chemiker der kein Physiker ist, ist gar nichts.*

A chemist who is not a physicist is nothing at all.

J. R. Partington (ed.), *A History of Chemistry* (1961), Vol. 4, 282.

3 ON **BUNSEN** As an investigator he was great, as a teacher he was greater, as a man and friend he was greatest.

Henry Roscoe on Bunsen. In J. R. Partington (ed.), *A History of Chemistry* (1961), Vol. 4, 282.

4 Working is beautiful and rewarding, but acquisition of wealth for its own sake is disgusting.

A comment Bunsen often told his students.
Quoted in R. Oesper, *The Human Side of Scientists* (1975), 28.

5 ON **BUNSEN** First I would like to wash Bunsen, and then I would like to kiss him because he is such a charming man.

Frau Fischer on meeting Bunsen for the first time.
Quoted in E. Fischer, *Aus meinem Leben* (1923). Trans. W. H. Brock.

Luis Buñuel 1900–83

Spanish director and film-maker

6 I find it [science] analytical, pretentious and superficial—largely because it does not address itself to dreams, chance, laughter, feelings, or paradox—in other words,—all the things I love the most.

My Last Sigh, trans. Abigail Israel (1983), 174.

John Bunyan 1628–88

British clergyman and writer

See **Osler, William** 472:4

7 *Atten.* Pray of what disease did Mr. Badman die, for now I perceive we are come up to his death? *Wise.* I cannot so properly say that he died of one disease, for there were many that had consented, and laid their heads together to bring him to his end. He was dropsical, he was consumptive, he was surfeited, was gouty, and, as some say, he had a tang of the pox in his bowels. Yet the captain of all these men of death that came against him to take him away, was the consumption, for it was that that brought him down to the grave.

The Life and Death of Mr Badman (1680). In *Grace Abounding & The Life and Death of Mr Badman* (1928), 282.

Jacob Christopher Burckhardt 1818–97

Swiss historian

8 History is the record of what one age finds worthy of note in another.

Judgements on History and Historians (1958), 11.

Edmund Burke 1729–97

British statesman, orator and politician

1 The age of chivalry is gone.—That of sophisters, economists, and calculators, has succeeded; and the glory of Europe is extinguished for ever.

Reflections on the Revolution in France (1790), 113.

2 Learning will be cast into the mire, and trodden down under the hoofs of a swinish multitude.

Reflections on the Revolution in France (1790), 117.

Frank Macfarlane Burnet 1899–1985

Australian immunologist and virologist

3 I like to think that when Medawar and his colleagues showed that immunological tolerance could be produced experimentally the new immunology was born. This is a science which to me has far greater potentialities both for practical use in medicine and for the better understanding of living process than the classical immunochemistry which it is incorporating and superseding.

'Immunological Recognition of Self', Nobel Lecture, 12 December 1960. In *Nobel Lectures Physiology or Medicine 1942–1962* (1964), 689.

4 I can see no practical application of molecular biology to human affairs . . . DNA is a tangled mass of linear molecules in which the informational content is quite inaccessible.

Immunological Surveillance (1970), 240–1.

Gilbert Burnet 1643–1715

British theologian and historian

5 Many physicians, and other ingenious men, went into the society for natural philosophy. But he who laboured most, at the greatest charge, and with the most success at experiments, was Robert Boyle, the earl of Cork's youngest son. He was looked on by all who knew him as a very perfect pattern. He was a very devout Christian, humble and modest, almost to a fault, of a most spotless and exemplary life in all respects. He was highly charitable; and was a mortified and self-denied man, that delighted in nothing so much as in the doing good. He neglected his person, despised the world, and lived abstracted from all pleasures, designs, and interests.

History of His Own Time (1724) (1833 edn.), Vol. 1, 351.

Thomas Burnet c.1635–1715

British cosmologist and theologian

6 'Tis a dangerous thing to ingage the authority of Scripture in disputes about the Natural World, in opposition to Reason; lest Time, which brings all things to light, should discover that to be evidently false which we had made Scripture to assert . . . We are not to suppose that any truth concerning the Natural World can be an Enemy to Religion; for Truth cannot be an Enemy to Truth, God is not divided against himself.

The Sacred Theory of the Earth (1691), reprint 1965, 16.

7 All I say, betwixt the first Chaos and the last Completion of Time and all Things temporary, This was given to the disquisitions of men; On either hand is Eternity, before the World and after, which is without our reach: But that little spot of Ground that lies betwixt those two great Oceans, this we are to cultivate, this we are Masters of, herein we are to exercise our thoughts, to understand and lay open the Treasuries of the Divine Wisdom and Goodness hid in this part of Nature and of Providence.

The Sacred Theory of the Earth (1691), reprint 1965, 25.

8 There is no Chase so pleasant, methinks, as to drive a Thought, by good conduct, from one end of the World to the other; and never to lose sight of it till it fall into Eternity, where all things are lost as to our knowledge.

The Sacred Theory of the Earth (1691), reprint 1965, 26.

1 We are almost the last Posterity of the first Men, and faln [sic] into the dying Age of the World; by what footsteps, or by what guide, we trace back our way to those first Ages, and the first order of things?

The Sacred Theory of the Earth (1691), reprint 1965, 27.

2 If this present state and form of the Earth had been from Eternity, it would have long ere this destroy'd it self, and chang'd it self: the Mountains sinking by degrees into the Valleys, and into the Sea, and the Waters rising above the Earth; which form it would certainly have come into sooner or later, and in it continu'd drown'd and uninhabitable, for all succeeding Generations . . . For whatsoever moulders or is washt away from them, is carried down into the lower grounds, and into the Sea, and nothing is ever brought back again by any circulation: Their losses are not repair'd, nor any proportionable recruits made from any other parts of Nature.

The Sacred Theory of the Earth (1691), reprint 1965, 44–5.

3 Oratours and Philosophers treat Nature after a very different manner . . . And as to this Earth in particular, if I was to describe it as an Oratour, I would suppose it a beautiful and regular Globe, and not only so, but that the whole Universe was made for its sake; that it was the darling and favourite of Heaven, that the Sun shin'd only to give it light, to ripen its Fruit, and make fresh its Flowers; And that the great Concave of the Firmament, and all the Stars in their several Orbs, were design'd only for a spangled Cabinet to keep this Jewel in. This *Idea* I would give of it as an Oratour; But a Philosopher that overheard me, would either think me in jest, or very injudicious, if I took the Earth for a body so regular in it self, or so considerable, if compar'd with the rest of the Universe. This, he would say, is to make the great World like one of the Heathen Temples, a beautiful and magnificent structure, and of the richest materials, yet built only for a little brute Idol, a Dog, or a Crocodile, or some deformed Creature, plac'd in a corner of it.

The Sacred Theory of the Earth (1691), reprint 1965, 90.

4 The truth is, the generality of people have not sence [sic] and curiosity enough to raise a question concerning these things [mountains], or concerning the Original of them. You may tell them that Mountains grow out of the earth like Fuzz-balls, or that there are Monsters under ground that throw up Mountains as Moles do Mole-hills; they will scarce raise one objection against your doctrine.

The Sacred Theory of the Earth (1691), reprint 1965, 110.

Edward Arthur Burroughs
(Bishop of Ripon) 1882–1934

British clergyman

5 The sum of human happiness, outside of scientific circles, would not necessarily be reduced if for, say ten years, every physical and chemical laboratory were closed and the patient and resourceful energy displayed in them transferred to recovering the lost art of getting together and finding a formula for making ends meet in the scale of human life.

Sermon to the British Association for the Advancement of Science, 1927. In J. Heilbron and Robert W. Seidel, *Lawrence and his Laboratory: A History of the Lawrence Berkeley Laboratory* (1989), Vol. 1, 30.

John Bagnell Bury 1861–1927
British classical scholar and historian

6 Evolution . . . does not necessarily mean applied to society, the movement of man to a desirable goal. It is a neutral, scientific conception, compatible either with optimism or with pessimism. According to different estimates it may appear to be a cruel sentence or a guarantee of steady amelioration. And it has been actually interpreted in both ways.

The Idea of Progress (1920), 335–6.

Vannevar Bush 1890–1974

American electrical engineer

1 Science—the Endless Frontier.

Science—The Endless Frontier: A Report to the President (1945).

2 Science—the Endless Expenditure.

Quoted in Thomas Hager, *Force of Nature: The Life of Linus Pauling* (1995), 290.

Leo W. Buss 1953–

American evolutionary biologist

3 The synthetic theory of evolution, with its emphasis on the individual as the unit of evolutionary modification, is frequently, and justly, criticized as a 'theory of adults'—one which has failed to address the diversity of ontogeny. Evolutionists, even today, seek to understand how development will illuminate patterns in evolution, not how evolution will illuminate the details of the developmental process.

The Evolution of Individuality (1987), 65.

Samuel Butler 1835–1902

British novelist, essayist and critic

4 There is a kind of plant that eats organic food with its flowers: when a fly settles upon the blossom, the petals close upon it and hold it fast till the plant has absorbed the insect into its system; but they will close on nothing but what is good to eat; of a drop of rain or a piece of stick they will take no notice. Curious! that so unconscious a thing should have such a keen eye to its own interest.

Erewhon (1872), 1966 edn, 142.

5 A pair of pincers set over a bellows and a stew pan and the whole thing fixed on stilts.

Definition of man.

H. Festing Jones (ed.), *The Notebooks of Samuel Butler* (1913), 18.

6 The idea of an indivisible, ultimate atom is inconceivable by the lay mind. If we can conceive of an idea of the atom at all, we can conceive it as capable of being cut in half; indeed, we cannot conceive it at all unless we so conceive it. The only true atom, the only thing which we cannot subdivide and cut in half, is the universe. We cannot cut a bit off the universe and put it somewhere else. Therefore the universe is a true atom and, indeed, is the smallest piece of indivisible matter which our minds can conceive; and they cannot conceive it any more than they can the indivisible, ultimate atom.

H. Festing Jones (ed.), *The Notebooks of Samuel Butler* (1913), 84.

7 I do not know whether my distrust of men of science is congenital or acquired, but I think I should have transmitted it to descendants.

Geoffrey Keynes and Brian Hill (eds.), *Samuel Butler's Notebooks* (1951), 32.

8 If it tends to thicken the crust of ice on which, as it were, we are skating, it is all right. If it tries to find, or professes to have found, the solid ground at the bottom of the water it is all wrong.

Geoffrey Keynes and Brian Hill (eds.), *Samuel Butler's Notebooks* (1951), 110.

9 There are two classes, those who want to know, and do not care whether others think they know or not, and those who do not much care about knowing, but care very greatly about being reputed as knowing.

Geoffrey Keynes and Brian Hill (eds.), *Samuel Butler's Notebooks* (1951), 119.

10 *The Athanasian Creed* is to me light and intelligible reading in comparison with much that now passes for science.

Geoffrey Keynes and Brian Hill (eds.), *Samuel Butler's Notebooks* (1951), 125.

1 Business should be like religion and science; it should know neither love nor hate.

Geoffrey Keynes and Brian Hill (eds.), *Samuel Butler's Notebooks* (1951), 144.

2 We no more deny the essential value of religion because we hold most religions false, and most professors of religion liars, than we deny that of science because we can see no great difference between men of science and theologians.

Geoffrey Keynes and Brian Hill (eds.), *Samuel Butler's Notebooks* (1951), 194.

3 Science is being daily more and more personified and anthromorphized into a god. By and by they will say that science took our nature upon him, and sent down his only begotten son, Charles Darwin, or Huxley, into the world so that those who believe in him, etc.; and they will burn people for saying that science, after all, is only an expression for our ignorance of our own ignorance.

Geoffrey Keynes and Brian Hill (eds.), *Samuel Butler's Notebooks* (1951), 233.

4 X-rays. Their moral is this—that a right way of looking at things will see through almost anything.

Geoffrey Keynes and Brian Hill (eds.), *Samuel Butler's Notebooks* (1951), 282.

5 A hen is only an egg's way of making another egg.

Attributed.

Aleksandr Mikhailovich Butlerov 1828–86

Russian chemist

6 *Die Chemische Struktur der Körper.*

The chemical structure of substances.

'Einiges Über Die Chemische Struktur der Körper', *Zeitschrift für Chemie und Pharmacie*, 1861, 549.

7 I would . . . change the accepted rule that the nature of a complex molecule is determined by the *nature, quantity, and position* of its elementary component parts, by the following statement: the chemical nature of a complex molecule is determined by the nature of its elementary component parts, their quantity and *chemical structure*.

'On the Chemical Structure of Substances' 1861.

Herbert Butterfield 1900–79

British historian

8 [The Whig interpretation of history] . . . is the tendency in many historians to write on the side of Protestants and Whigs, to praise revolutions provided they have been successful, to emphasise certain principles of progress in the past and to produce a story which is the ratification if not the glorification of the present.

The Whig Interpretation of History (1931), v.

9 The study of the past with one eye, so to speak, upon the present is the source of all sins and sophistries in history . . . It is the essence of what we mean by the word 'unhistorical'.

The Whig Interpretation of History (1931), 31–2.

10 The so-called 'scientific revolution', popularly associated with the sixteenth and seventeenth centuries, but reaching back in an unmistakably continuous line to a period much earlier still. Since that revolution overturned the authority in science not only of the middle ages but of the ancient world—since it ended not only in the eclipse of scholastic philosophy but in the destruction of Aristotelian physics—it outshines everything since the rise of Christianity and reduces the Renaissance and Reformation to the rank of mere episodes, mere internal displacements, within the system of medieval Christendom . . . It looms so large as the real origin of the modern world and of the modern mentality that our customary periodisation of European history has become an anachronism and an encumbrance.

The Origins of Modern Science (1949), viii.

1 Concerning alchemy it is more difficult to discover the actual state of things, in that the historians who specialise in this field seem sometimes to be under the wrath of God themselves; for, like those who write of the Bacon-Shakespeare controversy or on Spanish politics, they seem to become tinctured with the kind of lunacy they set out to describe.
The Origins of Modern Science (1949), 115.

Ada Byron see Lovelace, Augusta Ada King, Countess of

George Gordon, Lord Byron
1788–1824
British romantic poet

2 That knowledge is not happiness, and science
But an exchange of ignorance for that
Which is another kind of ignorance.
Manfred (1816), Act II, Scene IV, lines 61–3. In Jerome J. McGann (ed.), *Lord Byron: The Complete Poetical Works* (1986), Vol. 4, 83.

3 I had a dream, which was not all a dream.
The bright sun was extinguish'd, and the stars
Did wander darkling in the eternal space,
Rayless, and pathless, and the icy earth
Swung blind and blackening in the moonless air;
Morn came, and went—and came, and brought no day,
Darkness (1816), lines 1–6. In Jerome J. McGann (ed.), *Lord Byron: The Complete Poetical Works* (1986), Vol. 4, 40–1.

4 [My advice] will one day be found
With other relics of 'a former world,'
When this world shall be *former*, underground,
Thrown topsy-turvy, twisted, crisped, and curled,
Baked, fried or burnt, turned inside-out, or drowned,
Like all the worlds before, which have been hurled
First out of, and then back again to Chaos,
The Superstratum which will overlay us.
Don Juan (1821), Canto 9, Verse 37. In Jerome J. McGann (ed.), *Lord Byron: The Complete Poetical Works* (1986), Vol. 5, 420.

5 When Newton saw an apple fall, he found
In that slight startle from his contemplation—
'Tis *said* (for I'll not answer above ground
For any sage's creed or calculation)—
A mode of proving that the earth turn'd round
In a most natural whirl, called 'gravitation';
And this is the sole mortal who could grapple,
Since Adam, with a fall, or with an apple.
Don Juan (1821), Canto 10, Verse 1. In Jerome J. McGann (ed.), *Lord Byron: The Complete Poetical Works* (1986), Vol. 5, 437.

Pierre-Jean-Georges Cabanis
1757–1808
French physician and philosopher

6 In order to form for one's self a just notion of the operations which result in the production of thought, it is necessary to conceive of the brain as a peculiar organ, specially designed for the production thereof, just as the stomach is designed to effect digestion, the liver to filter the bile, the parotids and the maxillary and sublingual glands to prepare the salivary juices.
Rapports du Physique et du Moral de l'Homme (1805), 2nd edition, Vol. 1, 152–3. Translated in Robert M. Young, *Mind, Brain and Adaptation in the Nineteenth Century* (1970), 20.

Charles Caldwell 1772–1853
American physician and educator

7 I would not have it inferred . . . that I am, as yet, an advocate for the hypothesis of *chemical life*. The doctrine of the vitality of the blood, stands in no need of aid from that speculative source. If it did, I would certainly abandon it. For, notwithstanding the fashionableness of the hypothesis in Europe, and the ascendancy it has gained over some minds in this country [USA], it will require stubborn facts to convince me that man with all his corporeal and intellectual attributes is nothing but *hydro-phosphorated oxyde of azote* . . . When the chemist declares, that the same laws which direct the *crystallization* of spars, nitre and Glauber's salts, direct also the *crystallization* of man, he must pardon me if I neither understand him, *nor* believe him.

Medical Theses (1805), 391–2, footnote.

William H. Caldwell 1859–1941

British embryologist

1 Monotremes oviparous, ovum meroblastic.

The platypus and the echidna lay eggs which, having considerable yolk, lead to the development of the embryo above the yolkmass.

Cablegram sent to The British Association for the Advancement of Science in Montreal, 1884. Quoted in H. Burrell, *The Platypus* (1927), 45.

Norman Robert Campbell 1880–1949

British physicist and philosopher of science

2 Science would not be what it is if there had not been a Galileo, a Newton or a Lavoisier, any more than music would be what it is if Bach, Beethoven and Wagner had never lived. The world as we know it is the product of its geniuses—and there may be evil as well as beneficent genius—and to deny that fact, is to stultify all history, whether it be that of the intellectual or the economic world.

What is Science? (1921), 73.

Stanislao Cannizzaro 1826–1910

Italian chemist

3 Compare . . . the various quantities of the same element contained in the molecule of the free substance and in those of all its different compounds and you will not be able to escape the following law: The different quantities of the same element contained in different molecules are all whole multiples of one and the same quantity, which always being entire, has the right to be called an atom.

Sketch of a Course of Chemical Philosophy (1858), Alembic Club Reprint (1910), 11.

Annie Jump Cannon 1863–1941

American astronomer

4 A life spent in the routine of science need not destroy the attractive human element of a woman's nature.

Said of Williamina Paton Fleming 1857–1911, American Astronomer.

Obituary of Williamina Paton Fleming, *Science*, 1911, 33, 988.

Walter Bradford Cannon 1871–1945

American physiologist

5 Since the stomach gives no obvious external sign of its workings, investigators of gastric movements have hitherto been obliged to confine their studies to pathological subjects or to animals subjected to serious operative interference. Observations made under these necessarily abnormal conditions have yielded a literature which is full of conflicting statements and uncertain results. The only sure conclusion to be drawn from this material is that when the stomach receives food, obscure peristaltic contractions are set going, which in some way churn the food to a liquid chyme and force it into the intestines. How imperfectly this describes the real workings of the stomach will appear from the following account of the actions of the organ studied by a new method. The mixing of a small quantity of subnitrate of bismuth with the food allows not only the contractions of the gastric wall, but also the movements of the gastric contents to be seen with the

Röntgen rays in the uninjured animal during normal digestion.

'The Movements of the Stomach Studied by Means of the Röntgen Rays,' *American Journal of Physiology*, 1898, 1, 359–360.

1 'These changes in the body,' he wrote in the review paper he sent to the *American Journal of Physiology* late in 1913, 'are, each one of them, directly serviceable in making the organism more efficient in the struggle which fear or rage or pain may involve; for fear and rage are organic preparations for action, and pain is the most powerful known stimulus to supreme exertion. The organism which with the aid of increased adrenal secretion can best muster its energies, can best call forth sugar to supply the labouring muscles, can best lessen fatigue, and can best send blood to the parts essential in the run or the fight for life, is most likely to survive. Such, according to the view here propounded, is the function of the adrenal medulla at times of great emergency.'

Quoted in S. Benison, A. C. Barger and E. L. Wolfe, *Walter B Cannon: The Life and Times of a Young Scientist* (1987), 311.

2 I heard Professor Cannon lecture last night, going partly on your account, his subject was a physiological substitute for war—which is international sports and I suppose motorcycle races—to encourage the secretion of the adrenal glands!

Letter from James McKeen Cattell to his son, McKeen. In S. Benison, A. C. Barger and E. L. Wolfe, *Walter B Cannon: The Life and Times of a Young Scientist* (1987), 319.

3 The steady states of the fluid matrix of the body are commonly preserved by physiological reactions, i.e., by more complicated processes than are involved in simple physico-chemical equilibria. Special designations, therefore, are appropriate:—'homeostasis' to designate stability of the organism; 'homeostatic conditions,' to indicate details of the stability; and 'homeostatic reactions,' to signify means for maintaining stability.

'Physiological Regulation of Normal States: Some Tentative Postulates Concerning Biological Homeostatics', 1926. Reprinted in L. L. Langley (ed.), *Homeostasis: Origins of the Concept* (1973), 246.

4 The constant conditions which are maintained in the body might be termed *equilibria*. That word, however, has come to have fairly exact meaning as applied to relatively simple physico-chemical states, in closed systems, where known forces are balanced. The coordinated physiological processes which maintain most of the steady states in the organism are so complex and so peculiar to living beings—involving, as they may, the brain and nerves, the heart, lungs, kidneys and spleen, all working cooperatively—that I have suggested a special designation for these states, *homeostasis*. The word does not imply something set and immobile, a stagnation. It means a condition—a condition which may vary, but which is relatively constant.

The Wisdom of the Body (1932), 24.

5 Investigators are commonly said to be engaged in a search for the truth. I think they themselves would usually state their aims less pretentiously. What the experimenter is really trying to do is to learn whether facts can be established which will be recognized as facts by others and which will support some theory that in imagination he has projected. But he must be ingenuously honest. He must face facts as they arise in the course of experimental procedure, whether they are favourable to his idea or not. In doing this he must be ready to surrender his theory at any time if the facts are adverse to it.

The Way of an Investigator: A Scientist's Experiences in Medical Research (1945), 34.

6 These changes—the more rapid pulse, the deeper breathing, the increase of sugar in the blood, the secretion from the adrenal glands—were very diverse and seemed unrelated. Then, one wakeful night, after a considerable collection of these changes had been disclosed, the idea flashed through my mind that they could be nicely integrated if conceived as bodily preparations for supreme effort in flight or in fighting. Further investigation added to the collection and confirmed the general scheme suggested by the hunch.

The Way of an Investigator: A Scientist's Experiences in Medical Research (1945), 59–60.

1 Time . . . is an essential requirement for effective research. An investigator may be given a palace to live in, a perfect laboratory to work in, he may be surrounded by all the conveniences money can provide; but if his time is taken from him he will remain sterile.

Quoted in S. Benison, A. C. Barger and E. L. Wolfe, *Walter B Cannon: The Life and Times of a Young Scientist* (1987), 253.

Georg Cantor 1845–1918

German mathematician

2 The essence of mathematics lies precisely in its freedom.

Often misquoted.

Gesammelte Abhandlungen (1932), 182, trans. Ivor Grattan-Guinness.

3 The transfinite numbers are in a sense the *new irrationalities* [. . . they] stand or fall with the finite *irrational numbers*.

Gesammelte Abhandlungen (1932), 395, trans. Ivor Grattan-Guinness.

Karel Čapek 1890–1938

Czech writer and playwright

4 Rossum's Universal Robots.

R.U.R. (1920)

5 [Radius]: You will work. You will build . . . You will serve them . . . Robots of the world . . . The power of man has fallen . . . A new world has arisen. The rule of the Robots . . . March!

The play introduced the word 'robot' into the world's languages. The word robota *in Czech means compulsory labour: in the play it refers to a new working class of automatons.*

R.U.R. (1920), 89–90 in 1961 ed.

Girolamo Cardano 1501–76

Italian physician, mathematician and philosopher

6 By these pleasures it is permitted to relax the mind with play, in turmoils of the mind, or when our labors are light, or in great tension, or as a method of passing the time. A reliable witness is Cicero, when he says (*De Oratore*, 2): 'men who are accustomed to hard daily toil, when by reason of the weather they are kept from their work, betake themselves to playing with a ball, or with knucklebones or with dice, or they may also contrive for themselves some new game at their leisure.'

The Book of Games of Chance (1663), final sentences, trans. Sydney Henry Gould. In Oysten Ore, *The Gambling Scholar* (1953), 241.

Victoris Cardelini

7 Night and day, the ignorant as well as the learned give themselves over to the pleasure of making children. But no one knows how he has engendered his own progeny. If someone does understand it, he will not persuade others; because for thousands of years, those who study nature have been crossing swords with one another, and they will go on with the duel as long as the names of Hippocrates, Aristotle, and Galen continue to float upon men's lips.

De Origine Foetus Libri Duo (1628), 5. Quoted in Jacques Roger, *The Life Sciences in Eighteenth-Century French Thought*, Keith R. Benson (ed.) and trans. Robert Ellrich (1997), 37.

Sam Warren Carey 1911–2002

Australian geologist

8 Do not expect to be hailed as a hero when you make your great discovery. More likely you will be a ratbag—maybe failed by your examiners. Your statistics, or your observations, or your literature study, or your something else will be patently deficient. Do not doubt that in our enlightened age the really important advances are and will be rejected more often than acclaimed. Nor should we doubt that in our own professional lifetime we too will repudiate with like pontifical finality the most significant insight ever to reach our desk.

Theories of the Earth and Universe (1988), 365.

Thomas Carlyle 1795–1881

British historian and essayist

1 Statistics, one may hope, will improve gradually, and become good for something. Meanwhile, it is to be feared the crabbed satirist was partly right, as things go: 'A judicious man,' says he, 'looks at Statistics, not to get knowledge, but to save himself from having ignorance foisted on him.'

Chartism (1839), 1904, 125.

2 In all epochs of the world's history, we shall find the Great Man to have been the indispensable saviour of his epoch:—the lightning, without which the fuel never would have burnt. The History of the World, I said already, was the Biography of Great Men.

On Heroes, Hero-Worship and the Heroic in History (1841), 1903, 164.

3 Not a 'gay science,' I should say, like some we have heard of; no, a dreary, desolate, and indeed quite abject and distressing one; what we might call, by way of eminence, the *dismal science.*

Said of Social Science.

Occasional Discourse on The Nigger Question (1853), 9.

4 Man is a tool-using animal [*Handthierendes Tier*] . . . Without tools he is nothing, with tools he is all.

Sartor Resartus: The Life and Opinions of Herr Teufelsdröckh (1889), 36–7.

5 I don't pretend to understand the Universe—it's a great deal bigger than I am.

Letter to William Allingham, 28 December 1868. Quoted in H. Allingham and D. Radford (eds.), *William Allingham: A Diary* (1907), 196.

Rudolph Carnap 1891–1970

German-born American philosopher

6 Science is a system of statements based on direct experience, and controlled by experimental verification. Verification in science is not, however, of single statements but of the entire system of a sub-system of such statements.

The Unity of Science (1934), trans. M. Black, 42.

Andrew Carnegie 1835–1919

British-born American industrialist

7 While the law [of competition] may be sometimes hard for the individual, it is best for the race, because it insures the survival of the fittest in every department. We accept and welcome, therefore, as conditions to which we must accommodate ourselves, great inequality of environment, the concentration of business, industrial and commercial, in the hands of a few, and the law of competition between these, as being not only beneficial, but essential for the future progress of the race.

Wealth (1899), 655.

Lazare-Nicolas-Marguerite Carnot 1753–1823

French mathematician and engineer

8 When a body acts upon another one, it is always immediately or through some intermediate body; this intermediate body is in general what one calls a *machine.*

Essai sur les machines en général (1783), art. 8, trans. Ivor Grattan-Guinness.

Nicolas-Lèonard-Sadi Carnot 1796–1832

French engineer

9 The production of motion in the steam engine always occurs in circumstances which it is necessary to recognize, namely when the equilibrium of caloric is restored, or (to express this differently) when caloric passes from the body at one temperature to another body at a lower temperature.

'Réflexions sur la Puissance Motrice du Feu et sur les Machines Propres a Développer cette Puissance' (1824). Trans. Robert Fox, *Reflexions on the Motive Power of Fire* (1986), 64.

Alexis Carrel 1873–1944

French-born American surgeon

1 There is a strange disparity between the sciences of inert matter and those of life. Astronomy, mechanics, and physics are based on concepts which can be expressed, tersely and elegantly, in mathematical language. They have built up a universe as harmonious as the monuments of ancient Greece. They weave about it a magnificent texture of calculations and hypotheses. They search for reality beyond the realm of common thought up to unutterable abstractions consisting only of equations of symbols. Such is not the position of biological sciences. Those who investigate the phenomena of life are as if lost in an inextricable jungle, in the midst of a magic forest, whose countless trees unceasingly change their place and their shape. They are crushed under a mass of facts, which they can describe but are incapable of defining in algebraic equations.

Man the Unknown (1935), 1.

2 It seems that the increased number of scientific workers, their being split up into groups whose studies are limited to a small subject, and over-specialization have brought about a shrinking of intelligence. There is no doubt that the quality of any human group decreases when the number of the individuals composing this group increases beyond certain limits . . . The best way to increase the intelligence of scientists would be to decrease their number.

Man the Unknown (1935), 48–9.

Lewis Carroll 1832–98

British logician, mathematician and novelist

3 'When I use a word,' Humpty Dumpty said, in rather a scornful tone, 'it means just what I choose it to mean—neither more nor less.'

'The question is,' said Alice, 'whether you *can* make words mean so many different things.'

'The question is,' said Humpty Dumpty, 'which is to be master—that's all.'

Through the Looking Glass and what Alice Found there (1871). In Roger Lancelyn Green (ed.), *Alice's Adventures in Wonderland and Through the Looking-Glass, and what Alice Found there* (1971), 190.

4 'Can you do Addition?' the White Queen said. 'What's one and one and one and one and one and one and one and one and one and one?'

'I don't know', said Alice. 'I lost count'.

'She can't do Addition', the Red Queen interrupted.

Through the Looking Glass and what Alice Found there (1871). In Roger Lancelyn Green (ed.), *Alice's Adventures in Wonderland and Through the Looking-Glass, and what Alice Found there* (1971), 226.

William Herbert Carruth 1859–1924

American poet

5 Some call it Evolution,
And others call it God.

'Each in his Own Tongue', in *Each in his Own Tongue and Other Poems* (1909), 2.

Rachel Carson 1907–64

American biologist and writer

6 Over increasingly large areas of the United States, spring now comes unheralded by the return of the birds, and the early mornings are strangely silent where once they were filled with the beauty of bird song.

The Silent Spring (1962) ch. 8, 84.

Howard Carter 1874–1939 and George Herbert (5th Earl of Carnarvon) 1866–1923

British Egyptologists

1 LORD CARNARVON: Can you see anything?
HOWARD CARTER: Yes, wonderful things.

Lord Carnarvon and Howard Carter's dialogue upon Carter's first entry to the tomb of Tutankhamen.

Nancy Lynn Delaney Cartwright 1944–

American-born British philosopher of science

2 God may have written just a few laws and grown tired. We do not know whether we are in a tidy universe or an untidy one.

How the Laws of Physics Lie (1983), 49.

Carl Gustav Carus 1789–1869

German physiologist and psychologist

3 *Der Schlussel zur Erkenntnis vom Wesen des bewussten Seelenlebens liegt in der Region des Unbewusstseins.*

The key to the understanding of the character of the conscious lies in the region of the unconscious.

Psyche (1846), 1.

Paul Carus 1852–1919

German-born American philosopher

4 Science is wonderful at destroying metaphysical answers, but incapable of providing substitute ones. Science takes away foundations without providing a replacement. Whether we want to be there or not, science has put us in the position of having to live without foundations. It was shocking when Nietzsche said this, but today it is commonplace; *our* historical position—and no end to it is in sight—is that of having to philosophise without 'foundations'.

Quoted in Hilary Putnam (ed.), *The Many Faces of Realism: The Paul Carus Lectures* (1987), 29.

William Ernest Castle 1867–1962

American biologist

5 A moment's consideration of this case shows what a really great advance in the theory and practise of breeding has been obtained through the discovery of Mendel's law. What a puzzle this case would have presented to the biologist ten years ago! Agouti crossed with chocolate gives in the second filial generation (not in the first) four varieties, viz., agouti, chocolate, black and cinnamon. We could only have shaken our heads and looked wise (or skeptical).

Then we had no explanation to offer for such occurrences other than the 'instability of color characters under domestication,' the 'effects of inbreeding,' 'maternal impressions.' Serious consideration would have been given to the proximity of cages containing both black and cinnamon-agouti mice.

Now we have a simple, rational explanation, which any one can put to the test. We are able to predict the production of new varieties, and to produce them.

We must not, of course, in our exuberance, conclude that the powers of the hybridizer know no limits. The result under consideration consists, after all, only in the making of new combinations of unit characters, but it is much to know that these units exist and that all conceivable combinations of them are ordinarily capable of production. This valuable knowledge we owe to the discoverer and to the rediscoverers of Mendel's law.

'New Colour Variety of the Guinea Pig', *Science*, 1908, **28**, 250–252.

6 It is clear that in maize, seemingly blending is really segregating inheritance, but with entire absence of dominance, and it seems probably that the same will be found to be true among rabbits and other mammals; failure to observe it hitherto is probably due to the fact that the factors concerned are numerous. For the greater the number of factors concerned, the more nearly will the result obtained approximate a complete and permanent blend. As the number of factors approaches infinity, the result will become identical with a permanent blend.

Heredity: In Relation to Evolution and Animal Breeding (1911), 138–9.

Raymond Bernard Cattell

1905–98

British-born American psychologist

1 But psychology is a more tricky field, in which even outstanding authorities have been known to run in circles, 'describing things which everyone knows in language which no one understands'.

The Scientific Analysis of Personality (1965), 18.

Augustin-Louis Cauchy

1789–1857

French mathematician and physicist

2 As for methods I have sought to give them all the rigour that one requires in geometry, so as never to have recourse to the reasons drawn from the generality of algebra.

Cours d'analyse (1821), Preface, trans. Ivor Grattan-Guinness.

Henry Cavendish 1731–1810

British natural philosopher

3 A small bubble of air remained unabsorbed . . . if there is any part of the phlogisticated air [nitrogen] of our atmosphere which differs from the rest, and cannot be reduced to nitrous acid, we may safely conclude that it is not more than 1/120 part of the whole.

'Experiments on Air', read 2 June 1785, *Philosophical Transactions of the Royal Society*, 1785, 75, 382.

4 ON CAVENDISH His [Henry Cavendish's] Theory of the Universe seems to have been, that it consisted *solely* of a multitude of objects which could be weighed, numbered, and measured; and the vocation to which he considered himself called was, to weigh, number and measure as many of those objects as his allotted three-score years and ten would permit. This conviction biased all his doings, alike his great scientific enterprises, and the petty details of his daily life.

G. Wilson, *The Life of the Honourable Henry Cavendish: Including the Abstracts of his Important Scientific Papers* (1851), 186.

5 ON CAVENDISH Cavendish was a great Man with extraordinary singularities—His voice was squeaking his manner nervous He was afraid of strangers & seemed when embarrassed to articulate with difficulty—He wore the costume of our grandfathers. Was enormously rich but made no use of his wealth . . . He Cavendish lived latterly the life of a solitary, came to the Club dinner & to the Royal Society: but received nobody at his home. He was acute sagacious & profound & I think the most accomplished British Philosopher of his time.

Quoted in J. Z. Fullmer, 'Davy's Sketches of his Contemporaries', *Chymia*, 1967, 12, 133.

Margaret Lucas Cavendish

(Duchess of Newcastle) c.1624–74

British writer and natural philosopher

6 *Nature* is curious, and such *worke* may make,
That our dull *sense* can never finde, but scape.
For *Creatures*, small as *Atomes*, may be there,
If every *Atome a Creatures Figure* beare.
If foure *Atomes* a *World* can make, then see
What severall *Worlds* might in an *Eare-ring* bee:
For *Millions* of these *Atomes* may bee in
The *Head* of one *Small*, little, *Single Pin*.
And if thus *Small*, then *Ladies* may well weare

A *World* of *Worlds*, as *Pendents* in each
Eare.

From 'Of *Many* Worlds in *this* World', in *Poems and Fancies* (1653), 44–5.

Bennett Cerf 1898–1971

American editor and publisher

1 In a notable family called Stein
There were Gertrude, and Ep, and then
Ein.
Gert's writing was hazy,
Ep's statues were crazy,
And nobody understood Ein.

Out on a Limerick (1961), 76.

James Chadwick 1891–1974

British nuclear physicist

2 If we ascribe the ejection of the proton to a Compton recoil from a quantum of 52 × 106 electron volts, then the nitrogen recoil atom arising by a similar process should have an energy not greater than about 400,000 volts, should produce not more than about 10,000 ions, and have a range in the air at N.T.P. of about 1–3mm. Actually, some of the recoil atoms in nitrogen produce at least 30,000 ions. In collaboration with Dr. Feather, I have observed the recoil atoms in an expansion chamber, and their range, estimated visually, was sometimes as much as 3mm. at N.T.P.

These results, and others I have obtained in the course of the work, are very difficult to explain on the assumption that the radiation from beryllium is a quantum radiation, if energy and momentum are to be conserved in the collisions. The difficulties disappear, however, if it be assumed that the radiation consists of particles of mass 1 and charge 0, or neutrons. The capture of the α-particle by the Be9 nucleus may be supposed to result in the formation of a C12 nucleus and the emission of the neutron. From the energy relations of this process the velocity of the neutron emitted in the forward direction may well be about 3 × 109 cm. per sec. The collisions of this neutron with the atoms through which it passes give rise to the recoil atoms, and the observed energies of the recoil atoms are in fair agreement with this view. Moreover, I have observed that the protons ejected from hydrogen by the radiation emitted in the opposite direction to that of the exciting α-particle appear to have a much smaller range than those ejected by the forward radiation.

This again receives a simple explanation on the neutron hypothesis.

'Possible Existence of a Neutron', Letter to the Editor, *Nature*, 1932, **129**, 312.

Ernst Boris Chain 1906–79

German-born British biochemist

3 Science, as long as it limits itself to the descriptive study of the laws of nature, has no moral or ethical quality and this applies to the physical as well as the biological sciences.

'Social Responsibility and the Scientist', *New Scientist*, 22 October 1970, 166.

Thomas Chrowder Chamberlin 1843–1928

American geophysicist

4 In scientific study, or, as I prefer to phrase it, in creative scholarship, the truth is the single end sought; all yields to that. The truth is supreme, not only in the vague mystical sense in which that expression has come to be a platitude, but in a special, definite, concrete sense. Facts and the immediate and necessary inductions from facts displace all pre-conceptions, all deductions from general principles, all favourite theories. Previous mental constructions are bowled over as childish play-structures by facts as they come rolling into the mind. The dearest doctrines, the most fascinating hypotheses, the most cherished creations of the reason and of the imagination perish from a mind thoroughly inspired with the scientific spirit in the presence of incompatible facts. Previous intellectual affections are crushed without hesitation and without remorse. Facts are placed before reasonings and before ideals, even though the reasonings and the ideals be more beautiful, be seemingly more lofty, be seemingly better, be seemingly truer. The seemingly absurd and the seemingly impossible are sometimes true. The scientific disposition is to accept facts upon evidence, however absurd they may appear to our pre-conceptions.

The Ethical Functions of Scientific Study: An Address Delivered at the Annual Commencement of the University of Michigan, 28 June 1888, 7–8.

1 It will be observed that the distinction [between hypothesis and theory] is not such as to prevent a working hypothesis from gliding with the utmost ease into a ruling theory. Affection may as easily cling about a beloved intellectual child when named as a hypothesis as if named a theory, and its establishment in the one guise may become a ruling passion very much as in the other. The historical antecedents and the moral atmosphere associated with the working hypothesis lend some good influence however toward the preservation of its integrity. Conscientiously followed, the method of the working hypothesis is an incalculable advance upon the method of the ruling theory; but it has some serious defects. One of these takes concrete form, as just noted, in the ease with which the hypothesis becomes a controlling idea. To avoid this grave danger, the method of multiple working hypotheses is urged. It differs from the simple working hypothesis in that it distributes the effort and divides the affections . . . In developing the multiple hypotheses, the effort is to bring up into view every rational exploration of the phenomenon in hand and to develop every tenable hypothesis relative to its nature, cause or origin, and to give to all of these as impartially as possible a working form and a due place in the investigation. The investigator thus becomes the parent of a family of hypotheses; and by his parental relations to all is morally forbidden to fasten his affections unduly upon any one. In the very nature of the case, the chief danger that springs from affection is counteracted.

'Studies for Students. The Method of Multiple Working Hypotheses', *Journal of Geology*, 1897, 842–3.

Robert Chambers 1802–71

British naturalist and publisher

2 The hypothesis of the connexion of the first limestone beds with the commencement of organic life upon our planet is supported by the fact, that in these beds we find the first remains of the bodies of animated creatures.

Vestiges of the Natural History of Creation (1844), 57.

3 A comparatively small variety of species is found in the older rocks, although of some particular ones the remains are very abundant; . . . Ascending to the next group of rocks, we find the traces of life become more abundant, the number of species extended.

Vestiges of the Natural History of Creation (1844), 60–1.

4 This statistical regularity in moral affairs fully establishes their being under the presidency of law. Man is seen to be an enigma only as an individual: in the mass he is a mathematical problem.

Vestiges of the Natural History of Creation (1844), 331.

5 It is most interesting to observe into how small a field the whole of the mysteries of nature thus ultimately resolve themselves. The inorganic has one final comprehensive law, GRAVITATION. The organic, the other great department of mundane things, rests in like manner on one law, and that is,—DEVELOPMENT. Nor may even these be after all twain, but only branches of one still more comprehensive law, the expression of that unity which man's wit can scarcely separate from Deity itself.

Vestiges of the Natural History of Creation (1844), 360.

1 The organic creation, as we now see it . . . was not placed upon the earth at once;—it observed a PROGRESS.

Explanations (1845), 30.

2 Geology fully proves that organic creation passed through a series of stages before the highest vegetable and animal forms appeared.

Explanations (1845), 31.

Subrahmanyan Chandrasekhar 1910–95

Indian-born American astrophysicist

3 The pursuit of science has often been compared to the scaling of mountains, high and not so high. But who amongst us can hope, even in imagination, to scale the Everest and reach its summit when the sky is blue and the air is still, and in the stillness of the air survey the entire Himalayan range in the dazzling white of the snow stretching to infinity? None of us can hope for a comparable vision of nature and of the universe around us. But there is nothing mean or lowly in standing in the valley below and awaiting the sun to rise over Kinchinjunga.

Truth and Beauty: Aesthetics and Motivations in Science (1987), 26.

Charles Value Chapin 1856–1941

American physician and public health officer

4 Science can never be a closed book. It is like a tree, ever growing, ever reaching new heights. Occasionally the lower branches, no longer giving nourishment to the tree, slough off. We should not be ashamed to change our methods; rather we should be ashamed never to do so.

Papers of Charles V. Chapin, M.D.: A Review of Public Health Realities (1934), 55.

Erwin Chargaff 1905–2002

Austrian-born American biochemist

5 The results serve to disprove the tetranucleotide hypothesis. It is, however, noteworthy—whether this is more than accidental, cannot yet be said—that in all desoxypentose nucleic acids examined thus far the molar ratios of total purines to total pyrimidines, and also of adenine to thymine and of guanine to cytosine, were not far from 1.

'Chemical Specificity of Nucleic Acids and Mechanism of their Enzymatic Degradation', *Experientia*, 1950, 6, 206.

6 Molecular biology is essentially the practice of biochemistry without a license.

Essays on Nucleic Acids (1963), 176.

7 He [said of one or other eminent colleagues] is a very busy man, and half of what he publishes is true, but I don't know which half.

'Triviality in Science: A Brief Meditation on Fashions', *Perspectives in Biology and Medicine*, 1976, 19, 324.

8 If at one time or another I have brushed a few colleagues the wrong way, I must apologize: I had not realized that they were covered with fur.

Heraclitean Fire: Sketches from a Life before Nature (1978), Preface.

9 The double horror of two Japanese city names [Hiroshima and Nagasaki] grew for me into another kind of double horror; an estranging awareness of what the United States was capable of, the country that five years before had given me its citizenship; a nauseating terror at the direction the natural sciences were going. Never far from an apocalyptic vision of the world, I saw the end of the essence of mankind; an end brought nearer, or even made possible, by the profession to which I belonged. In my view, all natural sciences were as one; and if one science could no longer plead innocence, none could.

Heraclitean Fire: Sketches from a Life before Nature (1978), 3.

1 In 1945, therefore, I proved a sentimental fool; and Mr. Truman could safely have classified me among the whimpering idiots he did not wish admitted to the presidential office. For I felt that no man has the right to decree so much suffering, and that science, in providing and sharpening the knife and in upholding the ram, had incurred a guilt of which it will never get rid. It was at that time that the nexus between science and murder became clear to me. For several years after the somber event, between 1947 and 1952, I tried desperately to find a position in what then appeared to me as bucolic Switzerland,—but I had no success.

Heraclitean Fire: Sketches from a Life before Nature (1978), 4.

2 When the so-called think tanks began to replace the thought processes of human beings, I called them the aseptic tanks.

Heraclitean Fire: Sketches from a Life before Nature (1978), 5.

3 The narrow slit through which the scientist, if he wants to be successful, must view nature constructs, if this goes on for a long time, his entire character; and, more often than not, he ends up becoming what the German language so appropriately calls a *Fachidiot* (professional idiot).

Heraclitean Fire: Sketches from a Life before Nature (1978), 33.

4 I cannot serve as an example for younger scientists to follow. What I teach cannot be learned. I have never been a '100 percent scientist.' My reading has always been shamefully nonprofessional. I do not own an attaché case, and therefore cannot carry it home at night, full of journals and papers to read. I like long vacations, and a catalogue of my activities in general would be a scandal in the ears of the apostles of cost-effectiveness. I do not play the recorder, nor do I like to attend NATO workshops on a Greek island or a Sicilian mountain top; this shows that I am not even a molecular biologist. In fact, the list of what I have not got makes up the American Dream. Readers, if any, will conclude rightly that the *Gradus ad Parnassum* will have to be learned at somebody else's feet.

Heraclitean Fire: Sketches from a Life before Nature (1978), 7.

5 I came to biochemistry through chemistry; I came to chemistry, partly by the labyrinthine routes that I have related, and partly through the youthful romantic notion that the natural sciences had something to do with nature. What I liked about chemistry was its clarity surrounded by darkness; what attracted me, slowly and hesitatingly, to biology was its darkness surrounded by the brightness of the givenness of nature, the holiness of life. And so I have always oscillated between the brightness of reality and the darkness of the unknowable. When Pascal speaks of God in hiding, *Deus absconditus*, we hear not only the profound existential thinker, but also the great searcher for the reality of the world. I consider this unquenchable resonance as the greatest gift that can be bestowed on a naturalist.

Heraclitean Fire: Sketches from a Life before Nature (1978), 55.

6 Like all things of the mind, science is a brittle thing: it becomes absurd when you look at it too closely. It is designed for few at a time, not as a mass profession. But now we have megascience: an immense apparatus discharging in a minute more bursts of knowledge than humanity is able to assimilate in a lifetime. Each of us has two eyes, two ears, and, I hope, one brain. We cannot even listen to two symphonies at the same time. How do we get out of the horrible cacophony that assails our minds day and night? We have to learn, as others did, that if science is a machine to make more science, a machine to grind out so-called facts of nature, not all facts are equally worth knowing. Students, in other words, will have to learn to forget most of what they have learned. This process of forgetting must begin after each exam, but never before. The Ph.D. is essentially a license to start unlearning.

Voices In the Labyrinth: Nature, Man, and Science (1979), 2.

1 There is no question in my mind that we live in one of the truly bestial centuries in human history. There are plenty of signposts for the future historian, and what do they say? They say 'Auschwitz' and 'Dresden' and 'Hiroshima' and 'Vietnam' and 'Napalm.' For many years we all woke up to the daily body count on the radio. And if there were a way to kill people with the B Minor Mass, the Pentagon-Madison Avenue axis would have found it.

Voices in the Labyrinth: Nature, Man, and Science (1979), 2.

2 The modern version of Buridan's ass has a Ph.D., but no time to grow up as he is undecided between making a Leonardo da Vinci in the test tube or planting a Coca Cola sign on Mars.

Voices in the Labyrinth: Nature, Man, and Science (1979), 3.

3 In the last fifteen years we have witnessed an event that, I believe, is unique in the history of the natural sciences: their subjugation to and incorporation into the whirls and frenzies of disgusting publicity and propaganda. This is no doubt symptomatic of the precarious position assigned by present-day society to any form of intellectual activity. Such intellectual pursuits have at all times been both absurd and fragile; but they become ever more ludicrous when, as is now true of science, they become mass professions and must, as homeless pretentious parasites, justify their right to exist in a period devoted to nothing but the rapid consumption of goods and amusements. These sciences were always a *divertissement* in the sense in which Pascal used the word; but what is their function in a society living under the motto *lunam et circenses*? Are they only a band of court jesters in search of courts which, if they ever existed, have long lost their desire to be amused?

Voices in the Labyrinth: Nature, Man, and Science (1979), 27.

4 In science, attempts at formulating hierarchies are always doomed to eventual failure. A Newton will always be followed by an Einstein, a Stahl by a Lavoisier; and who can say who will come after us? What the human mind has fabricated must be subject to all the changes—which are not progress—that the human mind must undergo. The 'last words' of the sciences are often replaced, more often forgotten. Science is a relentlessly dialectical process, though it suffers continuously under the necessary relativation of equally indispensable absolutes. It is, however, possible that the ever-growing intellectual and moral pollution of our scientific atmosphere will bring this process to a standstill. The immense library of ancient Alexandria was both symptom and cause of the ossification of the Greek intellect. Even now I know of some who feel that we know too much about the wrong things.

Voices in the Labyrinth: Nature, Man, and Science (1979), 46.

5 One of the most insidious and nefarious properties of scientific models is their tendency to take over, and sometimes supplant, reality.

Quoted in J. J. Zuckerman, 'The Coming Renaissance of Descriptive Chemistry', *Journal of Chemical Education*, 1986, 63, 830.

Pierre Charron 1541–1603

French Roman Catholic theologian

6 *La vraye science et le vray étude de l'homme c'est l'homme.*

The true science and study of mankind is man.

De la Sagesse (1601), 1991 edn, Preface. See **Pope, Alexander** 502:1.

Geoffrey Chaucer c.1343–1400

English poet

1 With us ther was a DOCTOUR OF PHISIK;
In al the world ne was ther noon hym lik,
To speak of phisik and of surgerye,
For he was grounded in astronomye.
He kepte his pacient a ful greet deel
In houres by his magyk natureel.
Wel koude he fortunen the ascendent
Of his ymages for his pacient.
He knew the cause of everich maladye,
Were it of hoot, or cooled, or moyste, or drye,
And where they engendred, and of what humour.

Fragment I, General Prologue. In Larry D. Benson (ed.), *The Riverside Chaucer* (1988), 30.

2 I wol yow telle, as was me taught also,
The foure spirites and the bodies sevene,
By ordre, as ofte I herde my lord hem nevene.
The firste spirit quiksilver called is,
The second orpiment, the thridde, ywis,
Sal armoniak, and the firthe brimstoon.
The bodies sevene eek, lo! hem heer anoon:
Sol gold is, and *Luna* silver we threpe,
Mars yron, *Mercurie* quiksilver we clepe,
Saturnus leed, and *Jupiter* is tin,
And Venus coper, by my fader kin!

The Canon's Yeoman's Tale, lines 819–29. In Larry D. Benson (ed.), *The Riverside Chaucer* (1988), 273.

Anton Pavlovich Chekhov 1860–1904

Russian playwright and physician

3 A writer must be as objective as a chemist: he must abandon the subjective line; he must know that dung-heaps play a very reasonable part in a landscape, and that the evil passions are as inherent in life as good ones.

Letter to M. V. Kiselev, 14 January 1887. In L. S. Friedland (ed.), *Anton Chekov: Letters on the Short Story* (1967).

4 If a lot of cures are suggested for a disease, it means that the disease is incurable.

The Cherry Orchard (1904), Act 1. Trans. Elisaveta Fen.

5 There is no national science, just as there is no national multiplication table; what is national is no longer science.

The Note-Books of Anton Tchekhov (1967), trans. S. S. Koteliansky and L. Woolf, 4.

Edward Colin Cherry 1914–79

British electrical engineer

6 The human senses (above all, that of hearing) do not possess one set of constant parameters, to be measured independently, one at a time. It is even questionable whether the various 'senses' are to be regarded as separate, independent detectors. The human organism is one integrated whole, stimulated into response by physical signals; it is not to be thought of as a box, carrying various independent pairs of terminals labeled 'ears', 'eyes', 'nose', et cetera.

On Human Communication: A Review, A Survey and a Criticism (1957), 127–8.

Lord Cherwell see Lindemann, Frederick Alexander

Gilbert Keith Chesterton 1874–1936

British critic, poet and novelist

7 Men can construct a science with very few instruments, or with very plain instruments; but no one on earth could construct a science with unreliable instruments. A man might work out the whole of mathematics with a handful of pebbles, but not with a handful of clay which was always falling apart into new fragments, and falling together into new combinations. A man might measure heaven and earth with a reed, but not with a growing reed.

Heretics (1905), 146–7.

Hong-Yee Chiu 1932–

Chinese-born American astrophysicist

1 So far, the clumsily long name 'quasi-stellar radio sources' is used to describe these objects. Because the nature of these objects is entirely unknown, it is hard to prepare a short, appropriate nomenclature for them so that their essential properties are obvious from their name. For convenience, the abbreviated form '*quasar*' will be used throughout this paper.

'Gravitational Collapse', *Physics Today*, 1964, 17, 21.

Avram Noam Chomsky 1928–

American linguist

2 The fact that all normal children acquire essentially comparable grammars of great complexity with remarkable rapidity suggests that human beings are somehow specially designed to do this, with data-handling or 'hypothesis-formulating' ability of unknown character and complexity.

A review of B. F. Skinner, *Verbal Behavior* (1957). In *Language: Journal of the Linguistic Society of America*, 1959, 35, 57.

3 By a generative grammar I mean simply a system of rules that in some explicit and well-defined way assigns structural descriptions to sentences. Obviously, every speaker of a language has mastered and internalized a generative grammar that expresses his knowledge of his language. This is not to say that he is aware of the rules of the grammar or even that he can become aware of them, or that his statements about his intuitive knowledge of the language are necessarily accurate.

Aspects of the Theory of Syntax (1965), 8.

4 Hence, a generative grammar must be a system of rules that can iterate to generate an indefinitely large number of structures. This system of rules can be analyzed into the three major components of a generative grammar: the syntactic, phonological, and semantic components . . . the syntactic component of a grammar must specify, for each sentence, a *deep structure* that determines its semantic interpretation and a *surface structure* that determines its phonetic interpretation. The first of these is interpreted by the semantic component; the second, by the phonological component.

Aspects of the Theory of Syntax (1965), 15–6.

5 I think that in order to achieve progress in the study of language and human cognitive faculties in general it is necessary first to establish 'psychic distance' from the 'mental facts' to which Köhler referred, and then to explore the possibilities for developing explanatory theories . . . We must recognize that even the most familiar phenomena require explanation and that we have no privileged access to the underlying mechanisms, no more so than in physiology or physics.

Language and Mind (1972, enlarged edition), 26.

Chuang Tzu (Zhuangzi) c.369–286 BC

Chinese philosopher

6 There is no bandit so powerful as Nature.
The interaction of the positive and the negative principles, which produces the visible universe.
In the whole universe there is no escape from it.

Chuang Tzŭ: Mystic, Moralist and Social Reformer (1889), trans. Herbert A. Giles, 303.

7 Beneath multiple specific and individual distinctions, beneath innumerable and incessant transformations, at the bottom of the circular evolution without beginning or end, there hides a law, a unique nature participated in by all beings, in which this common participation produces a ground of common harmony.

A.W. Grabau, *Stratigraphy of China* (1928), title page.

Winston Leonard Spencer Churchill 1874–1965

British author and Prime Minister

1 Praise up the humanities, my boy. That will make them think that you are broad-minded.

Said to R. V. Jones in 'Science, Technology and Civilisation', *Bulletin of the Institute of Physics*, 1962, 13, 101.

2 Scientists should be on tap, but not on top.

Quoted in Randolph S. Churchill, *Twenty-One Years* (1964), 127.

3 The Dark Ages may return on the gleaming wings of Science.

Attributed.

Arthur Charles Clarke 1917–

British science-fiction writer

4 Any sufficiently advanced technology is indistinguishable from magic.

Clarke's Third Law. In *Profiles of the Future: An Enquiry into the Limits of the Possible* (1982), 36, footnote.

Rudolph Julius Emmanuel Clausius 1822–88

German physicist

5 In all cases where work is produced by heat, a quantity of heat proportional to the work done is expended; and inversely, by the expenditure of a like quantity of work, the same amount of heat may be produced.

'On the Moving Force of Heat, and the Laws regarding the Nature of Heat itself which are deducible therefrom', *Philosophical Magazine*, 1851, 2, 4.

6 Heat can never pass from a colder to a warmer body without some other change, connected therewith, occurring at the same time.

'On a Modified Form of the Second Fundamental Theorem in the Mechanical Theory of Heat', *Philosophical Magazine*, 1856, 12, 86.

7 The fundamental laws of the universe which correspond to the two fundamental theorems of the mechanical theory of heat.

1. *The energy of the universe is constant.*

2. *The entropy of the universe tends to a maximum.*

The Mechanical Theory of Heat (1867), 365.

Hans Cloos 1885–1951

German geologist

8 It was during my enchanted days of travel that the idea came to me, which, through the years, has come into my thoughts again and again and always happily—the idea that geology is the music of the earth.

Conversation with the Earth (1954), 3.

9 For a billion years the patient earth amassed documents and inscribed them with signs and pictures which lay unnoticed and unused. Today, at last, they are waking up, because man has come to rouse them. Stones have begun to speak, because an ear is there to hear them. Layers become history and, released from the enchanted sleep of eternity, life's motley, never-ending dance rises out of the black depths of the past into the light of the present.

Conversation with the Earth (1954), 4

10 The earth is large and old enough to teach us modesty.

Conversation with the Earth (1954), 8.

William Gemmell Cochran 1909–80

British mathematician and statistician

11 In 1905, a physicist measuring the thermal conductivity of copper would have faced, unknowingly, a very small systematic error due to the heating of his equipment and sample by the absorption of cosmic rays, then unknown to physics. In early 1946, an opinion poller, studying Japanese opinion as to who won the war, would have faced a very small systematic error due to the neglect of the 17 Japanese holdouts, who were discovered later north of Saipan. These cases are entirely parallel. Social, biological and physical scientists all need to remember that they have the same problems, the main difference being the decimal place in which they appear.

With Frederick Mosteller and John W. Tukey, 'Principles of Sampling', *Journal of the American Statistical Society*, 1954, **49**, 31.

I. Bernard Cohen 1914–2003

American historian of science

1 History without the history of science, to alter slightly an apothegm of Lord Bacon, resembles a statue of Polyphemus without his eye—that very feature being left out which most marks the spirit and life of the person. My own thesis is complementary: science taught . . . without a sense of history is robbed of those very qualities that make it worth teaching to the student of the humanities and the social sciences.

'The History of Science and the Teaching of Science', in I. Bernard Cohen and Fletcher G. Watson (eds.), *General Education in Science* (1952), 71.

2 All revolutionary advances in science may consist less of sudden and dramatic revelations than a series of transformations, of which the revolutionary significance may not be seen (except afterwards, by historians) until the last great step. In many cases the full potentiality and force of a most radical step in such a sequence of transformations may not even be manifest to its author.

The Newtonian Revolution (1980), 162.

Morris Raphael Cohen

1880–1947

American philosopher of science

3 It is interesting to note how many fundamental terms which the social sciences are trying to adopt from physics have as a matter of historical fact originated in the social field. Take, for instance, the notion of cause. The Greek *aitia* or the Latin *causa* was originally a purely legal term. It was taken over into physics, developed there, and in the 18th century brought back as a foreign-born kind for the adoration of the social sciences. The same is true of the concept of law of nature. Originally a strict anthropomorphic conception, it was gradually depersonalized or dehumanized in the natural sciences and then taken over by the social sciences in an effort to eliminate final causes or purposes from the study of human affairs. It is therefore not anomalous to find similar transformations in the history of such fundamental concepts of statistics as average and probability. The concept of average was developed in the Rhodian laws as to the distribution of losses in maritime risks. After astronomers began to use it in correcting their observations, it spread to other physical sciences; and the prestige which it thus acquired has given it vogue in the social field. The term probability, as its etymology indicates, originates in practical and legal considerations of probing and proving.

The Statistical View of Nature (1936), 327–8.

Samuel Taylor Coleridge

1772–1834

British poet and philosopher

4 Water, water, everywhere,
And all the boards did shrink;
Water, water everywhere,
Nor any drop to drink.
The very deep did rot: O Christ!
That ever this should be!

Yes, slimy things did crawl with legs
Upon the slimy sea.

'The Rime of the Ancient Mariner' (1798), Part 2.

1 Every subject in Davy's mind has the principle of Vitality. Living thoughts spring up like Turf under his feet.

Quoted in Joseph Cottle, *Reminiscences of Samuel Taylor Coleridge and Robert Southey* (1847), 329.

2 I should not think of devoting less than 20 years to an Epic Poem. Ten to collect materials and warm my mind with universal science. I would be a tolerable Mathematician, I would thoroughly know Mechanics, Hydrostatics, Optics, and Astronomy, Botany, Metallurgy, Fossilism, Chemistry, Geology, Anatomy, Medicine—then the *mind of man*—then the *minds of men*—in all Travels, Voyages and Histories. So I would spend ten years—the next five to the composition of the poem—and the five last to the correction of it. So I would write haply not unhearing of the divine and rightly-whispering Voice, which speaks to mighty minds of predestinated Garlands, starry and unwithering.

Letter to Joseph Cottle, early April 1797. In Earl Leslie Griggs (ed.), *The Collected Letters of Samuel Taylor Coleridge* (1956), Vol. 1, 320–1.

3 I shall attack Chemistry, like a Shark.

On his plans to set up a joint chemistry laboratory with Davy and Wordsworth in the Lake District.

Letter to Humphry Davy, 15 July 1800. In Earl Leslie Griggs (ed.), *The Collected Letters of Samuel Taylor Coleridge* (1956), Vol. 1, 605.

4 My Opinion is this—that deep Thinking is attainable only by a man of deep Feeling, and that all Truth is a species of Revelation. The more I understand of Sir Isaac Newton's works, the more boldly I dare utter to my own mind . . . that I believe the Souls of 500 Sir Isaac Newtons would go to the making up of a Shakspere [sic] or a Milton . . . *Mind* in his system is always *passive*—a lazy Looker-on on an external World. If the mind be not *passive*, if it be indeed made in God's Image, & that too in the sublimest sense—the image of the *Creator*—there is ground for suspicion, that any system built on the *passiveness* of the mind must be false, as a system.

Letter to Thomas Poole, 23 March 1801. In Earl Leslie Griggs (ed.), *The Collected Letters of Samuel Taylor Coleridge* (1956), Vol. 2, 709.

5 I attended Davy's lectures to renew my stock of metaphors.

In 1802 Coleridge attended an entire course of Humphry Davy's lectures at the Royal Institution, and took over 60 pages of notes.

Quoted in Sir Harold Hartley, *Humphry Davy* (1971), 45.

6 [Davy's] March of Glory, which he has run for the last six weeks—within which time by the aid and application of his own great discovery, of the identity of electricity and chemical attractions, he has placed all the elements and all their inanimate combinations in the power of man; having decomposed both the Alkalies, and three of the Earths, discovered as the base of the Alkalies a new metal . . . Davy supposes there is only one power in the world of the senses; which in particles acts as chemical attractions, in specific masses as electricity, & on matter in general, as planetary Gravitation . . . when this has been proved, it will then only remain to resolve this into some Law of vital Intellect—and all human knowledge will be Science and Metaphysics the only Science.

In November 1807 Davy gave his famous Second Bakerian Lecture at the Royal Society, in which he used Voltaic batteries to 'decompose, isolate and name' several new chemical elements, notably sodium and potassium.

Letter to Dorothy Wordsworth, 24 November 1807. In Earl Leslie Griggs (ed.), *The Collected Letters of Samuel Taylor Coleridge* (1956), Vol. 3, 38.

7 [Coleridge] selected an instance of what was called the sublime, in DARWIN, who imagined the creation of the universe to have taken place in a moment, by the explosion of a mass of matter in the womb, or centre of space. In one and the same instant of time, suns and planets shot into systems in every direction, and filled and spangled the illimitable void! He asserted this to be an intolerable degradation—referring, as it were, all the beauty and harmony of nature to something like the bursting of a *barrel of gunpowder*! that spit its combustible materials into a *pock-freckled* creation!

Report from Lectures (1812). In *The Collected Works of Samuel Taylor Coleridge: Lectures 1809–1819 on Literature* (1987), Vol. 5, 1, R. A. Foakes (ed.), 401.

1 What! Did Sir W[alter] R[aleigh] believe that a male and female ounce (and, if so, why not two tigers and lions, etc?) would have produced, in a course of generations, a cat, or a cat a lion? This is Darwinizing with a vengeance.

'Notes on Stillingfleet by S. T. Coleridge', *The Athenaeum*, no. 2474, 27 March 1875, 423.

2 I understood that you would take the Human Race in the concrete, have exploded the absurd notion of Pope's Essay on Man, [Erasmus] Darwin, and all the countless Believers—even (strange to say) among Xtians—of Man's having progressed from an Ouran Outang state—so contrary to all History, to all Religion, nay, to all Possibility—to have affirmed a Fall in some sense.

Letter to William Wordsworth, 30 May 1815. In Earl Leslie Griggs (ed.), *The Collected Letters of Samuel Taylor Coleridge* (1956), Vol. 4, 574–5.

3 This is, in truth, the first charm of chemistry, and the secret of the almost universal interest excited by its discoveries. The serious complacency which is afforded by the sense of truth, utility, permanence, and progression, blends with and ennobles the exhilarating surprise and the pleasurable sting of curiosity, which accompany the propounding and the solving of an Enigma . . . If in SHAKPEARE [sic] we find Nature idealized into Poetry, through the creative power of a profound yet observant meditation, so through the meditative observation of a DAVY, a WOOLLASTON [sic], or a HATCHETT; we find poetry, as if were, substantiated and realized in nature.

Essays on the Principle of Method, Essay VI (1818). In *The Collected Works of Samuel Taylor Coleridge: The Friend* (1969), Vol. 4, 1, Barbara E. Rooke (ed.), 471.

4 All Science is necessarily prophetic, so truly so, that the power of prophecy is the test, the infallible criterion, by which any presumed Science is ascertained to be actually & verily science. The Ptolemaic Astronomy was barely able to prognosticate a lunar eclipse; with Kepler and Newton came Science and Prophecy.

On the Constitution of the Church and State (1830). In *The Collected Works of Samuel Taylor Coleridge* (1976), John Colmer (ed.), Vol. 10, 118, footnote 1 on Coleridge's annotation.

5 Mr. Lyell's system of geology is just half the truth, and no more. He affirms a great deal that is true, and he denies a great deal which is equally true; which is the general characteristic of all systems not embracing the whole truth.

29 June 1833. Table Talk (1836). In *The Collected Works of Samuel Taylor Coleridge: Table Talk* (1990), Vol. 14, 2, Carl Woodring (ed.), 235.

6 The sublime discoveries of Newton, and, together with these, his not less fruitful than wonderful application, of the higher mathesis to the movement of the celestial bodies, and to the laws of light, gave almost religious sanction to the corpuscular system and mechanical theory. It became synonymous with philosophy itself. It was the sole portal at which truth was permitted to enter. The human body was treated as an hydraulic machine . . . In short, from the time of Kepler to that of Newton, and from Newton to Hartley, not only all things in external nature, but the subtlest mysteries of life, organization, and even of the intellect and moral being, were conjured within the magic circle of mathematical formulae.

Hints Towards the Formation of a more Comprehensive Theory of Life (1848). In *The Collected Works of Samuel Taylor Coleridge: Shorter Works and Fragments* (1995), H. J. Jackson and J. R. de J. Jackson (eds.), Vol. 11, 1, 498.

1 I must reject *fluids* and *ethers* of all kinds, magnetical, electrical, and universal, to whatever quintessential thinness they may be treble distilled, and (as it were) super-substantiated.

Hints Towards the Formation of a more Comprehensive Theory of Life (1848). In *The Collected Works of Samuel Taylor Coleridge: Shorter Works and Fragments* (1995), H. J. Jackson and J. R. de J. Jackson (eds.), Vol. 11, 1, 502.

John Norman Collie 1859–1942

British chemist

2 Organic Chemistry has become a vast rubbish heap of puzzling and bewildering compounds.

Preface to A. W. Stewart, *Recent Advances in Organic Chemistry* (1908), xiii.

3 The text-book is rare that stimulates its reader to ask, Why is this so? Or, How does this connect with what has been read elsewhere?

Preface to A. W. Stewart, *Recent Advances in Organic Chemistry* (1908), xiv.

Robin George Collingwood 1889–1943

British historian and philosopher

4 To the scientist, nature is always and merely a 'phenomenon,' not in the sense of being defective in reality, but in the sense of being a spectacle presented to his intelligent observation; whereas the events of history are never mere phenomena, never mere spectacles for contemplation, but things which the historian looks, not at, but through, to discern the thought within them.

The Idea of History (1946), 214.

5 To regard such a positive mental science [psychology] as rising above the sphere of history, and establishing the permanent and unchanging laws of human nature, is therefore possible only to a person who mistakes the transient conditions of a certain historical age for the permanent conditions of human life.

The Idea of History (1946), 224.

Francesco Maria Pompeo Colonna 1644–1726

French natural philosopher

6 There is no shame in admitting that the ultimate principle of all the wonders of the universe is unknown to us.

Les Principes de la Nature, ou de la Generation des Choses (1731), 165. Quoted in Jaques Roger, *The Life Sciences in Eighteenth-Century French Thought*, Keith R. Benson (ed.) and trans. Robert Ellrich (1997), 352.

Charles Caleb Colton c.1780–1832

British clergyman and writer

7 It is almost as difficult to make a man unlearn his errors, as his knowledge. Mal-information is more hopeless than non-information: for error is always more busy than ignorance. Ignorance is a blank sheet on which we may write; but error is a scribbled one on which we first erase. Ignorance is contented to *stand still* with her back to the truth; but error is more presumptuous, and *proceeds*, in the *same* direction. Ignorance has no light, but error follows a false one. The consequence is, that error, when she retraces her footsteps, has farther to go, before we can arrive at the truth, than ignorance.

Lacon: or Many things in Few Words; Addressed to Those Who Think (1820), Vol. 1, 15.

8 The study of the mathematics, like the Nile, begins in minuteness, but ends in magnificence.

Lacon: or Many things in Few Words; Addressed to Those Who Think (1820), Vol. 1, 162.

9 Professors in every branch of the sciences, prefer their own theories to truth: the reason is that their theories are *private property*, but truth is *common stock*.

Lacon: or Many things in Few Words; Addressed to Those Who Think (1820), Vol. 1, 169.

David E. Comings 1935–

American medical geneticist

1 One of the major goals when studying specific genetic diseases is to find the primary gene product, which in turn leads to a better understanding of the biochemical basis of the disorder. The bottom line often reads, 'This may lead to effective prenatal diagnosis and eventual eradication of the disease.' But we now have the ironic situation of being able to jump right to the bottom line without reading the rest of the page, that is, without needing to identify the primary gene product or the basic biochemical mechanism of the disease. The technical capability of doing this is now available. Since the degree of departure from our previous approaches and the potential of this procedure are so great, one will not be guilty of hyperbole in calling it the 'New Genetics'.

'Prenatal Diagnosis and the 'New Genetics', *The American Journal of Human Genetics*, 1980, 32:3, 453.

Arthur Holly Compton 1892–1962

American physicist

2 Yet is it possible in terms of the motion of atoms to explain how men can invent an electric motor, or design and build a great cathedral? If such achievements represent anything more than the requirements of physical law, it means that science must investigate the additional controlling factors, whatever they may be, in order that the world of nature may be adequately understood. For a science which describes only the motions of inanimate things but fails to include the actions of living organisms cannot claim universality.

The Human Meaning of Science (1940), 31.

3 The scientist who recognizes God knows only the God of Newton. To him the God imagined by Laplace and Comte is wholly inadequate. He feels that God is in nature, that the orderly ways in which nature works are themselves the manifestations of God's will and purpose. Its laws are his orderly way of working.

The Human Meaning of Science (1940), 69.

4 It is hard to think of fissionable materials when fashioned into bombs as being a source of happiness. However this may be, if with such destructive weapons men are to survive, they must grow rapidly in human greatness. A new level of human understanding is needed. The reward for using the atom's power towards man's welfare is great and sure. The punishment for its misuse would seem to be death and the destruction of the civilization that has been growing for a thousand years. These are the alternatives that atomic power, as the steel of Daedalus, presents to mankind. We are forced to grow to greater manhood.

Atomic Quest: A Personal Narrative (1956), xix.

John Lee Comstock 1789–1858

American surgeon and textbook compiler

5 The discovery of the laws of definite proportions is one of the most important and wonderful among the great and brilliant achievements of modern chemistry. It is sufficient of itself to convince any reasoning mind, that order and system pervade the universe, and that the minutest atoms of matter, and the vast orbs that move round the heavens are equally under the control of the invariable laws of the creator.

Elements of Chemistry (1845), 84.

Isidore Auguste Marie François Xavier Comte

1798–1857

French philosopher and mathematician

1 The law is this: that each of our leading conceptions—each branch of our knowledge—passes successively through three different theoretical conditions: the Theological, or fictitious: the Metaphysical, or abstract; and the Scientific, or positive.

The Positive Philosophy, trans. Harriet Martineau (1853), Vol. 1, 1–2.

2 In the final, the positive, state, the mind has given over the vain search after absolute notions, the origin and destination of the universe, and the causes of phenomena, and applies itself to the study of their laws—that is, their invariable relations of succession and resemblance. Reasoning and observation, duly combined, are the means of this knowledge. What is now understood when we speak of an explanation of facts is simply the establishment of a connection between single phenomena and some general facts.

The Positive Philosophy, trans. Harriet Martineau (1853), Vol. 1, 2.

3 The progress of the individual mind is not only an illustration, but an indirect evidence of that of the general mind. The point of departure of the individual and of the race being the same, the phases of the mind of a man correspond to the epochs of the mind of the race. Now, each of us is aware, if he looks back upon his own history, that he was a theologian in his childhood, a metaphysician in his youth, and a natural philosopher in his manhood. All men who are up to their age can verify this for themselves.

The Positive Philosophy, trans. Harriet Martineau (1853), Vol. 1, 3.

4 All good intellects have repeated, since Bacon's time, that there can be no real knowledge but that which is based on observed facts. This is incontestable, in our present advanced stage; but, if we look back to the primitive stage of human knowledge, we shall see that it must have been otherwise then. If it is true that every theory must be based upon observed facts, it is equally true that facts cannot be observed without the guidance of some theory. Without such guidance, our facts would be desultory and fruitless; we could not retain them: for the most part we could not even perceive them.

The Positive Philosophy, trans. Harriet Martineau (1853), Vol. 1, 3–4.

5 It needs scarcely be pointed out that in placing Mathematics at the head of Positive Philosophy, we are only extending the application of the principle which has governed our whole Classification. We are simply carrying back our principle to its first manifestation. Geometrical and Mechanical phenomena are the most general, the most simple, the most abstract of all,—the most irreducible to others, the most independent of them; serving, in fact, as a basis to all others. It follows that the study of them is an indispensable preliminary to that of all others. Therefore must Mathematics hold the first place in the hierarchy of the sciences, and be the point of departure of all Education whether general or special.

The Positive Philosophy, trans. Harriet Martineau (1853), Vol.1, 33.

6 In mathematics we find the primitive source of rationality; and to mathematics must the biologists resort for means to carry out their researches.

The Positive Philosophy, trans. Harriet Martineau (1853), Vol. 1, 388.

7 ON COMTE The only occasion when
Comte
Is known to have romped
Was when the multitude roared '*Vive La Philosophie Positive!*'

E. C. Bentley, *Biography for Beginners* (1905)

James Bryant Conant 1893–1978

American chemist and president of Harvard University

1 A conceptual scheme is never discarded merely because of a few stubborn facts with which it cannot be reconciled; a conceptual scheme is either modified or replaced by a better one, never abandoned with nothing left to take its place.

Science and Common Sense (1951), 173.

2 Even the development of the steam engine owed but little to the advancement of science.

Science and Common Sense (1951), 299–300.

Marie Jean Antoine Nicolas Caritat, Marquis de Condorcet 1743–94

French mathematician and natural philosopher

3 [All phenomena] are equally susceptible of being calculated, and all that is necessary, to reduce the whole of nature to laws similar to those which Newton discovered with the aid of the calculus, is to have a sufficient number of observations and a mathematics that is complex enough.

Unpublished Manuscript. Quoted in Frank E. Manuel, *The Prophets of Paris* (1962), 73.

Edwin Grant Conklin 1863–1952

American biologist

4 Heredity is to-day the central problem of biology. This problem may be approached from many sides—that of the breeder, the experimenter, the statistician, the physiologist, the embryologist, the cytologist—but the mechanism of heredity can be studied best by the investigation of the germ cells and their development.

'The Mechanism of Heredity', *Science*, 1905, 27, 89–90.

5 Life is not found in atoms or molecules or genes as such, but in organization; not in symbiosis but in synthesis.

'Cell and Protoplasm Concepts: Historical Account', *The Cell and the Protoplasm: Publication of the American Association of Science*, 1940, Number 114, 18.

John Constable 1776–1837

British landscape painter

6 Painting is a science, and should be pursued as an inquiry into the laws of nature. Why, then, may not landscape painting be considered as a branch of natural philosophy, of which pictures are but the experiments?

'The History of Landscape Painting', quoted in Charles Tomlinson, *Collected Poems* (1985), 33.

Anne Conway (Viscountess of Conway) 1631–79

British philosopher

7 For truly in nature there are many operations that are far more than mechanical. Nature is not simply an organic body like a clock, which has no vital principle of motion in it; but it is a living body which has life and perception, which are much more exalted than a mere mechanism or a mechanical motion.

The Principles of the Most Ancient and Modern Philosophy (1692), trans. and ed. Allison Coudert and Taylor Corse (1996), 64.

William Daniel Conybeare 1787–1857

British geologist and palaeontologist

8 Your admission of the late appearance of the great intellectual crown of the whole animal race [man] strikes me as perfectly fatal to the analogy of your system of a continually recurring series of identical terms.

Letter to Charles Lyell February 1841. In M.J.S. Rudwick, 'A Critique of Uniformitarian Geology: A Letter from W. D. Conybeare to Charles Lyell', *Proceedings of the American Philosophical Society*, 1967, 111, 282.

John Calvin Coolidge 1872–1933

American president

9 Wherever we look, the work of the chemist has raised the level of our civilisation and has increased the productive capacity of the nation.

White House lawn speech to American Chemical Society, April 1924. Quoted in H. Hale, *American Chemistry* (1928), 8.

William Cooper (Harry Summerfield Hoff) 1910–2002

British author and playwright

1 Dibdin said: 'I see you've put your own name at the top of your paper, Mr Woods.' His eyes looked sad and thoughtful. 'I always make it a matter of principle to put my name as well on every paper that comes out of the department.' 'Yours?' Albert said incredulously. 'Yes,' said Dibdin, still sad and thoughtful. 'I make it a matter of principle, Mr Woods. And I like my name to come first—it makes it easier for purposes of identification.' He rounded it off. 'First come, first served'.

The Struggles of Albert Woods (1952), 53.

Nicholas Copernicus 1473–1543

Polish astronomer

See ***Blackmore, Richard*** *68:3*

2 I have no doubt that certain learned men, now that the novelty of the hypotheses in this work has been widely reported—for it establishes that the Earth moves, and indeed that the Sun is motionless in the middle of the universe—are extremely shocked, and think that the scholarly disciplines, rightly established once and for all, should not be upset. But if they are willing to judge the matter thoroughly, they will find that the author of this work has committed nothing which deserves censure. For it is proper for an astronomer to establish a record of the motions of the heavens with diligent and skilful observations, and then to think out and construct laws for them, or rather hypotheses, whatever their nature may be, since the true laws cannot be reached by the use of reason; and from those assumptions the motions can be correctly calculated, both for the future and for the past. Our author has shown himself outstandingly skilful in both these respects. Nor is it necessary that these hypotheses should be true, nor indeed even probable, but it is sufficient if they merely produce calculations which agree with the observations . . . For it is clear enough that this subject is completely and simply ignorant of the laws which produce apparently irregular motions. And if it does work out any laws—as certainly it does work out very many—it does not do so in any way with the aim of persuading anyone that they are valid, but only to provide a correct basis for calculation. Since different hypotheses are sometimes available to explain one and the same motion (for instance eccentricity or an epicycle for the motion of the Sun) an astronomer will prefer to seize on the one which is easiest to grasp; a philosopher will perhaps look more for probability; but neither will grasp or convey anything certain, unless it has been divinely revealed to him. Let us therefore allow these new hypotheses also to become known beside the older, which are no more probable, especially since they are remarkable and easy; and let them bring with them the vast treasury of highly learned observations. And let no one expect from astronomy, as far as hypotheses are concerned, anything certain, since it cannot produce any such thing, in case if he seizes on things constructed for another other purpose as true, he departs from this discipline more foolish than he came to it.

Assumed by contemporary readers to be written by Copernicus himself, this preface suggested the earth's motion was merely a mathematical device and not to be taken seriously.

'To the Reader on the Hypotheses in this Work', Unsigned preface by Andreas Osiander to *Copernicus: On the Revolutions of the Heavenly Spheres* (1543), trans. A. M. Duncan (1976), 22–3.

3 I can well appreciate, Holy Father, that as soon as certain people realise that in these books which I have written about the Revolutions of the spheres of the universe I attribute certain motions to the globe of the Earth, they will at once clamour for me to be hooted off the stage with such an opinion.

'To His Holiness Pope Paul III', in *Copernicus: On the Revolutions of the Heavenly Spheres* (1543), trans. A.M. Duncan (1976), 23.

1 Those who knew that the judgements of many centuries had reinforced the opinion that the Earth is placed motionless in the middle of heaven, as though at its centre, if I on the contrary asserted that the Earth moves, I hesitated for a long time whether to bring my treatise, written to demonstrate its motion, into the light of day, or whether it would not be better to follow the example of the Pythagoreans and certain others, who used to pass on the mysteries of their philosophy merely to their relatives and friends, not in writing but by personal contact, as the letter of Lysis to Hipparchus bears witness. And indeed they seem to me to have done so, not as some think from a certain jealousy of communicating their doctrines, but so that their greatest splendours, discovered by the devoted research of great men, should not be exposed to the contempt of those who either find it irksome to waste effort on anything learned, unless it is profitable, or if they are stirred by the exhortations and examples of others to a high-minded enthusiasm for philosophy, are nevertheless so dull-witted that among philosophers they are like drones among bees.

'To His Holiness Pope Paul III', in *Copernicus: On the Revolutions of the Heavenly Spheres* (1543), trans. A. M. Duncan (1976), 24.

2 Therefore on long pondering this uncertainty of mathematical traditions on the deduction of the motions of the system of the spheres, I began to feel disgusted that no more certain theory of the motions of the mechanisms of the universe, which has been established for us by the best and most systematic craftsman of all, was agreed by the philosophers, who otherwise theorised so minutely with most careful attention to the details of this system. I therefore set myself the task of reading again the books of all philosophers which were available to me, to search out whether anyone had ever believed that the motions of the spheres of the universe were other than was supposed by those who professed mathematics in the schools.

'To His Holiness Pope Paul III', in *Copernicus: On the Revolutions of the Heavenly Spheres* (1543), trans. A. M. Duncan (1976), 25.

3 I therefore took this opportunity and also began to consider the possibility that the Earth moved. Although it seemed an absurd opinion, nevertheless because I knew that others before me had been granted the liberty of imagining whatever circles they wished to represent the phenomena of the stars, I thought that I likewise would readily be allowed to test whether, by assuming some motion of the Earth's, more dependable representations than theirs could be found for the revolutions of the heavenly spheres.

'To His Holiness Pope Paul III', in *Copernicus: On the Revolutions of the Heavenly Spheres* (1543), trans. A. M. Duncan (1976), 26.

4 In the first book I shall describe all the positions of the spheres, along with the motions which I attribute to the Earth, so that the book will contain as it were the general structure of the universe. In the remaining books I relate the motions of the remaining stars, and all the spheres, to the mobility of the Earth, so that it can be thence established how far the motions and appearances of the remaining stars and spheres can be saved, if they are referred to the motions of the Earth.

'To His Holiness Pope Paul III', in *Copernicus: On the Revolutions of the Heavenly Spheres* (1543), trans, A. M. Duncan (1976), 26.

5 Mathematics is written for mathematicians.

'To His Holiness Pope Paul III', in *Copernicus: On the Revolutions of the Heavenly Spheres* (1543), trans. A. M. Duncan (1976), 27.

1 Among the authorities it is generally agreed that the Earth is at rest in the middle of the universe, and they regard it as inconceivable and even ridiculous to hold the opposite opinion. However, if we consider it more closely the question will be seen to be still unsettled, and so decidedly not to be despised. For every apparent change in respect of position is due to motion of the object observed, or of the observer, or indeed to an unequal change of both.

'Book One. Chapter V. Whether Circular Motion is Proper to the Earth, and of its Place', in *Copernicus: On the Revolutions of the Heavenly Spheres* (1543), trans. A. M. Duncan (1976), 40.

2 Since, then, there is no objection to the mobility of the Earth, I think it must now be considered whether several motions are appropriate for it, so that it can be regarded as one of the wandering stars. For the fact that it is not the centre of all revolutions is made clear by the apparent irregular motion of the wandering stars, and their variable distances from the Earth, which cannot be understood in a circle having the same centre as the Earth.

'Book One. Chapter IX. Whether several motions can be attributed to the Earth, and on the centre of the universe', in *Copernicus: On the Revolutions of the Heavenly Spheres* (1543), trans. A. M. Duncan (1976), 46.

3 I myself consider that gravity is merely a certain natural inclination with which parts are imbued by the architect of all things for gathering themselves together into a unity and completeness by assembling into the form of a globe. It is easy to believe that the Sun, Moon and other luminaries among the wandering stars have this tendency also, so that by its agency they retain the rounded shape in which they reveal themselves, but nevertheless go round their orbits in various ways. If then the Earth also performs other motions, as for example the one about the centre, they must necessarily be like those which are similarly apparent in many external bodies in which we find an annual orbit.

'Book One. Chapter IX. Whether several motions can be attributed to the Earth, and on the centre of the universe', in *Copernicus: On the Revolutions of the Heavenly Spheres* (1543), trans. A. M. Duncan (1976), 46.

4 Then if the first argument remains secure (for nobody will produce a neater one, than the length of the periodic time is a measure of the size of the spheres), the order of the orbits follows this sequence, beginning from the highest: The first and highest of all is the sphere of the fixed stars, which contains itself and all things, and is therefore motionless. It is the location of the universe, to which the motion and position of all the remaining stars is referred. For though some consider that it also changes in some respect, we shall assign another cause for its appearing to do so in our deduction of the Earth's motion. There follows Saturn, the first of the wandering stars, which completes its circuit in thirty years. After it comes Jupiter which moves in a twelve-year long revolution. Next is Mars, which goes round biennially. An annual revolution holds the fourth place, in which as we have said is contained the Earth along with the lunar sphere which is like an epicycle. In fifth place Venus returns every nine months. Lastly, Mercury holds the sixth place, making a circuit in the space of eighty days. In the middle of all is the seat of the Sun. For who in this most beautiful of temples would put this lamp in any other or better place than the one from which it can illuminate everything at the same time? Aptly indeed is he named by some the lantern of the universe, by others the mind, by others the ruler. Trismegistus called him the visible God, Sophocles' Electra, the watcher over all things. Thus indeed the Sun as if seated on a royal throne governs his household of Stars as they circle around him. Earth also is by no means cheated of the Moon's attendance, but as Aristotle says in his book *On Animals* the Moon has the closest affinity with the Earth. Meanwhile the Earth conceives from the Sun, and is made pregnant with annual offspring. We find, then, in this arrangement the marvellous symmetry of the universe, and a sure linking together in harmony of the motion and size of the spheres, such as could be perceived in no other way. For here one may understand, by attentive observation, why Jupiter appears to have a larger progression and retrogression than Saturn, and smaller than Mars, and again why Venus has larger ones than Mercury; why such a doubling back appears more frequently in Saturn than in Jupiter, and still more rarely in Mars and Venus than in Mercury; and furthermore why Saturn, Jupiter and Mars are nearer to the Earth when in opposition than in the region of their occultation by the Sun and re-appearance. Indeed Mars in particular at the time when it is visible throughout the night seems to equal Jupiter in size, though marked out by its reddish colour; yet it is scarcely distinguishable among stars of the second magnitude, though recognized by those who track it with careful attention. All these phenomena proceed from the same course, which lies in the motion of the Earth. But the fact that none of these phenomena appears in the fixed stars shows their immense elevation, which makes even the circle of their annual motion, or apparent motion, vanish from our eyes.

'Book One. Chapter X. The Order of the Heavenly Spheres', in *Copernicus: On the Revolutions of the Heavenly Spheres* (1543), trans. A. M. Duncan (1976), 49–51.

1 The two revolutions, I mean the annual revolutions of the declination and of the centre of the Earth, are not completely equal; that is the return of the declination to its original value is slightly ahead of the period of the centre. Hence it necessarily follows that the equinoxes and solstices seem to anticipate their timing, not because the sphere of the fixed stars moves to the east, but rather the equatorial circle moves to the west, being at an angle to the plane of the ecliptic in proportion to the declination of the axis of the terrestrial globe.

'Book Three. Chapter I. The Precession of the equinoxes and solstices', in *Copernicus: On the Revolutions of the Heavenly Spheres* (1543), trans. A. M. Duncan (1976), 141.

2 After I had addressed myself to this very difficult and almost insoluble problem, the suggestion at length came to me how it could be solved with fewer and much simpler constructions than were formally used, if some assumptions (which are called axioms) were granted me. They follow in this order.

There is no one center of all the celestial circles or spheres.

The center of the earth is not the center of the universe, but only of gravity and of the lunar sphere.

All the spheres revolve about the sun as their mid-point, and therefore the sun is the center of the universe.

The ratio of the earth's distance from the sun to the height of the firmament is so much smaller than the ratio of the earth's radius to its distance from the sun that the distance from the earth to the sun is imperceptible in comparison with the height of the firmament.

Whatever motion appears in the firmament arises not from any motion of the firmament, but from the earth's motion. The earth together with its circumjacent elements performs a complete rotation on its fixed poles in a daily motion, while the firmament and highest heaven abide unchanged.

What appears to us as motions of the sun arise not from its motion but from the motion of the earth and our sphere, with which we revolve about the sun like any other planet. The earth has, then, more than one motion.

The apparent retrograde and direct motion of the planets arises not from their motion but from the earth's. The motion of the earth alone, therefore, suffices to explain so many apparent inequalities in the heavens.

'The Commentariolus', in *Three Copernican Treatises* (c.1510), trans. E. Rosen (1939), 58–9.

Carl Franz Joseph Erich Correns 1864–1933

German botanist and plant geneticist

1 In the gametes of an individual hybrid the Anlagen for each individual parental character are found in all possible combinations but never in a single gamete the Anlagen for a pair of characters. Each combination occurs with approximately the same frequency.

'Mendel's Regel über das Verhalten der Nachkommenschaft der Rassenbastarde', *Der Deutsche Botanische Gesellschaft*, 1900, **18**, 158–68. Trans. in Ernst Mayr, *The Growth of Biological Thought: Diversity, Evolution and Inheritance* (1982), 719.

Roger Cotes 1682–1716

British mathematician and astronomer

2 But shall gravity be therefore called an occult cause, and thrown out of philosophy, because the cause of gravity is occult and not yet discovered? Those who affirm this, should be careful not to fall into an absurdity that may overturn the foundations of all philosophy. For causes usually proceed in a continued chain from those that are more compounded to those that are more simple; when we are arrived at the most simple cause we can go no farther . . . These most simple causes will you then call occult and reject them? Then you must reject those that immediately depend on them.

Mathematical Principles (1729), 27.

Carl Bernhard von Cotta 1808–79

German geologist

3 In the course of the history of the earth innumerable events have occurred one after another, causing changes of states, all with certain lasting consequences. This is the basis of our developmental law, which, in a nutshell, claims that the diversity of phenomena is a necessary consequence of the accumulation of the results of all individual occurrences happening one after another . . . The current state of the earth, thus, constitutes the as yet most diverse final result, which of course represents not a real but only a momentary end-point.

Über das Entwicklung der Erde, (1867), 5–6.

Charles Augustin Coulomb 1736–1806

French physicist

4 Moreover, the sciences are monuments devoted to the public good; each citizen owes to them a tribute proportional to his talents. While the great men, carried to the summit of the edifice, draw and put up the higher floors, the ordinary artists scattered in the lower floors, or hidden in the obscurity of the foundations, must only seek to improve what cleverer hands have created.

Mémoires présentés par divers Savants à l'Académie des Sciences (1778), Introduction, trans. Ivor Grattan-Guinness.

Charles Alfred Coulson 1910–74

British theoretical chemist

5 One is almost tempted to say . . . at last I can almost see a bond. But that will never be, for a bond does not really exist at all: it is a most convenient fiction which, as we have seen, is convenient both to experimental and theoretical chemists.

'What is a Chemical Bond?', Coulson Papers, 25, Bodleian Library, Oxford. In Mary-Jo Nye, *From Chemical Philosophy to Theoretical Chemistry* (1993), 261.

6 [The chemical bond] First, it is related to the disposition of two electrons (remember, no one has ever seen an electron!): next, these electrons have their spins pointing in opposite directions (remember, no one can ever measure the spin of a particular electron!): then, the spatial distribution of these electrons is described analytically with some degree of precision (remember, there is no way of distinguishing experimentally the density distribution of one electron from another!): concepts like hybridization, covalent and ionic structures, resonance, all appear, not one of which corresponds to anything that is directly measurable. These concepts make a chemical bond seem so real, so life-like, that I can almost see it. Then I wake with a shock to the realization that a chemical bond does not exist; it is a figment of the imagination that we have invented, and no more real than the square root of −1. I will not say that the known is explained in terms of the unknown, for that is to misconstrue the sense of intellectual adventure. There is no explanation: there is form: there is structure: there is symmetry: there is growth: and there is therefore change and life.

Quoted in his obituary, *Biographical Memoirs of the Fellows of the Royal Society* 1974, 20, 96.

Victor Cousin 1792–1867

French philosopher, educational reformer and historian

1 Yes, gentlemen, give me the map of any country, its configuration, its climate, its waters, its winds, and the whole of its physical geography; give me its natural productions, its flora, its zoology, &c., and I pledge myself to tell you, *a priori*, what will be the quality of man in history;— not accidentally, but necessarily; not at any particular epoch, but in all; in short,— what idea he is called to represent.

Introduction to the History of Philosophy (1832), trans. by Henning Gotfried Linberg, 240.

Abraham Cowley 1618–67

British poet and essayist

2 Coy Nature, (which remain'd, though aged grown,
A beauteous virgin still, enjoy'd by none,
Nor seen unveil'd by any one),
When Harvey's violent passion she did see,
Began to tremble and to flee;
Took sanctuary, like Daphne, in a tree:
There Daphne's Lover stopped, and thought it much
The very leaves of her to touch:
But Harvey, our Apollo, stopp'd not so;
Into the Bark and Root he after her did go!

'Ode Upon Dr Harvey' (1663). In *The British Poets: Including Translations in One Hundred Volumes* (1822), Vol. 13, 245.

William Cowper 1731–1800

British poet

3 Some drill and bore
The solid earth, and from the strata there
Extract a register, by which we learn,
That he who made it, and reveal'd its date
To Moses, was mistaken in its age.

The Task and Other Poems, Book III, The Garden (1785). In John D. Baird and Charles Ryskamp (eds.), *The Poems of William Cowper* (1995), Vol. 2, 1782–1785, 166–7.

Donald James Cram 1919–2001

American chemist

4 Any chemist reading this book can see, in some detail, how I have spent most of my mature life. They can become familiar with the quality of my mind and imagination. They can make judgements about my research abilities. They can tell how well I have documented my claims of experimental results. Any scientist can redo my experiments to see if they still work—and this has happened! I know of no other field in which contributions to world culture are so clearly on exhibit, so cumulative, and so subject to verification.

From Design to Discovery (1990), 119–20.

1 I have always felt that I understood a phenomenon only to the extent that I could visualise it. Much of the charm organic chemical research has for me derives from structural formulae. When reading chemical journals, I look for formulae first.

From Design to Discovery (1990), 122.

2 Few scientists acquainted with the chemistry of biological systems at the molecular level can avoid being inspired. Evolution has produced chemical compounds exquisitely organized to accomplish the most complicated and delicate of tasks. Many organic chemists viewing crystal structures of enzyme systems or nucleic acids and knowing the marvels of specificity of the immune systems must dream of designing and synthesizing simpler organic compounds that imitate working features of these naturally occurring compounds.

'The Design of Molecular Hosts, Guests, and Their Complexes', Nobel Lecture, 8 December 1987. In *Nobel Lectures: Chemistry 1981–1990* (1992), 419.

Stephen Crane 1871–1900

American novelist and poet

3 A man said to the universe:
'Sir, I exist!'
'However,' replied the universe,
'The fact has not created in me
A sense of obligation.'

'A Man Said to the Universe', in *Poems by Stephen Crane* (1972), 139.

Francis Henry Compton Crick 1916–2004

British molecular biologist
See also ***Watson, James D.***

4 My own thinking (and that of many of my colleagues) is based on two general principles, which I shall call the Sequence Hypothesis and the Central Dogma. The direct evidence for both of them is negligible, but I have found them to be of great help in getting to grips with these very complex problems. I present them here in the hope that others can make similar use of them. Their speculative nature is emphasized by their names. It is an instructive exercise to attempt to build a useful theory without using them. One generally ends in the wilderness.

The Sequence Hypothesis

This has already been referred to a number of times. In its simplest form it assumes that the specificity of a piece of nucleic acid is expressed solely by the sequence of its bases, and that this sequence is a (simple) code for the amino acid sequence of a particular protein . . .

The Central Dogma

This states that once 'information' has passed into protein *it cannot get out again*. In more detail, the transfer of information from nucleic acid to nucleic acid, or from nucleic acid to protein may be possible, but transfer from protein to protein, or from protein to nucleic acid is impossible. Information means here the *precise* determination of sequence, either of bases in the nucleic acid or of amino acid residues in the protein. This is by no means universally held—Sir Macfarlane Burnet, for example, does not subscribe to it—but many workers now think along these lines. As far as I know it has not been *explicitly* stated before.

'On Protein Synthesis', *Symposia of the Society for Experimental Biology: The Biological Replication of Macromolecules*, 1958, **12**, 152–3.

5 Protein synthesis is a central problem for the whole of biology, and that it is in all probability closely related to gene action.

'On Protein Synthesis', *Symposia of the Society for Experimental Biology: The Biological Replication of Macromolecules*, 1958, **12**, 160.

6 One can say, looking at the papers in this symposium, that the elucidation of the genetic code is indeed a great achievement. It is, in a sense, the key to molecular biology because it shows how the great polymer languages, the nucleic acid language and the protein language, are linked together.

'The Genetic Code: Yesterday, Today, Tomorrow', *Cold Spring Harbour Symposium on Quantitative Biology*, 1966, 31, 9.

1 Finally one should add that in spite of the great complexity of protein synthesis and in spite of the considerable technical difficulties in synthesizing polynucleotides with defined sequences it is not unreasonable to hope that all these points will be clarified in the near future, and that the genetic code will be completely established on a sound experimental basis within a few years.

'On the Genetic Code', Nobel Lecture, 11 December 1962. In *Nobel Lectures: Physiology or Medicine 1942–1962* (1964), 808.

2 It is one of the striking generalizations of biochemistry—which surprisingly is hardly ever mentioned in the biochemical text-books—that the twenty amino acids and the four bases, are, with minor reservations, the same throughout Nature. As far as I am aware the presently accepted set of twenty amino acids was first drawn up by Watson and myself in the summer of 1953 in response to a letter of Gamow's.

'On the Genetic Code', Nobel Lecture, 11 December 1962. In *Nobel Lectures: Physiology or Medicine 1942–1962* (1964), 811.

3 Chance is the only source of true novelty.

Life Itself: Its Origin and Nature (1982), 58.

4 Almost all aspects of life are engineered at the molecular level, and without understanding molecules we can only have a very sketchy understanding of life itself.

What Mad Pursuit: A Personal View of Scientific Discovery (1988), 61.

5 While Occam's razor is a useful tool in the physical sciences, it can be a very dangerous implement in biology. It is thus very rash to use simplicity and elegance as a guide in biological research.

What Mad Pursuit: A Personal View of Scientific Discovery (1988), 138.

6 A busy life is a wasted life.

What Mad Pursuit: A Personal View of Scientific Discovery (1988), 145.

7 If you want to understand function, study structure. [I was supposed to have said in my molecular biology days.]

What Mad Pursuit: A Personal View of Scientific Discovery (1988), 150.

8 There is no form of prose more difficult to understand and more tedious to read than the average scientific paper.

The Astonishing Hypothesis: The Scientific Search for the Soul (1995), xiii.

9 When you start in science, you are brainwashed into believing how careful you must be, and how difficult it is to discover things. There's something that might be called the 'graduate student syndrome'; graduate students hardly believe they can make a discovery.

Quotation supplied by Professor Francis Crick.

Léon Croizat 1894–1982

Italian biogeographer, botanist and evolutionist

10 Let us suppose that we have laid on the table . . . [a] piece of glass . . . and let us homologize this glass to a whole order of plants or birds. Let us hit this glass a blow in such a manner as but to crack it up. The sectors circumscribed by cracks following the first blow may here be understood to represent families. Continuing, we may crack the glass into genera, species and subspecies to the point of finally having the upper right hand corner a piece about 4 inches square representing a sub-species.

Space, Time, Form: The Biological Synthesis (1962, issued 1964), 209.

James Croll 1821–90

British geologist

11 In geology the effects to be explained have almost all occurred already, whereas in these other sciences effects actually taking place have to be explained.

Climate and Time in their Geological Relations: A Theory of Secular Change of the Earth's Climate (1875), 4.

Axel Fredrik Cronstedt 1722–65

Swedish chemist and mineralogist

1 Sand in reality is nothing else than very small stones.

An Essay Towards a System of Mineralogy (1770), trans. G. Von Engestrom, xiv.

2 Meanwhile I flatter myself with so much success, that: students . . . will not be so easily mistaken in the subjects of the mineral kingdom, as has happened with me and others in following former systems; and I also hope to obtain some protectors against those who are so possessed with the *figuromania*, and *so addicted to the surface of things*, that they are shocked at the boldness of calling a *marble* a *limestone*, and of *placing the Porphyry amongst the Saxa*.

An Essay Towards a System of Mineralogy (1770), trans. G. Von Engestrom, xxi.

William Crookes 1832–1919

British chemist and physicist

3 The phenomena in these exhausted tubes reveal to physical science a new world—a world where matter may exist in a fourth state, where the corpuscular theory of light may be true, and where light does not always move in straight lines, but where we can never enter, and with which we must be content to observe and experiment from the outside.

'On the Illumination of Lines of Molecular Pressure and the Trajectory of Molecules', *Philosophical Transactions* 1879, 170, 164.

4 Probably our atomic weights merely represent a mean value around which the actual atomic weights of the atoms vary within certain narrow limits . . . when we say, the atomic weight of, for instance, calcium is 40, we really express the fact that, while the majority of calcium atoms have an actual atomic weight of 40, there are not but a few which are represented by 39 or 41, a less number by 38 or 42, and so on.

Presidential Address, 2 September 1886, Section B, Chemical Science. *Reports of the British Association for the Advancement of Science* (1886), 569.

5 England and all civilised nations stand in deadly peril of not having enough to eat. As mouths multiply, food resources dwindle. Land is a limited quantity, and the land that will grow wheat is absolutely dependent on difficult and capricious natural phenomena . . . I hope to point a way out of the colossal dilemma. It is the chemist who must come to the rescue of the threatened communities. It is through the laboratory that starvation may ultimately be turned into plenty . . . The fixation of atmospheric nitrogen is one of the great discoveries, awaiting the genius of chemists.

Presidential Address to the British Association for the Advancement of Science 1898. Published in *Chemical News*, 1898, 78, 125.

Roy Albert Crowson 1914–99

British taxonomist

6 There have been many authorities who have asserted that the basis of science lies in counting or measuring, i.e. in the use of mathematics. Neither counting nor measuring can however be the most fundamental processes in our study of the material universe—before you can do either to any purpose you must first select what you propose to count or measure, which presupposes a classification.

Classification and Biology (1970), 2.

Paul J. Crutzen 1933– and John W. Birks 1946–

Dutch and American chemists

7 In discussing the state of the atmosphere following a nuclear exchange, we point especially to the effects of the many fires that would be ignited by the thousands of nuclear explosions in cities, forests, agricultural fields, and oil and gas fields. As a result of these fires, the loading of the atmosphere with strongly light absorbing particles in the submicron size range (1 micron = 10^{-6} m) would increase so much that at noon solar radiation at the ground would be reduced by at least a factor of two and possibly a factor of greater than one hundred.

'The Atmosphere after a Nuclear War: Twilight at Noon', *Ambio*, 1982, 11, 115.

William Cullen 1710–90

Scottish physician and chemist

1 I would have you to observe that the difficulty & mystery which often appear in matters of science & learning are only owing to the terms of art used in them, & if many gentlemen had not been rebuted by the uncouth dress in which science was offered to them, we must believe that many of these who now shew an acute & sound judgement in the affairs of life would also in science have excelled many of those who are devoted to it & who were engaged in it only by necessity & a phlegmatic temper. This is particularly the case with respect to chemistry, which is as easy to be comprehended as any of the common affairs of life, but gentlemen have been kept from applying to it by the jargon in which it has been industriously involved.

Cullen MSS, No. 23, Glasgow University Library. In A. L. Donovan, *Philosophical Chemistry in the Scottish Enlightenment: The Doctrines and Discoveries of William Cullen and Joseph Black* (1975), 111.

2 Chemistry is an art that has furnished the world with a great number of useful facts, and has thereby contributed to the improvement of many arts; but these facts lie scattered in many different books, involved in obscure terms, mixed with many falsehoods, and joined to a great deal of false philosophy; so that it is not great wonder that chemistry has not been so much studied as might have been expected with regard to so useful a branch of knowledge, and that many professors are themselves but very superficially acquainted with it. But it was particularly to be expected, that, since it has been taught in universities, the difficulties in this study should have been in some measure removed, that the art should have been put into form, and a system of it attempted—the scattered facts collected and arranged in a proper order. But this has not yet been done; chemistry has not yet been taught but upon a very narrow plan. The teachers of it have still confined themselves to the purposes of pharmacy and medicine, and that comprehends a small branch of chemistry; and even that, by being a single branch, could not by itself be tolerably explained.

John Thomson, *An Account of the Life, Lectures and Writings of William Cullen, M.D.* (1832), Vol. 1, 40.

Richard Cumberland (Bishop of Peterborough) 1631–1718

English philosopher

3 All other men, being born of woman, have a navel, by reason of the umbilical vessels inserted into it, which from the placenta carry nourishment to children in the womb of their mothers; but it could not be so with our first parents. It cannot be believed that God gave them navels which would have been altogether useless.

A Treatise of Laws of Nature (1727).

Bruce Frederick Cummings (W. N. P. Barbellion) 1889–1919

British author and diarist

4 How I hate the man who talks about the 'brute creation', with an ugly emphasis on *Brute*. Only Christians are capable of it. As for me, I am proud of my close kinship with other animals. I take a jealous pride in my Simian ancestry. I like to think that I was once a magnificent hairy fellow living in the trees and that my frame has come down through geological time *via* sea jelly and worms and Amphioxus, Fish, Dinosaurs, and Apes. Who would exchange these for the pallid couple in the Garden of Eden?

W. N. P. Barbellion, *The Journal of a Disappointed Man* (1919), 27–8

Marie Curie (Marie Sklodowska) 1867–1934

Polish-born French physicist

1 We believe the substance we have extracted from pitch-blende contains a metal not yet observed, related to bismuth by its analytical properties. If the existence of this new metal is confirmed we propose to call it *polonium*, from the name of the original country of one of us.

In Eve Curie, *Madame Curie* (1938), 169.

2 In science we must be interested in things, not in persons.

In Eve Curie, *Madame Curie* (1938), 233.

3 Humanity certainly needs practical men, who get the most out of their work, and, without forgetting the general good, safeguard their own interests. But humanity also needs dreamers, for whom the disinterested development of an enterprise is so captivating that it becomes impossible for them to devote their care to their own material profit. Without the slightest doubt, these dreamers do not deserve wealth, because they do not desire it. Even so, a well-organised society should assure to such workers the efficient means of accomplishing their task, in a life freed from material care and freely consecrated to research.

In Eve Curie, *Madame Curie* (1938), 350.

4 Sometimes my courage fails me and I think I ought to stop working, live in the country and devote myself to gardening. But I am held by a thousand bonds, and I don't know when I shall be able to arrange things otherwise. *Nor do I know whether, even by writing scientific books, I could live without the laboratory*.

Letter to her sister Bronya, September 1927. In Eve Curie, *Madame Curie* (1938), 388.

5 My experiments proved that the radiation of uranium compounds can be measured with precision under determined conditions, and that this radiation is an atomic property of the element of uranium. Its intensity is proportional to the quantity of uranium contained in the compound, and depends neither on conditions of chemical combination, nor on external circumstances, such as light or temperature.

I undertook next to discover if there were other elements possessing the same property, and with this aim I examined all the elements then known, either in their pure state or in compounds. I found that among these bodies, thorium compounds are the only ones which emit rays similar to those of uranium. The radiation of thorium has an intensity of the same order as that of uranium, and is, as in the case of uranium, an atomic property of the element.

It was necessary at this point to find a new term to define this new property of matter manifested by the elements of uranium and thorium. I proposed the word radioactivity which has since become generally adopted; the radioactive elements have been called radio elements.

Marie Curie, *Pierre Curie, with the Autobiographical Notes of Marie Curie*, trans. Charlotte and Vernon Kellogg (1963), 45–6.

6 ON CURIE At my urgent request the Curie laboratory, in which radium was discovered a short time ago, was shown to me. The Curies themselves were away travelling. It was a cross between a stable and a potato-cellar, and, if I had not seen the worktable with the chemical apparatus, I would have thought it a practical joke.

Wilhelm Ostwald on seeing the Curie's laboratory facilities.
In R. Reid, *Marie Curie* (1974), 95.

1 ON **CURIE** It was my good fortune to be linked with Mme. Curie through twenty years of sublime and unclouded friendship. I came to admire her human grandeur to an ever growing degree. Her strength, her purity of will, her austerity toward herself, her objectivity, her incorruptible judgement—all these were of a kind seldom found joined in a single individual . . . The greatest scientific deed of her life—proving the existence of radioactive elements and isolating them—owes its accomplishment not merely to bold intuition but to a devotion and tenacity in execution under the most extreme hardships imaginable, such as the history of experimental science has not often witnessed.

Albert Einstein, *Out of My Later Years* (1950), 227–8.

Pierre Curie 1859–1906

French physicist

2 It can even be thought that radium could become very dangerous in criminal hands, and here the question can be raised whether mankind benefits from knowing the secrets of Nature, whether it is ready to profit from it or whether this knowledge will not be harmful for it. The example of the discoveries of Nobel is characteristic, as powerful explosives have enabled man to do wonderful work. They are also a terrible means of destruction in the hands of great criminals who lead the peoples towards war. I am one of those who believe with Nobel that mankind will derive more good than harm from the new discoveries.

'Radioactive Substances, Especially Radium', Nobel Lecture, 6 June 1905. In *Nobel Lectures: Physics 1901–1921* (1967), 78.

Georges Cuvier 1769–1832

French naturalist and palaeontologist

3 It is in this mutual dependence of the functions and the aid which they reciprocally lend one another that are founded the laws which determine the relations of their organs and which possess a necessity equal to that of metaphysical or mathematical laws, since it is evident that the seemly harmony between organs which interact is a necessary condition of existence of the creature to which they belong and that if one of these functions were modified in a manner incompatible with the modifications of the others the creature could no longer continue to exist.

Leçons d'anatomie comparée, Vol. 1, 47. Trans. William Coleman, *Georges Cuvier Zoologist: A Study in the History of Evolution Theory* (1964), 67–8.

4 Fortunately Nature herself seems to have prepared for us the means of supplying that want which arises from the impossibility of making certain experiments on living bodies. The different classes of animals exhibit almost all the possible combinations of organs: we find them united, two and two, three and three, and in all proportions; while at the same time it may be said that there is no organ of which some class or some genus is not deprived. A careful examination of the effects which result from these unions and privations is therefore sufficient to enable us to form probable conclusions respecting the nature and use of each organ, or form of organ. In the same manner we may proceed to ascertain the use of the different parts of the same organ, and to discover those which are essential, and separate them from those which are only accessory. It is sufficient to trace the organ through all the classes which possess it, and to examine what parts constantly exist, and what change is produced in the respective functions of the organ, by the absence of those parts which are wanting in certain classes.

Letter to Jean Claude Mertrud. In *Lectures on Comparative Anatomy* (1802), Vol. 1, xxiii-xxiv.

1 It is my object, in the following work, to travel over ground which has as yet been little explored and to make my reader acquainted with a species of Remains, which, though absolutely necessary for understanding the history of the globe, have been hitherto almost uniformly neglected.

'Preliminary discourse', to *Recherches sur les Ossemens Fossiles* (1812), trans. R. Kerr *Essay on the Theory of the Earth* (1813), 1.

2 As an antiquary of a new order, I have been obliged to learn the art of deciphering and restoring these remains, of discovering and bringing together, in their primitive arrangement, the scattered and mutilated fragments of which they are composed, of reproducing in all their original proportions and characters, the animals to which these fragments formerly belonged, and then of comparing them with those animals which still live on the surface of the earth; an art which is almost unknown, and which presupposes, what had scarcely been obtained before, an acquaintance with those laws which regulate the coexistence of the forms by which the different parts of organized being are distinguished.

'Preliminary discourse', to *Recherches sur les Ossemens Fossiles* (1812), trans. R. Kerr *Essay on the Theory of the Earth* (1813), 1–2.

3 If they [enlightened men] take any interest in examining, in the infancy of our species, the almost obliterated traces of so many nations that have become extinct, they will doubtless take a similar interest in collecting, amidst the darkness which covers the infancy of the globe, the traces of those revolutions which took place anterior to the existence of all nations.

'Preliminary discourse', to *Recherches sur les Ossemens Fossiles* (1812), trans. R. Kerr *Essay on the Theory of the Earth* (1813), 3.

4 Genius and science have burst the limits of space, and few observations, explained by just reasoning, have unveiled the mechanism of the universe. Would it not also be glorious for man to burst the limits of time, and, by a few observations, to ascertain the history of this world, and the series of events which preceded the birth of the human race?

'Preliminary discourse', to *Recherches sur les Ossemens Fossiles* (1812), trans. R. Kerr *Essay on the Theory of the Earth* (1813), 3–4.

5 Life, therefore, has been often disturbed on this earth by terrible events—calamities which, at their commencement, have perhaps moved and overturned to a great depth the entire outer crust of the globe, but which, since these first commotions, have uniformly acted at a less depth and less generally. Numberless living beings have been the victims of these catastrophes; some have been destroyed by sudden inundations, others have been laid dry in consequence of the bottom of the seas being instantaneously elevated. Their races even have become extinct, and have left no memorial of them except some small fragments which the naturalist can scarcely recognise.

'Preliminary discourse', to *Recherches sur les Ossemens Fossiles* (1812), trans. R. Kerr *Essay on the Theory of the Earth* (1813), 16–7.

6 It has been long considered possible to explain the more ancient revolutions on . . . [the Earth's] surface by means of these still existing causes; in the same manner as it is found easy to explain past events in political history, by an acquaintance with the passions and intrigues of the present day. But we shall presently see that unfortunately this is not the case in physical history:—the thread of operation is here broken, the march of nature is changed, and none of the agents that she now employs were sufficient for the production of her ancient works.

'Preliminary discourse', to *Recherches sur les Ossemens Fossiles* (1812), trans. R. Kerr *Essay on the Theory of the Earth* (1813), 24.

1 It is to them [fossils] alone that we owe the commencement of even a Theory of the Earth . . . By them we are enabled to ascertain, with the utmost certainty, that our earth has not always been covered over by the same external crust, because we are thoroughly assured that the organized bodies to which these fossil remains belong must have lived upon the surface before they came to be buried, as they now are, at a great depth.

'Preliminary discourse', to *Recherches sur les Ossemens Fossiles* (1812), trans. R. Kerr *Essay on the Theory of the Earth* (1813), 54–55.

2 I am of opinion, then, . . . that, if there is any circumstance thoroughly established in geology, it is, that the crust of our globe has been subjected to a great and sudden revolution, the epoch of which cannot be dated much farther back than five or six thousand years ago; that this revolution had buried all the countries which were before inhabited by men and by the other animals that are now best known; that the same revolution had laid dry the bed of the last ocean, which now forms all the countries at present inhabited; that the small number of individuals of men and other animals that escaped from the effects of that great revolution, have since propagated and spread over the lands then newly laid dry; and consequently, that the human race has only resumed a progressive state of improvement since that epoch, by forming established societies, raising monuments, collecting natural facts, and constructing systems of science and of learning.

'Preliminary discourse', to *Recherches sur les Ossemens Fossiles* (1812), trans. R. Kerr *Essay on the Theory of the Earth* (1813), 171–2.

3 Since nothing can exist that does not fulfil the conditions which render its existence possible, the different parts of each being must be co-ordinated in such a way as to render possible the existence of the being as a whole, not only in itself, but also in its relations with other beings, and the analysis of these conditions often leads to general laws which are as certain as those which are derived from calculation or from experiment.

Le Règne Animal distribué d'Après son Organisation (1817), 6. Translated in E. S. Russell, *Form and Function: A Contribution to the History of Animal Morphology* (1916), 34.

4 All organs of an animal form a single system, the parts of which hang together, and act and re-act upon one another; and no modifications can appear in one part without bringing about corresponding modifications in all the rest.

Histoire des Progrès des Sciences naturelles depuis (1789), Vol. 1, 310. Quoted in E. S. Russell, *Form and Function* (1916), 35.

5 In spite of what moralists say, the animals are scarcely less wicked or less unhappy than we are ourselves. The arrogance of the strong, the servility of the weak, low rapacity, ephemeral pleasure purchased by great effort, death preceded by long suffering, all belong to the animals as they do to men.

Recueil des Éloges Historiques 1819–27, Vol. 1, 91.

6 At the sight of a single bone, of a single piece of bone, I recognize and reconstruct the portion of the whole from which it would have been taken. The whole being to which this fragment belonged appears in my mind's eye.

Cited by Geoffroy Saint-Hilaire, *Comptes-Rendus de l'Académie des Sciences*. 1837, 7, 116. Trans. Franck Bourdier, 'Geoffroy Saint-Hilaire versus Cuvier: The Campaign for Paleontological Evolution (1825–1838)', Cecil J. Schneer (ed.), *Toward a History of Geology* (1969), 44.

7 ON CUVIER A famous anecdote concerning Cuvier involves the tale of his visitation from the devil—only it was not the devil but one of his students dressed up with horns on his head and shoes shaped like cloven hooves. This frightening apparition burst into Cuvier's bedroom when he was fast asleep and claimed:

'Wake up thou man of catastrophes. I am the Devil. I have come to devour you!'

Cuvier studied the apparition carefully and critically said, 'I doubt whether you can. You have horns and hooves. You eat only plants.'

Quoted in Glyn Daniel, *The Idea of Pre-History* (1962), 34.

1 ON **CUVIER** Cuvier had even in his address & manner the character of a superior Man, much general power & eloquence in conversation & great variety of information on scientific as well as popular subjects. I should say of him that he is the most distinguished man of *talents* I have ever known on the continent: but I doubt if He be entitled to the appellation of a Man of Genius.

J. Z. Fullmer, 'Davy's Sketches of his Contemporaries', *Chymia*, 1967, 12, 132.

Henry Hallett Dale 1875–1968

British physiologist

2 The question of a possible physiological significance, in the resemblance between the action of choline esters and the effects of certain divisions of the involuntary nervous system, is one of great interest, but one for the discussion of which little evidence is available. Acetyl-choline is, of all the substances examined, the one whose action is most suggestive in this direction. The fact that its action surpasses even that of adrenaline, both in intensity and evanescence, when considered in conjunction with the fact that each of these two bases reproduces those effects of involuntary nerves which are absent from the action of the other, so that the two actions are in many directions at once complementary and antagonistic, gives plenty of scope for speculation.

'The Action of Certain Esters and Ethers of Choline, and Their Relation to Muscarine', *The Journal of Pharmacology and Experimental Therapeutics*, 1914–15, 6, 188.

3 Hitherto the conception of chemical transmission at nerve endings and neuronal synapses, originating in Loewi's discovery, and with the extension that the work of my colleagues has been able to give to it, can claim one practical result, in the specific, though alas only short, alleviation of the condition of myasthenia gravis, by eserine and its synthetic analogues.

'Some recent extensions of the chemical transmission of the effects of nerve impulses', Nobel Lecture, 12 December 1936. In *Nobel Lectures: Physiology or Medicine 1922–1941* (1965), 412–3.

John Dalton 1766–1844

British chemist, physicist and meteorologist

See ***Bronowski, Jacob*** *93:3*

4 We should scarcely be excused in concluding this essay without calling the reader's attention to the beneficent and wise laws established by the author of nature to provide for the various exigencies of the sublunary creation, and to make the several parts dependent upon each other, so as to form one well-regulated system or whole.

'Experiments and Observations to Determine whether the Quantity of Rain and Dew is Equal to the Quantity of Water carried off by the Rivers and Raised by Evaporation', *Memoirs Manchester Literary and Philosophical Society*, 1803, 5, part 2, 372.

5 There are three distinctions in the kinds of bodies, or three states, which have more especially claimed the attention of philosophical chemists; namely, those which are marked by the terms *elastic fluids, liquids, and solids.* A very familiar instance is exhibited to us in water, of a body, which, in certain circumstances, is capable of assuming all the three states. In steam we recognise a perfectly elastic fluid, in water, a perfect liquid, and in ice of a complete solid. These observations have tacitly led to the conclusion which seems universally adopted, that all bodies of sensible magnitude, whether liquid or solid, are constituted of a vast number of extremely small particles, or atoms of matter bound together by a force of attraction.

A New System of Chemical Philosophy (1808), Vol. 1, 141.

1 The ultimate particles of all homogeneous bodies are perfectly alike in weight, figure &c.

A New System of Chemical Philosophy (1808), Vol. 1, 143.

2 Chemical analysis and synthesis go no farther than to the separation of particles one from another, and to their reunion. No new creation or destruction of matter is within the reach of chemical agency. We might as well attempt to introduce a new planet into the solar system, or to annihilate one already in existence, as to create or destroy a particle of hydrogen.

A New System of Chemical Philosophy (1808), Vol. 1, 212.

3 In all chemical investigations, it has justly been considered an important object to ascertain the relative *weights* of the simples which constitute a compound. But unfortunately the enquiry has terminated here; whereas from the relative weights in the mass, the relative weights of the ultimate particles or atoms of the bodies might have been inferred, from which their number and weight in various other compounds would appear, in order to assist and to guide future investigations, and to correct their results. Now it is one great object of this work, to shew the importance and advantage of ascertaining *the relative weights of the ultimate particles, both of simple and compound bodies, the number of simple elementary particles which constitute one compound particle, and the number of less compound particles which enter into the formation of one more compound particle.*

If there are two bodies, A and B, which are disposed to combine, the following is the order in which the combinations may take place, beginning with the most simple: namely,

1 atom of A + 1 atom of B = 1 atom of C, binary
1 atom of A + 2 atoms of B = 1 atom of D, ternary
2 atoms of A + 1 atom of B = 1 atom of E, ternary
1 atom of A + 3 atoms of B = 1 atom of F, quaternary
3 atoms of A and 1 atom of B = 1 atom of G, quaternary

A New System of Chemical Philosophy (1808), Vol. 1, 212–3.

4 When an element A has an affinity for another substance B, I see no mechanical reason why it should not take as many atoms of B as are presented to it, and can possibly come into contact with it (which may probably be 12 in general), *except so far as the repulsion of the atoms of B among themselves are more than a match for the attraction of an atom of A*. Now this repulsion begins with 2 atoms of B to 1 atom of A, in which case the 2 atoms of B are diametrically opposed; it increases with 3 atoms of B to 1 of A, in which case the atoms are only 120° asunder; with 4 atoms of B it is still greater as the distance is then only 90; and so on in proportion to the number of atoms. It is evident from these positions, that, as far as powers of attraction and repulsion are concerned (and we know of no other in chemistry), *binary* compounds must first be formed in the ordinary course of things, then *ternary* and so on, till the repulsion of the atoms of B (or A, whichever happens to be on the surface of the other), refuse to admit any more.

'Observations on Dr. Bostock's Review of the Atomic Principles of Chemistry', *Nicholson's Journal*, 1811, 29, 147.

5 I was introduced to Mr. Davy, who has rooms adjoining mine (in the Royal Institution); he is a very agreeable and intelligent young man, and we have interesting conversation in an evening; the principal failing in his character as a philosopher is that he does not smoke.

Letter to John Rothwell, January 1804. Quoted in J. P. Millington, *John Dalton* (1906), 141.

1 Berzelius' symbols are horrifying. A young student in chemistry might as soon learn Hebrew as make himself acquainted with them . . . They appear to me equally to perplex the adepts in science, to discourage the learner, as well as to cloud the beauty and simplicity of the atomic theory.

Quoted in H. E. Roscoe, 'Presidential Address', *Reports of the British Association for the Advancement of Science*, 57th report, 1887, 7.

2 **ON DALTON** Atoms are round balls of wood invented by Dr. Dalton.

Answer given by a pupil to a question on atomic theory, as reported by H. E. Roscoe.

Reports of the British Association for the Advancement of Science, 57th report, 1887, 7.

3 **ON DALTON** Mr. Dalton's aspect and manner were repulsive. There was no gracefulness belonging to him. His voice was harsh and brawling; his gait stiff and awkward; his style of writing and conversation dry and almost crabbed. In person he was tall, bony, and slender. He never could learn to swim: on investigating this circumstance he found that his spec. grav. as a mass was greater than that of water; and he mentioned this in his lectures on natural philosophy in illustration of the capability of different persons for attaining the art of swimming. Independence and simplicity of manner and originality were his best qualities. Though in comparatively humble circumstances he maintained the dignity of the philosophical character. As the first distinct promulgator of the doctrine that the elements of bodies unite in definite proportions to form chemical compounds, he has acquired an undying fame.

Dr John Davy's (brother of Humphry Davy) impressions of Dalton written in *c.*1830–1 in Malta. Quoted in W. C. Henry, *Memoirs of the Life and Scientific Researches of John Dalton* (1854), 217–8.

4 **ON DALTON** John Dalton was a very singular Man, a quaker by profession & practice: He has none of the manners or ways of the world. A tolerable mathematician He gained his livelihood I believe by teaching the mathematics to young people. He pursued science always with mathematical views. He seemed little attentive to the labours of men except when they countenanced or confirmed his own ideas . . . He was a very disinterested man, seemed to have no ambition beyond that of being thought a good Philosopher. He was a very coarse Experimenter & almost always found the results he required.— Memory & observation were subordinate qualities in his mind. He followed with ardour analogies & inductions & however his claims to originality may admit of question I have no doubt that he was one of the most original philosophers of his time & one of the most ingenious.

J. Z. Fullmer, 'Davy's Sketches of his Contemporaries', *Chymia*, 1967, **12**, 133–134.

Reginald Aldworth Daly

1871–1957

Canadian-American geologist

5 At bottom each 'exact' science is, and must be speculative, and its chief tool of research, too rarely used with both courage and judgement, is the regulated imagination.

Igneous Rocks and their Origin (1914), xxi.

6 This planet is essentially a body of crystallized and uncrystallized igneous material. The final philosophy of earth history will therefore be founded on igneous-rock geology.

Igneous Rocks and their Origin (1914), 1.

1 Our earth is very old, an old warrior that has lived through many battles. Nevertheless, the face of it is still changing, and science sees no certain limit of time for its stately evolution.

Our solid earth, apparently so stable, inert, and finished, is changing, mobile, and still evolving. Its major quakings are largely the echoes of that divine far-off event, the building of our noble mountains. The lava floods and intriguing volcanoes tell us of the plasticity, mobility, of the deep interior of the globe. The slow coming and going of ancient shallow seas on the continental plateaus tell us of the rhythmic distortion of the deep interior—deep-seated flow and changes of volume. Mountain chains prove the earth's solid crust itself to be mobile in high degree.

And the secret of it all—the secret of the earthquake, the secret of the 'temple of fire,' the secret of the ocean basin, the secret of the highland—is in the heart of the earth, forever invisible to human eyes.

Our Mobile Earth (1926), 320.

James Dwight Dana 1813–95

American geologist, mineralogist and naturalist

2 A map of the moon . . . should be in every geological lecture room; for no where can we have a more complete or more magnificent illustration of volcanic operations. Our sublimest volcanoes would rank among the smaller lunar eminences; and our Etnas are but spitting furnaces.

'On the Volcanoes of the Moon', *American Journal of Science*, 1846, 2 (2nd Series), 347.

3 *Are coral reefs growing from the depths of the oceans?* . . . [The] reply is a simple negative; and a single fact establishes its truth. The reef-forming coral zoophytes, as has been shown, cannot grow at greater depths than 100 or 120 feet; and therefore in seas deeper than this, the formation or growth of reefs over the bottom is impossible.

On Coral Reefs and Islands (1853), 138.

4 Science, while it penetrates deeply the system of things about us, sees everywhere, in the dim limits of vision, the word mystery.

Corals and Coral Islands, 3rd edition (1890), 17–18.

5 *The oceans have always been oceans.*

Corals and Coral Islands, 3rd edition (1890), 409.

6 Mr Hall's hypothesis [concerning the origin of mountains] has its cause for subsidence, but none for the lifting of the thickened sunken crust into mountains. It is a theory for the origin of mountains, with the origin of mountains left out.

'Observations on the Origin of Some of the Earth's Features', *American Journal of Science*, 1866, 42 (2nd Series), 210.

7 In using the present in order to reveal the past, we assume that the forces in the world are essentially the same through all time; for these forces are based on the very nature of matter, and could not have changed. The ocean has always had its waves, and those waves have always acted in the same manner. Running water on the land has ever had the same power of wear and transportation and mathematical value to its force. The laws of chemistry, heat, electricity, and mechanics have been the same through time. The plan of living structures has been fundamentally one, for the whole series belongs to one system, as much almost as the parts of an animal to the one body; and the relations of life to light and heat, and to the atmosphere, have ever been the same as now.

Manual of Geology (1867), 7.

Richard Henry Dana 1815–82

American lawyer and author

8 It is always observable that the physical and the exact sciences are the last to suffer under despotisms.

To Cuba and Back (1859), 192.

Dante Alighieri 1265–1321

Italian poet, philosopher and political thinker

1 *Il maestro di color che sanno.*

The master of those who know.

Of Aristotle.

Divina Commedia 'Inferno' canto 4, l.80.

2 *Considerate la vostra semenza:*
Fatti non foste a viver come bruti,
Ma per seguir virtute e conoscenza.

Consider your origins: you were not made to live as brutes, but to follow virtue and knowledge.

Divina Commedia 'Inferno', canto 26, l.118.

3 Man alone amongst the animals speaks and has gestures and expression which we call rational, because he alone has reason in him. And if anyone should say in contradiction that certain birds talk, as seems to be the case with some, especially the magpie and the parrot, and that certain beasts have expression or gestures, as the ape and some others seem to have, I answer that it is not true that they speak, nor that they have gestures, because they have no reason, from which these things need proceed; nor do they purpose to signify anything by them, but they merely reproduce what they see and hear.

Attributed.

Cyril Dean Darlington 1903–81

British botanist and geneticist

4 Cell genetics led us to investigate cell mechanics. Cell mechanics now compels us to infer the structures underlying it. In seeking the mechanism of heredity and variation we are thus discovering the molecular basis of growth and reproduction. The theory of the cell revealed the unity of living processes; the study of the cell is beginning to reveal their physical foundations.

Recent Advances in Cytology (1937), 562.

5 We are now witnessing, after the slow fermentation of fifty years, a concentration of technical power aimed at the essential determinants of heredity, development and disease. This concentration is made possible by the common function of nucleic acids as the molecular midwife of all reproductive particles. Indeed it is the nucleic acids which, in spite of their chemical obscurity, are giving to biology a unity which has so far been lacking, a chemical unity.

Nucleic Acid (1947), 266–7.

6 A large proportion of mankind, like pigeons and partridges, on reaching maturity, having passed through a period of playfulness or promiscuity, establish what they hope and expect will be a permanent and fertile mating relationship. This we call marriage.

Genetics And Man (1964), 298.

Clarence Darrow 1857–1938

American lawyer

7 If today you can take a thing like evolution and make it a crime to teach it in the public schools, tomorrow you can make it a crime to teach it in the private schools, and next year you can make it a crime to teach it to the hustings or in the church. At the next session you may ban books and the newspapers . . . Ignorance and fanaticism are ever busy and need feeding. Always feeding and gloating for more. Today it is the public school teachers; tomorrow the private. The next day the preachers and the lecturers, the magazines, the books, the newspapers. After a while, Your Honor, it is the setting of man against man and creed against creed until with flying banners and beating drums we are marching backward to the glorious ages of the sixteenth century when bigots lighted fagots to burn the men who dared to bring any intelligence and enlightenment and culture to the human mind.

Scopes Monkey Trial, Dayton, Tennessee, 1925. *The World's Most Famous Court Trial* (1925), Second Day's Proceedings, 87.

Karl Kelchner Darrow 1891–1982

American physicist

1 This 'truth for truth's sake' is a delusion of so-called savants.

'Industrial Physics', *The Review of Scientific Instruments with Physics News and Events*, 1934, 5(1), 60.

Raymond Arthur Dart 1893–1988

Australian-born South African anthropologist and anatomist

2 All fossil anthropoids found hitherto have been known only from mandibular or maxillary fragments, so far as crania are concerned, and so the general appearance of the types they represented had been unknown; consequently, a condition of affairs where virtually the whole face and lower jaw, replete with teeth, together with the major portion of the brain pattern, have been preserved, constitutes a specimen of unusual value in fossil anthropoid discovery. Here, as in *Homo rhodesiensis*, Southern Africa has provided documents of higher primate evolution that are amongst the most complete extant. Apart from this evidential completeness, the specimen is of importance because it exhibits an extinct race of apes *intermediate between living anthropoids and man* . . . Whether our present fossil is to be correlated with the discoveries made in India is not yet apparent; that question can only be solved by a careful comparison of the permanent molar teeth from both localities. It is obvious, meanwhile, that it represents a fossil group distinctly advanced beyond living anthropoids in those two dominantly human characters of facial and dental recession on one hand, and improved quality of the brain on the other. Unlike Pithecanthropus, it does not represent an ape-like man, a caricature of precocious hominid failure, but a creature well advanced beyond modern anthropoids in just those characters, facial and cerebral, which are to be anticipated in an extinct link between man and his simian ancestor. At the same time, it is equally evident that a creature with anthropoid brain capacity and lacking the distinctive, localised temporal expansions which appear to be concomitant with and necessary to articulate man, is no true man. It is therefore logically regarded as a man-like ape. I propose tentatively, then, that a new family of *Homo-simidæ* be created for the reception of the group of individuals which it represents, and that the first known species of the group be designated *Australopithecus africanus*, in commemoration, first, of the extreme southern and unexpected horizon of its discovery, and secondly, of the continent in which so many new and important discoveries connected with the early history of man have recently been made, thus vindicating the Darwinian claim that Africa would prove to be the cradle of mankind.

'*Australopithicus africanus*: The Man-Ape of South Africa', *Nature*, 1925, 115, 195.

3 From the time of Aristotle it had been said that man is a social animal: that human beings naturally form communities. I coudn't accept it. The whole of history and pre-history is against it. The two dreadful world wars we have recently been through, and the gearing of our entire economy today for defensive war belie it. Man's loathsome cruelty to man is his most outstanding characteristic; it is explicable only in terms of his carnivorous and cannibalistic origin. Robert Hartmann pointed out that both rude and civilised peoples show unspeakable cruelty to one another. We call it inhuman cruelty; but these dreadful things are unhappily truly human, because there is nothing like them in the animal world. A lion or tiger kills to eat, but the indiscriminate slaughter and calculated cruelty of human beings is quite unexampled in nature, especially among the apes. They display no hostility to man or other animals unless attacked. Even then their first reaction is to run away.

Africa's Place in the Emergence of Civilisation (1959), 41.

Charles Robert Darwin 1809–82

British naturalist

See ***Butler, Samuel*** *109:3;* ***Galton, Francis*** *240:1;* ***Hooker, Joseph Dalton*** *298:6*

1 To my deep mortification my father once said to me, 'You care for nothing but shooting, dogs, and rat-catching, and you will be a disgrace to yourself and all your family.'

'Recollections of the Development of my mind and Character' (written 1876). In *Autobiographies*, (eds.) Michael Neve and Sharon Messenger (2002), Penguin edn., 10.

2 The man who walks with Henslow.

'Recollections of the Development of my mind and Character' (written 1876). In *Autobiographies*, (eds.) Michael Neve and Sharon Messenger (2002), Penguin edn., 34.

3 The voyage of the *Beagle* has been by far the most important event in my life and has determined my whole career; yet it depended on so small a circumstance as my uncle offering to drive me 30 miles to Shrewsbury, which few uncles would have done, and on such a trifle as the shape of my nose.

'Recollections of the Development of my mind and Character' (written 1876). In *Autobiographies*, (eds.) Michael Neve and Sharon Messenger (2002), Penguin edn., 42.

4 But Geology carries the day: it is like the pleasure of gambling, speculating, on first arriving, what the rocks may be; I often mentally cry out 3 to 1 Tertiary against primitive; but the latter have hitherto won all the bets.

Letter to W. D. Fox, May 1832. In F. Burkhardt and S. Smith (eds.), *The Correspondence of Charles Darwin 1821–1836* (1985), Vol. 1, 232.

5 I shall always feel respect for every one who has written a book, let it be what it may, for I had no idea of the trouble which trying to write common English could cost one—And alas there yet remains the worst part of all correcting the press.

Letter to W. D. Fox, 7 July 1837, referring to his Journal of Researches. In F. Burkhardt and S. Smith (eds), *The Correspondence of Charles Darwin 1837–1843* (1986), Vol. 2, 29.

6 [Herschel and Humboldt] stirred up in me a burning zeal to add even the most humble contribution to the noble structure of Natural Science. No one or a dozen other books influenced me nearly so much as these two. I copied out from Humboldt long passages about Teneriffe and read them aloud on one of [my walking excursions].

Autobiographies, (eds.) Michael Neve and Sharon Messenger (2002), Penguin edn., 36.

7 A bad earthquake at once destroys the oldest associations: the world, the very emblem of all that is solid, has moved beneath our feet like a crust over a fluid; one second of time has conveyed to the mind a strange idea of insecurity, which hours of reflection would never have created.

Journal of Researches: into the Natural History and Geology of the Countries Visited During the Voyage of H.M.S. Beagle Round the World (1839), ch. XVI, 369.

8 I have stated, that in the thirteen species of ground-finches [in the Galapagos Islands], a nearly perfect gradation may be traced, from a beak extraordinarily thick, to one so fine, that it may be compared to that of a warbler. I very much suspect, that certain members of the series are confined to different islands; therefore, if the collection had been made on any *one* island, it would not have presented so perfect a gradation. It is clear, that if several islands have each their peculiar species of the same genera, when these are placed together, they will have a wide range of character. But there is not space in this work, to enter on this curious subject.

Journal of Researches: into the Natural History and Geology of the Countries Visited During the Voyage of H.M.S. Beagle Round the World (1839), ch. XIX, 475.

1 The theory which I would offer, is simply, that as the land with the attached reefs subsides very gradually from the action of subterranean causes, the coral-building polypi soon raise again their solid masses to the level of the water: but not so with the land; each inch lost is irreclaimably gone; as the whole gradually sinks, the water gains foot by foot on the shore, till the last and highest peak is finally submerged.

Journal of Researches: into the Natural History and Geology of the Countries Visited During the Voyage of H.M.S. Beagle Round the World (1839), ch. XXII, 557.

2 Hence, a traveller should be a botanist, for in all views plants form the chief embellishment.

Journal of Researches: into the Natural History and Geology of the Countries Visited During the Voyage of H.M.S. Beagle Round the World (1839), ch. XXIII, 604.

3 Among the scenes which are deeply impressed on my mind, none exceed in sublimity the primeval forests undefaced by the hand of man; whether those of Brazil, where the powers of Life are predominant, or those of Tierra del Fuego, where Death and Decay prevail. Both are temples filled with the varied productions of the God of Nature: no one can stand in these solitudes unmoved, and not feel that there is more in man than the mere breath of his body.

Journal of Researches: into the Natural History and Geology of the Countries Visited During the Voyage of H.M.S. Beagle Round the World (1839), ch. XXIII, 604–5.

4 Nothing can be more improving to a young naturalist, than a journey in a distant country.

Journal of Researches: into the Natural History and Geology of the Countries Visited During the Voyage of H.M.S. Beagle Round the World (1839), ch. XXIII, 607.

5 The natural history of these islands is eminently curious, and well deserves attention. Most of the organic productions are aboriginal creations, found nowhere else; there is even a difference between the inhabitants of the different islands; yet all show a marked relationship with those of America, though separated from that continent by an open space of ocean, between 500 and 600 miles in width. The archipelago is a little world within itself, or rather a satellite attached to America, whence it has derived a few stray colonists, and has received the general character of its indigenous productions. Considering the small size of these islands, we feel the more astonished at the number of their aboriginal beings, and at their confined range. Seeing every height crowned with its crater, and the boundaries of most of the lava-streams still distinct, we are led to believe that within a period, geologically recent, the unbroken ocean was here spread out. Hence, both in space and time, we seem to be brought somewhere near to that great fact—that mystery of mysteries—the first appearance of new beings on this earth.

Journal of Researches into the Natural History and Geology of the Countries Visited During the Voyage of H.M.S. Beagle Round the World, 2nd edn. (1845), 377–8.

6 What a scale of improvement is comprehended between the faculties of a Fuegian savage and a Sir Isaac Newton.

R. D. Keynes, *Darwin's Beagle Diary* (1988), 223.

7 I find in Geology a never failing interest, as [it] has been remarked, it creates the same gran[d] ideas respecting this world, which Astronomy do[es] for the universe.— We have seen much fine scenery that of the Tropics in its glory & luxuriance, exceeds even the language of Humboldt to describe. A Persian writer could alone do justice to it, & if he succeeded he would in England, be called the 'grandfather of all liars'.— But I have seen nothing, which more completely astonished me, than the first sight of a Savage; It was a naked Fuegian his long hair blowing about, his face besmeared with paint. There is in their countenances, an expression, which I believe to those who have not seen it, must be inconceivably wild. Standing on a rock he uttered tones & made gesticulations than which, the cries of domestic animals are far more intelligible.

Letter to Charles Whitley, 23 July 1834. In F. Burkhardt and S. Smith (eds.), *The Correspondence of Charles Darwin 1821–1836* (1985), Vol. 1, 397.

1 In July [1837] opened first note-book on Transmutation of Species. Had been greatly struck from about the month of previous March on character of South American fossils, and species on Galapagos Archipelago. These facts (especially latter), origin of all my views.

F. Darwin (ed.), *The Life and Letters of Charles Darwin, Including an Autobiographical Chapter* (1888), Vol. 1, 276.

2 In October 1838, that is, fifteen months after I had begun my systematic enquiry, I happened to read for amusement Malthus on *Population*, and being well prepared to appreciate the struggle for existence which everywhere goes on from long-continued observation of the habits of animals and plants, it at once struck me that under these circumstances favourable variations would tend to be preserved, and unfavourable ones to be destroyed. The result of this would be the formation of new species. Here, then, I had at last got a theory by which to work; but I was so anxious to avoid prejudice, that I determined not for some time to write even the briefest sketch of it.

Autobiographies, (eds.) Michael Neve and Sharon Messenger (2002), Penguin edn., 72–3.

3 The tree of life should perhaps be called the coral of life, base of branches dead; so that passages cannot be seen—this again offers contradiction to constant succession of germs in progress.

P. H. Barrett et al. (eds.), *Charles Darwin's Notebooks, 1836–1844: Geology, Transmutation of the Species, Metaphysical Enquiries* (1987), Notebook B, 25–6.

4 If we thus go very far back to the source of the Mammalian type of organisation; it is extremely improbable that any of [his relatives shall likewise] the successors of his relations now exist,—In same manner, if we take [a man from] any large family of 12 brothers & sisters [in a state which does not increase] it will be chances against any one [of them] having progeny living ten thousand years hence; because at present day many are relatives so that tracing back the [descen] fathers would be reduced to small percentage.—& [in] therefore the chances are excessively great against, any two of the 12. having progeny. after that distant period.

P. H. Barrett et al. (eds.), *Charles Darwin's Notebooks, 1836–1844: Geology, Transmutation of the Species and Metaphysical Enquiries* (1987), Notebook B, 40–1.

5 Origin of man now proved.—Metaphysics must flourish.—He who understands baboon [will] would do more towards metaphysics than Locke.

P. H. Barrett et al. (eds.), *Charles Darwin's Notebooks, 1836–1844: Geology, Transmutation of the Species and Metaphysical Enquiries* (1987), Notebook M, 84.

6 This is the question

Marry

Children—(if it Please God)—Constant companion (& friend in old age) who will feel interested in one—object to be beloved and played with—better than a dog anyhow. Home, & someone to take care of house—Charms of music and female chit-chat.—These things good for one's health.—*but terrible loss of time.*—

My God, it is intolerable to think of spending ones whole life, like a neuter bee, working, working—& nothing after all.—No, no, won't do. Imagine living all one's day solitary in smoky dirty

Not Marry

Freedom to go where one liked—choice of Society and *little of it*.—Conversation of clever men at clubs—Not forced to visit relatives, & to bend in every trifle.—to have the expense and anxiety of children—perhaps quarreling—**Loss of time.**—cannot read in the Evenings—fatness & idleness—Anxiety & responsibility—less money for books &c—if many children forced to gain one's bread.—(but then it is very bad for ones health to work too much)

Perhaps my wife wont like London; then the

London House.—Only picture to yourself a nice soft wife on a sofa with good fire, & books & music perhaps—Compare this vision with the dingy reality of Grt. Marlbro' Street.

sentence is banishment & degradation into indolent, idle fool.

Marry—Marry—Marry Q.E.D.
It being proved necessary to Marry
When? Soon or late?

Notes on Marriage, July 1838. In F. Burkhardt and S. Smith (eds.), *The Correspondence of Charles Darwin 1837–1843* (1986), Vol. 2, 444.

1 We may confidently come to the conclusion, that the forces which slowly and by little starts uplift continents, and that those which at successive periods pour forth volcanic matter from open orifices, are identical.

Journal of Researches into the Natural History and Geology of the Countries Visited During the Voyage of H.M.S. Beagle Round the World, 2nd edn. (1845), ch. XIV, 311.

2 I always feel as if my books came half out of Lyell's brain . . . & therefore that when seeing a thing never seen by Lyell, one yet saw it partially through his eyes.

Letter to Leonard Horner, 29 August 1844. In F. Burkhardt and S. Smith (eds.), *The Correspondence of Charles Darwin 1844–1846* (1987), Vol. 3, 55.

3 At last gleams of light have come, & I am almost convinced (quite contrary to opinion I started with) that species are not (it is like confessing a murder) immutable. Heaven forfend me from Lamarck nonsense of a 'tendency to progression', 'adaptations from the slow willing of animals', &c—but the conclusions I am led to are not widely different from his—though the means of change are wholly so—I think I have found out (here's presumption!) the simple way by which species become exquisitely adapted to various ends.

Letter to J. D. Hooker, 11 January 1844. In F. Burkhardt and S. Smith (eds.), *The Correspondence of Charles Darwin 1844–1846* (1987), Vol. 3, 2.

4 I have just finished my sketch of my species theory. If as I believe that my theory is true & if it be accepted even by one competent judge, it will be a considerable step in science. I therefore write this, in case of my sudden death, as my most solemn & last request, which I am sure you will consider the same as if legally entered in my will, that you will devote 400£ to its publication & further will yourself, or through Hensleigh [Wedgwood], take trouble in promoting it.

Letter to Emma Darwin, 5 July 1844. In F. Burkhardt and S. Smith (eds.), *The Correspondence of Charles Darwin 1844–1846* (1987), Vol. 3, 43.

5 What a book a Devil's chaplin might write on the clumsy, wasteful, blundering, low & horridly cruel works of nature!

Letter to J. D. Hooker, 13 July 1856. In F. Burkhardt and S. Smith (eds.), *The Correspondence of Charles Darwin 1844–1846* (1987), Vol. 6, 178.

6 I am a firm believer, that without speculation there is no good & original observation.

Letter to A. R. Wallace, 22 December 1857. In F. Burkhardt and S. Smith (eds.), *The Correspondence of Charles Darwin 1844–1846* (1987), Vol. 6, 514.

7 Your words have come true with a vengeance that I shd [should] be forestalled . . . I never saw a more striking coincidence. If Wallace had my M.S. sketch written out in 1842 he could not have made a better short abstract! Even his terms now stand as Heads of my Chapters.

Letter to Charles Lyell, 18 June 1858. In F. Burkhardt and S. Smith (eds.), *The Correspondence of Charles Darwin 1858–1859, Supplement 1821–1857* (1991), Vol. 7, 107.

8 The Struggle for Existence amongst all organic beings throughout the world, which inevitably follows from their high geometrical powers of increase, will be treated of. This is the doctrine of Malthus, applied to the whole animal and vegetable kingdoms. As many more individuals of each species are born than can possibly survive; and as, consequently, there is a frequently recurring struggle for existence, it follows that any being, if it vary however slightly in any manner profitable to itself, under the complex and sometimes varying conditions of life, will have a better chance of surviving, and thus be *naturally selected*. From the strong principle of inheritance, any selected variety will tend to propagate its new and modified form.

The Origin of Species (1859), Penguin edn, J. W. Burrow (ed.) (1968), 68.

1 All that we can do, is to keep steadily in mind that each organic being is striving to increase at a geometrical ratio; that each at some period of its life, during some season of the year, during each generation or at intervals, has to struggle for life, and to suffer great destruction. When we reflect on this struggle, we may console ourselves with the full belief, that the war of nature is not incessant, that no fear is felt, that death is generally prompt, and that the vigorous, the healthy, and the happy survive and multiply.

The Origin of Species (1859), Penguin edn, J. W. Burrow (ed.) (1968), 129.

2 It may be said that natural selection is daily and hourly scrutinising, throughout the world, every variation, even the slightest; rejecting that which is bad, preserving and adding up all that is good; silently and insensibly working, whenever and wherever opportunity offers, at the improvement of each organic being in relation to its organic and inorganic conditions of life.

The Origin of Species (1859), Penguin edn, J. W. Burrow (ed.) (1968), 133.

3 As buds give rise by growth to fresh buds, and these, if vigorous, branch out and overtop on all sides many a feebler branch, so by generation I believe it has been with the great Tree of Life, which fills with its dead and broken branches the crust of the earth, and covers the surface with its ever branching and beautiful ramifications.

The Origin of Species (1859), Penguin edn, J. W. Burrow (ed.) (1968), 172.

4 To suppose that the eye, with all its inimitable contrivances for adjusting the focus to different distances, for admitting different amounts of light, and for the correction of spherical and chromatic aberration, could have been formed by natural selection, seems, I freely confess, absurd in the highest possible degree. Yet reason tells me, that if numerous gradations from a perfect and complex eye to one very imperfect and simple, each grade being useful to its possessor, can be shown to exist; if further, the eye does vary ever so slightly, and the variations be inherited, which is certainly the case; and if any variation or modification in the organ be ever useful to an animal under changing conditions of life, then the difficulty of believing that a perfect and complex eye could be formed by natural selection, though insuperable by our imagination, can hardly be considered real.

The Origin of Species (1859), Penguin edn, J. W. Burrow (ed.) (1968), 217.

5 On the theory of natural selection we can clearly understand the full meaning of that old canon in natural history, 'Natura non facit saltum.' This canon, if we look only to the present inhabitants of the world, is not strictly correct, but if we include all those of past times, it must by my theory be strictly true.

The Origin of Species (1859), Penguin edn, J. W. Burrow (ed.), (1968), 233.

6 When the views entertained in this volume on the origin of species, or when analogous views are generally admitted, we can dimly forsee that there will be a considerable revolution in natural history.

The Origin of Species (1859), Penguin edn , J. W. Burrow (ed.) (1968), 455.

7 It is interesting to contemplate an entangled bank, clothed with many plants of many kinds, with birds singing on the bushes, with various insects flitting about, and with worms crawling through the damp earth, and to reflect that these elaborately constructed forms, so different from each other, and dependent on each other in so complex a manner, have all been produced by laws acting around us. These laws, taken in the largest sense, being Growth with Reproduction; Inheritance which is almost implied by reproduction; Variability from the indirect and direct action of the external conditions of life, and from use and disuse; a Ratio of Increase so high as to lead to a Struggle for Life, and as a consequence to Natural Selection, entailing Divergence of Character and the Extinction of less-improved forms. Thus, from the war of nature, from famine and death, the most exalted object which we are capable of conceiving, namely, the production of the higher animals, directly follows. There is grandeur in this view of life, with its several powers, having been originally breathed into a few forms or into one; and that, whilst this planet has gone cycling on according to the fixed law of gravity, from so simple a beginning endless forms most beautiful and most wonderful have been, and are being, evolved.

The Origin of Species (1859), Penguin edn, J. W. Burrow (ed.) (1968), 459–60.

1 I have heard by round about channel that Herschel says my book 'is the law of higgledy-pigglety.'

Letter to C. Lyell, 10 December 1859. In F. Burkhardt and S. Smith (eds.), *The Correspondence of Charles Darwin 1858–1859* (1991), Vol. 7, 423.

2 *Our* ancestor was an animal which breathed water, had a swim-bladder, a great swimming tail, an imperfect skull & undoubtedly was an hermaphrodite! Here is a pleasant genealogy for mankind.

Letter to C. Lyell, 10 January 1860. In F. Burkhardt and S. Smith (eds.), *The Correspondence of Charles Darwin 1860* (1993), Vol. 8, 29.

3 About weak points [of the *Origin*] I agree. The eye to this day gives me a cold shudder, but when I think of the fine known gradations, my reason tells me I ought to conquer the cold shudder.

Letter to Asa Gray, 8 or 9 February 1860. In F. Burkhardt and S. Smith (eds.), *The Correspondence of Charles Darwin 1860* (1993), Vol. 8, 75.

4 I have not [a] metaphysical Head.

Letter to C. Lyell, 23 February 1860. In F. Burkhardt and S. Smith (eds.), *The Correspondence of Charles Darwin 1860* (1993), Vol. 8, 103.

5 There seems to me too much misery in the world. I cannot persuade myself that a beneficent & omnipotent God would have designedly created the Ichneumonidae with the express intention of their feeding within the living bodies of caterpillars, or that a cat should play with mice.

Letter to Asa Gray, 22 May 1860. In F. Burkhardt and S. Smith (eds.), *The Correspondence of Charles Darwin 1860* (1993), Vol. 8, 224.

6 One more word on 'designed laws' and 'undesigned results.'—I see a bird which I want for food, take my gun and kill it, I do this *designedly*.—An innocent and good man stands under a tree and is killed by a flash of lightening. Do you believe (& I really should like to hear) that God *designedly* killed this man? Many or most persons do believe this; I can't and don't.—If you believe so, do you believe that when a swallow snaps up a gnat that God designed that that particular swallow should snap up that particular gnat at that particular instant? I believe that the man and the gnat are in the same predicament. If the death of neither man nor gnat are designed, I see no good reason to believe that their *first* birth or production should be necessarily designed.

Letter to Asa Gray, 3 July 1860. In F. Burkhardt and S. Smith (eds.), *The Correspondence of Charles Darwin 1860* (1993), Vol. 8, 275.

7 How odd it is that anyone should not see that all observation must be for or against some view, if it is to be of any service!

Letter to Henry Fawcett, 18 September 1861. In F. Burkhardt and S. Smith (eds.), *The Correspondence of Charles Darwin 1861* (1994), Vol. 9, 269.

1 It is mere rubbish thinking, at present, of origin of life; one might as well think of origin of matter.

Letter to J. D. Hooker, 29 March 1863. In F. Burkhardt and S. Smith (eds.), *The Correspondence of Charles Darwin 1863* (1999), Vol. 11, 278.

2 I fully agree with all that you say on the advantages of H. Spencer's excellent expression of 'the survival of the fittest.' This, however, had not occurred to me till reading your letter. It is, however, a great objection to this term that it cannot be used as a substantive governing a verb; and that this is a real objection I infer from H. Spencer continually using the words, natural selection.

Letter to A. R. Wallace July 1866. In F. Darwin (ed.), *The Life and Letters of Charles Darwin, Including an Autobiographical Chapter* (1887), Vol. 3, 45–6.

3 I hope you have not murdered too completely your own and my child.

Letter to A. R. Wallace, March 1869. In J. Marchant, *Alfred Russel Wallace: Letters and Reminiscences* (1916), Vol. 1, 240.

4 Many kinds of monkeys have a strong taste for tea, coffee and spirituous liqueurs.

The Descent of Man (1871), Vol. 1, 12.

5 Thus we have given to man a pedigree of prodigious length, but not, it may be said, of noble quality.

The Descent of Man (1871), Vol. 1, 213.

6 When the sexes differ in beauty, in the power of singing, or in producing what I have called instrumental music, it is almost invariably the male which excels the female.

The Descent of Man (1871), Vol. 2, 99.

7 With mammals the male appears to win the female much more through the law of battle than through the display of his charms.

The Descent of Man (1871), Vol. 2, 239.

8 If every one were cast in the same mould, there would be no such thing as beauty.

The Descent of Man (1871), Vol. 2, 354.

9 False facts are highly injurious to the progress of science, for they often long endure; but false views, if supported by some evidence, do little harm, as every one takes a salutary pleasure in proving their falseness; and when this is done, one path towards error is closed and the road to truth is often at the same time opened.

The Descent of Man (1871), Vol. 2, 385.

10 We thus learn that man is descended from a hairy quadruped, furnished with a tail and pointed ears, probably arboreal in its habits, and an inhabitant of the Old World.

The Descent of Man (1871), Vol. 2, 389.

11 The main conclusion arrived at in this work, namely that man is descended from some lowly-organised form, will, I regret to think, be highly distasteful to many persons. But there can hardly be a doubt that we are descended from barbarians.

The Descent of Man (1871), Vol. 2, 404.

12 For my own part I would as soon be descended from that heroic little monkey, who braved his dreaded enemy in order to save the life of his keeper; or from that old baboon, who, descending from the mountains, carried away in triumph his young comrade from a crowd of astonished dogs—as from a savage who delights to torture his enemies, offers up bloody sacrifices, practices infanticide without remorse, treats his wives like slaves, knows no decency, and is haunted by the grossest superstitions.

The Descent of Man (1871), Vol. 2, 404–5.

13 Man may be excused for feeling some pride at having risen, though not through his own exertions, to the very summit of the organic scale; and the fact of his having thus risen, instead of having been aboriginally placed there, may give him hopes for a still higher destiny in the distant future. But we are not here concerned with hopes or fears, only with the truth as far as our reason allows us to discover it. I have given the evidence to the best of my ability; and we must acknowledge, as it seems to me, that man with all his noble qualities, with sympathy which feels for the most debased, with benevolence which extends not only to other men but to the humblest living creature, with his god-like intellect which has penetrated into the movements and constitution of the solar system—with all these exalted powers—Man still bears in his bodily frame the indelible stamp of his lowly origin.

The Descent of Man (1871), Vol. 2, 405.

1 Blushing is the most peculiar and most human of all expressions.

The Expression of the Emotions in Man and Animals (1872), 310.

2 My mind seems to have become a kind of machine for grinding general laws out of large collections of facts.

Autobiographies, (eds.) Michael Neve and Sharon Messenger (2002), Penguin edn, 85.

3 Formerly Milton's *Paradise Lost* had been my chief favourite, and in my excursions during the voyage of the *Beagle*, when I could take only a single small volume, I always chose Milton.

Nora Barlow (ed.), *The Autobiography of Charles Darwin, 1809–1882* (1958), 85.

4 A surprising number [of novels] have been read aloud to me, and I like all if moderately good, and if they do not end unhappily—against which a law ought to be passed.

F. Darwin (ed.), *The Life and Letters of Charles Darwin, Including an Autobiographical Chapter* (1888), Vol. 1, 101.

5 The *Times* is getting more detestable (but that is too weak word) than ever.

Letter to Asa Gray, 23 February 1863. In F. Burkhardt and S. Smith (eds.), *The Correspondence of Charles Darwin 1863* (1999), Vol. 11, 167.

Erasmus Darwin 1731–1802

British physician, poet, and naturalist

6 Soon shall thy arm, UNCONQUER'D STEAM! afar
Drag the slow barge, or drive the rapid car;
Or on wide-waving wings expanded bear
The flying-chariot through the fields of air.

The Botanic Garden (1791) part 1, canto 1, lines 289–92, page 29.

7 The colours of insects and many smaller animals contribute to conceal them from the larger ones which prey upon them. Caterpillars which feed on leaves are generally green; and earth-worms the colour of the earth which they inhabit; butter-flies, which frequent flowers, are coloured like them; small birds which frequent hedges have greenish backs like the leaves, and light-coloured bellies like the sky, and are hence less visible to the hawk who passes under them or over them.

The Botanic Garden (1791), part 2, note to canto 1, line 375, page 38.

8 The great CREATOR of all things has infinitely diversified the works of his hands, but has at the same time stamped a certain similitude on the features of nature, that demonstrates to us, that *the whole is one family of one parent.*

Zoonomia (1794), Vol. 1, 1.

9 [Some] philosophers have been of opinion that our immortal part acquires during this life certain habits of action or of sentiment, which become forever indissoluble, continuing after death in a future state of existence . . . I would apply this ingenious idea to the generation, or production of the embryon, or new animal, which partakes so much of the form and propensities of the parent.

Zoonomia (1794), Vol. 1, 483–4.

1 So the horns of the stag are sharp to offend his adversary, but are branched for the purpose of parrying or receiving the thrusts of horns similar to his own, and have therefore been formed for the purpose of combating other stags for the exclusive possession of the females; who are observed, like the ladies in the times of chivalry, to attend to the car of the victor . . . The final cause of this contest amongst the males seems to be, that the strongest and most active animal should propagate the species, which should thence become improved.

Zoonomia (1794), Vol. 1, 507.

2 From thus meditating on the great similarity of the structure of the warm-blooded animals, and at the same time of the great changes they undergo both before and after their nativity; and by considering in how minute a portion of time many of the changes of animals above described have been produced; would it be too bold to imagine, that in the great length of time, since the earth began to exist, perhaps millions of ages before the commencement of the history of mankind, would it be too bold to imagine, that all warm-blooded animals have arisen from one living filament, which THE GREAT FIRST CAUSE endued with animality, with the power of acquiring new parts . . . and of delivering down those improvements by generation to its posterity, world without end!

Zoonomia (1794), Vol. 1, 509.

3 This compassion, or sympathy with the pains of others, ought also to extend to the brute creation, as far as our necessities will admit; for we cannot exist long without the destruction of other animal or vegetable beings either in their mature or embryon state. Such is the condition of mortality, that the first law of nature is 'eat, or be eaten.' Hence for the preservation of our existence we may be supposed to have a natural right to kill those brute creatures, which we want to eat, or which want to eat us; but to destroy even insects wantonly shows an unreflecting mind, or a depraved heart.

A Plan for the Conduct of Female Education in Boarding Schools (1797), 48.

4 From the sexual, or amatorial, generation of plants new varieties, or improvements, are frequently obtained; as many of the young plants from seeds are dissimilar to the parent, and some of them superior to the parent in the qualities we wish to possess . . . Sexual reproduction is the chef d'oeuvre, the master-piece of nature.

Phytologia (1800), 115, 103.

5 Such is the condition of organic nature! whose first law might be expressed in the words 'Eat or be eaten!' and which would seem to be one great slaughter-house, one universal scene of rapacity and injustice!

Phytologia (1800), 556.

6 We hence acquire this sublime and interesting idea; that all the calcareous mountains in the world, and all the strata of clay, coal, marl, sand, and iron, which are incumbent on them, are MONUMENTS OF THE PAST FELICITY OF ORGANIZED NATURE!

Phytologia (1800), 560.

7 By firm immutable immortal laws
Impress'd on Nature by the GREAT FIRST CAUSE,
Say, MUSE! how rose from elemental strife
Organic forms, and kindled into life;
How Love and Sympathy with potent charm
Warm the cold heart, the lifted hand disarm;
Allure with pleasures, and alarm with pains,
And bind Society in golden chains.

The Temple of Nature; or, The Origin of Society: A Poem, with Philosophical Notes (1803), 3.

8 ORGANIC LIFE beneath the shoreless waves
Was born and nurs'd in Ocean's pearly caves;

First, forms minute, unseen by spheric glass,
Move on the mud, or pierce the watery mass;
These, as successive generations bloom,
New powers acquire, and larger limbs assume;
Whence countless groups of vegetation spring,
And breathing realms of fin, and feet, and wing.
Thus the tall Oak, the giant of the wood,
Which bears Britannia's thunders on the flood;
The Whale, unmeasured monster of the main,
The lordly Lion, monarch of the plain,
The Eagle soaring in the realms of air,
Whose eye undazzled drinks the solar glare,
Imperious man, who rules the bestial crowd,
Of language, reason, and reflection proud,
With brow erect, who scorns this earthy sod,
And styles himself the image of his God;
Arose from rudiments of form and sense,
An embryon point, or microscopic ens!

The Temple of Nature (1803), canto 1, lines 295–314, pages 26–8.

1 It is often hazardous to marry an heiress, as she is not unfrequently the last of a diseased family.

The Temple of Nature (1803), notes, 45.

2 The Reproductions of the living Ens
From sires to sons, unknown to sex, commence . . .
Unknown to sex the pregnant oyster swells,
And coral-insects build their radiate shells . . .
Birth after birth the line unchanging runs,
And fathers live transmitted in their sons;
Each passing year beholds the unvarying kinds,
The same their manners, and the same their minds.

The Temple of Nature (1803), canto 2, lines 63–4, 89–90, 107–10, pages 48–52.

3 Each pregnant Oak ten thousand acorns forms
Profusely scatter'd by autumnal storms;
Ten thousand seeds each pregnant poppy sheds
Profusely scatter'd from its waving heads;
The countless Aphides, prolific tribe,
With greedy trunks the honey'd sap imbibe;
Swarm on each leaf with eggs or embryons big,
And pendent nations tenant every twig
. . .
—All these, increasing by successive birth,
Would each o'erpeople ocean, air, and earth.
So human progenies, if unrestrain'd,
By climate friended, and by food sustain'd,
O'er seas and soils, prolific hordes! would spread
Erelong, and deluge their terraqueous bed;
But war, and pestilence, disease, and dearth,
Sweep the superfluous myriads from the earth . . .
The births and deaths contend with equal strife,
And every pore of Nature teems with Life;
Which buds or breathes from Indus to the Poles,
And Earth's vast surface kindles, as it rolls!

The Temple of Nature (1803), canto 4, lines 347–54, 367–74, 379–82, pages 156–60.

4 So erst the Sage [Pythagoras] with scientific truth
In Grecian temples taught the attentive youth;
With ceaseless change how restless atoms pass
From life to life, a transmigrating mass;

How the same organs, which to-day compose
The poisonous henbane, or the fragrant rose,
May with to-morrow's sun new forms compile,
Frown in the Hero, in the Beauty smile.
Whence drew the enlighten'd Sage the moral plan,
That man should ever be the friend of man;
Should eye with tenderness all living forms,
His brother-emmets, and his sister-worms.

The Temple of Nature (1803), canto 4, lines 417–28, page 163.

1 I am sorry the infernal Divinities, who visit mankind with diseases, and are therefore at perpetual war with Doctors, should have prevented my seeing all you great Men at Soho to-day—Lord! what inventions, what wit, what rhetoric, metaphysical, mechanical and pyrotecnical, will be on the wing, bandy'd like a shuttlecock from one to another of your troop of philosophers! while poor I, I by myself I, imprizon'd in a post chaise, am joggled, and jostled, and bump'd, and bruised along the King's high road, to make war upon a pox or a fever!

Letter to Matthew Boulton, 5 April 1778. Quoted in Desmond King-Hele (ed.), *The Letters of Erasmus Darwin* (1981), 84.

2 Whilst I am writing to a Philosopher and a Friend, I can scarcely forget that I am also writing to the greatest Statesman of the present, or perhaps of any century, who spread the happy contagion of Liberty among his countrymen.

Letter to Benjamin Franklin, 29 May 1787. Quoted in Desmond King-Hele (ed.), *The Letters of Erasmus Darwin* (1981), 166.

3 I much condole with you on your late loss . . . pains and diseases of the mind are only cured by Forgetfulness;—Reason but skins the wound, which is perpetually liable to fester again.

Letter to Richard Lovell Edgeworth, 24 April 1790. Quoted in Desmond King-Hele (ed.), *The Letters of Erasmus Darwin* (1981), 201.

4 [Retirement] is a dangerous experiment, and generally ends in either drunkenness or hypochrondriacism.

Letter to Robert Darwin, October 1792. Quoted in Desmond King-Hele, *Erasmus Darwin: A Life of Unequalled Achievement* (1999), 274.

5 Life is a forced state! I am surprized that we live, rather than that our friends die.

Letter to James Watt, 21 June 1796. Quoted in Desmond King-Hele, *Erasmus Darwin: A Life of Unequalled Achievement* (1999), 305.

6 ON **DARWIN** A young man once asked him in, as he thought, an offensive manner, whether he did not find stammering very inconvenient. He answered, 'No, Sir, it gives me time for reflection, and saves me from asking impertinent questions.'

C. Darwin, *The Life of Erasmus Darwin* (1887), 40.

7 ON **DARWIN** He used to say that 'unitarianism was a feather-bed to catch a falling Christian.'

C. Darwin, *The Life of Erasmus Darwin* (1887), 44–5.

8 Man is an eating animal, a drinking animal, and a sleeping animal, and one placed in a material world, which alone furnishes all the human animal can desire. He is gifted besides with knowing faculties, practically to explore and to apply the resources of this world to his use. These are realities. All else is nothing; conscience and sentiment are mere figments of the imagination. Man has but five gates of knowledge, the five senses; he can know nothing but through them; all else is a vain fancy, and as for the being of a God, the existence of a soul, or a world to come, who can know anything about them? Depend upon it, my dear madam, these are only the bugbears by which men of sense govern fools.

Attributed to Darwin by Mary Anne Schimmelpennick. Quoted in Christiana C.

Hankin (ed.) *Life of Mary Anne Schimmelpenninck* (1858), Vol.1, 241–2.

1 Opium is the only drug to be rely'd on—all the boasted nostrums only take up time, and as the disease [is] often of short duration, or of small quantity, they have gain'd credit which they do not deserve.

Quoted in Desmond King-Hele, *Erasmus Darwin: A Life of Unequalled Achievement* (1999), 161.

2 *E conchis omnia.*
Everything from shells.

Motto on his bookplate, 1771, expressing his evolutionary beliefs.

Francis Darwin 1848–1925

British botanist

3 But in science the credit goes to the man who convinces the world, not to the man to whom the idea first occurs. Not the man who finds a grain of new and precious quality but to him who sows it, reaps it, grinds it and feeds the world on it.

'First Galton Lecture before the Eugenics Society', *Eugenics Review*, 1914, 6, 9.

John Davies 1569–1626

British poet

4 *But* how shall we this *union* well expresse?
Nought tyes the *soule*: her subtiltie is such
She moves the bodie, which she doth possesse.
Yet no part toucheth, but by *Vertue*'s touch.
Then dwels she not therein as in a tent;
Nor as a pilot in his Ship doth sit;
Nor as the spider in his web is pent;
Nor as the Waxe retaines the print in it;
Nor as a Vessell water doth containe;
Nor as one Liquor in another shed;
Nor as the heate doth in the fire remaine;
Nor as a voice throughout the ayre is spred;
But as the faire and cheerfull *morning light*,
Doth here, and there, her silver beames impart,
And in an instant doth her selfe unite
To the transparent Aire, in all, and part:
Still resting whole, when blowes the Aire devide;
Abiding pure, when th'Aire is most corrupted;
Throughout the Aire her beames dispersing wide,
And when the Aire is tost, not interrupted:
So doth the piercing *Soule* the body fill;
Being all in all, and all in part diffus'd;
Indivisible, incorruptible still,
Not forc't, encountred, troubled or confus'd.
And as the *Sunne* above the light doth bring,
Tough we behold it in the Aire below;
So from th'eternall light the *Soule* doth spring,
Though in the Bodie she her powers do show.

From 'Nosce Teipsum' (1599), in Claire Howard (ed.), *The Poems of Sir John Davies* (1941), 151–2.

5 Rudenesse it selfe she doth refine,
Euen like an Alchemist divine,
Grosse times of Iron turning
Into the purest forme of gold:
Not to corrupt, till heaven waxe old,
And be refin'd with burning.

From 'Hymnes of Astraea' (1599), Hymn I, in Clare Howard (ed.), *The Poems of John Davies* (1941), 197.

Kingsley Davis 1908–97

American sociologist and demographer

6 Most discussions of the population crisis lead logically to zero population growth as the ultimate goal, because *any* growth rate, if continued, will eventually use up the earth . . . Turning to the actual measures taken, we see that the very use of family planning as the means for implementing population policy poses serious but unacknowledged limits on the intended reduction in fertility. The family-planning movement, clearly devoted to the improvement and dissemination of contraceptive devices, states again and again that its purpose is that of enabling couples to have the number of children they want.

With the publication of this article 'zero population growth' and the acronym 'ZPG' came into general use.

'Population Policy: Will Current Programs Succeed?', *Science*, 1967, **158**, 732.

William Morris Davis 1850–1934

American geomorphologist

1 The meaning of geography is as much a sealed book to the person of ordinary intelligence and education as the meaning of a great cathedral would be to a backwoodsman, and yet no cathedral can be more suggestive of past history in its many architectural forms than is the land about us, with its innumerable and marvellously significant geographic forms. It makes one grieve to think of opportunity for mental enjoyment that is lost because of the failure of education in this respect.

'Geographic Methods in Geologic Investigation', *The National Geographic Magazine*, 1889, **1**, 23.

2 Once established, an original river advances through its long life, manifesting certain peculiarities of youth, maturity and old age, by which its successive stages of growth may be recognized without much difficulty.

'The Rivers and Valleys of Pennsylvania', *The National Geographic Magazine*, 1889, **1**, 203.

3 It is the relationship between the physical environment and the environed organism, between physiography and ontography (to coin a term), that constitutes the essential principles of geography today.

'Systematic Geography', read 3 April 1902. *Proceedings of the American Philosophical Society for Promoting Useful Knowledge*, 1902, **41**, 240.

4 The very foundation of our science is only an inference; for the whole of it rests on the unprovable assumption that, all through the inferred lapse of time which the inferred performance of inferred geological processes involves, they have been going on in a manner consistent with the laws of nature as we know them now.

'The Value of Outrageous Geological Hypotheses', *Science*, 1926, **63**, 465–466.

5 The more clearly the immensely speculative nature of geological science is recognized, the easier it becomes to remodel our concepts of any inferred terrestrial conditions and processes in order to make outrages upon them not outrageous.

'The Value of Outrageous Geological Hypotheses', *Science*, 1926, **63**, 466.

Humphry Davy 1778–1829

British chemist

*See **Coleridge, Samuel Taylor** 127:1*

6 The moment after, I began to respire 20 quarts of unmingled nitrous oxide. A thrilling, extending from the chest to the extremities, was almost immediately produced. I felt a sense of tangible extension highly pleasurable in every limb; my visible impressions were dazzling, and apparently magnified, I heard distinctly every sound in the room and was perfectly aware of my situation. By degrees, as the pleasurable sensations increased, I lost all connection with external things; trains of vivid visible images rapidly passed through my mind, and were connected with words in such a manner, as to produce perceptions perfectly novel. I existed in a world of newly connected and newly modified ideas. I theorised—I imagined that I made discoveries. When I was awakened from this semi-delirious trance by Dr. Kinglake, who took the bag from my mouth, indignation and pride were the first feelings produced by the sight of the persons about me. My emotions were enthusiastic and sublime; and for a minute I walked round the room, perfectly regardless of what was said to me. As I recovered my former state of mind, I felt an inclination to communicate the discoveries I had made during the experiment. I endeavoured to recall the ideas, they were feeble and indistinct; one collection of terms, however, presented itself: and with the most intense belief and prophetic manner, I exclaimed to Dr Kinglake, '*Nothing exists but thoughts!—the universe is composed of impressions, ideas, pleasures and pains!*'

Researches, Chemical and Philosophical (1800), in J. Davy (ed.), *The Collected Works of Sir Humphry Davy* (1839–40), Vol. 3, 289–90.

1 Geology, perhaps more than any other department of natural philosophy, is a science of contemplation. It requires no experience or complicated apparatus, no minute processes upon the unknown processes of matter. It demands only an enquiring mind and senses alive to the facts almost everywhere presented in nature. And as it may be acquired without much difficulty, so it may be improved without much painful exertion.

'Lectures on Geology, 1805 Lecture', in R. Siegfried and R. H. Dott (eds.), *Humphry Davy on Geology* (1980), 13.

2 Natural science is founded on minute critical views of the general order of events taking place upon our globe, corrected, enlarged, or exalted by experiments, in which the agents concerned are placed under new circumstances, and their diversified properties separately examined. The body of natural science, then, consists of *facts*; is analogy,—the relation of resemblance of facts by which its different parts are connected, arranged, and employed, either for popular use, or for new speculative improvements.

'Introductory Lecture to the Chemistry of Nature' (1807), in J. Davy (ed.), *The Collected Works of Sir Humphry Davy* (1839–40), Vol. 8, 167–8.

3 It will be a general expression of the facts that have been detailed, relating to the changes and transitions by electricity, in common philosophical language, to say, that hydrogen, the alkaline substances, the metals, and certain metallic oxides, are all attracted by negatively electrified metallic surfaces; and contrariwise, that oxygen and acid substances are attracted by positively electrified metallic surfaces and rejected by negatively electrified metallic surfaces; and these attractive and repulsive forces are sufficiently energetic to destroy or suspend the usual operation of elective affinity.

Bakerian Lecture, 'On Some Chemical Agencies of Electricity', *Philosophical Transactions of the Royal Society*, 1807, 97, 28–29.

4 In the present state of our knowledge, it would be useless to attempt to speculate on the remote cause of the electrical energy, or the reason why different bodies, after being brought into contact, should be found differently electrified; its relation to chemical affinity is, however, sufficiently evident. May it not be identical with it, and an essential property of matter?

Bakerian Lecture, 'On Some Chemical Agencies of Electricity', *Philosophical Transactions of the Royal Society*, 1807, 97, 39.

5 The progression of physical science is much more connected with your prosperity than is usually imagined. You owe to experimental philosophy some of the most important and peculiar of your advantages. It is not by foreign conquests chiefly that you are become great, but by a conquest of nature in your own country.

From an introductory lecture to a course on electro-chemical science in 1809, quoted in 'Extracts' in J. Davy (ed.), *The Collected Works of Sir Humphry Davy* (1839–40), Vol. 8, 358.

6 Nothing tends so much to the advancement of knowledge as the application of a new instrument.

Elements of Chemical Philosophy (1812), in J. Davy (ed.), *The Collected Works of Sir Humphry Davy* (1839–40), Vol. 4, 37.

7 It is contrary to the usual order of things, that events so harmonious as those of the system of the world, should depend on such diversified agents as are supposed to exist in our artificial arrangements; and there is reason to anticipate a great reduction in the number of undecompounded bodies, and to expect that the analogies of nature will be found conformable to the refined operations of art. The more the phenomena of the universe are studied, the more distinct their connection appears, and the more simple their causes, the more magnificent their design, and the more wonderful the wisdom and power of their Author.

Elements of Chemical Philosophy (1812), in J. Davy (ed.), *The Collected Works of Sir Humphry Davy* (1839–40), Vol. 4, 42.

1 And by the influence of heat, light, and electrical powers, there is a constant series of changes [in animal and vegetal substances]; matter assumes new forms, the destruction of one order of beings tends to the conservation of another, solution and consolidation, decay and renovation, are connected, and whilst the parts of the system, continue in a state of fluctuation and change, the order and harmony of the whole remain unalterable.

The Elements of Agricultural Chemistry (1813), in J. Davy (ed.) *The Collected Works of Sir Humphry Davy* (1839–40), Vol. 7, 182.

2 But if the two countries or governments are at war, the men of science are not. That would, indeed be a civil war of the worst description: we should rather, through the instrumentality of the men of science soften the asperities of national hostility.

Davy's remarks to Thomas Poole on accepting Napoleon's prize for the best experiment on Galvanism.

Quoted in Gavin de Beer, *The Sciences were Never at War* (1960), 204.

3 To me there never has been a higher source of honour or distinction than that connected with advances in science. I have not possessed enough of the eagle in my character to make a direct flight to the loftiest altitudes in the social world; and I certainly never endeavored to reach those heights by using the creeping powers of the reptile, who in ascending, generally chooses the dirtiest path, because it is the easiest.

Consolations in Travel (1830), Dialogue 5, The Chemical Philosopher, 225.

4 There are very few persons who pursue science with true dignity.

Consolations in Travel (1830), Dialogue 5, The Chemical Philosopher, 226.

5 When he saw the minute globules of potassium burst through the crust of potash, and take fire as they entered the atmosphere, he could not contain his joy—he actually bounded about the room in ecstatic delight.

Edmund Davy (Davy's cousin). Quoted in *Memoirs of the Life of Sir Humphry Davy*, in J. Davy (ed.), *The Collected Works of Sir Humphry Davy* (1839–40), Vol. 1, 109.

6 By science calmed, over the peaceful soul,
Bright with eternal Wisdom's lucid ray,
Peace, meek of eye, extends her soft control,
And drives the puny Passions far away.

Memoirs of the Life of Sir Humphry Davy, in J. Davy (ed.), *The Collected Works of Sir Humphrey Davy.* (1839–40), Vol. 1, 26.

7 Oh, most magnificent and noble Nature!
Have I not worshipped thee with such a love
As never mortal man before displayed?
Adored thee in thy majesty of visible creation,
And searched into thy hidden and mysterious ways
As Poet, as Philosopher, as Sage?

A late fragment, probably written when he knew he was dying, in *Fragmentary Remains* (1858), 14.

8 ON DAVY Mr Humphry Davy is a lively and talented man, and a thorough chemist; but if I might venture to give an opinion . . . he is rather too lively to fill the Chair of the Royal Society with that degree of gravity it is most becoming to assume.

Sir Joseph Banks, President of the Royal Society, on Davy. Quoted in John Barrow, *Sketches of the Royal Society* (1849), 52.

1 ON DAVY The poetic beauty of Davy's mind never seems to have left him. To that circumstance I would ascribe the distinguishing feature in his character, and in his discoveries,—a vivid imagination sketching out new tracts in regions unexplored, for the judgement to select those leading to the recesses of abstract truth.

Davies Gilbert, in the Presidential Address to the Royal Society on Davy's Death, 1829. Quoted in J. Davy, *Fragmentary Remains of Sir Humphry Davy* (1858), 314.

2 ON DAVY Too many men have often seen
Their talents underrated;
But Davy knows that his have been
Duly *Appreeciated.*

Davy married Mrs Apreece, a rich widow, in 1812.

Maria Edgeworth to Margaret Ruxton. Quoted in T. E. Thorpe, *Humphry Davy, Poet and Philosopher* (1896), 165.

3 ON DAVY Sir Humphrey [sic] Davy
Abominated gravy.
He lived in the odium
Of having discovered sodium.

'Sir Humphrey [sic] Davy', E. C. Bentley, *Biography for Beginners* (1905).

4 ON DAVY Davy was the type of all the jumped-up second-raters of all time.

Luard, a school chemistry teacher, speaking in C. P. Snow, *The Search* (1932), 21.

5 ON DAVY Sir H. Davy's greatest discovery was Michael Faraday.

'Michael Faraday', in Paul Harvey (ed.), *The Oxford Companion to English Literature* (1932), 279.

Richard Dawkins 1941–

British evolutionary biologist and science writer

6 We are survival machines, robot vehicles blindly programmed to preserve the selfish molecules known as genes. This is a truth which still fills me with astonishment.

The Selfish Gene (1976), Preface.

7 Intelligent life on a planet comes of age when it first works out the reason for its own existence.

The Selfish Gene (1976), 1.

8 Reductionism is a dirty word, and a kind of 'holistier than thou' self-righteousness has become fashionable.

The Extended Phenotype: The Gene as the Unit of Selection (1982), 113.

9 We animals are the most complicated things in the known universe.

The Blind Watchmaker (1986), 1.

10 All appearances to the contrary, the only watchmaker in nature is the blind forces of physics, albeit deployed in a very special way. A true watchmaker has foresight: he designs his cogs and springs, and plans their interconnections, with a future purpose in his mind's eye. Natural selection, the blind, unconscious, automatic process which Darwin discovered, and which we now know is the explanation for the existence and apparently purposeful form of all life, has no purpose in mind. It has no mind and no mind's eye. It does not plan for the future. It has no vision, no foresight, no sight at all. If it can be said to play the role of watchmaker in nature, it is the *blind* watchmaker.

The Blind Watchmaker (1986), 5.

11 It is raining DNA outside.

The Blind Watchmaker (1986), 111.

12 Nature is not cruel, only pitilessly indifferent. This is one of the hardest lessons for humans to learn. We cannot admit that things might be neither good nor evil, neither cruel nor kind, but simply callous—indifferent to all suffering, lacking all purpose.

River Out of Eden: A Darwinian View of Life (1995), 112.

1 DNA neither cares nor knows. DNA just is. And we dance to its music.

River Out of Eden: A Darwinian View of Life (1995), 133.

2 It is grindingly, creakingly, crashingly obvious that, if Darwinism were really a theory of chance, it couldn't work. You don't need to be a mathematician or physicist to calculate that an eye or a haemoglobin molecule would take from here to infinity to self-assemble by sheer higgledy-piggledy luck. Far from being a difficulty peculiar to Darwinism, the astronomic improbability of eyes and knees, enzymes and elbow joints and all the other living wonders is precisely the problem that *any* theory of life must solve, and that Darwinism uniquely *does* solve. It solves it by breaking the improbability up into small, manageable parts, smearing out the luck needed, going round the back of Mount Improbable and crawling up the gentle slopes, inch by million-year inch. Only God would essay the mad task of leaping up the precipice in a single bound.

Climbing Mount Improbable (1996), 67–8.

3 We are going to die, and that makes us the lucky ones. Most people are never going to die because they are never going to be born. The potential people who could have been here in my place but who will in fact never see the light of day outnumber the sand grains of Arabia. Certainly those unborn ghosts include greater poets than Keats, scientists greater than Newton. We know this because the set of possible people allowed by our DNA so massively outnumbers the set of actual people. In the teeth of these stupefying odds it is you and I, in our ordinariness, that are here.

Unweaving the Rainbow (1998), 1.

John William Dawson 1820–99

Canadian geologist

4 The science of the earth . . . invites us to be present at the origin of things, and to enter into the very worship of the Creator.

The Story of the Earth and Man (1887), vi.

Gavin Rylands de Beer

1899–1972

British zoologist and morphologist

5 A living organism must be studied from two distinct aspects. One of these is the causal-analytic aspect which is so fruitfully applicable to ontogeny. The other is the historical descriptive aspect which is unravelling lines of phylogeny with ever-increasing precision. Each of these aspects may make suggestions concerning the possible significance of events seen under the other, but does not explain or translate them into simpler terms.

'Embryology and Evolution', in G. R. de Beer (ed.), *Evolution: Essays on Aspects of Evolutionary Biology presented to Professor E. S. Goodrich on his Seventieth Birthday* (1938), 76–7.

6 But science is the collection of nature's answers.

Attributed.

James Dunwoody Brownson De Bow 1820–67

American statistician and editor

7 Statistics are far from being the barren array of figures ingeniously and laboriously combined into columns and tables, which many persons are apt to suppose them. They constitute rather the ledger of a nation, in which, like the merchant in his books, the citizen can read, at one view, all of the results of a year or of a period of years, as compared with other periods, and deduce the profit or the loss which has been made, in morals, education, wealth or power.

Statistical View of the United States (1854), 9.

Louis Victor de Broglie see Broglie, Louis de

Julius Wilhelm Richard Dedekind 1831–1916

German mathematician

1 That which is provable, ought not to be believed in science without proof.

Was sind und was sollen die Zahlen? (1888), Preface, trans. Ivor Grattan-Guinness.

John Dee 1527–1608

British mathematician

2 O comfortable allurement, O ravishing perswasion, to deal with a Science, whose subject is so Auncient, so pure, so excellent, so surmounting all creatures . . . By *Numbers* propertie . . . we may . . . arise, clime, ascend, and mount up (with Speculative winges) in spirit, to behold in the Glas of creation, the *Forme* of *Formes*, the *Exemplar Number* of all things Numerable . . . Who can remaine, therefore, unpersuaded, to love, allow, and honor the excellent science of Arithmatike?

'Mathematicall Preface', in H. Billingsley, trans. *The Elements of Geometry of the most Aunceint Philosopher Euclide of Megara* (1570), in J. L. Heilbron, *Weighing Imponderables and Other Quantitative Science around 1800* (1993), 2.

Braswell D. Deen Jr. 1925–97

American judge

3 This monkey mythology of Darwin is the cause of permissiveness, promiscuity, pills, prophylactics, perversions, abortions, pornography, pollution, poisoning, and proliferation of crimes of all types.

Georgia Court of Appeals, 1981. Quoted in K. M. Pierce, 'Putting Darwin back in the Dock', *Time*, 16 March 1981, 51–2.

Joseph-Marie Degerando 1772–1842

French philosopher

4 In an age of egoism, it is so difficult to persuade man that of all studies, the most important is that of himself. This is because egoism, like all passions, is blind. The attention of the egoist is directed to the immediate needs of which his senses give notice, and cannot be raised to those reflective needs that reason discloses to us; his aim is satisfaction, not perfection. He considers only his individual self; his species is nothing to him. Perhaps he fears that in penetrating the mysteries of his being he will ensure his own abasement, blush at his discoveries, and meet his conscience. True philosophy, always at one with moral science, tells a different tale. The source of useful illumination, we are told, like that of lasting content, is in ourselves. Our insight depends above all on the state of our faculties; but how can we bring our faculties to perfection if we do not know their nature and their laws? The elements of happiness are the moral sentiments; but how can we develop these sentiments without considering the principle of our affections, and the means of directing them? We become better by studying ourselves; the man who thoroughly knows himself is the wise man. Such reflection on the nature of his being brings a man to a better awareness of all the bonds that unite us to our fellows, to the re-discovery at the inner root of his existence of that identity of common life actuating us all, to feeling the full force of that fine maxim of the ancients: 'I am a man, and nothing human is alien to me.'

Considérations sur les diverses méthodes à suivre dans l'observation des peuples sauvages (1800). *The Observation of Savage Peoples*, trans. F. C. T. Moore (1969), 61.

Henry Thomas De la Beche 1796–1855

British geologist

5 The complacent manner in which geologists have produced their theories has been extremely amusing; for often, with knowledge (and that frequently

inaccurate) not extending beyond a given province, they have described the formation of a world with all the detail and air of eye-witnesses. That much good ensues, and that the science is greatly advanced, by the collision of various theories, cannot be doubted. Each party is anxious to support opinions by facts. Thus, new countries are explored, and old districts re-examined; facts come to light that do not suit either party; new theories spring up; and, in the end, a greater insight into the real structure of the earth's surface is obtained.

Sections and Views Illustrative of Geological Phenomena (1830), iii.

1 Generally speaking, geologists seem to have been much more intent on making little worlds of their own, than in examining the crust of that which they inhabit. It would be much more desirable that facts should be placed in the foreground and theories in the distance, than that theories should be brought forward at the expense of facts. So that, in after times, when the speculations of the present day shall have passed away, from a greater accumulation of information, the facts may be readily seized and converted to account.

Sections and Views Illustrative of Geological Phenomena (1830), iv.

2 It surely can be no offence to state, that the progress of science has led to new views, and that the consequences that can be deduced from the knowledge of a hundred facts may be very different from those deducible from five. It is also possible that the facts first known may be the exceptions to a rule and not the rule itself, and generalisations from these first-known facts, though useful at the time, may be highly mischievous, and impede the progress of the science if retained when it has made some advance.

Sections and Views Illustrative of Geological Phenomena (1830), viii.

Max Delbrück 1850–1919

German chemist

3 With the sword of Science and the armour of Practice German beer will encircle the world.

Bayerische Bierbauer, 1884, 19, 312.

Jean André Deluc 1727–1817

Swiss geologist

4 It will be contributing to bring forward the moment in which, seeing clearer into the nature of things, and having learnt to distinguish real knowledge from what has only the appearance of it, we shall be led to seek for exactness in every thing.

'An Essay on Pyrometry and Areometry, and on Physical Measures in General', *Philosophical Transactions*, 1778, 68, 493.

5 We may observe in some of the abrupt grounds we meet with, sections of great masses of *strata*, where it is as easy to read the history of the sea, as it is to read the history of *Man* in the *archives* of any nation.

'Geological Letters, Addressed to Professor Blumenbach, Letter 2', *The British Critic*, 1794, 226.

6 I am here tracing the *History* of the *Earth* itself, from its own *Monuments*.

'Geological Letters Addressed to Professor Blumenbach, Letter 3', *The British Critic*, 1794, 598.

7 According to the conclusion of Dr. Hutton, and of many other geologists, our continents are of definite antiquity, they have been peopled we know not how, and mankind are wholly unacquainted with their origin. According to my conclusions drawn from the same source, that of *facts*, our continents are of such small antiquity, that the memory of the revolution which gave them birth must still be preserved among men; and thus we are led to seek in the book of Genesis the record of the history of the human race from its origin. Can any object of importance superior to this be found throughout the circle of natural science?

An Elementary Treatise on Geology (1809), 82.

Democritus c.460–370 BC

Greek mathematician and philosopher

1 The first principles of the universe are atoms and empty space. Everything else is merely thought to exist. The worlds are unlimited. They come into being and perish. Nothing can come into being from that which is not nor pass away into that which is not. Further, the atoms are unlimited in size and number, and they are borne along in the whole universe in a vortex, and thereby generate all composite things—fire, water, air, earth. For even these are conglomerations of given atoms. And it is because of their solidarity that these atoms are impassive and unalterable. The sun and the moon have been composed of such smooth and spherical masses [i.e. atoms], and so also the soul, which is identical with reason.

Diogenes Laertius IX, 44. Trans. R. D. Hicks (1925), Vol. 2, 453–5.

2 All things happen by virtue of necessity.

Diogenes Laertius IX, 45. Trans. R. D. Hicks (1925), Vol. 2, 455.

3 Democritus sometimes does away with what appears to the senses, and says that none of these appears according to truth but only according to opinion: the truth in real things is that there are atoms and void. 'By convention sweet', he says, 'by convention bitter, by convention hot, by convention cold, by convention colour: but in reality atoms and void.'

Sextus Empiricus, *Against the Professors*, 7. 135. In G. S. Kirk, J. E. Raven and M. Schofield (eds.), *The Presocratic Philosophers: A Critical History with a Selection of Texts* (1983), p. 410.

4 These differences, they say, are three: shape, arrangement, and position; because they hold that what is differs only in *contour*, *inter-contact*, *inclination*.

Quoted in Aristotle, *Metaphysics* A 4 985b 13–16. Trans. Hugh Tredennick (1933), Vol. 1, 31.

5 I would rather discover one scientific fact than become King of Persia.

Dionysius cited by Eusebius, *Preparation for the Gospel*, 14.27.4. Democritus fr. 118. Diels-Kranz.

Augustus De Morgan 1806–71

British mathematician

6 It was long before I got at the maxim, that in reading an old mathematician you will not read his riddle unless you plough with his heifer; you must see with his light, if you want to know how much he saw.

Letter to W. R. Hamilton, 27 January 1853. In R. P. Graves (ed.), *A Life of Sir W. R. Hamilton* (1889), Vol. 3, 438.

7 Great fleas have little fleas upon their backs to bite 'em,
And little fleas have lesser fleas, and so *ad infinitum*.
And the great fleas themselves, in turn have, greater fleas to go on;
While these again have greater still, and greater still, and so on.

A Budget of Paradoxes (1915), first published 1872, Vol. 2, 191.

8 We know that mathematicians care no more for logic than logicians for mathematics. The two eyes of science are mathematics and logic; the mathematical set puts out the logical eye, the logical set puts out the mathematical eye; each believing that it sees better with one eye than with two.

Note De Morgan had only one eye himself.

Book review in the *Athenaeum*, 1868, Vol. 2, 71–3.

Daniel C. Dennett 1942–

American philosopher

9 Human consciousness is just about the last surviving mystery. A mystery is a phenomenon that people don't know how to think about—yet. There have been other great mysteries: the mystery of the origin of the universe, the mystery of life and reproduction, the mystery of the design to be found in nature, the mysteries of time, space, and gravity. These were not just areas of scientific ignorance, but of utter bafflement and wonder. We do not yet have the final answers to any of the questions of cosmology and particle physics, molecular genetics and evolutionary theory, but we do know how to think about them. The mysteries haven't vanished, but they have been tamed. They no longer overwhelm our efforts to think about the phenomena, because now we know how to tell the misbegotten questions from the right questions, and even if we turn out to be dead wrong about some of the currently accepted answers, we know how to go about looking for better answers. With consciousness, however, we are still in a terrible muddle. Consciousness stands alone today as a topic that often leaves even the most sophisticated thinkers tongue-tied and confused. And, as with all the earlier mysteries, there are many who insist—and hope—that there will never be a demystification of consciousness.

Consciousness Explained (1991), 21–22.

Christine de Pizan c.1364–1429

Italian-born writer and poet

1 If it were customary to send daughters to school like sons, and if they were then taught the natural sciences, they would learn as thoroughly and understand the subtleties of all the arts and sciences as well as sons. And by chance there happen to be such women, for, as I touched on before, just as women have more delicate bodies than men, weaker and less able to perform many tasks, so do they have minds that are freer and sharper whenever they apply themselves.

The Book of the City of Ladies (1405), part I, section 27. Trans. Earl Jeffrey Richards (1982), 63.

Thomas De Quincey 1785–1859

British essayist and critic

2 Mathematics has not a foot to stand upon which is not purely metaphysical.

'Kant in His Miscellaneous Essays', *Blackwood's Magazine*, 1830, **28**, 244–68.

René Du Perron Descartes 1596–1650

French philosopher and mathematician

3 As I considered the matter carefully it gradually came to light that all those matters only were referred to mathematics in which order and measurements are investigated, and that it makes no difference whether it be in numbers, figures, stars, sounds or any other object that the question of measurement arises. I saw consequently that there must be some general science to explain that element as a whole which gives rise to problems about order and measurement, restricted as these are to no special subject matter. This, I perceived was called 'universal mathematics'.

Rules for the Direction of the Mind (written 1628). In Haldane and Ross (eds.) *The Philosophical Works of Descartes* (1973), Vol. 1, 13.

4 I suppose the body to be just a statue or a machine made of earth.

The World and Other Writings (1633), trans. and ed. Stephen Gaukroger (1998), 99.

5 I was then in Germany, where I had been drafted because of the wars that are still going on there, and as I was returning to the army from the emperor's coronation, the arrival of winter delayed me in quarters where, finding no company to distract me and, luckily, having no cares or passions to trouble me, I used to spend the whole day alone in a room, that was heated by a stove, where I had plenty of time to concentrate on my own thoughts.

Discourse on Method in *Discourse on Method and Related Writings* (1637), trans. Desmond M. Clarke, Penguin edition (1999), Part 2, 11.

1 But from the time I was in college I learned that there is nothing one could imagine which is so strange and incredible that it was not said by some philosopher; and since that time, I have recognized through my travels that all those whose views are different from our own are not necessarily, for that reason, barbarians or savages, but that many of them use their reason either as much as or even more than we do. I also considered how the same person, with the same mind, who was brought up from infancy either among the French or the Germans, becomes different from what they would have been if they had always lived among the Chinese or among the cannibals, and how, even in our clothes fashions, the very thing that we liked ten years ago, and that we may like again within the next ten years, appears extravagant and ridiculous to us today. Thus our convictions result from custom and example very much more than from any knowledge that is certain . . . truths will be discovered by an individual rather than a whole people.

Discourse on Method in *Discourse on Method and Related Writings* (1637), trans. Desmond M. Clarke, Penguin edition (1999), Part 2, 14–5.

2 I believed that, instead of the multiplicity of rules that comprise logic, I would have enough in the following four, as long as I made a firm and steadfast resolution never to fail to observe them.

The first was never to accept anything as true if I did not know clearly that it was so; that is, carefully to avoid prejudice and jumping to conclusions, and to include nothing in my judgments apart from whatever appeared so clearly and distinctly to my mind that I had no opportunity to cast doubt upon it.

The second was to subdivide each of the problems I was about to examine into as many parts as would be possible and necessary to resolve them better.

The third was to guide my thoughts in an orderly way by beginning, as if by steps, to knowledge of the most complex, and even by assuming an order of the most complex, and even by assuming an order among objects in cases where there is no natural order among them.

And the final rule was: in all cases, to make such comprehensive enumerations and such general reviews that I was certain not to omit anything.

The long chains of inferences, all of them simple and easy, that geometers normally use to construct their most difficult demonstrations had given me an opportunity to think that all the things that can fall within the scope of human knowledge follow from each other in a similar way, and as long as one avoids accepting something as true which is not so, and as long as one always observes the order required to deduce them from each other, there cannot be anything so remote that it cannot be reached nor anything so hidden that it cannot be uncovered.

Discourse on Method in *Discourse on Method and Related Writings* (1637), trans. Desmond M. Clarke, Penguin edition (1999), Part 2, 16.

3 Finally, since I thought that we could have all the same thoughts, while asleep, as we have while we are awake, although none of them is true at that time, I decided to pretend that nothing that ever entered my mind was any more true than the illusions of my dreams. But I noticed, immediately afterwards, that while I thus wished to think that everything was false, it was necessarily the case that I, who was thinking this, was something. When I noticed that this truth 'I think, therefore I am' was so firm and certain that all the most extravagant assumptions of the sceptics were unable to shake it, I judged that I could accept it without scruple as the first principle of the philosophy for which I was searching. Then, when I was examining what I was, I realized that I could pretend that I had no body, and that there was no world nor any place in which I was present, but I could not pretend in the same way that I did not exist. On the contrary, from the very fact that I was thinking of doubting the truth of other things, it followed very evidently and very certainly that I existed; whereas if I merely ceased to think, even if all the rest of what I had ever imagined were true, I would have no reason to believe that I existed. I knew from this that I was a substance, the whole essence or nature of which was to think and which, in order to exist, has no need of any place and does not depend on anything material. Thus this self—that is, the soul by which I am what I am—is completely distinct from the body and is even easier to know than it, and even if the body did not exist the soul would still be everything that it is.

Discourse on Method in *Discourse on Method and Related Writings* (1637), trans. Desmond M. Clarke, Penguin edition (1999), Part 4, 24–5.

1 After that, I thought about what a proposition generally needs in order to be true and certain because, since I had just found one that I knew was such, I thought I should also know what this certainty consists in. Having noticed that there is nothing at all in the proposition 'I think, therefore I am' [*cogito ergo sum*] which convinces me that I speak the truth, apart from the fact that I see very clearly that one has to exist in order to think, I judged that I could adopt as a general rule that those things we conceive very clearly and distinctly are all true. The only outstanding difficulty is in recognizing which ones we conceive distinctly.

Discourse on Method in *Discourse on Method and Related Writings* (1637), trans. Desmond M. Clarke, Penguin edition (1999), Part 4, 25.

2 Then I had shown, in the same place, what the structure of the nerves and muscles of the human body would have to be in order for the animal spirits in the body to have the power to move its members, as one sees when heads, soon after they have been cut off, still move and bite the ground even though they are no longer alive; what changes must be made in the brain to cause waking, sleep and dreams; how light, sounds, odours, tastes, warmth and all the other qualities of external objects can impress different ideas on it through the senses; how hunger, thirst, and the other internal passions can also send their ideas there; what part of the brain should be taken as 'the common sense', where these ideas are received; what should be taken as the memory, which stores the ideas, and as the imagination, which can vary them in different ways and compose new ones and, by the same means, distribute the animal spirits to the muscles, cause the limbs of the body to move in as many different ways as our own bodies can move without the will directing them, depending on the objects that are present to the senses and the internal passions in the body. This will not seem strange to those who know how many different automata or moving machines can be devised by human ingenuity, by using only very few pieces in comparison with the larger number of bones, muscles, nerves, arteries, veins and all the other parts in the body of every animal. They will think of this body like a machine which, having been made by the hand of God, is incomparably better structured than any machine that could be invented by human beings, and contains many more admirable movements.

Discourse on Method in *Discourse on Method and Related Writings* (1637), trans. Desmond M. Clarke, Penguin edition (1999), Part 5, 39–40.

1 I specifically paused to show that, if there were such machines with the organs and shape of a monkey or of some other non-rational animal, we would have no way of discovering that they are not the same as these animals. But if there were machines that resembled our bodies and if they imitated our actions as much as is morally possible, we would always have two very certain means for recognizing that, none the less, they are not genuinely human. The first is that they would never be able to use speech, or other signs composed by themselves, as we do to express our thoughts to others. For one could easily conceive of a machine that is made in such a way that it utters words, and even that it would utter some words in response to physical actions that cause a change in its organs—for example, if someone touched it in a particular place, it would ask what one wishes to say to it, or if it were touched somewhere else, it would cry out that it was being hurt, and so on. But it could not arrange words in different ways to reply to the meaning of everything that is said in its presence, as even the most unintelligent human beings can do. The second means is that, even if they did many things as well as or, possibly, better than any one of us, they would infallibly fail in others. Thus one would discover that they did not act on the basis of knowledge, but merely as a result of the disposition of their organs. For whereas reason is a universal instrument that can be used in all kinds of situations, these organs need a specific disposition for every particular action.

Discourse on Method in *Discourse on Method and Related Writings* (1637), trans. Desmond M. Clarke, Penguin edition (1999), Part 5, 40.

2 *Et ainsi nous rendre maîtres et possesseurs de la nature.*

And thereby make ourselves, as it were, the lords and masters of nature.

Discourse on Method in *Discourse on Method and Related Writings* (1637), trans. Desmond M. Clarke, Penguin edition (1999), Part 6, 44.

3 Even the mind depends so much on temperament and the disposition of one's bodily organs that, if it is possible to find a way to make people generally more wise and more skilful than they have been in the past, I believe that we should look for it in medicine. It is true that medicine as it is currently practiced contains little of much use.

Discourse on Method in *Discourse on Method and Related Writings* (1637), trans. Desmond M. Clarke, Penguin edition (1999), Part 6, 44.

4 When someone says 'I am thinking, therefore I am, or I exist', he does not deduce existence from thought by means of a syllogism, but recognises it as something self-evident by a simple intuition of the mind. This is clear from the fact that if he were deducing it by means of a syllogism, he would have to have had previous knowledge of the major premiss 'Everything which thinks is, or exists'; yet in fact he learns it from experiencing in his own case that it is impossible that he should think without existing. It is in the nature of our mind to construct general propositions on the basis of our knowledge of particular ones.

Author's Replies to the Second set of Objections to Meditations on the First Philosophy (1641), in *The Philosophical Writings of Descartes* (1985), trans. J. Cottingham, R. Stoothoff and D. Murdoch, Vol. 2, 100.

5 The nature of matter, or body considered in general, consists not in its being something which is hard or heavy or coloured, or which affects the senses in any way, but simply in its being something which is extended in length, breadth and depth.

Principles of Philosophy (1644), trans. V. R. and R. P. Miller (1983), 40.

6 It must not be thought that it is ever possible to reach the interior earth by any perseverance in mining: both because the exterior earth is too thick, in comparison with human strength; and especially because of the intermediate waters, which would gush forth with greater impetus, the deeper the place in which their veins were first opened; and which would drown all miners.

Principles of Philosophy (1644), trans. V. R. and R. P. Miller (1983), 217–8.

1 And there are absolutely no judgments (or rules) in Mechanics which do not also pertain to Physics, of which Mechanics is a part or type: and it is as natural for a clock, composed of wheels of a certain kind, to indicate the hours, as for a tree, grown from a certain kind of seed, to produce the corresponding fruit. Accordingly, just as when those who are accustomed to considering automata know the use of some machine and see some of its parts, they easily conjecture from this how the other parts which they do not see are made: so, from the perceptible effects and parts of natural bodies, I have attempted to investigate the nature of their causes and of their imperceptible parts.

Principles of Philosophy (1644), trans. V. R. and R. P. Miller (1983), 285–6.

James Dewar 1842–1923

British chemist and physicist

2 ON DEWAR Professor Dewar
Is a better man than *you* are,
None of you asses
Can condense gases.

E. C. Bentley, *Biography for Beginners* (1905)

Michael J. S. Dewar 1918–97

British organic chemist

3 An aromatic compound may be defined as a cyclic compound with a large resonance energy where all the annular atoms take part in a single conjugated system.

Electronic Theory of Organic Chemistry (1949), 160.

John Dewey 1859–1952

American philosopher

4 Unless our laboratory results are to give us artificialities, mere scientific curiosities, they must be subjected to interpretation by gradual re-approximation to conditions of life.

'Psychology and Social Practice', *The Psychological Review*, 1900, 7, 119.

5 Every great advance in science has issued from a new audacity of imagination. What are now working conceptions, employed as a matter of course because they have withstood the tests of experiment and have emerged triumphant, were once speculative hypotheses.

The Quest for Certainty: A Study of the Relation of Knowledge and Action (1929), 294.

6 It has become a cheap intellectual pastime to contrast the infinitesimal pettiness of man with the vastnesses of the stellar universes. Yet all such comparisons are illicit. We cannot compare existence and meaning; they are disparate. The characteristic life of a man is itself the meaning of vast stretches of existences, and without it the latter have no value or significance. There is no common measure of physical existence and conscious experience because the latter is the only measure there is of the former. The significance of being, though not its existence, is the emotion it stirs, the thought it sustains.

Philosophy and Civilization (1931), reprinted in David Sidorsky (ed.), *John Dewey: The Essential Writings* (1977), 7.

7 The moment philosophy supposes it can find a final and comprehensive solution, it ceases to be inquiry and becomes either apologetics or propaganda.

Logic (1938), 42.

Charles John Huffam Dickens 1812–70

British novelist

8 'Now, what I want is, Facts. Teach these boys and girls nothing but Facts. Facts alone are wanted in life. Plant nothing else. And root out everything else. You can only form the minds of reasoning animals upon Facts: nothing else will ever be of any service to them. This is the principle on which I bring up my own children, and this is the principle on which I bring up these children. Stick to Facts, sir!'

Hard Times (1854). Penguin edition, ed. Kate Flint (1995), 1.

Emily Elizabeth Dickinson
1830–86

American poet

1 'Arcturus' is his other name—
I'd rather call him 'Star.'
It's very mean of Science
To go and interfere!

'Arcturus' (*c.*1859). *The Complete Poems of Emily Dickinson*, ed. Thomas H. Johnson (1970), 36.

Denis Diderot 1713–84

French philosopher and man of letters

2 I picture the vast realm of the sciences as an immense landscape scattered with patches of dark and light. The goal towards which we must work is either to extend the boundaries of the patches of light, or to increase their number. One of these tasks falls to the creative genius; the other requires a sort of sagacity combined with perfectionism.

Thoughts on the Interpretation of Nature and Other Philosophical Works (1753/4), ed. D. Adams (1999), Section XIV, 42.

3 We have three approaches at our disposal: the observation of nature, reflection, and experimentation. Observation serves to assemble the data, reflection to synthesise them and experimentation to test the results of the synthesis. The observation of nature must be assiduous, just as reflection must be profound, and experimentation accurate. These three approaches are rarely found together, which explains why creative geniuses are so rare.

Thoughts on the Interpretation of Nature and Other Philosophical Works (1753/4), ed. D. Adams (1999), section XV, 42.

4 Just as in the animal and vegetable kingdoms, an individual comes into being, so to speak, grows, remains in being, declines and passes on, will it not be the same for entire species? If our faith did not teach us that animals left the Creator's hands just as they now appear and, if it were permitted to entertain the slightest doubt as to their beginning and their end, may not a philosopher, left to his own conjectures, suspect that, from time immemorial, animal life had its own constituent elements, scattered and intermingled with the general body of matter, and that it happened when these constituent elements came together because it was possible for them to do so; that the embryo formed from these elements went through innumerable arrangements and developments, successively acquiring movement, feeling, ideas, thought, reflection, consciousness, feelings, emotions, signs, gestures, sounds, articulate sounds, language, laws, arts and sciences; that millions of years passed between each of these developments, and there may be other developments or kinds of growth still to come of which we know nothing; that a stationary point either has been or will be reached; that the embryo either is, or will be, moving away from this point through a process of everlasting decay, during which its faculties will leave it in the same way as they arrived; that it will disappear for ever from nature—or rather, that it will continue to exist there, but in a form and with faculties very different from those it displays at this present point in time? Religion saves us from many deviations, and a good deal of work. Had religion not enlightened us on the origin of the world and the universal system of being, what a multitude of different hypotheses we would have been tempted to take as nature's secret! Since these hypotheses are all equally wrong, they would all have seemed almost equally plausible. The question of why anything exists is the most awkward that philosophy can raise—and Revelation alone provides the answer.

Thoughts on the Interpretation of Nature and Other Philosophical Works (1753/4), ed. D. Adams (1999), Section LVIII, 75–6.

1 Man is merely a frequent effect, a monstrosity is a rare one, but both are equally natural, equally inevitable, equally part of the universal and general order. And what is strange about that? All creatures are involved in the life of all others, consequently every species . . . all nature is in a perpetual state of flux. Every animal is more or less a human being, every mineral more or less a plant, every plant more or less an animal . . . There is nothing clearly defined in nature.

D'Alembert's Dream (1769), in *Rameau's Nephew and D'Alembert's Dream*, trans. Leonard Tancock (Penguin edition 1966), 181.

2 To be born, to live and to die is merely to change forms . . . And what does one form matter any more than another? . . . Each form has its own sort of happiness and unhappiness. From the elephant down to the flea . . . from the flea down to the sensitive and living molecule which is the origin of all, there is not a speck in the whole of nature that does not feel pain or pleasure.

D'Alembert's Dream (1769), in *Rameau's Nephew and D'Alembert's Dream*, trans. Leonard Tancock (Penguin edition 1966), 182.

3 In England, philosophers are honoured, respected; they rise to public offices, they are buried with the kings . . . In France warrants are issued against them, they are persecuted, pelted with pastoral letters: Do we see that England is any the worse for it?

'Introduction aux Grands Principes ou Réception d'un Philosophe', in J. Assézat (ed.), *Oeuvres Complètes* (1875–7), Vol. 2, 80.

4 Without the English, reason and philosophy would still be in the most despicable infancy in France.

'Essai sur les Études en Russie', in J. Assézat (ed.), *Oeuvres Complètes* (1875–7), Vol. 3, 416. Quoted in Peter Gay, *The Enlightenment* (1966), Vol. 1, 12.

Eduard Jan Dijksterhuis

1892–1965

Dutch historian of science

5 The mechanization of the world picture.

The Mechanization of the World Picture, trans. C. Dikshoorn (1961), 39.

Paul Adrien Maurice Dirac

1902–84

British physicist

See ***Alvarez, Luis*** *10:1*

6 The only object of theoretical physics is to calculate results that can be compared with experiment . . . it is quite unnecessary that any satisfactory description of the whole course of the phenomena should be given.

The Principles of Quantum Mechanics (1930), 7.

7 When we make the photon meet a tourmaline crystal, we are subjecting it to an observation. We are observing whether it is polarised parallel or perpendicular to the optic axis. The effect of making the observation is to force the photon entirely into the state of perpendicular polarisation. It has to make a sudden jump from being partly in each of these two states to being entirely in one or other of them. Which of the two states it will jump into cannot be predicted, but is governed only by probability laws. If it jumps into the perpendicular state it passes through the crystal and appears on the other side preserving this state of polarisation.

The Principles of Quantum Mechanics (1930).

8 It is more important to have beauty in one's equations than to have them fit experiment . . . It seems that if one is working from the point of view of getting beauty in one's equations, and if one has really a sound insight, one is on a sure line of progress. If there is not complete agreement between the results of one's work and experiment, one should not allow oneself to be too discouraged, because the discrepancy may well be due to minor features that are not properly taken into account and that will get cleared up with further developments of the theory.

'The Evolution of the Physicist's Picture of Nature', *Scientific American*, May 1963, 208, 47.

1 Theoretical physicists accept the need for mathematical beauty as an act of faith . . . For example, the main reason why the theory of relativity is so universally accepted is its mathematical beauty.

'Methods in Theoretical Physics', *From A Life of Physics: Evening Lectures at the International Centre for Theoretical Physics, Trieste, Italy. A Special Supplement of the IAEA Bulletin* (1968), 22.

2 I found the best ideas usually came, not when one was actively striving for them, but when one was in a more relaxed state . . . I used to take long solitary walks on Sundays, during which I tended to review the current situation in a leisurely way. Such occasions often proved fruitful, even though (or perhaps, because) the primary purpose of the walk was relaxation and not research.

'Methods in Theoretical Physics', *From A Life of Physics: Evening Lectures at the International Centre for Theoretical Physics, Trieste, Italy. A Special Supplement of the IAEA Bulletin* (1968), 24.

Benjamin Disraeli

(Lord Beaconsfield) 1804–81

British Prime Minister and novelist

3 What Art was to the ancient world, Science is to the modern: the distinctive faculty. In the minds of men the useful has succeeded to the beautiful. Instead of the city of the Violet Crown, a Lancashire village has expanded into a mighty region of factories and warehouses. Yet, rightly understood, Manchester is as great a human exploit as Athens.

Coningsby or The New Generation (1844), Vol. 2, Book 4, Ch.1, 2.

4 You know, all is development. The principle is perpetually going on. First, there was nothing, then there was something; then—I forget the next—I think there were shells, then fishes; then we came—let me see—did we come next? Never mind that; we came at last. And at the next change there will be something very superior to us—something with wings. Ah! That's it: we were fishes, and I believe we shall be crows.

Tancred: or, The New Crusade (1847), 124.

5 What is the question now placed before society with a glib assurance the most astounding? The question is this—Is man an ape or an angel? My lord, I am on the side of the angels.

'Church and Queen', Five Speeches Delivered by the Rt. Hon. B. Disraeli, M.P. 1860–1864 (1865), 78.

6 There are three kinds of lies: lies, damned lies, and statistics.

Attributed to Disraeli in Mark Twain's *Autobiography* (1924), Vol. 1, 246. There is no further evidence beyond this attribution that Disraeli made this statement.

Carl Djerassi 1923–

Austrian-born American chemist and inventor

7 The vast outpourings of publications by Professor Djerassi and his cohorts marks him as one of the most prolific scientific writers of our day . . . a plot of N, the papers published by Professor Djerassi in a given year, against T, the year (starting with 1945, $T = 0$) gives a good straight-line relationship. This line follows the equation $N = 2.413T + 1.690$ . . . Assuming that the inevitable inflection point on the logistic curve is still some 10 years away, this equation predicts (a) a total of about 444 papers by the end of this year, (b) the average production of one paper per week or more every year beginning in 1966, and (c) the winning of the all-time productivity world championship in 10 years from now, in 1973. In that year Professor Djerassi should surpass the record of 995 items held by . . .

Steroids Made it Possible (1990), 11–12.

1 I have reviewed this work elsewhere under the title 'Natural Products Chemistry 1950 to 1980—A Personal View.' It is with some relish that I recall the flood of reprint requests prompted by the following footnote on the title page: 'Selected personal statements by the author were removed by the editor without Professor Djerassi's consent. An uncensored version of this paper can be obtained by writing to Professor C. Djerassi'.

Steroids Made it Possible (1990), 14.

Theodosius Dobzhansky

1900–75

Russian-born American geneticist

2 Mutations and chromosomal changes arise in every sufficiently studied organism with a certain finite frequency, and thus constantly and unremittingly supply the raw materials for evolution. But evolution involves something more than origin of mutations. Mutations and chromosomal changes are only the first stage, or level, of the evolutionary process, governed entirely by the laws of the physiology of individuals. Once produced, mutations are injected in the genetic composition of the population, where their further fate is determined by the dynamic regularities of the physiology of populations. A mutation may be lost or increased in frequency in generations immediately following its origin, and this (in the case of recessive mutations) without regard to the beneficial or deleterious effects of the mutation. The influences of selection, migration, and geographical isolation then mold the genetic structure of populations into new shapes, in conformity with the secular environment and the ecology, especially the breeding habits, of the species. This is the second level of the evolutionary process, on which the impact of the environment produces historical changes in the living population.

Genetics and Origin of Species (1937), 13.

3 Judged superficially, a progressive saturation of the germ plasm of a species with mutant genes a majority of which are deleterious in their effects is a destructive process, a sort of deterioration of the genotype which threatens the very existence of the species and can finally lead only to its extinction. The eugenical Jeremiahs keep constantly before our eyes the nightmare of human populations accumulating recessive genes that produce pathological effects when homozygous. These prophets of doom seem to be unaware of the fact that wild species in the state of nature fare in this respect no better than man does with all the artificality of his surroundings, and yet life has not come to an end on this planet. The eschatological cries proclaiming the failure of natural selection to operate in human populations have more to do with political beliefs than with scientific findings.

Genetics and Origin of Species (1937), 126.

4 Evolutionary plasticity can be purchased only at the ruthlessly dear price of continuously sacrificing some individuals to death from unfavourable mutations. Bemoaning this imperfection of nature has, however, no place in a scientific treatment of this subject.

Genetics and the Origin of Species (1937), 127.

5 In its essence, the theory of natural selection is primarily an attempt to give an account of the probable mechanism of the origin of the adaptations of the organisms to their environment, and only secondarily an attempt to explain evolution at large. Some modern biologists seem to believe that the word 'adaptation' has teleological connotations, and should therefore be expunged from the scientific lexicon. With this we must emphatically disagree. That adaptations exist is so evident as to be almost a truism, although this need not mean that ours is the best of all possible worlds. A biologist has no right to close his eyes to the fact that the precarious balance between a living being and its environment must be preserved by some mechanism or mechanisms if life is to endure.

Genetics and Origin of Species (1937), 150.

1 No evidence is powerful enough to force acceptance of a conclusion that is emotionally distasteful.

'Review of *Evolution, Creation and Science* by F. L. Marsh', *American Naturalist*, 79, 1945, 75.

2 The foundations of population genetics were laid chiefly by mathematical deduction from basic premises contained in the works of Mendel and Morgan and their followers. Haldane, Wright, and Fisher are the pioneers of population genetics whose main research equipment was paper and ink rather than microscopes, experimental fields, *Drosophila* bottles, or mouse cages. Theirs is theoretical biology at its best, and it has provided a guiding light for rigorous quantitative experimentation and observation.

'A Review of Some Fundamental Concepts and Problems of Population Genetics', *Cold Spring Harbor Symposia on Quantitative Biology*, 1955, 20, 13–14.

3 The process of mutation is the only known source of the raw materials of genetic variability, and hence of evolution. It is subject to experimental study, and considerable progress has been accomplished in this study in recent years. An apparent paradox has been disclosed. Although the living matter becomes adapted to its environment through formation of superior genetic patterns from mutational components, the process of mutation itself is not adaptive. On the contrary, the mutants which arise are, with rare exceptions, deleterious to their carriers, at least in the environments which the species normally encounters. Some of them are deleterious apparently in all environments. Therefore, the mutation process alone, not corrected and guided by natural selection, would result in degeneration and extinction rather than in improved adaptedness.

'On Methods of Evolutionary Biology and Anthropology', *American Scientist*, 1957, 45, 385.

4 Nature's stern discipline enjoins mutual help at least as often as warfare. The fittest may also be the gentlest.

Mankind Evolving (1962), 134.

5 Genetics is the first biological science which got in the position in which physics has been in for many years. One can justifiably speak about such a thing as theoretical mathematical genetics, and experimental genetics, just as in physics. There are some mathematical geniuses who work out what to an ordinary person seems a fantastic kind of theory. This fantastic kind of theory nevertheless leads to experimentally verifiable prediction, which an experimental physicist then has to test the validity of. Since the times of Wright, Haldane, and Fisher, evolutionary genetics has been in a similar position.

Oral history memoir. Columbia University, Oral History Research Office, New York, 1962. Quoted in William B. Provine, *Sewall Wright and Evolutionary Biology* (1989), 277.

6 Nothing in biology makes sense, except in the light of evolution.

'Nothing in Biology makes Sense except in the Light of Evolution', *The American Biology Teacher*, March 1973, 125.

Theodosius Dobzhansky 1900–75 and Gordon Allen 1919–

Russian-born American geneticist and American statistician

1 The frequent allegation that the selective processes in the human species are no longer 'natural' is due to persistence of the obsolete nineteenth-century concept of 'natural' selection. The error of this view is made clear when we ask its proponents such questions as, why should the 'surviving fittest' be able to withstand cold and inclement weather without the benefit of fire and clothing? Is it not ludicrous to expect selection to make us good at defending ourselves against wild beasts when wild beasts are getting to be so rare that it is a privilege to see one outside of a zoo? Is it necessary to eliminate everyone who has poor teeth when our dentists stand ready to provide us with artificial ones? Is it a great virtue to be able to endure pain when anaesthetics are available?

'Does Natural Selection Continue to Operate in Modern Mankind?', *American Anthropologist*, 1956, **58**, 595.

2 The outlook seems grim. Natural selection under civilized conditions may lead mankind to evolve towards a state of genetic overspecialization for living in gadget-ridden environments. It is certainly up to man to decide whether this direction of his evolution is or is not desirable. If it is not, man has, or soon will have, the knowledge requisite to redirect the evolution of his species pretty much as he sees fit. Perhaps we should not be too dogmatic about this choice of direction. We may be awfully soft compared to paleolithic men when it comes to struggling, unaided by gadgets, with climatic difficulties and wild beasts. Most of us feel most of the time that this is not a very great loss. If our remote descendants grow to be even more effete than we are, they may conceivably be compensated by acquiring genotypes conducive to kindlier dispositions and greater intellectual capacities than those prevalent in mankind today.

'Does Natural Selection Continue to Operate in Modern Mankind?', *American Anthropologist*, 1956, **58**, 599.

Charles Lutwidge Dodgson see Carroll, Lewis

William Richard Shaboe Doll 1912–2005 and Austin Bradford Hill 1897–1991

British epidemiologist, British medical statistician

3 The risk of developing carcinoma of the lung increases steadily as the amount smoked increases. If the risk among non-smokers is taken as unity and the resulting ratios in the three age groups in which a large number of patients were interviewed (ages 45 to 74) are averaged, the relative risks become 6, 19, 26, 49, and 65 when the number of cigarettes smoked a day are 3, 10, 20, 35, and, say, 60—that is, the mid-points of each smoking group. In other words, on the admittedly speculative assumptions we have made, the risk seems to vary in approximately simple proportion with the amount smoked.

'Smoking and Carcinoma of the Lung', *British Medical Journal*, 1950, **ii**, 746.

John Donne 1572–1631

British poet and divine

4 On a huge hill,
Cragged, and steep, Truth stands, and hee that will
Reach her, about must, and about must goe.

Satyre, III, l. 79–81. *The Works of John Donne* (Wordsworth edition 1994), 113.

5 And new philosophy calls all in doubt,
The Element of fire is quite put out;
The Sun is lost, and th'earth, and no mans wit
Can well direct him where to look for it.
And freely men confesse that this world's spent,

When in the Planets, and the Firmament
They seeke so many new; and then see that this
Is crumbled out againe to his Atomies.
'Tis all in pieces, all cohaerence gone;
All just supply, and all Relation;
Prince, Subject, Father, Sonne, are things forgot,
For every man alone thinkes he hath got
To be a phoenix, and that then can bee
None of that kinde, of which he is, but hee.

An Anatomie of the World, l. 205–18. *The Works of John Donne* (Wordsworth edition 1994), 177.

1 Poore soule, in this thy flesh what do'st thou know?
Thou know'st thy selfe so little, as thou know'st not.
How thou did'st die, nor how thou wast begot.
Thou neither know'st how thou at first camest in,
Nor how thou took'st the poyson of mans sin.
Nor dost thou, (though thou know'st, that thou art so)
By what way thou art made immortall, know.
Thou art too narrow, wretch, to comprehend
Even thy selfe; yea though thou wouldst but bend
To know thy body. Have not all soules thought
For many ages, that our body'is wrought
Of Ayre, and Fire, and other Elements?
And now they thinke of new ingredients,
And one soule thinkes one, and another way
Another thinkes, and 'tis an even lay.
Knowst thou but how the stone doth enter in
The bladder's Cave, and never breake the skin?
Knowst thou how blood, which to the hart doth flow,
Doth from one ventricle to th'other go?
And for the putrid stuffe, which thou dost spit,
Knowst thou how thy lungs have attracted it?
There are no passages, so that there is
(For aught thou knowst) piercing of substances.
And of those many opinions which men raise
Of Nailes and Haires, dost thou know which to praise?
What hope have we to know our selves, when wee
Know not the least things, which for our use bee?

Of the Progresse of the Soule. The Second Anniversarie, l. 254–280. *The Works of John Donne* (Wordsworth edition 1994), 196–7.

Margaret Mary Douglas 1921–

British anthropologist

2 Where there is dirt there is system. Dirt is the by-product of a systematic ordering and classification of matter.

Purity and Danger (1966), 35.

Arthur Conan Doyle 1859–1930

British doctor and writer

3 When you have eliminated the impossible, whatever remains, *however improbable*, must be the truth.

The Sign of Four (1890), Chapter 6. In *The Complete Sherlock Holmes* (Penguin edition 1981), 111.

4 I could not help laughing at the ease with which he explained his process of deduction. 'When I hear you give your reasons,' I remarked, 'the thing always appears to me to be so ridiculously simple that I could easily do it myself, though at each successive instance of your reasoning I am baffled, until you explain your process. And yet I believe that my eyes are as good as yours.'

'Quite so,' he answered, lighting a cigarette, and throwing himself down into an arm-chair. 'You see, but you do not observe. The distinction is clear. For example, you have frequently seen the steps which lead up from the hall to this room.'

'Frequently.'
'How often?'
'Well, some hundreds of times.'
'Then how many are there?'
'How many! I don't know.'
'Quite so! You have not observed. And yet you have seen. That is just my point. Now, I know that there seventeen steps, because I have both seen and observed.'

'A Scandal in Bohemia', *The Adventures of Sherlock Holmes* (1891). In *The Complete Sherlock Holmes* (Penguin edition 1981), 162–3.

1 'I had,' said he, 'come to an entirely erroneous conclusion which shows, my dear Watson, how dangerous it always is to reason from insufficient data.'

'The Adventure of the Speckled Band', *The Adventures of Sherlock Holmes* (1891). In *The Complete Sherlock Holmes* (Penguin edition 1981), 272.

2 ROSS: Is there any point to which you would wish to draw my attention?
HOLMES: To the curious incident of the dog in the night-time.
ROSS: The dog did nothing in the night-time.
HOLMES: That was the curious incident.

'Silver Blaze', *The Memoirs of Sherlock Holmes* (1894). In *The Complete Sherlock Holmes* (Penguin edition 1981), 347.

3 [Holmes]: The temptation to form premature theories upon insufficient data is the bane of our profession.

The Valley of Fear (1914–15), Chapter 2. In *The Complete Sherlock Holmes* (Penguin edition 1981), 779.

4 'There's no need for fiction in medicine,' remarks Foster . . . 'for the facts will always beat anything you fancy.'

'A Medical Document', in *Round the Red Lamp* (1894), 215.

5 'Men die of the diseases which they have studied most,' remarked the surgeon, snipping off the end of a cigar with all his professional neatness and finish. 'It's as if the morbid condition was an evil creature which, when it found itself closely hunted, flew at the throat of its pursuer. If you worry the microbes too much they may worry you. I've seen cases of it, and not necessarily in microbic diseases either. There was, of course, the well-known instance of Liston and the aneurism; and a dozen others that I could mention.'

'The Surgeon Talks', in *Round the Red Lamp* (1894), 316.

John William Draper 1811–82

American physician and amateur astronomer

6 As to Science, she has never sought to ally herself to civil power. She has never attempted to throw odium or inflict social ruin on any human being. She has never subjected any one to mental torment, physical torture, least of all to death, for the purpose of upholding or promoting her ideas. She presents herself unstained by cruelties and crimes. But in the Vatican—we have only to recall the Inquisition—the hands that are now raised in appeals to the Most Merciful are crimsoned. They have been steeped in blood!

History of the Conflict between Religion and Science (1875), xi.

Henry Drummond 1851–97

British theologian and evangelist

7 No man can run up the natural line of Evolution without coming to Christianity at the top.

The Ascent of Man (1894), 439.

John Dryden 1631–1700

British poet, critic and playwright

8 The longest tyranny that ever sway'd
Was that wherein our ancestors
betray'd

Their free-born *reason* to the *Stagirite* [Aristotle],
And made his torch their universal light.
So *truth*, while only one suppli'd the state,
Grew scarce, and dear, and yet sophisticate.

'To my Honour'd Friend, Dr Charleton' (1663), lines 1–6, in James Kinsley (ed.), *The Poems and Fables of John Dryden* (1962), 32.

1 *Gilbert* shall live, till *Load-stones* cease to draw,
Or *British* Fleets the boundless Ocean awe.

'To my Honour'd Friend, Dr Charleton' (1663), lines 25–6, in James Kinsley (ed.), *The Poems and Fables of John Dryden* (1962), 33.

2 Then we upon our globe's last verge shall go,
And view the ocean leaning on the sky:
From thence our rolling Neighbours we shall know,
And on the Lunar world securely pry.

'Annus Mirabilis The year of Wonders, 1666' (1667), lines 653–6, in James Kinsley (ed.), *The Poems and Fables of John Dryden* (1962), 81.

3 From Harmony, from heav'nly Harmony
This universal Frame began.

'A Song for St. Cecila's Day' (1687), lines 1–2, in James Kinsley (ed.), *The Poems and Fables of John Dryden* (1962), 422.

4 Is it not evident, in these last hundred years (when the Study of Philosophy has been the business of all the *Virtuosi* in *Christendome*) that almost a new Nature has been revealed to us? that more errours of the School have been detected, more useful Experiments in Philosophy have been made, more Noble Secrets in Opticks, Medicine, Anatomy, Astronomy, discover'd, than in all those credulous and doting Ages from *Aristotle* to us? So true it is that nothing spreads more fast than Science, when rightly and generally cultivated.

Of Dramatic Poesie (1684 edition), lines 258–67, in James T. Boulton (ed.) (1964), 44.

Charles Redway Dryer

1850–1927

American geographer

5 I shall take as a starting point for our flight into space two contrasted statements about geography. The first is that of a boy who said that the earth is a ball filled inside with dirt and worms and covered all over on the outside with nothing but geography.

'Genetic Geography: The Development of the Geographic Sense and Concept', *Annals of the Association of American Geographers*, 1920, 10, 4.

Marie Eugène François Thomas Dubois 1858–1940

Dutch anatomist and palaeoanthropologist

6 The Javanese *Anthropopithecus*, which in its skull is more human than any other known anthropoid ape, already had an upright, erect posture, which has always been considered to be the exclusive privilege of humans. Thus this ancient Pleistocene ape from our island is the first known transitional form linking Man more closely with his next of kin among the mammals. *Anthropopithecus erectus* Eug. Dubois, through each of its known skeletal elements, more closely approaches the human condition than any other anthropoid ape, especially in the femur—a fact that is totally in accord with what Lamarck proclaimed and which was explained later by Darwin and others: that the first step on the road to becoming human taken by our ancestors was acquiring upright posture. Consequently, the factual evidence is now in hand that, as some have already suspected, the East Indies was the cradle of mankind.

Report of 1892, quoted in Pat Shipman, *The Man Who Found the Missing Link: The Extraordinary Life of Eugène Dubois* (2001), 186.

René Jules Dubos 1901–82

French-born American microbiologist and environmentalist

1 Eradication of microbial disease is a will-o'-the-wisp; pursuing it leads into a morass of hazy biological concepts and half truths.

Man Adapting (1965), 381.

Lee Alvin DuBridge 1901–94

American physicist

2 The scientist or engineer—like every other human being—bears also the responsibility of being a useful member of his community . . . and should speak on issues which can be addressed with competence—including joining hands with other citizens when called to tasks of peace.

Quoted in Thomas Hager, *Force of Nature: The Life of Linus Pauling* (1995), 347.

Pierre-Maurice-Marie Duhem 1861–1916

French physicist and philosopher of science

3 If the aim of physical theories is to explain experimental laws, theoretical physics is not an autonomous science; it is subordinate to metaphysics.

The Aim and Structure of Physical Theory (1906), 2nd edition (1914), trans. Philip P. Wiener (1954), 10.

4 This whole theory of electrostatics constitutes a group of abstract ideas and general propositions, formulated in the clear and precise language of geometry and algebra, and connected with one another by the rules of strict logic. This whole fully satisfies the reason of a French physicist and his taste for clarity, simplicity and order.

The same does not hold for the Englishman. These abstract notions of material points, force, line of force, and equipotential surface do not satisfy his need to imagine concrete, material, visible, and tangible things. 'So long as we cling to this mode of representation,' says an English physicist, 'we cannot form a mental representation of the phenomena which are really happening.' It is to satisfy the need that he goes and creates a model.

The French or German physicist conceives, in the space separating two conductors, abstract lines of force having no thickness or real existence; the English physicist materializes these lines and thickens them to the dimensions of a tube which he will fill with vulcanised rubber. In place of a family of lines of ideal forces, conceivable only by reason, he will have a bundle of elastic strings, visible and tangible, firmly glued at both ends to the surfaces of the two conductors, and, when stretched, trying both to contact and to expand. When the two conductors approach each other, he sees the elastic strings drawing closer together; then he sees each of them bunch up and grow large. Such is the famous model of electrostatic action imagined by Faraday and admired as a work of genius by Maxwell and the whole English school.

The employment of similar mechanical models, recalling by certain more or less rough analogies the particular features of the theory being expounded, is a regular feature of the English treatises on physics. Here is a book [by Oliver Lodge] intended to expound the modern theories of electricity and to expound a new theory. In it are nothing but strings which move around pulleys, which roll around drums, which go through pearl beads, which carry weights; and tubes which pump water while others swell and contract; toothed wheels which are geared to one another and engage hooks. We thought we were entering the tranquil and neatly ordered abode of reason, but we find ourselves in a factory.

Footnote: O. Lodge, *Les Théories modernes* . . . p. 16.

[*Modern Views on Electricity* (1889)]
The Aim and Structure of Physical Theory (1906), 2nd edition (1914), trans. Philip P. Wiener (1954), 70–1.

1 The physicist can never subject an isolated hypothesis to experimental test, but only a whole group of hypotheses.
The Aim and Structure of Physical Theory (1906), 2nd edition (1914), trans. Philip P. Wiener (1954), 187.

2 The history of science alone can keep the physicist from the mad ambitions of dogmatism as well as the despair of Pyrrhonian scepticism.
The Aim and Structure of Physical Theory (1906), 2nd edition (1914), trans. Philip P. Wiener (1954), 270.

Jean-Baptiste-André Dumas

1800–84

French chemist

3 One of the most immediate consequences of the electrochemical theory is the necessity of regarding all chemical compounds as binary substances. It is necessary to discover in each of them the positive and negative constituents . . . No view was ever more fitted to retard the progress of organic chemistry. Where the theory of substitution and the theory of types assume similar molecules, in which some of the elements can be replaced by others without the edifice becoming modified either in form or outward behaviour, the electrochemical theory divides these same molecules, simply and solely, it may be said, in order to find in them two opposite groups, which it then supposes to be combined with each other in virtue of their mutual electrical activity . . . I have tried to show that in organic chemistry there exist types which are capable, without destruction, of undergoing the most singular transformations according to the nature of the elements.
Traité de Chemie Appliquée aux Arts, Vol. 1 (1828), 53. Trans. J. R. Partington, *A History of Chemistry*, Vol. 4, 366.

4 In chemistry, our theories are crutches; to show that they are valid, they must be used to walk . . . A theory established with the help of twenty facts must explain thirty, and lead to the discovery of ten more.
Leçons sur la Philosophie Chimique (1837), 60. Trans. S. Kapoor, 'Dumas and Organic Classification', *Ambix*, 1969, **16**, 4.

5 In inorganic chemistry the radicals are simple; in organic chemistry they are compounds—that is the sole difference.
Joint paper with Liebig, but written by Dumas, *Comptes Rendus* 1837, **5**, 567. Trans. J. R. Partington, *A History of Chemistry*, Vol. 4, 351.

6 In organic chemistry there exist certain types which are conserved even when, in place of hydrogen, equal volumes of chlorine, of bromine, etc. are introduced.
Comptes Rendus, 1839, **8**, 609–22. Trans. J. R. Partington, *A History of Chemistry*, Vol. 4, 364.

7 I have seen many phases of life; I have moved in imperial circles, I have been a Minister of State; but if I had to live my life again, I would always remain in my laboratory, for the greatest joy of my life has been to accomplish original scientific work, and, next to that, to lecture to a set of intelligent students.
Quoted in R. Oesper, *The Human Side of Scientists* (1975), 55.

8 The chemical compounds are comparable to a system of planets in that the atoms are held together by chemical affinity. They may be more or less numerous, simple or complex in composition, and in the constitution of the materials, they play the same role as Mars and Venus do in our planetary system, or the compound members such as our earth with its moon, or Jupiter with its satellites . . . If in such a system a particle is replaced by one of different character, the equilibrium can persist, and then the new compound will exhibit properties similar to those shown by the original substance.
Quoted in R. Oesper, *The Human Side of Scientists* (1975), 55.

Émile Durkheim 1858–1917

French sociologist

1 Science cannot describe individuals, but only types. If human societies cannot be classified, they must remain inaccessible to scientific description.

'Montesquieu's Contribution to the Rise of Social Science' (1892), in *Montesquieu and Rousseau. Forerunners of Sociology*, trans. Ralph Manheim (1960), 9.

2 Man is only a moral being because he lives in society, since morality consists in solidarity with the group, and varies according to that solidarity. Cause all social life to vanish, and moral life would vanish at the same time, having no object to cling to.

The Division of Labour in Society (1893), trans. W. D. Halls (1984), 331.

3 A social fact is every way of acting, fixed or not, capable of exercising on the individual an external constraint; or again, every way of acting which is general throughout a given society, while at the same time existing in its own right independent of its individual manifestations.

The Rules of Sociological Method (1895), 8th edition, trans. Sarah A. Solovay and John M. Mueller, ed. George E. G. Catlin (1938, 1964 edition), 13.

4 The first and most fundamental rule is: *Consider social facts as things.*

The Rules of Sociological Method (1895), 8th edition, trans. Sarah A. Solovay and John M. Mueller, ed. George E. G. Catlin (1938, 1964 edition), 14.

5 Even one well-made observation will be enough in many cases, just as one well-constructed experiment often suffices for the establishment of a law.

The Rules of Sociological Method (1895), 8th edition, trans. Sarah A. Solovay and John M. Mueller, ed. George E. G. Catlin (1938, 1964 edition), 80.

6 Society is not a mere sum of individuals. Rather, the system formed by their association represents a specific reality which has its own characteristics . . . The group thinks, feels, and acts quite differently from the way in which its members would were they isolated. If, then, we begin with the individual, we shall be able to understand nothing of what takes place in the group.

The Rules of Sociological Method (1895), 8th edition, trans. Sarah A. Solovay and John M. Mueller, ed. George E. G. Catlin (1938, 1964 edition), 103–4.

7 Sociological method as we practice it rests wholly on the basic principle that social facts must be studied as things, that is, as realities external to the individual. There is no principle for which we have received more criticism; but none is more fundamental. Indubitably for sociology to be possible, it must above all have an object all its own. It must take cognizance of a reality which is not in the domain of other sciences . . . *there can be no sociology unless societies exist, and that societies cannot exist if there are only individuals.*

Suicide: A Study in Sociology (1897), trans. J. A. Spaulding and G. Simpson (1952), 37–8.

8 An act cannot be defined by the end sought by the actor, for an identical system of behaviour may be adjustable to too many different ends without altering its nature.

Suicide: A Study in Sociology (1897), trans. J. A. Spaulding and G. Simpson (1952), 43.

9 For a long time it has been known that the first systems of representations with which men have pictured to themselves the world and themselves were of religious origin. There is no religion that is not a cosmology at the same time that it is a speculation upon divine things. If philosophy and the sciences were born of religion, it is because religion began by taking the place of the sciences and philosophy.

The Elementary Forms of the Religious Life (1912), trans. J. W. Swain (2nd edition 1976), 9.

10 It is only by historical analysis that we can discover what makes up man, since it is only in the course of history that he is formed.

'The Dualism of Human Nature and its Social Conditions' (1914), trans. Charles Blend, in

Kurt H. Wolff (ed.), *Emile Durkheim, 1858–1917* (1960), 325.

Alexander Logie Du Toit

1878–1948

South African geologist

1 Under the . . . new hypothesis [of Continental Drift] certain geological concepts come to acquire a new significance amounting in a few cases to a complete inversion of principles, and the inquirer will find it necessary to re-orient his ideas. For the first time he will get glimpses . . . of a pulsating restless earth, all parts of which are in greater or less degree of movement in respect to the axis of rotation, having been so, moreover, throughout geological time. He will have to leave behind him—perhaps reluctantly—the dumbfounding spectacle of the present continental masses, firmly anchored to a plastic foundation yet remaining fixed in space; set thousands of kilometres apart, it may be, yet behaving in almost identical fashion from epoch to epoch and stage to stage like soldiers, at drill; widely stretched in some quarters at various times and astoundingly compressed in others, yet retaining their general shapes, positions and orientations; remote from one another through history, yet showing in their fossil remains common or allied forms of terrestrial life; possessed during certain epochs of climates that may have ranged from glacial to torrid or pluvial to arid, though contrary to meteorological principles when their existing geographical positions are considered—to mention but a few such paradoxes!

Our Wandering Continents: An Hypothesis of Continental Drifting (1937), 3.

Clarence Edward Dutton

1841–1912

American geologist

2 Each volcano is an independent machine—nay, each vent and monticule is for the time being engaged in its own peculiar business, cooking as it were its special dish, which in due time is to be separately served. We have instances of vents within hailing distance of each other pouring out totally different kinds of lava, neither sympathizing with the other in any discernible manner nor influencing the other in any appreciable degree.

Report on the Geology of the High Plateaus of Utah (1880), 115.

George-Louis Duvernoy

1771–1855

French zoologist

3 As an organ is exercised, circulation is more particularly directed to it and is more readily performed in it. Consequently, all its secretions and excretions increase. The more an organ is exercised, the more it is nourished.

'Réflexions sur les corps organises et les sciences dont ils sont l'objet', *Magasin Encyclopédique*, III (1799), 471. Trans. Jonathan Mandelbaum in Pietro Corsi, *The Age of Lamarck* (1988), 75.

Freeman John Dyson 1923–

British-born American physicist

4 The reason why new concepts in any branch of science are hard to grasp is always the same; contemporary scientists try to picture the new concept in terms of ideas which existed before.

'Innovation in Physics', *Scientific American*, 1958, 199, 76.

5 Leaving aside genetic surgery applied to humans, I foresee that the coming century will place in our hands two other forms of biological technology which are less dangerous but still revolutionary enough to transform the conditions of our existence. I count these new technologies as powerful allies in the attack on Bernal's three enemies [the world, the flesh and the devil]. I give them the names 'biological engineering' and 'self-reproducing machinery'. Biological engineering means the artificial synthesis of living organisms designed to fulfil human purposes. Self-reproducing machinery means the imitation of the function and reproduction of a living organism with non-living materials, a computer-program imitating the function of DNA and a miniature factory imitating the functions of protein molecules. After we have attained a complete understanding of the principles of organization and development of a simple multicellular organism, both of these avenues of technological exploitation should be open to us.

The World, the Flesh and the Devil: 3rd J. D. Bernal Lecture, delivered at Birkbeck College London, 16th May 1972 (1972), 6

1 The technologies which have had the most profound effects on human life are usually simple. A good example of a simple technology with profound historical consequences is hay. Nobody knows who invented hay, the idea of cutting grass in the autumn and storing it in large enough quantities to keep horses and cows alive through the winter. All we know is that the technology of hay was unknown to the Roman Empire but was known to every village of medieval Europe. Like many other crucially important technologies, hay emerged anonymously during the so-called Dark Ages. According to the Hay Theory of History, the invention of hay was the decisive event which moved the center of gravity of urban civilization from the Mediterranean basin to Northern and Western Europe. The Roman Empire did not need hay because in a Mediterranean climate the grass grows well enough in winter for animals to graze. North of the Alps, great cities dependent on horses and oxen for motive power could not exist without hay. So it was hay that allowed populations to grow and civilizations to flourish among the forests of Northern Europe. Hay moved the greatness of Rome to Paris and London, and later to Berlin and Moscow and New York.

Infinite In All Directions (1988), 135.

2 The game of status seeking, organized around committees, is played in roughly the same fashion in Africa and in America and in the Soviet Union. Perhaps the aptitude for this game is a part of our genetic inheritance, like the aptitude for speech and for music. The game has had profound consequences for science. In science, as in the quest for a village water supply, big projects bring enhanced status; small projects do not. In the competition for status, big projects usually win, whether or not they are scientifically justified. As the committees of academic professionals compete for power and influence, big science becomes more and more preponderant over small science. The large and fashionable squeezes out the small and unfashionable. The space shuttle squeezes out the modest and scientifically more useful expendable launcher. The Great Observatory squeezes out the Explorer. The centralized adduction system squeezes out the village well. Fortunately, the American academic system is pluralistic and chaotic enough that first-rate small science can still be done in spite of the committees. In odd corners, in out-of the-way universities, and in obscure industrial laboratories, our Fulanis are still at work.

From Eros to Gaia (1992), 19.

3 The success of Apollo was mainly due to the fact that the project was conceived and honestly presented to the public as an international sporting event and not as a contribution to science. The order of priorities in Apollo was accurately reflected by the first item to be unloaded after each landing on the Moon's surface, the television camera. The landing, the coming and going of the astronauts, the exploring of the moon's surface, the gathering of Moon rocks and the earthward departure, all were expertly choreographed with the cameras placed in the right positions to make a dramatic show on television. This was to me the great surprise of the Apollo missions. There was nothing surprising in the fact that astronauts could walk on the Moon and bring home Moon rocks. There were no big scientific surprises in the chemistry of the Moon rocks or in the results of magnetic and seismic observations that the astronauts carried out. The big surprise was the quality of the public entertainment that the missions provided. I had never expected that we would see in real time astronauts hopping around in lunar gravity and driving their Rover down the Lincoln-Lee scarp to claim a lunar speed record of eleven miles per hour. Intensive television coverage was the driving force of Apollo. Von Braun had not imagined the possibilities of television when he decided that one kilohertz would be an adequate communication bandwidth for his Mars Project.

From Eros to Gaia (1992), 52.

1 Plasma seems to have the kinds of properties one would like for life. It's somewhat like liquid water—unpredictable and thus able to behave in an enormously complex fashion. It could probably carry as much information as DNA does. It has at least the potential for organizing itself in interesting ways.

Attributed.

2 I believe that life can go on forever. It takes a million years to evolve a new species, ten million for a new genus, one hundred million for a class, a billion for a phylum—and that's usually as far as your imagination goes. In a billion years, it seems, intelligent life might be as different from humans as humans are from insects. But what would happen in another ten billion years? It's utterly impossible to conceive of ourselves changing as drastically as that, over and over again. All you can say is, on that kind of time scale the material form that life would take is completely open. To change from a human being to a cloud may seem a big order, but it's the kind of change you'd expect over billions of years.

Attributed.

Edward Murray East 1879–1938

American plant geneticist

3 Genetics has enticed a great many explorers during the past two decades. They have labored with fruit-flies and guinea-pigs, with sweet peas and corn, with thousands of animals and plants in fact, and they have made heredity no longer a mystery but an exact science to be ranked close behind physics and chemistry in definiteness of conception. One is inclined to believe, however, that the unique magnetic attraction of genetics lies in the vision of potential good which it holds for mankind rather than a circumscribed interest in the hereditary mechanisms of the lowly species used as laboratory material. If man had been found to be sharply demarcated from the rest of the occupants of the world, so that his heritage of physical form, of physiological function, and of mental attributes came about in a superior manner setting him apart as lord of creation, interest in the genetics of the humbler organisms—if one admits the truth—would have flagged severely. Biologists would have turned their attention largely to the ways of human heredity, in spite of the fact that the difficulties encountered would have rendered progress slow and uncertain. Since this was not the case, since the laws ruling the inheritance of the denizens of the garden and the inmates of the stable were found to be applicable to prince and potentate as well, one could shut himself up in his laboratory and labor to his heart's content, feeling certain that any truth which it fell to his lot to discover had a real human interest, after all.

Mankind at the Crossroads (1923), v-vi.

Gregg Easterbrook 1953-

American author

1 Torture numbers, and they will confess to anything.

'Our Warming World', *New Republic*, 11 November 1999, Vol. 221, 42.

Amos Eaton 1776–1842

American botanist, geologist and lawyer

2 Geology has its peculiar difficulties, from which all other sciences are exempt. Questions in chemistry may be settled in the laboratory by experiment. Mathematical and philosophical questions may be discussed, while the materials for discussion are ready furnished by our own intellectual reflections. Plants, animals and minerals, may be arranged in the museum, and all questions relating to their intrinsic principles may be discussed with facility. But the relative positions, the shades of difference, the peculiar complexions, whether continuous or in subordinate beds, are subjects of enquiry in settling the character of rocks, which can be judged of while they are in situ only.

A Geological and Agricultural Survey of the District Adjoining the Erie Canal (1824), 8.

Hermann Ebbinghaus 1850–1909

German psychologist

3 What is true [in psychology] is alas not new, the new not true.

Über die hartmannsche Philosophie des Unbewussten (1873), 67.

4 Psychology has a long past, yet its real history is short.

Psychology (1885), 3.

Nicholas Eberstadt 1955-

American political economist and demographer

5 Though he may not always recognise his bondage, modern man lives under a tyranny of numbers.

The Tyranny of Numbers: Mismeasurement and Misrule (1995), 1.

Arthur Stanley Eddington 1882–1944

British astronomer and physicist

See **Fisher, Ronald Aylmer** *217:2*

6 Our ultimate analysis of space leads us not to a 'here' and a 'there', but to an extension such as that which relates 'here' and 'there'. To put the conclusion rather crudely—space is not a lot of points close together; it is a lot of distances interlocked.

The Mathematical Theory of Relativity (1923), 10.

7 The helium which we handle must have been put together at some time and some place. We do not argue with the critic who urges that the stars are not hot enough for this process; we tell him to go and find *a hotter place*.

The Internal Constitution of the Stars (1926), 301.

8 Let us draw an arrow arbitrarily. If as we follow the arrow we find more and more of the random element in the state of the world, then the arrow is pointing towards the future; if the random element decreases the arrow points towards the past . . . I shall use the phrase 'time's arrow' to express this one-way property of time which has no analogue in space.

Gifford Lectures (1927), *The Nature of The Physical World* (1928), 69.

9 If someone points out to you that your pet theory of the universe is in disagreement with Maxwell's equations—then so much the worse for Maxwell's equations. If it is found to be contradicted by observation—well these experimentalists do bungle things sometimes. But if your theory is found to be against the second law of thermodynamics I can give you no hope; there is nothing for it but to collapse in deepest humiliation.

Gifford Lectures (1927), *The Nature of the Physical World* (1928), 74.

1 So far as physics is concerned, time's arrow is a property of entropy alone.

Gifford Lectures (1927), *The Nature of the Physical World* (1928), 80.

2 The electron, as it leaves the atom, crystallises out of Schrödinger's mist like a genie emerging from his bottle.

Gifford Lectures (1927), *The Nature of the Physical World* (1928), 199.

3 Schrödinger's wave-mechanics is not a physical theory but a dodge—and a very good dodge too.

Gifford Lectures (1927), *The Nature of the Physical World,* (1928), 219.

4 Religious creeds are a great obstacle to any full sympathy between the outlook of the scientist and the outlook which religion is so often supposed to require . . . The spirit of seeking which animates us refuses to regard any kind of creed as its goal. It would be a shock to come across a university where it was the practice of the students to recite adherence to Newton's laws of motion, to Maxwell's equations and to the electromagnetic theory of light. We should not deplore it the less if our own pet theory happened to be included, or if the list were brought up to date every few years. We should say that the students cannot possibly realise the intention of scientific training if they are taught to look on these results as things to be recited and subscribed to. Science may fall short of its ideal, and although the peril scarcely takes this extreme form, it is not always easy, particularly in popular science, to maintain our stand against creed and dogma.

Swarthmore Lecture (1929), *Science and the Unseen World* (1929), 54–6.

5 An electron is no more (and no less) hypothetical than a star. Nowadays we count electrons one by one in a Geiger counter, as we count the stars one by one on a photographic plate.

Messenger Lectures (1934), *New Pathways in Science* (1935), 21.

6 But it is necessary to insist more strongly than usual that what I am putting before you is a *model*—the Bohr model atom—because later I shall take you to a profounder level of representation in which the electron instead of being confined to a particular locality is distributed in a sort of probability haze all over the atom.

Messenger Lectures (1934), *New Pathways in Science* (1935), 34.

7 There is no space without aether, and no aether which does not occupy space.

Messenger Lectures (1934), *New Pathways in Science* (1935), 39.

8 In the most modern theories of physics probability seems to have replaced aether as 'the nominative of the verb "to undulate"'.

Messenger Lectures (1934), *New Pathways in Science* (1935), 110.

9 The understanding between a non-technical writer and his reader is that he shall talk more or less like a human being and not like an Act of Parliament. I take it that the aim of such books must be to convey exact thought in inexact language . . . he can never succeed without the co-operation of the reader.

Messenger Lectures (1934), *New Pathways in Science* (1935), 279.

10 Just now nuclear physicists are writing a great deal about hypothetical particles called *neutrinos* supposed to account for certain peculiar facts observed in β-ray disintegration. We can perhaps best describe the neutrinos as little bits of spin-energy that have got detached. I am not much impressed by the neutrino theory. In an ordinary way I might say that I do not believe in neutrinos . . . But I have to reflect that a physicist may be an artist, and you never know where you are with artists. My old-fashioned kind of disbelief in neutrinos is scarcely enough. Dare I say that experimental physicists will not have sufficient ingenuity to *make* neutrinos? Whatever I may think, I am not going to be lured into a wager against the skill of experimenters under the impression that it is a wager against the truth of a theory. If they succeed in making neutrinos, perhaps even in developing industrial applications of them, I suppose I shall have to believe—though I may feel that they have not been playing quite fair.

Tarner Lectures (1938), *The Philosophy of Physical Science* (1939), 112.

1 I believe there are 15,747, 724,136, 275,002, 577,605, 653,961, 181,555, 468,044, 717,914, 527,116, 709,366, 231,025, 076,185, 631,031,296 protons in the universe, and the same number of electrons.

Tarner Lectures (1938), *The Philosophy of Physical Science* (1939), 170.

2 When an investigator has developed a formula which gives a complete representation of the phenomena within a certain range, he may be prone to satisfaction. Would it not be wiser if he should say 'Foiled again! I can find out no more about Nature along this line.'

Quoted in *Astro-Physical Journal*, 1945, 101, 133.

3 Science is one thing, wisdom is another. Science is an edged tool, with which men play like children, and cut their own fingers. If you look at the results which science has brought in its train, you will find them to consist almost wholly in elements of mischief. See how much belongs to the word 'Explosion' alone, of which the ancients knew nothing.

Attributed in Robert L. Weber, *More Random Walks in Science* (1982), 48.

4 On one occasion when Smart found him engrossed with his fundamental theory, he asked Eddington how many people he thought would understand what he was writing—after a pause came the reply, 'Perhaps seven.'

A. V. Douglas, *The Life of Arthur Stanley Eddington* (1956), 110.

Gerald Maurice Edelman 1929–

American immunologist and neurobiologist

5 Each nerve cell receives connections from other nerve cells at six sites called synapses. But here is an astonishing fact—there are about one million billion connections in the cortical sheet. If you were to count them, one connection (or synapse) per second, you would finish counting some thirty-two million years after you began. Another way of getting a feeling for the numbers of connections in this extraordinary structure is to consider that a large match-head's worth of your brain contains about a billion connections. Notice that I only mention counting connections. If we consider how connections might be variously combined, the number would be hyperastronomical—on the order of ten followed by millions of zeros. (There are about ten followed by eighty zero's worth of positively charged particles in the whole known universe!)

Bright and Brilliant Fire, On the Matters of the Mind (1992), 17.

6 What these two sciences of recognition, evolution and immunology, have in common is not found in nonbiological systems such as 'evolving' stars. Such physical systems can be explained in terms of energy transfer, dynamics, causes, and even 'information transfer'. But they do not exhibit repertoires of variants ready for interaction by selection to give a population response according to a hereditary principle. The application of a selective principle in a recognition system, by the way, does not *necessarily* mean that genes must be involved—it simply means that any state resulting after selection is highly correlated in structure with the one that gave rise to it and that the correlation continues to be propagated. Nor is it the case that selection cannot itself introduce variation. But a constancy or 'memory' of selected events *is* necessary. If changes occurred so fast that what was selected could not emerge in the population or was destroyed, a recognition system would not survive. Physics proper does not deal with recognition systems, which are by their nature biological and historical systems. But all the laws of physics nevertheless apply to recognition systems.

Bright and Brilliant Fire, On the Matters of the Mind (1992), 79.

1 There is no end of hypotheses about consciousness, particularly by philosophers. But most of these are not what we might call principled scientific theories, based on observables and related to the functions of the brain and body. Several theories of consciousness based on functionalism and on the machine model of the mind . . . have recently been proposed. These generally come in two flavors: one in which consciousness is assumed to be efficacious, and another in which it is considered an epiphenomenon. In the first, consciousness is likened to the executive in a computer systems program, and in the second, to a fascinating but more or less useless by-product of computation.

Bright and Brilliant Fire, On the Matters of the Mind (1992), 112.

Thomas Alva Edison 1847–1931

American inventor

2 I told [Kruesi] I was going to record talking, and then have the machine talk back. He thought it absurd. However, it was finished, the foil was put on; I then shouted 'Mary had a little lamb', etc. I adjusted the reproducer, and the machine reproduced it perfectly.

On first words spoken on a phonograph.
Byron M. Vanderbilt, *Thomas Edison, Chemist* (1971), 99.

3 Genius is two percent inspiration, ninety-eight percent perspiration.

Francis Arthur Jones, *The Life of Thomas Alva Edison: Sixty Years of an Inventor's Life* (1932), 371.

4 Oh these mathematicians make me tired! When you ask them to work out a sum they take a piece of paper, cover it with rows of A's, B's, and X's and Y's . . . scatter a mess of flyspecks over them, and then give you an answer that's all wrong!

Matthew Josephson, *Edison* (1959), 283.

Anthony William Fairbank Edwards 1935–

British statistician

5 What used to be called judgment is now called prejudice, and what used to be called prejudice is now called a null hypothesis.

Likelihood (1972), 180.

Paul Ehrenfest 1880–1933

Austrian physicist

6 Einstein, my upset stomach hates your theory [of General Relativity]—it almost hates you yourself! How am I to provide for my students? What am I to answer to the philosophers?!!

Letter to Albert Einstein, 20 November 1919. In Martin J. Klein, *Paul Ehrenfest: The Making of a Theoretical Physicist* (1970), Vol. 1, 315.

7 Don't be impatient with me. Bear in mind that I hop around among all of you big beasts like a harmless and helpless frog who is afraid of being squashed.

Letter to Albert Einstein, 16 August 1920. In Martin J. Klein, *Paul Ehrenfest: The Making of a Theoretical Physicist* (1970), Vol. 1, 319.

8 A thesis has to be presentable . . . but don't attach too much importance to it. If you do succeed in the sciences, you will do later on better things and then it will be of little moment. If you don't succeed in the sciences, it doesn't matter at all.

Quoted in *Leidraad* (periodical of the University of Leiden, Holland), 2, 1985.

Paul Ehrlich 1854–1915

German immunologist and pharmacologist

1 We may regard the cell quite apart from its familiar morphological aspects, and contemplate its constitution from the purely *chemical* standpoint. We are obliged to adopt the view, that the protoplasm is equipped with certain atomic groups, whose function especially consists in fixing to themselves food-stuffs, of importance to the cell-life. Adopting the nomenclature of organic chemistry, these groups may be designated side-chains. We may assume that the protoplasm consists of a special executive centre (Leistungs-centrum) in connection with which are nutritive side-chains . . . The relationship of the corresponding groups, *i.e.*, those of the food-stuff, and those of the cell, must be specific. They must be adapted to one another, as, *e.g.*, male and female screw (Pasteur), or as lock and key (E. Fischer).

Croonian Lecture, 'On Immunity with Special Reference to Cell Life', *Proceedings of the Royal Society of London*, 1900, **66**, 433–4.

2 The organism possesses certain contrivances by means of which the immunity reaction, so easily produced by all kinds of cells, is prevented from acting against the organism's own elements and so giving rise to autotoxins . . . so that one might be justified in speaking of a '*horror autotoxicus*' of the organism. These contrivances are naturally of the highest importance for the existence of the individual.

P. Ehrlich and J. Morgenroth, 'On Haemolysins', 1901. Reprinted in F. Himmelweit (ed.), *The Collected Papers of Paul Ehrlich* (1957), Vol. 2, 253.

3 The history of the knowledge of the phenomena of life and of the organized world can be divided into two main periods. For a long time anatomy, and particularly the anatomy of the human body, was the α and ω of scientific knowledge. Further progress only became possible with the discovery of the microscope. A long time had yet to pass until through Schwann the *cell* was established as the final biological unit. It would mean bringing coals to Newcastle were I to describe here the immeasurable progress which biology in all its branches owes to the introduction of this concept of the cell. For *this* concept is the axis around which the whole of the modern science of life revolves.

'Partial Cell Functions', Nobel Lecture, 11 December 1908. In *Nobel Lectures: Physiology or Medicine 1901–1921* (1967), 304.

4 In order to pursue chemotherapy successfully we must look for substances which possess a high affinity and high lethal potency in relation to the parasites, but have a low toxicity in relation to the body, so that it becomes possible to kill the parasites without damaging the body to any great extent. We want to hit the parasites as selectively as possible. In other words, we must *learn to aim and to aim in a chemical sense*. The way to do this is to synthesize by chemical means as many derivatives as possible of relevant substances.

'Ueber den jetzigen Stand der Chemotherapie'. *Berichte der Deutschen Chemischen Gesellschaft*, 1909, **42**, 17–47. Translated in B. Holmstedt and G. Liljestrand (eds.), *Readings in Pharmacology* (1963), 286.

5 It has been shown to be possible, by deliberately planned and chemotherapeutic approach, to discover curative agents which act specifically and aetiologically against diseases due to protozoal infections, and especially against the spirilloses, and amongst these against syphilis in the first place. Further evidence for the specificity of the action of dihydroxydiaminoarsenobenzene [Salvarsan '606'] is the disappearance of the Wasserman reaction, which reaction must . . . be regarded as indicative of a reaction of the organism to the constituents of the spirochaetes.

P. Ehrlich and S. Hata, 'Closing Notes to the Experimental Chemotherapy of Spirilloses', 1910. Reprinted in F. Himmelweit (ed.), *The*

Collected Papers of Paul Ehrlich (1957), Vol. 3, 302.

1 Success in research needs four Gs: Glück, Geduld, Geschick und Geld. Luck, patience, skill and money.
Quoted in M. Perutz, 'Rita and the Four Gs', *Nature*, 1988, 332, 791.

Albert Einstein 1879–1955

German-born American physicist

See **Curie, Marie** 144:1; **Freud, Sigmund** 231:4; **Reichenbach, Hans** 518:2; **Squire, John Collings** 563:5

2 Each ray of light moves in the coordinate system 'at rest' with the definite, constant velocity V independent of whether this ray of light is emitted by a body at rest or a body in motion.
Annalen der Physik, 1905, 17, 891–921. Trans. John Stachel et al (eds.), *The Collected Papers of Albert Einstein*, Vol. 2, (1989), Doc. 23, 143.

3 $E=MC^2$
E=energy, M=mass, C=velocity of light
Statement of the equivalency of mass and energy—energy equals mass times the speed of light squared—which opened up the atomic age.
The original statement was: 'If a body releases the energy L in the form of radiation, its mass is decreased by L/V^2'.
Annalen der Physik, 1905, 18, 639–641. Quoted in Alice Calaprice, *The Quotable Einstein* (1996), 165.

4 As far as the laws of mathematics refer to reality, they are not certain; and as far as they are certain, they do not refer to reality.
Sidelights on Relativity (1920), 28.

5 I was sitting in a chair in the patent office in Bern when all of a sudden a thought occurred to me: 'If a person falls freely he will not feel his own weight.' I was startled. This simple thought made a deep impression on me. It impelled me toward a theory of gravitation.
Kyoto Lecture (1922). Quoted in J. Ishiwara, *Einstein Koen-Roku* (1977).

6 If we consider that part of the theory of relativity which may nowadays in a sense be regarded as bone fide scientific knowledge, we note two aspects which have a major bearing on this theory. The whole development of the theory turns on the question of whether there are physically preferred states of motion in Nature (physical relativity problem). Also, concepts and distinctions are only admissible to the extent that observable facts can be assigned to them without ambiguity (stipulation that concepts and distinctions should have meaning). This postulate, pertaining to epistemology, proves to be of fundamental importance.
'Fundamental ideas and problems of the theory of relativity', Lecture delivered to the Nordic Assembly of Naturalists at Gothenburg, 11 July 1923. In *Nobel Lectures: Physics 1901–1921* (1998), 482.

7 Quantum mechanics is certainly imposing. But an inner voice tells me that this is not yet the real thing. The theory says a lot, but does not bring us any closer to the secrets of the 'old one'. I, at any rate, am convinced that *He* is not playing at dice.
Letter to Max Born, 4 December 1926. *The Born–Einstein Letters: Correspondence between Albert Einstein and Max and Hedwig Born from 1916–1955* (1971), 91.

8 I never think of the future. It comes soon enough.
Interview given on the liner *Belgenland*, December 1930.

9 Concern for man himself and his fate must always form the chief interest of all technical endeavours . . . in order that the creations of our minds shall be a blessing and not a curse to mankind. Never forget this in the midst of your diagrams and equations.
Address to students of the California Institute of Technology, February 1931. Quoted in Alice Calaprice, *The Quotable Einstein* (1996), 172.

10 Physical concepts are free creations of the human mind, and are not, however it may seem, uniquely determined by the external world. In our endeavour to understand reality we are somewhat like a man trying to understand the mechanism of a closed watch. He sees

the face and the moving hands, even hears its ticking, but he has no way of opening the case. If he is ingenious he may form some picture of a mechanism which could be responsible for all the things he observes, but he may never be quite sure his picture is the only one which could explain his observations. He will never be able to compare his picture with the real mechanism and he cannot even imagine the possibility or the meaning of such a comparison. But he certainly believes that, as his knowledge increases, his picture of reality will become simpler and simpler and will explain a wider and wider range of his sensuous impressions. He may also believe in the existence of the ideal limit of knowledge and that it is approached by the human mind. He may call this ideal limit the objective truth.

Albert Einstein and Leopold Infeld, *The Evolution of Physics* (1938), 33.

1 God does not care about our mathematical difficulties. He integrates empirically.

Leopold Infeld, *Quest* (1942), 222.

2 *Raffiniert ist der Herr Gott, aber boshaft ist er nicht.*

The Lord God is subtle, but malicious he is not.

Note in the Professor's lounge of the Mathematics Department at Princeton. 'God is slick, but he ain't mean'—Einstein's own translation to Derek Price, 1946.

Banesh Hoffmann, *Albert Einstein: Creator and Rebel* (1972), 146.

3 The unleashed power of the atom has changed everything save our modes of thinking and we thus drift toward unparalleled catastrophe.

Quoted in *New York Times*, 25 May 1946.

4 I have little patience with scientists who take a board of wood, look for its thinnest part and drill a great number of holes where drilling is easy.

P. Frank in 'Einstein's Philosophy of Science', *Reviews of Modern Physics* (1949).

5 If A is a success in life, then A equals x plus y plus z. Work is x; y is play; and z is keeping your mouth shut.

Quoted in *The Observer*, 15 January 1950.

6 Science is the attempt to make the chaotic diversity of our sense-experience correspond to a logically uniform system of thought.

Out of my Later Years (1950), 98.

7 While it is true that scientific results are entirely independent from religious and moral considerations, those individuals to whom we owe the great creative achievements of science were all of them imbued with the truly religious conviction that this universe of ours is something perfect and susceptible to the rational striving for knowledge. If this conviction had not been a strongly emotional one and if those searching for knowledge had not been inspired by Spinoza's *Amor Dei Intellectualis*, they would hardly have been capable of that untiring devotion which alone enables man to attain his greatest achievements.

'Religion and Science: Irreconcilable?' In *Ideas and Options* (1954), 52.

8 The conflict that exists today is no more than an old-style struggle for power, once again presented to mankind in semireligious trappings. The difference is that, this time, the development of atomic power has imbued the struggle with a ghostly character; for both parties know and admit that, should the quarrel deteriorate into actual war, mankind is doomed.

Unfinished address written shortly before his death in April 1955. In O. Nathan and H. Norden (eds.), *Einstein on Peace* (1960), 640.

9 The present theory of relativity is based on a division of physical reality into a metric field (gravitation) on the one hand and into an electromagnetic field and matter on the other hand. In reality space will probably be of a uniform character and the present theory will be valid only as a limiting case. For large densities of field and of matter, the field equations and even the field variables which enter into them will have no real significance. One may not therefore assume the validity of the equations for very high density of field and matter, and one may not conclude that the 'beginning of the expansion' must mean a singularity in the mathematical sense. All we have to realise is that the equations may not be continued over such regions.

The Meaning of Relativity (1956), 129.

1 During his Zurich stay the woman doctor, Paulette Brubacher, asked the whereabouts of his [Einstein's] laboratory. With a smile he took a fountain pen out of his breast pocket and said: 'here'.

C. Seelig, *Albert Einstein: A Documentary Biography* (1956), 154.

2 I have second thoughts. Maybe God *is* malicious.

Said to Valentine Bargmann. Quoted in J. Sayen, *Einstein in America* (1958), 51.

3 Culture in its higher forms is a delicate plant which depends on a complicated set of conditions and is wont to flourish only in a few places at any given time.

The World as I See it (1959), 74.

4 The most incomprehensible thing about the world is that it is comprehensible.

Banesh Hoffmann, *Albert Einstein: Creator and Rebel* (1972), 18.

5 Nationalism is an infantile sickness. It is the measles of the human race.

Helen Dukas and Banesh Hoffmann, *Albert Einstein, The Human Side* (1979), 38.

6 I used to wonder how it comes about that the electron is negative. Negative-positive—these are perfectly symmetric in physics. There is no reason whatever to prefer one to the other. Then why is the electron negative? I thought about this for a long time and at last all I could think was 'It won the fight!'

Quoted in G. Wald, *The Origin of Optical Activity* (1957), 352–68.

7 What I'm really interested in is whether God could have made the world in a different way; that is, whether the necessity of logical simplicity leaves any freedom at all.

Comment to Ernst Straus. Quoted in Gerald Holton, *The Scientific Imagination: Case Studies* (1978), xii.

8 If my theory of relativity is proven successful, Germany will claim me as a German and France will declare I am a citizen of the world. Should my theory prove untrue, France will say I am a German and Germany will declare I am a Jew.

Quoted in Alice Calaprice, *The Quotable Einstein* (1996), 8.

9 Development of Western science is based on two great achievements: the invention of the formal logical system (in Euclidean geometry) by the Greek philosophers, and the discovery of the possibility to find out causal relationships by systematic experiment (during the Renaissance).

Letter to J. S. Switzer, 23 April 1953, Einstein Archive 61–381. Quoted in Alice Calaprice, *The Quotable Einstein* (1996), 180.

10 It follows from the theory of relativity that mass and energy are both different manifestations of the same thing—a somewhat unfamiliar conception for the average man. Furthermore $E=MC^2$, in which energy is put equal to mass multiplied with the square of the velocity of light, showed that a very small amount of mass may be converted into a very large amount of energy . . . the mass and energy were in fact equivalent.

From the Einstein film, produced by Nova Television (1979). Quoted in Alice Calaprice, *The Quotable Einstein* (1996), 183.

11 Man has an intense desire for assured knowledge.

Quoted in P. A. Schilpp (ed.), *The Philosophy of Bertrand Russell* (1971), Vol. 1, 285.

12 Equations are more important to me, because politics is for the present, but an equation is something for eternity.

Quoted in Stephen Hawking, *A Brief History of Time* (1988), 178.

13 The aim of science is, on the one hand, as complete a comprehension as possible of the connection between perceptible experiences in their totality, and, on the other hand, the achievement of this aim *by employing a minimum of primary concepts and relations*.

H. Cuny, *Albert Einstein: The Man and his Theories* (1963), 128.

1 It is the theory which decides what we can observe.

Quoted in Werner Heisenberg, *Physics and Beyond: Encounters and Conversations* (1971), 77.

2 One thing I have learned in a long life: that all our science, measured against reality, is primitive and childlike—and yet is the most precious thing we have.

Banesh Hoffmann, *Albert Einstein: Creator and Rebel* (1972), Frontispiece.

3 To punish me for my contempt for authority, Fate made me an authority myself.

Banesh Hoffmann, *Albert Einstein: Creator and Rebel* (1972), 24.

4 If you are out to describe the truth, leave elegance to the tailor.

On being reproached that his formula of gravitation was longer and more cumbersome than Newton's.

Quoted in J. H. Mitchell, *Writing for Professional and Technical Journals* (1968), Introduction.

5 No amount of experimentation can ever prove me right; a single experiment can prove me wrong.

Attributed to Einstein. Quoted in Alice Calaprice, *The Quotable Einstein* (1996), 224.

6 Everything should be made as simple as possible, but not simpler.

Attributed.

7 ON **EINSTEIN** People complain that our generation has no philosophers. They are wrong. They now sit in another faculty. Their names are Max Planck and Albert Einstein.

Comment made by Adolf von Harnack in 1911 on being appointed the first president of the Kaiser Wilhelm Society founded in Berlin for the advancement of Science. Quoted in Carl Seelig, *Albert Einstein: A Documentary Biography* (1956), 45.

Albert Einstein 1879–1955 and **Leopold Infeld** 1898–1968

German-born American physicist, Polish physicist

8 Most of the fundamental ideas of science are essentially simple, and may, as a rule, be expressed in a language comprehensible to everyone.

The Evolution of Physics: The Growth of Ideas from the Early Concepts to Relativity and Quanta (1938), 29.

Niles Eldridge 1943– and **Stephen Jay Gould** 1941–2002

American palaeontologists

9 If new species arise very rapidly in small, peripherally isolated local populations, then the great expectation of insensibly graded fossil sequences is a chimera. A new species does not evolve in the area of its ancestors; it does not arise from the slow transformation of all its forbears.

'Punctuated Equilibria: An Alternative to Phyletic Gradualism', in Thomas J. M. Schopf (ed.), *Models in Paleobiology* (1972), 84.

10 The history of life is more adequately represented by a picture of 'punctuated equilibria' than by the notion of phyletic gradualism. The history of evolution is not one of stately unfolding, but a story of homeostatic equilibria, disturbed only 'rarely' (i.e. rather often in the fullness of time) by rapid and episodic events of speciation.

'Punctuated Equilibria: An Alternative to Phyletic Gradualism', in Thomas J. M. Schopf (ed.), *Models in Paleobiology* (1972), 84.

Norbert Elias 1897–1990

German sociologist

11 Scientific modes of thought cannot be developed and become generally accepted unless people renounce their primary, unreflecting, and spontaneous attempt to understand all their experience in terms of its purpose and meaning for themselves. The development that led to more adequate knowledge and increasing control of nature was therefore, considered from one aspect, also a development toward greater self-control by men.

The Civilizing Process: The Development of Manners—Changes in the Code of Conduct and Feeling in Early Modern Times (1939), trans. Edmund Jephcott (1978), 225. Originally published as *Über den Prozess der Zivilisation.*

Thomas Stearns Eliot 1888–1965

American-born British poet

1 Birth, and copulation, and death.
That's all the facts when you come to brass tacks:
Birth, and copulation, and death.
I've been born, and once is enough.

Sweeney Agonistes (1932), 24–5.

2 The wounded surgeon plies the steel
That questions the distempered part;
Beneath the bleeding hands we feel
The sharp compassion of the healer's art
Resolving the enigma of the fever chart.

'East Coker' (1940), Verse IV. Reprinted from the Easter Number of the *New English Weekly* (1940).

3 Where is the wisdom we have lost in knowledge?
Where is the knowledge we have lost in information?

The Rock (1934), part 1.

Thomas Renton Elliot 1877–1961

British physician and physiologist

4 Adrenalin does not excite sympathetic ganglia when applied to them directly, as does nicotine. Its effective action is localised at the periphery. The existence upon plain muscle of a peripheral nervous network, that degenerates only after section of both the constrictor and inhibitory nerves entering it, and not after section of either alone, has been described. I find that even after such complete denervation, whether of three days' or ten months' duration, the plain muscle of the dilatator pupillae will respond to adrenalin, and that with greater rapidity and longer persistence than does the iris whose nervous relations are uninjured. Therefore it cannot be that adrenalin excites any structure derived from, and dependent for its persistence on, the peripheral neurone. But since adrenalin does not evoke any reaction from muscle that has at no time of its life been innervated by the sympathetic, the point at which the stimulus of the chemical excitant is received, and transformed into what may cause the change of tension of the muscle fibre, is perhaps a mechanism developed out of the muscle cell in response to its union with the synapsing sympathetic fibre, the function of which is to receive and transform the nervous impulse. Adrenalin might then be the chemical stimulant liberated on each occasion when the impulse arrives at the periphery.

'On the Action of Adrenalin', Proceedings of the Physiological Society, 21 May 1904, in *The Journal of Physiology* 1904, 31, xxi.

Henry Havelock Ellis 1859–1939

British sexologist

5 I regard sex as the central problem of life. And now that the problem of religion has practically been settled, and that the problem of labor has at least been placed on a practical foundation, the question of sex—with the racial questions that rest on it—stands before the coming generations as the chief problem for solution. Sex lies at the root of life, and we can never learn to reverence life until we know how to understand sex.

Studies in the Psychology of Sex (1897), Vol. 1, xxx.

6 The second great channel through which the impulse towards the control of procreation for the elevation of the race is entering into practical life is by the general adoption, by the educated—of methods for the prevention of conception except when conception is deliberately desired.

Studies in the Psychology of Sex (1913), Vol. 4, 588.

1 What we call 'Progress' is the exchange of one Nuisance for another Nuisance.
Impressions and Comments (1914), 5.

2 Courtship, properly understood, is the process whereby both the male and the female are brought into that state of sexual tumescence which is a more or less necessary condition for sexual intercourse. The play of courtship cannot, therefore, be considered to be definitely brought to an end by the ceremony of marriage; it may more properly be regarded as the natural preliminary to every act of coitus.
Studies in the Psychology of Sex (1921), Vol. 3, 239.

3 Reproduction is so primitive and fundamental a function of vital organisms that the mechanism by which it is assured is highly complex and not yet clearly understood. It is not necessarily connected with sex, nor is sex necessarily connected with reproduction.
Psychology of Sex (1933), 7.

4 The modesty of women, which, in its most primitive form among animals, is based on sexual periodicity, is, with that periodicity, an essential condition of courtship.
Psychology of Sex (1933), 30.

5 'Auto erotism,' . . . spontaneous solitary sexual phenomena of which genital excitement during sleep may be said to be the type.
Psychology of Sex (1933), 91.

Walter Maurice Elsasser 1904–91

German-born American physicist

6 This irrelevance of molecular arrangements for macroscopic results has given rise to the tendency to confine physics and chemistry to the study of homogeneous systems as well as homogeneous classes. In statistical mechanics a great deal of labor is in fact spent on showing that homogeneous systems and homogeneous classes are closely related and to a considerable extent interchangeable concepts of theoretical analysis (Gibbs theory). Naturally, this is not an accident. The methods of physics and chemistry are ideally suited for dealing with homogeneous classes with their interchangeable components. But experience shows that the objects of biology are radically inhomogeneous both as systems (structurally) and as classes (generically). Therefore, the method of biology and, consequently, its results will differ widely from the method and results of physical science.
Atom and Organism: A New Approach to Theoretical Biology (1966), 34.

Ralph Waldo Emerson 1803–82

American philosopher and poet

7 A foolish consistency is the hobgoblin of little minds, adored by little statesmen and philosophers and divines.
'Self-Reliance', (1841), in *Essays and English Traits* (1910), 70.

8 So use all that is called Fortune. Most men gamble with her, and gain all, and lose all, as her wheel rolls. But do thou leave as unlawful these winnings, and deal with Cause and Effect, the Chancellors of God.
'Self-Reliance', (1841) in *Essays and English Traits* (1910), 88.

9 The dice of God are always loaded.
'Compensation', (1841), in *Essays and English Traits* (1910), 94.

10 All successful men have agreed to one thing,—they were *causationists*. They believed that things went not by luck, but by law; that there was not a weak or a cracked link in the chain that joins the first and last of things.
The Conduct of Life (1860), 48.

11 What is a weed? A plant whose virtues have not been discovered.
Fortune of the Republic (1878), 3.

12 If a man write a better book, preach a better sermon, or make a better mousetrap than his neighbour, tho' he build his house in the woods, the world will make a beaten path to his door.

Attributed to Emerson in Sarah S. B. Yule, *Borrowings* (1889). Mrs Yule states in *The Docket* February 1912 that she copied this in her handbook from a lecture delivered by Emerson; the quotation was the occasion of controversy owing to Elbert Hubbard's claim to its authorship.

1 Invention breeds invention.

Attributed.

Friedrich Engels 1820–95

German socialist

2 And what is impossible to science?

'Outlines of a Critique of Political Economy', in K. Marx (ed.), *Economic and Philosophic Manuscripts of 1844* (1844), 204.

3 The analysis of Nature into its individual parts, the grouping of the different natural processes and natural objects in definite classes, the study of the internal anatomy of organic bodies in their manifold forms—these were the fundamental conditions of the gigantic strides in our knowledge of Nature which have been made during the last four hundred years. But this method of investigation has also left us as a legacy the habit of observing natural objects and natural processes in their isolation, detached from the whole vast interconnection of things; and therefore not in their motion, but in their repose; not as essentially changing, but fixed constants; not in their life, but in their death.

Herr Eugen Dühring's Revolution in Science (Anti-Dühring), First Publication (1878). Trans. Emile Burns and ed. C. P. Dutt (1935), 27–8.

4 Without analysis, no synthesis.

Herr Eugen Dühring's Revolution in Science (Anti-Dühring), First Publication (1878). Trans. Emile Burns and ed. C. P. Dutt (1935), 52.

5 Just as Darwin discovered the law of evolution in organic nature, so Marx discovered the law of evolution in human history; he discovered the simple fact, hitherto concealed by an overgrowth of idealogy [sic], that mankind must first of all eat and drink, have shelter and clothing, before it can pursue politics, science, religion, art etc.

Engels' Speech over the Grave of Karl Marx, delivered at Highgate Cemetery, London, 17 March 1883. Quoted in *Karl Marx 1818–1883, for the Anniversary of his Death* (1942), 27.

6 The great basic thought that the world is not to be comprehended as a complex of ready-made *things*, but as a complex of *processes*, in which the things apparently stable no less than their mind-images in our heads, the concepts, go through an uninterrupted change of coming into being and passing away, in which, in spite of all seeming accidents and of all temporary retrogression, a progressive development asserts itself in the end—this great fundamental thought has, especially since the time of Hegel, so thoroughly permeated ordinary consciousness that in this generality it is scarcely ever contradicted.

Ludwig Feuerbach and the Outcome of Classical German Philosophy (1886). C. P. Dutt (ed.) (1934), 54.

7 One day we shall certainly 'reduce' thought experimentally to molecular and chemical motions in the brain; but does that exhaust the essence of thought?

Dialectics of Nature (1925), trans. Clemens Dutt (1940), 175.

Boris Ephrussi 1901–79

French biochemist and geneticist

See ***Beadle, George*** *48:5*

8 The admirable perfection of the adaptations of organisms and of their parts to the functions they perform has detracted attention from the fact that adaptedness does not consist of perfect fit, but capacity to fit or to adapt in a *variety* of ways: only in this sense is adaptedness a guarantee of further survival and evolutionary progress, for too perfect a fit is fatal to the species if not to the individual. This, I think, sets phylogeny and ontogeny in the correct perspective. It is the genotype which bears the marks of past experience of the species and defines the range of possible fits. What fit is actually chosen, what phenotype is actually evolved, is determined by the ever renewed individual history.

'The Interplay of Heredity and Environment in the Synthesis of Respiratory Enzymes in Yeast', *The Harvey Lectures: Delivered under the auspices of The Harvey Society of New York 1950–1951*, 1951, **156**, 45–6.

1 The ability of the genes to vary and, when they vary (mutate), to reproduce themselves in their new form, confers on these cell elements, as Muller has so convincingly pointed out, the properties of the building blocks required by the process of evolution. Thus, the cell, robbed of its noblest prerogative, was no longer the ultimate unit of life. This title was now conferred on the genes, subcellular elements, of which the cell nucleus contained many thousands and, more precisely, like Noah's ark, two of each kind.

Nucleo-cytoplasmic Relations in Micro-Organisms: Their Bearing on Cell Heredity and Differentiation (1953), 2–3.

2 The present knowledge of the biochemical constitution of the cell was achieved largely by the use of destructive methods. Trained in the tradition of the theory of solutions, many a biochemist tends, even today, to regard the cell as a 'bag of enzymes'. However, everyone realizes now that the biochemical processes studied *in vitro* may have only a remote resemblance to the events actually occurring in the living cell.

Nucleo-cytoplasmic Relations in Micro-Organisms: Their Bearing on Cell Heredity and Differentiation (1953), 108.

Epicurus 341–270 BC

Greek moral and natural philosopher

3 There are infinite worlds both like and unlike this world of ours. For the atoms being infinite in number . . . are borne on far out into space.

Letter to Herodotus, in *Epicurus: The Extant Remains* (1926), trans. C. Bailey, 25.

4 A world is a circumscribed portion of sky . . . it is a piece cut off from the infinite.

Letter to Pythocles, in *Epicurus: The Extant Remains* (1926), trans. C. Bailey, 59.

5 Earthquakes may be brought about because wind is caught up in the earth, so the earth is dislocated in small masses and is continually shaken, and that causes it to sway.

Letter to Pythocles, in *Epicurus: The Extant Remains* (1926), trans. C. Bailey, 71.

6 When someone admits one and rejects another which is equally in accordance with the appearances, it is clear that he has quitted all physical explanation and descended into myth.

Letter to Pythocles, 87. Trans. R. W. Sharples.

Erasistratus c.304–c.250 BC

Greek physician and physiologist

7 People who are unused to learning, learn little, and that slowly, while those more accustomed do much more and do it more easily. The same thing also happens in connection with research. Those who are altogether unfamiliar with this become blinded and bewildered as soon as their minds begin to work: they readily withdraw from the inquiry, in a state of mental fatigue and exhaustion, much like people who attempt to race without having been trained. He, on the other hand, who is accustomed to research, seeks and penetrates everywhere mentally, passing constantly from one topic to another; nor does he ever give up his investigation; he pursues it not merely for a matter of days, but throughout his whole life. Also by transferring his mind to other ideas which are yet not foreign to the questions at issue, he persists till he reaches the solution.

'On Paralysis'. Quoted in A. J. Brock, *Greek Medicine: Being Extracts Illustrative of Medical Writers from Hippocrates to Galen* (1929), 185.

Eratosthenes c.276–c.194 BC

Greek writer, astronomer and poet

1 In comparison with the great size of the earth the protrusion of mountains is not sufficient to deprive it of its spherical shape or to invalidate measurements based on its spherical shape. For Eratosthenes shows that the perpendicular distance from the highest mountain tops to the lowest regions is ten stades [*c.*5,000–5,500 feet]. This he shows with the help of dioptras which measure magnitudes at a distance.

Simplicius, *Commentary On Aristotle's De Caelo*, pp. 549.32–550.4 (Heiberg). Quoted in Morris R. Cohen and I. E. Drabkin, *A Sourcebook in Greek Science* (1948), 160 n.2.

2 The method of producing these numbers is called a sieve by Eratosthenes, since we take the odd numbers mingled and indiscriminate and we separate out of them by this method of production, as if by some instrument or sieve, the prime and incomposite numbers by themselves, and the secondary and composite numbers by themselves, and we find separately those that are mixed.

Nicomachus, *Introduction to Arithmetic*, 1.13.2. Quoted in Morris R. Cohen and I. E. Drabkin *A Sourcebook in Greek Science* (1948), 19–20.

3 Eratosthenes declares that it is no longer necessary to inquire as to the cause of the overflow of the Nile, since we know definitely that men have come to the sources of the Nile and have observed the rains there.

Proclus on Plato *Timaeus*, Vol. 1, 121.8–11 (Diehl). Quoted in Morris R. Cohen and I. E. Drabkin, *A Sourcebook in Greek Science* (1948), 383.

4 Eratosthenes contends that the aim of every poet is to entertain, not to instruct.

H. L. Jones (ed.), *The Geography of Strabo* (1917), Vol. 1, 55.

5 You [Eratosthenes] are wrong to declare that poetry is a fable-prating old wife, who has been permitted to 'invent' (as you call it) whatever she deems suitable for purposes of entertainment.

H. L. Jones (ed.), *The Geography of Strabo* (1917), Vol. 1, 59.

6 ON ERATOSTHENES [Eratosthenes] . . . is a mathematician among geographers, and yet a geographer among mathematicians; and consequently on both sides he offers his opponents occasions for contradiction.

H. L. Jones (ed.), *The Geography of Strabo* (1917), Vol. 1, 359–61.

John Eric Erichsen 1818–96

Danish-born British surgeon

7 There cannot always be fresh fields of conquest by the knife; there must be portions of the human frame that will ever remain sacred from its intrusions, at least in the surgeon's hands. That we have already, if not quite, reached these final limits, there can be little question. The abdomen, the chest, and the brain will be forever shut from the intrusion of the wise and humane surgeon.

Quoted in C. Cerf and V. Navasky (eds.), *I Wish I hadn't Said That: The Experts Speak and Get it Wrong!* (2000), 31.

Erik Homberger Erikson 1902–94

German-born American psychoanalyst

8 What was Freud's Galapagos, what species fluttered what kinds of wings before his searching eyes? It has often been pointed out derisively: his creative laboratory was the neurologist's office, the dominant species hysterical ladies.

The First Psychoanalyst (1957), 83.

Friedrich Ernst, Baron von Schlotheim 1764–1832

German geologist and palaeontologist

9 Where could the naturalist seek for more telling documents of the history of creation than in the fossils themselves?

Taschenbuch für Gesammte Mineralogie, 1813, 7, 2.

James Pollard Espy 1785–1860

American meteorologist

1 The astronomer is, in some measure, independent of his fellow astronomer; he can wait in his observatory till the star he wishes to observe comes to his meridian; but the meteorologist has his observations bounded by a very limited horizon, and can do little without the aid of numerous observers furnishing him contemporaneous observations over a wide-extended area.

Second Report on Meteorology to the Secretary of the Navy (1849), US Senate Executive Document 39, 31st Congress, 1st session. Quoted in J. R. Fleming, *Meteorology in America: 1800–1870* (1990), vii.

Euclid c.330–c.260 BC

Greek mathematician

2 That, if a straight line falling on two straight lines make the interior angles on the same side less than two right angles, the two straight lines, if produced indefinitely, meet on that side on which are the angles less than the two right angles.

The Thirteen Books of Euclid's Elements Translated from the Text of Heiberg, introduction and commentary by Sir T. L. Heath (1926), Vol. 1, 155.

3 In right-angled triangles the square on the side subtending the right angle is equal to the squares on the sides containing the right angle.

The Thirteen Books of Euclid's Elements Translated from the Text of Heiberg, introduction and commentary by Sir T. L. Heath (1926), Vol. 1, 349.

Leonhard Euler 1707–83

Swiss mathematician, astronomer and physicist

4 For since the fabric of the universe is most perfect and the work of a most wise creator, nothing at all takes place in the universe in which some rule of the maximum or minimum does not appear.

Methodus Inveniendi Lineas Curvas (1744), 1st addition, art. 1, trans. Ivor Grattan-Guinness.

5 $e^{\sqrt{-1}\pi}+1=0$

A special case of a formula published by Euler in his *Introductio ad analysin infinitorum* (1748), Vol. 1. However, he did not print it, either there or elsewhere. An early printing, maybe the first, is due to J. F. Français in *Annales des mathématique pures et appliquées* 1813–1814, **4**, 66. The formula was also highlighted by the American mathematician Benjamin Peirce around 1840. But its rise to 'fame' remains obscure.

Henry Eyring 1901–81

American physical chemist

6 The ingenuity and effective logic that enabled chemists to determine complex molecular structures from the number of isomers, the reactivity of the molecule and of its fragments, the freezing point, the empirical formula, the molecular weight, etc., is one of the outstanding triumphs of the human mind.

'Trends in Chemistry', *Chemical Engineering News*, 7 January 1963, 5.

7 A scientist's accomplishments are equal to the integral of his ability integrated over the hours of his effort.

J. O. Hirschfelder, in essay on Eyring, 'A Forecast for Theoretical Chemistry', *Journal of Chemical Education*, 1966, **45**, 457.

8 As I look back over my efforts, I would characterize my contributions as being largely in the realm of model building. . . . I perceive myself as rather uninhibited, with a certain mathematical facility and more interest in the broad aspect of a problem than the delicate nuances. I am more interested in discovering what is over the next rise than in assiduously cultivating the beautiful garden close at hand.

'Men, Mines and Molecules', *Annual Review of Physical Chemistry*, 1977, **28**, 13.

Kasimir Fajans 1887–1975

Polish-born American physical chemist

9 Saying that each of two atoms can attain closed electron shells by sharing a pair of electrons is equivalent to saying that husband and wife, by having a total of two dollars in a joint account and each having six dollars in individual bank accounts, have eight dollars apiece!

Quoted in Reynold E. Holmen, 'Kasimir Fajans (1887–1975): The Man and His Work', *Bulletin for the History of Chemistry*, 1990, 6, 7–8.

Michael Faraday 1791–1867

British chemist and physicist

See ***Davy, Humphry*** *168:5;* ***Whewell, William*** *617:2*

1 To day we made the grand experiment of burning the diamond and certainly the phenomena presented were extremely beautiful and interesting . . . The Duke's burning glass was the instrument used to apply heat to the diamond. It consists of two double convex lenses . . . The instrument was placed in an upper room of the museum and having arranged it at the window the diamond was placed in the focus and anxiously watched. The heat was thus continued for 3/4 of an hour (it being necessary to cool the globe at times) and during that time it was thought that the diamond was slowly diminishing and becoming opaque . . . On a sudden Sir H Davy observed the diamond to burn visibly, and when removed from the focus it was found to be in a state of active and rapid combustion. The diamond glowed brilliantly with a scarlet light, inclining to purple and, when placed in the dark, continued to burn for about four minutes. After cooling the glass heat was again applied to the diamond and it burned again though not for nearly so long as before. This was repeated twice more and soon after the diamond became all consumed. This phenomenon of actual and vivid combustion, which has never been observed before, was attributed by Sir H Davy to be the free access of air; it became more dull as carbonic acid gas formed and did not last so long.

From the journal kept by Faraday of his continental tour with Sir Humphry Davy. In Brian Bowers and Lenore Symons (eds.), *Curiosity Perfectly Satisfyed: Faraday's Travels in Europe 1813–1815* (1991), 75–6.

2 I am busy just now again on Electro-Magnetism and think I have got hold of a good thing but can't say; it may be a weed instead of a fish that after all my labour I may at last pull up.

Letter to Richard Phillips, 23 September 1831. In Frank A. J. L. James (ed.), *The Correspondence of Michael Faraday* (1991), Vol. 1, 579–60.

3 With respect to Committees as you would perceive I am very jealous of their formation. I mean *working committees*. I think business is always better done by few than by many. I think also the working few ought not to be embaras[s]ed by the idle many and further I think the idle many ought not to be honoured by association with the working few.

Letter to William Lubbock, 6 December 1833. In Frank A. J. L. James (ed.), *The Correspondence of Michael Faraday* (1993), Vol. 2, 160.

4 I wanted some new names to express my facts in Electrical science without involving more theory than I could help & applied to a friend Dr Nicholl, who has given me some that I intend to adopt for instance, a body decomposable by the passage of the Electric current, I call an '*electrolyte*' and instead of saying that water is *electro chemically decomposed* I say it is '*electrolyzed*'[.] The intensity above which a body is decomposed beneath which it conducts without decomposition I call the '*Electrolyte intensity*' &c &c. What have been called the poles of the battery I call the *electrodes* they are not merely surfaces of metal, but even of *water & air*, to which the term poles could hardly apply without receiving a new sense. *Electrolytes* must consist of two parts which during the *electrolization*, are determined the one in the one direction, and the other towards the poles where they are evolved; these evolved substances I call *zetodes*, which are therefore the direct constituents of electrolites.

Letter to William Whewell, 24 April 1834. In Frank A. J. L. James (ed.), *The Correspondence of Michael Faraday* (1993), Vol. 2, 176.

1 All your names I and my friend approve of or nearly all as to sense & expression, but I am frightened by their length & sound when compounded. As you will see I have taken *deoxide* and *skaiode* because they agree best with my natural standard East and West. I like Anode & Cathode better as to sound, but all to whom I have shewn them have supposed at first that by *Anode* I meant *No way.*

Letter to William Whewell, 3 May 1834. In Frank A. J. L. James (ed.), *The Correspondence of Michael Faraday* (1993), Vol. 2, 181.

2 I have taken your advice, and the names used are *anode cathode anions cations* and *ions*; the last I shall have but little occasion for. I had some hot objections made to them here and found myself very much in the condition of the man with his son and ass who tried to please every body; but when I held up the shield of your authority, it was wonderful to observe how the tone of objection melted away.

Letter to William Whewell, 15 May 1834. In Frank A. J. L. James (ed.), *The Correspondence of Michael Faraday* (1993), Vol. 2, 186.

3 I require a term to express those bodies which can pass to the *electrodes*, or, as they are usually called, the poles. Substances are frequently spoken of as being *electro-negative*, or *electro-positive*, according as they go under the supposed influence of a direct attraction to the positive or negative pole. But these terms are much too significant for the use to which I should have to put them; for though the meanings are perhaps right, they are only hypothetical, and may be wrong; and then, through a very imperceptible, but still very dangerous, because continual, influence, they do great injury to science, by contracting and limiting the habitual view of those engaged in pursuing it. I propose to distinguish these bodies by calling those *anions* which go to the *anode* of the decomposing body; and those passing to the *cathode*, *cations*; and when I have occasion to speak of these together, I shall call them *ions*.

Philosophical Transactions of the Royal Society of London, 1834, **124**, 79.

4 I have been so electrically occupied of late that I feel as if hungry for a little chemistry: but then the conviction crosses my mind that these things hang together under one law & that the more haste we make onwards each in his own path the sooner we shall arrive, and meet each other, at that state of knowledge of natural causes from which all varieties of effects may be understood & enjoyed.

Letter to Eilhard Mitscherlich, 24 January 1838. In Frank A. J. L. James (ed.), *The Correspondence of Michael Faraday* (1993), Vol. 2, 488.

5 But I must confess I am jealous of the term atom; for though it is very easy to talk of atoms, it is very difficult to form a clear idea of their nature, especially when compounded bodies are under consideration.

Experimental Researches in Electricity (1839–1855), Vol. 1, 195.

6 You can hardly imagine how I am struggling to exert my poetical ideas just now for the discovery of analogies & remote figures respecting the earth, Sun & all sorts of things—for I think it is the true way (corrected by judgement) to work out a discovery.

Letter to C. Schœnbein, 13th November, 1845. In Frank A. J. L. James (ed.), *The Correspondence of Michael Faraday* (1996), Vol. 3, 428.

7 What a delight it is to think that you are quietly & philosophically at work in the pursuit of science . . . rather than fighting amongst the crowd of black passions & motives that seem now a days to urge men every where into action. What incredible scenes every where, what unworthy motives ruled for the moment, under high sounding phrases and at the last what disgusting revolutions.

Letter to C. Schœnbein, 15 December 1848. In Frank A. J. L. James (ed.), *The Correspondence of Michael Faraday* (1996), Vol. 3, 742.

1 I have been driven to assume for some time, especially in relation to the gases, a sort of conducting power for magnetism. Mere space is Zero. One substance being made to occupy a given portion of space will cause more lines of force to pass through that space than before, and another substance will cause less to pass. The former I now call Paramagnetic & the latter are the diamagnetic. The former need not of necessity assume a polarity of particles such as iron has with magnetic, and the latter do not assume any such polarity either direct or reverse. I do not say more to you just now because my own thoughts are only in the act of formation, but this I may say: that the atmosphere has an extraordinary magnetic constitution, & I hope & expect to find in it the cause of the *annual* & *diurnal* variations, but *keep this to yourself* until I have time to see what harvest will spring from my growing ideas.

Letter to William Whewell, 22 August 1850. In L. Pearce Williams (ed.), *The Selected Correspondence of Michael Faraday* (1971), Vol. 2, 589.

2 What a weak, credulous, incredulous, unbelieving, superstitious, bold, frightened, what a ridiculous world ours is, as far as concerns the mind of man. How full of inconsistencies, contradictions and absurdities it is. I declare that taking the average of many minds that have recently come before me . . . I should prefer the obedience, affections and instinct of a dog before it.

Letter to C. Schœnbein, 25 July 1853. In Georg W. A. Kahlbaum and Francis Darbishire (eds.), *The Letters of Faraday and Schœnbein, 1836–1862* (1899), 215.

3 Magnetic *lines of force* convey a far better and purer idea than the phrase magnetic current or magnetic flood: it avoids the assumption of a current or of two currents and also of fluids or a fluid, yet conveys a full and useful pictorial idea to the mind.

Diary Entry for 10 September 1854. In Thomas Martin (ed.), *Faraday's Diary: Being the Various Philosophical Notes of Experimental Investigation* (1935), Vol. 6, 315.

4 I have never had any student or pupil under me to aid me with assistance; but have always prepared and made my experiments with my own hands, working & thinking at the same time. I do not think I could work in company, or think aloud, or explain my thoughts at the time. Sometimes I and my assistant have been in the Laboratory for hours & days together, he preparing some lecture apparatus or cleaning up, & scarcely a word has passed between us;—all this being a consequence of the *solitary & isolated* system of investigation; in contradistinction to that pursued by a Professor with his aids & pupils as in your Universities.

Letter to C. Hansteed, 16 December 1857. In L. Pearce Williams (ed.), *The Selected Correspondence of Michael Faraday* (1971), Vol. 2, 888.

5 The cases of action at a distance are becoming, in a physical point of view, daily more and more important. Sound, light, electricity, magnetism, gravitation, present them as a series. The nature of sound and its dependence on a medium we think we understand pretty well. The nature of light as dependent on a medium is now very largely accepted. The presence of a medium in the phenomena of electricity and magnetism becomes more and more probable daily. We employ ourselves, and I think rightly, in endeavouring to elucidate the physical exercise of these forces, or their sets of antecedents and consequents, and surely no one can find fault with the labours which eminent men have entered upon in respect of light, or into which they may enter as regards electricity and magnetism. Then what is there about gravitation that should exclude it from consideration also?

Newton did not shut out the physical view, but had evidently thought deeply of it; and if he thought of it, why should not we, in these advanced days, do so too?

Letter to E. Jones, 9 June 1857. In L. Pearce Williams (ed.), *The Selected Correspondence of Michael Faraday* (1971), Vol. 2, 870–1.

1 I could trust a fact and always cross-question an assertion.

Letter to De La Rive, 2 October 1858. Quotation supplied by Frank A. J. L. James.

2 Electricity is often called wonderful, beautiful; but it is so only in common with the other forces of nature. The beauty of electricity or of any other force is not that the power is mysterious, and unexpected, touching every sense at unawares in turn, but that it is under *law*, and that the taught intellect can even govern it largely. The human mind is placed above, and not beneath it, and it is in such a point of view that the mental education afforded by science is rendered super-eminent in dignity, in practical application and utility; for by enabling the mind to apply the natural power through law, it conveys the gifts of God to man.

Notes for a Friday Discourse at the Royal Institution (1858).

3 I purpose, in return for the honour you do us by coming to see what are our proceedings here, to bring before you, in the course of these lectures, the Chemical History of a Candle. I have taken this subject on a former occasion; and were it left to my own will, I should prefer to repeat it almost every year—so abundant is the interest that attaches itself to the subject, so wonderful are the varieties of outlet which it offers into the various departments of philosophy. There is not a law under which any part of this universe is governed which does not come into play, and is touched upon in these phenomena. There is no better, there is no more open door by which you can enter the study of natural philosophy, than by considering the physical phenomena of a candle.

A Course of Six Lectures on the Chemical History of a Candle (1861), 13–4.

4 Now I must take you to a very interesting part of our subject—to the relation between the combustion of a candle and that living kind of combustion which goes on within us. In every one of us there is a living process of combustion going on very similar to that of a candle, and I must try to make that plain to you. For it is not merely true in a poetical sense—the relation of the life of man to a taper; and if you will follow, I think I can make this clear.

A Course of Six Lectures on the Chemical History of a Candle (1861), 155–6.

5 Tyndall, . . . I must remain plain Michael Faraday to the last; and let me now tell you, that if I accepted the honour which the Royal Society desires to confer upon me, I would not answer for the integrity of my intellect for a single year.

On being offered the Presidency of the Royal Society. John Tyndall, *Faraday as a Discoverer* (1868), 157–8.

6 It is on record that when a young aspirant asked Faraday the secret of his success as a scientific investigator, he replied, 'The secret is comprised in three words—Work, Finish, Publish.'

J. H. Gladstone, *Michael Faraday* (1872), 122.

7 Why, sir, there is every probability that you will soon be able to tax it!

Said to William Gladstone, the Chancellor of the Exchequer, when he asked about the practical worth of electricity.

Quoted in R. A. Gregory, *Discovery, Or The Spirit and Service of Science* (1916), 3.

8 ON **FARADAY** To him [Faraday], as to all true philosophers, the main value of a fact was its position and suggestiveness in the general sequence of scientific truth.

John Tyndall, *Faraday as a Discoverer* (1868), 84.

1 ON FARADAY His [Faraday's] soul was above all littleness and proof to all egotism.

John Tyndall, *Faraday as a Discoverer* (1868), 104.

2 ON FARADAY His [Faraday's] third great discovery is the Magnetization of Light, which I should liken to the Weisshorn among mountains—high, beautiful, and alone.

John Tyndall, *Faraday as a Discoverer* (1868), 146.

3 ON FARADAY Taking him for all and all, I think it will be conceded that Michael Faraday was the greatest experimental philosopher the world has ever seen.

John Tyndall, *Faraday as a Discoverer* (1868), 147.

4 ON FARADAY The contemplation of Nature, and his own relation to her, produced in Faraday, a kind of spiritual exaltation which makes itself manifest here. His religious feeling and his philosophy could not be kept apart; there was an habitual overflow of the one into the other.

John Tyndall, *Faraday as a Discoverer* (1868), 152.

5 ON FARADAY He [Faraday] smells the truth.

F. W. G. Kohlrausch on Faraday. Quoted in John Tyndall, *Faraday as a Discoverer* (1868), 45.

Gustav Theodor Fechner

1801–87

German psychologist

6 *Vergleichende Anatomie der Engel.*

On the comparative anatomy of angels.

Vergleichende Anatomie der Engel: eine Skizze (1825). Written under the pseudonym of Dr Mises.

Enrico Fermi 1901–54

Italian physicist

See **Alvarez, Luis** 10:6

7 Young man, if I could remember the names of these particles, I would have been a botanist.

Quoted in Helge Kragh, *Quantum Generations* (1999), 321.

8 One might be led to question whether the scientists acted wisely in presenting the statesmen of the world with this appalling problem. Actually there was no choice. Once basic knowledge is acquired, any attempt at preventing its fruition would be as futile as hoping to stop the earth from revolving around the sun.

'Atomic Energy for Power', *Collected Papers (Note e Memorie)* (1939–1945), Vol. 2, 556.

9 Whatever Nature has in store for mankind, unpleasant as it may be, men must accept, for ignorance is never better than knowledge.

Quoted in Laura Fermi, *Atoms in the Family: My life with Enrico Fermi* (1954), 244.

10 The fact that no limits exist to the destructiveness of this weapon [i.e. the atomic bomb] makes its very existence and the knowledge of its construction a danger to humanity as a whole. It is necessarily an evil thing considered in any light. For these reasons, we believe it important for the President of the United States to tell the American public and the world what we think is wrong on fundamental ethical principles to initiate the development of such a weapon.

Enrico Fermi and I. I. Rabi, 'Minority Report of the General Advisory Committee', *United States Atomic Energy Commission: In the Matter of J. Robert Oppenheimer: Transcript of Hearing before Personnel Security Board, Washington, D.C. April 12th 1954—May 6th 1954* (1954), 79–80.

11 *When asked what he meant by a miracle:*

Oh, anything with a probability of less than 20%.

Attributed.

Jean François Fernel c.1497–1558

French physician

12 Anatomy is to physiology as geography is to history; it describes the theatre of events.

De Naturali Parte Medicinae Libri Septem (1542), Ch. 1.

Ludwig Andreas Feuerbach

1804–72

German philosopher and moralist

1 The doctrine of foods is of great ethical and political significance. Food becomes blood, blood becomes heart and brain, thoughts and mind stuff. Human fare is the foundation of human culture and thought. Would you improve a nation? Give it, instead of declamations against sin, better food. Man is what he eats [*Der Mensch ist, was er isst*].

Advertisement to Moleschott, *Lehre der Nahrungsmittel: Für das Volk* (1850).

2 Theology is Anthropology . . . [T]he distinction which is made, or rather supposed to be made, between the theological and anthropological predicates resolves itself into an absurdity.

The Essence of Christianity (1881), xi.

Paul K. Feyerabend 1924–94

American philosopher of science

3 Science is an essentially anarchic enterprise: theoretical anarchism is more humanitarian and more likely to encourage progress than its law-and-order alternatives.

Against Method: Outline of an Anarchistic Theory of Knowledge (1975), 9.

4 The only principle that does not inhibit progress is: *anything goes*.

Against Method: Outline of an Anarchistic Theory of Knowledge (1975), 23.

5 Given any rule, however 'fundamental' or 'necessary' for science, there are always circumstances when it is advisable not only to ignore the rule, but to adopt its opposite. For example, there are circumstances when it is advisable to introduce, elaborate and defend *ad hoc* hypotheses, or hypotheses which contradict well-established and generally accepted experimental results, or hypotheses whose content is smaller than the content of the existing and empirically adequate alternative, or self-inconsistent hypotheses, and so on.

Against Method: Outline of an Anarchistic Theory of Knowledge (1975), 23–4.

6 It is clear, then, that the idea of a fixed method, or of a fixed theory of rationality, rests on too naive a view of man and his social surroundings. To those who look at the rich material provided by history, and who are not intent on impoverishing it in order to please their lower instincts, their craving for intellectual security in the form of clarity, precision, 'objectivity', 'truth', it will become clear that there is only *one* principle that can be defended under *all* circumstances and in *all* stages of human development. It is the principle: *anything goes*.

Against Method: Outline of an Anarchistic Theory of Knowledge (1975), 27–8.

7 Unanimity of opinion may be fitting for a church, for the frightened or greedy victims of some (ancient, or modern) myth, or for the weak and willing followers of some tyrant. Variety of opinion is necessary for objective knowledge. And a method that encourages variety is also the only method that is comparable with a humanitarian outlook.

Against Method: Outline of an Anarchistic Theory of Knowledge (1975), 46.

8 No theory ever agrees with all the *facts* in its domain, yet it is not always the theory that is to blame. Facts are constituted by older ideologies, and a clash between facts and theories may be proof of progress. It is also a first step in our attempt to find the principles implicit in familiar observational notions.

Against Method: Outline of an Anarchistic Theory of Knowledge (1975), 55.

9 The separation of state and *church* must be complemented by the separation of state and *science*, that most recent, most aggressive, and most dogmatic religious institution.

Against Method: Outline of an Anarchistic Theory of Knowledge (1975), 295.

1 Everywhere science is enriched by unscientific methods and unscientific results, . . . the separation of science and non-science is not only artificial but also detrimental to the advancement of knowledge. If we want to understand nature, if we want to master our physical surroundings, then we must use *all* ideas, *all* methods, and not just a small selection of them.

Against Method: Outline of an Anarchistic Theory of Knowledge (1975), 305–6.

Richard Phillips Feynman

1918–88

American physicist

2 But the most impressive fact is that gravity is simple. It is simple to state the principles completely and not have left any vagueness for anybody to change the ideas of the law. It is simple, and therefore it is beautiful. It is simple in its pattern. I do not mean it is simple in its action—the motions of the various planets and the perturbations of one on the other can be quite complicated to work out, and to follow how all those stars in a globular cluster move is quite beyond our ability. It is complicated in its actions, but the basic pattern or the system beneath the whole thing is simple. This is common to all our laws; they all turn out to be simple things, although complex in their actual actions.

The Character of Physical Law (1967), 33–4.

3 For those who want some proof that physicists are human, the proof is in the idiocy of all the different units which they use for measuring energy.

The Character of Physical Law (1967), 75.

4 We have a habit in writing articles published in scientific journals to make the work as finished as possible, to cover up all the tracks, to not worry about the blind alleys or describe how you had the wrong idea first, and so on. So there isn't any place to publish, in a dignified manner, what you actually did in order to get to do the work, although, there has been in these days, some interest in this kind of thing.

'The Development of Space-time View of Quantum Electrodynamics', Nobel Lecture, 11 December 1965. In *Nobel Lectures: Physics 1963–1970* (1972), 155.

5 Philosophers have said that if the same circumstances don't always produce the same results, predictions are impossible and science will collapse. Here is a circumstance—identical photons are always coming down in the same direction to the piece of glass—that produces different results. We cannot predict whether a given photon will arrive at A or B. All we can predict is that out of 100 photons that come down, an average of 4 will be reflected by the front surface. Does this mean that physics, a science of great exactitude, has been reduced to calculating only the *probability* of an event, and not predicting exactly what will happen? Yes. That's a retreat, but that's the way it is: Nature permits us to calculate only probabilities. Yet science has not collapsed.

QED: The Strange Theory of Light and Matter (1985), 19.

6 Is no one inspired by our present picture of the universe? This value of science remains unsung by singers: you are reduced to hearing not a song or poem, but an evening lecture about it. This is not yet a scientific age.

Perhaps one of the reasons for this silence is that you have to know how to read music. For instance, the scientific article may say, 'The radioactive phosphorus content of the cerebrum of the rat decreases to one-half in a period of two weeks.' Now what does that mean?

It means that phosphorus that is in the brain of a rat—and also in mine, and yours—is not the same phosphorus as it was two weeks ago. It means the atoms that are in the brain are being replaced: the ones that were there before have gone away.

So what is this mind of ours: what

are these atoms with consciousness? Last week's potatoes! They now can *remember* what was going on in my mind a year ago—a mind which has long ago been replaced.

To note that the thing I call my individuality is only a pattern or dance, *that* is what it means when one discovers how long it takes for the atoms of the brain to be replaced by other atoms. The atoms come into my brain, dance a dance, and then go out—there are always new atoms, but always doing the same dance, remembering what the dance was yesterday.

'*What do You Care What Other People Think?' Further Adventures of a Curious Character* (1988), 244.

1 On the contrary, God was always invented to explain mystery. God is always invented to explain those things that you do not understand. Now when you finally discover how something works, you get some laws which you're taking away from God; you don't need him anymore. But you need him for the other mysteries. So therefore you leave him to create the universe because we haven't figured that out yet; you need him for understanding those things which you don't believe the laws will explain, such as consciousness, or why you only live to a certain length of time—life and death—stuff like that. God is always associated with those things that you do not understand. Therefore, I don't think that the laws can be considered to be like God because they have been figured out.

Quoted in P. C. W. Davies and Julian Brown (eds.), *Superstrings: A Theory of Everything?* (1988), 208–9.

Adolf Eugen Fick 1829–1901

German physiologist

2 A vital phenomenon can only be regarded as explained if it has been proven that it appears as the result of the material components of living organisms interacting according to the laws which those same components follow in their interactions outside of living systems.

Gesammelte Schriften (1904), Vol. 3, 767. Trans. Paul F. Cranefield, 'The Organic Physics of 1847 and the Biophysics of Today', *Journal of the History of Medicine and Allied Sciences*, 1957, **12**, 410.

Emil Hermann Fischer 1852–1919

German chemist

3 Once a molecule is asymmetric, its extension proceeds also in an asymmetrical sense. This concept completely eliminates the difference between natural and artificial synthesis. The advance of science has removed the last chemical hiding place for the once so highly esteemed *vis vitalis*.

'Synthesen in der Zuckergruppe', *Berichte der deutschen Chemischen Gesellschaft*, 1894, **27**, 3189.

4 Their specific effect on the glucosides might thus be explained by assuming that the intimate contact between the molecules necessary for the release of the chemical reaction is possible only with similar geometrical configurations. To give an illustration I will say that enzyme and glucoside must fit together like lock and key in order to be able to exercise a chemical action on each other. This concept has undoubtedly gained in probability and value for stereochemical research, after the phenomenon itself was transferred from the biological to the purely chemical field. It is an extension of the theory of asymmetry without being a direct consequence of it: for the conviction that the geometrical structure of the molecule even for optical isomers exercises such a great influence on the chemical affinities, in my opinion could only be gained by new actual observations.

'Einfluss der Configuration auf die wirkung der Enzyme', *Berichte der deutschen Chemischen Gesellschaft*, 1894, **27**, 2985–93. Trans. B. Holmstedt and G. Liljestrand (eds.) *Readings in Pharmacology* (1963), 251.

5 You are urgently warned against allowing yourself to be influenced in any way by theories or by other preconceived notions in the observation of phenomena, the performance of analyses and other determinations.

Laboratory Rules at Munich. Quoted by M. Bergmann, 'Fischer', in Bugge's *Das Buch der Grosse Chemiker*. Trans. Joseph S. Fruton, *Contrasts in Scientific Style: Research Groups in the Chemical and Biomedical Sciences* (1990), 172.

1 Modern warfare is in every respect so horrifying, that scientific people will only regret that it draws its means from the progress of the sciences. I hope that the present war [World War I] will teach the peoples of Europe a lasting lesson and bring the friends of peace into power. Otherwise the present ruling classes will really deserve to be swept away by socialism.

Fischer to Margaret Oppenheim, 14 December 1917. Fischer Papers, Bancroft Library, University of California. Quotation supplied by W. H. Brock.

2 **ON FISCHER** Emil Fischer represents a symbol of Germany's greatness.

C. Harries, 'Emil Fischer Wissenschaftliche Arbeiten', *Naturwissenschaften*, 1919, 7, 843. Trans. Joseph S. Fruton, *Contrasts in Scientific Style: Research Groups in the Chemical and Biomedical Sciences* (1990), 167.

Gustav Fischer 1870–1963

German agricultural engineer

3 If I wished to express the basic principle of my ideas in a somewhat strongly worded sentence, I would say that man, in his bodily development, is a primate fetus that has become sexually mature [*einen zur Geschlechtsreife gelangten Primatenfetus*].

Das Problem der Menschwerdung (1926), 44. Trans. Stephen Jay Gould, *Ontogeny and Phylogeny* (1977), 361.

Ronald Aylmer Fisher 1890–1962

British statistician and geneticist

4 When there are two independent causes of variability capable of producing in an otherwise uniform population distributions with standard deviations σ_1 and σ_2, it is found that the distribution, when both causes act together, has a standard deviation $\sqrt{\sigma_1^2 + \sigma_2^2}$. It is therefore desirable in analysing the causes of variability to deal with the square of the standard deviation as the measure of variability. We shall term this quantity the Variance of the normal population to which it refers, and we may now ascribe to the constituent causes fractions or percentages of the total variance which they together produce.

'The Correlation between Relatives on the Supposition of Mendelian Inheritance,' *Transactions of the Royal Society of Edinburgh*, 1918, 52, 399.

5 If one in twenty does not seem high enough odds, we may, if we prefer it, draw the line at one in fifty (the 2 per cent. point), or one in a hundred (the 1 per cent. point). Personally, the writer prefers to set a low standard of significance at the 5 per cent. point, and ignore entirely all results which fail to reach this level. A scientific fact should be regarded as experimentally established only if a properly designed experiment rarely fails to give this level of significance.

'The Arrangement of Field Experiments', *The Journal of the Ministry of Agriculture*, 1926, 33, 504.

6 No aphorism is more frequently repeated in connection with field trials, than that we must ask Nature few questions, or, ideally, one question, at a time. The writer is convinced that this view is wholly mistaken. Nature, he suggests, will best respond to a logical and carefully thought out questionnaire; indeed, if we ask her a single question, she will often refuse to answer until some other topic has been discussed.

'The Arrangement of Field Experiments', *The Journal of the Ministry of Agriculture*, 1926, 33, 511.

7 The neutral zone of selective advantage in the neighbourhood of zero is thus so narrow that changes in the environment, and in the genetic constitution of species, must cause this zone to be crossed and perhaps recrossed relatively rapidly in the course of evolutionary change, so that many possible gene substitutions may have a fluctuating history of advance and regression before the final balance of selective advantage is determined.

'The Distribution of Gene Ratios for Rare Mutations', *Proceedings of the Royal Society of Edinburgh*, 1930, 50, 219.

1 We may consequently state the fundamental theorem of Natural Selection in the form: The rate of increase in fitness of any organism at any time is equal to its genetic variance in fitness at that time.

The Genetical Theory of Natural Selection (1930), 35.

2 It will be noticed that the fundamental theorem proved above bears some remarkable resemblances to the second law of thermodynamics. Both are properties of populations, or aggregates, true irrespective of the nature of the units which compose them; both are statistical laws; each requires the constant increase of a measurable quantity, in the one case the entropy of a physical system and in the other the fitness, measured by *m*, of a biological population. As in the physical world we can conceive the theoretical systems in which dissipative forces are wholly absent, and in which the entropy consequently remains constant, so we can conceive, though we need not expect to find, biological populations in which the genetic variance is absolutely zero, and in which fitness does not increase. Professor Eddington has recently remarked that 'The law that entropy always increases—the second law of thermodynamics—holds, I think, the supreme position among the laws of nature'. It is not a little instructive that so similar a law should hold the supreme position among the biological sciences. While it is possible that both may ultimately be absorbed by some more general principle, for the present we should note that the laws as they stand present profound differences—(1) The systems considered in thermodynamics are permanent; species on the contrary are liable to extinction, although biological improvement must be expected to occur up to the end of their existence. (2) Fitness, although measured by a uniform method, is qualitatively different for every different organism, whereas entropy, like temperature, is taken to have the same meaning for all physical systems. (3) Fitness may be increased or decreased by changes in the environment, without reacting quantitatively upon that environment. (4) Entropy changes are exceptional in the physical world in being irreversible, while irreversible evolutionary changes form no exception among biological phenomena. Finally, (5) entropy changes lead to a progressive disorganization of the physical world, at least from the human standpoint of the utilization of energy, while evolutionary changes are generally recognized as producing progressively higher organization in the organic world.

The Genetical Theory of Natural Selection (1930), 36.

3 The statistician cannot excuse himself from the duty of getting his head clear on the principles of scientific inference, but equally no other thinking man can avoid a like obligation.

The Design of Experiments (1935), 2.

4 Inductive inference is the only process known to us by which essentially new knowledge comes into the world.

The Design of Experiments (1935), 8–9.

5 No isolated experiment, however significant in itself, can suffice for the experimental demonstration of any natural phenomenon; for the 'one chance in a million' will undoubtedly occur, with no less and no more than its appropriate frequency, however surprised we may be that it should occur to *us*.

The Design of Experiments (1935), 16.

1 To call in the statistician after the experiment is done may be no more than asking him to perform a postmortem examination: he may be able to say what the experiment died of.

Indian Statistical congress, Sankhya, *c.*1938.

2 The best causes tend to attract to their support the worst arguments, which seems to be equally true in the intellectual and in the moral sense.

Statistical Methods and Scientific Inference (1956), 31.

Walter M. Fitch 1929–

American evolutionary biologist

3 One major problem with any science is that people who don't know the conceptual history of their field go round re-inventing the elliptical wheel.

Telephone conversation with Susan Abrams 1983.

Gustave Flaubert 1821–80

French novelist

4 ARCHIMEDES. On hearing his name, shout 'Eureka!' Or else: 'Give me a fulcrum and I will move the world' There is also Archimedes' screw, but you are not expected to know what that is.

The Dictionary of Accepted Ideas (1881), trans. Jaques Barzun (1968), 15.

5 COLD. Healthier than heat.

The Dictionary of Accepted Ideas (1881), trans. Jaques Barzun (1968), 25.

6 DOCTOR. Always preceded by 'The good'. Among men, in familiar conversation, 'Oh! balls, doctor!' Is a wizard when he enjoys your confidence, a jack-ass when you're no longer on terms. All are materialists: 'you can't probe for faith with a scalpel.'

The Dictionary of Accepted Ideas (1881), trans. Jaques Barzun (1968), 30.

7 LITTRÉ. Snicker on hearing his name: 'the gentleman who thinks we are descended from the apes.'

The Dictionary of Accepted Ideas (1881), trans. Jaques Barzun (1968), 59.

Ludwik Fleck 1896–1961

Polish physician and microbiologist

8 If we define 'thought collective' as a community of persons mutually exchanging ideas or maintaining intellectual interaction, we will find by implication that it also provides the special 'carrier' for the historical development of any field of thought, as well as for the given stock of knowledge and level of culture. This we have designated thought style.

Genesis and the Development of a Scientific Fact (1935), 39.

9 The individual within the collective is never, or hardly ever, conscious of the prevailing thought style, which almost always exerts an absolutely compulsive force upon his thinking and with which it is not possible to be at variance.

Genesis and the Development of a Scientific Fact (1935), 41.

Alexander Fleming 1881–1955

British bacteriologist

10 While working with staphylococcus variants a number of culture-plates were set aside on the laboratory bench and examined from time to time. In the examinations these plates were necessarily exposed to the air and they became contaminated with various micro-organisms. It was noticed that around a large colony of a contaminating mould the staphylococcus colonies became transparent and were obviously undergoing lysis. Subcultures of this mould were made and experiments conducted with a view to ascertaining something of the properties of the bacteriolytic substance which had evidently been formed in the mould culture and which had diffused into the surrounding medium. It was found that broth in which the mould had been grown at room temperature for one or two weeks had acquired marked inhibitory, bacteriocidal and bacteriolytic properties to many of the more common pathogenic bacteria.

'On the Antibacterial Action of Cultures of a Penicillium, with Special Reference to their Use in the Isolation of B. Influenzae', *British Journal of Experimental Pathology*, 1929, 10, 226.

1 It has been demonstrated that a species of penicillium produces in culture a very powerful antibacterial substance which affects different bacteria in different degrees. Generally speaking it may be said that the least sensitive bacteria are the Gram-negative bacilli, and the most susceptible are the pyogenic cocci . . . In addition to its possible use in the treatment of bacterial infections penicillin is certainly useful . . . for its power of inhibiting unwanted microbes in bacterial cultures so that penicillin insensitive bacteria can readily be isolated.

'On the Antibacterial Action of Cultures of a Penicillium, with Special Reference to their Use in the Isolation of B. Influenzae', *British Journal of Experimental Pathology*, 1929, 10, 235–6.

2 In my first publication I might have claimed that I had come to the conclusion, as a result of serious study of the literature and deep thought, that valuable antibacterial substances were made by moulds and that I set out to investigate the problem. That would have been untrue and I preferred to tell the truth that penicillin started as a chance observation. My only merit is that I did not neglect the observation and that I pursued the subject as a bacteriologist. My publication in 1929 was the starting-point of the work of others who developed penicillin especially in the chemical field.

'Penicillin', Nobel Lecture, 11 December 1945. In *Nobel Lectures: Physiology or Medicine 1942–1962* (1964), 83.

Bernard Jacques Flürscheim

1874–1955

German-born British chemist

3 ON FLÜRSCHEIM [Flürscheim] was good at unanswerable arguments.

C. K. Ingold, in his Flürscheim obituary, *Journal of the Chemical Society*, 1956, 1087.

Robert William Fogel 1926– and Geoffrey Rudolph Elton

1921–94

American economist, British historian

4 Among the current discussions, the impact of new and sophisticated methods in the study of the past occupies an important place. The new 'scientific' or 'cliometric' history—born of the marriage contracted between historical problems and advanced statistical analysis, with economic theory as bridesmaid and the computer as best man—has made tremendous advances in the last generation.

Which Road to the Past? Two Views of History (1983), 2.

Bernard Le Bouyer (or Bovier) de Fontenelle 1657–1757

French natural philosopher

5 You say Beasts are Machines like Watches? But put the Machine called a Dog, and the Machine called a Bitch to one another for some time, and there may result another little Machine; whereas two Watches might be together all their Life-time, without ever producing a third Watch. Now Madam de B— and I think, according to our Philosophy, that all things which being but two, are endued with the Virtue of making themselves three, are of a much higher Nature than a Machine.

'Letter XII to Mr. C*****: Upon his studying the Philosophy of Descartes', *Letters of Gallantry* (1685), trans. Mr Ozell (1715), 25.

6 They will have the World to be in Large, what a Watch is in Small; which is very regular, and depends only upon the just disposing of the several Parts of the Movement.

Conversations on the Plurality of Worlds (1686), trans. William Gardiner (1715), 11.

1 As astronomy is the daughter of idleness, geometry is the daughter of property.

Conversations on the Plurality of Worlds (1686), trans. H. A. Hargreaves (1990), 13.

2 Since the princes take the Earth for their own, it's fair that the philosophers reserve the sky for themselves and rule there, but they should never permit the entry of others.

Conversations on the Plurality of Worlds (1686), trans. H. A. Hargreaves (1990), 51.

3 Let us be well assured of the Matter of Fact, before we trouble our selves with enquiring into the Cause. It is true, that this Method is too slow for the greatest part of Mankind, who run naturally to the Cause, and pass over the Truth of the Matter of Fact.

The History of Oracles. In two Dissertations (1687), trans. S. Whatley (1750), 20.

4 Nature is never so admired as when she is understood.

Prefacé sur l'utilité des mathématiques et de la physique (1733), in *Oeuvres*, Vol. 5, 11. Trans. John Heilbron, *Electricity in the 17th and 18th Centuries: A Study of Early Modern Physics* (1979), 43.

Edward Forbes 1815–54

British naturalist

5 Who that has ever visited the borders of this classic sea, has not felt at the first sight of its waters a glow of reverent rapture akin to devotion, and an instinctive sensation of thanksgiving at being permitted to stand before these hallowed waves? All that concerns the Mediterranean is of the deepest interest to civilized man, for the history of its progress is the history of the development of the world; the memory of the great men who have lived and died around its banks; the recollection of the undying works that have come thence to delight us for ever; the story of patient research and brilliant discoveries connected with every physical phenomenon presented by its waves and currents, and with every order of creatures dwelling in and around its waters. The science of the Mediterranean is the epitome of the science of the world.

On Early Explorations in the Mediterranean.

In George Wilson and Archibald Geikie, *Memoir of Edward Forbes F.R.S.* (1861), 279.

6 [Geology] may be looked upon as the history of the earth's changes during its preparation for the reception of organized beings, a history, which has all the character of a great epic.

Paper read to the Linnean Society, 7 February 1843. In George Wilson and Archibald Geikie, *Memoir of Edward Forbes F.R.S.* (1861), 343.

7 As to giving credit to whom credit is due, rest assured the best way to do good to one's-self is to do justice to others. There is plenty for everybody in science, and more than can be consumed in our time. One may get a fair name by suppressing references, but the Jewish maxim is true, 'He who seeks a name loses fame.'

Note to George Wilson, 1844. In George Wilson and Archibald Geikie, *Memoir of Edward Forbes F.R.S.* (1861), 366.

8 People without independence have no business to meddle with science. It should never be linked with lucre.

In George Wilson and Archibald Geikie, *Memoir of Edward Forbes F.R.S.* (1861), 392.

9 In truth, *ideas* and *principles* are independent of men; the application of them and their illustration is man's duty and merit. The time will come when the author of a view shall be set aside, and the view only taken cognizance of. This will be the millennium of Science.

Notes of hints to Mr Ramsey, Professor of Geology, University College London, 1847. In George Wilson and Archibald Geikie, *Memoir of Edward Forbes F.R.S.* (1861), 429.

10 How many and how curious problems concern the commonest of the sea-snails creeping over the wet sea-weed!

In how many points of view may its history be considered! There are its origin and development, the mystery of its generation, the phenomena of its growth, all concerning each apparently insignificant individual; there is the history of the species, the value of its distinctive marks, the features which link it with the higher and lower creatures, the reason why it takes its stand where we place it in the scale of creation, the course of its distribution, the causes of its diffusion, its antiquity or novelty, the mystery (deepest of mysteries) of its first appearance, the changes of the outline of continents and of oceans which have taken place since its advent, and their influence on its own wanderings.

On the Natural History of European Seas. In George Wilson and Archibald Geikie, *Memoir of Edward Forbes F.R.S.* (1861), 547–8.

Edmund Briscoe Ford 1901–88

British geneticist

1 It is evident that certain genes which either initially or ultimately have beneficial effects may at the same time produce characters of a non-adaptive type, which will therefore be established with them. Such characters may sometimes serve most easily to distinguish different races or species; indeed, they may be the only ones ordinarily available, when the advantages with which they are associated are of a physiological nature. Further, it may happen that the chain of reactions which a gene sets going is of advantage, while the end-product to which this gives rise, say a character in a juvenile or the adult stage, is of no adaptive significance.

Mendelism and Evolution (1931), 78–9.

Henry Ford 1863–1947

American industrialist

2 History is more or less bunk. It's tradition. We want to live in the present and the only history that is worth a tinker's damn is the history we make today.

Often misquoted as 'History is bunk'.

From a 1916 interview with Charles Wheeler from the *Chicago Tribune*. Quoted in C. Gelderman, *Henry Ford* (1981), 177.

Alberto Fortis 1741–1803

Italian geologist

3 It seems to me that the physical constitution of the valley, on which I am reporting, must cast doubt in the minds of those who may have accepted the assumptions of any of the geologic systems hitherto proposed; and that those who delight in science would do better to enrich themselves with empirical facts than take upon themselves the burden of defending and applying general hypotheses.

Della valle vulcanico-marina di Roncà nel Territorio Veronese (1778), trans. Ezio Vaccari, vii–viii.

Dian Fossey 1932–85

American zoologist

4 The more you learn about the dignity of the gorilla, the more you want to avoid people.

W. E. Smith, 'The Case of the Gorilla Lady Murder', *Time*, 1 September 1986, 18.

Michael Foster 1836–1907

British physiologist

5 The determining cause of most wars in the past has been, and probably will be of all wars in the future, the uncertainty of the result; war is acknowledged to be a challenge to the Unknown, it is often spoken of as an appeal to the God of Battles. The province of science is to foretell; this is true of every department of science. And the time must come—how soon we do not know—when the real science of war, something quite different from the application of science to the means of war, will make it possible to foresee with certainty the issue of a projected war. That will mark the end of battles; for however strong the spirit of contention, no nation will spend its money in a fight in which it knows it must lose.

Times Literary Supplement, 28 November 1902, 353–4.

Jean Bernard Léon Foucault

1819–68

French physicist

1 The observations, so numerous and so important, of the pendulum as object are especially relevant to the length of its oscillations. Those that I propose to make known to the [Paris] Academy [of Sciences] are principally addressed to the direction of the plane of its oscillation, which, moving gradually from east to west, provides evidence to the senses of the diurnal movement of the terrestrial globe.

'Demonstration Physique du mouvement de Rotation de la Terre', 3 Février 1851. In C. M. Gariel and J. Bertrand (eds.), *Recueil des Travaux Scientifiques de Léon Foucault* (1878), Vol. 2, 378. Trans. Harold Burstyn.

2 Science gains from it [the pendulum] more than one can expect. With its huge dimensions, the apparatus presents qualities that one would try in vain to communicate by constructing it on a small [scale], no matter how carefully. Already the regularity of its motion promises the most conclusive results. One collects numbers that, compared with the predictions of theory, permit one to appreciate how far the true pendulum approximates or differs from the abstract system called 'the simple pendulum'.

'Demonstration Expérimentale du movement de Rotation de la Terre', 31 Mai 1851. In C. M. Gariel and J. Bertrand (eds.), *Recueil des Travaux Scientifiques de Léon Foucault* (1878), Vol. 2, 527. Trans. Harold Burstyn.

Michel Paul Foucault 1926–84

French philosopher

3 To seek in the great accumulation of the already-said the text that resembles "in advance" a later text, to ransack history in order to rediscover the play of anticipations or echoes, to go right back to the first seeds or to go forward to the last traces, to reveal in a work its fidelity to tradition or its irreducible uniqueness, to raise or lower its stock of originality, to say that the Port-Royal grammarians invented nothing, or to discover that Cuvier had more predecessors than one thought, these are harmless enough amusements for historians who refuse to grow up.

The Archaeology of Knowledge (1969), trans. A. M. Sheridan Smith (1972), 144.

4 It might be said that all knowledge is linked to the essential forms of cruelty.

Mental Illness and Psychology (1976), trans. Alan Sheridan, 73.

5 Truth is not by nature free—nor error servile—its production is thoroughly imbued with relations of power.

History of Sexuality (1976), trans Robert Hurley (1978), Vol. 1, 60.

Antoine François de Fourcroy

1755–1809

French chemist

6 Each man who receives a liberal education today counts chemistry as one of the indispensable parts of his studies.

Systéme des connaissances chimiques (1801), Vol. 1, xviii, trans. W. A. Smeaton.

7 The serum, when subjected to heat, coagulates and hardens like egg. This property is one of its striking characteristics; it is attributed to a particular substance which is thereby readily recognizable, and which is called *albumine*, because it is the one present in egg white, termed *albumen*.

Systéme des connaissances chimiques (1801), Vol. 5, 117. Trans. Joseph S. Fruton, *Proteins, Enzymes, Genes: The Interplay of Chemistry and Biology* (1999), 161.

8 *Compounds formed by chemical attraction, possess new properties different from those of their component parts* . . . chemists have long believed that the contrary took place in their combination. They

thought, in fact, that the compounds possessed properties intermediate between those of their component parts; so that two bodies, very coloured, very sapid, or insapid, soluble or insoluble, fusible or infusible, fixed or volatile, assumed in chemical combination, a shade or colour, or taste, solubility or volatility, intermediate between, and in some sort composed of, the same properties which were considered in their principles. This is an illusion or error which modern chemistry is highly interested to overthrow.

Quoted in *A General System of Chemical Knowledge* (1804), Vol. 1, trans. W. Nicholson, 102–3.

Jean Baptiste Joseph Fourier
1768–1830

French mathematician and physicist

1 Profound study of nature is the most fertile source of mathematical discoveries.

Théorie analytique de la chaleur (1822), Preface, trans. Ivor Grattan-Guinness.

2 The integrals which we have obtained are not only general expressions which satisfy the differential equation, they represent in the most distinct manner the natural effect which is the object of the phenomenon . . . when this condition is fulfilled, the integral is, properly speaking, *the equation of the phenomenon*; it expresses clearly the character and progress of it, in the same manner as the finite equation of a line or curved surface makes known all the properties of those forms.

Théorie analytique de la chaleur (1822), Art. 428, trans. Ivor Grattan-Guinness.

Patrick William Fowler 1956–
British chemist

3 Theoretical chemistry is a peculiar subject. It is based on an equation that can hardly ever be solved.

'Orbital Update', review of R. McWenny, *Methods of Molecular Quantum Mechanics* (1989). *Nature*, 1990, 343, 222.

Cornelius Benjamin Fox
1839–c.1884

British chemist

4 To the Philosopher, the Physician, the Meteorologist, and the Chemist, there is perhaps no subject more attractive than that of Ozone.

Ozone and Antozone (1873), 1.

Philipp Frank 1884–1966
Austrian-American physicist, mathematician and philosopher

5 If in the description of an experimental arrangement the expression 'position of a particle' can be used, then in the description of the same arrangement the expression 'velocity of a particle' can *not* be used, and vice versa. Experimental arrangements, one of which can be described with the help of the expression 'position of a particle' and the other with the help of the expression 'velocity' or, more exactly, 'momentum', are called *complementary* arrangements, and the descriptions are referred to as *complementary* descriptions.

Modern Science and its Philosophy (1949), 163–4.

Edward Frankland 1825–99
British chemist

6 When the formulæ of inorganic chemical compounds are considered, even a superficial observer is struck with the general symmetry of their construction; the compounds of nitrogen, phosphorus, antimony and arsenic especially exhibit the tendency of the elements to form compounds containing 3 or 5 equivs. of other elements, and it is in these proportions that their affinities are best satisfied; thus in the ternal group we have NO_3, NH_3, NI_3, NS_3, PO_3, PH_3, PCl_3, SbO_3, SbH_3, $SbCl_3$, AsO_3, AsH_3, $AsCl_3$, &c; and in the five-atom group NO_5, NH_4O, NH_4I, PO_5, PH_4I, &c. Without offering any hypothesis regarding the cause of this symmetrical grouping of atoms, it is sufficiently evident, from the examples just given, that such a tendency or law prevails, and that, no matter what the character of the uniting atoms may be, the combining power of the attracting element, if I may be allowed the term, is always satisfied by the same number of these atoms.

'On a New Series of Organic Bodies Containing Metals', *Philosophical Transactions of the Royal Society of London*, 1852, 141:2, 440.

1 ON FRANKLAND On coming down the stairs at dinner Tris [Trismegistus = Frankland] who walked before me seemed impressed by a mechanical impulse which impelled him along the corridor with a fervid velocity. On reaching the stair bottom I discovered the cause of the attraction. Miss Edmondson, like a pure planet, had checked his gravitating tendencies and lo! He stood radiant with smiles dropping joysparkes from his eyes as he clasped her hand. His countenance became a transparency through which the full proportions of his soul shone manifest; his blood tingled from his eyebrows to his finger ends, and wealthy with rich emotions his face became the avenue of what he felt.

Journals of John Tyndall, 18 January 1848. Royal Institution Archives.

Benjamin Franklin 1706–90

American natural philosopher

2 And we daily in our experiments electrise bodies *plus* or *minus*, as we think proper. [These terms we may use till your Philosophers give us better.] To electrise *plus* or *minus*, no more needs to be known than this, that the parts of the Tube or Sphere, that are rubb'd, do, in the Instant of Friction, attract the Electrical Fire, and therefore take it from the Thin rubbing; the same parts immediately, as the Friction upon them ceases, are disposed to give the fire they have received, to any Body that has less.

Letter 25 May 1747. Quoted in I. B. Cohen, *Franklin and Newton: An Enquiry into Speculative Newtonian Experimental Science and Franklin's Work in Electricity as an Example Thereof* (1956), 439.

3 In going on with these Experiments, how many pretty systems do we build, which we soon find ourselves oblig'd to destroy! If there is no other Use discover'd of Electricity, this, however, is something considerable, that it may *help to make a vain Man humble*.

Letter to Peter Collinson, 14 August 1747. In I. Bernard Cohen (ed.), *Benjamin Franklin's Experiments* (1941), 63.

4 Dangerous, therefore, is it to take shelter under a tree, during a thunder-gust. It has been fatal to many, both men and beasts.

Letter to Dr John Mitchel F.R.S., 29 April 1749. In I. Bernard Cohen (ed.), *Benjamin Franklin's Experiments* (1941), 209.

5 Chagrined a little that we have been hitherto able to produce nothing in this way of use to mankind; and the hot weather coming on, when electrical experiments are not so agreeable, it is proposed to put an end to them for this season, somewhat humorously, in a party of pleasure, on the banks of *Skuylkil*. Spirits, at the same time, are to be fired by a spark sent from side to side through the river, without any other conductor that the water; an experiment which we some time since performed, to the amazement of many. A turkey is to be killed for our dinner by the *electrified bottle*: when the healths of all the famous electricians in *England*, *Holland*, *France*, and *Germany* are to be drank in *electrified bumpers*, under the discharge of guns from the *electrical battery*.

Letter to Peter Collinson, 29 April 1749. In I. Bernard Cohen (ed.), *Benjamin Franklin's Experiments* (1941), 199–200.

6 The electrical matter consists of particles extremely subtile, since it can permeate common matter, even the densest metals, with such ease and freedom as not to receive any perceptible resistance.

If any one should doubt whether the

electrical matter passes through the substance of bodies, or only over along their surfaces, a shock from an electrified large glass jar, taken through his own body, will probably convince him.

Electrical matter differs from common matter in this, that the parts of the latter mutually attract, those of the former mutually repel each other.

'Opinions and Conjectures, Concerning the Properties and Effects of the Electrical Matter, arising from Experiments and Observations, made at Philadelphia, 1749.' In I. Bernard Cohen (ed.), *Benjamin Franklin's Experiments* (1941), 213.

1 In New England they once thought *blackbirds* useless, and mischievous to the corn. They made efforts to destroy them. The consequence was, the blackbirds were diminished; but a kind of worm, which devoured their grass, and which the blackbirds used to feed on, increased prodigiously; then, finding their loss in grass much greater than their saving in corn, they wished again for their blackbirds.

Letter to Richard Jackson, 5 May 1753. In Albert Henry Smyth, *The Writings of Benjamin Franklin* (1905), Vol. 3, 135.

2 Let the experiment be made.

Letter to Dr L—, 18 March 1755. In I. Bernard Cohen (ed.), *Benjamin Franklin's Experiments* (1941), 334.

3 What signifies Philosophy that does not apply to some Use? May we not learn from hence, that black Clothes are not so fit to wear in a hot Sunny Climate or Season, as white ones; because in such Cloaths the Body is more heated by the Sun when we walk abroad, and are at the same time heated by the Exercise, which double Heat is apt to bring on putrid dangerous Fevers? The Soldiers and Seamen, who must march and labour in the Sun, should in the East or West Indies have an Uniform of white?

Letter to Miss Mary Stevenson, 20 September 1761. In Albert Henry Smyth (ed.), *The Writings of Benjamin Franklin* (1906), Vol. 4, 115.

4 An iron rod being placed on the outside of a building from the highest part continued down into the moist earth, in any direction strait or crooked, following the form of the roof or other parts of the building, will receive the lightening at its upper end, attracting it so as to prevent it's striking any other part; and, affording it a good conveyance into the earth, will prevent its damaging any part of the building.

'Of Lightening, and the Method (now used in America) of securing Buildings and Persons from its mischievous Effects', Paris 1767. In I. Bernard Cohen (ed.), *Benjamin Franklin's Experiments* (1941), 390.

5 I always rejoice to hear of your being still employed in experimental researches into nature, and of the success you meet with. The rapid progress *true* science now makes, occasions my regretting sometimes that I was born so soon: it is impossible to imagine the height to which may be carried, in a thousand years, the power of man over matter; we may perhaps learn to deprive large masses of their gravity, and give them absolute levity for the sake of easy transport. Agriculture may diminish its labour and double its produce; all diseases may by sure means be prevented or cured (not excepting even that of old age), and our lives lengthened at pleasure even beyond the antediluvian standard. Oh! that moral science were in as fair a way of improvement; that men would cease to be wolves to one another; and that human beings would at length learn what they now improperly call humanity!

Letter to Dr Priestley, 8 February 1780. In *Memoirs of Benjamin Franklin* (1845), Vol. 2, 152.

6 Our new Constitution is now established, and has an appearance that promises permanency; but in this world nothing can be said to be certain, except death and taxes.

Letter to Jean Baptiste Le Roy, 13 November 1789. Quoted in Albert Henry Smyth (ed.) *The Writings of Benjamin Franklin* (1907), vol. 10, 69.

1 What is the use of a new-born child?

When asked of the use of a new invention.

In I. Bernard Cohen, *Benjamin Franklin's Science* (1990), 38.

2 [Franklin always found it a] pleasure . . . to see good workmen handle their tools.

Autobiography.

3 ON FRANKLIN *Eripuit coelo fulmen sceptrumque tyrannis.*

He snatched the lightning from the sky and the sceptre from tyrants.

Line applied to Franklin by French economist Anne-Robert-Jacques Turgot. In I. Bernard Cohen, *Benjamin Franklin's Experiments* (1941), xxvii.

Rosalind Elsie Franklin 1920–58

British biophysicist

4 You look at science (or at least talk of it) as some sort of demoralising invention of man, something apart from real life, and which must be cautiously guarded and kept separate from everyday existence. But science and everyday life cannot and should not be separated. Science, for me, gives a partial explanation for life. In so far as it goes, it is based on fact, experience and experiment.

Letter to Ellis Franklin, no date, possibly summer 1940 whilst Rosalind was an undergraduate at Cambridge. Cited in Brenda Maddox, *The Dark Lady of DNA* (2002), 60–1.

5 Conclusion: Big helix in several chains, phosphates on outside, phosphate-phosphate inter-helical bonds disrupted by water. Phosphate links available to proteins.

Lecture Notes of Franklin. Headed 'Colloquium November 1951', the report is typewritten dated 7 February 1952, in A. Sayre, *Rosalind Franklin and DNA* (1975), 128.

6 The results suggest a helical structure (which must be very closely packed) containing probably 2, 3 or 4 coaxial nucleic acid chains per helical unit and having the phosphate groups near the outside.

Official Report, submitted in February 1952. In A. Klug, 'Rosalind Franklin and the Discovery of the Structure of DNA', *Nature*, 1968, **219**, 843.

7 We wish to discuss a structure for the salt of deoxyribose nucleic acid (D.N.A.). This structure has novel features which are of considerable biologic interest.

Rosalind Franklin and R. G. Gosling, 'Molecular Structures of Nucleic Acids: A Structure for Deoxyribose Nucleic Acid', *Nature*, 1953, **171**, 737.

8 While the biological properties of deoxypentose nucleic acid suggest a molecular structure containing great complexity, X-ray diffraction studies described here . . . show the basic molecular configuration has great simplicity.

Rosalind Franklin and R. G. Gosling, 'Molecular Structures of Nucleic Acids: A Structure for Deoxyribose Nucleic Acid', *Nature*, 1953, **171**, 741.

9 ON FRANKLIN Her [Rosalind Franklin's] photographs are among the most beautiful X-ray photographs of any substance ever taken.

Obituary of Rosalind Franklin, *Nature*, 1958, **182**, 154.

10 ON FRANKLIN She would have solved it, but it would have come out in stages. For the feminists, however, she has become a doomed heroine, and they have seized upon her as an icon, which is not, of course, her fault. Rosalind was not a feminist in the ordinary sense, but she was determined to be treated equally like anybody else.

A. Klug, 'Rosalind Franklin and the Discovery of the Structure of DNA'. Cited in Brenda Maddox, *The Dark Lady of DNA* (2002), 326.

11 ON FRANKLIN She discovered in a series of beautifully executed researches the fundamental distinction between carbons that turned on heating into graphite and those that did not. Further she related this difference to the chemical constitution of the molecules from which carbon was made. She was already a recognized authority in industrial physico-chemistry when she chose to abandon this work in favour of the far more difficult and more exciting fields of biophysics.

Comment of J. D. Bernal in *The Times*, 19 April 1958, shortly after Franklin's death. In Jenifer Glynn, 'Rosalind Franklin', in E. Shils and C. Blacker (eds.), *Cambridge Women: Twelve Portraits* (1996), 206.

1 ON **FRANKLIN** Our dark lady is leaving us next week.

Letter from Maurice Wilkins to Francis Crick, 7 March 1953. Cited in Brenda Maddox, *The Dark Lady of DNA* (2002), xvii.

James George Frazer 1854–1941

British anthropologist

2 The position of the anthropologist of to-day resembles in some sort the position of classical scholars at the revival of learning. To these men the rediscovery of ancient literature came like a revelation, disclosing to their wondering eyes a splendid vision of the antique world, such as the cloistered of the Middle Ages never dreamed of under the gloomy shadow of the minster and within the sound of its solemn bells. To us moderns a still wider vista is vouchsafed, a greater panorama is unrolled by the study which aims at bringing home to us the faith and the practice, the hopes and the ideals, not of two highly gifted races only, but of all mankind, and thus at enabling us to follow the long march, the slow and toilsome ascent, of humanity from savagery to civilization. And as the scholar of the Renaissance found not merely fresh food for thought but a new field of labour in the dusty and faded manuscripts of Greece and Rome, so in the mass of materials that is steadily pouring in from many sides—from buried cities of remotest antiquity as well as from the rudest savages of the desert and the jungle—we of to-day must recognise a new province of knowledge which will task the energies of generations of students to master.

'Author's Introduction' (1900). In Dr Theodor H. Gaster (ed.), *The New Golden Bough* (1959), xxv-xxvi.

Edward Augustus Freeman

1823–92

British historian

3 To write about history or language is supposed to be within the reach of every man. To write about natural science is allowed to be within the reach only of those who have mastered the subjects on which they write.

The Methods of Historical Study (1886), 91.

Francis Arthur Freeth 1884–1970

British industrial chemist

4 I am Freeth, and I have come to apply the phase-rule to the ammonia-soda process.

First words on joining the Brunner-Mond Company in 1907.

W. F. L. Dick, *A Hundred Years of Alkali in Cheshire* (1973), 38.

Friedrich Ludwig Gottlob Frege 1848–1925

German logician

5 The aim of scientific work is *truth*. While we internally *recognise* something *as true*, we *judge*, and while we utter judgements, we *assert*.

Frege m.s., after 1879 (Manuskcripte edition 2), trans. Ivor Grattan-Guinness.

6 I compare arithmetic with a tree that unfolds upwards in a multitude of techniques and theorems while the root drives into the depths.

Grundgesetze der Arithmetik (1893), xiii, trans. Ivor Grattan-Guinness.

7 It really is worth the trouble to invent a new symbol if we can thus remove not a few logical difficulties and ensure the rigour of the proofs. But many mathematicians seem to have so little feeling for logical purity and accuracy that they will use a word to mean three or four different things, sooner than make the frightful decision to invent a new word.

Grundgesetze der Arithmetik (1893). Vol. 2, Section 60. In P. Greach and M. Black (eds.), *Translations from the Philosophical Writings of Gottlob Frege* (1952), 144.

1 There is more danger of numerical sequences continued indefinitely than of trees growing up to heaven. Each will some time reach its greatest length.

Grundgesetze der Arithmetik (1893), Vol. 2, Section 128. In P. Greach and M. Black (eds.), *Translations from the Philosophical Writings of Gottlob Frege* (1952), 204.

2 'Facts, facts, facts,' cries the scientist if he wants to emphasize the necessity of a firm foundation for science. What is a fact? A fact is a thought that is true. But the scientist will surely not recognize something which depends on men's varying states of mind to be the firm foundation of science.

Attributed.

Carl Remigius Fresenius 1818–97

German analytical chemist

3 Knowledge and ability must be combined with ambition as well as with a sense of honesty and a severe conscience. Every analyst occasionally has doubts about the accuracy of his results, and also there are times when he knows his results to be incorrect. Sometimes a few drops of the solution were spilt, or some other slight mistake made. In these cases it requires a strong conscience to repeat the analysis and to make a rough estimate of the loss or apply a correction. Anyone not having sufficient will-power to do this is unsuited to analysis no matter how great his technical ability or knowledge. A chemist who would not take an oath guaranteeing the authenticity, as well as the accuracy of his work, should never publish his results, for if he were to do so, then the result would be detrimental not only to himself, but to the whole of science.

Anleitung zur quantitativen Analyse (1847), preface. F. Szabadvary, *History of Analytical Chemistry* (1966), trans. Gyula Svehla, 176.

Sigmund Freud 1856–1939

Austrian psychoanalyst

See ***Erikson, Erik Homberger*** *206:8;* ***Jones, Ernest*** *329:5;* ***Trilling, Lionel*** *587:1*

4 In matters of sexuality we are at present, every one of us, ill or well, nothing but hypocrites.

Sexuality in the Aetiology of the Neuroses (1898), in James Strachey (ed.), *The Standard Edition of the Complete Psychological Works of Sigmund Freud* (1952), Vol. 3, 266.

5 Being in love with the one parent and hating the other are among the essential constituents of the stock of psychical impulses which is formed at that time and which is of such importance in determining the symptoms of the later neurosis . . . This discovery is confirmed by a legend that has come down to us from classical antiquity: a legend whose profound and universal power to move can only be understood if the hypothesis I have put forward in regard to the psychology of children has an equally universal validity. What I have in mind is the legend of King Oedipus and Sophocles' drama which bears his name.

The Interpretation of Dreams (1900), in James Strachey (ed.), *The Standard Edition of the Complete Psychological Works of Sigmund Freud* (1953), Vol. 4, 260–1.

6 The interpretation of dreams is the royal road to a knowledge of the unconscious activities of the mind.

The Interpretation of Dreams (1900), in James Strachey (ed.), *The Standard Edition of the Complete Psychological Works of Sigmund Freud* (1953), Vol. 5, 608.

7 The unconscious is the true psychical reality; *in its innermost nature it is as much unknown to us as the reality of the external world, and it is as incompletely presented by the data of consciousness as is the external world by the communications of our sense organs.*

The Interpretation of Dreams (1900), in James Strachey (ed.), *The Standard Edition of the Complete Psychological Works of Sigmund Freud* (1953), Vol. 5, 613.

8 Sexuality is the key to the problem of the psychoneuroses and of the neuroses in general. No one who disdains the key will ever be able to unlock the door.

Fragment of an Analysis of a Case of Hysteria (1905), in James Strachey (ed.), *The Standard Edition of the Complete Psychological Works of Sigmund Freud* (1953), Vol. 7, 115.

1 No one who has seen a baby sinking back satiated from the breast and falling asleep with flushed cheeks and a blissful smile can escape the reflection that this picture persists as a prototype of the expression of sexual satisfaction in later life.

Three Essays on Sexuality: Infantile Sexuality (1905), in James Strachey (ed.), *The Standard Edition of the Complete Psychological Works of Sigmund Freud* (1953), Vol. 7, 182.

2 In human beings pure masculinity or femininity is not to be found either in a psychological or biological sense.

Three Essays on the Theory of Sexuality (1905), in James Strachey (ed.), *The Standard Edition of the Complete Psychological Works of Sigmund Freud* (1953), Vol. 7, 220, fn 1.

3 A layman will no doubt find it hard to understand how pathological disorders of the body and mind can be eliminated by 'mere' words. He will feel that he is being asked to believe in magic. And he will not be so very wrong, for the words which we use in our everyday speech are nothing other than watered-down magic. But we shall have to follow a roundabout path in order to explain how science sets about restoring to words a part at least of their former magical power.

Psychical (or Mental) Treatment (1905), in James Strachey (ed.), *The Standard Edition of the Complete Psychological Works of Sigmund Freud* (1953), Vol. 7, 283.

4 [The child] takes his play very seriously and he expends large amounts of emotion on it. The opposite of play is not what is serious but what is real.

Creative Writers and Day-Dreaming (1906), in James Strachey (ed.), *The Standard Edition of the Complete Psychological Works of Sigmund Freud* (1959), Vol. 9, 144.

5 We may lay it down that a happy person never phantasises, only an unsatisfied one . . . The motive forces of phantasies are unsatisfied wishes, and every single phantasy is the fulfilment of a wish, a correction of unsatisfying reality. These motivating wishes vary according to the sex, character and circumstances of the person who is having the phantasy; but they fall naturally into two main groups. They are either ambitious wishes, which serve to elevate the subject's personality; or they are erotic ones. It was shocking when Nietzsche said this, but today it is commonplace; *our* historical position—and no end to it is in sight—is that of having to philosophise without 'foundations'.

Creative Writers and Day-Dreaming (1908), in James Strachey (ed.), *The Standard Edition of the Complete Psychological Works of Sigmund Freud* (1959), Vol. 9, 146–7.

6 The excremental is all too intimately and inseparably bound up with the sexual; the position of the genitals—*inter urinas et faeces*—remains the decisive and unchangeable factor. One might say here, varying a well-known saying of the great Napoleon: 'Anatomy is destiny'.

On the Universal Tendency to Debasement in the Sphere of Love (*Contributions to the Psychology of Love II*) (1912), in James Strachey (ed.), *The Standard Edition of the Complete Psychological Works of Sigmund Freud* (1957), Vol. 11, 189.

7 In the course of centuries the *naïve* self-love of men has had to submit to two major blows at the hands of science. The first was when they learnt that our earth was not the centre of the universe but only a tiny fragment of a cosmic system of scarcely imaginable vastness . . . the second blow fell when biological research destroyed man's supposedly privileged place in creation and proved his descent from the animal kingdom and his ineradicable animal nature . . . But human megalomania will have suffered its third and most wounding blow from the psychological research of the present time which seeks to prove to the ego that it is not even master in its own house, but must content itself with scanty information of what is going on unconsciously in its mind.

Introductory Lectures on Psychoanalyis (1916), in James Strachey (ed.), *The Standard Edition of the Complete Psychological Works of Sigmund Freud* (1963), Vol. 16, 284–5.

1 The ego is not master in its own house.

From the History of an Infantile Neuroses (1918), in James Strachey (ed.), *The Standard Edition of the Complete Psychological Works of Sigmund Freud* (1955), Vol. 17, 143.

2 Neurosis is the result of a conflict between the ego and its id, whereas psychosis is the analogous outcome of a similar disturbance in the relation between the ego and the external world.

Neurosis and Psychosis (1924), in James Strachey (ed.), *The Standard Edition of the Complete Psychological Works of Sigmund Freud* (1961), Vol. 19, 149.

3 We know less about the sexual life of little girls than of boys. But we need not feel ashamed of this distinction; after all, the sexual life of adult women is a 'dark continent' for psychology.

The Question of Lay Analysis (1926), in James Strachey (ed.), *The Standard Edition of the Complete Psychological Works of Sigmund Freud* (1959), Vol. 20, 212.

4 Religion is comparable to a childhood neurosis.

The Future of an Illusion (1927), 53.

5 No, our science is no illusion. But an illusion it would be to suppose that what science cannot give us we can get elsewhere.

The Future of an Illusion (1927), in James Strachey (ed.), *The Standard Edition of the Complete Psychological Works of Sigmund Freud* (1961), Vol. 21, 56.

6 The price we pay for our advance in civilization is a loss of happiness through the heightening of the sense of guilt.

Civilization and Its Discontents (1930), in James Strachey (ed.), *The Standard Edition of the Complete Psychological Works of Sigmund Freud* (1978), Vol. 21, 134.

7 Analogies, it is true, decide nothing but they can make one feel more at home.

New Introductory Lectures on Psycho-Analysis (1933), in James Strachey (ed.), *The Standard Edition of the Complete Psychological Works of Sigmund Freud* (1964), Vol. 22, 72.

8 You may take it as an instance of male injustice if I assert that envy and jealousy play an even greater part in the mental life of women than of men. It is not that I think these characteristics are absent in men or that I think they have no other roots in women than envy for the penis; but I am inclined to attribute their greater amount in women to this latter influence.

New Introductory Lectures on Psycho-Analysis (1933), in James Strachey (ed.), *The Standard Edition of the Complete Psychological Works of Sigmund Freud* (1964), Vol. 22, 125.

9 The reproaches against science for not having yet solved the problems of the universe are exaggerated in an unjust and malicious manner; it has truly not had time enough yet for these great achievements. Science is very young—a human activity which developed late.

The Question of a Weltanschauung? (1932), in James Strachey (ed.), *The Standard Edition of the Complete Psychological Works of Sigmund Freud* (1964), Vol. 22, 173.

10 Smoking is indispensable if one has nothing to kiss.

Letter to Martha Bernays, 22 January 1884. Quoted in P. Gay, *Freud: A Life For Our Time* (1988) 39.

11 For I am actually not at all a man of science, not an observer, nor an experimenter, not a thinker. I am by temperament nothing but a conquistador—an adventurer . . . with all the curiosity, daring, and tenacity characteristic of a man of this sort.

Letter to Wilhelm Fliess, 1 February 1900. In Jeffrey Moussaieff Masson (ed.), *The Complete Letters of Sigmund Freud to Wilhelm Fliess* (1985), 398.

12 In this House on July 24, 1895 the Secret of Dreams was revealed to Dr. Sigmund Freud.

Letter to Wilhelm Fliess, 20 June 1900. Quoted in Ernst L. Freud (ed.), *Letters of Sigmund Freud 1873–1939* (1961), 250. A plaque was placed at Bellevue on 6 May 1977—a house on the slopes of the

Wienerwald, where the Freud family spent their summers.

1 In view of the kind of matter we work with, it will never be possible to avoid little laboratory explosions.

Letter to Carl Jung, 18 June 1909. Quoted in William McGuire (ed.), *The Freud-Jung Letters: The Correspondence between Sigmund Freud and C. G. Jung* (1974), 235.

2 I cannot face with comfort the idea of life without work; work and the free play of the imagination are for me the same thing, I take no pleasure in anything else.

Letter to Oskar Pfister, 3 June 1910. Quoted in H. Meng and E. Freud (eds.), *Psycho-Analysis and Faith: The Letters of Sigmund Freud and Oskar Pfister* (1963), 146.

3 It is unreasonable to expect science to produce a system of ethics—ethics are a kind of highway code for traffic among mankind—and the fact that in physics atoms which were yesterday assumed to be square are now assumed to be round is exploited with unjustified tendentiousness by all who are hungry for faith; so long as physics extends our dominion over nature, these changes ought to be a matter of complete indifference to you.

Letter to Oskar Pfister, 24 February 1928. Quoted in H. Meng and E. Freud (eds.), *Psycho-Analysis and Faith: The Letters of Sigmund Freud and Oscar Pfister* (1963), 123.

4 I no longer count as one of my merits that I always tell the truth as much as possible; it has become my métier.

Letter to Albert Einstein, 8 December 1932. Quoted in P. Gay, *Freud: A Life for Our Time* (1988), xvii.

5 The great question that has never been answered and which I have not been able to answer, despite my thirty years of research into the feminine soul, is 'What does a woman want?'

Freud once said to Marie Bonaparte. Quoted in Ernest Jones (ed.), *Sigmund Freud: Life and Work* (1955), Vol. 2, 468.

6 Sometimes a cigar is just a cigar.

Attributed.

7 Wit is the best safety valve modern man has evolved; the more civilization, the more repression, the more the need there is for wit.

Attributed.

8 When someone abuses me I can defend myself, but against praise I am defenceless.

Attributed.

Milton Friedman 1912–

American economist

9 In both social and natural sciences, the body of positive knowledge grows by the failure of a tentative hypothesis to predict phenomena the hypothesis professes to explain; by the patching up of that hypothesis until someone suggests a new hypothesis that more elegantly or simply embodies the troublesome phenomena, and so on ad infinitum. In both, experiment is sometimes possible, sometimes not (witness meteorology). In both, no experiment is ever completely controlled, and experience often offers evidence that is the equivalent of controlled experiment. In both, there is no way to have a self-contained closed system or to avoid interaction between the observer and the observed. The Gödel theorem in mathematics, the Heisenberg uncertainty principle in physics, the self-fulfilling or self-defeating prophecy in the social sciences all exemplify these limitations.

Inflation and Unemployment (1976), 348.

Otto Robert Frisch 1904–79

Austrian-born British physicist

10 We [Frisch and Lise Meitner] walked up and down in the snow, I on skis and she on foot (she said and proved that she could get along just as fast that way), and gradually the idea took shape that this was no chipping or cracking of the nucleus but rather a process to be explained by Bohr's idea that the nucleus was like a liquid drop;

such a drop might elongate and divide itself.

Otto Frisch and John A. Wheeler, 'The Discovery of Fission', *Physics Today*, November 1967, 20, 47.

1 Today we no longer ask what really goes on in an atom; we ask what is likely to be observed—and with what likelihood—when we subject atoms to any specified influences such as light or heat, magnetic fields or electric currents.

What Little I Remember (1979), 20.

2 Scientists have one thing in common with children: curiosity. To be a good scientist you must have kept this trait of childhood, and perhaps it is not easy to retain just one trait. A scientist *has* to be curious like a child; perhaps one can understand that there are other childish features he hasn't grown out of.

What Little I Remember (1979), 86.

Robert Frost 1874–1963

American poet

3 For, dear me, why abandon a belief
Merely because it ceases to be true.
Cling to it long enough, and not a doubt
It will turn true again, for so it goes.
Most of the change we think we see in life
Is due to truths being in and out of favour.

'The Black Cottage'. In Edward Connery Latham (ed.), *The Poetry of Robert Frost* (1971), 77.

4 Sarcastic Science, she would like to know,
In her complacent ministry of fear,
How we propose to get away from here
When she has made things so we have to go
Or be wiped out. Will she be asked to show
Us how by rocket we may hope to steer
To some star off there, say, a half light-year
Through temperature of absolute zero?
Why wait for Science to supply the how
When any amateur can tell it now?
The way to go away should be the same
As fifty million years ago we came—
If anyone remembers how that was
I have a theory, but it hardly does.

'Why Wait for Science?' In Edward Connery Latham (ed.), *The Poetry of Robert Frost: The Collected Poems, Complete and Unabridged* (1979), 395.

5 As a confirmed astronomer
I'm always for a better sky.

'The Objection to being Stepped on'. In Edward Connery Latham (ed.), *The Poetry of Robert Frost* (1971), 451.

6 And one of the three great things in the world is gossip, you know. First there's religion; and then there's science; and there's—and then there's friendly gossip. Those are the three—the three great things.

From the *Claremont Quarterly*, Spring 1958. Transcript of a taped conversation between Robert Frost and British author Cecil Day Lewis which was broadcast on the BBC on 13 September 1957.

James Anthony Froude 1818–94

British historian and biographer

7 The superstition of science scoffs at the superstition of faith.

Attributed.

Harold Shipley Fry 1878–1949

American chemist

8 The structural formula of the organic chemist is not the canvas on which the cubist artist should impose his drawings which he alone can interpret.

'A Pragmatic System of Notation for Electronic Valance Conceptions in Chemical Formulas', *Chemical Reviews*, 1928, 5, 558–9.

Elizabeth Fulhame fl. 1780–1810

British chemist

9 I publish this Essay in its present imperfect state, in order to prevent the furacious attempts of the prowling plagiary, and the insidious pretender to chymistry, from arrogating to themselves, and assuming my invention, in plundering silence: for there are those, who, if they can not be chymical, never fail by stratagem, and mechanical means, to deprive industry of the fruits, and fame of her labours.

Preface to *An Essay on Combustion with a View to a New Art of Dyeing and Painting* (1794), vii-viii.

Thomas Fuller 1654–1734

British physician

1 Nature, in the first compounding and forming of us, hath laid into the Substance and Constitution of each something equivalent to Ovula, of various distinct Kinds, productive of all the contagious, venomous Fevers, we can possibly have as long as we live.

Exanthematologia: Or, An Attempt to Give a Rational Account of Eruptive Fevers, Especially of the Measles and SmallPox (1730), 175.

Yuri Alekseyevich Gagarin 1934–68

Russian cosmonaut

2 *After two years of highly secret training, on 12 April 1961 Yuri Gagarin climbed into his Vostok ('East') spacecraft. As the engines fired on the Baikonur launchpad in Kazakhstan, he shouted:*
Poyekhali!

Let's go!

Attributed.

John Kenneth Galbraith 1908–

American economist

3 Inventions that are not made, like babies that are not born, are rarely missed. In the absence of new developments, old ones may seem very impressive for quite a long while.

The Affluent Society (1958), 127.

4 Faced with the choice between changing one's mind and proving that there is no need to do so, almost everyone gets busy with the proof.

Quotation supplied by John Kenneth Galbraith.

5 She is a reflection of comfortable middle-class values that do not take seriously the continuing unemployment. What I particularly regret is that she does not take seriously the intellectual decline. Having given up the Empire and the mass production of industrial goods, Britain's future lay in its scientific and artistic pre-eminence. Mrs Thatcher will be long remembered for the damage she has done.

On Mrs Margaret H. Thatcher.

The Guardian, 15 October 1988.

Galen 129–c.216

Graeco-Roman physician

6 The best physician is also a philosopher.

Title of one of Galen's works.

7 The plexus called rectiform [rete mirabile] by anatomists, is the most wonderful of the bodies located in this region. It encircles the gland [the hypophysis] itself and extends far to the rear; for nearly the whole base of the encephalon has this plexus lying beneath it. It is not a simple network but [looks] as if you had taken several fisherman's nets and superimposed them. It is characteristic of this net of Nature's, however, that the meshes of one layer are always attached to those of another, and it is impossible to remove any one of them alone; for, one after another, the rest follow the one you are removing, because they are all attached to one another successively.

On the Usefulness of the Parts of the Body, Book IX, 4. Trans. Margaret Tallmadge May (1968), Vol. 1, 430–1.

8 Every animal is sad after coitus except the human female and the rooster.

Attributed.

Galileo Galilei 1564–1642

Italian physicist and astronomer

1 I accepted the Copernican position several years ago and discovered from thence the causes of many natural effects which are doubtless inexplicable by the current theories. I have written up many reasons and refutations on the subject, but I have not dared until now to bring them into the open, being warned by the fortunes of Copernicus himself, our master, who procured for himself immortal fame among a few but stepped down among the great crowd (for this is how foolish people are to be numbered), only to be derided and dishonoured. I would dare publish my thoughts if there were many like you; but since there are not, I shall forbear.

Letter to Johannes Kepler, 4 August 1597. Quoted in G. de Santillana, *Crime of Galileo* (1955), 11.

2 But what exceeds all wonders, I have discovered four new planets and observed their proper and particular motions, different among themselves and from the motions of all the other stars; and these new planets move about another very large star [Jupiter] like Venus and Mercury, and perchance the other known planets, move about the Sun. As soon as this tract, which I shall send to all the philosophers and mathematicians as an announcement, is finished, I shall send a copy to the Most Serene Grand Duke, together with an excellent spyglass, so that he can verify all these truths.

Letter to the Tuscan Court, 30 January 1610. Quoted in Albert van Helden (ed.), *Siderius Nuncius or The Sidereal Messenger* (1989), 18.

3 About ten months ago [1609] a report reached my ears that a certain Fleming [Hans Lippershey] had constructed a spyglass, by means of which visible objects, though very distant from the eye of the observer, were distinctly seen as if nearby . . . Of this truly remarkable effect several experiences were related, to which some persons gave credence while others denied them. A few days later the report was confirmed to me in a letter from a noble Frenchman at Paris, Jacques Badovere, which caused me to apply myself wholeheartedly to enquire into the means by which I might arrive at the invention of a similar instrument. This I did shortly afterwards, my basis being the theory of refraction. First I prepared a tube of lead, at the ends of which I fitted two glass lenses, both plane on one side while on the other side one was spherically convex and the other concave.

The Starry Messenger (1610), trans. Stillman Drake, *Discoveries and Opinions of Galileo* (1957), 28–9.

4 When the moon is ninety degrees away from the sun it sees but half the earth illuminated (the western half). For the other (the eastern half) is enveloped in night. Hence the moon itself is illuminated less brightly from the earth, and as a result its secondary light appears fainter to us.

The Starry Messenger (1610), trans. Stillman Drake, *Discoveries and Opinions of Galileo* (1957), 45.

5 For the holy Bible and the phenomena of nature proceed alike from the divine Word, the former as the dictate of the Holy Ghost and the latter as the observant executrix of God's commands. It is necessary for the Bible, in order to be accommodated to the understanding of every man, to speak many things which appear to differ from the absolute truth so far as the bare meaning of the words is concerned. But Nature, on the other hand, is inexorable and immutable; she never transgresses the laws imposed upon her, or cares a whit whether her abstruse reasons and methods of operation are understandable to men. For that reason it appears that nothing physical which sense-experience sets before our eyes, or which necessary demonstrations prove to us, ought to be called in question (much less condemned) upon the testimony of biblical passages which may have some different meaning beneath their words.

Letter to Madame Christina of Lorraine, Grand Duchess of Tuscany: Concerning the Use of Biblical Quotations in Matters of Science (1615), trans. Stillman Drake, *Discoveries and Opinions of Galileo* (1957), 182–3.

1 But I do not feel obliged to believe that that same God who has endowed us with senses, reason, and intellect has intended to forgo their use and by some other means to give us knowledge which we can attain by them.

Letter to Madame Christina of Lorraine, Grand Duchess of Tuscany: (1615). In *Discoveries and Opinions of Galileo*, trans. Stillman Drake (1957), 183.

2 I would say here something that was heard from an ecclesiastic of the most eminent degree [Cardinal Baronius (1538–1607).]: 'That the intention of the holy ghost is to teach us how one goes to heaven, not how heaven goes.'

Letter to Madame Christina of Lorraine, Grand Duchess of Tuscany: concerning the Use of Biblical Quotations in Matters of Science (1615), trans. Stillman Drake, *Discoveries and Opinions of Galileo* (1957), 186.

3 Two truths cannot contradict one another.

Letter to Madame Christina of Lorraine, Grand Duchess of Tuscany: concerning the Use of Biblical Quotations in Matters of Science (1615) , trans. Stillman Drake, *Discoveries and Opinions of Galileo* (1957), 186.

4 Philosophy is written in this grand book, the universe, which stands continually open to our gaze. But the book cannot be understood unless one first learns to comprehend the language and read the letters in which it is composed. It is written in the language of mathematics, and its characters are triangles, circles, and other geometric figures without which it is humanly impossible to understand a single word of it; without these, one wanders about in a dark labyrinth.

'The Assayer' (1623), trans. Stillman Drake, *Discoveries and Opinions of Galileo* (1957), 237–8.

5 I cannot but be astonished that Sarsi should persist in trying to prove by means of witnesses something that I may see for myself at any time by means of experiment. Witnesses are examined in doutbful matters which are past and transient, not in those which are actual and present. A judge must seek by means of witnesses to determine whether Peter injured John last night, but not whether John was injured, since the judge can see that for himself.

'The Assayer' (1623), trans. Stillman Drake, *Discoveries and Opinions of Galileo* (1957), 271.

6 If experiments are performed thousands of times at all seasons and in every place without once producing the effects mentioned by your philosophers, poets, and historians, this will mean nothing and we must believe their words rather our own eyes? But what if I find for you a state of the air that has all the conditions you say are required, and still the egg is not cooked nor the lead ball destroyed? Alas! I should be wasting my efforts . . . for all too prudently you have secured your position by saying that 'there is needed for this effect violent motion, a great quantity of exhalations, a highly attenuated material and whatever else conduces to it.' This 'whatever else' is what beats me, and gives you a blessed harbor, a sanctuary completely secure.

'The Assayer' (1623), trans. Stillman Drake, *Discoveries and Opinions of Galileo* (1957), 273.

7 As to what Simplicius said last, that to contend whether the parts of the Sun, Moon, or other celestial body, separated from their whole, should naturally return to it, is a vanity, for that the case is impossible, it being clear by the demonstrations of Aristotle that the celestial bodies are impassible, impenetrable, unpartable, etc., I answer that none of the conditions whereby Aristotle distinguishes the celestial bodies from the elementary has any foundation other than what he deduces from the diversity of their natural motions; so that, if it is denied that the circular motion is peculiar to celestial bodies, and affirmed instead that it is agreeable to all naturally moveable bodies, one is led by necessary confidence to say either that the attributes of generated or ungenerated, alterable or unalterable, partable or unpartable, etc., equally and commonly apply to all bodies, as well to the celestial as to the elementary, or that Aristotle has badly and erroneously deduced those from the circular motion which he has assigned to celestial bodies.

Dialogue on the Great World Systems (1632). Revised and Annotated by Giorgio De Santillana (1953), 45.

1 Nor need you doubt that Pythagoras, a long time before he found the demonstration for the Hecatomb, had been certain that the square of the side subtending the right angle in a rectangular triangle was equal to the square of the other two sides; the certainty of the conclusion helped not a little in the search for a demonstration. But whatever was the method of Aristotle, and whether his arguing a priori preceded sense a posteriori, or the contrary, it is sufficient that the same Aristotle (as has often been said) put sensible experiences before all discourses. As to the arguments a priori, their force has already been examined.

Dialogue on the Great World Systems (1632). Revised and Annotated by Giorgio De Santillana (1953), 60.

2 [Simplicio] is much puzzled and perplexed. I think I hear him say, 'To whom then should we repair for the decision of our controversies if Aristotle were removed from the choir? What other author should we follow in the schools, academies, and studies? What philosopher has written all the divisions of Natural Philosophy, and so methodically, without omitting as much as a single conclusion? Shall we then overthrow the building under which so many voyagers find shelter? Shall we destroy that sanctuary, that Prytaneum, where so many students find commodious harbour; where without exposing himself to the injuries of the air, with only the turning over of a few leaves, one may learn all the secrets of Nature.'

Dialogue on the Great World Systems (1632). Revised and Annotated by Giorgio De Santillana (1953), 66.

3 Their vain presumption of knowing all can take beginning solely from their never having known anything; for if one has but once experienced the perfect knowledge of one thing, and truly tasted what it is to know, he shall perceive that of infinite other conclusions he understands not so much as one.

Dialogue on the Great World Systems (1632). Revised and Annotated by Giorgio De Santillana (1953), 112.

4 We see only the simple motion of descent, since that other circular one common to the Earth, the tower, and ourselves remains imperceptible. There remains perceptible to us only that of the stone, which is not shared by us; and, because of this, sense shows it as by a straight line, always parallel to the tower, which is built upright and perpendicular upon the terrestrial surface.

Dialogue on the Great World Systems (1632). Revised and Annotated by Giorgio De Santillana (1953), 177.

5 Take note, theologians, that in your desire to make matters of faith out of propositions relating to the fixity of sun and earth you run the risk of eventually having to condemn as heretics those who would declare the earth to stand still and the sun to change position—eventually, I say, at such a time as it might be physically or logically proved that the earth moves and the sun stands still.

Note added by Galileo in the preliminary

leaves of his copy of *Dialogue on the Great World Systems* (1632).

1 It seems to me that your Reverence and Signor Galileo act prudently when you content yourselves with speaking hypothetically and not absolutely, as I have always understood that Copernicus spoke. To say that on the supposition of the Earth's movement and the Sun's quiescence all the celestial appearances are explained better than by the theory of eccentrics and epicycles is to speak with excellent good sense and to run no risk whatsoever. Such a manner of speaking is enough for a mathematician. But to want to affirm that the Sun, in very truth, is at the center of the universe and only rotates on its axis without going from east to west, is a very dangerous attitude and one calculated not only to arouse all Scholastic philosophers and theologians but also to injure our holy faith by contradicting the Scriptures.

Roberto Bellarmino, Cardinal Inquisitor of the Holy Office, letter to Paolo Antonio Foscarini, 12 April 1615. Quoted in Giorgio De Santillana, *The Crime of Galileo* (1955), 99.

2 I, Galileo Galilei, son of the late Vincenzio Galilei of Florence, aged seventy years, being brought personally to judgment, and kneeling before you, Most Eminent and Most Reverend Lords Cardinals, General Inquisitors of the Universal Christian Commonwealth against heretical depravity, having before my eyes the Holy Gospels which I touch with my own hands, swear that I have always believed, and, with the help of God, will in future believe, every article which the Holy Catholic and Apostolic Church of Rome holds, teaches, and preaches. But because I have been enjoined, by the Holy Office, altogether to abandon the false opinion which maintains that the Sun is the center and immovable, and forbidden to hold, defend, or teach, the said false doctrine in any manner . . . I am willing to remove from the minds of your Eminences, and of every Catholic Christian, this vehement suspicion rightly entertained towards me, therefore, with a sincere heart and unfeigned faith, I abjure, curse, and detest said errors and heresies, and generally every other error and sect contrary to the said Holy Church; and I swear that I will never more in future say, or assert anything, verbally or in writing, which may give rise to a similar suspicion of me; but that if I shall know any heretic, or anyone suspected of heresy, I will denounce him to this Holy Office, or to the Inquisitor and Ordinary of the place in which I may be. I swear, moreover, and promise that I will fulfil and observe fully all the penances which have been or shall be laid on me by this Holy Office. But if it shall happen that I violate any of my said promises, oaths, and protestations (which God avert), I subject myself to all the pains and punishments which have been decreed and promulgated by the sacred canons and other general and particular constitutions against delinquents of this description. So, may God help me, and his Holy Gospels, which I touch with my own hands, I, the above named Galileo Galilei, have abjured, sworn, promised, and bound myself as above; and, in witness thereof, with my own hand have subscribed this present writing of my abjuration, which I have recited word for word.

The terms of Galileo's Abjuration. Quoted in J. J. Fahie, *Galileo, His Life and Work* (1903), 319–21.

3 *Eppur si muove.*

And yet it does move.

Apocryphal words to himself after making his abjuration of heliocentricity.

A painting of 1643 by Murillo depicts Galileo pointing at these words on the wall of his cell, so it was an early tradition. See Stillman Drake, *Galileo at Work* (1978), 356–7.

4 In questions of science the authority of a thousand is not worth the humble reasoning of a single individual.

Attributed by F. Arago.

1 ON GALILEI It took Galileo 16 years to master the universe. You have one night. It seems unfair. The genius had all that time. While you have a few short hours to learn sun spots from your satellites before the dreaded astronomy exam. If Galileo had used Vivarin, maybe he could have mastered the solar system faster, too.

Advert for stimulant in student newspaper, *Daily Philadelphian*, 6 December 1990. Quotation supplied by W. H. Brock.

Franz Joseph Gall 1758–1828

German anatomist and phrenologist

2 Whoever would not remain in complete ignorance of the resources which cause him to act; whoever would seize, at a single philosophical glance, the nature of man and animals, and their relations to external objects; whoever would establish, on the intellectual and moral functions, a solid doctrine of mental diseases, of the general and governing influence of the brain in the states of health and disease, should know, that it is indispensable, that the study of the organization of the brain should march side by side with that of its functions.

On the Organ of the Moral Qualities and Intellectual Faculties, and the Plurality of the Cerebral Organs (1835), 45–6.

3 The fate of the physiology of the brain is independent of the truth and falsity of my assertions relative to the laws of the organization of the nervous system, in general, and of the brain in particular, just as the knowledge of the functions of a sense is independent of the knowledge of the structure of its apparatus.

Critical Review of Some Anatomical-Physiological Works; With an Explanation of a New Philosophy of the Moral Qualities and Intellectual Faculties (1835), 237–8.

Arthur William Galston 1920–

American plant biologist

4 In my view, the only recourse for a scientist concerned about the social consequences of his work is to remain involved with it to the end.

'Science and Social Responsibility: A Case Study', *Annals of the New York Academy of Science*, 1972, **196**, 223.

Francis Galton 1822–1911

British statistician and psychologist

See ***Webb, Beatrice*** *610:6.*

5 Exercising the right of occasional suppression and slight modification, it is truly absurd to see how plastic a limited number of observations become, in the hands of men with preconceived ideas.

Meteorographica (1863), 5.

6 We shall therefore take an appropriately correct view of the origin of our life, if we consider our own embryos to have sprung immediately from those embryos whence our parents were developed, and these from the embryos of *their* parents, and so on for ever. We should in this way look on the nature of mankind, and perhaps on that of the whole animated creation, as one Continuous System, ever pushing out new branches in all directions, that variously interlace, and that bud into separate lives at every point of interlacement.

'Hereditary Talent and Character', *Macmillan's Magazine*, 1865, **12**, 322.

7 Characteristics cling to families.

Hereditary Genius (1869), v.

8 It is notorious that the same discovery is frequently made simultaneously and quite independently, by different persons. Thus, to speak of only a few cases in late years, the discoveries of photography, of electric telegraphy, and of the planet Neptune through theoretical calculations, have all their rival claimants. It would seem, that discoveries are usually made when the time is ripe for them—that is to say, when the ideas from which they naturally flow are fermenting in the minds of many men.

Hereditary Genius (1869), 192.

1 A *primâ facie* argument in favour of the efficacy of prayer is therefore to be drawn from the very general use of it. The greater part of mankind, during all the historic ages, have been accustomed to pray for temporal advantages. How vain, it may be urged, must be the reasoning that ventures to oppose this mighty consensus of belief! Not so. The argument of universality either proves too much, or else it is suicidal.

'Statistical Inquiries into the Efficacy of Prayer', *Fortnightly Review*, 1872, 12, 126.

2 The phrase 'nature and nurture' is a convenient jingle of words, for it separates under two distinct heads the innumerable elements of which personality is composed. Nature is all that a man brings with himself into the world; nurture is every influence without that affects him after his birth.

English Men of Science: Their Nature and Nurture (1874), 12.

3 The processes concerned in simple descent are those of Family Variability and Reversion. It is well to define these words clearly. By family variability is meant the departure of the children of the same or similarly descended families from the ideal mean type of all of them. Reversion is the tendency of that ideal mean type to depart from the parent type, 'reverting' towards what may be roughly and perhaps fairly described as the average ancestral type. If family variability had been the only process in simple descent, the dispersion of the race would indefinitely increase with the number of the generations, but reversion checks this increase, and brings it to a standstill.

Typical Laws of Heredity (1877), 513.

4 We greatly want a brief word to express the science of improving stock, which is by no means confined to questions of judicious mating, but which, especially in the case of man, takes cognisance of all influences that tend in however remote a degree to give to the more suitable races or strains of blood a better chance of prevailing speedily over the less suitable than they otherwise would have had. The word *eugenics* would sufficiently express the idea; it is at least a neater word and a more generalised one than *viviculture*, which I once ventured to use.

First use of the term Eugenics.

Inquiries into Human Faculty and its Development (1883), 25, footnote.

5 My method consists in allowing the mind to play freely for a very brief period, until a couple or so of ideas have passed through it, and then, while the traces or echoes of those ideas are still lingering in the brain, to turn the attention upon them with a sudden and complete awakening; to arrest, to scrutinise them, and to record their exact appearance . . . The general impression they have left upon me is like that which many of us have experienced when the basement of our house happens to be under thorough sanitary repairs, and we realise for the first time the complex system of drains and gas and water pipes, flues, bell-wires, and so forth, upon which our comfort depends, but which are usually hidden out of sight, and with whose existence, so long as they acted well, we had never troubled ourselves.

Inquiries into Human Faculty and its Development (1883), 185–6.

6 *The Charms of Statistics.*—It is difficult to understand why statisticians commonly limit their inquiries to Averages, and do not revel in more comprehensive views. Their souls seem as dull to the charm of variety as that of the native of one of our flat English counties, whose retrospect of Switzerland was that, if its mountains could be thrown into its lakes, two nuisances would be got rid of at once. An Average is but a solitary fact, whereas if a single other fact be added to it, an entire Normal Scheme, which nearly corresponds to the observed one, starts potentially into existence. Some people hate the very name of statistics, but I find them full of beauty and interest. Whenever they are not brutalised, but delicately handled by the higher methods, and are warily interpreted, their power of dealing with complicated phenomena is extraordinary. They are the only tools by which an opening can be cut through the formidable thicket of difficulties that bars the path of those who pursue the Science of man.

Natural Inheritance (1889), 62–3.

1 The publication in 1859 of the *Origin of Species* by Charles Darwin made a marked epoch in my own mental development, as it did in that of human thought generally. Its effect was to demolish a multitude of dogmatic barriers by a single stroke, and to arouse a spirit of rebellion against all ancient authorities whose positive and unauthenticated statements were contradicted by modern science.

Memories of My Life (1908), 287.

2 Whenever you can, count.

Quoted in James R. Newman, *Commentary on Sir Francis Galton* (1956), 1169.

3 **ON GALTON** Francis Galton, whose mission it seems to be to ride other men's hobbies to death, has invented the felicitous expression 'structureless germs'.

Letter from James Clerk Maxwell to Professor Lewis Campbell, 26th September 1874. Quoted in Lewis Campbell and William Garnett (eds.), *The Life of James Clerk Maxwell* (1884), 299.

George Gamow 1904–68

Russian-born American physicist

4 Adam, the first man, didn't know anything about the nucleus but Dr. George Gamow, visiting professor from George Washington University, pretends he does. He says for example that the nucleus is 0.000000000000003 feet in diameter. Nobody believes it, but that doesn't make any difference to him.

He also says that the nuclear energy contained in a pound of lithium is enough to run the United States Navy for a period of three years. But to get this energy you would have to heat a mixture of lithium and hydrogen up to 50,000,000 degrees Fahrenheit. If one has a little stove of this temperature installed at Stanford, it would burn everything alive within a radius of 10,000 miles and broil all the fish in the Pacific Ocean.

If you could go as fast as nuclear particles generally do, it wouldn't take you more than one ten-thousandth of a second to go to Miller's where you could meet Gamow and get more details.

'Gamow interviews Gamow', *Stanford Daily*, 25 June 1936. In Helge Kragh, *Cosmology and Controversy: The Historical Development of Two Theories of the Universe* (1996), 90.

5 If and when all the laws governing physical phenomena are finally discovered, and all the empirical constants occurring in these laws are finally expressed through the four independent basic constants, we will be able to say that physical science has reached its end, that no excitement is left in further explorations, and that all that remains to a physicist is either tedious work on minor details or the self-educational study and adoration of the magnificence of the completed system. At that stage physical science will enter from the epoch of Columbus and Magellan into the epoch of the National Geographic Magazine!

'Any Physics Tomorrow?', *Physics Today*, January 1949, 2, 17.

6 If the expansion of the space of the universe is uniform in all directions, an observer located in any one of the galaxies will see all other galaxies running away from him at velocities proportional to their distances from the observer.

The Creation of the Universe (1952), 31.

7 It took less than an hour to make the atoms, a few hundred million years to make the stars and planets, but five billion years to make man!

The Creation of the Universe (1952), 139.

8 Twinkle, twinkle, quasi-star
Biggest puzzle from afar

How unlike the other ones
Brighter than a billion suns
Twinkle, twinkle, quasi-star
How I wonder what you are.

Originally published in *Newsweek*, 25 May 1964. In Ivor Robinson, Alfred Schild, and E. L. Schuckling (eds.), *Quasi-Stellar Sources and Gravitational Collapse* (1965), 472.

Martin Gardner 1914-

American science writer

1 Politicians, real-estate agents, used-car salesmen, and advertising copy-writers are expected to stretch facts in self-serving directions, but scientists who falsify their results are regarded by their peers as committing an inexcusable crime. Yet the sad fact is that the history of science swarms with cases of outright fakery and instances of scientists who unconsciously distorted their work by seeing it through lenses of passionately held beliefs.

Science Good, Bad and Bogus (1981), 123.

Archibald Garrod 1857–1936

British physician and biochemist

2 All the more recent work on alkaptonuria has . . . strengthened the belief that the homogentisic acid excreted is derived from tyrosin, but why alkaptonuric individuals pass the benzene ring of their tyrosin unbroken and how and where the peculiar chemical change from tyrosin to homogentisic acid is brought about, remain unsolved problems.

'The Incidence of Alkaptonuria: A Study in Chemical Individuality', *The Lancet*, 1902, 2, 1616.

3 Inborn errors of metabolism.

Inborn Errors of Metabolism. The Croonian Lectures delivered before the Royal College of Physicians of London, in June, 1908 (1909).

4 Nor can it be supposed that the diversity of chemical structure and process stops at the boundary of the species, and that within that boundary, which has no real finality, rigid uniformity reigns. Such a conception is at variance with any evolutionary conception of the nature and origin of species. The existence of chemical individuality follows of necessity from that of chemical specificity, but we should expect the differences between individuals to be still more subtle and difficult of detection. Indications of their existence are seen, even in man, in the various tints of skin, hair, and eyes, and in the quantitative differences in those portions of the end-products of metabolism which are endogenous and are not affected by diet, such as recent researches have revealed in increasing numbers. Even those idiosyncrasies with regard to drugs and articles of food which are summed up in the proverbial saying that what is one man's meat is another man's poison presumably have a chemical basis.

Inborn Errors of Metabolism. The Croonian Lectures delivered before the Royal College of Physicians of London, in June, 1908 (1909), 2–3.

5 Nevertheless, scientific method is not the same as the scientific spirit. The scientific spirit does not rest content with applying that which is already known, but is a restless spirit, ever pressing forward towards the regions of the unknown, and endeavouring to lay under contribution for the special purpose in hand the knowledge acquired in all portions of the wide field of exact science. Lastly, it acts as a check, as well as a stimulus, sifting the value of the evidence, and rejecting that which is worthless, and restraining too eager flights of the imagination and too hasty conclusions.

'The Scientific Spirit in Medicine: Inaugural Sessional Address to the Abernethian Society', *St. Bartholomew's Hospital Journal*, 1912, 20, 19.

6 Science is not, as so many seem to think, something apart, which has to do with telescopes, retorts, and test-tubes, and especially with nasty smells, but it is a way of searching out by observation, trial and classification; whether the phenomena investigated be the outcome of human activities, or of the more direct workings of nature's laws. Its methods admit of nothing untidy or slip-shod; its keynote is accuracy and its goal is truth.

The University of Utopia (1918), 17.

Walter Holbrook Gaskell

1847–1914

British physiologist

1 The inhibitory nerves are of as fundamental importance in the economy of the body as the motor nerves. No evidence exists that the same nerve fibre is sometimes capable of acting as a motor nerve, sometimes as a nerve of inhibition, but on the contrary the latter nerves form a separate and complete nervous system subject to as definite anatomical and histological laws as the former.

'On the Structure, Distribution and Function of the Nerves which Innervate the Visceral and Vascular Systems', *The Journal of Physiology*, 1886, 7, 40.

Herbert Spencer Gasser

1888–1963 and

Joseph Erlanger 1874–1965

American physiologists

2 We have seen that the cytoplasm of nerve has a fluid consistency. Hence its molecules are free to move. According to the thermodynamic principle known as the Gibbs-Thompson rule, any substance in the interior of a liquid which will reduce the free energy of the surface of the liquid, will be concentrated in the surface. The composition of the surface is, therefore, determined by the composition of the fluid from which it is formed; and as the rule is one having universal application, it must hold also for the cytoplasm of nerve. We must think of the surface membrane, then, as a structure which is in equilibrium with the interior of the axon, or at least as one which deviates from equilibrium only because, for dynamic reasons, equilibrium cannot be attained.

Joseph Erlanger and Herbert S. Gasser (eds.), *Electrical Signs of Nervous Activity* (1937), 136.

Marc Antoine Augustin Gaudin

1804–80

French crystallographer and chemist

3 For us, an *atom* shall be a small, spherical, homogeneous body or an essentially indivisible, material point, whereas a *molecule* shall be a separate group of atoms in any number and of any nature.

Annales de Chimie 1833, 52, 133. Trans. W. H. Brock.

Joseph Louis Gay-Lussac

1778–1850

French chemist and physicist

4 Compounds of gaseous substances with each other are always formed in very simple ratios, so that representing one of the terms by unity, the other is 1, or 2, or at most 3 . . . The apparent *contraction* of volume suffered by gases on combination is also very simply related to the volume of one of them.

Mémoires de la Société d'Arcueil, 1809, 2, 233–4. Trans. *Foundations of the Molecular Theory*, Alembic Club Reprint, no. 4 (1950), 24.

5 I have not chosen a career that will lead me to a great fortune, but that is not my principal ambition.

Letter to his father, 1803. Ironic in view of his later large income through applied science. Maurice Crosland, *Gay-Lussac, Scientist, Bourgeois* (1978), 1.

6 At Arcueil . . . I dined in distinguished company . . . There was a lot of very interesting discussion. It is these gatherings which are the joy of life.

Maurice Crosland, *Gay-Lussac, Scientist and Bourgeois* (1978), 21.

7 If one were not animated with the desire to discover laws, they would escape the most enlightened attention.

Maurice Crosland, *Gay-Lussac, Scientist and Bourgeois* (1978), 54.

8 In the natural sciences, and particularly in chemistry, generalities must come after the detailed knowledge of each fact and not before it.

Maurice Crosland, *Gay-Lussac, Scientist and Bourgeois* (1978), 69.

Archibald Geikie 1835–1924

British geologist

1 This boulder seemed like a curious volume, regularly paged, with a few extracts from older works. Bacon tells us that 'some books are to be tasted, others to be swallowed, and some few to be chewed and digested.' Of the last honour I think the boulder fully worthy.

The Story of a Boulder (1858), 4.

2 Geologists have not been slow to admit that they were in error in assuming that they had an eternity of past time for the evolution of the earth's history. They have frankly acknowledged the validity of the physical arguments which go to place more or less definite limits to the antiquity of the earth. They were, on the whole, disposed to acquiesce in the allowance of 100 millions of years granted to them by Lord Kelvin, for the transaction of the whole of the long cycles of geological history. But the physicists have been insatiable and inexorable. As remorseless as Lear's daughters, they have cut down their grant of years by successive slices, until some of them have brought the number to something less than ten millions. In vain have the geologists protested that there must somewhere be a flaw in a line of argument which tends to results so entirely at variance with the strong evidence for a higher antiquity, furnished not only by the geological record, but by the existing races of plants and animals. They have insisted that this evidence is not mere theory or imagination, but is drawn from a multitude of facts which become hopelessly unintelligible unless sufficient time is admitted for the evolution of geological history. They have not been able to disapprove the arguments of the physicists, but they have contended that the physicists have simply ignored the geological arguments as of no account in the discussion.

'Twenty-five years of Geological Progress in Britain', *Nature*, 1895, 51, 369.

3 Apart from its healthful mental training as a branch of ordinary education, geology as an open-air pursuit affords an admirable training in habits of observation, furnishes a delightful relief from the cares and routine of everyday life, takes us into the open fields and the free fresh face of nature, leads us into all manner of sequestered nooks, whither hardly any other occupation or interest would be likely to send us, sets before us problems of the highest interest regarding the history of the ground beneath our feet, and thus gives a new charm to scenery which may be already replete with attractions.

Outlines of Field-Geology (1900), 251–2.

4 If all history is only an amplification of biography, the history of science may be most instructively read in the life and work of the men by whom the realms of Nature have been successively won.

The Founders of Geology (1905), 4.

5 The present is the key to the past.

The Founders of Geology (1905), 299.

Murray Gell-Mann 1929–

American physicist

6 In 1963, when I assigned the name 'quark' to the fundamental constituents of the nucleon, I had the sound first, without the spelling, which could have been 'kwork.' Then, in one of my occasional perusals of *Finnegans Wake*, by James Joyce, I came across the word 'quark' in the phrase 'Three quarks for Muster Mark.' Since 'quark' (meaning, for one thing, the cry of a gull) was clearly intended to rhyme with 'Mark,' as well as 'bark' and other such words, I had to find an excuse to pronounce it as 'kwork.' But the book represents the dreams of a publican named Humphrey Chimpden Earwicker. Words in the text are typically drawn from several sources at once, like the 'portmanteau words' in *Through the Looking Glass.* From time to time, phrases occur in the book that are partially determined by calls for drinks at the bar. I argued, therefore, that perhaps one of the multiple sources of the cry 'Three quarks for Muster Mark' might be pronunciation for 'Three quarts for Mister Mark,' in which case the pronunciation 'kwork' would not be totally unjustified. In any case, the number three fitted perfectly the way quarks occur in nature.

The Quark and the Jaguar (1994), 180.

Ernest André Gellner 1925–95

Czech-British philosopher and social anthropologist

1 Roughly, science is the mode of cognition of industrial society, and industry is the ecology of science.

Thought and Change (1964), 179.

Etienne Geoffroy Saint-Hilaire 1772–1844

French zoologist

2 We know that nature invariably uses the same materials in its operations. Its ingeniousness is displayed only in the variation of form. Indeed, as if nature had voluntarily confined itself to using only a few basic units, we observe that it generally causes the same elements to reappear, in the same number, in the same circumstances, and in the same relationships to one another. If an organ happens to grow in an unusual manner, it exerts a considerable influence on adjacent parts, which as a result fail to reach their standard degree of development.

'Considérations sur les pieces de la tête osseuse des animaux vertebras, et particulièrement sur celle du crane des oiseaux', *Annales du Museum d'Histoire Naturelle*, 1807, 10, 343. Trans. J. Mandelbaum. Quoted in Pietro Corsi, *The Age of Lamarck* (1988), 232.

3 Nature . . . tends to repeat the same organs in the same number and in the same relations, and varies to infinity only their form. In accordance with this principle I shall have to draw my conclusions, in the determining the bones of the fish's skull, not from a consideration of their form, but from a consideration of their connections.

'Considérations sur les pieces de la tête osseuse des animaux vertebras, et particulièrement sur celle du crane des oiseaux', *Annales du Museum d'Histoire Naturelle*, 1807, 10, 344. Trans. E. S. Russell, *Form and Function* (1916), 71.

George III 1738–1820

British monarch

4 I spend money on war because it is necessary, but to spend it on science, that is pleasant to me. This object costs no tears; it is an honour to humanity.

Said to Lalande. Quoted in R. A. Gregory, *Discovery, Or the Spirit and Service of Science* (1916), 47–8.

Charles Frédéric Gerhardt 1816–56

French chemist

5 In deriving a body from the water type I intend to express that to this body, considered as an oxide, there corresponds a chloride, a bromide, a sulphide, a nitride, etc., susceptible of double compositions, or resulting from double decompositions, analogous to those presented by hydrochloric acid, hydrobromic acid, sulphuretted hydrogen, ammonia etc., or which give rise to the same compounds. The type is thus the unit of comparison for all the bodies which, like it, are susceptible of similar changes or result from similar changes.

Traité de Chimie Organique, 1856, 4, 587. Trans. J. R. Partington, *A History of Chemistry* (1970), Vol. 4, 456.

Edward Gibbon 1737–94

British historian

1 The laws of probability, so true in general, so fallacious in particular.

Memoirs of my Life (1774) ed. G. A. Bonnard (1966), 188.

Michael Gibbons 1939– and **Philip Gummett** 1947–

British social scientists

2 During the eighteenth and nineteenth centuries we can see the emergence of a tension that has yet to be resolved, concerning the attitude of scientists towards the usefulness of science. During this time, scientists were careful not to stress too much their relationships with industry or the military. They were seeking autonomy for their activities. On the other hand, to get social support there had to be some perception that the fruits of scientific activity could have useful results. One resolution of this dilemma was to assert that science only contributed at the discovery stage; others, industrialists for example, could apply the results. . . . Few noted the obvious paradox of this position; that, if scientists were to be distanced from the 'evil' effects of the applications of scientific ideas, so too should they receive no credit for the 'good' or socially beneficial, effects of their activities.

Science, Technology and Society Today (1984), Introduction, 4.

Josiah Willard Gibbs 1839–1903

American theoretical physicist and chemist

3 The laws of thermodynamics, as empirically determined, express the approximate and probable behavior of systems of a great number of particles, or, more precisely, they express the laws of mechanics for such systems as they appear to beings who have not the fineness of perception to enable them to appreciate quantities of the order of magnitude of those which relate to single particles, and who cannot repeat their experiments often enough to obtain any but the most probable results.

Elementary Principles in Statistical Mechanics (1902), Preface, viii.

4 A mathematician may say anything he pleases, but a physicist must be at least partially sane.

Attributed. Cited in R. B. Lindsay, 'On the Relation of Mathematics and Physics', *The Scientific Monthly*, December 1944, 59, 456.

Grove Karl Gilbert 1843–1918

American geologist and geomorphologist

5 The conflict of theories, leading, as it eventually must, to the survival of the fittest, is advantageous.

'The Origin of Hypotheses, illustrated by the Discussion of a Topographical Problem', *Science*, 1896, 3, 2.

6 Knowledge of Nature is an account at bank, where each dividend is added to the principal and the interest is ever compounded; and hence it is that human progress, founded on natural knowledge, advances with ever increasing speed.

'The Origin of Hypotheses, illustrated by the Discussion of a Topographical Problem', *Science*, 1896, 3, 13.

William Gilbert 1544–1603

British natural philosopher

See ***Dryden, John*** *186:1*

7 Since the discovery of secret things and in the investigation of hidden causes, stronger reasons are obtained from sure experiments and demonstrated arguments than from probable conjectures and the opinions of philosophical speculators of the common sort; therefore to the end that the noble substance of that great loadstone, our common mother (the earth), still quite unknown, and also the forces extraordinary and exalted of this globe may the better be understood, we have decided first to begin with the common stony and ferruginous matter, and magnetic bodies, and the parts of the earth that we may handle and may perceive with the senses; then to proceed with plain magnetic experiments, and to penetrate to the inner parts of the earth.

On the Loadstone and Magnetic Bodies and on the Great Magnet the Earth: A New Physiology, Demonstrated with many Arguments and Experiments (1600), trans. P. Fleury Mottelay (1893), Author's Preface, xlvii.

1 In like manner, the loadstone has from nature its two poles, a northern and a southern; fixed, definite points in the stone, which are the primary termini of the movements and effects, and the limits and regulators of the several actions and properties. It is to be understood, however, that not from a mathematical point does the force of the stone emanate, but from the parts themselves; and all these parts in the whole—while they belong to the whole—the nearer they are to the poles of the stone the stronger virtues do they acquire and pour out on other bodies. These poles look toward the poles of the earth, and move toward them, and are subject to them. The magnetic poles may be found in very loadstone, whether strong and powerful (male, as the term was in antiquity) or faint, weak, and female; whether its shape is due to design or to chance, and whether it be long, or flat, or four-square, or three-cornered or polished; whether it be rough, broken-off, or unpolished: the loadstone ever has and ever shows its poles.

On the Loadstone and Magnetic Bodies and on the Great Magnet the Earth: A New Physiology, Demonstrated with many Arguments and Experiments (1600), trans. P. Fleury Mottelay (1893), 23.

George Robert Gissing

1857–1903

British novelist

2 I hate and fear 'science' because of my conviction that, for long to come if not for ever, it will be the remorseless enemy of mankind. I see it destroying all simplicity and gentleness of life, all the beauty of the world; I see it restoring barbarism under a mask of civilization; I see it darkening men's minds and hardening their hearts.

The Private Papers of Henry Ryecroft (1903), 268–9.

William Ewart Gladstone

1809–98

British Prime Minister

3 It is difficult to see anything but infatuation in the destructive temperament which leads to the action . . . that each of us is to rejoice that our several units are to be distinguished at death into countless millions of organisms; for such, it seems, is the latest revelation delivered from the fragile tripod of a modern Delphi.

'Dawn and the Creation of Worship', *The Nineteenth Century*, 1885, **18**, 706.

Joseph Glanvill 1636–80

British natural philosopher

4 And for *mathematical sciences*, he that doubts their certainty hath need of a dose of *Hellebore*.

The Vanity of Dogmatizing (1661), ch. xxi, 209.

5 We cannot conceive how the *Foetus* is form'd in the *Womb*, nor as much as how a *Plant* springs from the *Earth* we tread on . . . And if we are ignorant of the most *obvious* things about us, and the most *considerable within* our selves, 'tis then no wonder that we know not the *constitution* and *powers* of the *creatures*, to whom we are such strangers.

Saducismus Triumphatus or Full and Plain Evidence Concerning Witches and Apparitions (1689), 72–3.

Sheldon Lee Glashow 1932–

American theoretical physicist

6 Tapestries are made by many artisans working together. The contributions of separate workers cannot be discerned in the completed work, and the loose and false threads have been covered over. So it is in our picture of particle physics.

'Towards a Unified Theory—Threads in a Tapestry', Nobel Lecture, 8 December 1979. In *Nobel Lectures: Physics 1971–1980* (1992), 494.

Henry Allan Gleason 1882–1975

American botanist

1 Every species of plant is a law unto itself.

'The Individualistic Concept of the Plant Association', *Bulletin of the Torrey Botanical Club*, 1926, 53, 26.

Herman Max Gluckman 1911–75

South African social anthropologist

2 A science is any discipline in which the fool of this generation can go beyond the point reached by the genius of the last generation.

Politics, Law and Ritual in Tribal Society (1965), 32.

Kurt Gödel 1906–78

Austrian-born American logician and mathematician

3 The development of mathematics toward greater precision has led, as is well known, to the formalization of large tracts of it, so that one can prove any theorem using nothing but a few mechanical rules . . . One might therefore conjecture that these axioms and rules of inference are sufficient to decide *any* mathematical question that can at all be formally expressed in these systems. It will be shown below that this is not the case, that on the contrary there are in the two systems mentioned relatively simple problems in the theory of integers that cannot be decided on the basis of the axioms.

'On Formally Undecidable Propositions of *Principia Mathematica* and Related Systems I' (1931), in S. Feferman (ed.), *Kurt Gödel Collected Works: Publications 1929–1936* (1986), Vol. 1, 145.

4 Classes and concepts may, however, also be conceived as real objects, namely classes as 'pluralities of things' or as structures consisting of a plurality of things and concepts as the properties and relations of things existing independently of our definitions and constructions. It seems to me that the assumption of such objects is quite as legitimate as the assumption of physical bodies and there is quite as much reason to believe in their existence. They are in the same sense necessary to obtain a satisfactory system of mathematics as physical bodies are necessary for a satisfactory theory of our sense perceptions . . .

'Russell's Mathematical Logic', in P. A. Schilpp (ed.), *The Philosophy of Bertrand Russell* (1944), Vol. 1, 137.

5 Non-standard analysis frequently simplifies substantially the proofs, not only of elementary theorems, but also of deep results. This is true, e.g., also for the proof of the existence of invariant subspaces for compact operators, disregarding the improvement of the result; and it is true in an even higher degree in other cases. This state of affairs should prevent a rather common misinterpretation of non-standard analysis, namely the idea that it is some kind of extravagance or fad of mathematical logicians. Nothing could be farther from the truth. Rather, there are good reasons to believe that non-standard analysis, in some version or other, will be the analysis of the future.

'Remark on Non-standard Analysis' (1974), in S. Feferman (ed.), *Kurt Gödel Collected Works: Publications 1938–1974* (1990), Vol. 2, 311.

Johann Wolfgang von Goethe 1749–1832

German naturalist, philosopher and writer

6 The Primal Plant is going be the strangest creature in the world, which Nature herself must envy me. With this model and the key to it, it will be possible to go on for ever inventing plants and know that their existence is logical; that is to say, if they do not actually exist, they could, for they are not the shadowy phantoms of a vain imagination, but possess an inner necessity and truth. The same law will be applicable to all other living organisms.

To Herder, 17 May 1787. *Italian Journey* (1816–17), trans. W. H. Auden and Elizabeth Mayer (1970), 310-11.

1 *Zweck sein selbst ist jegliches Tier.*

Each animal is an end in itself.

'Metamorphose der Tiere' (1806), in David Luke (ed.), *Goethe* (1964), 152.

2 *Blut ist ein ganz besondrer Saft.*

Blood is a very special juice.

'Faust I' (1808), Faust's Study III, l. 1740, in *Faust I & II*, trans. Stuart Atkins (1984), 45.

3 *Die Wahlverwandtschaften.*

Elective affinities.

Title of a novel, 1809.

4 Nature goes her own way, and all that to us seems an exception is really according to order.

Thursday 9 December 1824. Johann Peter Eckermann, *Conversations with Goethe*, ed. J. K. Moorhead and trans. J. Oxenford (1971), 75.

5 I could never have known so well how paltry men are, and how little they care for really high aims, if I had not tested them by my scientific researches. Thus I saw that most men only care for science so far as they get a living by it, and that they worship even error when it affords them a subsistence.

Wednesday 12 October 1825. Johann Peter Eckermann, *Conversations with Goethe*, ed. J. K. Moorhead and trans. J. Oxenford (1971), 119–20.

6 Man is born, not to solve the problems of the universe, but to find out where the problem applies, and then to restrain himself within the limits of the comprehensible.

Wednesday 12 October 1825. Johann Peter Eckermann, *Conversations with Goethe*, ed. J. K. Moorhead and trans. J. Oxenford (1971), 120.

7 Without my attempts in natural science, I should never have learned to know mankind such as it is. In nothing else can we so closely approach pure contemplation and thought, so closely observe the errors of the senses and of the understanding, the weak and strong points of character.

Friday 13 February 1829. Johann Peter Eckermann, *Conversations with Goethe*, ed. J. K. Moorhead and trans. J. Oxenford (1971), 293.

8 As for what I have done as a poet, I take no pride in whatever. Excellent poets have lived at the same time with me, poets more excellent lived before me, and others will come after me. But that in my country I am the only person who knows the truth in the difficult science of colors—of that, I say, I am not a little proud, and here have a consciousness of superiority to many.

Wednesday 18 February 1829. Johann Peter Eckermann, *Conversations with Goethe*, ed. J. K. Moorhead and trans. J. Oxenford (1971), 302.

9 Only he who finds empiricism irksome is driven to method.

Maxims and Reflections (1998), trans. E. Stopp, 154.

10 Hypotheses are scaffoldings erected in front of a building and then dismantled when the building is finished. They are indispensable for the workman; but you mustn't mistake the scaffolding for the building.

Maxims and Reflections (1998), trans. E. Stopp, 154.

11 Mathematicians are like a certain type of Frenchman: when you talk to them they translate it into their own language, and then it soon turns into something completely different.

Maxims and Reflections (1998), trans. E. Stopp, 162.

12 *Dauer in Wechsel.*

Duration in change.

Favourite expression.

13 Someday someone will write a pathology of experimental physics and bring to light all those swindles which subvert our reason, beguile our judgement and, what is worse, stand in the way of any practical progress. The phenomena must be freed once and for all from their grim torture chamber of empiricism, mechanism, and dogmatism; they must be brought before the jury of man's common sense.

Jeremy Naydler (ed.), *Goethe On Science: An Anthology of Goethe's Scientific Writings* (1996), 31.

1 *Wilst du ins Unendliche schreiten,*
Geh nur im Endlichen nach allen Seiten.

If you want to reach the infinite,
explore every aspect of the finite.

Jeremy Naydler (ed.), *Goethe On Science: An Anthology of Goethe's Scientific Writings* (1996), 37.

2 Nothing is more consonant with Nature than that she puts into operation in the smallest detail that which she intends as a whole.

Jeremy Naydler (ed.), *Goethe On Science: An Anthology of Goethe's Scientific Writings* (1996), 59.

3 Whatever Nature undertakes, she can only accomplish it in a sequence. She never makes a leap. For example she could not produce a horse if it were not preceded by all the other animals on which she ascends to the horse's structure as if on the rungs of a ladder. Thus every one thing exists for the sake of all things and all for the sake of one; for the one is of course the all as well. Nature, despite her seeming diversity, is always a unity, a whole; and thus, when she manifests herself in any part of that whole, the rest must serve as a basis for that particular manifestation, and the latter must have a relationship to the rest of the system.

Jeremy Naydler (ed.), *Goethe On Science: An Anthology of Goethe's Scientific Writings* (1996), 60.

4 What is the universal?
The single case.
What is the particular?
Millions of cases.

Jeremy Naydler (ed.), *Goethe On Science: An Anthology of Goethe's Scientific Writings* (1996), 92.

5 No one can take from us the joy of the first becoming aware of something, the so-called discovery. But if we also demand the honor, it can be utterly spoiled for us, for we are usually not the first. What does discovery mean, and who can say that he has discovered this or that? After all it's pure idiocy to brag about priority; for it's simply unconscious conceit, not to admit frankly that one is a plagiarist.

Epigraph to Lancelot Law Whyte, *The Unconscious before Freud* (1960).

6 Science has been seriously retarded by the study of what is not worth knowing, and of what is not knowable.

Attributed.

Thomas Gold 1920–2004

Austrian-born American astrophysicist

7 In no subject is there a rule, compliance with which will lead to new knowledge or better understanding. Skilful observations, ingenious ideas, cunning tricks, daring suggestions, laborious calculations, all these may be required to advance a subject. Occasionally the conventional approach in a subject has to be studiously followed; on other occasions it has to be ruthlessly disregarded. Which of these methods, or in what order they should be employed is generally unpredictable. Analogies drawn from the history of science are frequently claimed to be a guide; but, as with forecasting the next game of roulette, the existence of the best analogy to the present is no guide whatever to the future. The most valuable lesson to be learnt from the history of scientific progress is how misleading and strangling such analogies have been, and how success has come to those who ignored them.

'Cosmology', in Arthur Beer (ed.), *Vistas in Astronomy* (1956), Vol. 2, 1722.

Victor Moritz Goldschmidt
1853–1947

Swiss-born Norwegian geochemist

1 This [cyanide] poison is for professors of chemistry only. You, as a professor of mechanics, will have to use the rope. *Said during the Nazi occupation of Norway.*

Quoted in Kaufman, *Industrial Chemist and Chemical Manufacturer*, January 1988.

Camillo Golgi 1843–1926

Italian histologist and pathologist

2 I am delighted that I have found a new reaction to demonstrate even to the blind the structure of the interstitial stroma of the cerebral cortex. I let the silver nitrate react with pieces of brain hardened in potassium dichromate. I have already obtained magnificent results and hope to do even better in the future.

Letter to Nicolò Manfredi, 16 February 1873. Archive source. Quoted in Paolo Mazzarello, *The Hidden Structure: A Scientific Biography of Camillo Golgi*, trans. and ed. Henry A. Buchtel and Aldo Badiani (1999), 63.

3 Taking advantage of the method, found by me, of the black staining of the elements of the brain, staining obtained by the prolonged immersion of the pieces, previously hardened with potassium or ammonium bichromate, in a 0.50 or 1.0% solution of silver nitrate, I happened to discover some facts concerning the structure of the cerebral gray matter that I believe merit immediate communication.

'On the Structure of the Gray Matter of the Brain', *Gazetta Medica Italiana*, 2 August 1873. Trans. Maurizio Santini (ed.), *Golgi Centennial Symposium: Perspectives in Neurobiology* (1975), 647.

4 I have never had reason, up to now, to give up the concept which I have always stressed, that nerve cells, instead of working individually, act together, so that we must think that several groups of elements exercise a cumulative effect on the peripheral organs through whole bundles of fibres. It is understood that this concept implies another regarding the opposite action of sensory functions. However opposed it may seem to the popular tendency to individualize the elements, I cannot abandon the idea of a unitary action of the nervous system, without bothering if, by that, I approach old conceptions.

'The Neuron Doctrine—Theory and Facts', Nobel Lecture 11 December 1906. In *Nobel Lectures: Physiology or Medicine 1901–1921* (1967), 216.

Edmond and Jules de Goncourt 1822–96, 1830–70

French writers

5 There has never been an age so full of humbug. Humbug everywhere, even in science. For years now the scientists have been promising us every morning a new miracle, a new element, a new metal, guaranteeing to warm us with copper discs immersed in water, to feed us with nothing, to kill us at no expense whatever on a grand scale, to keep us alive indefinitely, to make iron out of heaven knows what. And all this fantastic, scientific humbugging leads to membership of the Institut, to decorations, to influence, to stipends, to the respect of serious people. In the meantime the cost of living rises, doubles, trebles; there is a shortage of raw materials; even death makes no progress—as we saw at Sebastopol, where men cut each other to ribbons—and the cheapest goods are still the worst goods in the world.

Diary entry, 7 January 1857. In R. Baldick (ed. & trans.), *Pages from the Goncourt Journal* (1978), 24.

Nelson Goodman 1906–98

American philosopher

6 For if as scientists we seek simplicity, then obviously we try the simplest surviving theory first, and retreat from it only when it proves false. Not this course, but any other, requires explanation. If you want to go somewhere quickly, and several alternate routes are equally likely to be open, no one asks why you take the shortest. The simplest theory is to be chosen not because it is the most likely to be true but because it is scientifically the most rewarding among equally likely alternatives. We aim at simplicity and hope for truth.

Problems and Projects (1972), 352.

John Goodsir 1814–67

British anatomist

1 A nutritive centre, anatomically considered, is merely a cell, the nucleus of which is the permanent source of successive broods of young cells, which from time to time fill the cavity of their parent, and carrying with them the cell wall of the parent, pass off in certain directions, and under various forms, according to the texture or organ of which their parent forms a part.

Anatomical and Pathological Observations (1845), 2.

Philip Henry Gosse 1810–88

British naturalist

2 That alone is worthy to be called *Natural History*, which investigates and records the condition of living things, of things in a state of nature; if animals, of *living* animals:— which tells of their 'sayings and doings,' their varied notes and utterances, songs and cries; their actions, in ease and under the pressure of circumstances; their affections and passions, towards their young, towards each other, towards other animals, towards man: their various arts and devices, to protect their progeny, to procure food, to escape from their enemies, to defend themselves from attacks; their ingenious resources for concealment; their stratagems to overcome their victims; their modes of bringing forth, of feeding, and of training, their offspring; the relations of their structure to their wants and habits; the countries in which they dwell; their connexion with the intimate world around them, mountain or plain, forest or field, barren heath or bushy dell, open savanna or wild hidden glen, river, lake, or sea:— this would be indeed *zoology*, *i.e.* the science of *living* creatures.

A Naturalist's Sojourn in Jamaica (1851), vi-vii.

3 I assume that each organism which the Creator educed was stamped with an indelible specific character, which made it what it was, and distinguished it from everything else, however near or like. I assume that such character has been, and is, indelible and immutable; that the characters which distinguish species *now*, were as definite at the first instant of their creation as now and are as distinct now as they were then. If any choose to maintain . . . that species were gradually bought to their present maturity from humbler forms . . . he is welcome to his hypothesis, but I have nothing to do with it.

Omphalos: An Attempt to Untie the Geological Knot (1857), 111.

4 Admit for a moment, as a hypothesis, that the Creator had before his mind a projection of the whole life-history of the globe, commencing with any point which the geologist may imagine to have been a fit commencing point, and ending with some unimaginable acme in the indefinitely distant future. He determines to call this idea into actual existence, not at the supposed commencing point, but at some stage or other of its course. It is clear, then, that at the selected stage it appears, exactly as it would have appeared at that moment of its history, if all the preceding eras of its history had been real.

Omphalos: An Attempt to Untie the Geological Knot (1857), 351.

William Sealy Gosset (Student) 1876–1937

British statistician

5 Any experiment may be regarded as forming an individual of a 'population' of experiments which might be performed under the same conditions. A series of experiments is a sample drawn from this population.

Now any series of experiments is only of value in so far as it enables us to form a judgment as to the statistical constants of the population to which the experiments belong. In a great number of cases the question finally turns on the value of a mean, either directly, or as the mean difference between the two qualities.

If the number of experiments be very large, we may have precise information as to the value of the mean, but if our sample be small, we have two sources of uncertainty:— (1) owing to the 'error of random sampling' the mean of our series of experiments deviates more or less widely from the mean of the population, and (2) the sample is not sufficiently large to determine what is the law of distribution of individuals.

'The Probable Error of a Mean', *Biometrika*, 1908, 6, 1.

1 *Conclusions*

I. A curve has been found representing the frequency distribution of standard deviations of samples drawn from a normal population.

II. A curve has been found representing the frequency distribution of values of the means of such samples, when these values are measured from the mean of the population in terms of the standard deviation of the sample . . .

IV. Tables are given by which it can be judged whether a series of experiments, however short, have given a result which conforms to any required standard of accuracy or whether it is necessary to continue the investigation.

'The Probable Error of a Mean', *Biometrika*, 1908, 6, 25.

2 I fancy you give me credit for being a more systematic sort of cove than I really am in the matter of limits of significance. What would actually happen would be that I should make out Pt (normal) and say to myself that would be about 50:1; pretty good but as it may not be normal we'd best not be too certain, or 100:1; even allowing that it may not be normal it seems good enough and whether one would be content with that or would require further work would depend on the importance of the conclusion and the difficulty of obtaining suitable experience.

Letter to E. S. Pearson, 18 May 1929.
E. S. Pearson, '"Student" as Statistician', *Biometrika*, 1939, 30, 244.

Stephen Jay Gould 1941–2002

American palaeontologist

See **Eldridge, Niles and Gould, Stephen Jay**

3 Is uniformitarianism necessary?

'Is Uniformitarianism Necessary?', *American Journal of Science*, 1965, 263, 223.

4 Without a commitment to science and rationality in its proper domain, there can be no solution to the problems that engulf us. Still, the Yahoos never rest.

Ever Since Darwin (1980), 146.

5 Biological determinism is, in its essence, a *theory of limits*. It takes the current status of groups as a measure of where they should and must be . . . We inhabit a world of human differences and predilections, but the extrapolation of these facts to theories of rigid limits is ideology.

The Mismeasure of Man (1981), 28–9.

6 In the great debates of early-nineteenth century geology, catastrophists followed the stereotypical method of objective science—empirical literalism. They believed what they saw, interpolated nothing, and read the record of the rocks directly.

'The Stinkstones of Oeningen', in *Hen's Teeth and Horse's Toes* (1983), 105.

7 In science 'fact' can only mean 'confirmed to such a degree that it would be perverse to withhold provisional assent'.

'Evolution as Fact and Theory', in *Hen's Teeth and Horse's Toes* (1983), 255.

1 The median isn't the message.

'The Median Isn't the Message', *Discover*, June 1985, 40.

2 No Geologist worth anything is permanently bound to a desk or laboratory, but the charming notion that true science can only be based on unbiased observation of nature in the raw is mythology. Creative work, in geology and anywhere else, is interaction and synthesis: half-baked ideas from a bar room, rocks in the field, chains of thought from lonely walks, numbers squeezed from rocks in a laboratory, numbers from a calculator riveted to a desk, fancy equipment usually malfunctioning on expensive ships, cheap equipment in the human cranium, arguments before a road cut.

An Urchin in the Storm (1988), 98.

3 Life is a copiously branching bush, continually pruned by the grim reaper of extinction, not a ladder of predictable progress.

Wonderful Life (1989), 35.

4 Taxonomy (the science of classification) is often undervalued as a glorified form of filing—with each species in its folder, like a stamp in its prescribed place in an album; but taxonomy is a fundamental and dynamic science, dedicated to exploring the causes of relationships and similarities among organisms. Classifications are theories about the basis of natural order, not dull catalogues compiled only to avoid chaos.

Wonderful Life (1989), 98.

5 Asian *Homo erectus* died without issue and does not enter our immediate ancestry (for we evolved from African populations); Neanderthal people were collateral cousins, perhaps already living in Europe while we emerged in Africa . . . In other words, we are an improbable and fragile entity, fortunately successful after precarious beginnings as a small population in Africa, not the predictable end result of a global tendency. We are a thing, an item of history, not an embodiment of general principles.

Wonderful Life (1989), 319.

6 Run the tape again, and let the tiny twig of *Homo sapiens* expire in Africa. Other hominids may have stood on the threshold of what we know as human possibilities, but many sensible scenarios would never generate our level of mentality. Run the tape again, and this time Neanderthal perishes in Europe and *Homo erectus* in Asia (as they did in our world). The sole surviving human stock, *Homo erectus* in Africa, stumbles along for a while, even prospers, but does not speciate and therefore remains stable. A mutated virus then wipes *Homo erectus* out, or a change in climate reconverts Africa into inhospitable forest. One little twig on the mammalian branch, a lineage with interesting possibilities that were never realized, joins the vast majority of species in extinction. So what? Most possibilities are never realized, and who will ever know the difference? Arguments of this form lead me to the conclusion that biology's most profound insight into human nature, status, and potential lies in the simple phrase, the embodiment of contingency: *Homo sapiens* is an entity, not a tendency.

Wonderful Life (1989), 320.

7 Some beliefs may be subject to such instant, brutal and unambiguous rejection. For example: no left-coiling periwinkle has ever been found among millions of snails examined. If I happen to find one during my walk on Nobska beach tomorrow morning, a century of well nurtured negative evidence will collapse in an instant.

'A Foot Soldier for Evolution', in *Eight Little Piggies* (1994), 452.

1 If one small and odd lineage of fishes had not evolved fins capable of bearing weight on land (though evolved for different reasons in lakes and seas,) terrestrial vertebrates would never have arisen. If a large extraterrestrial object—the ultimate random bolt from the blue—had not triggered the extinction of dinosaurs 65 million years ago, mammals would still be small creatures, confined to the nooks and crannies of a dinosaur's world, and incapable of evolving the larger size that brains big enough for self-consciousness require. If a small and tenuous population of protohumans had not survived a hundred slings and arrows of outrageous fortune (and potential extinction) on the savannas of Africa, then *Homo sapiens* would never have emerged to spread throughout the globe. We are glorious accidents of an unpredictable process with no drive to complexity, not the expected results of evolutionary principles that yearn to produce a creature capable of understanding the mode of its own necessary construction.

Full House: The Spread of Excellence from Plato to Darwin (1996), 216.

Stephen Jay Gould 1941–2002 and R. C. Lewontin 1929–

American palaeontologist and American biologist

2 The spandrels of San Marco and the Panglossian Paradigm.

'The Spandrels of San Marco and the Panglossian Paradigm: A Critique of the Adaptionist Programme', *Proceedings of the Royal Society*, 1979, **205**, 581–98.

Amadeus William Grabau 1870–1946

American geologist and palaeontologist

3 Even more difficult to explain, than the breaking-up of a single mass into fragments, and the drifting apart of these blocks to form the foundations of the present-day continents, is the explanation of the original production of the single mass, or PANGAEA, by the concentration of the former holosphere of granitic sial into a hemisphere of compressed and crushed gneisses and schists. Creep and the effects of compression, due to shrinking or other causes, have been appealed to, but this is hardly a satisfactory explanation. The earth could no more shrug itself out of its outer rock-shell unaided, than an animal could shrug itself out of its hide, or a man wriggle out of his skin, or even out of his closely buttoned coat, without assistance either of his own hands or those of others.

The Rhythm of Ages (1940), 9–10.

John Graunt 1620–74

British demographer and statistician

4 Now having (I know not by what accident) engaged my thoughts upon the *Bills of Mortality*, and so far succeeded therein, as to have reduced several great confused *Volumes* into a few perspicuous *Tables*, and abridged such *Observations* as naturally flowed from them, into a few succinct *Paragraphs*, without any long Series of *multiloquious Deductions*, I have presumed to sacrifice these my small, but first publish'd, *Labours* unto your Lordship, as unto whose benign acceptance of some other of my *Papers*, even the birth of these is due; hoping (if I may without vanity say it) they may be of as much use to persons in your Lordships place, as they are of little or none to me, which is no more than the fairest *Diamonds* are to the *Journeyman Jeweller* that works them, or the poor *Labourer* that first digg'd them from the Earth.

Natural and Political Observations Mentioned in a Following Index and made upon Bills of Mortality (1662), 5th edition (1676), the Epistle Dedicatory. In Charles Henry Hull (ed.), *The Economic Writings of Sir William*

Petty Together with The Observations upon the Bills of Mortality by John Graunt (1899), Vol. 2, 320.

Asa Gray 1810–88

American botanist

1 The longest-domesticated of all species.

On human beings.

'Notice of Dr. Hooker's Flora of New Zealand', *American Journal of Science*, 1854, 17, 336.

2 The beginning of things must needs lie in obscurity, beyond the bounds of proof, though within those of conjecture or of analogical inference. Why not hold fast to the customary view, that all species were directly, instead of indirectly, created after their respective kinds, as we now behold them,—and that in a manner which, passing our comprehension, we intuitively refer to the supernatural? Why this continual striving after 'the unattained and dim,'—these anxious endeavors, especially of late years, by naturalists and philosophers of various schools and different tendencies, to penetrate what one of them calls 'the mystery of mysteries,' the origin of species? To this, in general, sufficient answer may be found in the activity of the human intellect, 'the delirious yet divine desire to know,' stimulated as it has been by its own success in unveiling the laws and processes of inorganic Nature,—in the fact that the principal triumphs of our age in physical science have consisted in tracing connections where none were known before, in reducing heterogenous phenomena to a common cause or origin, in a manner quite analogous to that of the reduction of supposed independently originated species to a common ultimate origin,—thus, and in various other ways, largely and legitimately extending the domain of secondary causes. Surely the scientific mind of an age which contemplates the solar system as evolved from a common, revolving, fluid mass,—which, through experimental research, has come to regard light, heat, electricity, magnetism, chemical affinity, and mechanical power as varieties or derivative and convertible forms of one force, instead of independent species,—which has brought the so-called elementary kinds of matter, such as the metals, into kindred groups, and raised the question, whether the members of each group may not be mere varieties of one species,—and which speculates steadily in the direction of the ultimate unity of matter, of a sort of prototype or simple element which may be to the ordinary species of matter what the *protozoa* or component cells of an organism are to the higher sorts of animals and plants,—the mind of such an age cannot be expected to let the old belief about species pass unquestioned.

'Darwin on the Origin of Species', *The Atlantic Monthly*, July 1860, 112–3.

3 Natural selection is not the wind which propels the vessel, but the rudder which, by friction, now on this side and now on that, shapes the course.

Quoted in A. Hunter Dupree, *Asa Gray: American Botanist, Friend of Darwin* (1988), 367.

4 This view, as a rounded whole and in all its essential elements, has very recently disappeared from science. It died a royal death with Agassiz.

The older view that species had no genetic connection.

'Scientific Beliefs': Two Lectures delivered to the Theological School of Yale College. In *Natural Science and Religion* (1880), 35.

5 We have really, that I know of, no philosophical basis for high and low. Moreover, the vegetable kingdom does not culminate, as the animal kingdom does. It is not a kingdom, but a common-wealth; a democracy, and therefore puzzling and unaccountable from the former point of view.

Letter to Charles Darwin, 27 January 1863. In *Letters of Asa Gray* (1893), Vol. 2, 496.

John Grebe 1900–84

American chemist

1 If you cannot measure it you cannot control it.

Ray Boundy and J. Laurence Amos (eds.), *A History of the Dow Chemical Physics Lab, The Freedom to Be Creative* (1990), 53.

2 Fine, fine; don't do anything to patch it up. The way things are going, gangrene will set in. Then we can amputate and clean up the problem once and for all.

Ray Boundy and J. Laurence Amos (eds.), *A History of the Dow Chemical Physics Lab, The Freedom to Be Creative* (1990), 180.

Edward Greenly 1861–1951

British geologist

3 The surveyor ought to work in solitude. He must no more admit company while mapping than a writer admits visitors to his study while writing. This applies even to geological company, nay, even to the company of a skilled fellow-surveyor . . . The two authors of this book [Edward Greenly and Howell Williams] once thought that it would be pleasant to have a day's mapping together, and decided to break through their rule. The result was a ludicrous paralysis. The commonest operation seemed a mountain of difficulty. Next day the senior author (whose ground it was) swept an india-rubber over every line and went out again, when, hey presto! all was clear.

Edward Greenly and Howell Williams, *Methods in Geological Surveying* (1930), 375–6.

Alan Gregg 1890–1957

American medical researcher and science administrator

4 Most of the knowledge and much of the genius of the research worker lie behind his selection of what is worth observing. It is a crucial choice, often determining the success or failure of months of work, often differentiating the brilliant discoverer from an otherwise admirable but entirely unproductive plodder.

The Furtherance of Medical Research (1941), 8.

William King Gregory 1876–1970

American palaeontologist

5 Perhaps the majority of paleontologists of the present time, who believe in orthogenesis, the irreversibility of evolution and the polyphyletic origin of families, will assume that a short molar must keep on getting shorter, that it can never get longer and then again grow relatively shorter and therefore that *Propliopithecus* with its extremely short third molar and *Dryopithecus* with its long m3 are alike excluded from the ancestry of the Gorilla, in which there is a slight retrogression in length of m3. After many years reflection and constant study of the evolution of the vertebrates however, I conclude that 'orthogenesis' should mean solely that structures and races evolve in a certain direction, or toward a certain goal, only until the direction of evolution shifts toward some other goal. I believe that the 'irreversibility of evolution' means only that past changes irreversibly limit and condition future possibilities, and that, as a matter of experience, if an organ is once lost the same (homogenous) organ can never be regained, although nature is fertile in substituting imitations. But this does not mean, in my judgement, that if one tooth is smaller than its fellows it will in all cases continue to grow smaller.

'Studies on the Evolution of the Primates', *Bulletin of the American Museum of Natural History*, 1916, **35**, 307.

6 As long as museums and universities send out expeditions to bring to light new forms of living and extinct animals and new data illustrating the interrelations of organisms and their environments, as long as anatomists desire a broad comparative basis for human anatomy, as long as even a few students feel a strong curiosity to learn about the course of evolution and the relationships of animals, the old problems of taxonomy, phylogeny and evolution will gradually reassert themselves even in competition with brilliant and highly fruitful laboratory studies in cytology, genetics and physiological chemistry.

'Genetics Versus Paleontology', *The American Naturalist*, 1917, 51, 623.

1 'Planning' is simply the result of experience read backward and projected into the future. To me the 'purposive' action of a beehive is simply the summation and integration of its units, and Natural Selection has put higher and higher premiums on the most 'purposeful' integration. It is the same way (to me) in the evolution of the middle ear, the steps in the Cynodonts (clearly shown by me in 1910 and by you later in Oudenodon) make it easier to see how such a wonderful device as the middle ear could arise without any predetermination or human-like planning, and in fact in the good old Darwinian way, if only we admit that as the 'twig is bent the tree's inclined' and that each stage conserves the advantages of its predecessors . . . The simple idea that planning is only experience read backward and combined by selection in suitable or successful combinations takes the mystery out of Nature and out of men's minds.

Letter to Robert Broom [1933]. In Ronald Rainger, *An Agenda for Antiquity* (1991), 238.

Thomas Gresham 1518–79

British merchant, financier and founder of Royal Exchange

2 [Gresham's Law]: Bad money drives out good money.

Quoted in C. Alexander Harris, *Gresham's Law* (1896), 262.

Nehemiah Grew 1641–1712

British botanist

3 First, by what means it is that a *Plant*, or any *Part* of it, comes to *Grow*, a *Seed* to put forth a *Root* and *Trunk* . . . How the Aliment by which a *Plant* is fed, is duly prepared in its several *Parts* . . . How not only their *Sizes*, but also their *Shapes* are so exceedingly various . . . Then to inquire, What should be the reason of their various *Motions*; that the *Root* should *descend*; that its descent should sometimes be *perpendicular*, sometimes more *level*: That the *Trunk* doth *ascend*, and that the ascent thereof, as to the space of *Time* wherein it is made, is of different measures . . . Further, what may be the *Causes* as of the *Seasons* of their *Growth*; so of the *Periods* of their *Lives*; some being *Annual*, others *Biennial*, others *Perennial* . . . what manner the *Seed* is prepared, formed and fitted for *Propagation*.

'An Idea of a Philosophical History of Plants', in *The Anatomy of Plants With an Idea of a Philosophical History of Plants and Several Other Lectures Read Before the Royal Society* (1682), 3–4.

4 [We need not think] that there is any Contradiction, when *Philosophy* teaches that to be done by *Nature*; which *Religion*, and the Sacred *Scriptures*, teach us to be done by *God*: no more, than to say, That the balance of a *Watch* is moved by the next *Wheel*, is to deny that *Wheel*, and the rest, to be moved by the *Spring*; and that both the *Spring*, and all the other *Parts*, are caused to move together by the *Maker* of them. So God may be truly the *Cause* of *This Effect*, although a Thousand other *Causes* should be supposed to intervene: For all Nature is as one Great *Engine*, made by, and held in His Hand.

'An Idea of a Philosophical History of Plants', in *The Anatomy of Plants With an Idea of a Philosophical History of Plants and Several Other Lectures Read Before the Royal Society* (1682), 80.

John Joseph Griffin 1802–77

British chemist

5 Imagine a school-boy who has outgrown his clothes. Imagine the *repairs* made on the vestments where the enlarged frame had burst the narrow limits of the enclosure. Imagine the *additions* made where the projecting limbs had fairly and far emerged beyond the confines of the garment. Imagine the boy *still growing*, and the *clothes*, mended all over, now *more than ever in want of mending*—such is chemistry, and such is nomenclature.

Chemical Recreations (1834), 206, footnote.

1 I appeal to the contemptible speech made lately by Sir Robert Peel to an applauding House of Commons. '*Orders of merit*,' said he, '*were the proper rewards of the military*' (the desolators of the world in all ages). '*Men of science are better left to the applause of their own hearts.*' Most learned Legislator! Most *liberal* cotton-spinner! Was *your title* the proper reward of military prowess? Pity you hold not the dungeon-keys of an English Inquisition! Perhaps Science, like creeds, would flourish best under a little persecution.

Chemical Recreations (1834), 232.

François Auguste Victor Grignard 1871–1935

French chemist

2 On the terrace of the Pépinière, the 150 pupils of the Institut Chemique talk chemistry as they leave the auditoria and the laboratory. The echoes of the magnificent public garden of the city of Nancy make the words reverberate; coupling, condensation, grignardization. Moreover, their clothes stay impregnated with strong and characteristic odours; we follow the initiates of Hermes by their scent. In such an environment, how is it possible not to be productive?

Charles Courtot, 'Notice sur la vie de Victor Grignard', *Bulletin Societé Chimie*, 1936, 3, 1445. Trans. in Mary Jo Nye, *Science in the Provinces* (1986), 184.

Robert Grosseteste c.1168–1253

British natural philosopher

3 Now, all causes of natural effects must be expressed by means of lines, angles and figures, for otherwise it is impossible to grasp their explanation. This is evident as follows. A natural agent multiplies its power from itself to the recipient, whether it acts on sense or on matter. This power is sometimes called species, sometimes a likeness, and it is the same thing whatever it may be called; and the agent sends the same power into sense and into matter, or into its own contrary, as heat sends the same thing into the sense of touch and into a cold body. For it does not act by deliberation and choice, and therefore it acts in a single manner whatever it encounters, whether sense or something insensitive, whether something animate or inanimate. But the effects are diversified by the diversity of the recipient, for when this power is received by the senses, it produces an effect that is somehow spiritual and noble; on the other hand, when it is received by matter, it produces a material effect. Thus the sun produces different effects in different recipients by the same power, for it cakes mud and melts ice.

De Lineis, Angulis et Figuris seu Fractionibus et Reflexionibus Radiorum (*On Lines, Angles and Figures or On the Refraction and Reflection of Rays*) [1230/31], trans. D. C. Lindberg, quoted in E. Grant (ed.), *A Source Book in Medieval Science* (1974), 385–6.

4 It is not possible for form to do without matter because it is not separable, nor can matter itself be purged of form.

De Luce seu De Inchoatione Formarum (*On Light or On The Beginning of Forms*) [1220], trans. A. C. Crombie, quoted in A. C. Crombie, *Robert Grosseteste and the Origins of Experimental Science, 1100–1700* (1953), 106.

Guang Zhong

Chinese natural philosopher

5 Where there is cinnabar above, yellow gold will be found below. Where there is lodestone above, copper and gold will be found below. Where there is calamine above, lead, tin, and red copper will be found below. Where there is haematite above, iron will be found below. Thus it can be seen that mountains are full of riches.

From Guo Me-ruo et al., *Collections of Rectifications of the Book of Guang Zi* (1956), 146–7. Trans. Yang Jing-Yi.

Jürgen Habermas 1929–

German philosopher

1 Positivism stands or falls with the principle of scientism, that is that the meaning of knowledge is defined by what the sciences do and can thus be adequately explicated through the methodological analysis of scientific procedures.

Knowledge and Human Interests (1968), 67.

Ernst Heinrich Philip August Haeckel 1834–1919

German zoologist

2 The nucleus has to take care of the inheritance of the heritable characters, while the surrounding cytoplasm is concerned with accommodation or adaptation to the environment.

Generelle Morphologie (1866), Vol. 1, 287–8. Trans. Ernst Mayr, *The Growth of Biological Thought: Diversity, Evolution and Inheritance* (1982), 672.

3 Phylogeny and ontogeny are, therefore, the two coordinated branches of morphology. Phylogeny is the developmental history [*Entwickelungsgeschichte*] of the abstract, genealogical individual; ontogeny, on the other hand, is the developmental history of the concrete, morphological individual.

Allgemeine Entwickelungsgeschichte der Organismen (1866), Vol. 1, 60. Trans. Stephen Jay Gould, *Ontogeny and Phylogeny* (1977), 80.

4 In the course of individual development, inherited characters appear, in general, earlier than adaptive ones, and the earlier a certain character appears in ontogeny, the further back must lie in time when it was acquired by its ancestors.

Allgemeine Entwickelungsgeschichte der Organismen (1866), Vol. 2, 298. Trans. Stephen Jay Gould, *Ontogeny and Phylogeny* (1977), 81.

5 *Die Phylogenese ist die mechanische Ursache der Ontogenese.*

Phylogenesis is the mechanical cause of ontogenesis.

Anthropogenie oder Entwickelungsgeschichte des Menschen (1874), 7.

6 The cell never acts; it *re*acts.

Generelle Morphologie (1866).

7 In consequence of Darwin's reformed Theory of Descent, we are now in a position to establish scientifically the groundwork of a *non-miraculous history of the development of the human race* . . . If any person feels the necessity of conceiving the coming into existence of this matter as the work of a supernatural creative power, of the creative force of something outside of matter, we have nothing to say against it. But we must remark, that thereby not even the smallest advantage is gained for a scientific knowledge of nature. Such a conception of an immaterial force, which as the first creates matter, is an article of faith which has nothing whatever to do with human science. *Where faith commences, science ends.*

The History of Creation (1876), Vol. 1, 6–9.

8 Ontogeny is a short and quick repetition, or recapitulation, of Phylogeny, determined by the laws of Inheritance and Adaptation.

The History of Creation (1876), Vol. 2, 33.

9 An irrefutable proof that such single-celled primæval animals really existed as the direct ancestors of Man, is furnished according to the fundamental law of biogeny by the fact that the human egg is nothing more than a simple cell.

Natürliche Schöpfungsgeschichte, trans. E. R. Lankester, *The History of Creation* (1892), Vol. 2, 381.

10 We may now give the following more precise expression to our chief law of biogeny:— The evolution of the foetus (or *ontogenesis*) is a condensed and abbreviated recapitulation of the evolution of the stem (or *phylogenesis*); and this recapitulation is the more complete in proportion as the original development (or *palingenesis*) is preserved by a constant heredity; on the other hand, it becomes less complete in proportion as a varying adaptation to new conditions increases the disturbing factors in the development (or *cenogenesis*).

The Evolution of Man. Translated from the 5th edition of *Anthropogenie* by Joseph McCabe (1910), 8.

1 Nothing is constant but change! All existence is a perpetual flux of 'being and becoming!' That is the broad lesson of the evolution of the world.

Attributed.

Peter Haggett 1933–

British geographer

2 The sound of progress is perhaps the sound of plummeting hypotheses.

Locational Analysis in Human Geography (1965), 277.

Otto Hahn 1879–1968 and Fritz Strausmann 1902–80

German chemists

3 As chemists, we must rename [our] scheme and insert the symbols Ba, La, Ce in place of Ra, Ac, Th. As nuclear chemists closely associated with physics, we cannot yet convince ourselves to make this leap, which contradicts all previous experience in nuclear physics.

'Über den nachweis und das Verhalten der bei der Bestrahlung des Urans mittels Neutronen entstehenden Erdalkalimetalle', *Die Naturwissenschaften*, 1939, **27**, 11–15. Trans. J. Heilbron and Robert W. Seidel, *Lawrence and his Laboratory: A History of the Lawrence Berkeley Laboratory* (1989), Vol. 1, 436–7.

Yrjö Haila 1947–

Finnish environmental scientist

4 Science is bound to language.

'On the Semiotic Dimension of Ecological Theory: The Case of Island Biogeography', *Biology and Philosophy*, 1986, **1**, 378.

John Burden Sanderson Haldane 1892–1964

British geneticist, physiologist and biochemist

5 Capitalism, though it may not always give the scientific worker a living wage, will always protect him, as being one of the geese which produce golden eggs for its table.

Daedalus or Science and the Future (1924), 6.

6 I have tried to show why I believe that the biologist is the most romantic figure on earth at the present day. At first sight he seems to be just a poor little scrubby underpaid man, groping blindly amid the mazes of the ultra-microscopic, engaging in bitter and lifelong quarrels over the nephridia of flatworms, waking perhaps one morning to find that someone whose name he has never heard has demolished by a few crucial experiments the work which he had hoped would render him immortal.

Daedalus or Science and the Future (1924), 77.

7 Man armed with science is like a baby with a box of matches.

Daedalus or Science and the Future (1924), 82.

8 Quantitative work shows clearly that natural selection is a reality, and that, among other things, it selects Mendelian genes, which are known to be distributed at random through wild populations, and to follow the laws of chance in their distribution to offspring. In other words, they are an agency producing variation of the kind which Darwin postulated as the raw material on which selection acts.

'Natural Selection', *Nature*, 1929, **124**, 444.

9 An attempt to study the evolution of living organisms without reference to cytology would be as futile as an account of stellar evolution which ignored spectroscopy.

'Foreword', in C. D. Darlington, *Recent Advances in Cytology* (1937), v.

1 I'd lay down my life for two brothers or eight cousins.

Quipped in a pub conversation.
'Accidental Career', *New Scientist*, 8 August 1974, 325.

2 Coming to the question of life being found on other planets, Professor Haldane apologized for discoursing, as a mere biologist, on a subject on which we had been expecting a lecture by a physicist [J. D. Bernal]. He mentioned three hypotheses:

(a) That life had a supernatural origin,

(b) That it originated from inorganic materials, and

(c) That life is a constituent of the Universe and can only arise from pre-existing life.

The first hypothesis, he said, should be taken seriously, and he would proceed to do so. From the fact that there are 400,000 species of beetle on this planet, but only 8,000 species of mammals, he concluded that the Creator, if he exists, has a special preference for beetles, and so we might be more likely to meet them than any other type of animal on a planet which would support life.

In Mark Williamson, 'Haldane's Special Preference', *The Linnean*, 1992, 8, 14.

Stephen Hales 1677–1761

British plant physiologist

3 The farther researches we make into this admirable scene of things, the more beauty and harmony we see in them: And the stronger and clearer convictions they give us, of the being, power and wisdom of the divine Architect, who has made all things to concur with a wonderful conformity, in carrying on, by various and innumerable combinations of matter, such a circulation of causes, and effects, as was necessary to the great ends of nature. And since we are assured that the all-wise Creator has observed the most exact proportions, *of number, weight and measure*, in the make of all things; the most likely way therefore, to get any insight into the nature of those parts of the creation, which come within our observation, must in all reason be to number, weigh and measure. And we have much encouragement to pursue this method, of searching into the nature of things, from the great success that has attended any attempts of this kind.

Vegetable Staticks (1727), xxxi.

Granville Stanley Hall 1844–1924

American psychologist

4 All possible truth is practical. To ask whether our conception of chair or table corresponds to the real chair or table apart from the uses to which they may be put, is as utterly meaningless and vain as to inquire whether a musical tone is red or yellow. No other conceivable relation than this between ideas and things can exist. The unknowable is what I cannot react upon. The active part of our nature is not only an essential part of cognition itself, but it always has a voice in determining what shall be believed and what rejected.

The Muscular Perception of Space (1878), 446.

Marshall Hall 1790–1857

British physiologist

5 Unhappily for the physiologist, the subjects of the principal department of his science, that of animal physiology, are sentient beings; and every experiment, every new or unusual situation of such a being, is necessarily attended by pain or suffering of a bodily or mental kind.

A Critical and Experimental Essay on the Circulation of the Blood (1831), 1.

6 It is distinctly proved, by this series of observations, that the reflex function exists in the medulla independently of the brain; in the medulla oblongata independently of the medulla spinalis; and in the spinal marrow of the anterior extremities, of the posterior extremities, and of the tail, independently of that of each other of these parts, respectively. There is still a more interesting and satisfactory mode of performing the experiment: it is to divide the spinal marrow between the nerves of the superior and inferior extremities. We have then two modes of animal life: the first being the assemblage of the voluntary and respiratory powers with those of the reflex function and irritability; the second, the two latter powers only: the first are those which obtain in the perfect animal, the second those which animate the foetus. The phenomena are precisely what might have been anticipated. If the spinal marrow be now destroyed, the irritability alone remains,—all the other phenomena having ceased.

'On the Reflex Function of the Medulla Oblongata and Medulla Spinalis,' *Philosophical Transactions of the Royal Society*, 1833, 123, 650.

1 The true-spinal system consists of a series of nerves passing principally from the cutaneous surface, and the surface of the mucous membranes, to the spinal marrow; and of another series of nerves passing from the spinal marrow to a series of muscles, destined to be moved simultaneously. The former, thence designated the *incident* nerves; the latter, *reflex* nerves: the spinal marrow is their common *centre*.

On the Mutual Relations between Anatomy, Physiology, Pathology and Therapeutics, and the Practice of Medicine. Being the Gulstonian Lectures for 1842 (1842), 32.

Victor Albrecht von Haller

1708–77

Swiss physiologist and poet

2 Nature never jests.

Réflexions sur le systême de la Génération de M. de Buffon (1751), 50. Quoted in Jacques Roger, *The Life Sciences in Eighteenth-Century French Thought*, edited by Keith R. Benson, trans. Robert Ellrich (1997), 494.

3 The result of all these experiments has given place to a new division of the parts of the human body, which I shall follow in this short essay, by distinguishing those which are susceptible of Irritability and Sensibility, from those which are not. But the theory, why some parts of the human body are endowed with these properties, while others are not, I shall not at all meddle with. For I am persuaded that the source of both lies concealed beyond the reach of the knife and microscope, beyond which I do not chuse to hazard many conjectures, as I have no desire of teaching what I am ignorant of myself. For the vanity of attempting to guide others in paths where we find ourselves in the dark, shews, in my humble opinion, the last degree of arrogance and ignorance.

'A Treatise on the Sensible and Irritable Parts of Animals' (Read 1752). Trans. 1755 and reprinted in *Bulletin of the Institute of the History of Medicine*, 1936, 4(2), 657–8.

4 I call that part of the human body irritable, which becomes shorter upon being touched; very irritable if it contracts upon a slight touch, and the contrary if by a violent touch it contracts but little. I call that a sensible part of the human body, which upon being touched transmits the impression of it to the soul; and in brutes, in whom the existence of a soul is not so clear, I call those parts sensible, the Irritation of which occasions evident signs of pain and disquiet in the animal. On the contrary, I call that insensible, which being burnt, tore, pricked, or cut till it is quite destroyed, occasions no sign of pain nor convulsion, nor any sort of change in the situation of the body.

'A Treatise on the Sensible and Irritable Parts of Animals' (Read 1752). Trans. 1755 and reprinted in *Bulletin of the Institute of the History of Medicine*, 1936, 4(2), 658–9.

5 The ovary of an ancestress will contain not only her daughter, but also her granddaughter, her great-grand-daughter, and her great-great-granddaughter, and if it is once proved that an ovary can contain many generations, there is no absurdity in saying that it contains them all.

Attributed.

Edmond Halley c.1656–1743

British astronomer and geophysicist

1 In the year 1456 . . . a Comet was seen passing Retrograde between the Earth and the sun . . . Hence I dare venture to foretell, that it will return again in the year 1758.

A Synopsis of the Astronomy of Comets (1705), 22.

2 This incomparable Author having at length been prevailed upon to appear in public, has in this Treatise given a most notable instance of the extent of the powers of the Mind; and has at once shown what are the Principles of Natural Philosophy, and so far derived from them their consequences, that he seems to have exhausted his Argument, and left little to be done by those that shall succeed him.

Attributed.

3 Aristotle's opinion . . . that comets were nothing else than sublunary vapors or airy meteors . . . prevailed so far amongst the Greeks, that this sublimest part of astronomy lay altogether neglected; since none could think it worthwhile to observe, and to give an account of the wandering and uncertain paths of vapours floating in the Ether.

Attributed.

William Hamilton 1730–1803

British diplomat and antiquary

4 If I was to establish a system, it would be, that *Mountains are produced by Volcanoes, and not Volcanoes by Mountains*.

Observations on Mount Vesuvius, Mount Etna, and other Volcanoes (1774), 52.

5 May not subterraneous fire be considered as the great plough (if I may be allowed the expression) which Nature makes use of to turn up the bowels of the earth?

Observations on Mount Vesuvius, Mount Etna, and other Volcanoes (1774), 161.

William Rowan Hamilton 1805–65

Irish mathematician

6 $i^2 = j^2 = k^2 = ijk = -1$

These formulae were conceived on 16th October 1843, and carved by Hamilton, apparently in this form on a stone of Brougham Bridge, over the Royal Canal, Dublin, at the time. It has now worn away.

The first version that was written down in a note-book is reproduced in the Frontispiece of Hamilton's *Mathematical Papers* (1967), Vol. 3.

Louis Plack Hammett 1894–1987

American chemist

7 One of the most striking evidences of the reliability of the organic chemist's methods of determining molecular structure is the fact that he has never been able to derive satisfactory structures for supposed molecules which are in fact nonexistent.

Physical Organic Chemistry: Reaction Rates, Equilibria, and Mechanisms (1940), 38.

8 To many physical chemists in the 1920's and early 1930's, the organic chemist was a grubby artisan engaged in an unsystematic search for new compounds, a search which was strongly influenced by the profit motive.

'Physical Organic Chemistry in Retrospect', *Journal of Chemical Education*, 1966, 43, 464.

Norwood Russell Hanson 1924–67

American philosopher of science

9 Seeing is an experience. A retinal reaction is only a physical state . . . People, not their eyes, see. Cameras, and eye-balls, are blind . . . there is more to seeing than meets the eyeball.

Patterns of Discovery (1958), 6–7.

10 There is no logical staircase running from the physics of 10^{-28} cm. to the physics of 10^{28} light-years.

Patterns of Discovery (1958), 157.

Garrett Hardin 1915–2003

American ecologist and microbiologist

1 Ecological differentiation is the necessary condition for coexistence.

'The Competitive Exclusion Principle', *Science*, 1960, 131, 1296.

Godfrey Harold Hardy 1877–1947

British mathematician

See ***Ramanujan, Srinivasa***

2 It is a melancholy experience for a professional mathematician to find himself writing about mathematics. The function of a mathematician is to do something, to prove new theorems, to add to mathematics, and not to talk about what he or other mathematicians have done.

A Mathematician's Apology (1940), 1.

3 No mathematician should ever allow himself to forget that mathematics, more than any other art or science, is a young man's game.

A Mathematician's Apology (1940), 10.

4 Archimedes will be remembered when Aeschylus is forgotten, because languages die and mathematical ideas do not. 'Immortality' may be a silly word, but probably a mathematician has the best chance of whatever it may mean.

A Mathematician's Apology (1940), 21.

5 The mathematician's patterns, like the painter's or the poet's, must be *beautiful*; the ideas, like the colours or the words, must fit together in a harmonious way. Beauty is the first test; there is no permanent place in the world for ugly mathematics.

A Mathematician's Apology (1940), 25.

6 I believe that mathematical reality lies outside us, that our function is to discover or *observe* it, and that the theorems which we prove, and which we describe grandiloquently as our 'creations', are simply our notes of our observations.

A Mathematician's Apology (1940), 63–4.

7 I have never done anything 'useful'. No discovery of mine has made, or is likely to make, directly or indirectly, for good or ill, the least difference to the amenity of the world . . . Judged by all practical standards, the value of my mathematical life is nil; and outside mathematics it is trivial anyhow. I have just one chance of escaping a verdict of complete triviality, that I may be judged to have created something worth creating. And that I have created something is undeniable: the question is about its value.

A Mathematician's Apology (1940), 90–1.

Thomas Hardy 1840–1928

British novelist and poet

8 I am the family face:
Flesh perishes, I live on,
Projecting trait and trace
Through time to times anon,
And leaping from place to place
Over oblivion.

'Heredity'. In James Gibson (ed.), *The Complete Poems of Thomas Hardy* (1976), 434.

9 Thy shadow, Earth, from Pole to Central Sea,
Now steals along upon the Moon's meek shine
In even monochrome and curving line
Of imperturbable serenity.
How shall I link such sun-cast symmetry
With the torn troubled form I know as thine,
That profile, placid as a brow divine,
With continents of moil and misery?

'At a Lunar Eclipse'. In James Gibson (ed.), *The Complete Poems of Thomas Hardy* (1976), 116.

10 She had been made to break an accepted social law, but no law known to the environment in which she fancied herself such an anomaly.

Tess of the d'Urbervilles (1891), 146.

Alfred Harker 1859–1939

British petrologist

1 The laws of physics and chemistry must be the same in a crucible as in the larger laboratory of Nature.

The Natural History of Igneous Rocks (1909), 282.

John Lander Harper 1925–

British botanist and ecologist

2 The theory of evolution by natural selection is an ecological theory—founded on ecological observation by perhaps the greatest of all ecologists. It has been adopted by and brought up by the science of genetics, and ecologists, being modest people, are apt to forget their distinguished parenthood.

'A Darwinian Approach to Plant Ecology', *Journal of Ecology*, 1967, **55**, 247.

Robert Harrington 1751–1837

British physician and chemist

3 Such pretensions to *nicety* in experiments of this nature, are truly laughable! They will be telling us some day of the WEIGHT of the MOON, even to *drams*, *scruples* and *grains*—nay, *to the very fraction of a grain!*—I wish there were infallible experiments to ascertain the *quantum* of *brains* each man possesses, and every man's *integrity* and *candour*:—This is a *desideratum* in science which is most of all wanted.

The Death Warrant of the French Theory of Chemistry (1804), 217.

Edward Robert Harrison 1919–

British astronomer and physicist

4 'By convention there is color, by convention sweetness, by convention bitterness, but in reality there are atoms and the void,' announced Democritus. The universe consists only of atoms and the void; all else is opinion and illusion. If the soul exists, it also consists of atoms.

Masks of the Universe (1985), 55.

David Hartley 1705–57

British psychologist

5 The influence of Association over our Opinions and Affections, and its Use in explaining those Things in an accurate and precise Way, which are commonly referred to the Power of Habit and Custom, is a general and indeterminate one.

Observations on Man, His Frame, His Duty, and His Expectations (1749), part 1, 5–6.

6 The White medullary Substance of the Brain is also the immediate Instrument, by which Ideas are presented to the Mind: Or, in other Words, whatever Changes are made in this Substance, corresponding Changes are made in our Ideas; and vice versa.

Observations on Man, His Frame, His Duty, and His Expectations (1749), part 1, 8.

7 When external objects are impressed on the sensory nerves, they excite vibrations in the aether residing in the pores of these nerves . . . Thus it seems that light affects both the optic nerve and the aether and . . . the affections of the aether are communicated to the optic nerve, and vice versa. And the same may be observed of frictions of the skin, taste, smells and sounds . . . Vibrations in the aether will agitate the small particles of the medullary substance of the sensory nerves with synchronous vibrations . . . up to the brain . . . These vibrations are motions backwards and forwards of small particles, of the same kind with the oscillations of pendulums, and the tremblings of the particles of the sounding bodies (but) exceedingly short and small, so as not to have the least efficacy to disturb or move the whole bodies of the nerves . . . That the nerves themselves should vibrate like musical strings is highly absurd.

Observations on Man, His Frame, His Duty, and His Expectations (1749), part 1, 11–22.

1 It is often in our Power to obtain an Analogy where we cannot have an Induction.

Observations on Man, His Frame, His Duty, and His Expectations (1749).

William Harvey 1578–1657

British physician, physiologist, anatomist and embryologist

2 The animal's heart is the basis of its life, its chief member, the sun of its microcosm; on the heart all its activity depends, from the heart all its liveliness and strength arise. Equally is the king the basis of his kingdoms, the sun of his microcosm, the heart of the state; from him all power arises and all grace stems.

De Motu Cordis (1628), *The Circulation of the Blood and Other Writings*, trans. Kenneth J. Franklin (1957), Dedication to the King, 3.

3 I profess to learn and to teach anatomy not from books but from dissections, not from the tenets of Philosophers but from the fabric of Nature.

De Motu Cordis (1628), *The Circulation of the Blood and Other Writings*, trans. Kenneth J. Franklin (1957), Dedication to Doctor Argent, 7.

4 When in many dissections, carried out as opportunity offered upon living animals, I first addressed my mind to seeing how I could discover the function and offices of the heart's movement in animals through the use of my own eyes instead of through the books and writings of others, I kept finding the matter so truly hard and beset with difficulties that I all but thought, with Fracastoro, that the heart's movement had been understood by God alone.

De Motu Cordis (1628), *The Circulation of the Blood and Other Writings*, trans. Kenneth J. Franklin (1957), Chapter 1, author's motives for writing, 23.

5 In attempting to discover how much blood passes from the veins into the arteries I made dissections of living animals, opened up arteries in them, and carried out various other investigations. I also considered the symmetry and size of the ventricles of the heart and of the vessels which enter and leave them (since Nature, who does nothing purposelessly, would not purposelessly have given these vessels such relatively large size). I also recalled the elegant and carefully contrived valves and fibres and other structural artistry of the heart; and many other points. I considered rather often and with care all this evidence, and took correspondingly long trying to assess how much blood was transmitted and in how short a time. I also noted that the juice of the ingested food could not supply this amount without our having the veins, on the one hand, completely emptied and the arteries, on the other hand, brought to bursting through excessive inthrust of blood, unless the blood somehow flowed back again from the arteries into the veins and returned to the right ventricle of the heart. In consequence, I began privately to consider that it had a movement, as it were, in a circle.

De Motu Cordis (1628), *The Circulation of the Blood and Other Writings*, trans. Kenneth J. Franklin (1957), Chapter 8, 57–8.

6 This organ deserves to be styled the starting point of life and the sun of our microcosm just as much as the sun deserves to be styled the heart of the world. For it is by the heart's vigorous beat that the blood is moved, perfected, activated, and protected from injury and coagulation. The heart is the tutelary deity of the body, the basis of life, the source of all things, carrying out its function of nourishing, warming, and activating body as a whole. But we shall more fittingly speak of these matters when we consider the final cause of this kind of movement.

De Motu Cordis (1628), *The Circulation of the Blood and Other Writings*, trans. Kenneth J. Franklin (1957), Chapter 8, 59.

7 In man, then, let us take the amount that is extruded by the individual beats, and that cannot return into the heart because of the barrier set in its way by the valves, as half an ounce, or three drachms, or at least one drachm. In half an hour the heart makes over a thousand beats; indeed, in some individuals, and on occasion, two, three, or four thousand. If you multiply the drachms per beat by the number of beats you will see that in half an hour either a thousand times three drachms or times two drachms, or five hundred ounces, or other such proportionate quantity of blood has been passed through the heart into the arteries, that is, in all cases blood in greater amount than can be found in the whole of the body. Similarly in the sheep or the dog. Let us take it that one scruple passes in a single contraction of the heart; then in half an hour a thousand scruples, or three and a half pounds of blood, do so. In a body of this size, as I have found in the sheep, there is often not more than four pounds of blood.

In the above sort of way, by calculating the amount of blood transmitted [at each heart beat] and by making a count of the beats, let us convince ourselves that the whole amount of the blood mass goes through the heart from the veins to the arteries and similarly makes the pulmonary transit.

Even if this may take more than half an hour or an hour or a day for its accomplishment, it does nevertheless show that the beat of the heart is continuously driving through that organ more blood than the ingested food can supply, or all the veins together at any time contain.

De Motu Cordis (1628), *The Circulation of the Blood and Other Writings*, trans. Kenneth J. Franklin (1957), Chapter 9, 62–3.

1 *Ex ovo omnia.*

All out of the egg.

Disputations Touching the Generation of Animals (1651), Legend to Title Page Illustration.

2 Everyone admits that the male is the primary efficient cause in generation, as being that in whom the species or form resides, and they further assert that his genitures emitted in coitus causes the egg both to exist and to be fertile. But how the semen of the cock produces the chick from the egg, neither the philosophers nor the physicians of yesterday or today have satisfactorily explained, or solved the problem formulated by Aristotle.

Disputations Touching the Generation of Animals (1651), trans. Gweneth Whitteridge (1981), Chapter 47, 214.

3 And so I conclude that blood lives and is nourished of itself and in no way depends on any other part of the body as being prior to it or more excellent . . . So that from this we may perceive the causes not only of life in general . . . but also of longer or shorter life, of sleeping and waking, of skill, of strength and so forth.

Disputations Touching the Generation of Animals (1651), trans. Gweneth Whitteridge (1981), Chapter 52, 247.

4 It is, however, an argument of no weight to say that natural bodies are first generated or compounded out of those things into which they are at the last broken down or dissolved.

Disputations Touching the Generation of Animals (1651), trans. Gweneth Whitteridge (1981), Chapter 72, 389.

5 ON **HARVEY** He [William Harvey] bid me to goe to the Fountain-head, and read Aristotle, Cicero, Avicenna, and did call the Neoteriques shitt-breeches.

John Aubrey, *Brief Lives* (1680), edited by Oliver Lawson Dick (1949), 129.

6 ON **HARVEY** I have heard him [William Harvey] say, that after his Booke of the *Circulation of the Blood* came-out, that he fell mightily in his Practize, and that 'twas beleeved by the vulgar that he was crack-brained.

John Aubrey, *Brief Lives* (1680), edited by Oliver Lawson Dick (1949), 131.

7 ON **HARVEY** He did not care for chymistrey, and was wont to speake against them with an undervalue.

John Aubrey, *Brief Lives* (1680), edited by Oliver Lawson Dick (1949), 132.

René Juste Haüy 1743–1822

French mineralogist

1 A casual glance at crystals may lead to the idea that they were *pure sports of nature*, but this is simply an elegant way of declaring one's ignorance. With a thoughtful examination of them, we discover laws of arrangement. With the help of these, calculation portrays and links up the observed results. How variable and at the same time how precise and regular are these laws! How simple they are ordinarily, without losing anything of their significance! The theory which has served to develop these laws is based entirely on a fact, whose existence has hitherto been vaguely discerned rather than demonstrated. This fact is that in all minerals which belong to the same species, these little solids, which are the crystal elements and which I call their *integrant molecules*, have an invariable form, in which the faces lie in the direction of the natural fracture surfaces corresponding to the mechanical division of the crystals. Their angles and dimensions are derived from calculations combined with observation.

Traité de minéralogie . . . Publié par le conseil des mines (1801), Vol. 1, xiii-iv, trans. Albert V. and Marguerite Carozzi.

2 The exterior form and the chemical composition are each other's image.

Traité de minéralogie . . . Publié par le conseil des mines, (1801), Vol. 1, xxxi, trans. Albert V. and Marguerite Carozzi.

Stephen W. Hawking 1942–

British theoretical physicist and cosmologist

3 Let me describe briefly how a black hole might be created. Imagine a star with a mass 10 times that of the sun. During most of its lifetime of about a billion years the star will generate heat at its center by converting hydrogen into helium. The energy released will create sufficient pressure to support the star against its own gravity, giving rise to an object with a radius about five times the radius of the sun. The escape velocity from the surface of such a star would be about 1,000 kilometers per second. That is to say, an object fired vertically upward from the surface of the star with a velocity of less than 1,000 kilometers per second would be dragged back by the gravitational field of the star and would return to the surface, whereas an object with a velocity greater than that would escape to infinity.

When the star had exhausted its nuclear fuel, there would be nothing to maintain the outward pressure, and the star would begin to collapse because of its own gravity. As the star shrank, the gravitational field at the surface would become stronger and the escape velocity would increase. By the time the radius had got down to 30 kilometers the escape velocity would have increased to 300,000 kilometers per second, the velocity of light. After that time any light emitted from the star would not be able to escape to infinity but would be dragged back by the gravitational field. According to the special theory of relativity nothing can travel faster than light, so that if light cannot escape, nothing else can either.

The result would be a black hole: a region of space-time from which it is not possible to escape to infinity.

'The Quantum Mechanics of Black Holes', *Scientific American*, 1977, 236, 34–40.

4 Consideration of particle emission from black holes would seem to suggest that God not only plays dice, but also sometimes throws them where they cannot be seen.

'The Quantum Mechanics of Black Holes', *Scientific American*, 1977, 236, 40.

5 Hubble's observations suggested that there was a time, called the big bang, when the universe was infinitesimally small and infinitely dense. Under such conditions all the laws of science, and therefore all ability to predict the future, would break down. If there were events earlier than this time, then they could not affect what happens at the present time. Their existence can be ignored because it would have no observational consequences. One may say that time had a beginning at the big bang, in the sense that earlier times simply would not be defined. It should be emphasized that this beginning in time is very different from those that had been considered previously. In an unchanging universe a beginning in time is something that has to be imposed by some being outside the universe; there is no physical necessity for a beginning. One can imagine that God created the universe at literally any time in the past. On the other hand, if the universe is expanding, there may be physical reasons why there had to be a beginning. One could still imagine that God created the universe at the instant of the big bang, or even afterwards in just such a way as to make it look as though there had been a big bang, but it would be meaningless to suppose that it was created *before* the big bang. An expanding universe does not preclude a creator, but it does place limits on when he might have carried out his job!

A Brief History of Time: From the Big Bang to Black Holes (1988), 8–9.

1 Today scientists describe the universe in terms of two basic partial theories—the general theory of relativity and quantum mechanics. They are the great intellectual achievements of the first half of this century. The general theory of relativity describes the force of gravity and the large-scale structure of the universe, that is, the structure on scales from only a few miles to as large as a million million million million (1 with twenty-four zeros after it) miles, the size of the observable universe. Quantum mechanics, on the other hand, deals with phenomena on extremely small scales, such as a millionth of a millionth of an inch. Unfortunately, however, these two theories are known to be inconsistent with each other—they cannot both be correct.

A Brief History of Time: From the Big Bang to Black Holes (1988), 11–2.

2 There are something like ten million million million million million million million million million million million million million million (1 with eighty zeroes after it) particles in the region of the universe that we can observe. Where did they all come from? The answer is that, in quantum theory, particles can be created out of energy in the form of particle/antiparticle pairs. But that just raises the question of where the energy came from. The answer is that the total energy of the universe is exactly zero. The matter in the universe is made out of positive energy. However, the matter is all attracting itself by gravity. Two pieces of matter that are close to each other have less energy than the same two pieces a long way apart, because you have to expend energy to separate them against the gravitational force that is pulling them together. Thus, in a sense, the gravitational field has negative energy. In the case of a universe that is approximately uniform in space, one can show that this negative gravitational energy exactly cancels the positive energy represented by the matter. So the total energy of the universe is zero.

A Brief History of Time: From the Big Bang to Black Holes (1988), 129.

3 The quantum theory of gravity has opened up a new possibility, in which there would be no boundary to space-time and so there would be no need to specify the behaviour at the boundary. There would be no singularities at which the laws of science broke down and no edge of space-time at which one would have to appeal to God or some new law to set the boundary conditions for space-time. One could say: 'The boundary condition of the universe is that it has no boundary.' The universe would be completely self-contained and not affected by anything outside itself.

It would neither be created nor destroyed. It would just BE.

A Brief History of Time: From the Big Bang to Black Holes (1988), 136.

Benjamin Robert Haydon 1786–1846

British painter and writer

1 Newton's health, and confusion to mathematics.

28 December 1817, in Tom Taylor (ed.), *The Autobiography and Memoirs of Benjamin Robert Haydon (1786–1846)*, intro. Aldous Huxley (1926), Vol. 1, 269.

William Hazlitt 1778–1830

British writer

2 You shall yourself be judge. Reason, with most people, means their own opinion.

Essay XVII. 'A New School of Reform: A Dialogue between a Rationalist and a Sentimentalist', in A.R. Waller and A. Glover (eds.), *The Collected Works of William Hazlitt* (1903), Vol. 7, 188.

3 The origin of all science is in the desire to know causes; and the origin of all false science and imposture is in the desire to accept false causes rather than none; or, which is the same thing, in the unwillingness to acknowledge our own ignorance.

'Burke and the Edinburgh Phrenologists', *The Atlas*, 15 February 1829.

Oliver Heaviside 1850–1925

British physicist

4 Now, in the development of our knowledge of the workings of Nature out of the tremendously complex assemblage of phenomena presented to the scientific inquirer, mathematics plays in some respects a very limited, in others a very important part. As regards the limitations, it is merely necessary to refer to the sciences connected with living matter, and to the ologies generally, to see that the facts and their connections are too indistinctly known to render mathematical analysis practicable, to say nothing of the complexity. Facts are of not much use, considered as facts. They bewilder by their number and their apparent incoherency. Let them be digested into theory, however, and brought into mutual harmony, and it is another matter. Theory is the essence of facts. Without theory scientific knowledge would be only worthy of the madhouse.

Electromagnetic Theory (1893), Vol. 1, 12.

5 Ohm found that the results could be summed up in such a simple law that he who runs may read it, and a schoolboy now can predict what a Faraday then could only guess at roughly. By Ohm's discovery a large part of the domain of electricity became annexed by Coulomb's discovery of the law of inverse squares, and completely annexed by Green's investigations. Poisson attacked the diffcult problem of induced magnetisation, and his results, though differently expressed, are still the theory, as a most important first approximation. Ampère brought a multitude of phenomena into theory by his investigations of the mechanical forces between conductors supporting currents and magnets. Then there were the remarkable researches of Faraday, the prince of experimentalists, on electrostatics and electrodynamics and the induction of currents. These were rather long in being brought from the crude experimental state to a compact system, expressing the real essence. Unfortunately, in my opinion, Faraday was not a mathematician. It can scarely be doubted that had he been one, he would have anticipated much later work. He would, for instance, knowing Ampère's theory, by his own results have readily been led to Neumann's theory, and the connected work of Helmholtz and Thomson. But it is perhaps too much to expect a man to be both the prince of experimentalists and a competent mathematician.

Electromagnetic Theory (1893), Vol. 1, 14.

Georg Wilhelm Friedrich Hegel 1770–1831

German philosopher

1 In history an additional result is commonly produced by human actions beyond that which they aim at and obtain—that which they immediately recognize and desire. They gratify their own interest; but something further is thereby accomplished, latent in the actions in question, though not present to their consciousness, and not included in their design . . . This may be called the cunning of reason [*List der Vernunft*].

The Philosophy of History (1837), 27–33.

2 The History of the World is nothing but the development of the Idea of Freedom.

The Philosophy of History (1837), 456.

3 Reason is just as cunning as she is powerful. Her cunning consists principally in her mediating activity, which, by causing objects to act and re-act on each other in accordance with their own nature, in this way, without any direct interference in the process, carries out reason's intentions.

Die Logik (1840), 382.

4 To him who looks at the world rationally the world looks rationally back.

Quoted in *Reason in History: A General Introduction to the Philosophy of History* (1953), trans. Robert S. Hartmann, 13.

5 When philosophy paints its grey in grey, then has a shape of life grown old. By philosophy's grey in grey it cannot be rejuvenated but only understood. The owl of Minerva spreads its wings only with the falling of the dusk.

Hegel's Philosophy of Right (1821), trans. T. M. Knox (1942), 13.

Martin Heidegger 1889–1976

German philosopher

6 We are too late for the gods and too
Early for the being. Being's a poem,
Just begun, is man.

Poetry, Language, Thought (1971), trans. Albert Hofstadter, 4.

Ian Morris Heilbron 1886–1959

British chemist

7 And so, after many years, victory has come, and the romance of exploration, of high hopes and bitter disappointment, will in a few years simply be recorded in the text-books of organic chemistry in a few terse sentences.

'Recent Developments in the Vitamin A Field', The Pedlar Lecture, December 4th 1947, *Journal of the Chemical Society*, Part 1 (1948), 393.

Werner Heisenberg 1901–76

German physicist

8 The basic idea is to shove all fundamental difficulties onto the neutron and to do quantum mechanics in the nucleus.

Letter to Niels Bohr, 20 June 1932. Quoted in David C. Cassidy, *Uncertainty: The Life and Science of Werner Heisenberg* (1992), 292.

9 It is not surprising that our language should be incapable of describing the processes occurring within the atoms, for, as has been remarked, it was invented to describe the experiences of daily life, and these consists only of processes involving exceedingly large numbers of atoms. Furthermore, it is very difficult to modify our language so that it will be able to describe these atomic processes, for words can only describe things of which we can form mental pictures, and this ability, too, is a result of daily experience. Fortunately, mathematics is not subject to this limitation, and it has been possible to invent a mathematical scheme—the quantum theory—which seems entirely adequate for the treatment of atomic processes; for visualization, however, we must content ourselves with two incomplete analogies—the wave picture and the corpuscular picture.

The Physical Principles of the Quantum Theory, trans. Carl Eckart and Frank C. Hoyt (1949), 11.

1 Thus one becomes entangled in contradictions if one speaks of the probable position of the electron without considering the experiment used to determine it . . . It must also be emphasized that the statistical character of the relation depends on the fact that the influence of the measuring device is treated in a different manner than the interaction of the various parts of the system on one another. This last interaction also causes changes in the direction of the vector representing the system in the Hilbert space, but these are completely determined. If one were to treat the measuring device as a part of the system—which would necessitate an extension of the Hilbert space—then the changes considered above as indeterminate would appear determinate. But no use could be made of this determinateness unless our observation of the measuring device were free of indeterminateness. For these observations, however, the same considerations are valid as those given above, and we should be forced, for example, to include our own eyes as part of the system, and so on. The chain of cause and effect could be quantitatively verified only if the whole universe were considered as a single system—but then physics has vanished, and only a mathematical scheme remains. The partition of the world into observing and observed system prevents a sharp formulation of the law of cause and effect. (The observing system need not always be a human being; it may also be an inanimate apparatus, such as a photographic plate.)

The Physical Principles of the Quantum Theory, trans. Carl Eckart and Frank C. Hoyt (1949), 58.

2 According to Democritus, atoms had lost the qualities like colour, taste, etc., they only occupied space, but geometrical assertions about atoms were admissible and required no further analysis. In modern physics, atoms lose this last property, they possess geometrical qualities in no higher degree than colour, taste, etc. The atom of modern physics can only be symbolized by a partial differential equation in an abstract multidimensional space. Only the experiment of an observer forces the atom to indicate a position, a colour and a quantity of heat. All the qualities of the atom of modern physics are derived, it has no *immediate and direct* physical properties at all, i.e. every type of visual conception we might wish to design is, *eo ipso*, faulty. An understanding of 'the first order' is, I would almost say by definition, impossible for the world of atoms.

Philosophic Problems of Nuclear Science, trans. F. C. Hayes (1952), 38.

3 What we observe is not nature itself but nature exposed to our method of questioning. Our scientific work in physics consists in asking questions about nature in the language that we possess and trying to get an answer from experiment by the means that are at our disposal.

Physics and Philosophy: The Revolution in Modern Science (1958), 78.

4 *The incomplete knowledge of a system must be an essential part of every formulation in quantum theory.* Quantum theoretical laws must be of a statistical kind. To give an example: we know that the radium atom emits alpha-radiation. Quantum theory can give us an indication of the probability that the alpha-particle will leave the nucleus in unit time, but it cannot predict at what precise point in time the emission will occur, for this is uncertain in principle.

The Physicist's Conception of Nature (1958), 41.

1 Can quantum mechanics represent the fact that an electron finds itself approximately in a given place and that it moves approximately with a given velocity, and can we make these approximations so close that they do not cause experimental difficulties?

Physics and Beyond: Encounters and Conversations, trans. Arnold J. Pomerans (1971), 78.

2 It seems sensible to discard all hope of observing hitherto unobservable quantities, such as the position and period of the electron . . . Instead it seems more reasonable to try to establish a theoretical quantum mechanics, analogous to classical mechanics, but in which only relations between observable quantities occur.

In Helge Kragh, *Quantum Generations: A History of Physics in the Twentieth Century* (1999), 161.

Joseph Heller 1923–99

American author

3 There was only one catch and that was Catch-22, which specified that a concern for one's own safety in the face of dangers that were real and immediate was the process of a rational mind . . . Orr would be crazy to fly more missions and sane if he didn't, but if he was sane he had to fly them. If he flew them he was crazy and didn't have to; but if he didn't want to he was sane and had to.

Catch-22 (1964), 54.

Hermann Ludwig Ferdinand von Helmholtz 1821–94

German physicist and physiologist

4 The relationships of free and latent heat set forth in the language of the materialistic theory remain the same if in place of the quantity of matter we put the constant quantity of motion in accordance with the laws of mechanics. The only difference enters where it concerns the generations of heat through other motive forces and where it concerns the equivalent of heat that can be produced by a particular quantity of a mechanical or electrical force.

'Wärme, physiologisch', *Handwörterbuch der medicinischen Wissenschaften* (1845). In Timothy Lenoir, *The Strategy of Life* (1982), 203.

5 There is deposited in them [plants] an enormous quantity of potential energy [Spannkräfte], whose equivilent is provided to us as heat in the burning of plant substances. So far as we know at present, the only living energy [lebendige Kraft] absorbed during plant growth are the chemical rays of sunlight . . . Animals take up oxygen and complex oxidizable compounds made by plants, release largely as combustion products carbonic acid and water, partly as simpler reduced compounds, thus using a certain amount of chemical potential energy to produce heat and mechanical forces. Since the latter represent a relatively small amount of work in relation to the quantity of heat, the question of the conservation of energy reduces itself roughly to whether the combustion and transformation of the nutritional components yields the same amount of heat released by animals.

Wissenschaftliche Abhandlungen (1847), 66. Trans. Joseph S. Fruton, *Proteins, Enzymes, Genes: The Interplay of Chemistry and Biology* (1999), 247.

6 I have found that a measurable period of time elapses before the stimulus applied to the iliac plexus of the frog is transmitted to the insertion of the crural nerve into the gastrocnemius muscle by a brief electric current. In large frogs, in which the nerves were from 50–60 mm. in length, and which were preserved at a temperature of 2–6° C, although the temperature of the observation chanber was between 11° and 15° C, the elapsed time was 0.0014 to 0.0020 of a second.

'Vorläufiger Bericht über die Fortpflanzungsgeschwindigkeit der Nervenreizung' (1850). Trans. Edwin Clarke and C. D. O'Malley, *The Human Brain and Spinal Cord* (1968), 207.

1 In all cases of the motion of free material points under the influence of their attractive and repulsive forces, whose intensity depends solely upon distance, the loss in tension is always equal to the gain in *vis viva*, and the gain in the former equal to the loss in the latter. Hence *the sum of the existing tensions and vires vivae is always constant*. In this most general form we can distinguish our law as *the principle of the conservation of force*.

'On the Conservation of Force; a Physical Memoir'. In John Tyndall and William Francis (eds.), *Scientific Memoirs: Natural Philosophy* (1853), 121.

2 The total quantity of all the forces capable of work in the whole universe remains eternal and unchanged throughout all their changes. All change in nature amounts to this, that force can change its form and locality without its quantity being changed. The universe possesses, once for all, a store of force which is not altered by any change of phenomena, can neither be increased nor diminished, and which maintains any change which takes place on it.

The Conservation of Energy, from a Lecture, 1863. Trans. Edmund Blair Bolles (ed.), *Galileo's Commandment: An Anthology of Science Writing* (2000), 407.

3 If we accept the hypothesis that the elementary substances are composed of atoms, we cannot avoid concluding that electricity also, positive as well as negative, is divided into definite elementary portions, which behave like atoms of electricity.

'On the Modern Development of Faraday's Conception of Electricity', *Journal of the Chemical Society* 1881, 39, 290.

4 I think the facts leave no doubt that the very mightiest among the chemical forces are of electric origin. The atoms cling to their electric charges, and opposite electric charges cling to each other.

'On the Modern Development of Faraday's Conception of Electricity', *Journal of the Chemical Society* 1881, 39, 302.

5 Isolated facts and experiments have in themselves no value, however great their number may be. They only become valuable in a theoretical or practical point of view when they make us acquainted with the *law* of a series of uniformly recurring phenomena, or, it may be, only give a negative result showing an incompleteness in our knowledge of such a law, till then held to be perfect.

'The Aim and Progress of Physical Science' (1869). Trans. E. Atkinson, *Popular Lectures on Scientific Subjects* (1873), 369.

6 A metaphysical conclusion is either a false conclusion or a concealed experimental conclusion.

'On Thought in Medicine' (1877). Trans. E. Atkinson, *Popular Lectures on Scientific Subjects* (1881), 234.

7 Just as a physicist has to examine the telescope and galvanometer with which he is working; has to get a clear conception of what he can attain with them, and how they may deceive him; so, too, it seemed to me necessary to investigate likewise the capabilities of our power of thought.

'An Autobiographical Sketch' (1891). Trans. E. Atkinson, *Popular Lectures on Scientific Subjects*, Second Series, New Edition (1895), 284–5.

Johannes Baptista van Helmont 1579–1644

Belgian natural philosopher, chemist and physician

8 *Archeus*, the Workman and Governour of generation, doth cloath himself presently with a bodily cloathing: For in things soulified he walketh thorow all the Dens and retiring places of his Seed, and begins to transform the matter, according to the perfect act of his own Image.

Oriatrike: Or, Physick Refined, trans. John Chandler (1662), 35.

1 I call this Spirit, unknown hitherto, by the new name of Gas, which can neither be constrained by Vessels, nor reduced into a visible body, unless the feed being first extinguished. But Bodies do contain this Spirit, and do sometimes wholly depart into such a Spirit, not indeed, because it is actually in those very bodies (for truly it could not be detained, yea the whole composed body should flie away at once) but it is a Spirit grown together, coagulated after the manner of a body, and is stirred up by an attained ferment, as in Wine, the juyce of unripe Grapes, bread, hydromel or water and Honey.

Oriatrike: Or, Physick Refined, trans. John Chandler (1662), 106.

2 For I took an Earthen Vessel, in which I put 200 pounds of Earth that had been dried in a Furnace, which I moystened with Rain-water, and I implanted therein the Trunk or Stem of a Willow Tree, weighing five pounds: and about three ounces: But I moystened the Earthen Vessel with Rain-water, or distilled water (alwayes when there was need) and it was large, and implanted into the Earth, and leaft of the Vessel, with an Iron-Plate covered with Tin, and easily passable with many holes. I computed not the weight of the leaves that fell off in the four Autumnes. At length, I again dried the Earth of the Vessel, and there were found the same 200 pounds, wanting about two ounces. Therefore 164 pounds of Wood, Barks, and Roots, arose out of water onely.

Oriatrike: Or, Physick Refined, trans. John Chandler (1662), 109.

3 That Mettals, Small Stones, Rocky-Stones, Sulphurs, Salts, and so the whole rank of Minerals, do find their Seeds in the Matrix or Womb of the Waters, which contain the Reasons, Gifts, Knowledges, Progresses, Appointments, Offices, and Durations of the same.

Oriatrike: Or, Physick Refined, trans. John Chandler (1662), 693.

Philip Showalter Hench

1896–1965

American physician

4 These hormones still belong to the physiologist and to the clinical investigator as much as, if not more than, to the practicing physician. But as Professor Starling said many years ago, 'The physiology of today is the medicine of tomorrow'.

'The Reversibility of Certain Rheumatic and Non-rheumatic Conditions by the use of Cortisone or of the Pituitary Adrenocorticotropic Hormone', Nobel Lecture, 11 December 1950. In *Nobel Lectures: Physiology or Medicine 1942–1962* (1964), 334.

Lawrence Joseph Henderson

1878–1942

American physiologist

5 Historically the most striking result of Kant's labors was the rapid separation of the thinkers of his own nation and, though less completely, of the world, into two parties;—the philosophers and the scientists.

The Order of Nature: An Essay (1917), 69.

6 The concept of an *independent system* is a pure creation of the imagination. For no material system is or can ever be perfectly isolated from the rest of the world. Nevertheless it completes the mathematician's 'blank form of a universe' without which his investigations are impossible. It enables him to introduce into his geometrical space, not only masses and configurations, but also physical structure and chemical composition. Just as Newton first conclusively showed that this is a world of masses, so Willard Gibbs first revealed it as a world of systems.

The Order of Nature: An Essay (1917), 126.

7 A . . . difference between most system-building in the social sciences and systems of thought and classification of the natural sciences is to be seen in their evolution. In the natural sciences both theories and descriptive systems grow by adaptation to the increasing knowledge and experience of the scientists. In the social sciences, systems often issue fully formed from the mind of one man. Then they may be much discussed if they attract attention, but progressive adaptive modification as a result of the concerted efforts of great numbers of men is rare.

The Study of Man (1941), 19–20.

1 Science owes more to the steam engine than the steam engine owes to science.

Attributed.

2 I have always had the feeling that organic chemistry is a very peculiar science, that organic chemists are unlike other men, and there are few occupations that give more satisfactions [sic] than masterly experimentation along the old lines of this highly specialised science.

Henderson's memories, unpublished typescript, 85–6, Harvard University Archives 4450.7.2. Quoted in J. S. Fruton, *Contrasts in Scientific Style* (1990), 194.

3 ON **HENDERSON** Somewhere between 1900 and 1912 in this country, according to one sober medical scientist [Henderson] a random patient, with a random disease, consulting a doctor chosen at random had, for the first time in the history of mankind, a better than fifty-fifty chance of profiting from the encounter.

Quoted in *New England Journal of Medicine*, 1964, 270, 449.

Friedrich Gustav Jacob Henle

1809–85

German anatomist and pathologist

4 To choose a rough example, think of a thorn which has stuck in a finger and produces an inflammation and suppuration. Should the thorn be discharged with the pus, then the finger of another individual may be pricked with it, and the disease may be produced a second time. In this case it would not be the disease, not even its product, that would be transmitted by the thorn, but rather the stimulus which engendered it. Now supposing that the thorn is capable of multiplying in the sick body, or that every smallest part may again become a thorn, then one would be able to excite the same disease, inflammation and suppuration, in other individuals by transmitting any of its smallest parts. The disease is not the parasite but the thorn. Diseases resemble one another, because their causes resemble each other. The contagion in our sense is therefore not the germ or seed of the disease, but rather the cause of the disease. For example, the egg of a taenia is not the product of a worm disease even though the worm disease may have been the cause, which first gave rise to the taenia in the intestinal contents—nor of the individual afflicted with the worm disease, but rather of the parasitic body, which, no matter how it may have come into the world at first, now reproduces itself by means of eggs, and produces the symptoms of the worm disease, at least in part. It is not the seed of the disease; the latter multiplies in the sick organism, and is again excreted at the end of the disease.

'On Miasmata and Contagia', trans. G. Rosen, *Bulletin of the Institute of the History of Medicine*, 1938, 6, 924.

5 I have shown first in the course of this investigation that the infective matter is independently animate, and further that one could think of the independently animate matter as either animal and plant organisms or elementary parts of animals, which have achieved a relative individuality.

'On Miasmata and Contagia', trans. G. Rosen, *Bulletin of the Institute of the History of Medicine*, 1938, 6, 957.

6 I, however, believe that for the ripening of experience the light of an intelligent theory is required. People are amused by the witticism that the man with a theory forces from nature that answer to his question which he wishes to have but nature never answers unless she is questioned, or to speak more accurately, she is always talking to us and with a thousand tongues but we only catch the answer to our own question.

Quoted in Major Greenwood, *Epidemiology Historical and Experimental. The Herter Lectures for 1931* (1932), 13.

Joseph Henry 1797–1878

American geophysicist

1 Meteorology has ever been an apple of contention, as if the violent commotions of the atmosphere induced a sympathetic effect on the minds of those who have attempted to study them.

'Meteorology in its Connection with Agriculture', *US Patent Office Annual Report Agricultural*, 1858. In J .R. Fleming, *Meteorology in America: 1800–1870* (1990), 23.

2 The seeds of great discoveries are constantly floating around us, but they only take root in minds well-prepared to receive them.

In C. Guy Suits (ed.), *The Collected Works of Irving Langmuir* (1962), Vol. 12, 80.

Heraclitus c.550–c.475 BC

Greek moral and natural philosopher

3 Things of which there is sight, hearing, apprehension, these I prefer.

Heraclitus, fr. 55. Trans. R. W. Sharples.

4 Nature is accustomed to hide itself.

Heraclitus, fr. 123. Trans. R. W. Sharples.

5 Eyes and ears are bad witnesses for men who have barbarian souls.

Heraclitus, fr. 107. Trans. R. W. Sharples.

6 Heraclitus son of Bloson (or, according to some, of Herakon) of Ephesus. This man was at his prime in the 69th Olympiad. He grew up to be exceptionally haughty and supercilious, as is clear also from his book, in which he says: 'Learning of many things does not teach intelligence; if so it would have taught Hesiod and Pythagoras, and again Xenophanes and Hecataeus.' . . . Finally he became a misanthrope, withdrew from the world, and lived in the mountains feeding on grasses and plants. However, having fallen in this way into a dropsy he came down to town and asked the doctors in a riddle if they could make a drought out of rainy weather. When they did not understand he buried himself in a cow-stall, expecting that the dropsy would be evaporated off by the heat of the manure; but even so he failed to effect anything, and ended his life at the age of sixty.

Diogenes Laertius 9.1. In G. S. Kirk, J. E. Raven, and M. Schofield (eds.), *The Presocratic Philosophers: A Critical History with a Selection of Texts* (1983), 181.

7 Heraclitus somewhere says that all things are in process and nothing stays still, and likening existing things to the stream of a river he says that you would not step twice into the same river.

Plato, *Cratylus* 402A. In G. S. Kirk, J. E. Raven, and M. Schofield (eds.), *The Presocratic Philosophers: A Critical History with a Selection of Texts* (1983), 195.

Johann Gottfried Herder 1744–1803

German philosopher

8 So says the most ancient book of the Earth; thus it is written on its leaves of marble, lime, sand, slate, and clay: . . . that our Earth has fashioned itself, from its chaos of substances and powers, through the animating warmth of the creative spirit, to a peculiar and original whole, by a series of preparatory revolutions, till at last the crown of its creation, the exquisite and tender creature man, was enabled to appear.

Outlines of a Philosophy of the History of Man (1803). Translated from 1784 Original, Vol. 1, Book 10, 465–6.

Robert Herrick 1591–1674

British clergyman and poet

9 Attempt the end and never stand to
doubt;

Nothing's so hard, but search will find it out.

'Seeke and Finde', *Hesperides: Or, the Works Both Humane and Divine of Robert Herrick Esq.* (1648). In J. Max Patrick (ed.), *The Complete Poetry of Robert Herrick* (1963), 411.

Caroline Lucretia Herschel

1750–1848

German-born British astronomer

1 My dear nephew was only in his sixth year when I came to be detached from the family circle. But this did not hinder *John and I* from remaining the most affectionate friends, and many a half or whole holiday he was allowed to spend with me, was dedicated to making experiments in chemistry, where generally all boxes, tops of tea-canisters, pepper-boxes, teacups, &c., served for the necessary vessels, and the sand-tub furnished the matter to be analysed. I only had to take care to exclude water, which would have produced havoc on my carpet.

Letter to Lady Herschel, 6 September 1833. Quoted in *Memoir and Correspondence of Caroline Herschel* (1876), 259.

John Frederick William Herschel 1792–1871

British astronomer

See ***Darwin, Charles*** *158:1*

2 Speculations apparently the most unprofitable have almost invariably been those from which the greatest practical applications have emanated.

Preliminary Discourse on the Study of Natural Philosophy (1831), 11.

3 A mind which has once imbibed a taste for scientific enquiry, and has learnt the habit of applying its principles readily to the cases which occur, has within itself an inexhaustible source of pure and exciting contemplations.

Preliminary Discourse on the Study of Natural Philosophy (1831), 14–5.

4 Geology, in the magnitude and sublimity of the objects of which it treats, undoubtedly ranks, in the scale of the sciences, next to astronomy.

Preliminary Discourse on the Study of Natural Philosophy (1831), 287.

5 Every student who enters upon a scientific pursuit, especially if at a somewhat advanced period of life, will find not only that he has much to learn, but much also to unlearn.

Outlines of Astronomy (1871), 11th edn., 1.

6 Words are to the Anthropologist what rolled pebbles are to the Geologist—Battered relics of past ages often containing within them indelible records capable of intelligible interpretion—and when we see what amount of change 2000 years has been able to produce in the languages of Greece & Italy or 1000 in those of Germany, France & Spain we naturally begin to ask how long a period must have lapsed since the Chinese, the Hebrew, the Delaware & the Malesass had a point in common with the German & Italian & each other.—Time! Time! Time!—we must not impugn the Scripture Chronology, but we *must* interpret it in accordance with *whatever* shall appear on fair enquiry to be the *truth* for there cannot be two truths. And really there is scope enough: for the lives of the Patriarchs may as reasonably be extended to 5000 or 50000 years apiece as the days of Creation to as many thousand millions of years.

Letter to Charles Lyell, 20 February 1836, in Walter F. Cannon, 'The Impact of Uniformitarianism', *Proceedings of the American Philosophical Society*, 1961, 105, 308.

7 According to this view of the matter, there is nothing casual in the formation of Metamorphic Rocks. All strata, once buried deep enough, (and due TIME allowed!!!) must assume that state,—none can escape. All records of former worlds must ultimately perish.

Letter to Mr Murchison, in explanation of the views expressed in his previous letter to Mr Lyell, 15 November 1836. Quoted in the

Appendix to Charles Babbage, *The Ninth Bridgewater Treatise: A Fragment* (1838), 240.

William Herschel 1738–1822

German-born British astronomer

*See **Walpole, Horace** 607:1*

1 It will be found that those contained in one article [class of nebulae], are so closely allied to those in the next, that there is perhaps not so much difference between them, if I may use the comparison, as there would be in an annual description of the human figure, were it given from the birth of a child till he comes to be a man in his prime.

'Astronomical Observations . . . ', *Philosophical Transactions*, 1811, 101, 271.

2 We may also draw a very important additional conclusion from the gradual dissolution of the milky way; for the state into which the incessant action of the clustering power [presumably, gravity] has brought it at present, is a kind of chronometer that may be used to measure the time of its past and future existence; and although we do not know the rate of going of this mysterious chronometer, it is nevertheless certain, that since the breaking up of the parts of the milky way affords a proof that it cannot last for ever, it equally bears witness that its past duration cannot be admitted to the infinite.

'Astronomical Observations . . . ', *Philosophical Transactions*, 1814, 104, 284.

3 These instruments have play'd me so many tricks that I have at last found them out in many of their humours.

On the problems with telescopes.
Quoted in C. A. Lubbock, *The Herschel Chronicle* (1933), 102.

4 I have *looked further into space than ever human being did before me.* I have observed stars of which the light, it can be proved, must take two million years to reach the earth.

Quoted in C. A. Lubbock, *The Herschel Chronicle* (1933), 336.

Wilhelm August Oscar Hertwig 1849–1922

German biologist

5 The cell, this elementary keystone of living nature, is far from being a peculiar chemical giant molecule or even a living protein and as such is not likely to fall prey to the field of an advanced chemistry. The cell is itself an organism, constituted of many small units of life.

Quoted in Joseph S. Fruton, *Proteins, Enzymes, Genes: The Interplay of Chemistry and Biology* (1999), 59.

Heinrich Rudolf Hertz 1857–94

German physicist

6 Maxwell's theory is Maxwell's system of equations.

Electric Waves (1893), 21.

Alexander Herzen 1812–70

Russian social philosopher

7 It seems to me that you are solving a problem which goes beyond the limits of physiology in too simple a way. Physiology has realized its problem with fortitude, breaking man down into endless actions and counteractions and reducing him to a crossing, a vortex of reflex acts. Let it now permit sociology to restore him as a whole. Sociology will wrest man from the anatomical theatre and return him to history.

Letter to his son, Alexander, July-August 1868. Trans. Roger Smith, *Inhibition: History and Meaning in the Sciences of Mind and Brain* (1992), 223.

Hesiod fl. 800 BC

Greek poet

8 Tell me these things, Olympian Muses, tell
From the beginning, which first came to be?
Chaos was first of all, but next appeared
Broad-bosomed Earth, Sure standing-place for all

The gods who live on snowy Olympus' peak,
And misty Tartarus, in a recess
Of broad-pathed earth, and Love, most beautiful
Of all the deathless gods. He makes men weak,
He overpowers the clever mind, and tames
The spirit in the breasts of men and gods.
From Chaos came black Night and Erebos.
And Night in turn gave birth to Day and Space
Whom she conceived in love to Erebos.
And Earth bore starry Heaven, first, to be
An equal to herself, to cover her
All over, and to be a resting-place,
Always secure, for all the blessed gods.

Theogony, l. 114–28. In *Hesiod and Theognis*, trans. Dorothea Wender (1973), 26–7.

Harry Hammond Hess 1906–69

American geologist

1 I shall consider this paper an essay in geopoetry. In order not to travel any further into the realm of fantasy than is absolutely necessary I shall hold as closely as possibly to a uniformitarian approach; even so, at least one great catastrophe will be required early in the Earth's history.

'History of Ocean Basins', in A. E. J. Engel, H. L. James and B. F. Leonard (eds.), *Petrologic Studies: A Volume to Honour F. Buddington* (1962), 599–600.

Jack H. Hexter 1910–96

American historian

2 If physicists could not quote in the text, they would not feel that much was lost with respect to advancement of knowledge of the natural world. If historians could not quote, they would deem it a disastrous impediment to the communication of knowledge about the past. A luxury for physicists, quotation is a necessity for historians, indispensable to historiography.

Historiography (1968), 385.

William Mitchinson Hicks

1850–1934

British physicist

3 While, on the one hand, the end of scientific investigation is the discovery of laws, on the other, science will have reached its highest goal when it shall have reduced ultimate laws to one or two, the necessity of which lies outside the sphere of our cognition. These ultimate laws—in the domain of physical science at least—will be the dynamical laws of the relations of matter to number, space, and time. The ultimate data will be number, matter, space, and time themselves. When these relations shall be known, all physical phenomena will be a branch of pure mathematics.

'Address to the section of Mathematical and Physical Science', *Reports of the British Association for the Advancement of Science* (1895), 595.

David Hilbert 1862–1943

German mathematician

4 Geometry is the most complete science.

Hilbert lecture course on Geometry, unpublished. Quotation supplied by Ivor Grattan-Guinness.

5 Besides it is an error to believe that rigour is the enemy of simplicity. On the contrary we find it confirmed by numerous examples that the rigorous method is at the same time the simpler and the more easily comprehended.

Bulletin of the American Mathematical Society, 1902, 8, 442.

6 He who seeks for methods without having a definite problem in mind seeks for the most part in vain.

Bulletin of the American Mathematical Society, 1902, 8, 445.

7 *Meine Herren*, I do not see that the sex of the candidate is an argument against her admission as a Privatdozent. After all, the Senate is not a bathhouse.

Remark made by Hilbert about the University of Göttingen's decision not to accept Emmy Noether as a female faculty member.

Quoted in C. Reid, *Hilbert: With an appreciation of Hilbert's Mathematical Work by Hermann Weyl* (1970), 143.

1 *Wir mussen wissen. Wir werden wissen.*

We must know. We will know.

Now on his tomb in Göttingen.

Lecture at Königsberg, 1930. *Gesammelte Abhandlungen*, Vol. 3, 387, trans. Ivor Grattan-Guinness.

2 One hears a lot of talk about the hostility between scientists and engineers. I don't believe in any such thing. In fact I am quite certain it is untrue . . . There cannot possibly be anything in it because neither side has anything to do with the other.

Quoted in A. Rosenfeld, *Langmuir: The Man and the Scientist* (1962), 57.

Joel Henry Hildebrand 1881–1983

American chemist

3 Chemistry is fun!

In Joseph Hirschfelder, *Annual Review of Physical Chemistry*, 1983, 3, 1.

Archibald Vivian Hill 1886–1977

British physiologist

4 To prove to an indignant questioner on the spur of the moment that the work I do was useful seemed a thankless task and I gave it up. I turned to him with a smile and finished, 'To tell you the truth we don't do it because it is useful but because it's amusing.' The answer was thought of and given in a moment: it came from deep down in my soul, and the results were as admirable from my point of view as unexpected. My audience was clearly on my side. Prolonged and hearty applause greeted my confession. My questioner retired shaking his head over my wickedness and the newspapers next day, with obvious approval, came out with headlines 'Scientist Does It Because It's Amusing!' And if that is not the best reason why a scientist should do his work, I want to know what is. Would it be any good to ask a mother what practical use her baby is? That, as I say, was the first evening I ever spent in the United States and from that moment I felt at home. I realised that all talk about science purely for its practical and wealth-producing results is as idle in this country as in England. Practical results will follow right enough. No real knowledge is sterile. The most useless investigation may prove to have the most startling practical importance: Wireless telegraphy might not yet have come if Clerk Maxwell had been drawn away from his obviously 'useless' equations to do something of more practical importance. Large branches of chemistry would have remained obscure had Willard Gibbs not spent his time at mathematical calculations which only about two men of his generation could understand. With this faith in the ultimate usefulness of all real knowledge a man may proceed to devote himself to a study of first causes without apology, and without hope of immediate return.

Quoted in Larry R. Squire (ed.), *The History of Neuroscience in Autobiography* (1996), Vol. 1, 351.

Austin Bradford Hill 1897–1991

British statistician

5 The statistical method is required in the interpretation of figures which are at the mercy of numerous influences, and its object is to determine whether individual influences can be isolated and their effects measured. The essence of the method lies in the determination that we are really comparing like with like, and that we have not overlooked a relevant factor which is present in Group A and absent from Group B. The variability of human beings in their illnesses and in their reactions to them is a fundamental reason *for* the planned clinical trial and not *against it*.

Principles of Medical Statistics (1971), 13.

James Hilton 1900–54

British novelist

1 And I believe that the Binomial Theorem and a Bach Fugue are, in the long run, more important than all the battles of history.

This Week Magazine (1937).

Cyril Hinshelwood 1897–1967

British chemist

2 The natural sciences are sometimes said to have no concern with values, nor to seek morality and goodness, and therefore belong to an inferior order of things. Counter-claims are made that they are the only living and dynamic studies . . . Both contentions are wrong. Language, Literature and Philosophy express, reflect and contemplate the world. But it is a world in which men will never be content to stay at rest, and so these disciplines cannot be cut off from the great searching into the nature of things without being deprived of life-blood.

Presidential Address to Classical Association, 1959. In E. J. Bowen's obituary of Hinshelwood, *Chemistry in Britain* (1967), Vol. 3, 536.

3 A common fallacy in much of the adverse criticism to which science is subjected today is that it claims certainty, infallibility and complete emotional objectivity. It would be more nearly true to say that it is based upon wonder, adventure and hope.

Quoted in E. J. Bowen's obituary of Hinshelwood, *Chemistry in Britain* (1967), Vol. 3, 536.

4 Chemistry: that most excellent child of intellect and art.

Quoted without reference in Mary-Jo Nye, *From Chemical Philosophy to Theoretical Chemistry* (1993), 1.

Hippocrates c.460–c.370 BC

Greek physician

The Hippocratic Writings were produced by a number of individuals.

5 But medicine has long had all its means to hand, and has discovered both a principle and a method, through which the discoveries made during a long period are many and excellent, while full discovery will be made, if the inquirer be competent, conduct his researches with knowledge of the discoveries already made, and make them his starting-point. But anyone who, casting aside and rejecting all these means, attempts to conduct research in any other way or after another fashion, and asserts that he has found out anything, is and has been victim of deception.

Ancient Medicine, in *Hippocrates*, trans. W. H. S. Jones (1923), Vol. 1, 15.

6 And in Europe is a Scythian race, dwelling round Lake Maeotis, which differs from the other races. Their name is Sauromatae. Their women, so long as they are virgins, ride, shoot, throw the javelin while mounted, and fight with their enemies. They do not lay aside their virginity until they have killed three of their enemies, and they do not marry before they have performed the traditional sacred rites. A woman who takes herself a husband no longer rides, unless she is compelled to do so by a general expedition. They have no right breast; for while they are yet babies their mothers make a red-hot bronze instrument constructed for this very purpose and apply it to the right breast and cauterise it, so that its growth is arrested, and all its strength and bulk are diverted to the right shoulder and right arm.

Airs Waters Places, in *Hippocrates*, trans. W. H. S. Jones (1923), Vol. 1, 117–9.

7 As to diseases, make a habit of two things—to help, or at least to do no harm.

Epidemics, in *Hippocrates*, trans. W. H. S. Jones (1923), Vol.1, 165.

8 Into whatsoever houses I enter, I will enter to help the sick, and I will abstain from all intentional wrong-doing and harm, especially from abusing the bodies of man or woman, bond or free. And whatsoever I shall see or hear in the course of my profession, as well as outside my profession in my intercourse with men, if it be what should not be published abroad, I will never divulge, holding such things to be holy secrets.

Oath, in *Hippocrates*, trans. W. H. S. Jones (1923), Vol. 1, 301.

1 In acute diseases the physician must conduct his inquiries in the following way. First he must examine the face of the patient, and see whether it is like the faces of healthy people, and especially whether it is like its usual self. Such likeness will be the best sign, and the greatest unlikeness will be the most dangerous sign. The latter will be as follows. Nose sharp, eyes hollow, temples sunken, ears cold and contracted with their lobes turned outwards, the skin about the face hard and tense and parched, the colour of the face as a whole being yellow or black.

Prognostic, in *Hippocrates*, trans. W. H. S. Jones (1923), Vol. 2, 9.

2 I am about to discuss the disease called 'sacred'. It is not, in my opinion, any more divine or more sacred that other diseases, but has a natural cause, and its supposed divine origin is due to men's inexperience, and to their wonder at its peculiar character.

The Sacred Disease, in *Hippocrates*, trans. W. H. S. Jones (1923), Vol. 2, 139.

3 The brain of man, like that of all animals is double, being parted down its centre by a thin membrane. For this reason pain is not always felt in the same part of the head, but sometimes on one side, sometimes on the other, and occasionally all over.

The Sacred Disease, in *Hippocrates*, trans. W. H. S. Jones (1923), Vol. 2, 153.

4 Men ought to know that from the brain, and from the brain only, arise our pleasures, joys, laughter and jests, as well as our sorrows, pains, griefs and tears. Through it, in particular, we think, see, hear, and distinguish the ugly from the beautiful, the bad from the good, the pleasant from the unpleasant, in some cases using custom as a test, in others perceiving them from their utility. It is the same thing which makes us mad or delirious, inspires us with dread or fear, whether by night or by day, brings sleeplessness, inopportune mistakes, aimless anxieties, absent-mindedness, and acts that are contrary to habit. These things that we suffer all come from the brain, when it is not healthy, but becomes abnormally hot, cold, moist, or dry, or suffers any other unnatural affection to which it was not accustomed. Madness comes from its moistness.

The Sacred Disease, in *Hippocrates*, trans. W. H. S. Jones (1923), Vol. 2, 175.

5 There are some arts which to those that possess them are painful, but to those that use them are helpful, a common good to laymen, but to those that practise them grievous. Of such arts there is one which the Greeks call medicine. For the medical man sees terrible sights, touches unpleasant things, and the misfortunes of others bring a harvest of sorrows that are peculiarly his; but the sick by means of the art rid themselves of the worst of evils, disease, suffering, pain and death.

Breaths, in *Hippocrates*, trans. W. H. S. Jones (1923), Vol. 2, 227.

6 There are in fact two things, science and opinion; the former begets knowledge, the latter ignorance.

Law sect. 4, in *Hippocrates*, trans. W. H. S. Jones (1923), Vol. 2, 265.

7 And if incision of the temple is made on the left, spasm seizes the parts on the right, while if the incision is on the right, spasm seizes the parts on the left.

On Wounds in the Head, in *Hippocrates*, trans. E. T. Withington (1927), Vol. 3, 33.

8 I have clearly recorded this: for one can learn good lessons also from what has been tried but clearly has not succeeded, when it is clear why it has not succeeded.

On Joints, 47. Trans. R. W. Sharples.

9 The forms of diseases are many and the healing of them is manifold.

Nature of Man, in *Hippocrates*, trans. W. H. S. Jones (1931), Vol. 4, 7.

1 The body of man has in itself blood, phlegm, yellow bile and black bile; these make up the nature of this body, and through these he feels pain or enjoys health. Now he enjoys the most perfect health when these elements are duly proportioned to one another in respect of compounding, power and bulk, and when they are perfectly mingled.

Nature of Man, in *Hippocrates*, trans. W. H. S. Jones (1931), Vol. 4, 11.

2 Life is short, the Art long, opportunity fleeting, experience treacherous, judgment difficult. The physician must be ready, not only to do his duty himself, but also to secure the co-operation of the patient, of the attendants and of externals.

Aphorisms, in *Hippocrates*, trans. W. H. S. Jones (1931), Vol. 4, 99.

3 Even when all is known, the care of a man is not yet complete, because eating alone will not keep a man well; he must also take exercise. For food and exercise, while possessing opposite qualities, yet work together to produce health.

Regimen, in *Hippocrates*, trans. W. H. S. Jones (1931), Vol. 4, 229.

4 Through seven figures come sensations for a man; there is hearing for sounds, sight for the visible, nostril for smell, tongue for pleasant or unpleasant tastes, mouth for speech, body for touch, passages outwards and inwards for hot or cold breath. Through these come knowledge or lack of it.

Regimen, in *Hippocrates*, trans. W. H. S. Jones (1931), Vol. 4, 261.

5 How twins are born my discourse will explain thus. The cause is chiefly the nature of the womb in woman. For if it has grown equally on either side of its mouth, and if it opens equally, and also dries equally after menstruation, it can give nourishment, if it conceive the secretion of the man so that it immediately divides into both parts of the womb equally. Now if the seed secreted from both parents be abundant and strong, it can grow in both places, as it masters the nourishment that reaches it. In all other cases twins are not formed. Now when the secretion from both parents is male, of necessity boys are begotten in both places; but when from both it is female, girls are begotten. But when one secretion is female and the other male, whichever masters the other gives the embryo its sex. Twins are like one another for the following reasons. First, the places are alike in which they grow; then they were secreted together; then they grow by the same nourishment, and at birth they reach together the light of day.

Regimen, in *Hippocrates*, trans. W. H. S. Jones (1931), Vol. 4, 273.

6 Any man who is intelligent must, on considering that health is of the utmost value to human beings, have the personal understanding necessary to help himself in diseases, and be able to understand and to judge what physicians say and what they administer to his body, being versed in each of these matters to a degree reasonable for a layman.

Affections, in *Hippocrates*, trans. P. Potter (1988), Vol. 5, 7.

7 About medications that are drunk or applied to wounds it is worth learning from everyone; for people do not discover these by reasoning but by chance, and experts not more than laymen.

Affections, in *Hippocrates*, trans. P. Potter (1988), Vol. 5, 69. Littré VI, 254.

8 Correct is to recognize what diseases are and whence they come; which are long and which are short; which are mortal and which are not; which are in the process of changing into others; which are increasing and which are diminishing; which are major and which are minor; to treat the diseases that can be treated, but to recognize the ones that cannot be, and to know why they cannot be; by treating patients with the former, to give them the benefit of treatment as far as it is possible.

Diseases, in *Hippocrates*, trans. P. Potter (1988), Vol. 5, 113.

1 And if this were so in all cases, the principle would be established, that sometimes conditions can be treated by things opposite to those from which they arose, and sometimes by things like to those from which they arose.

Places in Man, in *Hippocrates*, trans. P. Potter (1995), Vol. 8, 87.

2 Medicine in its present state is, it seems to me, by now completely discovered, insofar as it teaches in each instance the particular details and the correct measures. For anyone who has an understanding of medicine in this way depends very little upon good luck, but is able to do good with or without luck. For the whole of medicine has been established, and the excellent principles discovered in it clearly have very little need of good luck.

Places in Man, in *Hippocrates*, trans. P. Potter (1995), Vol. 8, 93.

3 An insolent reply from a polite person is a bad sign.

Prorrhetic, in *Hippocrates*, trans. P. Potter (1995), Vol. 8, 181.

Wilhelm His 1831–1904

Swiss anatomist and embryologist

4 I defend the following postulate as an indisputable principle: *that each nerve fibre originates as a process from a single cell. This is its genetic, nutritive, and functional center; all other connections of the fibre are either indirect or secondary.*

'Zur Geschichte des menschlichen Rückenmarkes und der Nervenwurzeln' (1887). Trans. Edwin Clarke and C. D. O'Malley, *The Human Brain and Spinal Cord* (1968), 103.

Edward Hitchcock 1793–1864

American geologist and clergyman

5 Why may we not add Geology to the list of poetical sciences? Why shall not that science, which is the second science in eras and magnitudes, and the first, in affording scope for the imagination, be brought into favor with the Muses and afford themes for the Poet?

'The Poetry of Geology', *The Indicator*, 1849, 109.

Adolf Hitler 1889–1945

German dictator

6 Our national policies will not be revoked or modified, even for scientists. If the dismissal of Jewish scientists means the annihilation of contemporary German science, then we shall do without science for a few years.

Said to Max Planck as President of the Kaiser Wilhelm Society for the Advancement of Science as he petitioned the Führer to stop the dismissal of scientists on political grounds.

In E. Y. Hartshorne, *The German Universities and National Socialism* (1937), 112.

Sarah Hoare 1777–1856

British poet

7 Science, illuminating ray!
Fair mental beam, extend thy sway,
And shine from pole to pole!
From thy accumulated store,
O'er every mind thy riches pour,
Excite from low desires to soar,
And dignify the soul.

'Botany', 1. From *Poems on Conchology and Botany* (1831), 176.

Thomas Hobbes 1588–1679

British political and moral philosopher

8 Why may we not say, that all *Automata* (Engines that move themselves by springs and wheeles as doth a watch) have an artificiall life? For what is the *Heart*, but a *Spring*; and the *Nerves*, but so many *Strings*; and the *Joynts*, but so many *Wheeles*, giving motion to the whole Body, such as was intended by the Artificer? *Art* goes yet further, imitating the rationall and most excellent worke of Nature, *Man*. For by

Art is created the great LEVIATHAN called a COMMON-WEALTH, or STATE, (in latine CIVITAS) which is but an Artificiall Man; though of greater stature and strength than the Naturall, for whose protection and defence it was intended; and in which, the *Soveraignty* is an Artificiall *Soul*, as giving life and motion to the whole body.

Leviathan (1651), ed. C. B. Macpherson (1968), Part 1, Introduction, 81.

1 The most noble and profitable invention of all other, was that of SPEECH, consisting of *Names* or *Appellations*, and their Connexion; whereby men register their Thoughts; recall them when they are past; and also declare them one to another for mutuall utility and conversation; without which, there had been amongst men, neither Common-wealth, nor Society, nor Contract, nor Peace, no more than amongst Lyons, Bears, and Wolves.

Leviathan (1651), ed. C. B. Macpherson (1968), Part 1, Chapter 4, 100.

2 In Geometry, (which is the only Science that it hath pleased God hitherto to bestow on mankind,) men begin at settling the significations of their words; which settling of significations, they call *Definitions*; and place them in the beginning of their reckoning.

Leviathan (1651), ed. C. B. Macpherson (1968), Part 1, Chapter 4, 105.

3 For between true Science, and erroneous Doctrines, Ignorance is in the middle. Naturall sense and imagination, are not subject to absurdity. Nature it selfe cannot erre: and as men abound in copiousnesses of language; so they become more wise, or more mad than ordinary. Nor is it possible without Letters for any man to become either excellently wise, or (unless his memory be hurt by disease, or ill constitution of organs) excellently foolish. For words are wise men's counters, they do but reckon by them; but they are the money of fools that value them by the authority of an *Aristotle*, a *Cicero*, or a *Thomas*, or any other Doctor whatsoever, if but a man.

Leviathan (1651), ed. C. B. Macpherson (1968), Part 1, Chapter 4, 106.

4 *Science* is the knowledge of Consequences, and dependance of one fact upon another.

Leviathan (1651), ed. C. B. Macpherson (1968), Part 1, Chapter 5, 115.

5 And as to the faculties of the mind, setting aside the arts grounded upon words, and especially that skill of proceeding upon generall, and infallible rules, called Science; which very few have, and but in few things; as being not a native faculty, born within us; nor attained, (as Prudence,) while we look after somewhat else.

Leviathan (1651), ed. C. B. Macpherson (1968), Part 1, Chapter 13, 183.

6 Whatsoever therefore is consequent to a time of Warre, where every man is Enemy to every man; the same is consequent to the time, wherein men live without other security, than what their own strength, and their own invention shall furnish them withall. In such condition, there is no place for Industry; because the fruit thereof is uncertain: and consequently no Culture of the Earth; no Navigation, nor use of the commodities that may be imported by Sea; no commodious Building; no Instruments of moving, and removing, such things as require much force; no Knowledge of the face of the Earth; no account of Time; no Arts; no Letters; no Society; and which is worst of all, continual fear, and danger of violent death; And the life of man, solitary, poore, nasty, brutish, and short.

Leviathan (1651), ed. C. B. Macpherson (1968), Part 1, Chapter 13, 186.

7 That wee have of Geometry, which is the mother of all Naturall Science, wee are not indebted for it to the Schools.

Leviathan (1651), ed. C. B. Macpherson (1968), Part 4, Chapter 46, 686.

8 Whatsoever accidents or qualities our sense make us think there be in the world, they are not there, but are seemings and apparitions only. The things that really are in the world without us, are those motions by which these seemings are caused. And this is the great deception of sense, which also is by sense to be corrected. For as sense telleth me, when I see directly, that the colour seemeth to be in the object; so also sense telleth me, when I see by reflection, that colour is not in the object.

The Elements of Law: Natural and Politic (1640), Ferdinand Tönnies edn. (1928), Part 1, Chapter 2, 6.

1 By this we may understand, there be two sorts of knowledge, whereof the one is nothing else but sense, or knowledge original (as I have said at the beginning of the second chapter), and remembrance of the same; the other is called science or knowledge of the truth of propositions, and how things are called, and is derived from understanding.

The Elements of Law: Natural and Politic (1640), Ferdinand Tönnies edn. (1928), Part 1, Chapter 6, 18–9.

2 Mr. Hobbes told me that the cause of his Lordship's [Francis Bacon's] death was trying an Experiment: viz. as he was taking the aire in a Coach with Dr. Witherborne (a Scotchman, Physitian to the King) towards High-gate, snow lay on the ground, and it came into my Lord's thoughts, why flesh might not be preserved in snow, as in Salt. They were resolved they would try the Experiment presently. They alighted out of the Coach and went into a poore woman's house at the bottom of Highgate hill, and bought a Hen, and made the woman exenterate it, and then stuffed the body with Snow, and my Lord did help to doe it himselfe. The Snow so chilled him that he immediately fell so extremely ill, that he could not return to his Lodging.

John Aubrey, *Brief Lives* (1680), edited by Oliver Lawson Dick (1949), 16.

3 ON **HOBBES** He was 40 yeares old before he looked on Geometry; which happened accidentally. Being in a Gentleman's Library, Euclid's Elements lay open, and 'twas the 47 *El. libri* I [Pythagoras' Theorem]. He read the proposition. *By* G—, sayd he (he would now and then sweare an emphaticall Oath by way of emphasis) *this is impossible!* So he reads the Demonstration of it, which referred him back to such a Proposition; which proposition he read. That referred him back to another, which he also read. *Et sic deinceps* [and so on] that at last he was demonstratively convinced of that trueth. This made him in love with Geometry.

Of Thomas Hobbes, in 1629.
John Aubrey, *Brief Lives* (1680), edited by Oliver Lawson Dick (1949), 150.

William Herbert Hobbs

1864–1953

American geologist and Arctic explorer

4 Anyone who has examined into the history of the theories of earth evolution must have been astounded to observe the manner in which the unique and the difficultly explainable has been made to take the place of the common and the natural in deriving the framework of these theories.

Earth Evolution and Facial Expression (1921), 174.

Alan Lloyd Hodgkin 1914–98

British physiologist

See also ***Hodgkin, A. L. and Huxley, A. F.***

5 One of the many useful properties of giant nerve fibres is that samples of protoplasm or axoplasm as it is usually called can be obtained by squeezing out the contents from a cut end . . . As in many other cells there is a high concentration of potassium ions and relatively low concentration of sodium and chloride ions. This is the reverse of the situation in the animals' blood or in sea water, where sodium and chloride are the dominant ions and potassium is relatively dilute.

The Conduction of the Nervous Impulse (1964), 27.

1 The zoologist is delighted by the differences between animals, whereas the physiologist would like all animals to work in fundamentally the same way.

Chance and Design: Reminiscences of Science in Peace and War (1992), 66.

Alan Lloyd Hodgkin 1914–98 and Andrew Fielding Huxley 1917–

British physiologists
See also ***Hodgkin, Alan***

2 Nervous messages are invariably associated with an electrical change known as the action potential. This potential is generally believed to arise at a membrane which is situated between the axoplasm and the external medium. If this theory is correct, it should be possible to record the action potential between an electrode inside a nerve fibre and the conducting fluid outside it. Most nerve fibres are too small for this to be tested directly, but we have recently succeeded in inserting micro-electrodes into the giant axons of squids (*Loligo forbesi*).

'Action Potentials Recorded from Inside a Nerve Fibre', *Nature*, 1939, **144**, 710.

Dorothy Crowfoot Hodgkin

1910–94
British chemist and crystallographer

3 From the intensity of the spots near the centre, we can infer that the protein molecules are relatively dense globular bodies, perhaps joined together by valency bridges, but in any event separated by relatively large spaces which contain water. From the intensity of the more distant spots, it can be inferred that the arrangement of atoms inside the protein molecule is also of a perfectly definite kind, although without the periodicities characterising the fibrous proteins. The observations are compatible with oblate spheroidal molecules of diameters about 25 A. and 35 A., arranged in hexagonal screw-axis. . . . At this stage, such ideas are merely speculative, but now that a crystalline protein has been made to give X-ray photographs, it is clear that we have the means of checking them and, by examining the structure of all crystalline proteins, arriving at a far more detailed conclusion about protein structure than previous physical or chemical methods have been able to give.

'X-Ray Photographs of Crystalline Pepsin', *Nature*, 1934, **133**, 795.

4 A great advantage of X-ray analysis as a method of chemical structure analysis is its power to show some totally unexpected and surprising structure with, at the same time, complete certainty.

'X-ray Analysis of Complicated Molecules', Nobel Lecture 11 December 1964. In *Nobel Lectures: Chemistry 1942–1962* (1964), 83.

5 I used to say the evening that I developed the first x-ray photograph I took of insulin in 1935 was the most exciting moment of my life. But the Saturday afternoon in late July 1969, when we realized that the insulin electron density map was interpretable, runs that moment very close.

'X Rays and the Structure of Insulin', *British Medical Journal*, 1971, **4**, 449.

Jacobus Henricus van't Hoff

1852–1911
Dutch chemist

6 It is sometimes easier to *circumvent* prevailing difficulties [in science] rather than to *attack* them.

Imagination in Science, inaugural lecture (1878), trans. G. F. Springer (1967), 8.

7 A famous name has this peculiarity that it becomes gradually smaller especially in natural sciences where each succeeding discovery invariably overshadows what precedes.

H. S. Van Klooster, 'Van't Hoff (1852–1911)

in Retrospect', *Journal of Chemical Education*, 1952, **29**, 376.

1 Whereas the chemico-chemists always find in industry a beautiful field of gold-laden soil, the physico-chemists stand somewhat farther off, especially those who seek only the greatest dilution, for in general there is little to make with watery solutions.

Attributed.

2 ON **HOFF** A Dr van't Hoff of the veterinary college at Utrecht, appears to have no taste for exact chemical investigation. He finds it a less arduous task to mount Pegasus (evidently borrowed from the veterinary school) and to proclaim in his *La Chemie dans l'espace* how, during his bold fight to the top of the chemical Parnassus, the atoms appeared to him to have grouped themselves together throughout universal space . . . I should have taken no notice of this matter had not Wislicenus oddly enough written a preface to the pamphlet, and not by way of a joke but in all seriousness recommended it a worthwhile performance.

Herman Kolbe, 'Signs of the Times', *Journal für Praktische Chemie*, **15**, 473. Trans. W. H. Brock.

Banesh Hoffmann 1906–86

American scientist and writer

3 It did not cause anxiety that Maxwell's equations did not apply to gravitation, since nobody expected to find any link between electricity and gravitation at that particular level. But now physics was faced with an entirely new situation. The same entity, light, was at once a wave and a particle. How could one possibly imagine its proper size and shape? To produce interference it must be spread out, but to bounce off electrons it must be minutely localized. This was a fundamental dilemma, and the stalemate in the wave-photon battle meant that it must remain an enigma to trouble the soul of every true physicist. It was intolerable that light should be two such contradictory things. It was against all the ideals and traditions of science to harbor such an unresolved dualism gnawing at its vital parts. Yet the evidence on either side could not be denied, and much water was to flow beneath the bridges before a way out of the quandary was to be found. The way out came as a result of a brilliant counterattack initiated by the wave theory, but to tell of this now would spoil the whole story. It is well that the reader should appreciate through personal experience the agony of the physicists of the period. They could but make the best of it, and went around with woebegone faces sadly complaining that on Mondays, Wednesdays, and Fridays they must look on light as a wave; on Tuesdays, Thursdays, and Saturdays, as a particle. On Sundays they simply prayed.

The Strange Story of the Quantum (1947), 42.
See **Bragg, William** 83:6.

4 'Daddy,' she says, 'which came first, the chicken or the egg?'

Steadfastly, even desperately, we have been refusing to commit ourselves. But our questioner is insistent. The truth alone will satisfy her. Nothing less. At long last we gather up courage and issue our solemn pronouncement on the subject:

'Yes!'

So it is here.

'Daddy, is it a wave or a particle?'

'Yes.'

'Daddy, is the electron here or is it there?'

'Yes.'

'Daddy, do scientists really know what they are talking about?'

'Yes!'

The Strange Story of the Quantum (1947), 156–7.

August Wilhelm von Hofmann 1818–92

German organic chemist

5 I will on this occasion . . . select my illustrations from that most delightful of games, *croquet*. Let the croquet balls represent our atoms, and let us distinguish the atoms of different elements by different colours. The white balls are hydrogen, the green ones chlorine atoms; the atoms of fiery oxygen are red, those of nitrogen, blue; the carbon atoms, lastly, are naturally represented by black balls.

'On the Combining Power of Atoms', *Proceedings of the Royal Institution* (1865), 416.

1 In the benzene nucleus we have been given a soil out of which we can see with surprise the already-known realm of organic chemistry multiply, not once or twice but three, four, five or six times just like an equivalent number of trees. What an amount of work had suddenly become necessary, and how quickly were busy hands found to carry it out! First the eye moves up the six stems opening out from the tremendous benzene trunk. But already the branches of the neighbouring stems have become intertwined, and a canopy of leaves has developed which becomes more spacious as the giant soars upwards into the air. The top of the tree rises into the clouds where the eye cannot yet follow it. And to what an extent is this wonderful benzene tree thronged with blossoms! Everywhere in the sea of leaves one can spy the slender hydroxyl bud: hardly rarer is the forked blossom [Gabelblüte] which we call the amine group, the most frequent is the beautiful cross-shaped blossom we call the methyl group. And inside this embellishment of blossoms, what a richness of fruit, some of them shining in a wonderful blaze of color, others giving off an overwhelming fragrance.

A. W. Hofmann, after-dinner speech at Kekulé Benzolfest in March 1890. Trans. in W. H. Brock, O. Theodor Benfrey and Susanne Stark, 'Hofmann's Benzene Tree at the Kekulé Festivities', *Journal of Chemical Education*, 1991, **68**, 887–8.

2 People have wracked their brains for an explanation of benzene and how the celebrated man [Kekulé] managed to come up with the concept of the benzene theory. With regard to the last point especially, a friend of mine who is a farmer and has a lively interest in chemistry has asked me a question which I would like to share with you. My 'agricultural friend' apparently believes he has traced the origins of the benzene theory. 'Has Kekulé,' so ran the question, 'once been a bee-keeper? You certainly know that bees too build hexagons; they know well that they can store the greatest amount of honey that way with the least amount of wax. I always liked it,' my agricultural friend went on, 'When I received a new issue of the *Berichte*; admittedly, I don't read the articles, but I like the pictures very much. The patterns of benzene, naphthalene and especially anthracene are indeed wonderful. When I look at the pictures I always have to think of the honeycombs of my bee hives.'

A. W. Hofmann, after-dinner speech at Kekulé Benzolfest in March 1890. Trans. in W. H. Brock, O. Theodor Benfrey and Susanne Stark, 'Hofmann's Benzene Tree at the Kekulé Festivities' *Journal of Chemical Education*, 1991, **68**, 888.

3 I would trade all my experimental works for the single idea of the benzene theory.

Quoted by B. L. Lepsius in 'Hofmann und die Deutsche Chemische Gesellschaft', *Berichte der Deutschen Chemischen Gesellschaft*, 1918, **51**, 51.

Douglas R. Hofstadter 1945–

American cognitive and computer scientist

4 *Hofstadter's Law*: It always takes longer than you expect, even when you take into account Hofstadter's Law.

Gödel, Escher, Bach: An Eternal Golden Braid (1979), 152.

Richard Hofstadter 1916–70

American historian

5 Such biological ideas as the 'survival of the fittest,' whatever their doubtful value in natural science, are utterly useless in attempting to understand society . . . The life of a man in society, while it is incidentally a biological fact, has characteristics that are not reducible to biology and must be explained in the distinctive terms of a cultural analysis . . . the physical well-being of men is a result of their social organization and not vice versa . . . Social improvement is a product of advances in technology and social organization, not of breeding or selective elimination . . . Judgments as to the value of competition between men or enterprises or nations must be based upon social and not allegedly biological consequences; and . . . there is nothing in nature or a naturalistic philosophy of life to make impossible the acceptance of moral sanctions that can be employed for the common good.

Social Darwinism in American Thought 1860–1915 (1945), 176.

Lancelot Hogben 1895–1975

British zoologist and geneticist

1 With full responsibility for my words as a professional biologist, I do not hesitate to say that all existing and genuine knowledge about the way in which the physical characteristics of human communities are related to their cultural capabilities can be written on the back of a postage stamp.

Preface on Prejudices (1937), 9.

2 *Science for the Citizen* is . . . also written for the large and growing number of adolescents, who realize that they will be the first victims of the new destructive powers of science misapplied.

Science for the Citizen: A Self-Educator based on the Social Background of Scientific Discovery (1938), Author's Confessions, 9.

Paul Henri Thiry, Baron D'Holbach 1723–89

German-born French philosopher of science

3 If ignorance of nature gave birth to the Gods, knowledge of nature is destined to destroy them.

Système de la Nature (1770), Part 2, Chapter 1.

4 The universe, that vast assemblage of every thing that exists, presents only matter and motion: the whole offers to our contemplation, nothing but an immense, an uninterrupted succession of causes and effects.

The System of Nature (1770), trans. Samuel Wilkinson (1820), Vol. 1, 12–3.

5 With respect to those who may ask why Nature does not produce new beings? We may enquire of them in turn, upon what foundation they suppose this fact? What it is that authorizes them to believe this sterility in Nature? Know they if, in the various combinations which she is every instant forming, Nature be not occupied in producing new beings, without the cognizance of these observers? Who has informed them that this Nature is not actually assembling, in her immense elaboratory, the elements suitable to bring to light, generations entirely new, that will have nothing in common with those of the species at present existing? What absurdity then, or what want of just inference would there be, to imagine that the man, the horse, the fish, the bird will be no more? Are these animals so indispensably requisite to Nature, that without them she cannot continue her eternal course? Does not all change around us? Do we not ourselves change? . . . Nature contains no one constant form.

The System of Nature (1770), trans. Samuel Wilkinson (1820), Vol. 1, 94–5.

Josiah Gilbert Holland 1819–81

American physician, editor, and poet

6 Geology gives us a key to the patience of God.

Attributed.

Thomas Henry Holland 1868–1947

British geologist

7 The intensity and quantity of polemical literature on scientific problems frequently varies inversely as the number of direct observations on which the discussions are based: the number and variety of theories concerning a subject thus often form a coefficient of our ignorance. Beyond the superficial observations, direct and indirect, made by geologists, not extending below about one two-hundredth of the Earth's radius, we have to trust to the deductions of mathematicians for our ideas regarding the interior of the Earth; and they have provided us successively with every permutation and combination possible of the three physical states of matter—solid, liquid, and gaseous.

'Address delivered by the President of Section [Geology] at Sydney, on Friday, August 21', *Report of the Eighty-Fourth Meeting of the British Association for the Advancement of Science: Australia 1914*, 1915, 345.

Oliver Wendell Holmes

1809–94

American writer, physician and physiologist

1 I was just going to say, when I was interrupted, that one of the many ways of classifying minds is under the heads of arithmetical and algebraical intellects. All economical and practical wisdom is an extension or variation of the following arithmetical formula: 2+2=4. Every philosophical proposition has the more general character of the expression a+b=c. We are mere operatives, empirics, and egotists, until we learn to think in letters instead of figures.

The Autocrat of the Breakfast Table (1858), 1.

2 Science ... —in other words, knowledge,—is not the enemy of religion; for, if so, then religion would mean ignorance. But it is often the antagonist of school-divinity.

The Professor at the Breakfast-Table (1860), 100.

3 Heads I win, tails you lose.

The Professor at the Breakfast-Table (1860), 176.

4 Science is a first-rate piece of furniture for a man's upper chamber, if he has common sense on the ground-floor. But if a man hasn't got plenty of good common sense, the more science he has the worse for his patient.

The Poet at the Breakfast Table (1872), 126.

5 Throw out opium, which the Creator himself seems to prescribe, for we often see the scarlet poppy growing in the cornfields, as if it were foreseen that wherever there is hunger to be fed there must also be a pain to be soothed; throw out a few specifics which our art did not discover, and it is hardly needed to apply; throw out wine, which is a food, and the vapors which produce the miracle of anaesthesia, and I firmly believe that if the whole materia medica, *as now used*, could be sunk to the bottom of the sea, it would be all the better for mankind,—and all the worse for the fishes.

'Currents and Counter-Currents in Medical Science', Address to Massachusetts Medical Society, 30 May 1860. In *Medical Essays 1842–1882* (1891), 202–3.

6 I would never use a long word, even, where a short one would answer the purpose. I know there are professors in this country who 'ligate' arteries. Other surgeons only tie them, and it stops the bleeding just as well.

'Scholastic and Bedside Teaching', Introductory Lecture to the Medical Class of Harvard University, 6 November 1867. In *Medical Essays 1842–1882* (1891), 302.

7 Go on, fair Science; soon to thee
Shall Nature yield her idle boast;
Her vulgar fingers formed a tree,
But thou hast trained it to a post.

'The meeting of the Dryads' (1830), *Poems* (1891), 152.

8 A mind that is stretched by a new idea can never go back to its original dimensions.

Attributed.

Gerald Holton 1922–

American physicist and historian of science

9 As a result of the phenomenally rapid change and growth of physics, the men and women who did their great work one or two generations ago may be our distant predecessors in terms of the state of the field, but they are our close neighbors in terms of time and tastes. This may be an unprecedented state of affairs among professionals; one can perhaps be forgiven if one characterizes it epigrammatically with a disastrously mixed metaphor; in the sciences, we are now uniquely privileged to sit side-by-side with the giants on whose shoulders we stand.

'On the Recent Past of Physics', *American Journal of Physics*, 1961, **29**, 807.

1 The flight of most members of a profession to the high empyrean, where they can work peacefully on purely scientific problems, isolated from the turmoil of real life, was perhaps quite appropriate at an earlier stage of science; but in today's world it is a luxury we cannot afford.

The Scientific Imagination: Case Studies (1978), 250.

2 To this day, we see all around us the Promethean drive *to omnipotence through technology and to omniscience through science*. The effecting of all things possible and the knowledge of all causes are the respective primary imperatives of technology and of science. But the motivating imperative of society continues to be the very different one of its physical and spiritual survival. It is now far less obvious than it was in Francis Bacon's world how to bring the three imperatives into harmony, and how to bring all three together to bear on problems where they superpose.

'Science, Technology and the Fourth Discontinuity' (1982). Reprinted in *The Advancement of Science, and its Burdens* (1986), 183.

Robert Hooke 1635–1703

British physicist and microscopist

3 The Designe of the Royall Society being the Improvement of Naturall knowledge all ways and meanes that tend thereunto ought to be made use of in the prosecution thereof. Naturall knowledge then being the thing sought for, we are to consider by what meanes it may soonest easiest and most certainly attaind. These meanes we shall the sooner find if we consider where tis to be had to wit in three places. first in bookes, 2dly in men. 3ly in the things themselves. and these three point us out the search of books. the converse & correspondence with men the Experimenting and Examining the things themselves. under each of these there is a multitude of businesse to be done but the first hath the Least [and is] the most easily attained, the 2d hath a great Deal and requires much en[deavour] and Industry; and the 3d is infinite and the difficultest of all.

'Proposals for advancement of the R[oyal] S[ociety]' (*c.*1700), quoted in Michael Hunter, *Establishing the New Science: The Experience of the Early Royal Society* (1989), 232.

4 The other experiment (which I shall hardly, I confess, make again, because it was cruel) was with a dog, which, by means of a pair of bellows, wherewith I filled his lungs, and suffered them to empty again, I was able to preserve alive as long as I could desire, after I had wholly opened the thorax, and cut off all the ribs, and opened the belly. Nay, I kept him alive above an hour after I had cut off the pericardium and the mediastinum, and had handled and turned his lungs and heart and all the other parts of its body, as I pleased. My design was to make some enquiries into the nature of respiration. But though I made some considerable discovery of the necessity of fresh air, and the motion of the lungs for the continuance of the animal life, yet I could not make the least discovery in this of what I longed for, which was, to see, if I could by any means discover a passage of the air of the lungs into either the vessels or the heart; and I shall hardly be induced to make any further trials of this kind, because of the torture of this creature: but certainly the enquiry would be very noble, if we could any way find a way so to stupify the creature, as that it might not be sensible.

Letter from Robert Hooke to Robert Boyle, 10 November 1664. In M. Hunter, A. Clericuzio and L. M. Principe (eds.), *The Correspondence of Robert Boyle* (2001), Vol. 2, 399.

1 The Reason of making Experiments is, for the Discovery of the Method of Nature, in its Progress and Operations.

Whosoever, therefore doth rightly make Experiments, doth design to enquire into some of these Operations; and, in order thereunto, doth consider what Circumstances and Effects, in the Experiment, will be material and instructive in that Enquiry, whether for the confirming or destroying of any preconceived Notion, or for the Limitation and Bounding thereof, either to this or that Part of the Hypothesis, by allowing a greater Latitude and Extent to one Part, and by diminishing or restraining another Part within narrower Bounds than were at first imagin'd, or hypothetically supposed.

The Method therefore of making Experiments by the *Royal Society* I conceive should be this.

First, To propound the Design and Aim of the Curator in his present Enquiry.

Secondly, To make the Experiment, or Experiments, leisurely, and with Care and Exactness.

Thirdly, To be diligent, accurate, and curious, in taking Notice of, and shewing to the Assembly of Spectators, such Circumstances and Effects therein occurring, as are material, or at least, as he conceives such, in order to his Theory.

Fourthly, After finishing the Experiment, to discourse, argue, defend, and further explain, such Circumstances and Effects in the preceding Experiments, as may seem dubious or difficult: And to propound what new Difficulties and Queries do occur, that require other Trials and Experiments to be made, in order to their clearing and answering: And farther, to raise such Axioms and Propositions, as are thereby plainly demonstrated and proved.

Fifthly, To register the whole Process of the Proposal, Design, Experiment, Success, or Failure; the Objections and Objectors, the Explanation and Explainers, the Proposals and Propounders of new and farther Trials; the Theories and Axioms, and their Authors; and, in a Word the history of every Thing and Person, that is material and circumstantial in the whole Entertainment of the said Society; which shall be prepared and made ready, fairly written in a bound Book, to be read at the Beginning of the Sitting of the Society: The next Day of their Meeting, then to be read over and further discoursed, augmented or diminished, as the Matter shall require, and then to be sign'd by a certain Number of the Persons present, who have been present, and Witnesses of all the said Proceedings, who, by Subscribing their names, will prove undoubted testimony to Posterity of the whole History.

'Dr Hooke's Method of Making Experiments' (1664–5). In W. Derham (ed.), *Philosophical Experiments and Observations Of the Late Eminent Dr. Robert Hooke, F.R.S. And Geom. Prof. Gresh. and Other Eminent Virtuoso's in his Time* (1726), 26–8.

2 For the Members of the Assembly having before their eyes so many fatal Instances of the errors and falshoods, in which the greatest part of mankind has so long wandred, because they rely'd upon the strength of humane Reason alone, have begun anew to correct all Hypotheses by sense, as Seamen do their dead Reckonings by Coelestial Observations; and to this purpose it has been their principal indeavour to enlarge and strengthen the Senses by Medicine, and by such outward Instruments as are proper for their particular works.

Micrographia, or some Physiological Descriptions of Minute Bodies made by Magnifying Glasses with Observations and Inquiries thereupon (1665), preface sig.

1 The next care to be taken, in respect of the Senses, is a supplying of their infirmities with Instruments, and, as it were, the adding of artificial Organs to the natural; this in one of them has been of late years accomplisht with prodigious benefit to all sorts of useful knowledge, by the invention of Optical Glasses. By the means of Telescopes, there is nothing so far distant but may be represented to our view; and by the help of Microscopes, there is nothing so small, as to escape our inquiry; hence there is a new visible World discovered to the understanding. By this means the Heavens are open'd, and a vast number of new Stars, and new Motions, and new Productions appear in them, to which all the ancient Astronomers were utterly Strangers. By this the Earth it self, which lyes so neer us, under our feet, shews quite a new thing to us, and in every little particle of its matter, we now behold almost as great a variety of creatures as we were able before to reckon up on the whole Universe it self.

Micrographia, or some Physiological Descriptions of Minute Bodies made by Magnifying Glasses with Observations and Inquiries thereupon (1665), preface, sig. A2v.

2 I took a good clear piece of Cork and with a Pen-knife sharpen'd as keen as a Razor, I cut a piece of it off, and thereby left the surface of it exceeding smooth, then examining it very diligently with a *Microscope*, me thought I could perceive it to appear a little porous; but I could not so plainly distinguish them, as to be sure that they were pores, much less what Figure they were of: But judging from the lightness and yielding quality of the Cork, that certainly the texture could not be so curious, but that possibly, if I could use some further diligence, I might find it to be discernable with a *Microscope*, I with the same sharp Penknife, cut off from the former smooth surface an exceeding thin piece of it with a deep *plano-convex Glass*, I could exceedingly plainly perceive it to be all perforated and porous, much like a Honey-comb, but that the pores of it were not regular; yet it was not unlike a Honey-comb in these particulars.

First, in that it had a very little solid substance, in comparison of the empty cavity that was contain'd between, . . . for the *Interstitia* or walls (as I may so call them) or partitions of those pores were neer as thin in proportion to their pores as those thin films of Wax in a Honey-comb (which enclose and constitute the *sexangular* cells) are to theirs.

Next, in that these pores, or cells, were not very deep, but constituted of a great many little Boxes, separated out of one continued long pore, by certain *Diaphragms* . . .

I no sooner discerned these (which were indeed the first *microscopical* pores I ever saw, and perhaps, that were ever seen, for I had not met with any Writer or Person, that had made any mention of them before this) but me thought I had with the discovery of them, presently hinted to me the true and intelligible reason of all the *Phœnomena of Cork.*

Micrographia, or some Physiological Descriptions of Minute Bodies made by Magnifying Glasses with Observations and Inquiries thereupon (1665), 112–6.

3 We have the opportunity of observing her [Nature] through these delicate and pellucid teguments of the bodies of Insects acting according to her usual course and way, undisturbed, whereas when we endeavour to pry into her secrets by breaking open the doors upon her, and dissecting and mangling creatures whil'st there is life yet within them, we find her indeed at work, but put into such disorder by the violence offer'd, as it may easily be imagin'd how differing a thing we should find, if we could, as we can with a *Microscope*, in these smaller creatures, quietly peep in at the windows, without frighting her out of her usual byas.

Micrographia, or some Physiological Descriptions of Minute Bodies made by Magnifying Glasses with Observations and Inquiries thereupon (1665), 186.

1 Most of these Mountains and Inland places whereon these kind of Petrify'd Bodies and Shells are found at present, or have been heretofore, were formerly under the Water, and that either by the descending of the Waters to another part of the Earth by the alteration of the Centre of Gravity of the whole bulk, or rather by the Eruption of some kind of Subterraneous Fires or Earthquakes, great quantities of Earth have been deserted by the Water and laid bare and dry.

Lectures and Discourses of Earthquakes (1668). In *The Posthumous Works of Robert Hooke, containing his Cutlerian Lectures and other Discourses read at the Meetings of the Illustrious Royal Society* (1705), 320–1.

2 I do . . . humbly conceive (tho' some possibly may think there is too much notice taken of such a trivial thing as a rotten Shell, yet) that Men do generally rally too much slight and pass over without regard these Records of Antiquity which Nature have left as Monuments and Hieroglyphick Characters of preceding Transactions in the like duration or Transactions of the Body of the Earth, which are infinitely more evident and certain tokens than any thing of Antiquity that can be fetched out of Coins or Medals, or any other way yet known, since the best of those ways may be counterfeited or made by Art and Design, as may also Books, Manuscripts and Inscriptions, as all the Learned are now sufficiently satisfied, has often been actually practised; but those Characters are not to be Counterfeited by all the Craft in the World, nor can they be doubted to be, what they appear, by any one that will impartially examine the true appearances of them: And tho' it must be granted, that it is very difficult to read them, and to raise *a Chronology* out of them, and to state the intervalls of the Times wherein such, or such Catastrophies and Mutations have happened; yet 'tis not impossible, but that, by the help of those joined to other means and assistances of Information, much may be done even in that part of Information also.

Lectures and Discourses of Earthquakes (1668). In *The Posthumous Works of Robert Hooke, containing his Cutlerian Lectures and other Discourses read at the Meetings of the Illustrious Royal Society* (1705), 411.

3 If the finding of Coines, Medals, Urnes, and other Monuments of famous Persons, or Towns, or Utensils, be admitted for unquestionable Proofs, that such Persons or things have, in former Times, had a being, certainly those Petrifactions may be allowed to be of equal Validity and Evidence, that there have been formerly such Vegetables or Animals. These are truly Authentick Antiquity not to be counterfeited, the Stamps, and Impressions, and Characters of Nature that are beyond the Reach and Power of Humane Wit and Invention, and are true universal Characters legible to all rational Men.

Lectures and Discourses of Earthquakes (1668). In *The Posthumous Works of Robert Hooke, containing his Cutlerian Lectures and other Discourses read at the Meetings of the Illustrious Royal Society* (1705), 449.

4 I shall explain a System of the World differing in many particulars from any yet known, answering in all things to the common Rules of Mechanical Motions: This depends upon three Suppositions. First, That all Cœlestial Bodies whatsoever, have an attraction or gravitating power towards their own Centers, whereby they attract not only their own parts, and keep them from flying from them, as we may observe the Earth to do, but that they do also attract all the other Cœlestial bodies that are within the sphere of their activity; and consequently that not only the Sun and Moon have an influence upon the body and motion of the Earth, and the Earth upon them, but that Mercury also Venus, Mars,

Saturn and Jupiter by their attractive powers, have a considerable influence upon its motion in the same manner the corresponding attractive power of the Earth hath a considerable influence upon every one of their motions also. The second supposition is this, That all bodies whatsoever that are put into a direct and simple motion, will continue to move forward in a streight line, till they are by some other effectual powers deflected and bent into a Motion, describing a Circle, Ellipse, or some other more compounded Curve Line. The third supposition is, That these attractive powers are so much the more powerful in operating, by how much the nearer the body wrought upon is to their own Centers. Now what these several degrees are I have not yet experimentally verified; but it is a notion, which if fully prosecuted as it ought to be, will mightily assist the Astronomer to reduce all the Cœlestial Motions to a certain rule, which I doubt will never be done true without it. He that understands the nature of the Circular Pendulum and Circular Motion, will easily understand the whole ground of this Principle, and will know where to find direction in Nature for the true stating thereof. This I only hint at present to such as have ability and opportunity of prosecuting this Inquiry, and are not wanting of Industry for observing and calculating, wishing heartily such may be found, having myself many other things in hand which I would first compleat and therefore cannot so well attend it. But this I durst promise the Undertaker, that he will find all the Great Motions of the World to be influenced by this Principle, and that the true understanding thereof will be the true perfection of Astronomy.

An Attempt to Prove the Motion of the Earth from Observations (1674), 27–8. Based on a Cutlerian Lecture delivered by Hooke at the Royal Society four years earlier.

1 ON HOOKE He [Robert Hooke] is but of midling stature, something crooked, pale faced, and his face but little belowe, but his head is lardge; his eie full and popping, and not quick; a grey eie. He haz a delicate head of haire, browne, and of an excellent moist curle. He is and ever was very temperate, and moderate in dyet, etc. As he is of prodigious inventive head, so is a person of great vertue and goodnes. Now when I have sayd his Inventive faculty is so great, you cannot imagine his Memory to be excellent, for they are like two Bucketts, as one goes up, the other goes downe. He is certainly the greatest Mechanick this day in the World.

John Aubrey, *Brief Lives* (1680), edited by Oliver Lawson Dick (1949), 165.

2 ON HOOKE Mr Hooke sent, in his next letter [to Sir Isaac Newton] the whole of his Hypothesis, *scil.* that the gravitation was reciprocall to the square of the distance: . . . This is the greatest Discovery in Nature that ever was since the World's Creation. It was never so much as hinted by any man before. I wish he had writt plainer, and afforded a little more paper.

John Aubrey, *Brief Lives* (1680), edited by Oliver Lawson Dick (1949), 166–7.

Joseph Dalton Hooker 1817–1911

British botanist

3 It is difficult to conceive a grander mass of vegetation:—the straight shafts of the timber-trees shooting aloft, some naked and clean, with grey, pale, or brown bark; others literally clothed for yards with a continuous garment of epiphytes, one mass of blossoms, especially the white Orchids *Caelogynes*, which bloom in a profuse manner, whitening their trunks like snow. More bulky trunks were masses of interlacing climbers, *Araliaceae*, *Leguminosae*, Vines, and *Menispermeae*, Hydrangea, and Peppers, enclosing a hollow, once filled by the now strangled supporting tree, which has long ago decayed away. From the sides and summit of these, supple branches hung forth, either leafy or naked; the latter resembling cables flung from one tree to another, swinging in the breeze, their rocking motion increased by the weight of great bunches of ferns or Orchids, which were perched aloft in the loops. Perpetual moisture nourishes this dripping forest: and pendulous mosses and lichens are met with in profusion.

Himalayan Journals (1854), vol. 1. 110–1.

1 The modern system of elevating every minor group, however trifling the characters by which it is distinguished, to the rank of genus, evinces, we think, a want of appreciation of the true value of classification. The genus is the group which, in consequence of our system of nomenclature, is kept most prominently before the mind, and which has therefore most importance attached to it . . . The rashness of some botanists is productive of still more detrimental effects to the science in the case of species; for though a beginner may pause before venturing to institute a genus, it rarely enters into his head to hesitate before proposing a new species.

(With Thomas Thomson) *Flora Indica: A Systematic Account of the Plants of British India* (1855), 10–11.

2 Plants, in a state of nature, are always warring with one another, contending for the monopoly of the soil,—the stronger ejecting the weaker,—the more vigorous overgrowing and killing the more delicate. Every modification of climate, every disturbance of the soil, every interference with the existing vegetation of an area, favours some species at the expense of others.

(With Thomas Thomson) *Flora Indica: A Systematic Account of the Plants of British India* (1855), 41.

3 I am above the forest region, amongst grand rocks & such a torrent as you see in Salvator Rosa's paintings vegetation all a scrub of rhodods. with Pines below me as thick & bad to get through as our Fuegian Fagi on the hill tops, & except the towering peaks of P. S. [perpetual snow] that, here shoot up on all hands there is little difference in the mt scenery—here however the blaze of Rhod. flowers and various colored jungle proclaims a differently constituted region in a naturalists eye & twenty species here, to one there, always are asking me the vexed question, where do we come from?

Letter to Charles Darwin, 24 June 1849. Quoted in *The Correspondence of Charles Darwin* (1988), Vol. 4 1847–1850, 242.

4 These parsons are so in the habit of dealing with the abstractions of doctrines as if there was no difficulty about them whatever, so confident, from the practice of having the talk all to themselves for an hour at least every week with no one to gainsay a syllable they utter, be it ever so loose or bad, that they gallop over the course when their field is Botany or Geology as if we were in the pews and they in the pulpit . . . There is a story somewhere of an Englishman, Frenchman, and German being each called on to describe a camel. The Englishman immediately embarked for Egypt, the Frenchman went to the Jardin des Plantes, and the German shut himself up in his study and thought it out!

Letter to Asa Gray, 29 March 1857. Quoted in Leonard Huxley, *Life and Letters of Sir Joseph Dalton Hooker* (1918), Vol. 1, 477.

5 All I ever aim to do is to put the Development hypothesis in the same coach as the creation one. It will only be a question of who is to ride outside & who in after all.

Letter to Asa Gray, 31 May 1859. Quoted in A. Hunter Dupree, *Asa Gray: American Botanist, Friend of Darwin* (1988), 265. Originally published as *Asa Gray: 1810–1888* (1959).

6 I expect to think that I would rather be author of your book [*The Origin of Species*] than of any other on Nat. Hist. Science.

Letter to Darwin, 12 December 1859. Quoted in Leonard Huxley, *Life and Letters of Sir Joseph Dalton Hooker* (1918), Vol. 1, 511.

Albert Earnest Hooton
1887–1954
American anthropologist

1 Nothing is more detestable to the physical anthropologist than . . . [the] wretched habit of cremating the dead. It involves not only a prodigal waste of costly fuel and excellent fertilizer, but also the complete destruction of physical historical data. On the other hand, the custom of embalming and mummification is most praiseworthy and highly to be recommended.

Up From the Ape (1931), 531.

2 In order to survive, an animal must be born into a favoring or at least tolerant environment. Similarly, in order to achieve preservation and recognition, a specimen of fossil man must be discovered in intelligence, attested by scientific knowledge, and interpreted by evolutionary experience. These rigorous prerequisites have undoubtedly caused many still-births in human palaeontology and are partly responsible for the high infant mortality of discoveries of geologically ancient man.

Apes, Men and Morons (1938), 106.

Reijer Hooykaas 1906–94
Dutch historian of science

3 Both history of nature and history of humanity are 'historical' and yet cannot dispense with uniformity. In both there is 'uniformity' ('science') as well as non-uniformity ('history'); in both 'history respects itself' and 'history does not repeat itself'. But, as even the history of humanity has its uniformitarian features, uniformity can still less be dispensed with in 'history' of nature, which, being one of the natural sciences, is less historical and, consequently, more uniformitarian.

Natural Law and Divine Miracle: The Principle of Uniformity in Geology, Biology and Theology (1963), 151.

4 The student of palaetiological sciences is a scientist *and* a historian. The former tries to be as uniformitarian as possible, the latter has to recognize the contingency of events which will ever be a 'skandalon' to the scientist. Verily, the geologist 'lives in a divided world'.

Natural Law and Divine Miracle: The Principle of Uniformity in Geology, Biology and Theology (1963), 151.

5 The history of science shows so many examples of the 'irrational' notions and theories of to-day becoming the 'rational' notions and theories of to-morrow, that it seems largely a matter of being *accustomed* to them whether they are considered rational or not, natural or not.

Natural Law and Divine Miracle: The Principle of Uniformity in Geology, Biology and Theology (1963), 167.

Frederick Gowland Hopkins
1861–1947
British biochemist

6 But, further, no animal can live upon a mixture of pure protein, fat and carbohydrate, and even when the necessary inorganic material is carefully supplied, the animal still cannot flourish. The animal body is adjusted to live either upon plant tissues or the tissues of other animals, and these contain countless substances other than the proteins, carbohydrates and fats . . . In diseases such as rickets, and particularly in scurvy, we have had for long years knowledge of a dietetic factor; but though we know how to benefit these conditions empirically, the real errors in the diet are to this day quite obscure. They are, however, certainly of the kind which comprises these minimal qualitative factors that I am considering.

'The Analyst and the Medical Man', *The Analyst*, 1906, 31, 395–6.

7 My main thesis will be that in the study of the intermediate processes of metabolism we have to deal not with complex substances which elude ordinary chemical methods, but with the simple substances undergoing comprehensible reactions . . . I intend also to emphasise the fact that it is not alone with the separation and identification of products from the animal that our present studies deal; but with their reactions in the body; with the dynamic side of biochemical phenomena.

'The Dynamic Side of Biochemistry', Address delivered on 11 September 1913. *Report on the 83rd Meeting of the British Association for the Advancement of Science* (1914), 653.

1 One reason which has led the organic chemist to avert his mind from the problems of Biochemistry is the obsession that the really significant happenings in the animal body are concerned in the main with substances of such high molecular weight and consequent vagueness of molecular structure as to make their reactions impossible of study by his available and accurate methods. There remains, I find, pretty widely spread, the feeling—due to earlier biological teaching—that, apart from substances which are obviously excreta, all the simpler products which can be found in cells or tissues are as a class mere objects, already too remote from the fundamental biochemical events to have much significance. So far from this being the case, recent progress points in the clearest way to the fact that the molecules with which a most important and significant part of the chemical dynamics of living tissues is concerned are of a comparatively simple character.

'The Dynamic Side of Biochemistry', Address delivered on 11 September 1913. *Report on the 83rd Meeting of the British Association for the Advancement of Science* (1914), 657–8.

2 The cell, too, has a geography, and its reactions occur in colloidal apparatus, of which the form, and the catalytic activity of its manifold surfaces, must efficiently contribute to the due guidance of chemical reactions.

'Some Aspects of Biochemistry', *The Irish Journal of Medical Science*, 1932, 79, 344.

3 A cell has a history; its structure is inherited, it grows, divides, and, as in the embryo of higher animals, the products of division differentiate on complex lines. Living cells, moreover, transmit all that is involved in their complex heredity. I am far from maintaining that these fundamental properties may not depend upon organisation at levels above any chemical level; to understand them may even call for different methods of thought; I do not pretend to know. But if there be a hierarchy of levels we must recognise each one, and the physical and chemical level which, I would again say, may be the level of self-maintenance, must always have a place in any ultimate complete description.

'Some Aspects of Biochemistry', *The Irish Journal of Medical Science*, 1932, 79, 346.

4 As a progressive discipline [biochemistry] belongs to the present century. From the experimental physiologists of the last century it obtained a charter, and, from a few pioneers of its own, a promise of success; but for the furtherance of its essential aim that century left it but a small inheritance of facts and methods. By its essential or ultimate aim I myself mean an adequate and acceptable description of molecular dynamics in living cells and tissues.

'Some Chemical Aspects of Life', Address delivered in September 1933. *Report on the 103rd Meeting of the British Association for the Advancement of Science* (1933), 3.

5 It is an old saying, abundantly justified, that where sciences meet there growth occurs. It is true moreover to say that in scientific borderlands not only are facts gathered that [are] often new in kind, but it is in these regions that wholly new concepts arise. It is my own faith that just as the older biology from its faithful studies of external forms provided a new concept in the doctrine of evolution, so the new biology is yet fated to furnish entirely new fundamental concepts of science, at which physics and chemistry when concerned with the non-living alone could never arrive.

'Biological Thought and Chemical Thought: A Plea for Unification', Linacre Lecture, *Lancet*, 1938, 2, 1204.

1 ON **HOPKINS** The growth curves of the famous Hopkins' rats are familiar to anyone who has ever opened a textbook of physiology. One recalls the proud ascendant curve of the milk-fed group which suddenly turns downwards as the milk supplement is removed, and the waning curve of the other group taking its sudden milk-assisted upward spring, until it passes its fellow now abruptly on the decline. 'Feeding experiments illustrating the importance of accessory factors in normal dietaries', *Jour. Physiol.*, 1912, xliv, 425, ranks aesthetically beside the best stories of H. G. Wells.

W. R. Aykroyd, *Vitamins and Other Dietary Essentials* (1933), 46.

2 ON **HOPKINS** In trying to evaluate Hopkins' unique contribution to biochemistry it may perhaps be said that he alone amongst his contemporaries succeeded in *formulating* the subject. Among others whose several achievements in their own fields may have surpassed his, no one has ever attempted to unify and correlate biochemical knowledge so as to form a comprehensible picture of the cell and its relation to life, reproduction and function.

M. Stephenson, 'Sir F G Hopkins' Teaching and Scientific Influence'. In J. Needham and E. Baldwin (eds.), *Hopkins and Biochemistry, 1861–1947* (1949), 36.

3 ON **HOPKINS** But nothing ever put 'Hoppy' in the shade. No one could fail to recognize in the little figure . . . the authentic gold of intellectual inspiration, the *Fundator et Primus Abbas* of biochemistry in England.

J. and D. M. Needham, 'F. G. Hopkins' Personal Influence'. In J. Needham and E. Baldwin (eds.), *Hopkins and Biochemistry, 1861–1947* (1949), 115.

Harold Horace Hopkins 1918–94 and **Narinder Singh Kapany**

British physicist and Indian-born American physicist

4 An optical unit has been devised which will convey optical images along a flexible axis. The unit comprises a bundle of fibres of glass, or other transparent material, and it therefore appears appropriate to introduce the term 'fibrescope' to denote it.

'A Flexible Fibrescope, using Static Scanning', *Nature*, 1954, 173, 39.

Horace 65–8 BC

Roman poet

5 *Nos numeros sumus et fruges consumere nati.*

We are but ciphers, born to consume earth's fruits.

Sometimes translated as 'We are just statistics, born to consume resources'.

Epistles bk. I, no. 2, l. 27. In *Satires, Epistles and Ars Poetica*, trans. H. Rushton Fairclough (1926), 264–5.

6 *Sapere aude.*

Dare to be wise.

More often translated as 'Dare to know'.

Epistles bk. I, no. 2, l. 40. In *Satires, Epistles and Ars Poetica*, trans. H. Rushton Fairclough (1926), 264–5.

7 *Decus et pretium recte petit experiens vir.*

The man who makes the attempt justly aims at honour and reward.

Epistles bk. I, no. 17, l. 42. In *Satires, Epistles and Ars Poetica*, trans. H. Rushton Fairclough (1926), 364–5.

8 *Dulce et decorum est pro patria mori.*

Lovely and honourable it is to die for one's country.

Odes bk. 3, no. 2, l. 13. In *Odes and Epodes*, trans. C. E. Bennett (1914, 1995 printing), 174–5.

Leonard Horner 1785–1864

British geologist

9 How does it arise that, while the statements of geologists that other organic bodies existed millions of years ago are tacitly accepted, their conclusions as to man having existed many thousands of years ago should be received with hesitation by some geologists, and be altogether repudiated by a not inconsiderable number among the other educated classes of society?

'Anniversary Address of the Geological Society of London', *Proceedings of the Geological Society of London*, 1861, 17, lxvii.

Norman Horowitz 1915–

American biochemist and geneticist

1 One gene one enzyme.

'The one-gene-one-enzyme hypothesis', *Genetics*, 1948, 33, 612–3.

Neale Eltinge Howard 1911–

American science writer

2 Astronomers work always with the past; because light takes time to move from one place to another, they see things as they *were*, not as they *are*.

The Telescope Handbook and Star Atlas (1967), 33.

Fred Hoyle 1915 –2001 and Raymond Arthur Lyttleton 1911–95

British astrophysicist and British astronomer

3 It is often held that scientific hypotheses are constructed, and are to be constructed, only after a detailed weighing of all possible evidence bearing on the matter, and that then and only then may one consider, and still only tentatively, any hypotheses. This traditional view however, is largely incorrect, for not only is it absurdly impossible of application, but it is contradicted by the history of the development of any scientific theory. What happens in practice is that by intuitive insight, or other inexplicable inspiration, the theorist decides that certain features seem to him more important than others and capable of explanation by certain hypotheses. Then basing his study on these hypotheses the attempt is made to deduce their consequences. The sucessful pioneer of theoretical science is he whose intuitions yield hypotheses on which satisfactory theories can be built, and conversely for the unsuccessful (as judged from a purely scientific standpoint).

'The Internal Constitution of the Stars', *Occasional Notes of the Royal Astronomical Society*, 1948, 12, 90.

Elbert Green Hubbard 1856–1915

American author and editor

4 One machine can do the work of fifty ordinary men. No machine can do the work of one extraordinary man.

Attributed.

Marion King Hubbert 1903–89

American geophysicist

5 [The] first postulate of the Principle of Uniformity, namely, that the laws of nature are invariant with time, is not peculiar to that principle or to geology, but is a common denominator of all science. In fact, instead of being an assumption or an *ad hoc* hypothesis, it is simply a succinct summation of the totality of all experimental and observational evidence.

'Critique of the Principle of Uniformity', in C. C. Albritton (ed.), *Uniformity and Simplicity* (1967), 29.

6 History, human or geological, represents our hypothesis, couched in terms of past events, devised to explain our present-day observations.

'Critique of the Principle of Uniformity', in C. C. Albritton (ed.), *Uniformity and Simplicity* (1967), 30.

7 Historical chronology, human or geological, depends . . . upon comparable impersonal principles. If one scribes with a stylus on a plate of wet clay two marks, the second crossing the first, another person on examining these marks can tell unambiguously which was made first and which second, because the latter event irreversibly disturbs its predecessor. In virtue of the fact that most of the rocks of the earth contain imprints of a succession of such irreversible events, an unambiguous working out of the chronological sequence of these events becomes possible.

'Critique of the Principle of Uniformity', in C. C. Albritton (ed.), *Uniformity and Simplicity* (1967), 31.

Edwin Powell Hubble 1889–1953

American astronomer

1 Science is the one human activity that is truly progressive. The body of positive knowledge is transmitted from generation to generation.

The Realm of the Nebulae (1936), 1.

2 The history of astronomy is a history of receding horizons.

The Realm of the Nebulae (1936), 21.

3 With increasing distance, our knowledge fades, and fades rapidly. Eventually, we reach the dim boundary—the utmost limits of our telescopes. There, we measure shadows, and we search among ghostly errors of measurement for landmarks that are scarcely more substantial. The search will continue. Not until the empirical resources are exhausted, need we pass on to the dreamy realms of speculation.

The Realm of the Nebulae (1936), 202.

4 Past time is finite, future time is infinite.

The Observational Approach to Cosmology (1937), 62.

5 Positive, objective knowledge is public property. It can be transmitted directly from one person to another, it can be pooled, and it can be passed on from one generation to the next. Consequently, knowledge accumulates through the ages, each generation adding its contribution. Values are quite different. By values, I mean the standards by which we judge the significance of life. The meaning of good and evil, of joy and sorrow, of beauty, justice, success—all these are purely private convictions, and they constitute our store of wisdom. They are peculiar to the individual, and no methods exist by which universal agreement can be obtained. Therefore, wisdom cannot be readily transmitted from person to person, and there is no great accumulation through the ages. Each man starts from scratch and acquires his own wisdom from his own experience. About all that can be done in the way of communication is to expose others to vicarious experience in the hope of a favorable response.

The Nature of Science and other Lectures (1954), 7.

David H. Hubel 1926–

Canadian physiologist

6 Those who think 'Science is Measurement' should search Darwin's works for numbers and equations.

'David H. Hubel', in Larry R. Squire (ed.), *The History of Neuroscience in Autobiography* (1996), Vol. 1, 313.

Kurt Hübner 1921–

German philosopher

7 Factual assertions and fundamental principles are . . . merely parts of theories: they are given within the framework of a theory; they are chosen and valid within this framework; and subsequently they are dependent upon it. This holds for all empirical sciences—for the natural sciences as well as those pertaining to history.

Critique of Scientific Reason (1983), 106.

Maurice Loyal Huggins 1897–1981

American chemist

8 The most stable arrangement for an assemblage of molecules is one in which the component atoms and groups are packed together so that (*a*) the distances between neighbors are close to the equilibrium distance, (*b*) each atom or group has as many close neighbors as possible, and (*c*) there are no large unoccupied regions. In other words, each structure tends to be as 'close-packed' as possible, consistent with the 'sizes' of its component atoms or groups.

'The Structure of Fibrous Proteins', *Chemical Reviews*, 1943, **32**, 198.

William Huggins 1824–1910

British astrophysicist

1 One important object of this original spectroscopic investigation of the light of the stars and other celestial bodies, namely to discover whether the same chemical elements as those of our earth are present throughout the universe, was most satisfactorily settled in the affirmative.

Scientific Papers of Sir William Huggins, (ed.) Sir William Huggins & Lady Huggins (1909), 49, footnote [added in 1909 to 'On the Spectra of some of the Fixed Stars' (1864)]

Victor Hugo 1802–85

French novelist, poet and dramatist

2 Science says the first word on everything, and the last word on nothing.

Things of the Infinite: Intellectual Autobiography, trans. L O'Rourke (1907).

3 An invasion of armies can be resisted; an invasion of ideas cannot be resisted.

Histoire d'un Crime (written 1851–2, published 1877), conclusion, chap. 10. Trans. T. H. Joyce and Arthur Locker (1886), 413.

4 Science is continually correcting what it has said. Fertile corrections . . . science is a ladder . . . poetry is a winged flight . . . An artistic masterpiece exists for all time . . . Dante does not efface Homer.

Quoted in Pierre Biquard, *Frédéric Joliot-Curie: The Man and his Theories* (1961), trans. Geoffrey Strachan (1965), 168.

Friedrich Wilhelm Heinrich Alexander von Humboldt

1769–1859

German naturalist

5 I shall collect plants and fossils, and with the best of instruments make astronomic observations. Yet this is not the main purpose of my journey. I shall endeavor to find out how nature's forces act upon one another, and in what manner the geographic environment exerts its influence on animals and plants. *In short, I must find out about the harmony in nature.*

Letter to Karl Freiesleben, June 1799. In Helmut de Terra, *Humboldt: The Life and Times of Alexander von Humboldt 1769–1859* (1955), 87.

6 That which we call the Atlantic Ocean is only a valley excavated by the force of the waters; the form of the seacoast, the salient and re-entrant angles of America, of Africa, and of Europe proclaim this catastrophe.

'Esquisse d'un tableau géologique de L'amerique maridonale', *Journal de Physique, de Chemie, d'Histoire Naturelle*, 1801, **53**, 33.

7 The application of botanical and zoological evidence to determine the relative age of rocks—this chronometry of the earth's surface which was already present to the lofty mind of Hooke—indicates one of the most glorious epochs of modern geognosy, which has finally, on the Continent at least, been emancipated from the way of Semitic doctrines. Palaeontological investigations have imparted a vivifying breath of grace and diversity to the science of the solid structure of the earth.

Cosmos: A Sketch of a Physical Description of the Universe (1845–62), trans. E. C. Otté (1849), Vol. 1, 272.

8 While we maintain the unity of the human species, we at the same time repel the depressing assumption of superior and inferior races of men. There are nations more susceptible of cultivation, more highly civilized, more ennobled by mental cultivation than others—but none in themselves nobler than others. All are in like degree designed for freedom.

Cosmos: A Sketch of a Physical Description of the Universe (1845–62), trans. E. C. Otté (1849), Vol 1, 368.

1 The philosophical study of nature rises above the requirements of mere delineation, and does not consist in the sterile accumulation of isolated facts. The active and inquiring spirit of man may therefore be occasionally permitted to escape from the present into the domain of the past, to conjecture that which cannot yet be clearly determined, and thus to revel amid the ancient and ever-recurring myths of geology.

Views of Nature: Or Contemplation of the Sublime Phenomena of Creation (1850), trans. E. C. Otté and H. G. Bohn, 375.

Wilhelm von Humboldt

1767–1835

German naturalist and educationalist

2 Regardless of communication between man and man, speech is a necessary condition for the thinking of the individual in solitary seclusion. In appearance, however, language develops only *socially*, and man understands himself only once he has tested the intelligibility of his words by trial upon others.

On Language (1836), trans. Peter Heath (1988), 56.

3 Words well up freely from the breast, without necessity or intent, and there may well have been no wandering horde in any desert that did not already have its own songs. For man, as a species, is a singing creature, though the notes, in his case, are also coupled with thought.

On Language (1836), trans. Peter Heath (1988), 60.

David Hume 1711–76

Scottish philosopher

4 All the sciences have a relation, greater or less, to human nature; and . . . however wide any of them may seem to run from it, they still return back by one passage or another. Even *Mathematics, Natural Philosophy, and Natural Religion*, are in some measure dependent on the science of MAN; since they lie under the cognizance of men, and are judged of by their powers and faculties.

A Treatise on Human Nature (1739–40), ed. L. A. Selby-Bigge (1888), introduction, xix.

5 We have no other notion of cause and effect, but that of certain objects, which have *always conjoin'd* together, and which in all past instances have been found inseparable. We cannot penetrate into the reason of the conjunction. We only observe the thing itself, and always find that from the constant conjunction the objects acquire an union in the imagination.

A Treatise on Human Nature (1739–40), ed. L. A. Selby-Bigge (1888), book 1, part 3, section 6, 93.

6 To consider the matter aright, reason is nothing but a wonderful and unintelligible instinct in our souls, which carries us along a certain train of ideas, and endows them with particular qualities, according to their particular situations and relations. This instinct, 'tis true, arises from past observation and experience; but can any one give the ultimate reason, why past experience and observation produces such an effect, any more than why nature alone shou'd produce it?

A Treatise on Human Nature (1739–40), ed. L. A. Selby-Bigge (1888), book 1, part 3, section 16, 179.

7 All knowledge degenerates into probability.

A Treatise on Human Nature (1739–40), ed. L. A. Selby-Bigge (1888), book 1, part 4, section 1, 180.

8 We need only reflect on what has been prov'd at large, that we are never sensible of any connexion betwixt causes and effects, and that 'tis only by our experience of their constant conjunction, we can arrive at any knowledge of this relation.

A Treatise on Human Nature (1739–40), ed. L. A. Selby-Bigge (1888), book 1, part 4, section 165, 247.

1 Generally speaking, the errors in religion are dangerous; those in philosophy only ridiculous.

A Treatise on Human Nature (1739–40), ed. L. A. Selby-Bigge (1888), book 1, part 4, section 7, 272.

2 Reason is, and ought only to be the slave of the passions, and can never pretend to any other office than to serve and obey them.

A Treatise on Human Nature (1739–40), ed. L. A. Selby-Bigge (1888), book 2, part 3, section 3, 415.

3 'Tis evident that all reasonings concerning *matter of fact* are founded on the relation of cause and effect, and that we can never infer the existence of one object from another, unless they be connected together, either mediately or immediately . . . Here is a billiard ball lying on the table, and another ball moving toward it with rapidity. They strike; and the ball which was formerly at rest now acquires a motion. This is as perfect an instance of the relation of cause and effect as any which we know, either by sensation or reflection.

An Abstract of a Treatise on Human Nature (1740), ed. John Maynard Keynes and Piero Sraffa (1938), 11.

4 What is possible can never be demonstrated to be false; and 'tis possible the course of nature may change, since we can conceive such a change. Nay, I will go farther, and assert, that he could not so much as prove by any *probable* arguments, that the future must be conformable to the past. All probable arguments are built on the supposition, that there is this conformity betwixt the future and the past, and therefore can never prove it. This conformity is a *matter of fact*, and if it must be proved, will admit of no proof but from experience. But our experience in the past can be a proof of nothing for the future, but upon a supposition, that there is a resemblance betwixt them. This therefore is a point, which can admit of no proof at all, and which we take for granted without any proof.

An Abstract of a Treatise on Human Nature (1740), ed. John Maynard Keynes and Piero Sraffa (1938), 15.

5 There is nothing, in itself, valuable or despicable, desirable or hateful, beautiful or deformed; but that these attributes arise from the particular constitution and fabric of human sentiment and affection.

The Skeptic (1742). In T. H. Green and T. H. Grose (eds.), *The Philosophical Works of David Hume* (1874), Vol. 3, 224.

6 Be a philosopher; but, amidst all your philosophy, be still a man.

An Enquiry Concerning Human Understanding (1748), ed. L. A. Selby-Bigge (1894), section 1, 9.

7 *That the sun will not rise tomorrow* is no less intelligible a proposition and implies no more contradiction than the affirmation *that it will rise*. We should in vain, therefore, attempt to demonstrate its falsehood.

An Enquiry Concerning Human Understanding (1748), ed. L. A. Selby-Bigge (1894), section 4, part 1, 26.

8 In reality, all arguments from experience are founded on the similarity which we discover among natural objects, and by which we are induced to expect effects similar to those which we have found to follow from such objects. And though none but a fool or madman will ever pretend to dispute the authority of experience, or to reject that great guide of human life, it may surely be allowed a philosopher to have so much curiosity at least as to examine the principle of human nature, which gives this mighty authority to experience, and makes us draw advantage from that similarity which nature has placed among different objects. From causes which appear *similar* we expect similar effects.

This is the sum of our experimental conclusions.

An Enquiry Concerning Human Understanding (1748), ed. L. A. Selby-Bigge (1894), section 4, part 2, 36.

1 If . . . the past may be no rule for the future, all experience becomes useless and can give rise to no inference or conclusion.

An Enquiry Concerning Human Understanding (1748), ed. L. A. Selby-Bigge (1894), section 4, part 2, 37–8.

2 Though there be no such thing as *chance* in the world; our ignorance of the real cause of any event has the same influence on the understanding, and begets a like species of belief or opinion.

An Enquiry Concerning Human Understanding (1748), ed. L. A. Selby-Bigge (1894), section 6, 56.

3 It seems to me, that the only objects of the abstract sciences or of demonstration are quantity and number, and that all attempts to extend this more perfect species of knowledge beyond these bounds are mere sophistry and illusion.

An Enquiry Concerning Human Understanding (1748), ed. L. A. Selby-Bigge (1894), section 7, part 3, 163.

4 A miracle is a violation of the laws of nature; and as a firm and unalterable experience has established these laws, the proof against a miracle, from the very nature of the fact, is as entire as any argument from experience can possibly be imagined. Why is it more than probable, that all men must die; that lead cannot, of itself, remain suspended in the air; that fire consumes wood, and is extinguished by water; unless it be, that these events are found agreeable to the laws of nature, and there is required a violation of these laws, or in other words, a miracle to prevent them? Nothing is esteemed a miracle, if it ever happen in the common course of nature . . . There must, therefore, be a uniform experience against every miraculous event, otherwise the event would not merit that appellation. And as a uniform experience amounts to a proof, there is here a direct and full *proof*, from the nature of the fact, against the existence of any miracle; nor can such a proof be destroyed, or the miracle rendered credible, but by an opposite proof, which is superior.

An Enquiry Concerning Human Understanding (1748), ed. L. A. Selby-Bigge (1894), section 10, part 1, 114–5.

5 If we take in our hand any volume; of divinity or school metaphysics, for instance; let us ask, *Does it contain any abstract reasoning concerning quantity or number?* No. *Does it contain any experimental reasoning concerning matter of fact and existence?* No. Commit it then to the flames: for it can contain nothing but sophistry and illusion.

An Enquiry Concerning Human Understanding (1748), ed. L. A. Selby-Bigge (1894), section 12, part 3, 165.

6 While Newton seemed to draw off the veil from some of the mysteries of nature, he showed at the same time the imperfections of the mechanical philosophy; and thereby restored her ultimate secrets to that obscurity, in which they ever did and ever will remain.

The History of England (1754–62) (1926 edition), Vol. 8, 294.

7 But the life of a man is of no greater importance to the universe than that of an oyster.

'On Suicide' (written 1755, published 1777), in Stephen Copley and Andrew Edgar (eds.), *David Hume: Selected Essays* (1993), 319.

8 Look round the world, contemplate the whole and every part of it: you will find it to be nothing but one great machine, subdivided into an infinite number of lesser machines, which again admit of subdivisions to a degree beyond what human senses and faculties can trace and explain. All these various machines, and even their most minute parts, are adjusted to each other with an accuracy which ravishes into admiration all men who have ever contemplated them. The curious adapting of means to ends, throughout all nature, resembles exactly, though it much exceeds, the productions of human contrivance—of human design, thought, wisdom, and intelligence.

Dialogues Concerning Natural Religion (1779), ed. Norman Kemp Smith (1935), 176–7.

1 Look round this universe. What an immense profusion of beings, animated and organized, sensible and active! You admire this prodigious variety and fecundity. But inspect a little more narrowly these living existences, the only beings worth regarding. How hostile and destructive to each other! How insufficient all of them for their own happiness! How contemptible or odious to the spectator! The whole presents nothing but the idea of a blind nature, inpregnated by a great vivifying principle, and pouring forth from her lap, without discernment or parental care, her maimed and abortive children.

Dialogues Concerning Natural Religion (1779), ed. Norman Kemp Smith (1935), 259–60.

John Hunter 1728–93

British anatomist and surgeon

2 I thank you for your Expt on the Hedge Hog; but why do you ask me such a question, by way of solving it. I think your solution is just; but why think, why not try the Expt.

Usually quoted as 'But why think, why not try the experiment?'

Letter to Edward Jenner, 2 August 1775. In A. J. Harding Rains (ed.), *Letters From the Past: From John Hunter to Edward Jenner* (1976), 9.

3 Some physiologists will have it that the stomach is a mill; others, that it is a fermenting vat; others, again that it is a stew-pan; but in my view of the matter, it is neither a mill, a fermenting vat nor a stew-pan, but a stomach gentlemen, a stomach.

'MS Note From his Lectures', in J. A. Paris, *A Treatise on Diet* (1824), epigraph.

4 If we were capable of following the progress of increase of the number of the parts of the most perfect animal, as they first formed in succession, from the very first to its state of full perfection, we should probably be able to compare it with some one of the incomplete animals themselves, of every order of animals in the Creation, being at no stage different from some of the inferior orders; or, in other words, if we were to take a series of animals, from the more imperfect to the perfect, we should probably find an imperfect animal, corresponding with some stage of the most perfect.

R. Owen (ed.), *John Hunter's Observations on Animal Development* (1841), 14.

James Hutton 1726–97

British philosopher and geologist

See **Lyell, Charles** *408:5;* **Playfair, John**

5 Nature, everywhere the most amazingly and outstandingly remarkable producer of living bodies, being most carefully arranged according to physical, mechanical, and chemical laws, does not give even the smallest hint of its extraordinary and tireless workings and quite clearly points to its work as being alone worthy of a benign and omnipotent God; and it carries this bright quality in all of its traces, in that, just as all of its general mechanisms rejoice, so also do all of their various smallest component parts rejoice in the depth of wisdom, in the height of perfection, and in the lofty arrangement of forms and qualities, which lie far beyond every investigation of the human mind.

'Inaugural Physico-Medical Dissertation on the Blood and the Circulation of the Microcosm' (1749). Trans. Arthur Donovan and Joseph Prentiss, *James Hutton's Medical Dissertation* (1980), 29.

6 As there is not in human observation proper means for measuring the waste of land upon the globe, it is hence inferred, that we cannot estimate the duration of what we see at present, nor calculate the period at which it had begun; so that, with respect to human observation, this world has neither a beginning nor an end.

Abstract of a Dissertation . . . Concerning the System of the Earth, its Duration, and Stability (1785), 28.

1 When we trace the part of which this terrestrial system is composed, and when we view the general connection of those several parts, the whole presents a machine of a peculiar construction by which it is adapted to a certain end. We perceive a fabric, erected in wisdom, to obtain a purpose worthy of the power that is apparent in the production of it.

'Theory of the Earth', *Transactions of the Royal Society of Edinburgh*, 1788, 1, 209.

2 Time, which measures everything in our idea, and is often deficient to our schemes, is to nature endless and as nothing; it cannot limit that by which alone it had existence; and as the natural course of time, which to us seems infinite, cannot be bounded by any operation that may have an end, the progress of things upon this globe, that is, the course of nature, cannot be limited by time, which must proceed in a continual succession.

'Theory of the Earth', *Transactions of the Royal Society of Edinburgh*, 1788, 1, 215.

3 Error, never can be consistent, nor can truth fail of having support from the accurate examination of every circumstance.

'Theory of the Earth', *Transactions of the Royal Society of Edinburgh*, 1788, 1, 259.

4 Man is made for science; he reasons from effects to causes, and from causes to effects; but he does not always reason without error. In reasoning, therefore, from appearances which are particular, care must be taken how we generalize; we should be cautious not to attribute to nature, laws which may perhaps be only of our own invention.

'Theory of the Earth', *Transactions of the Royal Society of Edinburgh*, 1788, 1, 273.

5 We are not to suppose, that there is any violent exertion of power, such as is required in order to produce a great event in little time; in nature, we find no deficiency in respect of time, nor any limitation with regard to power. But time is not made to flow in vain; nor does there ever appear the exertion of superfluous power, or the manifestation of design, not calculated in wisdom to effect some general end.

'Theory of the Earth', *Transactions of the Royal Society of Edinburgh*, 1788, 1, 294.

6 We have the satisfaction to find, that in nature there is wisdom, system and consistency. For having, in the natural history of this earth, seen a succession of worlds, we may from this conclude that, there is a system in nature; in like manner as, from seeing revolutions of the planets, it is concluded, that there is a system by which they are intended to continue those revolutions. But if the succession of worlds is established in the system of nature, it is vain to look for anything higher in the origin of the earth. The result, therefore, of our present enquiry is, that we find no vestige of a beginning,—no prospect of an end.

'Theory of the Earth', *Transactions of the Royal Society of Edinburgh*, 1788, 1, 304.

7 With such wisdom has nature ordered things in the economy of this world, that the destruction of one continent is not brought about without the renovation of the earth in the production of another.

Theory of the Earth, with Proofs and Illustrations, Vol. 1 (1795), 183.

8 A rock or stone is not a subject that, of itself, may interest a philosopher to study; but, when he comes to see the necessity of those hard bodies, in the constitution of this earth, or for the permanency of the land on which we dwell, and when he finds that there are means wisely provided for the renovation of this necessary decaying part, as well as that of every other, he then, with pleasure, contemplates this manifestation of design, and thus connects the mineral system of this earth with that by which the heavenly bodies are made to move perpetually in their orbits.

Theory of the Earth, with Proofs and Illustrations, Vol. 1 (1795), 276.

1 [It] is the little causes, long continued, which are considered as bringing about the greatest changes of the earth.

Theory of the Earth with Proofs and Illustrations, Vol. 2 (1795), 205.

2 In matters of science, curiosity gratified begets not indolence, but new desires.

Theory of the Earth, with Proofs and Illustrations, Vol. 3, ed. Archibald Geikie (1899), 16.

Aldous Leonard Huxley 1894–1963

British novelist and critic

3 A million million spermatozoa,
All of them alive:
Out of their cataclysm but one poor
Noah
Dare hope to survive.
And among that billion minus one
Might have chanced to be
Shakespeare, another Newton, a new
Donne—
But the One was Me.

'Fifth Philosopher's Song', *Leda* (1920), 33.

4 Everything's incredible, if you can skin off the crust of obviousness our habits put on it.

Point Counter Point (1928), 407.

5 The ductless glands secrete among other things our moods, our aspirations, our philosophy of life.

'And Wanton Optics Roll the Melting Eye', in *Music at Night and Other Essays* (1931), 26.

6 We have learned that there is an endocrinology of elation and despair, a chemistry of mystical insight, and, in relation to the autonomic nervous system, a meteorology and even . . . an astro-physics of changing moods.

Literature and Science (1963), 90.

7 Even if I could be Shakespeare I think that I should still choose to be Faraday.

In 1925, attributed. Walter M. Elsasser *Memoirs of a Physicist in the Atomic Age* (1978), epigraph.

Andrew Fielding Huxley 1917–

British physiologist

See also ***Hodgkin, A. L. and Huxley, A. F.***

Julian Huxley 1887–1975

British biologist and writer

8 To speculate without facts is to attempt to enter a house of which one has not the key, by wandering aimlessly round and round, searching the walls and now and then peeping through the windows. Facts are the key.

'Heredity I: The Behaviour of the Chromosomes', in *Essays in Popular Science* (1926), 1–2.

9 By death the moon was gathered in
Long ago, ah long ago;
Yet still the silver corpse must spin
And with another's light must glow.
Her frozen mountains must forget
Their primal hot volcanic breath,
Doomed to revolve for ages yet,
Void amphitheatres of death.
And all about the cosmic sky,
The black that lies beyond our blue,
Dead stars innumerable lie,
And stars of red and angry hue
Not dead but doomed to die.

'Cosmic Death' (1923), in *The Captive Shrew and Other Poems of a Biologist* (1932), 30.

10 Evolution: At the Mind's Cinema
I turn the handle and the story starts:
Reel after reel is all astronomy,
Till life, enkindled in a niche of sky,
Leaps on the stage to play a million
parts.
Life leaves the slime and through all
ocean darts;
She conquers earth, and raises wings to
fly;
Then spirit blooms, and learns how not
to die,—
Nesting beyond the grave in others'
hearts.
I turn the handle: other men like me
Have made the film: and now I sit and
look

In quiet, privileged like Divinity
To read the roaring world as in a book.
If this thy past, where shall they future climb,
O Spirit, built of Elements and Time?

'Evolution: At the Mind's Cinema' (1922), in *The Captive Shrew and Other Poems of a Biologist* (1932), 55.

1 Evolution: The Modern Synthesis.

Evolution: The Modern Synthesis (1942).

2 The proof given by Wright, that non-adaptive differentiation will occur in small populations owing to 'drift', or the chance fixation of some new mutation or recombination, is one of the most important results of mathematical analysis applied to the facts of neo-mendelism. It gives accident as well as adaptation a place in evolution, and at one stroke explains many facts which puzzled earlier selectionists, notably the much greater degree of divergence shown by island than mainland forms, by forms in isolated lakes than in continuous river-systems.

Evolution: The Modern Synthesis (1942), 199–200.

3 The scientific doctrine of progress is destined to replace not only the myth of progress, but all other myths of human earthly destiny. It will inevitably become one of the cornerstones of man's theology, or whatever may be the future substitute for theology, and the most important external support for human ethics.

New Bottles for New Wine (1957), 21.

4 Operationally, God is beginning to resemble not a ruler, but the last fading smile of a cosmic Cheshire Cat.

Religion without Revelation (1957), 58.

5 Evolution . . . is the most powerful and the most comprehensive idea that has ever arisen on Earth.

'Education and Humanism', in *Essays of a Humanist* (1964), 125.

Julian Huxley 1887–1975 and **Alfred Cort Haddon** 1855–1940

British biologist and writer and British anthropologist

6 The popular and scientific views of 'race' no longer coincide. The word 'race', as applied scientifically to human groupings, has lost any sharpness of meaning. To-day it is hardly definable in scientific terms, except as an abstract concept which *may*, under certain conditions, very different from those now prevalent, have been realized approximately in the past and *might*, under certain other but equally different conditions, be realized in the distant future.

We Europeans (1935), 107.

Thomas Henry Huxley 1825–95

British biologist
See ***Darwin, Charles***

7 Living things have no inertia, and tend to no equilibrium.

'On the Educational Value of the Natural History Sciences' (1854). In *Collected Essays* (1893), Vol. 3, 40.

8 Science is, I believe, nothing but *trained and organised common sense*, differing from the latter only as a veteran may differ from a raw recruit: and its methods differ from those of common sense only so far as the guardsman's cut and thrust differ from the manner in which a savage wields his club.

'On the Educational Value of the Natural History Sciences' (1854). In *Collected Essays* (1893), Vol. 3, 45.

9 Deduction, which takes us from the general proposition to facts again—teaches us, if I may so say, to anticipate from the ticket what is inside the bundle.

'On the Educational Value of the Natural History Sciences' (1854). In *Collected Essays* (1893), Vol. 3, 52.

10 To a person uninstructed in natural history, his country or sea-side stroll is a walk through a gallery filled with wonderful works of art, nine-tenths of which have their faces turned to the wall. Teach him something of natural history, and you place in his hands a catalogue of those which are worth turning around. Surely our innocent pleasures are not so abundant in this life, that we can afford to despise this or any other source of them.

'On the Educational Value of the Natural History Sciences' (1854). In *Collected Essays* (1893), Vol. 3, 63.

1 Every philosophical thinker hails it [*The Origin of Species*] as a veritable Whitworth gun in the armoury of liberalism.

'The Origin of Species' (1860). In *Collected Essays* (1893), Vol. 2, 23.

2 Extinguished theologians lie about the cradle of every science, as the strangled snakes besides that of Hercules.

'The Origin of Species' (1860). In *Collected Essays* (1893), Vol. 2, 52.

3 Unity of plan everywhere lies hidden under the mask of diversity of structure—the complex is everywhere evolved out of the simple.

'A Lobster; or, the Study of Zoology' (1861). In *Collected Essays* (1894), Vol. 8, 205–6.

4 Man's Place in Nature.

Title of an 1863 Volume, reprinted in *Collected Essays* (1894), Vol. 7.

5 The question of questions for mankind—the problem which underlies all others, and is more deeply interesting than any other—is the ascertainment of the place which Man occupies in nature and of his relations to the universe of things.

'On the Relations of Man to the Lower Animals' (1863). In *Collected Essays* (1894), Vol. 7, 77.

6 It is not I who seek to base Man's dignity upon his great toe, or insinuate that we are lost if an Ape has a hippocampus minor. On the contrary, I have done my best to sweep away this vanity. I have endeavoured to show that no absolute structural line of demarcation, wider than that between the animals which immediately succeed us in the scale, can be drawn between the animal world and ourselves; and I may add the expression of my belief that the attempt to draw a physical distinction is equally futile, and that even the highest faculties of feeling and of intellect begin to germinate in lower forms of life. At the same time, no one is more strongly convinced than I am of the vastness of the gulf between civilized man and the brutes; or is more certain that whether *from* them or not, he is assuredly not *of* them.

'On the Relations of Man to the Lower Animals' (1863). In *Collected Essays* (1894), Vol. 7, 152–3.

7 Where, then, must we look for primæval Man? Was the oldest *Homo sapiens* pliocene or miocene, or yet more ancient? In still older strata do the fossilized bones of an ape more anthropoid, or a Man more pithecoid, than any yet known await the researches of some unborn paleontologist?

'On some Fossil Remains of Man' (1863). In *Collected Essays* (1894), Vol. 7, 208.

8 The man of science has learned to believe in justification, not by faith, but by verification.

'On the Advisableness of Improving Natural knowledge' (1866). In *Collected Essays* (1893), Vol. 1, 41.

9 Yet it is a very plain and elementary truth, that the life, the fortune, and the happiness of every one of us, and, more or less, of those who are connected with us, do depend upon our knowing something of the rules of a game infinitely more difficult and complicated than chess. It is a game which has been played for untold ages, every man and woman of us being one of the two players in a game of his or her own. The chess-board is the world, the pieces are the phenomena of the universe, the rules of the game are what we call the laws of Nature. The player on the other side is hidden from us. We know that his play is always fair, just and patient. But also we know, to our cost, that he never overlooks a mistake, or makes the smallest allowance for ignorance.

To the man who plays well, the highest stakes are paid, with that sort of overflowing generosity with which the strong shows delight in strength. And one who plays ill is checkmated—without haste, but without remorse.

'A Liberal Education and Where to Find it' (1868). In *Collected Essays* (1893), Vol. 3, 82.

1 Mathematics may be compared to a mill of exquisite workmanship, which grinds you stuff of any degree of fineness; but, nevertheless, what you get out depends upon what you put in; and as the grandest mill in the world will not extract wheat-flour from peascods, so pages of formulae will not get a definite result out of loose data.

'Geological Reform' (1869). In *Collected Essays* (1894), Vol. 8, 333.

2 If some great Power would agree to make me always think what is true and do what is right, on condition of being turned into a sort of clock and wound up every morning before I got out of bed, I should instantly close with the offer.

'On Descartes' "Discourse Touching the Method of Using One's Reason Rightly and of Seeking Scientific Truth"' (1870). In *Collected Essays* (1893), Vol. 1, 192–3.

3 The great tragedy of Science—the slaying of a beautiful hypothesis by an ugly fact.

'Biogenesis and Abiogenesis'(1870), In *Collected Essays* (1894), Vol. 8, 244.

4 What men of science want is only a fair day's wages for more than a fair day's work.

'Administrative Nihilism' (1871). In *Collected Essays* (1893), Vol. 1, 287.

5 If I may paraphrase Hobbes's well-known aphorism, I would say that 'books are the money of Literature, but only the counters of Science.'

'Universities: Actual and Ideal' (1874). In *Collected Essays* (1893), Vol. 3, 213.

6 I venture to maintain, that, if the general culture obtained in the Faculty of Arts were what it ought to be, the student would have quite as much knowledge of the fundamental principles of Physics, of Chemistry, and of Biology, as he needs, before he commenced his special medical studies. Moreover, I would urge, that a thorough study of Human Physiology is, in itself, an education broader and more comprehensive than much that passes under that name. There is no side of the intellect which it does not call into play, no region of human knowledge into which either its roots, or its branches, do not extend; like the Atlantic between the Old and the New Worlds, its waves wash the shores of the two worlds of matter and of mind; its tributary streams flow from both; through its waters, as yet unfurrowed by the keel of any Columbus, lies the road, if such there be, from the one to the other; far away from that North-west Passage of mere speculation, in which so many brave souls have been hopelessly frozen up.

'Universities: Actual and Ideal' (1874). In *Collected Essays* (1893), Vol. 3, 220.

7 The great end of life is not knowledge but action.

Marx was saying much the same thing at the time.

'Technical Education' (1877). In *Collected Essays* (1893), Vol. 3, 422.

8 History warns us . . . that it is the customary fate of new truths to begin as heresies and to end as superstitions.

'The Coming of Age of the "Origin of Species"' (1880). In *Collected Essays* (1893), Vol. 2, 229.

9 The scientific spirit is of more value than its products, and irrationally held truths may be more harmful than reasoned errors.

'The Coming of Age of the "Origin of Species"' (1880). In *Collected Essays* (1893), Vol. 2, 229.

10 Very few, even among those who have taken the keenest interest in the progress of the revolution in natural knowledge set afoot by the publication of the 'Origin of Species'; and who have watched, not without astonishment, the rapid and complete change which has been effected both inside and outside the boundaries of the scientific world in the attitude of men's minds towards the doctrines which are expounded in that great work, can have been prepared for the extraordinary manifestation of affectionate regard for the man, and of profound reverence for the philosopher, which followed the announcement, on Thursday last, of the death of Mr Darwin.

'Obituary [of Charles Darwin]' (1882). In *Collected Essays* (1893), Vol. 2, 244.

1 Science . . . commits suicide when it adopts a creed.

'The Darwin Memorial' (1885). In *Collected Essays* (1893), Vol. 2, 252.

2 If there is anything in the world which I do firmly believe in, it is the universal validity of the law of causation.

'Science and Morals' (1886). In *Collected Essays* (1994), Vol. 9, 121.

3 The publication of the Darwin and Wallace papers in 1858, and still more that of the 'Origin' in 1859, had the effect upon them of the flash of light, which to a man who has lost himself in a dark night, suddenly reveals a road which, whether it takes him straight home or not, certainly goes his way. That which we were looking for, and could not find, was a hypothesis respecting the origin of known organic forms, which assumed the operation of no causes but such as could be proved to be actually at work. We wanted, not to pin our faith to that or any other speculation, but to get hold of clear and definite conceptions which could be brought face to face with facts and have their validity tested. The 'Origin' provided us with the working hypothesis we sought.

'On the Reception of the "Origin of Species"'. In F. Darwin (ed.), *The Life and Letters of Charles Darwin, Including an Autobiographical Chapter* (1888), Vol 2, 197.

4 My reflection, when I first made myself master of the central idea of the 'Origin', was, 'How extremely stupid not to have thought of that!'

'On the Reception of the "Origin of Species"'. In F. Darwin (ed.), *The Life and Letters of Charles Darwin, Including an Autobiographical Chapter* (1888), Vol. 2, 197.

5 If one of these people, in whom the chance-worship of our remoter ancestors thus strangely survives, should be within reach of the sea when a heavy gale is blowing, let him betake himself to the shore and watch the scene. Let him note the infinite variety of form and size of the tossing waves out at sea; or against the curves of their foam-crested breakers, as they dash against the rocks; let him listen to the roar and scream of the shingle as it is cast up and torn down the beach; or look at the flakes of foam as they drive hither and thither before the wind: or note the play of colours, which answers a gleam of sunshine as it falls upon their myriad bubbles. Surely here, if anywhere, he will say that chance is supreme, and bend the knee as one who has entered the very penetralia of his divinity. But the man of science knows that here, as everywhere, perfect order is manifested; that there is not a curve of the waves, not a note in the howling chorus, not a rainbow-glint on a bubble, which is other than a necessary consequence of the ascertained laws of nature; and that with a sufficient knowledge of the conditions, competent physico-mathematical skill could account for, and indeed predict, every one of these 'chance' events.

'On the Reception of the "Origin of Species"'. In F. Darwin (ed.), *The Life and Letters of Charles Darwin, Including an Autobiographical Chapter* (1888), Vol. 2, 200–1.

6 The known is finite, the unknown infinite; intellectually we stand on an islet in the midst of an illimitable ocean of inexplicability. Our business in every generation is to reclaim a little more land, to add something to the extent and the solidity of our possessions. And even a cursory glance at the history of the biological sciences during the last quarter of a century is sufficient to justify the assertion, that the most potent instrument for the extension of the realm of natural knowledge which has come into men's hands, since the publication of Newton's 'Principia', is Darwin's 'Origin of Species'.

'On the Reception of the "Origin of Species"'. In F. Darwin (ed.), *The Life and Letters of Charles Darwin, Including an Autobiographical Chapter* (1888), Vol. 2, 204.

1 It was badly received by the generation to which it was first addressed, and the outpouring of angry nonsense to which it gave rise is sad to think upon. But the present generation will probably behave just as badly if another Darwin should arise, and inflict upon them that which the generality of mankind most hate—the necessity of revising their convictions. Let them, then, be charitable to us ancients; and if they behave no better than the men of my day to some new benefactor, let them recollect that, after all, our wrath did not come to much, and vented itself chiefly in the bad language of sanctimonious scolds. Let them as speedily perform a strategic right-about-face, and follow the truth wherever it leads.

'On the Reception of the "Origin of Species"'. In F. Darwin (ed.), *The Life and Letters of Charles Darwin, Including an Autobiographical Chapter* (1888), Vol. 2, 204.

2 It is an error to imagine that evolution signifies a constant tendency to increased perfection. That process undoubtedly involves a constant remodeling of the organism in adaptation to new conditions; but it depends on the nature of those conditions whether the direction of the modifications effected shall be upward or downward.

'The Struggle for Existence in Human Society' (1888). In Collected Essays (1894), Vol. 9, 199.

3 Every variety of philosophical and theological opinion was represented there [The Metaphysical Society], and expressed itself with entire openness; most of my colleages were *-ists* of one sort or another; and, however kind and friendly they might be, I, the man without a rag of a label to cover himself with, could not fail to have some of the uneasy feelings which must have beset the historical fox when, after leaving the trap in which his tail remained, he presented himself to his normally elongated companions. So I took thought, and invented what I conceived to be the appropriate title of 'agnostic'.

'Agnosticism' (1889). In *Collected Essays* (1894), Vol. 5, 239.

4 I know no study which is so unutterably saddening as that of the evolution of humanity, as it is set forth in the annals of history. Out of the darkness of prehistoric ages man emerges with the marks of his lowly origin strong upon him. He is a brute, only more intelligent than the other brutes, a blind prey to impulses, which as often as not led him to destruction; a victim to endless illusions, which make his mental existence a terror and a burden, and fill his physical life with barren toil and battle.

'Agnosticism' (1889). In Collected Essays (1894), Vol. 5, 256.

5 Cosmic evolution may teach us how the good and evil tendencies of man may have come about; but, in itself, it is incompetent to furnish any better reason why what we call good is preferable to what we call evil than we had before. Some day, I doubt not, we shall arrive at an understanding of the evolution of the aesthetic faculty; but all the understanding in the world will neither increase nor diminish the force of the intuition that this is beautiful and that is ugly.

'Evolution and Ethics' (1893). In *Collected Essays* (1894), Vol. 9, 80.

6 The practice of that which is ethically best—what we call goodness or virtue—involves a course of conduct which, in all respects, is opposed to that which leads to success in the cosmic struggle for existence. In place of ruthless self-assertion it demands self-restraint; in place of thrusting aside, or treading down, all competitors, it requires that the individual shall not merely respect, but shall help his fellows . . . It repudiates the gladiatorial theory of existence . . . Laws and moral precepts are directed to the end of curbing the cosmic process.

'Evolution and Ethics' (1893). In *Collected Essays* (1894), Vol. 9, 81–2.

1 Not only is the state of nature hostile to the state of art of the garden; but the principle of the horticultural process, by which the latter is created and maintained, is antithetic to that of the cosmic process. The characteristic feature of the latter is the intense and unceasing competition of the struggle for existence. The characteristic of the former is the elimination of that struggle, by the removal of the conditions which give rise to it. The tendency of the cosmic process is to bring about the adjustment of the forms of plant life to the current conditions; the tendency of the horticultural process is the adjustment of the conditions to the needs of the forms of plant life which the gardener desires to raise.

'Evolution and Ethics—Prolegomena' (1894). In *Collected Essays* (1894), Vol. 9, 13.

2 Some experience of popular lecturing had convinced me that the necessity of making things plain to uninstructed people, was one of the very best means of clearing up the obscure corners in one's own mind.

'Preface'. In *Man's Place in Nature and Other Anthropological Essays. Collected Essays* (1894), Vol. 7, ix.

3 *Asked by Samuel Wilberforce, Bishop of Oxford, whether he traced his descent from an ape on his mother's or his father's side:*

If then, said I, the question is put to me would I rather have a miserable ape for a grandfather or a man highly endowed by nature and possessing great means and influence and yet who employs those faculties for the mere purpose of introducing ridicule into a grave scientific discussion—I unhesitatingly affirm my preference for the ape.

Commemorated by a plate in the University Museum.

Letter to Dr. Dyster, 9 September 1860. Huxley Papers, Imperial College of Science and Technology. Quoted in D.J. Foskett, 'Wilberforce and Huxley on Evolution' Letter to *Nature*, 1953, **172**, 920.

4 Science seems to me to teach in the highest and strongest manner the great truth which is embodied in the Christian conception of entire surrender to the will of God. Sit down before fact as a little child, be prepared to give up every preconceived notion, follow humbly wherever and to whatever abysses nature leads, or you shall learn nothing. I have only begun to learn content and peace of mind since I have resolved at all risks to do this.

Letter to Charles Kingsley, 23 September 1860. In L. Huxley, *The Life and Letters of Thomas Henry Huxley* (1903), Vol. 1, 316.

5 As I stood behind the coffin of my little son the other day, with my mind bent on anything but disputation, the officiating minister read, as part of his duty, the words, 'If the dead rise not again, let us eat and drink, for to-morrow we die.' I cannot tell you how inexpressibly they shocked me. Paul had neither wife nor child, or he must have known that his alternative involved a blasphemy against all that well best and noblest in human nature. I could have laughed with scorn. What! Because I am face to face with irreparable loss, because I have given back to the source from whence it came, the cause of a great happiness, still retaining through all my life the blessings which have sprung and will spring from that cause, I am to renounce my manhood, and, howling, grovel in bestiality? Why, the very apes know better, and if you shoot their young, the poor brutes grieve their grief out and do not immediately seek distraction in a gorge.

Letter to Charles Kingsley, 23 September 1860. In L. Huxley, *The Life and Letters of Thomas Henry Huxley* (1903), Vol. 1, 318.

1 In fact a favourite problem of [Tyndall] is—Given the molecular forces in a mutton chop, deduce Hamlet or Faust therefrom. He is confident that the Physics of the Future will solve this easily.

Letter to Herbert Spencer, 3 August 1861. In L. Huxley, *The Life and Letters of Thomas Henry Huxley* (1903), Vol. 1, 333.

2 There is no absurdity in theology so great that you cannot parallel it by a greater absurdity in Nature.

Letter to Charles Kingsley, 5 May 1863. In L. Huxley, *The Life and Letters of Thomas Henry Huxley* (1903), Vol. 1, 347.

3 I am too much of a sceptic to deny the possibility of anything.

Letter to Herbert Spencer, 22 March 1886. In L. Huxley, *The Life and Letters of Thomas Henry Huxley* (1903), Vol. 2, 443.

Christiaan Huygens 1629–95

Dutch physicist and astronomer

4 I believe that we do not know anything for certain, but everything probably.

Letter to Pierre Perrault, 'Sur la préface de M. Perrault de son traité de l'Origine des fontaines' [1763], *Oeuvres Complètes de Christiaan Huygens* (1897), Vol. 7, 298. Quoted in Jacques Roger, *The Life Sciences in Eighteenth-Century French Thought*, ed. Keith R. Benson and trans. Robert Ellrich (1997), 163.

5 One may conceive light to spread successively, by spherical waves.

Attributed.

Libbie Henrietta Hyman 1888–1969

American invertebrate zoologist

6 The whole aim of comparative anatomy is to discover what structures are homologous.

A Laboratory Manual for Comparative Vertebrate Anatomy (1922), 3.

Ibn al-Haitham (Alhazan) 965–1040

Arabic natural philosopher

7 The sun's rays proceed from the sun along straight lines and are reflected from every polished object at equal angles, i.e. the reflected ray subtends, together with the line tangential to the polished object which is in the plane of the reflected ray, two equal angles. Hence it follows that the ray reflected from the spherical surface, together with the circumference of the circle which is in the plane of the ray, subtends two equal angles. From this it also follows that the reflected ray, together with the diameter of the circle, subtends two equal angles. And every ray which is reflected from a polished object to a point produces a certain heating at that point, so that if numerous rays are collected at one point, the heating at that point is multiplied: and if the number of rays increases, the effect of the heat increases accordingly.

In H. J. J. Winter, 'A Discourse of the Concave Spherical Mirror by Ibn Al-Haitham', *Journal of the Royal Asiatic Society of Bengal*, 1950, 16, 2.

Ibn al-Nafīs d. 1288

Arab physician

8 This is the right cavity of the two cavities of the heart. When the blood in this cavity has become thin, it must be transferred into the left cavity, where the pneuma is generated. But there is no passage between these two cavities, the substance of the heart there being impermeable. It neither contains a visible passage, as some people have thought, nor does it contain an invisible passage which would permit the passage of blood, as Galen thought. The pores of the heart there are compact and the substance of the heart is thick. It must, therefore, be that when the blood has become thin, it is passed into the arterial vein [pulmonary artery] to the lung, in order to be dispersed inside the substance of the lung, and to mix with the air. The finest parts of the blood are then strained, passing into the venous artery [pulmonary vein] reaching the left of the two cavities of the heart, after mixing with the air and becoming fit for the generation of pneuma.

Albert Z. Iskandar, 'Ibn al-Nafīs', in Charles Coulston Gillispie (ed.), *Dictionary of Scientific Biography* (1974), Vol. 9, 603.

Ibn Khaldun 1332–1406

Arab historian

1 Untruth naturally afflicts historical information. There are various reasons that make this unavoidable. One of them is partisanship for opinions and schools . . . Another reason making untruth unavoidable in historical information is reliance upon transmitters . . . Another reason is unawareness of the purpose of an event . . . Another reason is unfounded assumption as to the truth of a thing . . . Another reason is ignorance of how conditions conform with reality . . . Another reason is the fact that people as a rule approach great and high-ranking persons with praise and encomiums . . . Another reason making untruth unavoidable—and this one is more powerful than all the reasons previously mentioned—is ignorance of the nature of the various conditions arising in civilization. Every event (or phenomenon), whether (it comes into being in connection with some) essence or (as the result of an) action, must inevitably possess a nature peculiar to its essence as well as to the accidental conditions that may attach themselves to it.

The Muqaddimah. An Introduction to History, trans. Franz Rosenthal, 2nd edition (1967), Vol. 1, 71–2.

Ibn Sina see Avicenna

Henrik Johan Ibsen 1828–1906

Norwegian playwright

2 I propose to raise a revolution against the lie that the majority has the monopoly of the truth.

An Enemy of the People (1882), Act IV. In *Ghosts and two Other Plays* (1911), 218.

Ikhwan al-Safa

Secret Arab brotherhood, Basra, Iraq (10th century onwards)

3 Know, oh Brother (May God assist thee and us by the Spirit from Him) that God, Exalted Be His Praise, when He created all creatures and brought all things into being, arranged them and brought them into existence by a process similar to the process of generation of numbers from one, so that the multiplicity [of numbers] should be a witness to His Oneness, and their classification and order an indication of the perfection of His wisdom in creation. And this would be a witness to the fact, too, that they [creatures] are related to Him who created them, in the same way as the numbers are related to the One which is prior to two, and which is the principle, origin and source of numbers, as we have shown in our treatise on arithmetic.

Rasā'il. In Seyyed Hossein Nasr, *Science and Civilisation in Islam* (1968), 155–6.

Leopold Infeld 1898–1968

Polish physicist

4 Einstein uses his concept of God more often than a Catholic priest. Once I asked him:

'Tomorrow is Sunday. Do you want me to come to you, so we can work?'

'Why not?'

'Because I thought perhaps you would like to rest on Sunday.'

Einstein settled the question by saying with a loud laugh: 'God does not rest on Sunday either.'

Quest: The Evolution of a Scientist (1941), 247.

Robert Green Ingersoll 1833–99

American lawyer and politician

5 We must remember that in nature there are neither rewards nor punishments—there are consequences.

'Some Reasons Why' (1896) in *Lectures and Essays* (1907), Series 3, 31.

Christopher Kelk Ingold

1893–1970

British chemist

1 Reagents are regarded as acting by virtue of a constitutional affinity either for electrons or for nuclei . . . the terms *electrophilic* (electron-seeking) and *nucleophilic* (nucleus-seeking) are suggested . . . and the organic molecule, in the activation necessary for reaction, is therefore required to develop at the seat of attack either a high or low electron density as the case may be.

'Significance of Tautomerism and of the Reactions of Aromatic Compounds in the Electronic Theory of Organic Relations', *Journal of the Chemical Society*, 1933, 136, 1121, fn.

2 Since it is proposed to regard chemical reactions as electrical transactions in which reagents act by reason of a constitutional affinity either for electrons or for atomic nuclei, it is important to be able to recognize which type of reactivity any given reagent exhibits.

'Principles of an Electronic Theory of Organic Reactions', *Chemical Reviews*, 1934, 15, 265.

3 You ask whether I am going over to the history of science . . . no, I am not as old as that.

Said on his retirement *c.*1969.

St. Isidore of Seville *c.*560–636

Spanish theologian and bishop

4 *Tolle numerum omnibus rebus et omnia pereunt.*

Take from all things their number and all shall perish.

Etymologies [*c.*600], Book III, chapter 4, quoted in E. Grant (ed.), *A Source Book in Medieval Science* (1974), trans. E. Brehaut (1912), revised by E. Grant, 5.

5 Number is divided into even and odd. Even number is divided into the following: evenly even, evenly uneven, and unevenly uneven. Odd number is divided into the following: prime and incomposite, composite, and a third intermediate class (*mediocris*) which in a certain way is prime and incomposite but in another way secondary and composite.

Etymologies [*c.*600], Book III, chapter 5, quoted in E. Grant (ed.), *A Source Book in Medieval Science* (1974), trans. E. Brehaut (1912), revised by E. Grant, 5.

6 It is agreed that all sound which is the material of music is of three sorts. First is *harmonica*, which consists of vocal music; second is *organica*, which is formed from the breath; third is *rhythmica*, which receives its numbers from the beat of the fingers. For sound is produced either by the voice, coming through the throat; or by the breath, coming through the trumpet or tibia, for example; or by touch, as in the case of the cithara or anything else that gives a tuneful sound on being struck.

Etymologies [*c.*600], Book III, chapter 19, quoted in E. Grant (ed.), *A Source Book in Medieval Science* (1974), trans. E. Brehaut (1912), revised by E. Grant, 10.

7 The name of medicine is thought to have been given from 'moderation', *modus*, that is, from a due proportion, which advises that things be done not to excess, but 'little by little', *paulatim*. For nature is pained by surfeit but rejoices in moderation. Whence also those who take drugs and antidotes constantly, or to the point of saturation, are sorely vexed, for every immoderation brings not health but danger.

Etymologies [*c.*600], Book IV, chapter 2, quoted in E. Grant (ed.), *A Source Book in Medieval Science* (1974), trans. W. D. Sharpe (1964), 701.

Charles Ian Jackson 1935–

Canadian geographer and environmentalist

8 Whether or not you agree that trimming and cooking are likely to lead on to downright forgery, there is little to support the argument that trimming and cooking are less reprehensible and more forgivable. Whatever the rationalization is, in the last analysis one can no more than be a bit dishonest than one can be a little bit pregnant. Commit any of these three sins and your scientific career is in jeopardy and deserves to be.

Honour in Science (1984), 14.

John Hughlings Jackson

1835–1911

British neurologist

1 For in disease the most voluntary or most special movements, faculties, etc., suffer first and most, that is in an order the exact opposite of evolution. Therefore I call this the principle of Dissolution.

'On the Anatomical and Physiological Localisation of Movements in the Brain' (1875), Preface. In James Taylor (ed.), *Selected Writings of John Hughlings Jackson*, Vol. 1 (1931), 38.

2 It has been said that he who was the first to abuse his fellow-man instead of knocking out his brains without a word, laid thereby the basis of civilisation.

'On affections of Speech from the Disease of the Brain' (1878). In James Taylor (ed.), *Selected Writings of John Hughlings Jackson*, Vol. 2 (1932), 179.

3 The doctrine of evolution implies the passage from the most organised to the least organised, or, in other terms, from the most general to the most special. Roughly, we say that there is a gradual 'adding on' of the more and more special, a continual adding on of new organisations. But this 'adding on' is at the same time a 'keeping down'. The higher nervous arrangements evolved out of the lower keep down those lower, just as a government evolved out of a nation controls as well as directs that nation.

'Evolution and Dissolution of the Nervous System', *British Medical Journal*, 1884, i, 662.

François Jacob 1920–

French biologist

4 It is natural selection that gives direction to changes, orients chance, and slowly, progressively produces more complex structures, new organs, and new species. Novelties come from previously unseen association of old material. To create is to recombine.

'Evolution and Tinkering', *Science*, 1977, 196, 1163.

5 Myths and science fulfill a similar function: they both provide human beings with a representation of the world and of the forces that are supposed to govern it. They both fix the limits of what is considered as possible.

The Possible and the Actual (1982), 9.

Jean Jacques

French chemist and philosopher

6 Chemistry . . . is like the maid occupied with daily civilisation; she is busy with fertilisers, medicines, glass, insecticides . . . for she dispenses the recipes.

Les Confessions d'un Chimiste Ordinaire (1981), 5. Trans. W. H. Brock.

Jalal ad-Din ar-Rumi 1207–73

Persian writer and poet

7 I died as mineral and became a plant,
I died as plant and rose to animal,
I died as animal and I became man.
Why should I fear? When was I less by dying?

'I Died as a Mineral', trans. A. J. Arberry. http://www.iles.umn.edu/faculty/bashiri/Poets%20folder/Rumi.html

Sayyid Jamal ad-Din 1838–97

Afghani philosopher

8 The strangest thing of all is that our ulama these days have divided science into two parts. One they call Muslim science, and one European science. Because of this they forbid others to teach some of the useful sciences. They have not understood that science is that noble thing that has no connection with any nation, and is not distinguished by anything but itself. Rather, everything that is known is known by science, and every nation that becomes renowned becomes renowned through science. Men must be related to science, not science to men. How very strange it is that the Muslims study those sciences that are ascribed to Aristotle with the greatest delight, as if Aristotle were one of the pillars of the Muslims. However, if the discussion relates to Galileo, Newton, and Kepler, they consider them infidels. The father and mother of science is proof, and proof is neither Aristotle nor Galileo. The truth is where there is proof, and those who forbid science and knowledge in the belief that they are safeguarding the Islamic religion are really the enemies of that religion.

Lecture on Teaching and Learning (1882). In Nikki R. Keddie, *An Islamic Response to Imperialism* (1983), 107.

Henry James 1843–1916

American author

1 Huxley is a very genial, comfortable being—yet with none of the noise and windy geniality of some folks here.

Letter to William James, 29 March 1877. In Percy Lubbock (ed.), *The Letters of Henry James* (1920), 52.

William James 1842–1910

American psychologist

2 A Beethoven string-quartet is truly, as some one has said, a scraping of horses' tails on cats' bowels, and may be exhaustively described in such terms; but the application of this description in no way precludes the simultaneous applicability of an entirely different description.

The Sentiment of Rationality (1882), 76.

3 'Facts' are the bounds of human knowledge, set for it, not by it.

'On Some Hegelisms' (1882). In *The Will to Believe and Other Essays in Popular Philosophy* (1897), 271.

4 All our scientific and philosophic ideals are altars to unknown gods.

'The Dilemma of Determinism' (1884). In *The Will to Believe and Other Essays in Popular Philosophy* (1897), 147.

5 The traditional psychology talks like one who should say a river consists of nothing but pailsful, spoonsful, quartpotsful, barrelsful, and other moulded forms of water. Even were the pails and the pots all actually standing in the stream, still between them the free water would continue to flow. It is just this free water of consciousness that psychologists resolutely overlook. Every definite image in the mind is steeped and dyed in the free water that flows round it. With it goes the sense of its relations, near and remote, the dying echo of whence it came to us, the dawning sense of whither it is to lead.

'On Some Omissions of Introspective Psychology', *Mind*, 1884, 9, 16.

6 Consciousness . . . does not appear to itself chopped up in bits. Such words as 'chain' or 'train' do not describe it fitly as it presents itself in the first instance. It is nothing jointed; it flows. A 'river' or a 'stream' are the metaphors by which it is most naturally described. *In talking of it hereafter, let us call it the stream of thought, of consciousness, or of subjective life.*

The origin of the phrase 'stream of consciousness'.

The Principles of Psychology (1890), Vol. 1, 239.

7 Our natural way of thinking about these coarser emotions is that the mental perception of some fact excites the mental affection called the emotion, and that this latter state of mind gives rise to the bodily expression. My theory, on the contrary, is that *the bodily changes follow directly the perception of the exciting fact, and that our feeling of the same changes as they occur IS the emotion.* Common-sense says, we lose our fortune, are sorry and weep; we meet a bear, are frightened and run; we are insulted by a rival, are angry and strike. The hypothesis here to be defended says that this order of sequence is incorrect, that the one mental state is not immediately induced by the other, that the bodily manifestations must first be interposed between, and that the more rational statement is that we feel sorry because we cry, angry because we strike, afraid because we tremble, and not that we cry, strike, or tremble, because we are sorry, angry, or fearful, as the case may be. Without the bodily states following on the perception, the latter would be purely cognitive in form, pale, colorless, destitute of emotional warmth. We might then see the bear, and judge it best to run, receive the insult and deem it right to strike, but we should not actually *feel* afraid or angry.

The Principles or Psychology (1890), Vol. 2, 449–50.

1 Science as such assuredly has no authority, for she can only say what is, not what is not.

'Is Life Worth Living?' (1895). In *The Will to Believe and Other Essays in Popular Philosophy* (1897), 56.

2 Objective evidence and certitude are doubtless very fine ideals to play with, but where on this moonlit and dream-visited planet are they found?

'The Will to Believe' (1896). In *The Will to Believe and Other Essays in Popular Philosophy* (1897), 14.

3 Science can tell us what exists; but to compare the *worths*, both of what exists and of what does not exist, we must consult not science, but what Pascal calls our heart.

'The Will to Believe' (1896). In *The Will to Believe and Other Essays in Popular Philosophy* (1897), 22.

4 The first thing the intellect does with an object is to class it along with something else. But any object that is infinitely important to us and awakens our devotion feels to us also as if it must be *sui generis* and unique. Probably a crab would be filled with a sense of personal outrage if it could hear us class it without ado or apology as a crustacean, and thus dispose of it. 'I am no such thing,' it would say; 'I am MYSELF, MYSELF alone.'

The Varieties of Religious Experience: A Study in Human Nature (1902), 9.

5 The God whom science recognizes must be a God of universal laws exclusively, a God who does a wholesale, not a retail business. He cannot accommodate his processes to the convenience of individuals.

The Varieties of Religious Experience: A Study in Human Nature (1902), 493–5.

6 The history of philosophy is to a great extent that of a certain clash of human temperaments ... I will write these traits down in two columns. I think you will practically recognize the two types of mental make-up that I mean if I head the columns by the titles 'tender-minded' and 'tough-minded' respectively.

THE TENDER-MINDED.	THE TOUGH-MINDED.
Rationalistic (going by 'principles'),	Empiricist (going by 'facts'),
Intellectualistic,	Sensationalistic,
Idealistic,	Materialistic,
Optimistic,	Pessimistic,
Religious,	Irreligious,
Free-willist,	Fatalistic,
Monistic,	Pluralistic,
Dogmatical.	Sceptical.

'The Present Dilemma in Philosophy', in *Pragmatism: A New Way for Some Old Ways of Thinking, Popular Lectures on Philosophy* (1907), 6, 12.

7 First ... a new theory is attacked as absurd; then it is admitted to be true, but obvious and insignificant; finally it is seen to be so important that its adversaries claim that they themselves discovered it.

'Pragmatism's Conception of Truth', in *Pragmatism: A New Way for some Old Ways of Thinking, Popular Lectures on Philosophy* (1907), 198.

Robert Jameson 1774–1854

British naturalist and geologist

8 According to the common law of nature, deficiency of power is supplied by duration of time.

'Geological Illustrations', Appendix to G. Cuvier, *Essay on the Theory of the Earth*, trans. R. Jameson (1827), 430.

Karl Jaspers 1883–1969

German philosopher and psychiatrist

1 On the question of the world as a whole, science founders. For scientific knowledge the world lies in fragments, the more so the more precise our scientific knowledge becomes.

Kleine Schule des philosophischen Denkens (1965), trans. R. F. C. Hull and G. Wels, *Philosophy is for Everyman: A Short Course in Philosophical Thinking* (1969), 8.

James Hopwood Jeans 1877–1946

British physicist and mathematician

2 Humanity is at the very beginning of its existence—a new-born babe, with all the unexplored potentialities of babyhood; and until the last few moments its interest has been centred, absolutely and exclusively, on its cradle and feeding bottle.

EOS: Or the Wider Aspects of Cosmology (1928), 12.

3 Taking a very gloomy view of the future of the human race, let us suppose that it can only expect to survive for two thousand millions years longer, a period about equal to the past age of the earth. Then, regarded as a being destined to live for three-score years and ten, humanity although it has been born in a house seventy years old, is itself only three days old. But only in the last few minutes has it become conscious that the whole world does not centre round its cradle and its trappings, and only in the last few ticks of the clock has any adequate conception of the size of the external world dawned upon it. For our clock does not tick seconds, but years; its minutes are the lives of men.

EOS: Or the Wider Aspects of Cosmology (1928), 12–3.

4 Life exists in the universe only because the carbon atom possesses certain exceptional properties.

The Mysterious Universe (1930), 8.

5 The tendency of modern physics is to resolve the whole material universe into waves, and nothing but waves. These waves are of two kinds: bottled-up waves, which we call matter, and unbottled waves, which we call radiation or light. If annihilation of matter occurs, the process is merely that of unbottling imprisoned wave-energy and setting it free to travel through space. These concepts reduce the whole universe to a world of light, potential or existent, so that the whole story of its creation can be told with perfect accuracy and completeness in the six words: 'God said, "Let there be light"'.

The Mysterious Universe (1930), 37–8.

6 We have already considered with disfavour the possibility of the universe having been planned by a biologist or an engineer; from the intrinsic evidence of his creation, the Great Architect of the Universe now begins to appear as a pure mathematician.

The Mysterious Universe (1930), 134.

7 ON JEANS Sir James Jeans
Always says what he means
He is really perfectly serious
About the Universe being Mysterious.

E. C. Bentley, *Baseless Biography* (1939), 44.

Thomas Jefferson 1743–1826

American President

See ***Kennedy, John F.*** *338:3*

8 It is well known, that on the Ohio, and in many parts of America further north, tusks, grinders, and skeletons of unparalleled magnitude are found in great numbers, some lying on the surface of the earth, and some a little below it . . . But to whatever animal we ascribe these remains, it is certain that such a one has existed in America, and that it has been the largest of all terrestrial beings.

Notes on the State of Virginia (1782), 71, 77.

9 The ocean . . . like the air, is the common birth-right of mankind.

'Reply to the Society of Tammany, or Columbian Order, No. 1, of the City of New

York', 29 February 1808. In H. A. Washington (ed.), *The Writings of Thomas Jefferson*, Vol. 8 (1854), 127.

1 For I agree with you that there is a natural aristocracy among men. The grounds of this are virtue and talents.

Letter to John Adams, 28 October 1813. In Paul Wilstach (ed.), *Correspondence of John Adams and Thomas Jefferson 1812–1826* (1925), 92.

Lord Francis Jeffrey 1773–1850

British lawyer and literary critic

2 Damn the Solar System. Bad light; planets too distant; pestered with comets; feeble contrivance; could make a better myself.

Attributed.

Harold Jeffreys 1891–1989

British astronomer and geophysicist

3 What the use of P [the significance level] implies, therefore, is that a hypothesis that may be true may be rejected because it has not predicted observable results that have not occurred.

Theory of Probability (1939), 316.

4 The real difficulty about vulcanism is not to see how it can start, but how it can stop.

Earthquakes and Mountains, 2nd edition (1950), 187.

Herbert Spencer Jennings 1868–1947

American zoologist

5 A cell of a higher organism contains a thousand different substances, arranged in a complex system. This great organized system was not discovered by chemical or physical methods; they are inadequate to its refinement and delicacy and complexity.

'The Cell in Relation to its Environment', *Journal of the Maryland Academy of Sciences*, 1931, 2, 25.

Glenn L. Jepsen 1904–74

American geologist

6 *Why Become Extinct?* Authors with varying competence have suggested that dinosaurs disappeared because the climate deteriorated (became suddenly or slowly too hot or cold or dry or wet), or that the diet did (with too much food or not enough of such substances as fern oil; from poisons in water or plants or ingested minerals; by bankruptcy of calcium or other necessary elements). Other writers have put the blame on disease, parasites, wars, anatomical or metabolic disorders (slipped vertebral discs, malfunction or imbalance of hormone and endocrine systems, dwindling brain and consequent stupidity, heat sterilization, effects of being warm-blooded in the Mesozoic world), racial old age, evolutionary drift into senescent overspecialization, changes in the pressure or composition of the atmosphere, poison gases, volcanic dust, excessive oxygen from plants, meteorites, comets, gene pool drainage by little mammalian egg-eaters, overkill capacity by predators, fluctuation of gravitational constants, development of psychotic suicidal factors, entropy, cosmic radiation, shift of Earth's rotational poles, floods, continental drift, extraction of the moon from the Pacific Basin, draining of swamp and lake environments, sunspots, God's will, mountain building, raids by little green hunters in flying saucers, lack of standing room in Noah's Ark, and palaeoweltschmerz.

'Riddles of the Terrible Lizards', *American Scientist*, 1964, 52, 231.

Niels Jerne 1911–94

Danish immunologist

7 A cis-immunologist will sometimes speak to a trans-immunologist; but the latter rarely answers.

'Summary: Waiting for the end', *Cold Spring Harbor Symposia on Quantitative Biology*, 1967, 32, 591.

1 An immune system of enormous complexity is present in all vertebrate animals. When we place a population of lymphocytes from such an animal in appropriate tissue culture fluid, and when we add an antigen, the lymphocytes will produce specific antibody molecules, in the absense of any nerve cells. I find it astonishing that the immune system embodies a degree of complexity which suggests some more or less superficial though striking analogies with human language, and that this cognitive system has evolved and functions without assistance of the brain.

'The Generative Grammar of the Immune System', Nobel Lecture, 8 December 1984. In *Nobel Lectures: Physiology or Medicine 1981–1990* (1993), 223.

2 It seems a miracle that young children easily learn the language of any environment into which they were born. The generative approach to grammar, pioneered by Chomsky, argues that this is only explicable if certain deep, universal features of this competence are innate characteristics of the human brain. Biologically speaking, this hypothesis of an inheritable capability to learn any language means that it must somehow be encoded in the DNA of our chromosomes. Should this hypothesis one day be verified, then lingusitics would become a branch of biology.

'The Generative Grammar of the Immune System', Nobel Lecture, 8 December 1984. In *Nobel Lectures: Physiology or Medicine 1981–1990* (1993), 223.

3 More about the selection theory: Jerne meant that the Socratic idea of learning was a fitting analogy for 'the logical basis of the selective theories of antibody formation': Can the truth (the capability to synthesize an antibody) be learned? If so, it must be assumed not to pre-exist; to be learned, it must be acquired. We are thus confronted with the difficulty to which Socrates calls attention in Meno [. . .] namely, that it makes as little sense to search for what one does not know as to search for what one knows; what one knows, one cannot search for, since one knows it already, and what one does not know, one cannot search for, since one does not even know what to search for. Socrates resolves this difficulty by postulating that learning is nothing but recollection. The truth (the capability to synthesize an antibody) cannot be brought in, but was already inherent.

'The Natural Selection Theory', in John Cairns, Gunther S. Stent, and James D. Watson (eds.) *Phage and the Origins of Molecular Biology* (1966), 301.

Otto Jespersen 1860–1943

Danish linguist

4 Man is a classifying animal: in one sense it may be said that the whole process of speaking is nothing but distributing phenomena, of which no two are alike in every respect, into different classes on the strength of perceived similarities and dissimilarities. In the name-giving process we witness the same ineradicable and very useful tendency to see likenesses and to express similarity in the phenomena through similarity in name.

Language: Its Nature, Development and Origin (1922), 388–9.

William Stanley Jevons 1835–82

British philosopher and economist

5 It seems perfectly clear that Economy, if it is to be a science at all, must be a mathematical science. There exists much prejudice against attempts to introduce the methods and language of mathematics into any branch of the moral sciences. Most persons appear to hold that the physical sciences form the proper sphere of mathematical method, and that the moral sciences demand some other method—I know not what.

The Theory of Political Economy (1871), 3.

6 Science arises from the discovery of Identity amid Diversity.

The Principles of Science: A Treatise on Logic and Scientific Method (1874), 1.

1 I am convinced that it is impossible to expound the methods of induction in a sound manner, without resting them upon the theory of probability. Perfect knowledge alone can give certainty, and in nature perfect knowledge would be infinite knowledge, which is clearly beyond our capacities. We have, therefore, to content ourselves with partial knowledge—knowledge mingled with ignorance, producing doubt.

The Principles of Science: A Treatise on Logic and Scientific Method, 2nd edition (1877), 197.

Wilhelm Ludvig Johannsen

1857–1927

Danish botanist

2 No hypothesis concerning the nature of this 'something' shall be advanced thereby or based thereon. Therefore it appears as most simple to use the last syllable 'gen' taken from Darwin's well-known word pangene since it alone is of interest to use, in order thereby to replace the poor, more ambiguous word, 'Anlage'. Thus, we will say for 'das pangene' and 'die pangene' simply 'Das Gen' and 'Die Gene,' The word Gen is fully free from every hypothesis; it expresses only the safely proved fact that in any case many properties of organisms are conditioned by separable and hence independent 'Zustände,' 'Grundlagen,' 'Anlagen'—in short what we will call 'just genes'—which occur specifically in the gametes.

Elemente der Exakten Erblichkeitslehre (1909), 124. Trans. G. E. Allen and quoted in G. E. Allen, *Thomas Hunt Morgan: The Man and His Science* (1978), 209–10 (Footnote 79).

3 This constitution we designate by the word genotype. The word is entirely independent of any hypothesis; it is *fact*, not hypothesis that different zygotes arising by fertilisation can thereby have different qualities, that, even under quite similar conditions of life, phenotypically diverse individuals can develop.

Elemente der Exakten Erblichkeitslehre (1909), 165–70. Trans. in Ernst Mayr, *The Growth of Biological Thought: Diversity, Evolution and Inheritance* (1982), 782.

4 The genotypic constitution of a gamete or a zygote may be parallelized with a complicated chemico-physical structure. This reacts exclusively *in consequence of its realized state*, but not in consequence of the history of its creation. So it may be with the genotypical constitution of gametes and zygotes: its history is without influence upon its reactions, which are determined exclusively by its actual nature. The genotype-conception is thus an 'ahistoric' view of the reactions of living beings—of course only as far as true heredity is concerned. This view is an analog to the chemical view, as already pointed out; chemical compounds have no compromising ante-act, H_2O is always H_2O, and reacts always in the same manner, whatsoever may be the 'history' of its formation or the earlier states of its elements. I suggest that it is useful to emphasize this 'radical' ahistoric genotype-conception of heredity in its strict antagonism to the transmission—or phenotype—view.

'The Genotype Conception of Heredity', *The American Naturalist*, 1911, **45**, 129.

5 The science of genetics is in a transition period, becoming an exact science just as the chemistry in the times of Lavoisier, who made the balance an indispensable implement in chemical research.

'The Genotype Conception of Heredity', *The American Naturalist*, 1911, **45**, 131.

Douglas Wilson Johnson

1878–1944

American geologist

6 If an explanation is so vague in its inherent nature, or so unskillfully molded in its formulation, that specific deductions subject to empirical verification or refutation can not be based upon it, then it can never serve as a working hypothesis. A hypothesis with which one can not work is not a working hypothesis.

'Role of Analysis in Scientific Investigation', *Bulletin of the Geological Society of America*, 1933, **44**, 479.

George Johnson 1952–

American science writer

1 Artificial intelligence is based on the assumption that the mind can be described as some kind of formal system manipulating symbols that stand for things in the world. Thus it doesn't matter what the brain is made of, or what it uses for tokens in the great game of thinking. Using an equivalent set of tokens and rules, we can do thinking with a digital computer, just as we can play chess using cups, salt and pepper shakers, knives, forks, and spoons. Using the right software, one system (the mind) can be mapped onto the other (the computer).

Machinery of the Mind: Inside the New Science of Artificial Intelligence (1986), 250.

Samuel Johnson 1709–84

British author and lexicographer

See ***Boswell, James*** *78:4*

2 No place affords a more striking conviction of the vanity of human hopes than a publick library; for who can see the wall crouded on every side by mighty volumes, the works of laborious meditation, and accurate inquiry, now scarcely known but by the catalogue, and preserved only to encrease the pomp of learning, without considering how many hours have been wasted in vain endeavours, how often imagination has anticipated the praises of futurity, how many statues have risen to the eye of vanity, how many ideal converts have elevated zeal, how often wit has exulted in the eternal infamy of his antagonists, and dogmatism has delighted in the gradual advances of his authority, the immutability of his decrees, and the perpetuity of his power.

Non unquam dedit
Documenta fors majora, quam fragili loco
Starent superbi.

Seneca, Troades, II, 4–6

Insulting chance ne'er call'd with louder voice,
On swelling mortals to be proud no more.

Of the innumerable authors whose performances are thus treasured up in magnificent obscurity, most are forgotten, because they never deserved to be remembered, and owed the honours which they have once obtained, not to judgment or to genius, to labour or to art, but to the prejudice of faction, the stratagem of intrigue, or the servility of adulation.

Nothing is more common than to find men whose works are now totally neglected, mentioned with praises by their contemporaries, as the oracles of their age, and the legislators of science. Curiosity is naturally excited, their volumes after long enquiry are found, but seldom reward the labour of the search. Every period of time has produced these bubbles of artificial fame, which are kept up a while by the breath of fashion and then break at once and are annihilated. The learned often bewail the loss of ancient writers whose characters have survived their works; but perhaps if we could now retrieve them we should find them only the Granvilles, Montagus, Stepneys, and Sheffields of their time, and wonder by what infatuation or caprice they could be raised to notice.

It cannot, however, be denied, that many have sunk into oblivion, whom it were unjust to number with this despicable class. Various kinds of literary fame seem destined to various measures of duration. Some spread into exuberance with a very speedy growth, but soon wither and decay; some rise more slowly, but last long. Parnassus has its flowers of transient fragrance as well as its oaks of towering height, and its laurels of eternal verdure.

The Rambler, Number 106, 23 March 1751. In W. J. Bate and Albrecht B. Strauss (eds.), *The Rambler* (1969), Vol. 2, 200–1.

1 Nothing has more retarded the advancement of learning than the disposition of vulgar minds to ridicule and vilify what they cannot comprehend.

The Rambler, Number 117, 30 April 1751. In W. J. Bate and Albrecht B. Strauss (eds.), *The Rambler* (1969), Vol. 2, 258–9.

2 There prevails among men of letters, an opinion, that all appearance of science is particularly hateful to Women; and that therefore whoever desires to be well received in female assemblies, must qualify himself by a total rejection of all that is serious, rational, or important; must consider argument or criticism as perpetually interdicted; and devote all his attention to trifles, and all his eloquence to compliment.

The Rambler, Number 173, 12 November 1751. In W. J. Bate and Albrecht B. Strauss (eds.), *The Rambler* (1969), Vol. 3, 152–3.

3 I am not yet so lost in lexicography, as to forget that *words are the daughters of the earth, and that things are the sons of heaven.* Language is only the instrument of science, and words are but the signs of ideas: I wish, however, that the instrument might be less apt to decay, and that signs might be permanent, like the things which they denote.

'Preface', *A Dictionary of the English Language* (1755), Vol. 1.

4 No, Sir, I am not a botanist; and (alluding, no doubt, to his near sightedness) should I wish to become a botanist, I must first turn myself into a reptile.

Entry for Tuesday 20 July 1762. In George Birkbeck-Hill (ed.), *Boswell's Life of Johnson* (1934–50), Vol 1, 377, fn2.

5 Knowledge is of two kinds. We know a subject ourselves, or we know where we can find information upon it.

Entry for Tuesday 18 April 1775. In George Birkbeck-Hill (ed.), *Boswell's Life of Johnson* (1934–50), Vol. 2, 7.

6 People have now a-days got a strange opinion that every thing should be taught by lectures. Now, I cannot see that lectures can do as much good as reading the books from which the lectures are taken.

Entry for February 1776. In George Birkbeck-Hill (ed.), *Boswell's Life of Johnson* (1934–50), Vol. 2, 7.

7 He sometimes employed himself in chymistry, sometimes in watering and pruning a vine, and sometimes in small experiments, at which those who may smile, should recollect that there are moments which admit of being soothed only by trifles.

Entry for Thursday 9 December 1779. In George Birkbeck-Hill (ed.), *Boswell's Life of Johnson* (1934–50), Vol. 3, 398.

8 [Boswell]: Sir Alexander Dick tells me, that he remembers having a thousand people in a year to dine at his house: that is, reckoning each person as one, each time that he dined there. [Johnson]: That, Sir, is about three a day. [Boswell]: How your statement lessens the idea. [Johnson]: That, Sir, is the good of counting. It brings every thing to a certainty, which before floated in the mind indefinitely.

Entry for Friday 18 April 1783. In George Birkbeck-Hill (ed.), *Boswell's Life of Johnson* (1934–50), Vol. 4, 204.

Frédéric Jean Joliot 1900–58 and **Irène Joliot-Curie** 1897–1956

French chemists

9 The farther an experiment is from theory, the closer it is to the Nobel Prize.

Variously attributed to each.

Alison Jolly 1937–

American evolutionary biologist

10 Primates stand at a turning point in the course of evolution. Primates are to the biologist what viruses are to the biochemist. They can be analysed and partly understood according to the rules of a simpler discipline, but they also present another level of complexity: viruses are living chemicals, and primates are animals who love and hate and think.

'The Evolution of Primate Behavior: A survey

of the primate order traces the progressive development of intelligence as a way of life', *American Scientist*, 1985, 73, 288.

John Joly 1857–1933

Irish geologist and physicist

1 It is . . . indisputable that the orogenic movements which uplift the hills have been at the basis of geological history. To them the great accumulation of sediments which now form so large a part of continental land are mainly due. There can be no doubt of the fact that these movements have swayed the entire history, both inorganic and organic, of the world in which we live.

Radioactivity and Geology (1909), 115–6.

2 With an interest almost amounting to anxiety, geologists will watch the development of researches which may result in timing the strata and the phases of evolutionary advance; and may even—going still further back—give us reason to see in the discrepancy between denudative and radioactive methods, glimpses of past aeons, beyond that day of regeneration which at once ushered in our era of life, and, for all that went before, was 'a sleep and a forgetting'.

Radioactivity and Geology (1909), 250–1.

3 We can . . . find no beginning of the world. We trace back events and come to barriers which close our vistas—barriers which, for all we know, may for ever close it. They stand like the gates of ivory and of horn; portals from which only dreams proceed; and Science cannot as yet say of this or that dream if it proceeds from the gate of horn or from that of ivory.

The Birth-Time of the World (1915), 2.

4 Only mountains can beget mountains.

The Birth-Time of the World (1915), 141.

Ernest Jones 1879–1958

British psychoanalyst

5 [Freud's] great strength, though sometimes also his weakness, was the quite extraordinary respect he had for the singular fact . . . When he got hold of a simple but significant fact he would feel, and know, that it was an example of something general or universal, and the idea of collecting statistics on the matter was quite alien to him.

The Life and Work of Sigmund Freud (1953), Vol.1, 96–7.

Frederick Wood Jones 1879–1954

British anatomist and anthropologist

6 Man is no new-begot child of the ape, bred of a struggle for existence upon brutish lines—nor should the belief that such is his origin, oft dinned into his ears by scientists, influence his conduct. Were he to regard himself as an extremely ancient type, distinguished chiefly by the qualities of his mind, and to look upon the existing Primates as the failures of his line, as his misguided and brutish collaterals, rather than as his ancestors, I think it would be something gained for the ethical outlook of Homo—and also it would be consistent with present knowledge.

The Origin of Man (1918), a pamphlet published by The Society for the Promotion of Christian Knowledge, reprinted in Arthur Dendy (ed.), *Animal Life and Human Progress* (1919), 131.

John Stephen Jones 1944–

British geneticist and popular science writer

7 The language of the genes has a simple alphabet, not with twenty-six letters, but just four. These are the four different DNA bases—adenine, guanine, cytosine and thymine (A, G, C and T for short). The bases are arranged in words of three letters such as CGA or TGG. Most of the words code for different amino acids, which themselves are joined together to make proteins, the building blocks of the body.

The Language of the Genes: Biology, History and the Evolutionary Future (1993), 3.

1 Philosophy is to science as pornography is to sex: it is cheaper, easier and some people prefer it.

Review of Simon Pinker *How the Mind Works* (1997). In *New York Review of Books*, 6 November 1997.

2 Evolution triumphs because it turns to natural selection, the plodding accumulation of error.

Almost Like a Whale: The Origin of Species Updated (1999), 9.

3 Genetics is to biology what atomic theory is to physics. Its principle is clear: that inheritance is based on particles and not on fluids. Instead of the essence of each parent mixing, with each child the blend of those who made him, information is passed on as a series of units. The bodies of successive generations transport them through time, so that a long-lost character may emerge in a distant descendant. The genes themselves may be older than the species that bear them.

Almost Like a Whale: The Origin of Species Updated (1999), 115.

William Jones 1726–1800

British physician

4 An experiment in nature, like a text in the Bible, is capable of different interpretations, according to the preconceptions of the interpreter.

Physiological Disquisitions (1781), 148.

David Starr Jordan 1851–1931

American ichthyologist

5 The process of natural selection has been summed up in the phrase 'survival of the fittest'. This, however, tells only part of the story. 'Survival of the existing' in many cases covers more of the truth. For in hosts of cases the survival of characters rests not on any special usefulness or fitness, but on the fact that individuals possessing these characters have inhabited or invaded a certain area. The principle of utility explains survivals among competing structures. It rarely accounts for qualities associated with geographic distribution.

The nature of animals which first colonize a district must determine what the future fauna will be. From their specific characters, which are neither useful nor harmful, will be derived for the most part the specific characters of their successors.

It is not essential to the meadow lark that he should have a black blotch on the breast or the outer tail-feather white. Yet all meadow larks have these characters just as all shore larks have the tiny plume behind the ear. Those characters of the parent stock, which may be harmful in the new relations, will be eliminated by natural selection. Those especially helpful will be intensified and modified, but the great body of characters, the marks by which we know the species, will be neither helpful nor hurtful. These will be meaningless streaks and spots, variations in size of parts, peculiar relations of scales or hair or feathers, little matters which can neither help nor hurt, but which have all the persistence heredity can give.

Foot-notes to Evolution. A Series of Popular Addresses on the Evolution of Life (1898), 218.

Edward Jorden 1569–1632

British physician

6 [The] seminary spirit of minerals hath its proper wombs where it resides, and is like a Prince or Emperour, whose prescripts both Elements and matter must obey; and it is never idle, but always in action, producing and maintaining natural substances, untill they have fulfilled their destiny.

A Discourse of Natural Bathes, and Mineral Waters (1669), 104.

Christian Klixbull Jørgensen

1931–2000

Danish spectroscopist

7 *On the future of Chemistry:*

Chemistry is not the preservation hall

of old jazz that it sometimes looks like. We cannot know what may happen tomorrow. Someone may oxidize mercury (II), francium (I), or radium (II). A mineral in Nova Scotia may contain an unsaturated quark per 1020 nucleons. (This is still 6000 per gram.) We may pick up an extraterrestrial edition of *Chemical Abstracts*. The universe may be a 4-dimensional soap bubble in an 11-dimensional space as some supersymmetry theorists argued in May of 1983. Who knows?

George B. Kaufmann, 'Interview with Jannik Bjerrum and Christian Klixbull Jørgensen', *Journal of Chemical Education*, 1985, **62**, 1005.

Joseph Joubert 1754–1824

French philosopher

1 *Combien de gens se font abstraits pour paraître profonds! La plupart des termes abstraits sont des ombres qui cachent des vides.*

How many people become abstract in order to appear profound! Most abstract terms are shadows that conceal a void.

Quoted in M. Paul De Raynal, *Pensées de J. Joubert* (1862), 456.

James Prescott Joule 1818–89

British physicist

2 The most convincing proof of the conversion of heat into living force [*vis viva*] has been derived from my experiments with the electro-magnetic engine, a machine composed of magnets and bars of iron set in motion by an electrical battery. I have proved by actual experiment that, in exact proportion to the force with which this machine works, heat is abstracted from the electrical battery. You see, therefore, that living force may be converted into heat, and that heat may be converted into living force, or its equivalent attraction through space.

'On Matter, Living Force, and Heat' (1847). In *The Scientific Papers of James Prescott Joule* (1884), Vol. 1, 270–1.

3 The animal frame, though destined to fulfill so many other ends, is as a machine more perfect than the best contrived steam-engine—that is, is capable of more work with the same expenditure of fuel.

'On Matter, Living Force, and Heat' (1847). In *The Scientific Papers of James Prescott Joule* (1884), Vol. 1, 271.

4 The earth in its rapid motion round the sun possesses a degree of living force so vast that, if turned into the equivalent of heat, its temperature would be rendered at least one thousand times greater than that of red-hot iron, and the globe on which we tread would in all probability be rendered equal in brightness to the sun itself.

'On Matter, Living Force, and Heat' (1847). In *The Scientific Papers of James Prescott Joule* (1884), Vol. 1, 271.

5 ON JOULE You'll be thought cool
If you call it the joule.
But there'll be a howl
If you call it the jowl.

Anonymous. Quotation supplied by W. H. Brock.

Horace Freeland Judson 1931–

American historian

6 Science is our century's art.

The Search for Solutions (1980), 10.

Carl Gustav Jung 1875–1961

Swiss analytic psychologist and psychiatrist

7 Dream analysis stands or falls with [the hypothesis of the unconscious]. Without it the dream appears to be merely a freak of nature, a meaningless conglomerate of memory-fragments left over from the happenings of the day.

Dream Analysis in its Practical Application (1930), 1–2.

8 Complexes are psychic contents which are outside the control of the conscious mind. They have been split off from consciousness and lead a separate existence in the unconscious, being at all times ready to hinder or to reinforce the conscious intentions.

A Psychological Theory of Types (1931), 79.

1 Man's unconscious . . . contains all the patterns of life and behaviour inherited from his ancestors, so that every human child, prior to consciousness, is possessed of a potential system of adapted psychic functioning.

The Basic Postulates of Analytical Psychology (1931), 186.

2 I can still recall vividly how Freud said to me, 'My dear Jung, promise me never to abandon the sexual theory. That is the most essential thing of all. You see, we must make a dogma of it, an unshakable bulwark' . . . In some astonishment I asked him, 'A bulwark—against what?' To which he replied, 'Against the black tide of mud'—and here he hesitated for a moment, then added—'of occultism'.

Memories, Dreams, Reflections (1963), 147–8.

Antoine de Jussieu 1686–1758

French physician and botanist

3 I observed on most collected stones the imprints of innumerable plant fragments which were so different from those which are growing in the Lyonnais, in the nearby provinces, and even in the rest of France, that I felt like collecting plants in a new world . . . The number of these leaves, the way they separated easily, and the great variety of plants whose imprints I saw, appeared to me just as many volumes of botany representing in the same quarry the oldest library of the world.

'Examen des causes des Impressions des Plantes marquées sur certaines Pierres des environs de Saint-Chaumont dans le Lionnais', *Mémoires de l'Académie Royale des Sciences* (1718), 364. Trans. Albert V. and Marguerite Carozzi.

Leo P. Kadanoff 1937–

American physicist

4 Science is in low regard.

'Hard Times', *Physics Today*, October 1992, 45, 9.

5 We are fast approaching a situation in which nobody will believe anything we [physicists] say in any matter that touches upon our self-interest. Nothing we do is likely to arrest our decline in numbers, support or social value.

'Hard Times', *Physics Today*, October 1992, 45, 9.

6 Today when the public thinks of the products of science it is likely to think about environmental problems, an unproductive armament industry, careless or dishonest 'scientific' reports, Livermore cheers for 'nukes forever' and a huge amount of self-serving noise on every subject from global warming to 'the face of God'.

'Hard Times', *Physics Today*, October 1992, 45, 9.

Louis Albrecht Kahlenberg 1870–1941

American physical chemist

7 Professor, how can you bring yourself to enter this chemical building that has Ionic columns?

Kahlenberg opposed the ionic theory.

Quoted in R. Oesper, *The Human Side of Scientists* (1975), 106.

8 We already have anions and cations and now the biochemists and nutritionists are speaking of rat-ions.

Quoted in R. Oesper, *The Human Side of Scientists* (1975), 106.

9 Harvard never produced anyone of great originality.

Quoted in R. Oesper, *The Human Side of Scientists* (1975), 107.

10 Philosophers go the way in which scientists shove them.

Quoted in R. Oesper, *The Human Side of Scientists* (1975), 107.

Michio Kaku 1947–

American physicist

11 In fact, it is often stated that of all the theories proposed in this century, the silliest is quantum theory. Some say that the only thing that quantum theory has going for it, in fact, is that it is unquestionably correct.

Hyperspace: A Scientific Odyssey Through

Parallel Universes, Time Warps, and The Tenth Dimension (1994), 262.

Heike Kamerlingh Onnes

1853–1926

Dutch physicist

1 According to my views, aiming at quantitative investigations, that is at establishing relations between measurements of phenomena, should take first place in the experimental practice of physics. By measurement to knowledge [*door meten tot weten*] I should like to write as a motto above the entrance to every physics laboratory.

'The Significance of Quantitative Research in Physics', Inaugural Address at the University of Leiden (1882). In Hendrik Casimar, *Haphazard Reality: Half a Century of Science* (1983), 160–1.

2 The experiment left no doubt that, as far as accuracy of measurement went, the resistance disappeared. At the same time, however, something unexpected occurred. The disappearance did not take place gradually but *abruptly*. From 1/500 the resistance at 4.2K, it could be established that the resistance had become less than a thousand-millionth part of that at normal temperature. Thus the mercury at 4.2K has entered a new state, which, owing to its particular electrical properties, can be called the state of superconductivity.

'Investigations into the Properties of Substances at low Temperatures, which have led, amongst other Things, to the Preparation of Liquid Helium', Nobel Lecture, 11 December 1913. In *Nobel Lectures in Physics 1901–1921* (1967), 333.

Immanuel Kant 1724–1804

German philosopher

3 Give me matter, and I will construct a world out of it!

'Universal Natural History and Theory of the Heavens' (1755), preface. In W. Hastie (ed. and trans.), *Kant's Cosmogony: As in his Essay on the Retardation of the Rotation of the Earth and his Natural History and Theory of the Heavens* (1900), 29.

4 We come no nearer the infinitude of the creative power of God, if we enclose the space of its revelation within a sphere described with the radius of the Milky Way, than if we were to limit it to a ball an inch in diameter. All that is finite, whatever has limits and a definite relation to unity, is equally far removed from the infinite . . . Eternity is not sufficient to embrace the manifestations of the Supreme Being, if it is not combined with the infinitude of space.

'Universal Natural History and Theory of the Heavens' (1755), part 2, ch.7. In W. Hastie (ed. and trans.), *Kant's Cosmogony: As in his Essay on the Retardation of the Rotation of the Earth and his Natural History and Theory of the Heavens* (1900), 139–40.

5 When Galileo caused balls, the weights of which he had himself previously determined, to roll down an inclined plane; when Torricelli made the air carry a weight which he had calculated beforehand to be equal to that of a definite volume of water; or in more recent times, when Stahl changed metal into lime, and lime back into metal, by withdrawing something and then restoring it, a light broke upon all students of nature. They learned that reason has insight only into that which it produces after a plan of its own, and that it must not allow itself to be kept, as it were, in nature's leading-strings, but must itself show the way with principles of judgement based upon fixed laws, constraining nature to give answer to questions of reason's own determining. Accidental observations, made in obedience to no previously thought-out plan, can never be made to yield a necessary law, which alone reason is concerned to discover.

Critique of Pure Reason (1781), trans. Norman Kemp Smith (1929), 20.

6 Our knowledge springs from two fundamental sources of the mind; the first is the capacity of receiving representations (receptivity for impressions), the second is the power of knowing an object through these representations (spontaneity [in the production] of concepts).

Critique of Pure Reason (1781), trans. Norman Kemp Smith (1929), 92.

1 Thoughts without content are empty, intuitions without concepts are blind . . . The understanding can intuit nothing, the senses can think nothing. Only through their union can knowledge arise.

Critique of Pure Reason (1781), trans. Norman Kemp Smith (1929), 93.

2 The ideal of the supreme being is nothing but a *regulative principle* of reason which directs us to look upon all connection in the world *as if* it originated from an all-sufficient necessary cause.

Critique of Pure Reason (1781), trans. Norman Kemp Smith (1929), 517.

3 *Enlightenment is man's emergence from his self-incurred immaturity. Immaturity* is the inability to use one's own understanding without the guidance of another. This immaturity is *self-incurred* if its cause is not lack of understanding, but lack of resolution and courage to use it without the guidance of another. The motto of enlightenment is therefore: *Sapere aude!* Have courage to use your *own* understanding!

'An Answer to the Question: What is Enlightenment?', (1784). In Hans Reiss (ed.), *Kant: Political Writings*, trans. H. B. Nisbet (1970), 54.

4 If it were possible for us to have so deep an insight into a man's character as shown both in inner and in outer actions, that every, even the least, incentive to these actions and all external occasions which affect them were so known to us that his future conduct could be predicted with as great a certainty as the occurrence of a solar or lunar eclipse, we could nevertheless still assert that the man is free.

Critique of Practical Reason (1788). In L. W. Beck (ed. & trans.), *Critique of Practical Reason and Other Writings in Moral Philosophy* (1949), 204–5.

5 Two things fill the mind with ever new and increasing admiration and awe, the oftener and more steadily they are reflected on: the starry heavens above me and the moral law within me.

Critique of Practical Reason (1788). In L. W. Beck (ed. and trans.), *Critique of Practical Reason and Other Writings in Moral Philosophy* (1949), 258.

6 In scientific matters . . . the greatest discoverer differs from the most arduous imitator and apprentice only in degree, whereas he differs in kind from someone whom nature has endowed for fine art. But saying this does not disparage those great men to whom the human race owes so much in contrast to those whom nature has endowed for fine art. For the scientists' talent lies in continuing to increase the perfection of our cognitions and on all the dependent benefits, as well as in imparting that same knowledge to others; and in these respects they are far superior to those who merit the honour of being called geniuses. For the latter's art stops at some point, because a boundary is set for it beyond which it cannot go and which has probably long since been reached and cannot be extended further.

The Critique of Judgement (1790), trans. J. C. Meredith (1991), 72.

Peter Leonidovich Kapitsa

1894–1984

Russian physicist

7 The year that Rutherford died (1938 [sic]) there disappeared forever the happy days of free scientific work which gave us such delight in our youth. Science has lost her freedom. Science has become a productive force. She has become rich but she has become enslaved and part of her is veiled in secrecy. I do not know whether Rutherford would continue to joke and laugh as he used to.

'Notes from Here and There', *Science Policy News*, 1969, 1, No 2, 33.

Aharon Katchalsky-Katzir

1913–72

Israeli chemist

1 Whether we like it or not, the ultimate goal of every science is to become trivial, to become a well-controlled apparatus for the solution of schoolbook exercises or for practical application in the construction of engines.

'Nonequilibrium Thermodynamics', *International Science and Technology*, October 1963, 44.

John Keats 1795–1821

British romantic poet

2 I am certain of nothing but the holiness of the Heart's affections and the truth of Imagination—What the imagination seizes as Beauty must be truth—whether it existed before or not.

Letter to Benjamin Bailey, 22 November 1817. In H. E. Rollins (ed.), *Letters of John Keats* (1958), Vol. 1, 184.

3 *Negative Capability*, that is when man is capable of being in uncertainties, Mysteries, doubts, without any irritable reaching after fact & reason—Coleridge, for instance, would let go by a fine isolated verisimilitude caught from the Penetralium of mystery, from being incapable of remaining content with half knowledge.

Letter to George and Thomas Keats, 21 December 1817. In H. E. Rollins (ed.), *Letters of John Keats* (1958), Vol. 1, 193–4.

4 Who, of men, can tell
That flowers would bloom, or that green fruit would swell
To melting pulp, that fish would have bright mail,
The earth its dower of river, wood, and vale,
The meadows runnels, runnels pebble-stones,
The seed its harvest, or the lute its tones,
Tones ravishment, or ravishment its sweet,
If human souls did never kiss and greet?

Endymion (1818), bk. 1, l. 835–842. In John Barnard (ed.), *John Keats. The Complete Poems* (1973), 129.

5 Do not all charms fly
At the mere touch of cold philosophy?
There was an awful rainbow once in heaven:
We know her woof, her texture; she is given
In the dull catalogue of common things.
Philosophy will clip an Angel's wings,
Conquer all mysteries by rule and line,
Empty the haunted air, and gnomèd mine
Unweave a rainbow.

Lamia 1820, II, lines 229–37. In John Barnard (ed.), *John Keats. The Complete Poems* (1973), 431.

Arthur Keith 1866–1955

British anthropologist and anatomist

6 Religious leaders and men of science have the same ideals; they want to understand and explain the universe of which they are part; they both earnestly desire to solve, if a solution be ever possible, that great riddle: Why are we here?

Concerning Man's Origin (1927), viii.

Friedrich August Kekulé von Stradonitz 1829–96

German chemist

See ***Hofmann, August*** *290:2*

7 Carbon is, as may easily be shown and as I shall explain in greater detail later, tetrabasic or tetratomic, that is 1 atom of carbon = Ͼ = 12 is equivalent to 4 At. H.

'On the so-called Copulated Compounds and the Theory of Polyatomic Radicals', *Annalen*, 1857, **4**, 133. Trans. in J. R. Partington, *A History of Chemistry* (1972), Vol. 4, 536.

8 When the simplest compounds of this element are considered (marsh gas, chloride of carbon, chloroform, carbonic acid, phosgene, sulphide of carbon, hydrocyanic acid , etc.) it is seen that the quantity of carbon which chemists have recognised as the smallest possible, that is, as an atom, always unites with 4 atoms of a monatomic or with two atoms of a diatomic element; that in general, the sum of the chemical units of the elements united with one atom of carbon is 4. This leads us to the view that carbon is tetratomic or tetrabasic. In the cases of substances which contain several atoms of carbon, it must be assumed that at least some of the atoms are in some way held in the compound by the affinity of carbon, and that the carbon atoms attach themselves to one another, whereby a part of the affinity of the one is naturally engaged with an equal part of the affinity of the other. The simplest and therefore the most probable case of such an association of carbon atoms is that in which one affinity unit of one is bound by one of the other. Of the 2 x 4 affinity units of the two carbon atoms, two are used up in holding the atoms together, and six remain over, which can be bound by atoms of other elements.

'Ueber die Konstitution und die Metamorphosen der chemischen Verbindungen', *Annalen*, 1858, 5, 106. Trans. in J. R. Partington, *A History of Chemistry* (1972), Vol. 4, 536.

1 We define organic chemistry as the chemistry of carbon compounds.

Lehrbuch der Organischen Chemie (1861), Vol. 1, 11. Trans. W. H. Brock.

2 If we wish to give an account of the atomic constitution of the aromatic compounds, we are bound to explain the following facts:

1) All aromatic compounds, even the most simple, are relatively richer in carbon than the corresponding compounds in the class of fatty bodies.

2) Among the aromatic compounds, as well as among the fatty bodies, a large number of homologous substances exist.

3) The most simple aromatic compounds contain at least six atoms of carbon.

4) All the derivatives of aromatic substances exhibit a certain family likeness; they all belong to the group of 'Aromatic compounds'. In cases where more vigorous reactions take place, a portion of the carbon is often eliminated, but the chief product contains at least six atoms of carbon . . .

These facts justify the supposition that all aromatic compounds contain a common group, or, we may say, a common *nucleus* consisting of six atoms of carbon. Within this *nucleus* a more intimate combination of the carbon atoms takes place; they are more compactly placed together, and this is the cause of the aromatic bodies being relatively rich in carbon. Other carbon atoms can be joined to this *nucleus* in the same way, and according to the same law, as in the case of the group of fatty bodies, and in this way the existence of homologous compounds is explained.

Bulletin de la Societé Chimique de France, 1865, 1, 98. Trans. W. H. Brock.

3 The question whether atoms exist or not . . . belongs rather to metaphysics. In chemistry we have only to decide whether the assumption of atoms is an hypothesis adapted to the explanation of chemical phenomena . . . whether a further development of the atomic hypothesis promises to advance our knowledge of the mechanism of chemical phenomena . . . I rather expect that we shall some day find, for what we now call atoms, a mathematico-mechanical explanation, which will render an account of atomic weight, of atomicity, and of numerous other properties of the so-called atoms.

Laboratory, 1867, 1, 303.

4 The separate atoms of a molecule are not connected all with all, or all with one, but, on the contrary, each one is connected only with one or with a few neighbouring atoms, just as in a chain link is connected with link.

'The Scientific Aims and Achievements of Chemistry', *Nature*, 18, 1878, 212.

1 During my stay in London I resided for a considerable time in Clapham Road in the neighbourhood of Clapham Common . . . One fine summer evening I was returning by the last bus 'outside' as usual, through the deserted streets of the city, which are at other times so full of life. I fell into a reverie (Träumerei), and lo, the atoms were gambolling before my eyes! Whenever, hitherto, these diminutive beings had appeared to me, they had always been in motion: but up to that time I had never been able to discern the nature of their motion. Now, however, I saw how, frequently, two smaller atoms united to form a pair: how the larger one embraced the two smaller ones: how still larger ones kept hold of three or even four of the smaller: whilst the whole kept whirling in a giddy dance. I saw how the larger ones formed a chain, dragging the smaller ones after them but only at the ends of the chain. I saw what our past master, Kopp, my highly honoured teacher and friend has depicted with such charm in his *Molekular-Welt*: but I saw it long before him. The cry of the conductor 'Clapham Road', awakened me from my dreaming: but I spent part of the night in putting on paper at least sketches of these dream forms. This was the origin of the 'Structural Theory'.

Kekulé at Benzolfest in *Berichte*, 1890, 23, 1302.

2 I was sitting writing at my textbook but the work did not progress; my thoughts were elsewhere. I turned my chair to the fire and dozed. Again the atoms were gambolling before my eyes. This time the smaller groups kept modestly in the background. My mental eye, rendered more acute by the repeated visions of the kind, could now distinguish larger structures of manifold confirmation: long rows, sometimes more closely fitted together all twining and twisting in snake like motion. But look! What was that? One of the snakes had seized hold of its own tail, and the form whirled mockingly before my eyes. As if by a flash of lightning I awoke; and this time also I spent the rest of the night in working out the rest of the hypothesis. Let us learn to dream, gentlemen, then perhaps we shall find the truth . . . But let us beware of publishing our dreams till they have been tested by waking understanding.

Kekulé at Benzolfest in *Berichte*, 1890, 23, 1302.

3 Originally a pupil of Liebig, I became a pupil of Dumas, Gerhardt and Williamson: I no longer belonged to any school.

J. R. Partington, *A History of Chemistry* (1970), Vol. 4, 533.

4 ON **KEKULÉ** The structural theory of Kekulé has been the growth hormone of organic chemistry.

G. E. K. Branch and M. Calvin, *The Theory of Organic Chemistry* (1941), Preface, v.

Lord Kelvin see **Thomson, William Thomas**

May Kendall 1861–1931

British poet

5 I abide in a goodly Museum,
Frequented by sages profound:
'Tis a kind of strange mausoleum,
Where the beasts that have vanished abound.
There's a bird of the ages Triassic,
With his antediluvian beak,
And many a reptile Jurassic,
And many a monster antique.

'Ballad of the Ichthyosaurus', *Dreams to Sell* (1887), 14.

6 Ay, driven no more by passion's gale,
Nor impulse unforeseen,
Humanity shall faint and fail,
And on her ruins will prevail
The Conquering Machine!
Responsibility begone!
Let Freedom's flag be furled;

Oh, coming ages, hasten on,
And bring the true Automaton,
The monarch of the world.

'The Conquering Machine', *Dreams to Sell* (1887), 29–30.

John Cowdrey Kendrew et al.

1917–97

British biochemists

1 In describing a protein it is now common to distinguish the primary, secondary and tertiary structures. The *primary structure* is simply the order, or sequence, of the amino-acid residues along the polypeptide chains. This was first determined by Sanger using chemical techniques for the protein insulin, and has since been elucidated for a number of peptides and, in part, for one or two other small proteins. The *secondary structure* is the type of folding, coiling or puckering adopted by the polypeptide chain: the α-helix structure and the pleated sheet are examples. Secondary structure has been assigned in broad outline to a number of fibrous proteins such as silk, keratin and collagen; but we are ignorant of the nature of the secondary structure of any globular protein. True, there is suggestive evidence, though as yet no proof, that α-helices occur in globular proteins, to an extent which is difficult to gauge quantitatively in any particular case. The *tertiary structure* is the way in which the folded or coiled polypeptide chains are disposed to form the protein molecule as a three-dimensional object, in space. The chemical and physical properties of a protein cannot be fully interpreted until all three levels of structure are understood, for these properties depend on the spatial relationships between the amino-acids, and these in turn depend on the tertiary and secondary structures as much as on the primary. Only X-ray diffraction methods seem capable, even in principle, of unravelling the tertiary and secondary structures.

Kendrew, J. C., Bodo, G., Dintzis, H. M., Parrish, R. G., Wyckoff, H. and Phillips, D. C., 'A Three-Dimensional Model of the Myoglobin Molecule Obtained by X-ray Analysis', *Nature*, 1958, **181**, 662.

John Fitzgerald Kennedy

1917–63

American President

2 I believe that this Nation should commit itself to achieving the goal, before this decade is out, of landing a man on the moon and returning him safely to earth.

'Freedom's Cause: These are Extraordinary Times' (25 May 1961). In *Vital Speeches of the Day* (15 June 1961), Vol. 27, No. 17, 518–9.

3 The President described the dinner [for Nobel Prizewinners] as 'probably the greatest concentration of talent and genius in this house except for perhaps those times when Thomas Jefferson ate alone.'

'49 Nobel Prize Winners Honored at White House', by M. Hunter. In *New York Times* (30 April 1962), 1.

4 The language of science is universal, and perhaps scientists have been the most international of all professions in their outlook . . . Every time you scientists make a major invention, we politicians have to invent a new institution to cope with it—and almost invariably, these days, it must be an international institution.

'Science as a Guide of Public Policy', Address to the National Academy of Science, Washington D. C., 22 October 1963.

5 In the years since man unlocked the power stored up within the atom, the world has made progress, halting, but effective, toward bringing that power under human control. The challenge may be our salvation. As we begin to master the destructive potentialities of modern science, we move toward a new era in which science can fulfill its creative promise and help bring into existence the happiest society the world has ever known.

'Science as a Guide of Public Policy', Address to the National Academy of Science, Washington D. C., 22 October 1963.

1 Science contributes to our culture in many ways, as a creative intellectual activity in its own right, as the light which has served to illuminate man's place in the universe, and as the source of understanding of man's own nature.

'Science as a Guide of Public Policy', Address to the National Academy of Science, Washington D. C., 22 October 1963.

Anthony John Patrick Kenny

1931–

British philosopher

2 It is characteristic of our age to endeavour to replace virtues by technology. That is to say, wherever possible we strive to use methods of physical or social engineering to achieve goals which our ancestors thought attainable only by the training of character. Thus, we try so far as possible to make contraception take the place of chastity, and anaesthetics to take the place of fortitude; we replace resignation by insurance policies and munificence by the Welfare State. It would be idle romanticism to deny that such techniques and institutions are often less painful and more efficient methods of achieving the goods and preventing the evils which unaided virtue once sought to achieve and avoid. But it would be an equal and opposite folly to hope that the take-over of virtue by technology may one day be complete, so that the necessity for the laborious acquisition of the capacity for rational choice by individuals can be replaced by the painless application of the fruits of scientific discovery over the whole field of human intercourse and enterprise.

'Mental Health in Plato's Republic', in *The Anatomy of the Soul: Historical Essays in the Philosophy of Mind* (1973), 26.

Johannes Kepler 1571–1630

German astronomer and physicist

See **Blackmore, Richard** 68:3

3 Some of what these pamphlets [of astrological forecasts] say will turn out to be true, but most of it time and experience will expose as empty and worthless. The latter part will be forgotten [literally: written on the winds] while the former will be carefully entered in people's memories, as is usual with the crowd.

On giving astrology sounder foundations, *De fundamentis astrologiae certioribus*, (1602), Thesis 2, *Johannes Kepler Gesammelte Werke* (1937–), Vol. 4, 12, trans. J. V. Field, in *Archive for History of Exact Sciences*, 1984, 31, 229–72.

4 However, before we come to [special] creation, which puts an end to all discussion: I think we should try everything else.

De Stella Nova, *On the New Star* (1606), Chapter 22, in *Johannes Kepler Gesammelte Werke* (1937–), Vol. 1, 257, ll. 23–4.

5 After the birth of printing books became widespread. Hence everyone throughout Europe devoted himself to the study of literature . . . Every year, especially since 1563, the number of writings published in every field is greater than all those produced in the past thousand years. Through them there has today been created a new theology and a new jurisprudence; the Paracelsians have created medicine anew and the Copernicans have created astronomy anew. I really believe that at last the world is alive, indeed seething, and that the stimuli of these remarkable conjunctions did not act in vain.

De Stella Nova, *On the New Star* (1606), *Johannes Kepler Gesammelte Werke* (1937–), Vol. 1, 330–2. Quoted in N. Jardine, *The Birth of History and Philosophy of Science: Kepler's A Defence of Tycho Against Ursus With Essays on its Provenance and Significance* (1984), 277–8.

6 I myself, a professional mathematician, on re-reading my own work find it strains my mental powers to recall to mind from the figures the meanings of the demonstrations, meanings which I myself originally put into the figures and the text from my mind. But when I attempt to remedy the obscurity of the material by putting in extra words, I see myself falling into the opposite fault of becoming chatty in something mathematical.

Astronomia Nova, New Astronomy, (1609), Introduction, second paragraph.

1 And from this such small difference of eight minutes [of arc] it is clear why Ptolemy, since he was working with bisection [of the linear eccentricity], accepted a fixed equant point. . . . For Ptolemy set out that he actually did not get below ten minutes [of arc], that is a sixth of a degree, in making observations. To us, on whom Divine benevolence has bestowed the most diligent of observers, Tycho Brahe, from whose observations this eight-minute error of Ptolemy's in regard to Mars is deduced, it is fitting that we accept with grateful minds this gift from God, and both acknowledge and build upon it. So let us work upon it so as to at last track down the real form of celestial motions (these arguments giving support to our belief that the assumptions are incorrect). This is the path I shall, in my own way, strike out in what follows. For if I thought the eight minutes in [ecliptic] longitude were unimportant, I could make a sufficient correction (by bisecting the [linear] eccentricity) to the hypothesis found in Chapter 16. Now, because they could not be disregarded, these eight minutes alone will lead us along a path to the reform of the whole of Astronomy, and they are the matter for a great part of this work.

Astronomia Nova, New Astronomy (1609), ch. 19, 113–4, *Johannes Kepler Gesammelte Werke* (1937–), Vol. 3, 177–8.

2 Geometry is one and eternal shining in the mind of God. That share in it accorded to men is one of the reasons that Man is the image of God.

Conversation with the Sidereal Messenger [an open letter to Galileo Galilei], *Dissertatio cum Nuncio Sidereo* (1610), in *Johannes Kepler Gesammelte Werke* (1937–), Vol. 4, 308, ll. 9–10.

3 Truth is the daughter of time, and I feel no shame in being her midwife.

Account of personal observations of the four moving satellites of Jupiter . . . , Narratio de observatis a se quatuor Jovis satellitibus erronibus (1611), first words of text, in *Johannes Kepler Gesammelte Werke* (1937–), Vol. 4, 317.

4 Geometry, which before the origin of things was coeternal with the divine mind and is God himself (for what could there be in God which would not be God himself?), supplied God with patterns for the creation of the world, and passed over to Man along with the image of God; and was not in fact taken in through the eyes.

Harmonice Mundi, The Harmony of the World (1619), book IV, ch. 1. Trans. E. J. Aiton, A. M. Duncan, and J. V. Field (1997), 304.

5 I am stealing the golden vessels of the Egyptians to build a tabernacle to my God from them, far far away from the boundaries of Egypt. If you forgive me, I shall rejoice; if you are enraged with me, I shall bear it. See, I cast the die, and I write the book. Whether it is to be read by the people of the present or of the future makes no difference: let it await its reader for a hundred years, if God himself has stood ready for six thousand years for one to study him.

Harmonice Mundi, The Harmony of the World (1619), end of Introduction to Book V. Trans. E. J. Aiton, A. M. Duncan, and J. V. Field (1997), 391.

6 Yet in this my stars were not Mercury as morning star in the angle of the seventh house, in quartile with Mars, but they were Copernicus, they were Tycho Brahe, without whose books of observations everything which has now been brought by me into the brightest daylight would lie buried in darkness.

Harmonice Mundi, The Harmony of the World (1619), book IV, Epilogue on Sublunary Nature. Trans. E. J. Aiton, A. M. Duncan, and J. V. Field (1997), 377.

7 And if you want the exact moment in time, it was conceived mentally on 8th March in this year one thousand six hundred and eighteen, but submitted to calculation in an unlucky way, and therefore rejected as false, and finally returning on the 15th of May and adopting a new line of attack, stormed the darkness of my mind. So strong was the support from the combination of my labour of seventeen years on the observations of Brahe and the present study, which conspired together, that at first I believed I was dreaming, and assuming my conclusion among my basic premises. But it is absolutely certain and exact that the proportion between the periodic times of any two planets is precisely the sesquialterate proportion of their mean distances.

Harmonice Mundi, The Harmony of the World (1619), book V, ch. 3. Trans. E. J. Aiton, A. M. Duncan, and J. V. Field (1997), 411.

1 The cause of the six-sided shape of a snowflake is none other than that of the ordered shapes of plants and of numerical constants; and since in them nothing occurs without supreme reason—not, to be sure, such as discursive reasoning discovers, but such as existed from the first in the Creators's design and is preserved from that origin to this day in the wonderful nature of animal faculties, I do not believe that even in a snowflake this ordered pattern exists at random.

Di Nive Sexangula, On the Six-Cornered Snowflake (1611), K18, l. 6–12. Trans. and ed. Colin Hardie (1966), 33.

2 If this [the Mysterium cosmographicum] is published, others will perhaps make discoveries I might have reserved for myself. But we are all ephemeral creatures (and none more so than I). I have, therefore, for the Glory of God, who wants to be recognized from the book of Nature, that these things may be published as quickly as possible. The more others build on my work the happier I shall be.

Letter to Michael Maestlin, 3 October 1595. *Johannes Kepler Gesammelte Werke* (1937–), Vol. 13, letter 23, l. 251, p. 39–40.

3 I wanted to become a theologian; for a long time I was unhappy. Now, behold, God is praised by my work even in astronomy.

Letter to Michael Maestlin, 3 October 1595. *Johannes Kepler Gesammelte Werke* (1937–), Vol. 13, letter 23, l. 256–7, p. 40.

4 I am a Lutheran astrologer, I throw away the nonsense and keep the hard kernel.

Letter to Michael Maestlin, 15 March 1598. *Johannes Kepler Gesammelte Werke* (1937–), Vol. 13, letter 68, l. 177, p.184.

5 My aim is to say that the machinery of the heavens is not like a divine animal but like a clock (and anyone who believes a clock has a soul gives the work the honour due to its maker) and that in it almost all the variety of motions is from one very simple magnetic force acting on bodies, as in the clock all motions are from a very simple weight.

Letter to J. G. Herwart von Hohenburg, 16 February 1605. *Johannes Kepler Gesammelte Werke* (1937–), Vol. 15, letter 325, l. 57–61, p. 146.

6 So, Fabricius, I already have this: that the most true path of the planet [Mars] is an ellipse, which Dürer also calls an oval, or certainly so close to an ellipse that the difference is insensible.

Letter to David Fabricius, 11 October 1605. *Johannes Kepler Gesammelte Werke* (1937–), Vol. 15, letter 358, l. 390–92, p. 249.

7 Repudiating the sensible world, which he neither sees himself nor believes from those who have, the Peripatetic joins combat by childish quibbling in a world on paper, and denies the Sun shines because he himself is blind.

Letter to Galileo Galilei, 28 March 1611. *Johannes Kepler Gesammelte Werke* (1937–), Vol. 16, letter 611, l. 17–20, p. 372.

8 I also ask you my friends not to condemn me entirely to the mill of mathematical calculations, and allow me time for philosophical speculations, my only pleasures.

Letter to Vincenzo Bianchi, 17 February 1619. *Johannes Kepler Gesammelte Werke* (1937–), Vol. 17, letter 827, l. 249–51, p. 327.

9 I used to measure the Heavens, now I measure the shadows of Earth. The mind belonged to Heaven, the body's shadow lies here.

Kepler's epitaph for himself (*Johannes Kepler Gesammelte Werke* (1937–), vol. 19, p. 393).

Thomas Hewitt Key 1799–1875

British philosopher and classical scholar

1 What is Matter?—Never mind.
What is Mind?—No Matter.

A Short Cut to Metaphysics (1855), 19.

John Maynard Keynes 1883–1946

British economist and mathematician

2 This long run is a misleading guide to current affairs. In the long run we are all dead. Economists set themselves too easy, too useless a task if in tempestuous seasons they can only tell us that when the storm is long past the ocean is flat again.

A Tract on Monetary Reform (1923), 80.

3 The study of economics does not seem to require any specialised gifts of an unusually high order. Is it not, intellectually regarded, a very easy subject compared with the higher branches of philosophy and pure science? Yet good, or even competent, economists are the rarest of birds. An easy subject, at which very few excel! The paradox finds its explanation, perhaps, in that the master-economist must possess a rare *combination* of gifts. He must reach a high standard in several different directions and must combine talents not often found together. He must be mathematician, historian, statesman, philosopher—in some degree. He must understand symbols and speak in words. He must contemplate the particular in terms of the general, and touch abstract and concrete in the same flight of thought. He must study the present in the light of the past for the purposes of the future. No part of man's nature or his institutions must lie entirely outside his regard. He must be purposeful and disinterested in a simultaneous mood; as aloof and incorruptible as an artist, yet sometimes as near the earth as a politician.

'Alfred Marshall: 1842–1924' (1924). In Geoffrey Keynes (ed.), *Essays in Biography*, (1933), 170.

4 Professor [Max] Planck, of Berlin, the famous originator of the Quantum Theory, once remarked to me that in early life he had thought of studying economics, but had found it too difficult! Professor Planck could easily master the whole corpus of mathematical economics in a few days. He did not mean that! But the amalgam of logic and intuition and the wide knowledge of facts, most of which are not precise, which is required for economic interpretation in its highest form is, quite truly, overwhelmingly difficult for those whose gift mainly consists in the power to imagine and pursue to their furthest points the implications and prior conditions of comparatively simple facts which are known with a high degree of precision.

'Alfred Marshall: 1842–1924' (1924). In Geoffrey Keynes (ed.), *Essays in Biography*, (1933), 191–2 fn.

5 Newton was not the first of the age of reason. He was the last of the magicians, the last of the Babylonians and Sumerians . . . Isaac Newton, a posthumous child born with no father on Christmas Day, 1642, was the last wonder child to whom the Magi could do sincere and appropriate homage . . . Why do I call him a magician? Because he looked on the whole universe and all that is in it *as a riddle*, as a secret which could be read by applying pure thought to certain evidence, certain mystic clues which God had laid about the world to allow a sort of philosopher's treasure hunt to the esoteric brotherhood . . . He regarded the Universe as a cryptogram set by the Almighty—just as he himself wrapt the discovery of the calculus in a cryptogram when he communicated with Leibniz. By pure thought, by concentration of mind, the riddle, he believed, would be revealed to the initiate.

'Newton, the Man' (1946). In Geoffrey Keynes (ed.), *Essays in Biography*, 2nd edition (1951), 311–4.

Cassius Jackson Keyser

1862–1947

American mathematician

1 The validity of mathematical propositions is independent of the actual world—the world of existing subject-matters—is logically prior to it, and would remain unaffected were it to vanish from being.

The Pastures of Wonder: The Realm of Mathematics and the Realm of Science (1929), 99.

Charles Kingsley 1819–75

British clergyman, novelist and naturalist

2 You are literally filled with the fruit of your own devices, with rats and mice and such small deer, paramecia, and entomostraceæ, and kicking things with horrid names, which you see in microscopes at the Polytechnic, and rush home and call for brandy—without the water—stone, and gravel, and dyspepsia, and fragments of your own muscular tissue tinged with your own bile.

'The Water Supply of London', *North British Review*, 1851, **15**, 246

3 'But a water-baby is contrary to nature.'

Well, but, my dear little man, you must learn to talk about such things, when you grow older, in a very different way from that. You must not talk about 'ain't' and 'can't' when you speak of this great wonderful world round you, of which the wisest man knows only the very smallest corner, and is, as the great Sir Isaac Newton said, only a child picking up pebbles on the shore of a boundless ocean.

You must not say that this cannot be, or that that is contrary to nature. You do not know what nature is, or what she can do; and nobody knows; not even Sir Roderick Murchison, or Professor Owen, or Professor Sedgwick, or Professor Huxley, or Mr. Darwin, or Professor Faraday, or Mr. Grove, or any other of the great men whom good boys are taught to respect. They are very wise men; and you must listen respectfully to all they say: but even if they should say, which I am sure they never would, 'That cannot exist. That is contrary to nature,' you must wait a little, and see; for perhaps even they may be wrong.

The Water-Babies (1863), Ch. 2, 71–2.

4 For science is, I verily believe, like virtue, its own exceeding great reward. I can conceive few human states more enviable than that of the man to whom, panting in the foul laboratory, or watching for his life under the tropic forest, Isis shall for a moment lift her sacred veil, and show him, once and for ever, the thing he dreamed not of; some law, or even mere hint of a law, explaining one fact; but explaining with it a thousand more, connecting them all with each other and with the mighty whole, till order and meaning shoots through some old Chaos of scattered observations.

Health and Education (1874), 289.

Alfred Charles Kinsey 1894–1956

American zoologist and sexologist

5 The male's difficulties in his sexual relations after marriage include a lack of facility, of ease, or of suavity in establishing rapport in a sexual situation.

Sexual Behavior in the Human Male (1948), 545.

6 Few males achieve any real freedom in their sexual relations even with their wives. Few males realise how badly inhibited they are on these matters.

Sexual Behavior in the Human Male (1948), 545.

7 The heterosexuality or homosexuality of many individuals is not an all-or-none proposition.

Sexual Behavior in the Human Male (1948), 638.

1 Males do not represent two discrete populations, heterosexual and homosexual.

Sexual Behavior in the Human Male (1948), 639.

2 Among all types of sexual activity, masturbation is . . . the one in which the female most frequently reaches orgasm.

Sexual Behavior in the Human Female (1953), 132.

3 There is a tendency to consider anything in human behavior that is unusual, not well known, or not well understood, as neurotic, psychopathic, immature, perverse, or the expression of some other sort of psychologic disturbance.

Sexual Behavior in the Human Female (1953), 195.

4 The range of variation in the female far exceeds the range of variation in the male.

Sexual Behavior in the Human Female (1953), 537–8.

Joseph Rudyard Kipling 1865–1936

British novelist and poet

5 The motto of all the mongoose family is, 'Run and find out'.

'Rikki-Tikki-Tavi', *The Jungle Book* (1894), 124.

6 Doctors have been exposed—you always will be exposed—to the attacks of those persons who consider their own undisciplined emotions more important than the world's most bitter agonies—the people who would limit and cripple and hamper research because they fear research may be accompanied by a little pain and suffering.

Doctors (1908), 28–9.

7 The female of the species is more deadly than the male.

'The female of the species' (1911). In *Rudyard Kipling's Verse, Inclusive Edition 1885–1918* (1919), Vol. 2, 166.

Clifford Kirkpatrick 1898–1971

American sociologist

8 Science recognizes no personal powers in the universe responsive to the prayers and needs of men. Belief in mysterious powers which constitutes, according to our definition, the conceptual aspect of religion is usually an animistic belief in personal powers. Science in effect denies the existence of spiritual beings which religion affirms.

Religion in Human Affairs (1929), 470.

Richard Kirwan 1733–1812

British chemist and geologist

9 Geological facts being of an historical nature, all attempts to deduce a complete knowledge of them merely from their still, subsisting consequences, to the exclusion of unexceptionable testimony, must be deemed as absurd as that of deducing the history of ancient Rome solely from the medals or other monuments of antiquity it still exhibits, or the scattered ruins of its empire, to the exclusion of a Livy, a Sallust, or a Tacitus.

Geological Essays (1799), 5.

10 We have seven or eight geological facts, related by Moses on the one part, and on the other, deduced solely from the most exact and best verified geological observations, and yet agreeing perfectly with each other, not only in *substance*, but in the order of their succession . . . That two accounts derived from sources totally distinct from and independent on each other should agree not only in the substance but in the order of succession of two events only, is already highly improbable, if these facts be not true, both substantially and as to the order of their succession. Let this improbability, as to the substance of the facts, be represented only by 1/10. Then the improbability of their agreement as to seven events is 1.7/10.7 that is, as one to ten million, and would be much

higher if the *order* also had entered into the computation.

Geological Essays (1799), 52–3.

David Burlingame Kitts 1923–

American geologist

1 In terms of the way a geologist operates, there is no past until after the assumption of uniformity has been made.

'The Theory of Geology', in C. C. Albritton (ed.), *The Fabric of Geology* (1963), 63.

Christian Felix Klein 1849–1925

German mathematician

2 The teacher manages to get along still with the cumbersome algebraic analysis, in spite of its difficulties and imperfections, and avoids the smooth infinitesimal calculus, although the eighteenth century shyness toward it had long lost all point.

Elementary Mathematics From an Advanced Standpoint (1908). 3rd edition (1924), trans. E. R. Hedrick and C. A. Noble (1932), Vol. 1, 155.

Morris Kline 1908–92

American mathematician

3 The stone that Dr. Johnson once kicked to demonstrate the reality of matter has become dissipated in a diffuse distribution of mathematical probabilities. The ladder that Descartes, Galileo, Newton, and Leibniz erected in order to scale the heavens rests upon a continually shifting, unstable foundation.

Mathematics in Western Culture (1953), 382.

Clyde Ray Maben Kluckhohn 1905–60 and **Henry Murray** 1893–1988

American anthropologist and American psychologist

4 Without the discovery of uniformities there can be no concepts, no classifications, no formulations, no principles, no laws; and without these no science can exist.

'Personality Formation: the Determinants'. In Clyde Kluckhohn and Henry A. Murray (eds.), *Personality in Nature, Society, and Culture* (1949), 37–8.

Heinrich Hermann Robert Koch 1843–1910

German bacteriologist

5 However, on many occasions, I examined normal blood and normal tissues and there was no possibility of overlooking bacteria or confusing them with granular masses of equal size. I never found organisms. Thus, I conclude that bacteria do not occur in healthy human or animal tissues.

'Investigations of the Etiology of Wound Infections' (1878), *Essays of Robert Koch* (1987), trans. K. Codell Carter, 27.

6 The pure culture is the foundation for all research on infectious disease.

'Zur Untersuchungen von Pathologen Organismen', *Mittheilungen aus dem Kaiserlichen Gesundheitsamte*, 1881, 1, 1–48. Quoted in English in Thomas D. Brock, *Robert Koch: A Life in Medicine and Bacteriology* (1988), 94.

7 The facts obtained in this study may possibly be sufficient proof of the causal relationship, that only the most sceptical can raise the objection that the discovered microorganism is not the cause but only an accompaniment of the disease . . . It is necessary to obtain a perfect proof to satisfy oneself that the parasite and the disease are . . . actually causally related, and that the parasite is the . . . direct cause of the disease. This can only be done by completely separating the parasite from the diseased organism [and] introducing the isolated parasite into healthy organisms and induce the disease anew with all its characteristic symptoms and properties.

Berliner Klinische Wochenschrift (1882), 393. Quoted in Edward J. Huth and T. Jock Murray (eds.), *Medicine in Quotations: Views of Health and Disease Through the Ages* (2000), 52.

8 From my numerous observations, I conclude that these tubercle bacilli occur in all tuberculous disorders, and that they are distinguishable from all other microorganisms.

'The Etiology of Tuberculosis' (1882), *Essays of Robert Koch* (1987), trans. K. Codell Carter, 87.

1 To prove that tuberculosis is caused by the invasion of bacilli, and that it is a parasitic disease primarily caused by the growth and multiplication of bacilli, it is necessary to isolate the bacilli from the body, to grow them in pure culture until they are freed from every disease product of the animal organism, and, by introducing isolated bacilli into animals, to reproduce the same morbid condition that is known to follow from inoculation with spontaneously developed tuberculous material.

'The Etiology of Tuberculosis' (1882), *Essays of Robert Koch* (1987), trans. K. Codell Carter, 87.

2 It is safe to say that the little pamphlet which was left to find its way through the slow mails to the English scientist outweighed in importance and interest for the human race all the press dispatches which have been flashed under the channel since the delivery of the address—March 24. The rapid growth of the Continental capitals, the movements of princely noodles and fat, vulgar Duchesses, the debates in the Servian Skupschina, and the progress or receding of sundry royal gouts are given to the wings of lightening; a lumbering mail-coach is swift enough for the news of one of the great scientific discoveries of the age. Similarly, the gifted gentlemen who daily sift out for the American public the pith and kernel of the Old World's news; leave Dr. KOCH and his *bacilli* to chance it in the ocean mails, while they challenge the admiration of every gambler and jockey in this Republic by the fullness and accuracy of their cable reports of horse-races.

New York Times, 3 May 1882. Quoted in Thomas D. Brock, *Robert Koch* (1988), 131.

3 Our studies have shown that all cases of typhoid of this type have arisen by contact, that is, carried directly from one person to another. There was no trace of a connection to drinking water.

'Die Bekämpfing des Typhus', *Veröffentlichungen aus dem Gebiete des Militär-Sanitätswesens*, 1903, **21**. Quoted in English in Thomas D. Brock, *Robert Koch: A Life in Medicine and Bacteriology* (1988), 256.

4 In the first papers concerning the aetiology of tuberculosis I have already indicated the dangers arising from the spread of the bacilli-containing excretions of consumptives, and have urged moreover that prophylactic measures should be taken against the contagious disease. But my words have been unheeded. It was still too early, and because of this they still could not meet with full understanding. It shared the fate of so many similar cases in medicine, where a long time has also been necessary before old prejudices were overcome and the new facts were acknowledged to be correct by the physicians.

'The current state of the struggle against tuberculosis', Nobel Lecture, 12 December 1905. In *Nobel Lectures: Physiology or Medicine 1901–1921* (1967), 169.

5 If my efforts have led to greater success than usual, this is due, I believe, to the fact that during my wanderings in the field of medicine, I have strayed onto paths where the gold was still lying by the wayside. It takes a little luck to be able to distinguish gold from dross, but that is all.

'Robert Koch', *Journal of Outdoor Life*, 5, 1908, 164–9.

Arthur Koestler 1905–83

Hungarian-born British novelist

6 Darkness at noon.

Darkness at Noon (1940).

7 As modern physics started with the Newtonian revolution, so modern philosophy starts with what one might call the Cartesian Catastrophe. The catastrophe consisted in the splitting up of the world into the realms of matter and mind, *and* the identification of 'mind' with conscious thinking. The result of this identification was the shallow rationalism of *l'esprit Cartesien*, and an impoverishment of psychology which it took three centuries to remedy even in part.

The Act of Creation (1964), 148.

1 God seems to have left the receiver off the hook, and time is running out.

The Ghost in the Machine (1967), 339.

Charles Atwood Kofoid

1865–1947

American zoologist

2 The Mecca of the biological world.

The Naples Biological Station—one of the foremost marine laboratories in the world.

The Biological Stations of Europe (1910), 9.

Wolfgang Köhler 1887–1967

German-born American psychologist

3 If we wish to imitate the physical sciences, we must not imitate them in their contemporary, most developed form; we must imitate them in their historical youth, when their state of development was comparable to our own at the present time. Otherwise we should behave like boys who try to copy the imposing manners of full-grown men without understanding their *raison d'être*, also without seeing that in development one cannot jump over intermediate and preliminary phases.

Gestalt Psychology (1929), 32.

Adolph Wilhelm Hermann Kolbe 1818–84

German chemist

4 The weeds of a seemingly learned and brilliant but actually trivial and empty philosophy of Nature which, after having been replaced some 50 years ago by the exact sciences, is now once more dug up by pseudo scientists from the lumber room of human fallacies, and like a trollop, newly attired in elegant dress and make-up, is smuggled into respectable company, to which she does not belong.

'Sign of the Times', *Journal für Praktische Chemie*, 1877, 15, 473, trans. W. H. Brock.

5 ON KOLBE Not one of them [formulae] can be shown to have any existence, so that the formula of one of the simplest of organic bodies is confused by the introduction of unexplained symbols for imaginary differences in the mode of combination of its elements . . . It would be just as reasonable to describe an oak tree as composed of blocks and chips and shavings to which it may be reduced by the hatchet, as by Dr Kolbe's formula to describe acetic acid as containing the products which may be obtained from it by destructive influences. A Kolbe botanist would say that half the chips are united with some of the blocks by the force *parenthesis*; the other half joined to this group in a different way, described by a *buckle*; shavings stuck on to these in a third manner, *comma*; and finally, a compound of shavings and blocks united together by a fourth force, *juxtaposition*, is joined to the main body by a fifth force, *full stop*.

Alexander Williamson, 'On Dr. Kolbe's Additive Formulae', *Quarterly Journal of the Chemical Society*, 1855, 7, 133–4.

Konrad of Megenberg

c.1180–1233

German papal inquisitor

6 *Buch der Nature.*

Book of Nature.

Title of book *c.*1350

Hermann Kopp 1817–92

German chemist

7 The alchemists of past centuries tried hard to make the elixir of life: . . . Those efforts were in vain; it is not in our power to obtain the experiences and the views of the future by prolonging our lives forward in this direction. However, it is well possible in a certain sense to prolong our lives backwards by acquiring the experiences of those who existed before us and by learning to know their views as well as if we were their contemporaries. The means for doing this is also an elixir of life.

Foreword to *Die Entwicklung der Chemie in der neueren Zeit* (1873), trans. W. H. Brock.

Karl Martin Leonhard Albrecht Kossel 1853–1927

German biochemist

1 I should like to compare this rearrangement which the proteins undergo in the animal or vegetable organism to the making up of a railroad train. In their passage through the body parts of the whole may be left behind, and here and there new parts added on. In order to understand fully the change we must remember that the proteins are composed of Bausteine united in very different ways. Some of them contain Bausteine of many kinds. The multiplicity of the proteins is determined by many causes, first through the differences in the nature of the constituent Bausteine; and secondly, through differences in the arrangement of them. The number of Bausteine which may take part in the formation of the proteins is about as large as the number of letters in the alphabet. When we consider that through the combination of letters an infinitely large number of thoughts may be expressed, we can understand how vast a number of the properties of the organism may be recorded in the small space which is occupied by the protein molecules. It enables us to understand how it is possible for the proteins of the sex-cells to contain, to a certain extent, a complete description of the species and even of the individual. We may also comprehend how great and important the task is to determine the structure of the proteins, and why the biochemist has devoted himself with so much industry to their analysis.

'The Chemical Composition of the Cell', *The Harvey Lectures*, 1911, 7, 45.

Alexander Onufrievich Kovalevsky 1840–1901

Russian embryologist

2 Darwin's theory was received in Russia with profound sympathy. While in Western Europe it met firmly established old traditions which it had first to overcome, in Russia its appearance coincided with the awakening of our society after the Crimean War and here it immediately received the status of full citizenship and ever since has enjoyed widespread popularity.

Quoted in Thomas F. Glick (ed.), *The Comparative Reception of Darwinism* (1988), 229–30.

Alexandre Koyré 1892–1964

Russian-born French philosopher and historian of science

3 What the founders of modern science, among them Galileo, had to do, was not to criticize and to combat certain faulty theories, and to correct or to replace them by better ones. They had to do something quite different. They had to destroy one world and to replace it by another. They had to reshape the framework of our intellect itself, to restate and to reform its concepts, to evolve a new approach to Being, a new concept of knowledge, a new concept of science—and even to replace a pretty natural approach, that of common sense, by another which is not natural at all.

Galileo and Plato (1943), 405.

4 It is possible that the deepest meaning and aim of Newtonianism, or rather, of the whole scientific revolution of the seventeenth century, of which Newton is the heir and the highest expression, is just to abolish the world of the 'more or less', the world of qualities and sense perception, the world of appreciation of our daily life, and to replace it by the (Archimedean) universe of precision, of exact measures, of strict determination . . . This revolution [is] one of the deepest, if not the deepest, mutations and transformations accomplished—or suffered—by the human mind since the invention of the cosmos by the Greeks, two thousand years before.

'The Significance of the Newtonian Synthesis' (1950). In *Newtonian Studies* (1965), 4–5.

Richard Freiherr von Krafft-Ebing 1840–1902

German sexologist and psychiatrist

1 The weakness of men in comparison with women lies in the great intensity of their sexual desires. Man becomes dependent upon woman, and the more, the weaker and more sensual he becomes; and this just in proportion as he becomes neuropathic.

Psychopathia Sexualis: With Special Reference to Contrary Sexual Instinct: A Medico-Legal Study (1886), trans. Charles Gilbert Chaddock (1892), 14.

2 With respect of the development of physiological love, it is probable that its nucleus is always to be found in an individual fetich (charm) which a person of one sex exercises over a person of the opposite sex.

Psychopathia Sexualis: With Special Reference to Contrary Sexual Instinct: A Medico-Legal Study (1886), trans. Charles Gilbert Chaddock (1892), 17.

3 Sexual instinct—as emotion, idea, and impulse—is a function of the cerebral cortex. Thus far no definite region of the cortex has been proved to be exclusively the seat of sexual sensations and impulses.

Psychopathia Sexualis: With Special Reference to Contrary Sexual Instinct: A Medico-Legal Study (1886), trans. Charles Gilbert Chaddock (1892), 24.

4 Too early and perverse sexual satisfaction injures not merely the mind, but also the body; inasmuch as it induces neuroses of the sexual apparatus (irritable weakness of the centres governing erection and ejaculation; defective pleasurable feeling in coitus), while, at the same time, it maintains the imagination and libido in continuous excitement.

Psychopathia Sexualis: With Special Reference to Contrary Sexual Instinct: A Medico-Legal Study (1886), trans. Charles Gilbert Chaddock (1892), 189.

5 While up to this time contrary sexual instinct has had but an anthropological, clinical, and forensic interest for science, now, as a result of the latest investigations, there is some thought of therapy in this incurable condition, which so heavily burdens its victims, socially, morally, and mentally. A preparatory step for the application of therapeutic measures is the exact differentiation of the acquired from the congenital cases; and among the latter again, the assignment of the concrete case to its proper position in the categories that have been established empirically.

Psychopathia Sexualis: With Special Reference to Contrary Sexual Instinct: A Medico-Legal Study (1886), trans. Charles Gilbert Chaddock (1892), 319.

Hans Adolf Krebs 1900–81

German-born British biochemist

6 What, then, is it in particular that can be learned from teachers of special distinction? Above all, what they teach is high standards. We measure everything, including ourselves, by comparisons; and in the absence of someone with outstanding ability there is a risk that we easily come to believe that we are excellent and much better than the next man. Mediocre people may appear big to themselves (and to others) if they are surrounded by small circumstances. By the same token, big people feel dwarfed in the company of giants, and this is a most useful feeling. So what the giants of science teach us is to see ourselves modestly and not to overrate ourselves.

Reminiscences and Reflections (1981), 175.

Ernst Kretschmer 1888–1964

German psychiatrist

7 The psychopaths are always around. In calm times we study them, but in times of upheaval, they rule over us.

Gestalten und Gedanken (1963), 94, trans. P. Lerner.

Leopold Kronecker 1823–91

German mathematician

1 *Die ganzen Zahlen hat der liebe Gott gemacht, alles andere ist Menschenwerk.*
The dear God has made the whole numbers, all the rest is man's work.

Speech at the Berlin meeting of the Society of German Scientists and Doctors in 1886, published in *Jahresbericht der Deutschen Mathematiker-Vereinigung*. Trans. obituary of Kronecker by H. E. Weber, *Year Book of the German Mathematics Association*, 1893, 19.

Petr Alekseevich Kropotkin 1842–1921

Russian geographer

2 In the animal world we have seen that the vast majority of species live in societies, and that they find in association the best arms for the struggle for life: understood, of course, in its wide Darwinian sense—not as a struggle for the sheer means of existence, but as a struggle against all natural conditions unfavourable to the species. The animal species, in which individual struggle has been reduced to its narrowest limits, and the practice of mutual aid has attained the greatest development, are invariably the most numerous, the most prosperous, and the most open to further progress. The mutual protection which is obtained in this case, the possibility of attaining old age and of accumulating experience, the higher intellectual development, and the further growth of sociable habits, secure the maintenance of the species, its extension, and its further progressive evolution. The unsociable species, on the contrary, are doomed to decay.

Mutual Aid: A Factor of Evolution (1902), 293.

Joseph Wood Krutch 1893–1970

American naturalist and conservationist

3 We must be part not only of the human community, but of the whole community; we must acknowledge some sort of oneness not only with our neighbors, our countrymen and our civilization but also some respect for the natural as well as for the man-made community. Ours is not only 'one world' in the sense usually implied by that term. It is also 'one earth'. Without some acknowledgement of that fact, men can no more live successfully than they can if they refuse to admit the political and economic interdependency of the various sections of the civilized world. It is not a sentimental but a grimly literal fact that unless we share this terrestrial globe with creatures other than ourselves, we shall not be able to live on it for long.

The Voice of the Desert (1956), 194–5.

Phillip H. Kuenen 1902–76

Dutch geologist

4 Experimental geology has this in common with all other branches of our science, petrology and palaeontology included, that in the long run it withers indoors.

'Experiments in Geology', *Transactions of the Geological Society of Glasgow*, 1958, 23, 25.

5 On the basis of the results recorded in this review, it can be claimed that the average sand grain has taken many hundreds of millions of years to lose 10 per cent. of its weight by abrasion and become subangular. It is a platitude to point to the slowness of geological processes. But much depends on the way things are put. For it can also be said that a sand grain travelling on the bottom of a river loses 10 million molecules each time it rolls over on its side and that representation impresses us with the high rate of this loss. The properties of quartz have led to the concentration of its grains on the continents, where they could now form a layer averaging several hundred metres thick. But to my mind the most astounding numerical estimate that follows from the present evaluations, is that during each and every second of the incredibly long geological past the number of quartz grains on earth has increased by 1,000 million.

'Sand—its Origin, Transportation, Abrasion and Accumulation', *The Geological Society of South Africa*, 1959, Annexure to Volume 62, 31.

Thomas S. Kuhn 1922–96

American historian and philosopher of science

1 History, if viewed as a repository for more than anecdote or chronology, could produce a decisive transformation in the image of science by which we are now possessed.

The Structure of Scientific Revolutions (1962), 1.

2 'Normal science' means research firmly based upon one or more past scientific achievements, achievements that some particular scientific community acknowledges for a time as supplying the foundation for its further practice.

The Structure of Scientific Revolutions (1962), 10.

3 The success of the paradigm . . . is at the start largely a promise of success . . . Normal science consists in the actualization of that promise . . . Mopping up operations are what engage most scientists throughout their careers. They constitute what I am here calling normal science . . . That enterprise seems an attempt to force nature into the preformed and relatively inflexible box that the paradigm supplies. No part of the aim of normal science is to call forth new sorts of phenomena; indeed those that will not fit the box are often not seen at all. Nor do scientists normally aim to invent new theories, and they are often intolerant of those invented by others.

The Structure of Scientific Revolutions (1962), 23–4.

4 Research under a paradigm must be a particularly effective way of inducing paradigm change.

The Structure of Scientific Revolutions (1962), 52.

5 The transition from a paradigm in crisis to a new one from which a new tradition of normal science can emerge is far from a cumulative process, one achieved by an articulation or extension of the old paradigm. Rather it is a reconstruction of the field from new fundamentals, a reconstruction that changes some of the field's most elementary theoretical generalizations as well as many of its paradigm methods and applications. During the transition period there will be a large but never complete overlap between the problems that can be solved by the old and by the new paradigm. But there will also be a decisive difference in the modes of solution. When the transition is complete, the profession will have changed its view of the field, its methods, and its goals.

The Structure of Scientific Revolutions (1962), 84–5.

6 Almost always the men who achieve these fundamental inventions of a new paradigm have been either very young or very new to the field whose paradigm they change.

The Structure of Scientific Revolutions (1962), 89–90.

7 As in political revolutions, so in paradigm choice—there is no standard higher than the assent of the relevant community . . . this issue of paradigm choice can never be unequivocally settled by logic and experiment alone.

The Structure of Scientific Revolutions (1962), 93.

8 Each paradigm will be shown to satisfy more or less the criteria that it dictates for itself and to fall short of a few of those dictated by its opponent.

The Structure of Scientific Revolutions (1962), 108–9.

9 Though the world does not change with a change of paradigm, the scientist afterward works in a different world . . . I am convinced that we must learn to make sense of statements that at least resemble these. What occurs during a scientific revolution is not fully reducible to a re-interpretation of individual and stable data. In the first place, the data are not unequivocally stable.

The Structure of Scientific Revolutions (1962), 120.

1 What chemists took from Dalton was not new experimental laws but a new way of practicing chemistry (he himself called it the 'new system of chemical philosophy'), and this proved so rapidly fruitful that only a few of the older chemists in France and Britain were able to resist it.

The Structure of Scientific Revolutions (1962), 133.

2 We may . . . have to relinquish the notion, explicit or implicit, that changes of paradigm carry scientists and those who learn from them closer and closer to the truth . . . The developmental process described in this essay has been a process of evolution *from* primitive beginnings—a process whose successive stages are characterized by an increasingly detailed and refined understanding of nature. But nothing that has been or will be said makes it a process of evolution *toward* anything.

The Structure of Scientific Revolutions (1962), 169–70.

3 The resolution of revolutions is selection by conflict within the scientific community of the fittest way to practice future science. The net result of a sequence of such revolutionary selections, separated by periods of normal research, is the wonderfully adapted set of instruments we call modern scientific knowledge.

The Structure of Scientific Revolutions (1962), 171.

4 Later scientific theories are better than earlier ones for solving puzzles in the often quite different environments to which they are applied. That is not a relativist's position, and it displays the sense in which I am a convinced believer in scientific progress.

The Structure of Scientific Revolutions, 2nd edition (1970), 206.

5 To turn Karl [Popper]'s view on its head, it is precisely the abandonment of critical discourse that marks the transition of science. Once a field has made the transition, critical discourse recurs only at moments of crisis when the bases of the field are again in jeopardy. Only when they must choose between competing theories do scientists behave like philosophers.

'Logic of Discovery or Psychology of Research', in I. Lakatos and A. Musgrave (eds.), *Criticism and the Growth of Knowledge* (1970), 6–7.

6 Scientific development depends in part on a process of non-incremental or revolutionary change. Some revolutions are large, like those associated with the names of Copernicus, Newton, or Darwin, but most are much smaller, like the discovery of oxygen or the planet Uranus. The usual prelude to changes of this sort is, I believed, the awareness of anomaly, of an occurrence or set of occurrences that does not fit existing ways of ordering phenomena. The changes that result therefore require 'putting on a different kind of thinking-cap', one that renders the anomalous lawlike but that, in the process, also transforms the order exhibited by some other phenomena, previously unproblematic.

The Essential Tension (1977), xvii.

7 Concerned to reconstruct past ideas, historians must approach the generation that held them as the anthropologist approaches an alien culture. They must, that is, be prepared at the start to find that natives speak a different language and map experience into different categories from those they themselves bring from home. And they must take as their object the discovery of those categories and the assimilation of the corresponding language.

'Revisiting Planck', *Historical Studies in the Physical Sciences*, 1984, 14, 246.

8 Groups do not have experiences except insofar as all their members do. And there are no experiences . . . that all the members of a scientific community must share in the course of a [scientific] revolution. Revolutions should be described not in terms of group experience but in terms of the varied experiences of individual group members. Indeed, that variety itself turns out to play an essential role in the evolution of scientific knowledge.

Thomas S. Kuhn's Foreword to Paul Hoyningen-Huene, *Reconstructing Scientific Revolutions: Thomas S Kuhn's Philosophy of Science* (1993), xiii.

Frank Kyte 1949–

American geologist and geochemist

1 You will never convince some palaeontologists that an impact killed the dinosaurs unless you find a dinosaur skeleton with a crushed skull and a ring of iridium round the hole.

Quoted in 'Extinction Wars' by Stefi Weisburd, *Science News*, 1 February 1986, 77.

Jacques Marie Émile Lacan 1901–81

French psychoanalyst

2 *C'est à vous d'être lacaniens, si vous voulez. Moi, je suis freudien.*

You can be Lacanians, if you want. As for me, I'm a Freudian.

'Le séminaire de Caracas' 12 July 1980, 30–1. Transcription printed in *L'Ane*, 1981. Trans. O. Zentner (ed.), *Papers of the Freudian School of Melbourne* (1981).

David Lambert Lack 1910–73

British ornithologist

3 Rumour has it that the gardens of natural history museums are used for surreptitious burial of those intermediate forms between species which might disturb the orderly classifications of the taxonomist.

Darwin's Finches (1947), 23.

4 On consideration, it is not surprising that Darwin's finches should recognize their own kind primarily by beak characters. The beak is the only prominent specific distinction, and it features conspicuously both in attacking behaviour, when the birds face each other and grip beaks, and also in courtship, when food is passed from the beak of the male to the beak of the female. Hence though the beak differences are primarily correlated with differences in food, secondarily they serve as specific recognition marks, and the birds have evolved behaviour patterns to this end.

Darwin's Finches (1947), 54.

5 The fundamental problem in the origin of species is not the origin of differences in appearance, since these arise at the level of the geographical race, but the origin of genetic segregation. The test of species-formation is whether, when two forms meet, they interbreed and merge, or whether they keep distinct.

Darwin's Finches (1947), 129.

Théophile-Réne-Hyacinthe Laennec 1781–1826

French physician

6 ON LAENNEC Laennec—this Herschel of the human thorax.

Carl Pfeufer in his Zurich speech of 1840, 'Ueber den gegenwärtigen Zustand der Medizin', 402. Trans. Arleen M. Tuchman, 'From the Lecture to the Laboratory: the Institutionalization of Scientific Medicine at the University of Heidelberg', in William Coleman and Frederic L. Holmes (eds.), *The Investigative Enterprise: Experimental Physiology in Nineteenth Century Medicine* (1988), 90.

Joseph Louis Lagrange 1736–1813

Italian-French astronomer and mathematician

7 I regarded as quite useless the reading of large treatises of pure analysis: too large a number of methods pass at once before the eyes. It is in the works of application that one must study them; one judges their utility there and appraises the manner of making use of them.

As reported by J. F. Maurice in *Moniteur Universel* (1814), 228.

1 ON LAGRANGE Then one day Lagrange took out of his pocket a paper which he read at the *Académe*, and which contained a demonstration of the famous *Postulatum* of Euclid, relative to the theory of parallels. This demonstration rested on an obvious paralogism, which appeared as such to everybody; and probably Lagrange also recognised it such during his lecture. For, when he had finished, he put the paper back in his pocket, and spoke no more of it. A moment of universal silence followed, and one passed immediately to other concerns.

Lagrange at a meeting of the class of mathematical and physical sciences at the Institut de France, 3 February 1806. Quoted in J. B. Biot, *Journal des Savants* (1837), 84, trans. Ivor Grattan-Guinness.

Ronald David Laing 1927–89

British psychiatrist

2 The experience and behaviour that gets labelled schizophrenic is a special strategy that a person invents in order to live in an unlivable situation.

Politics of Experience (1967), 95.

3 Madness need not be all breakdown. It may also be break-through. It is potential liberation and renewal as well as enslavement and existential death.

Politics of Experience (1967), 110.

Imre Lakatos 1922–74

Hungarian-born British philosopher

4 No experimental result can ever kill a theory: any theory can be saved from counterinstances either by some auxiliary hypothesis or by a suitable reinterpretation of its terms.

'Falsification and the Methodology of Scientific Research Programmes', in I. Lakatos and A. Musgrave (eds.), *Criticism and the Growth of Knowledge: Proceedings of the International Colloquium in the Philosophy of Science, London 1965* (1970), Vol. 4, 116.

5 It is not that we propose a theory and Nature may shout NO; rather, we propose a maze of theories, and Nature may shout INCONSISTENT.

'Falsification and the Methodology of Scientific Research Programmes', in I. Lakatos and A. Musgrave (eds.), *Criticism and the Growth of Knowledge: Proceedings of the International Colloquium in the Philosophy of Science, London 1965* (1970), Vol. 4, 130.

6 Creative imagination is likely to find corroborating novel evidence even for the most 'absurd' programme, if the search has sufficient drive. This look-out for *new confirming evidence* is perfectly permissible. Scientists dream up phantasies and then pursue a highly selective hunt for new facts which fit these phantasies. This process may be described as 'science creating its own universe' (as long as one remembers that 'creating' here is used in a provocative-idiosyncratic sense). A brilliant school of scholars (backed by a rich society to finance a few well-planned tests) might succeed in pushing any fantastic programme ahead, or alternatively, if so inclined, in overthrowing any arbitrarily chosen pillar of 'established knowledge'.

'Falsification and the Methodology of Scientific Research Programmes', in I. Lakatos and A. Musgrave (eds.), *Criticism and the Growth of Knowledge: Proceedings of the International Colloquium in the Philosophy of Science, London 1965* (1970), Vol. 4, 187–8.

Jean Baptiste Pierre Antoine de Monet de Lamarck 1744–1829

French naturalist

7 It is not always the magnitude of the differences observed between species that must determine specific distinctions, but the constant preservation of those differences in reproduction.

'Espèce', *Encyclopédie Méthodique Botanique* (1773–1789), Vol. 2, 396. In Pietro Corsi, *The Age of Lamarck: Evolutionary Theories in France 1790–1830*, trans. J. Mandelbaum, (1988), 43.

8 One must believe that every living thing whatsoever must change insensibly in its organization and in its form . . . One must therefore never expect to find among living species all those which are found in the fossil state, and yet one may not assume that any species has really been lost or rendered extinct.

Système des Animaux sans Vertèbres, (1801) trans. D. R. Newth, in *Annals of Science*, 1952, 8, 253–4.

1 A sound *Physics of the Earth* should include all the primary considerations of the earth's atmosphere, of the characteristics and continual changes of the earth's external crust, and finally of the origin and development of living organisms. These considerations naturally divide the physics of the earth into three essential parts, the first being a theory of the atmosphere, or *Meteorology*, the second, a theory of the earth's external crust, or *Hydrogeology*, and the third, a theory of living organisms, or *Biology*.

Hydrogéologie (1802), trans. A. V. Carozzi (1964), 18.

2 On our planet, all objects are subject to continual and inevitable changes which arise from the essential order of things. These changes take place at a variable rate according to the nature, condition, or situation of the objects involved, but are nevertheless accomplished within a certain period of time. Time is insignificant and never a difficulty for Nature. It is always at her disposal and represents an unlimited power with which she accomplishes her greatest and smallest tasks.

Hydrogéologie (1802), trans. A. V. Carozzi (1964), 61.

3 After having produced aquatic animals of all ranks and having caused extensive variations in them by the different environments provided by the waters, nature led them little by little to the habit of living in the air, first by the water's edge and afterwards on all the dry parts of the globe. These animals have in course of time been profoundly altered by such novel conditions; which so greatly influenced their habits and organs that the regular gradation which they should have exhibited in complexity of organisation is often scarcely recognisable.

Hydrogéologie (1802), trans. A. V. Carozzi (1964), 69–70.

4 The great age of the earth will appear greater to man when he understands the origin of living organisms and the reasons for the gradual development and improvement of their organization. This antiquity will appear even greater when he realizes the length of time and the particular conditions which were necessary to bring all the living species into existence. This is particularly true since man is the latest result and present *climax* of this development, the ultimate limit of which, if it is ever reached, cannot be known.

Hydrogéologie (1802), trans. A. V. Carozzi (1964), 77.

5 What nature does in the course of long periods we do every day when we suddenly change the environment in which some species of living plant is situated.

Philosophie Zoologique (1809), Vol. 1, 226, trans. Hugh Elliot (1914), 109.

6 First Law

In every animal which has not passed the limit of its development, a more frequent and continuous use of any organ gradually strengthens, develops and enlarges that organ, and gives it a power proportional to the length of time it has been so used; while the permanent disuse of any organ imperceptibly weakens and deteriorates it, and progressively diminishes its functional capacity, until it finally disappears.

Philosophie Zoologique (1809), Vol. 1, 235, trans. Hugh Elliot (1914), 113.

7 Second Law

All the acquisitions or losses wrought by nature on individuals, through the influence of the environment in which their race has long been placed, and hence through the influence of the predominant use or permanent disuse of any organ; all these are preserved by reproduction to the new individuals which arise, provided that the acquired modifications are common to both sexes, or at least to the individuals which produce the young.

Philosophie Zoologique (1809), Vol. 1, 235, trans. Hugh Elliot (1914), 113.

1 The bird which is drawn to the water by its need of finding there the prey on which it lives, separates the digits of its feet in trying to strike the water and move about on the surface. The skin which unites these digits at their base acquires the habit of being stretched by these continually repeated separations of the digits; thus in course of time there are formed large webs which unite the digits of ducks, geese, etc., as we actually find them. In the same way efforts to swim, that is to push against the water so as to move about in it, have stretched the membranes between the digits of frogs, sea-tortoises, the otter, beaver, etc.

On the other hand, a bird which is accustomed to perch on trees and which springs from individuals all of whom had acquired this habit, necessarily has longer digits on its feet and differently shaped from those of the aquatic animals that I have just named. Its claws in time become lengthened, sharpened and curved into hooks, to clasp the branches on which the animal so often rests.

We find in the same way that the bird of the water-side which does not like swimming and yet is in need of going to the water's edge to secure its prey, is continually liable to sink into the mud. Now this bird tries to act in such a way that its body should not be immersed in the liquid, and hence makes its best efforts to stretch and lengthen its legs. The long-established habit acquired by this bird and all its race of continually stretching and lengthening its legs, results in the individuals of this race becoming raised as though on stilts, and gradually obtaining long, bare legs, denuded of feathers up to the thighs and often higher still.

Philosophie Zoologique (1809), Vol. 1, 249–50, trans. Hugh Elliot (1914), 119–20.

2 It is interesting to observe the result of habit in the peculiar shape and size of the giraffe (*Camelo-pardalis*): this animal, the largest of the mammals, is known to live in the interior of Africa in places where the soil is nearly always arid and barren, so that it is obliged to browse on the leaves on the trees and to make constant efforts to reach them. From this habit long maintained in all its race, it has resulted that the animal's fore-legs have become longer than its hind legs, and that its neck is lengthened to such a degree that the giraffe, without standing up on its hind legs, attains a height of six metres (nearly 20 feet).

Philosophie Zoologique (1809), Vol. 1, 256, trans. Hugh Elliot (1914), 122.

3 If some race of quadrumanous animals, especially one of the most perfect of them, were to lose, by force of circumstances or some other cause, the habit of climbing trees and grasping the branches with its feet in the same way as with its hands, in order to hold on to them; and if the individuals of this race were forced for a series of generations to use their feet only for walking, and to give up using their hands like feet; there is no doubt, according to the observations detailed in the preceding chapter, that these quadrumanous animals would at length be transformed into bimanous, and that the thumbs on their feet would cease to be separated from the other digits, when they only used their feet for walking.

Philosophie Zoologique (1809), Vol. 1, 349, trans. Hugh Elliot (1914), 170.

4 It is not enough to discover and prove a useful truth previously unknown, but that it is necessary also to be able to propagate it and get it recognized.

Philosophie Zoologique (1809), Vol. 2, 450, trans. Hugh Elliot (1914), 404

1 The plan followed by nature in producing animals clearly comprises a predominant prime cause. This endows animal life with the power to make organization gradually more complex, and to bring increasing complexity and perfection not only to the total organization but also to each individual apparatus when it comes to be established by animal life. This progressive complication of organisms was in effect accomplished by the said principal cause in all existing animals. Occasionally a foreign, accidental, and therefore variable cause has interfered with the execution of the plan, without, however, destroying it. This has created gaps in the series, in the form either of terminal branches that depart from the series in several points and alter its simplicity, or of anomalies observable in specific apparatuses of various organisms.

Histoire Naturelle des Animaux sans Vertèbres (1815–22), Vol. 1, 133. In Pietro Corsi *The Age of Lamarck: Evolutionary Theories in France 1790–1830*, trans. J. Mandelbaum (1988), 189.

2 Now this circumscribed power, which we have scarcely examined, scarcely studied, this power to whose actions we nearly always attribute an intention and a goal, this power, finally, that always does necessarily the same things in the same circumstances and nevertheless does so many and such admirable ones, is what we call 'nature'.

Histoire Naturelle des Animaux sans Vertèbres (1815), Vol. 1, 312, trans. M. H. Shank and quoted in Madeleine Barthélemy-Madaule, *Lamarck the Mythical Precursor: A Study of the Relations between Science and Ideology* (1982), 28–9.

3 Life, in a body whose order and state of affairs can make it manifest, is assuredly, as I have said, a real power that gives rise to numerous phenomena. This power has, however, neither goal nor intention. It can do only what it does; it is only a set of acting causes, not a particular being. I was the first to establish this truth at a time when *life* was still thought to be a *principle*, an *archeia*, a *being* of some sort.

'Système Analytique des Connaissances Positives de l'Homme, restreintes a celles qui proviennent directement ou indirectement de l'observation' (1820), trans. M. H. Shank and quoted in Madeleine Barthélemy-Madaule, *Lamarck the Mythical Precursor: A Study of the Relations between Science and Ideology* (1982), 102.

4 It is not the organs—that is, the character and form of the animal's bodily parts—that have given rise to its habits and particular structures. It is the habits and manner of life and the conditions in which its ancestors lived that have in the course of time fashioned its bodily form, its organs and qualities.

Attributed.

5 All known living bodies are sharply divided into two special kingdoms, based upon the essential differences which distinguish animals from plants, and in spite of what has been said, I am convinced that these two kingdoms do not really merge into one another at any point.

Attributed.

Charles Lamb 1775–1834

British essayist and critic

6 Science has succeeded to poetry, no less in the little walks of children than with men. Is there no possibility of averting this sore evil?

Letter to Coleridge, 23 October 1802. In Edwin W. Marrs, Jr. (ed.), *The Letters of Charles and Mary Anne Lamb* (1976), Vol. 2, 81–2.

7 Nothing puzzles me more than time and space; and yet nothing puzzles me less, for I never think about them.

Letter to Thomas Manning, 2 January 1810. In Edwin W. Marrs, Jr. (ed.), *The Letters of Charles and Mary Anne Lamb* (1978), Vol. 3, 36.

8 In every thing that relates to *science*, I am a whole Encyclopaedia behind the rest of the world.

'The Old and the New Schoolmaster', in *Elia* (1823), 111.

1 Alas! can we ring the bells backward? Can we unlearn the arts that pretend to civilize, and then burn the world? There is a march of science; but who shall beat the drums for its retreat?

Letter to George Dyer, 20 December 1830. In Alfred Ainger (ed.), *The Letters of Charles Lamb* (1904), Vol. 2, 277.

William Lamb (Lord Melbourne)

1779–1848

British Prime Minister

2 I look upon the whole system of giving pensions to literary and scientific people as a piece of gross humbug. It is not done for any good purpose; it ought never to have been done. It is gross humbug from beginning to end.

Words attributed to Melbourne in *Fraser's Magazine*, 1835, 12, 707.

Julien Offray de La Mettrie

1709–51

French physician and philosopher

3 We can see that there is only one substance in the universe and that man is the most perfect one. He is to the ape and the cleverest animals what Huygens's planetary clock is to one of Julien Leroy's watches. If it took more instruments, more cogs, more springs to show or tell the time, if it took Vaucanson more artistry to make his flautist than his duck, he would have needed even more to make a speaking machine, which can no longer be considered impossible, particularly at the hands of a new Prometheus. Thus, in the same way, nature needed more artistry and machinery to construct and maintain a machine which could continue for a whole century to tell all the beats of the heart and the mind; for we cannot tell the time from the pulse, it is at least the barometer of heat and liveliness, from which we can judge the nature of the soul.

Machine Man (1747), in Ann Thomson (ed.), *Machine Man and Other Writings* (1996), 33–4.

4 Break the chains of your prejudices and take up the torch of experience, and you will honour nature in the way she deserves, instead of drawing derogatory conclusions from the ignorance in which she has left you. Simply open your eyes and ignore what you cannot understand, and you will see that a labourer whose mind and knowledge extend no further than the edges of his furrow is no different essentially from the greatest genius, as would have been proved by dissecting the brains of Descartes and Newton; you will be convinced that the imbecile or the idiot are animals in human form, in the same way as the clever ape is a little man in another form; and that, since everything depends absolutely on differences in organisation, a well-constructed animal who has learnt astronomy can predict an eclipse, as he can predict recovery or death when his genius and good eyesight have benefited from some time at the school of Hippocrates and at patients' bedsides.

Machine Man (1747), in Ann Thomson (ed.), *Machine Man and Other Writings* (1996), 38.

5 Let him who so wishes take pleasure in boring us with all the wonders of nature: let one spend his life observing insects, another counting the tiny bones in the hearing membrane of certain fish, even in measuring, if you will, how far a flea can jump, not to mention so many other wretched objects of study; for myself, who am curious only about philosophy, who am sorry only not to be able to extend its horizons, active nature will always be my sole point of view; I love to see it from afar, in its breadth and its entirety, and not in specifics or in little details, which, although to some extent necessary in all the sciences, are generally the mark of little genius among those who devote themselves to them.

'L'Homme Plante', in *Oeuvres Philosophiques de La Mettrie* (1796), Vol. 2, 70–1. Jacques Roger, *The Life Sciences in Eighteenth-Century French Thought*, edited by Keith R. Benson and trans. Robert Ellrich (1997), 377.

Karl Gottfried Lamprecht

1856–1915

German historian and psychologist

1 History is primarily a socio-psychological science. In the conflict between the old and the new tendencies in historical investigation . . . we are at the turn of the stream, the parting of the ways in historical science.

Historical Development and Present Character of the Science of History (1906), 111.

Peter Theodore Landsberg

1922–

British physicist

2 It has been suggested that thermodynamic irreversibility is due to cosmological expansion.

'Thermodynamics, Cosmology, and the Physical Constants', in J. T. Fraser (ed.), *The Study of Time III* (1973), 117–8.

3 In a sense cosmology contains all subjects because it is the story of everything, including biology, psychology and human history. In that single sense it can be said to contain an explanation also of time's arrow. But this is not what is meant by those who advocate the cosmological explanation of irreversibility. They imply that in some way the time arrow of cosmology imposes its sense on the thermodynamic arrow. I wish to disagree with this view. The explanation assumes that the universe is expanding. While this is current orthodoxy, there is no certainty about it. The red-shifts might be due to quite different causes. For example, when light passes through the expanding clouds of gas it will be red-shifted. A large number of such clouds might one day be invoked to explain these red shifts. It seems an odd procedure to attempt to 'explain' everyday occurrences, such as the diffusion of milk into coffee, by means of theories of the universe which are themselves less firmly established than the phenomena to be explained. Most people believe in explaining one set of things in terms of others about which they are more certain, and the explanation of normal irreversible phenomena in terms of the cosmological expansion is not in this category.

'Thermodynamics, Cosmology, and the Physical Constants', in J. T. Fraser (ed.), *The Study of Time III* (1973), 117–8.

4 Do these models give a pointer to God? The steady-state universe, the Hawking model . . . and the infinitely oscillating model decidedly do not. One might almost regard them as models manufactured for a Society of Atheists.

'From Entropy to God', in K. Martinás, L. Ropolyi and P. Szegedi (eds.) *Thermodynamics: History and Philosophy: Facts, Trends, Debates* (1991), 386.

Karl Landsteiner 1868–1943

Austrian-American immunologist and pathologist

5 I have recently observed and stated that the serum of normal people is capable of clumping the red cells of other healthy individuals . . . As commonly expressed, it can be said that in these cases at least two different kinds of agglutinins exist, one kind in A, the other in B, both together in C. The cells are naturally insensitive to the agglutinins in their own serum.

'Ueber Agglutinationserscheinungen normalen menschlichen Blutes', *Wiener klinische Wochenschrift* (1901), **14**, 1132–1134. Trans. Pauline M. H. Mazumdar.

6 A single kind of red cell is supposed to have an enormous number of different substances on it, and in the same way there are substances in the serum to react with many different animal cells. In addition, the substances which match each kind of cell are different in each kind of serum. The number of hypothetical different substances postulated makes this conception so uneconomical that the question must be asked whether it is the only one possible. . . . We ourselves hold that another, simpler, explanation is possible.

Landsteiner and Adriano Sturli, 'Hämagglutinine normaler Sera', *Wiener klinische Wochenschrift* (1902), **15**, 38–40. Trans. Pauline M. H. Mazumdar.

1 There is no sharp boundary line separating the reactions of the immune bodies from chemical processes between crystalloids, just as in nature there exists every stage between crystalloid and colloid. The nearer the colloid particle approximates to the normal electrolyte, the nearer its compounds must obviously come to conforming to the law of simple stoichiometric proportions, and the compounds themselves to simple chemical compounds. At this point, it should be recalled that Arrhenius has shown that the quantitative relationship between toxin and antitoxin is very similar to that between acid and base.

Landsteiner and Nicholas von Jagic, 'Über Reaktionen anorganischer Kolloide und Immunkörper', *Münchener medizinischer Wochenschrift* (1904), **51**, 1185–1189. Trans. Pauline M. H. Mazumdar.

2 According to the older view, for every single effect of a serum, there was a separate substance, or at least a particular chemical group . . . A normal serum contained as many different haemagglutinins as it agglutinated different cells. The situation was undoubtedly made much simpler if, to use the Ehrlich terminology . . . the separate haptophore groups can combine with an extremely large number of receptors in stepwise differing quantities as a stain does with different animal tissues, though not always with the same intensity. A normal serum would therefore visibly affect such a large number of different blood cells . . . not because it contained countless special substances, but because of the colloids of the serum, and therefore of the agglutinins by reason of their chemical constitution and the electrochemical properties resulting from it. That this manner of representation is a considerable simplification is clear; it also opens the way to direct experimental testing by the methods of structural chemistry.

'Die Theorien der Antikörperbildung . . . ', *Wiener klinische Wochenschrift* (1909), **22**, 1623–1631. Trans. Pauline M. H. Mazumdar.

3 The morphological characteristics of plant and animal species form the chief subject of the descriptive natural sciences and are the criteria for their classification. But not until recently has it been recognized that in living organisms, as in the realm of crystals, chemical differences parallel the variation in structure.

The Specificity of Serological Reactions (1936), 3.

4 On the whole, at least in the author's experience, the preparation of species-specific antiserum fractions and the differentiation of closely related species with precipitin sera for serum proteins does not succeed so regularly as with agglutinins and lysins for blood cells. This may be due to the fact that in the evolutional scale the proteins undergo continuous variations whereas cell antigens are subject to sudden changes not linked by intermediary stages.

The Specificity of Serological Reactions (1936), 12–3.

5 The reactions follow a pattern, which is valid for the blood of all humans . . . Basically, in fact, there are four different types of human blood, the so-called blood groups. The number of the groups follows from the fact that the erythrocytes evidently contain substances (iso-agglutinogens) with two different structures, of which both may be absent, or one or both present, in the erythrocytes of a person. This alone would still not explain the reactions; the active substances of the sera, the iso-agglutinins, must also be present in a specific distribution. This is actually the case, since every serum contains those agglutinins which react with the agglutinogens not present in the cells—a remarkable phenomenon, the cause of which is not yet known for certain.

'On Individual Differences in Human Blood', Nobel Lecture, 11 December 1930. In *Nobel Lectures: Physiology or Medicine 1922–1941* (1965), 235.

Andrew Lang 1844–1912

British scholar and man of letters

1 He uses statistics as a drunken man uses lamp-posts—for support rather than illumination.

Attributed.

Susanne Knauth Langer 1895–1985

American philosopher

2 The faith of scientists in the power and truth of mathematics is so implicit that their work has gradually become less and less observation, and more and more calculation. The promiscuous collection and tabulation of data have given way to a process of assigning possible meanings, merely supposed real entities, to mathematical terms, working out the logical results, and then staging certain crucial experiments to check the hypothesis against the actual empirical results. But the facts which are accepted by virtue of these tests are not actually *observed* at all. With the advance of mathematical technique in physics, the tangible results of experiment have become less and less spectacular; on the other hand, their *significance* has grown in inverse proportion. The men in the laboratory have departed so far from the old forms of experimentation—typified by Galileo's weights and Franklin's kite—that they cannot be said to observe the actual objects of their curiosity at all; instead, they are watching index needles, revolving drums, and sensitive plates. No psychology of 'association' of sense-experiences can relate these data to the objects they signify, for in most cases the objects have never been experienced. Observation has become almost entirely indirect; and *readings* take the place of genuine witness.

Philosophy in a New Key; A Study in the Symbolism of Reason, Rite, and Art (1942), 19–20.

John Newport Langley 1852–1925

British physiologist

3 I propose to substitute the word 'autonomic'. The word implies a certain degree of independent action, but exercised under control of a higher power. The 'autonomic' nervous system means the nervous system of the glands and of the involuntary muscle; it governs the 'organic' functions of the body.

'On the Union of Cranial Autonomic (Visceral) Fibres with the Nerve Cells of the Superior Cervical Ganglion', *The Journal of Physiology*, 1898–99, 23, 241.

4 Those who have occasion to enter into the depths of what is oddly, if generously, called the literature of a scientific subject, alone know the difficulty of emerging with an unsoured disposition. The multitudinous facts presented by each corner of Nature form in large part the scientific man's burden to-day, and restrict him more and more, willy-nilly, to a narrower and narrower specialism. But that is not the whole of his burden. Much that he is forced to read consists of records of defective experiments, confused statement of results, wearisome description of detail, and unnecessarily protracted discussion of unnecessary hypotheses. The publication of such matter is a serious injury to the man of science; it absorbs the scanty funds of his libraries, and steals away his poor hours of leisure.

'Physiology, including Experimental Pathology and Experimental Physiology', *Reports of the British Association for the Advancement of Science*, 1899, 891–2.

Irving Langmuir 1881–1957

American chemist and physicist

1 A chemist who does not know mathematics is seriously handicapped.

Quoted in Albert Rosenfeld, *Langmuir: The Man and the Scientist* (1962), 293.

2 ON LANGMUIR Langmuir is the most convincing lecturer that I have ever heard. I have heard him talk to an audience of chemists when I knew they did not understand more than one-third of what he was saying; but they thought they did. Its very easy to be swept off one's feet by Langmuir. You remember in [Kipling's novel] *Kim* that the water jar was broken and Lurgan Sahib was trying to hypnotise Kim into seeing it whole again. Kim saved himself by saying the multiplication table [so] I have heard Langmuir lecture when I knew he was wrong, but I had to repeat to myself: 'He is wrong; I know he is wrong; he is wrong', or I should have believed like the others.

Wilder D. Bancroft, 'How to Ripen Time', *Journal of Physical Chemistry*, 1931, 35, 1917.

3 ON LANGMUIR Langmuir is a regular thinking machine. Put in facts, and you get out a theory.

Comment by Saul Dushman. In C. Guy Suits (ed.), *The Collected Works of Irving Langmuir* (1962), Vol. 12, 6.

Edwin Ray Lankester 1847–1929

British zoologist

4 The chemical differences among various species and genera of animals and plants are certainly as significant for the history of their origins as the differences in form. If we could define clearly the differences in molecular constitution and functions of different kinds of organisms, there would be possible a more illuminating and deeper understanding of question of the evolutionary reactions of organisms than could ever be expected from morphological considerations.

'Uber das Vorkommen von Haemoglobin in den Muskeln der Mollusken und die Verbreitung desselben in den lebenden Organismen', *Pflügers Archiv für die gesamte Physiologie des Menschen und der Tiere*, 1871, 4, 318–9. Trans. Joseph S. Fruton, *Proteins, Enzymes, Genes: The Interplay of Chemistry and Biology* (1999), 270.

5 Through it [Science] we believe that man will be saved from misery and degradation, not merely acquiring new material powers, but learning to use and to guide his life with understanding. Through Science he will be freed from the fetters of superstition; through faith in Science he will acquire a new and enduring delight in the exercise of his capacities; he will gain a zest and interest in life such as the present phase of culture fails to supply.

'Biology and the State', *The Advancement of Science: Occasional Essays & Addresses* (1890), 108–9.

6 It appears, nevertheless, that all such simple solutions of the problem of vertebrate ancestry are without warrant. They arise from a very common tendency of the mind, against which the naturalist has to guard himself,—a tendency which finds expression in the very widespread notion that the existing anthropoid apes, and more especially the gorilla, must be looked upon as the ancestors of mankind, if once the doctrine of the descent of man from ape-like forefathers is admitted. A little reflexion suffices to show that any given living form, such as the gorilla, cannot possibly be the ancestral form from which man was derived, since *ex-hypothesi* that ancestral form underwent modification and development, and in so doing, ceased to exist.

'Vertebrata', entry in *Encyclopaedia Britannica*, 9th edition (1899), Vol. 24, 180.

Isaac de la Peyrère 1594–1676

French scholar and writer

7 But as Geographers use to place Seas upon that place of the Globe which they know not: so chronologers, who are near of kin to them, use to blot out ages past, which they know not. They drown those Countries which they know not: These with cruel pen kill the times they heard not of, and deny which they know not.

Prae-Adamitae (1655), trans. *Men Before Adam* (1656), 164, published anonymously.

Pierre Simon, Marquis de Laplace 1749–1827

French mathematician and physicist

See **Bonaparte, Napoléon** *74:4;* **Bowditch, Nathaniel** *80:1*

1 If an event can be produced by a number *n* of different causes, the probabilities of the existence of these causes, given the event (*prises de l'événement*), are to each other as the probabilities of the event, given the causes: and the probability of each cause is equal to the probability of the event, given that cause, divided by the sum of all the probabilities of the event, given each of the causes.

'Mémoire sur la Probabilité des Causes par les Événements' (1774). In *Oeuvres complètes de Laplace*, 14 Vols. (1843–1912), Vol. 8, 29, trans. Charles Coulston Gillispie, *Pierre-Simon Laplace 1749–1827: A Life in Exact Science* (1997), 16.

2 The present state of the system of nature is evidently a consequence of what it was in the preceding moment, and if we conceive of an intelligence that at a given instant comprehends all the relations of the entities of this universe, it could state the respective position, motions, and general affects of all these entities at any time in the past or future. Physical astronomy, the branch of knowledge that does the greatest honor to the human mind, gives us an idea, albeit imperfect, of what such an intelligence would be. The simplicity of the law by which the celestial bodies move, and the relations of their masses and distances, permit analysis to follow their motions up to a certain point; and in order to determine the state of the system of these great bodies in past or future centuries, it suffices for the mathematician that their position and their velocity be given by observation for any moment in time. Man owes that advantage to the power of the instrument he employs, and to the small number of relations that it embraces in its calculations. But ignorance of the different causes involved in the production of events, as well as their complexity, taken together with the imperfection of analysis, prevents our reaching the same certainty about the vast majority of phenomena. Thus there are things that are uncertain for us, things more or less probable, and we seek to compensate for the impossibility of knowing them by determining their different degrees of likelihood. So it was that we owe to the weakness of the human mind one of the most delicate and ingenious of mathematical theories, the science of chance or probability.

'Recherches, 1°, sur l'Intégration des Équations Différentielles aux Différences Finies, et sur leur Usage dans la Théorie des Hasards' (1773, published 1776). In *Oeuvres complètes de Laplace*, 14 Vols. (1843–1912), Vol. 8, 144–5, trans. Charles Coulston Gillispie, *Pierre-Simon Laplace 1749–1827: A Life in Exact Science* (1997), 26.

3 The word 'chance' then expresses only our ignorance of the causes of the phenomena that we observe to occur and to succeed one another in no apparent order. Probability is relative in part to this ignorance, and in part to our knowledge.

'Mémoire sur les Approximations des Formules qui sont Fonctions de Très Grands Nombres' (1783, published 1786). In *Oeuvres complète de Laplace*, 14 Vols. (1843–1912), Vol. 10, 296, trans. Charles Coulston Gillispie, *Pierre-Simon Laplace 1749–1827: A Life in Exact Science* (1997), 91.

4 [It] may be laid down as a general rule that, if the result of a long series of precise observations approximates a simple relation so closely that the remaining difference is undetectable by observation and may be attributed to the errors to which they are liable, then this relation is probably that of nature.

'Mémoire sur les Inégalités Séculaires des Planètes et des Satellites' (1785, published 1787). In *Oeuvres complètes de Laplace*, 14 Vols. (1843–1912), Vol. 11, 57, trans. Charles Coulston Gillispie, *Pierre-Simon Laplace 1749–1827: A Life in Exact Science* (1997), 130.

1 Thus the system of the world only oscillates around a mean state from which it never departs except by a very small quantity. By virtue of its constitution and the law of gravity, it enjoys a stability that can be destroyed only by foreign causes, and we are certain that their action is undetectable from the time of the most ancient observations until our own day. This stability in the system of the world, which assures its duration, is one of the most notable among all phenomena, in that it exhibits in the heavens the same intention to maintain order in the universe that nature has so admirably observed on earth for the sake of preserving individuals and perpetuating species.

'Sur l'Équation Séculaire de la Lune' (1786, published 1788). In *Oeuvres complètes de Laplace*, 14 Vols. (1843–1912), Vol. 11, 248–9, trans. Charles Coulston Gillispie, *Pierre-Simon Laplace 1749–1827: A Life in Exact Science* (1997), 145.

2 [Science] dissipates errors born of ignorance about our true relations with nature, errors the more damaging in that the social order should rest only on those relations. TRUTH! JUSTICE! Those are the immutable laws. Let us banish the dangerous maxim that it is sometimes useful to depart from them and to deceive or enslave mankind to assure its happiness.

Exposition du Système du Monde (1796), 2, 312, trans. Charles Coulston Gillispie, *Pierre-Simon Laplace 1749–1827: A Life in Exact Science* (1997), 175.

3 However, the small probability of a similar encounter [of the earth with a comet], can become very great in adding up over a huge sequence of centuries. It is easy to picture to oneself the effects of this impact upon the Earth. The axis and the motion of rotation changed; the seas abandoning their old position to throw themselves toward the new equator; a large part of men and animals drowned in this universal deluge, or destroyed by the violent tremor imparted to the terrestrial globe.

Exposition du Système du Monde, 2nd edition (1799), 208, trans. Ivor Grattan-Guinness.

4 It is remarkable that a science which began with the consideration of games of chance, should have become the most important object of human knowledge.

Théorie Analytique des Probabilités (1812). Quoted in A. S. Eddington, *New Pathways in Science* (1935), 110.

5 Here I shall present, without using Analysis [mathematics], the principles and general results of the *Théorie*, applying them to the most important questions of life, which are indeed, for the most part, only problems in probability. One may even say, strictly speaking, that almost all our knowledge is only probable; and in the small number of things that we are able to know with certainty, in the mathematical sciences themselves, the principal means of arriving at the truth—induction and analogy—are based on probabilities, so that the whole system of human knowledge is tied up with the theory set out in this essay.

Philosophical Essay on Probabilities (1814), 5th edition (1825), trans. Andrew I. Dale (1995), 1.

6 We ought then to consider the present state of the universe as the effect of its previous state and as the cause of that which is to follow. An intelligence that, at a given instant, could comprehend all the forces by which nature is animated and the respective situation of the beings that make it up, if moreover it were vast enough to submit these data to analysis, would encompass in the same formula the movements of the greatest bodies of the universe and those of the lightest atoms. For such an intelligence nothing would be uncertain, and the future, like the past, would be open to its eyes.

Philosophical Essay on Probabilities (1814), 5th edition (1825), trans. Andrew I. Dale (1995), 2.

1 Without any doubt, the regularity which astronomy shows us in the movements of the comets takes place in all phenomena. The trajectory of a simple molecule of air or vapour is regulated in a manner as certain as that of the planetary orbits; the only difference between them is that which is contributed by our ignorance. Probability is relative in part to this ignorance, and in part to our knowledge.

Philosophical Essay on Probabilities (1814), 5th edition (1825), trans. Andrew I. Dale (1995), 3.

2 The theory of probabilities is basically only common sense reduced to a calculus. It makes one estimate accurately what right-minded people feel by a sort of instinct, often without being able to give a reason for it.

Philosophical Essay on Probabilities (1814), 5th edition (1825), trans. Andrew I. Dale (1995), 124.

3 The simplicity of nature is not to be measured by that of our conceptions. Infinitely varied in its effects, nature is simple only in its causes, and its economy consists in producing a great number of phenomena, often very complicated, by means of a small number of general laws.

Philosophical Essay on Probabilities (1825), trans. Andrew I. Dale (1995), book 1, chap. 14.

4 I am particularly concerned to determine the probability of causes and results, as exhibited in events that occur in large numbers, and to investigate the laws according to which that probability approaches a limit in proportion to the repetition of events. That investigation deserves the attention of mathematicians because of the analysis required. It is primarily there that the approximation of formulas that are functions of large numbers has its most important applications. The investigation will benefit observers in identifying the mean to be chosen among the results of their observations and the probability of the errors still to be apprehended. Lastly, the investigation is one that deserves the attention of philosophers in showing how in the final analysis there is a regularity underlying the very things that seem to us to pertain entirely to chance, and in unveiling the hidden but constant causes on which that regularity depends. It is on the regularity of the main outcomes of events taken in large numbers that various institutions depend, such as annuities, tontines, and insurance policies. Questions about those subjects, as well as about inoculation with vaccine and decisions of electoral assemblies, present no further difficulty in the light of my theory. I limit myself here to resolving the most general of them, but the importance of these concerns in civil life, the moral considerations that complicate them, and the voluminous data that they presuppose require a separate work.

Philosophical Essay on Probabilities (1825), trans. Andrew I. Dale (1995), Introduction.

5 I see with much pleasure that you are working on a large work on the integral Calculus [. . .] The reconciliation of the methods which you are planning to make, serves to clarify them mutually, and what they have in common contains very often their true metaphysics; this is why that metaphysics is almost the last thing that one discovers. The spirit arrives at the results as if by instinct; it is only on reflecting upon the route that it and others have followed that it succeeds in generalising the methods and in discovering its metaphysics.

Letter to S. F. Lacroix, 1792. Quoted in S. F. Lacroix, *Traité du calcul differentiel et du calcul integral* (1797), Vol. 1, xxiv, trans. Ivor Grattan-Guinness.

1 NAPOLEON: You have written this huge book on the system of the world without once mentioning the author of the universe.
LAPLACE: Sire, I had no need of that hypothesis.
Later, when told by Napoleon about the incident, Lagrange commented: 'Ah, but that is a fine hypothesis. It explains so many things'.
Quoted in Augustus De Morgan, *A Budget of Paradoxes* (1872), Vol. 2, 2.

2 ON LAPLACE The genius of Laplace was a perfect sledge hammer in bursting purely mathematical obstacles; but, like that useful instrument, it gave neither finish nor beauty to the results. In truth, in truism if the reader please, Laplace was neither Lagrange nor Euler, as every student is made to feel. The second is power and symmetry, the third power and simplicity; the first is power without either symmetry or simplicity. But, nevertheless, Laplace never attempted investigation of a subject without leaving upon it the marks of difficulties conquered: sometimes clumsily, sometimes indirectly, always without minuteness of design or arrangement of detail; but still, his end is obtained and the difficulty is conquered.
Augustus de Morgan, 'Review of "Théorie Analytique des Probabilités" par M. le Marquis de Laplace, 3ème édition. Paris. 1820', *Dublin Review*, 1837, 2, 348.

Charles Lapworth 1842–1920

British geologist

3 Darwin was a biological evolutionist, because he was first a uniformitarian geologist. Biology is pre-eminent to-day among the natural sciences, because its younger sister, Geology, gave it the means.
Presidential Address to the Geology Section, *Report of the British Association for the Advancement of Science*, 1892, 696.

4 Uniformity and Evolution are one.
Presidential Address to the Geology Section, *Report of the British Association for the Advancement of Science*, 1892, 707.

5 The course of the line we indicated as forming our grandest terrestrial fold [along the shores of Japan] returns upon itself. It is an endless fold, an endless band, the common possession of two sciences. It is geological in origin, geographical in effect. It is the wedding ring of geology and geography, uniting them at once and for ever in indissoluble union.
Presidential Address to the Geology Section, *Report of the British Association for the Advancement of Science*, 1892, 705.

6 Nothing perhaps has so retarded the reception of the higher conclusions of Geology among men in general, as . . . [the] instinctive parsimony of the human mind in matters where time is concerned.
Proceedings of the Geological Society of London, 1903, **59**, lxx.

7 Far be it from me to suggest that geologists should be reckless in their drafts upon the bank of Time; but nothing whatever is gained, and very much is lost, by persistent niggardliness in this direction.
Proceedings of the Geological Society of London, 1903, **59**, lxxii.

8 All that comes above the surface [of the globe] lies within the province of Geography; all that comes below that surface lies inside the realm of Geology. The surface of the earth is that which, so to speak, divides them and at the same time 'binds them together in indissoluble union.' We may, perhaps, put the case metaphorically. The relationships of the two are rather like that of man and wife. Geography, like a prudent woman, has followed the sage advice of Shakespeare and taken unto her 'an elder than herself'; but she does not trespass on the domain of her consort, nor could she possibly maintain the respect of her children were she to flaunt before the world the assertion that she is 'a woman with a past.'

Proceedings of the Geological Society of London, 1903, **59**, lxxviii.

1 Astronomy concerns itself with the whole of the visible universe, of which our earth forms but a relatively insignificant part; while Geology deals with that earth regarded as an individual. Astronomy is the oldest of the sciences, while Geology is one of the newest. But the two sciences have this in common, that to both are granted a magnificence of outlook, and an immensity of grasp denied to all the rest.

Proceedings of the Geological Society of London, 1903, **59**, lxviii.

François de La Rochefoucauld 1613–80

French classical author

2 Death, like the sun, cannot be looked at steadily.

Maxims (1678), no. 26, trans. F. G. Stevens (1939), 11.

3 It is easier to understand mankind in general than any individual man.

Maxims (1678), no. 436, trans. F. G. Stevens (1939), 137.

Harold Dwight Lasswell 1902–78

American political scientist

4 Political science without biography is a form of taxidermy.

Psychopathology and Politics (1930), 1.

Peter Mere Latham 1789–1875

British physician

5 Unfortunately, where there is no experiment of exact science to settle the matter, it takes as much time and trouble to pull down a falsehood as to build up a truth.

Robert Martin (ed.), 'General Remarks on the Practice of Medicine', *The Collected Works of Dr. P. M. Latham* (1873), Vol. II, 398.

6 The practice of physic is jostled by quacks on the one side, and by science on the other.

William B. Bean (ed.), *Aphorisms from Latham* (1962), 22.

Max von Laue 1879–1960

German physicist

7 With crystals we are in a situation similar to an attempt to investigate an optical grating merely from the spectra it produces . . . But a knowledge of the positions and intensities of the spectra does not suffice for the determination of the structure. The phases with which the diffracted waves vibrate relative to one another enter in an essential way. To determine a crystal structure on the atomic scale, one must know the phase differences between the different interference spots on the photographic plate, and this task may certainly prove to be rather difficult.

Physikalische Zeitschrift, 1913, **14**. Translated in Walter Moore, *Schrödinger. Life and Thought* (1989), 73.

William L. Laurence 1888–1977

American journalist

8 The Atomic Age began at exactly 5.30 Mountain War Time on the morning of July 15, 1945, on a stretch of semi-desert land about 50 airline miles from Alamogordo, New Mexico. And just at that instance there rose from the bowels of the earth a light not of this world, the light of many suns in one . . . At first it was a giant column that soon took the shape of a supramundane mushroom.

On the first atomic explosion in New Mexico, 16 July 1945.

In *New York Times*, 26 September 1945.

Auguste Laurent 1807–53

French chemist

9 The chemists who uphold dualism are far from being agreed among themselves; nevertheless, all of them in maintaining their opinion, rely upon the phenomena of chemical reactions. For a long time the uncertainty of this method has been pointed out: it has been shown repeatedly, that the atoms put into movement during a reaction take at that time a new arrangement, and that it is impossible to deduce the old arrangement from the new one. It is as if, in the middle of a game of chess, after the disarrangement of all the pieces, one of the players should wish, from the inspection of the new place occupied by each piece, to determine that which it originally occupied.

Chemical Method (1855), 18.

1 But experiments went for nothing,—dualism had sworn to uphold its position.

Chemical Method (1855), 203.

2 I was an impostor, the worthy associate of a brigand, &c., &c., and all this for an atom of chlorine put in the place of an atom of hydrogen, for the simple correction of a chemical formula!

Chemical Method (1855), 203.

3 From this time everything was copulated. Acetic, formic, butyric, margaric, &c., acids, alkaloids, ethers, amides, anilides, all became copulated bodies. So that to make acetanilide, for example, they no longer employed acetic acid and aniline, but they re-copulated a copulated oxalic acid with a copulated ammonia. I am inventing nothing—altering nothing. Is it my fault if, when writing history, I appear to be composing a romance?

Chemical Method (1855), 204.

4 ON LAURENT Both died, ignored by most; they neither sought nor found public favour, for high roads never lead there. Laurent and Gerhardt never left such roads, were never tempted to peruse those easy successes which, for strongly marked characters, offer neither allure nor gain. Their passion was for the search for truth; and, preferring their independence to their advancement, their convictions to their interests, they placed their love for science above that of their worldly goods; indeed above that for life itself, for death was the reward for their pains. Rare example of abnegation, sublime poverty that deserves the name nobility, glorious death that France must not forget!

'Éloge de Laurent et Gerhardt', *Moniteur Scientifique*, 1862, 4, 473–83, trans. Alan J. Rocke.

Antoine-Laurent Lavoisier

1743–94

French chemist

5 Perhaps . . . some day the precision of the data will be brought so far that the mathematician will be able to calculate at his desk the outcome of any chemical combination, in the same way, so to speak, as he calculates the motions of celestial bodies.

Oeuvres (1862), Vol. 2, 550–1. Trans. John Heilbron, *Weighing Imponderables and Other Quantitative Science around 1800* (1993), 14.

6 About eight days ago I discovered that sulfur in burning, far from losing weight, on the contrary, gains it; it is the same with phosphorus; this increase of weight arises from a prodigious quantity of air that is fixed during combustion and combines with the vapors. This discovery, which I have established by experiments, that I regard as decisive, has led me to think that what is observed in the combustion of sulfur and phosphorus may well take place in the case of all substances that gain in weight by combustion and calcination; and I am persuaded that the increase in weight of metallic calxes is due to the same cause . . . This discovery seems to me one of the most interesting that has been made since Stahl and since it is difficult not to disclose something inadvertently in conversation with friends that could lead to the truth I have thought it necessary to make the present deposit to the Secretary of the

Academy to await the time I make my experiments public.

Sealed note deposited with the Secretary of the French Academy 1 November 1772. *Oeuvres de Lavoisier, Correspondance*, Fasc. II 1770–75 (1957), 389–90. Adapted from translation by A. N. Meldrum, *The Eighteenth-Century Revolution in Science* (1930), 3.

1 In every combustion there is disengagement of the matter of fire or of light. A body can burn only in pure air [oxygen]. There is no destruction or decomposition of pure air and the increase in weight of the body burnt is exactly equal to the weight of air destroyed or decomposed. The body burnt changes into an acid by addition of the substance that increases its weight. Pure air is a compound of the matter of fire or of light with a base. In combustion the burning body removes the base, which it attracts more strongly than does the matter of heat, which appears as flame, heat and light.

'Mémoire sur la combustion en général', *Mémoires de l'Académie des Sciences*, 1777, 592. Reprinted in *Oeuvres de Lavoisier* (1864), Vol. 2, 225–33, trans. M. P. Crosland.

2 If everything in chemistry is explained in a satisfactory manner without the help of phlogiston, it is by that reason alone infinitely probable that the principle does not exist; that it is a hypothetical body, a gratuitous supposition; indeed, it is in the principles of good logic, not to multiply bodies without necessity.

'Réflexions sur le phlogistique', *Mémoires de l'Académie des Sciences*, 1783, 505–38. Reprinted in *Oeuvres de Lavoisier* (1864), Vol. 2, 623, trans. M. P. Crosland.

3 Chemists have made of phlogiston a vague principle which is not at all rigorously defined, and which, in consequence, adapts itself to all explanations in which it is wished it shall enter; sometimes it is free fire, sometimes it is fire combined with the earthy element; sometimes it passes through the pores of vessels, sometimes they are impenetrable to it; it explains both the causticity and non-causticity, transparency and opacity, colours and absence of colours. It is a veritable Proteus which changes its form every instant. It is time to conduct chemistry to a more rigorous mode of reasoning . . . to distinguish fact and observation from what is systematic and hypothetical.

'Réflexions sur le phlogistique', *Mémoires de l'Académie des Sciences*, 1783, 505–38. Reprinted in *Oeuvres de Lavoisier* (1864), Vol. 2, 640, trans. M. P. Crosland.

4 The impossibility of separating the nomenclature of a science from the science itself, is owing to this, that every branch of physical science must consist of three things; the series of facts which are the objects of the science, the ideas which represent these facts, and the words by which these ideas are expressed. Like three impressions of the same seal, the word ought to produce the idea, and the idea to be a picture of the fact.

Elements of Chemistry (1790), trans. R. Kerr, Preface, xiv.

5 As ideas are preserved and communicated by means of words, it necessarily follows that we cannot improve the language of any science, without at the same time improving the science itself; neither can we, on the other hand, improve a science without improving the language or nomenclature which belongs to it.

Elements of Chemistry (1790), trans. R. Kerr, Preface, xiv-v.

6 Imagination, on the contrary, which is ever wandering beyond the bounds of truth, joined to self-love and that self-confidence we are so apt to indulge, prompt us to draw conclusions which are not immediately derived from facts.

Elements of Chemistry (1790), trans. R. Kerr, Preface, xvii.

7 We must trust to nothing but facts: These are presented to us by Nature, and cannot deceive. We ought, in every instance, to submit our reasoning to the test of experiment, and never to search for truth but by the natural road of experiment and observation.

Elements of Chemistry (1790), trans. R. Kerr, Preface, xviii.

1 All that can be said upon the number and nature of elements is, in my opinion, confined to discussions entirely of a metaphysical nature. The subject only furnishes us with indefinite problems, which may be solved in a thousand different ways, not one of which, in all probability, is consistent with nature. I shall therefore only add upon this subject, that if, by the term *elements*, we mean to express those simple and indivisible atoms of which matter is composed, it is extremely probable we know nothing at all about them; but, if we apply the term *elements*, or *principles of bodies*, to express our idea of the last point which analysis is capable of reaching, we must admit, as elements, all the substances into which we are capable, by any means, to reduce bodies by decomposition.

Elements of Chemistry (1790), trans. R. Kerr, Preface, xxiv.

2 Chemistry affords two general methods of determining the constituent principles of bodies, the method of analysis, and that of synthesis. When, for instance, by combining water with alkohol, we form the species of liquor called, in commercial language, brandy or spirit of wine, we certainly have a right to conclude, that brandy, or spirit of wine, is composed of alkohol combined with water. We can produce the same result by the analytical method; and in general it ought to be considered as a principle in chemical science, never to rest satisfied without both these species of proofs. We have this advantage in the analysis of atmospherical air, being able both to decompound it, and to form it a new in the most satisfactory manner.

Elements of Chemistry (1790), trans. R. Kerr, 33.

3 We may lay it down as an incontestible axiom, that, in all the operations of art and nature, nothing is created; an equal quantity of matter exists both before and after the experiment; the quality and quantity of the elements remain precisely the same; and nothing takes place beyond changes and modifications in the combination of these elements. Upon this principle the whole art of performing chemical experiments depends: We must always suppose an exact equality between the elements of the body examined and those of the products of its analysis.

Elements of Chemistry (1790), trans. R. Kerr, 130–1.

4 I have had a fairly long life, above all a very happy one, and I think that I shall be remembered with some regrets and perhaps leave some reputation behind me. What more could I ask? The events in which I am involved will probably save me from the troubles of old age. I shall die in full possession of my faculties, and that is another advantage that I should count among those that I have enjoyed. If I have any distressing thoughts, it is of not having done more for my family; to be unable to give either to them or to you any token of my affection and my gratitude is to be poor indeed.

Letter to Augez de Villiers, undated. Quoted in D. McKie, *Antoine Lavoisier: Scientist, Economist, Social Reformer* (1952), 303.

5 This theory [the oxygen theory] is not as I have heard it described, that of the French chemists, it is *mine* (elle est *la mienne*); it is a property which I claim from my contemporaries and from posterity.

Mémoires de Chimie (1805), Vol. 2, 87, trans. M. P. Crosland.

6 ON **LAVOISIER** To Monsieur Lavoisier by appointment. Madame Lavoisier, a lively, sensible, scientific lady, had prepared a *dejuné Anglois* of tea and coffee, but her conversation on Mr. Kirwan's Essay on Phlogiston, which she is translating from the English, and on other subjects, which a woman of understanding, that works with her husband in his laboratory, knows how to adorn, was the best repast.

Entry for 16 October 1787. In Arthur Young, *Travels in France During the Years, 1787, 1788 and 1789* (1792), 64.

1 ON **LAVOISIER** I have witnessed a most remarkable drama here, one which to me as a German was very unexpected, and quite shocking. I saw the famous M. Lavoisier hold a ceremonial auto-da-fe of phlogiston in the Arsenal. His wife . . . served as the sacrificial priestess, and Stahl appeared as the *advocatus diaboli* to defend phlogiston. In the end, poor phlogiston was burned on the accusation of oxygen. Do you not think I have made a droll discovery? Everything is literally true. I will not say whether the cause of phlogiston is now irretrievably lost, or what I think about the issue. But I am glad that this spectacle was not presented in my fatherland.

Lorenz Crell, in a letter to *Chemische Annalen*, 1789, 1, 519. Quoted (in English) in K. Hufbauer, *The Formation of the German Chemical Community* (1982), 96.

2 ON **LAVOISIER** *Le mur murant Paris rend Paris murmurant.*

The wall surrounding Paris is making Paris grumble.

Parisian saying after the Farmers-General of taxes, acting on a proposal by Lavoisier, erected a customs wall around Paris.

Quoted in D. McKie, *Antoine Lavoisier: Scientist, Economist, Social Reformer* (1952), 136.

3 ON **LAVOISIER** I denounce to you the Coryphaeus—the leader of the chorus—of charlatans, Sieur Lavoisier, son of a land-grabber, apprentice-chemist, pupil of the Genevan stock-jobber [Necker], a Farmer-General, Commisioner for Gunpowder and Saltpetre, Governor of the Discount Bank, Secretary to the King, Member of the Academy of Sciences.

Marat's denunciation of 1791.

Jean Paul Marat, *L'Ami du Peuple*, 27 January 1791. Trans. D. McKie, *Antoine Lavoisier, Scientist, Economist, Social Reformer* (1952), 242.

4 ON **LAVOISIER** The republic has no need of scientists [*savants*].

The judge's reply to Lavoisier when he appealed at his trial for more time to complete his scientific work.

Apocryphal remark. Cited in H. Guerlac, 'Lavoisier', in Charles Gillispie (ed.), *Dictionary of Scientific Biography* (1973), Vol. 8, 85.

5 ON **LAVOISIER** Only a moment to cut off that head and a hundred years may not give us another like it.

J. L. Lagrange to Delambre on Lavoisier's execution, 8 May 1794.

Quoted in D. McKie, *Antoine Lavoisier: Scientist, Economist, Social Reformer* (1962), 309.

William Lawrence 1783–1867

British surgeon

6 It is strongly suspected that a NEWTON or SHAKESPEARE excels other mortals only by a more ample development of the anterior cerebral lobes, by having an extra inch of brain in the right place.

Lectures on Physiology, Zoology, and the Natural History of Man (1819), 110.

7 I consider the differences between man and animals in propensities, feelings, and intellectual faculties, to be the result of the same cause as that which we assign for the variations in other functions, *viz.* difference of organization; and that the superiority of man in rational endowments is not greater than the more exquisite, complicated, and perfectly developed structure of his brain, and particularly of his ample cerebral hemispheres, to which the rest of the animal kingdom offers no parallel, nor even any near approximation, is sufficient to account for.

Lectures on Physiology, Zoology, and the Natural History of Man (1819), 237.

Richard Erskine Leakey 1944– and **Roger Amos Lewin** 1946–

Kenyan anthropologist and palaeontologist and American science writer

8 When out fossil hunting, it is very easy to forget that rather than telling you *how* the creatures *lived*, the remains you find indicate only *where* they became *fossilized.*

Origins: What New Discoveries Reveal about the Emergence of our Species and its Possible Future (1977), 96.

1 For three million years we were hunter-gatherers, and it was through the evolutionary pressures of that way of life that a brain so adaptable and so creative eventually emerged. Today we stand with the brains of hunter-gatherers in our heads, looking out on a modern world made comfortable for some by the fruits of human inventiveness, and made miserable for others by the scandal of deprivation in the midst of plenty.

Origins: What New Discoveries Reveal about the Emergence of our Species and its Possible Future (1977), 249.

2 An evolutionary perspective of our place in the history of the earth reminds us that *Homo sapiens sapiens* has occupied the planet for the tiniest fraction of that planet's four and a half thousand million years of existence. In many ways we are a biological accident, the product of countless propitious circumstances. As we peer back through the fossil record, through layer upon layer of long-extinct species, many of which thrived far longer than the human species is ever likely to do, we are reminded of our mortality as a species. There is no law that declares the human animal to be different, as seen in this broad biological perspective, from any other animal. There is no law that declares the human species to be immortal.

Origins: What New Discoveries Reveal about the Emergence of our Species and its Possible Future (1977), 256.

Mary Leapor 1722–46

British poet

3 How near one Species to the next is join'd,
The due Gradations please a thinking Mind;
And there are Creatures which no eye can see,
That for a Moment live and breathe like me:
Whom a small Fly in bulk as far exceeds,
As yon tall Cedar does the waving Reeds:
These we can reach—and may we not suppose
There still are Creatures more minute than those.

'The Enquiry'. In *Poems Upon Several Occasions* (1748), 198.

Gustave Le Bon 1841–1931

French social psychologist

4 Under certain given circumstances, and only under those circumstances, an agglomeration of men presents new characteristics very different from those of the individuals composing it. The sentiments and ideas of all the persons in the gathering take one and the same direction, and their conscious personality vanishes. A collective mind is formed, doubtless transitory, but presenting very clearly defined characteristics. The gathering has thus become what, in the absence of a better expression, I will call an organized crowd, or, if the term is considered preferable, a psychological crowd. It forms a single being and is subject to the law of the mental unity of crowds.

The Crowd (1895), 58–9.

Pierre Lecomte du Nouy 1883–1947

French biophysicist

5 The man of science who cannot formulate a hypothesis is only an accountant of phenomena.

The Road to Reason (1948), 77.

Daniel Lednicer 1929–

American science writer

6 To LIBIDO. Without which the work described in this volume would have been neither possible or necessary.

Dedication to his *Contraception: The Chemical Control of Fertility* (1969).

Tsung-Dao Lee 1926–

Chinese-born American physicist

1 The progress of science has always been the result of a close interplay between our concepts of the universe and our observations on nature. The former can only evolve out of the latter and yet the latter is also conditioned greatly by the former. Thus in our exploration of nature, the interplay between our concepts and our observations may sometimes lead to totally unexpected aspects among already familiar phenomena.

'Weak Interactions and Nonconservation of Parity', Nobel Lecture, 11 December 1957. In *Nobel Lectures: Physics 1942–1962* (1964), 417.

2 Since the beginning of physics, symmetry considerations have provided us with an extremely powerful and useful tool in our effort to understand nature. Gradually they have become the backbone of our theoretical formulation of physical laws.

Particle Physics and an Introduction to Field Theory (1981), 177.

Yuan Tseh Lee 1936–

Taiwanese-born American chemist

3 Chemistry is the study of material transformations. Yet a knowledge of the rate, or time dependence, of chemical change is of critical importance for the successful synthesis of new materials and for the utilization of the energy generated by a reaction. During the past century it has become clear that all macroscopic chemical processes consist of many elementary chemical reactions that are themselves simply a series of encounters between atomic or molecular species. In order to understand the time dependence of chemical reactions, chemical kineticists have traditionally focused on sorting out all of the elementary chemical reactions involved in a macroscopic chemical process and determining their respective rates.

'Molecular Beam Studies of Elementary Chemical Processes', Nobel Lecture, 8 December 1986. In *Nobel Lectures: Chemistry 1981–1990* (1992), 320.

Antoni van Leeuwenhoek 1632–1723

Dutch naturalist and microscopist

4 I have divers times endeavoured to see and to know, what parts the *Blood* consists of; and at length I have observ'd, taking some Blood out of my own hand, that it consists of small round *globuls* [sic] driven through a Crystalline humidity or water.

Letter to H. Oldenburg, 7 April 1674. In *The Collected Letters of Antoni van Leeuwenhoek* (1957), Vol. 1, 75.

5 The 31*th* of *May*, I perceived in the same water more of those Animals, as also some that were somewhat bigger. And I imagine, that [ten hundred thousand] of these little Creatures do not equal an ordinary grain of Sand in bigness: And comparing them with a Cheese-mite (which may be seen to move with the naked eye) I make the proportion of one of these small Water-creatures to a Cheese-mite, to be like that of a Bee to a Horse: For, the circumference of one of these little Animals in water, is not so big as the thickness of a hair in a Cheese-mite.

Letter to H. Oldenburg, 9 October 1676. In *The Collected Letters of Antoni van Leeuwenhoek* (1957), Vol. 2, 75.

6 The 4*th* sort of creatures . . . which moved through the 3 former sorts, were incredibly small, and so small in my eye that I judged, that if 100 of them lay [stretched out] one by another, they would not equal the *length* of a grain of course Sand; and according to this estimate, ten hundred thousand of them could not equal the dimensions of a *grain* of such course Sand. There was discover'd by me a fifth sort, which had near the thickness of the former, but they were almost twice as long.

The first microscopic observation of bacteria.

Letter to H. Oldenburg, 9 October 1676. In *The Collected Letters of Antoni van Leeuwenhoek* (1957), Vol. 2, 95.

1 I now saw very distinctly that these were little eels or worms . . . Lying huddled together and wriggling, just as if you saw with your naked eye a whole tubful of very little eels and water, the eels moving about in swarms; and the whole water seemed to be alive with the multitudinous animalcules. For me this was among all the marvels that I have discovered in nature the most marvellous of all, and I must say that, for my part, no more pleasant sight has yet met my eye than this of so many thousands of living creatures in one small drop of water, all huddling and moving, but each creature having its own motion.

Letter to H. Oldenburg, 9 October 1676. In *The Collected Letters of Antoni van Leeuwenhoek* (1957), Vol. 2, 115.

2 I have divers times examined the same matter (human semen) from a healthy man . . . not from a sick man . . . nor spoiled by keeping . . . for a long time and not liquefied after the lapse of some time . . . but immediately after ejaculation before six beats of the pulse had intervened; and I have seen so great a number of living animalcules . . . in it, that sometimes more than a thousand were moving about in an amount of material the size of a grain of sand . . . I saw this vast number of animalcules not all through the semen, but only in the liquid matter adhering to the thicker part.

Letter to W. Brouncker, President of the Royal Society, undated, November 1677. In *The Collected Letters of Antoni van Leeuwenhoek* (1957), Vol. 2, 283–4.

3 But many of our imaginations and investigations of nature are futile, especially when we see little living animals and see their legs and must judge the same to be ten thousand times thinner than a hair of my beard, and when I see animals living that are more than a hundred times smaller and am unable to observe any legs at all, I still conclude from their structure and the movements of their bodies that they do have legs . . . and therefore legs in proportion to their bodies, just as is the case with the larger animals upon which I can see legs . . . Taking this number to be about a hundred times smaller, we therefore find a million legs, all these together being as thick as a hair from my beard, and these legs, besides having the instruments for movement, must be provided with vessels to carry food.

Letter to N. Grew, 27 September 1678. In *The Collected Letters of Antoni van Leeuwenhoek* (1957), Vol. 2, 391.

4 But in my opinion we can now be assured sufficiently that no animals, however small they may be, take their origin in putrefaction, but exclusively in procreation . . . For seeing that animals, from the largest down to the little despised animal, the flea, have animalcules in their semen, seeing also that some of the vessels of the lungs of horses and cows consist of rings and that these rings can occur on the flea's veins, why cannot we come to the conclusion that as well as the male sperm of that large animal the horse and similar animals, and of all manner of little animals, the flea included, is furnished with animalcules (and other intestines, for I have often been astonished when I beheld the numerous vessels in a flea), why, I say should not the male sperm of the smallest animals, smaller than a flea, nay even the very smallest animalcules have the perfection that we find in a flea.

Letter to Robert Hooke, 12 November 1680. In *The Collected Letters of Antoni van Leeuwenhoek* (1957), Vol. 3, 329.

5 If we reflect that a small creature such as this is provided, not only with external members, but also with intestines and other organs, we have no reason to doubt that a like creature, even if a thousand million times smaller, may already be provided with all its external and internal organs . . . though they may be hidden from our eyes. For, if we consider the external and internal organs of animalcules which are so small that a thousand million of them together would amount to the size of a coarse grain of sand, it may well be, however incomprehensible and unsearchable it may seem to us, that an animalcule from the male seed of whatever members of the animal kingdom, contains within itself . . . all the limbs and organs which an animal has when it is born.

Letter to the Gentlemen of the Royal Society, 30 March 1685. In *The Collected Letters of Antoni van Leeuwenhoek* (1957), Vol. 5, 185.

Julien Jean César Legallois

1770–1814

French physiologist

1 For, every time a certain portion is destroyed, be it of the brain or of the spinal cord, a function is compelled to cease suddenly, and before the time known beforehand when it would stop naturally, it is certain that this function depends upon the area destroyed. It is in this way that I have recognized that the prime motive power of respiration has its seat in that part of the medulla oblongata that gives rise to the nerves of the eighth pair [vagi]; and it is by this method that up to a certain point it will be possible to discover the use of certain parts of the brain.

Expériences sur le Principe de la Vie, Notamment sur celui des Mouvements du Coeur, et sur le Siège de ce Principe (1812), 148–149. Translated in Edwin Clarke and L. S. Jacyna, *Nineteenth Century Origins of Neuroscientific Concepts* (1987), 247.

Tom Lehrer 1928–

American satirist

2 There's antimony, arsenic, aluminium, selenium,
And hydrogen and oxygen and nitrogen and rhenium,
And nickel, neodymium, neptunium, germanium,
And iron, americium, ruthenium, uranium,
Europium, zirconium, lutetium, vanadium,
And lanthanum and osmium and astatine and radium,
And gold and protactinium and indium and gallium,
And iodine and thorium and thulium and thallium.
There's yttrium, ytterbium, actinium, rubidium,
And boron, gadolinium, niobium, iridium,
And strontium and silicon and silver and samarium,
And bismuth, bromine, lithium, beryllium and barium.
There's holmium and helium and hafnium and erbium,
And phosphorus and francium and fluorine and terbium,
And manganese and mercury, molybdenum, magnesium,
Dysprosium and scandium and cerium and cesium,
And lead, praseodymium and platinum, plutonium,
Palladium, promethium, potassium, polonium,
And tantalum, technetium, titanium, tellurium,
And cadmium and calcium and chromium and curium.
There's sulfur, californium and fermium, berkelium,
And also mendelevium, einsteinium, nobelium,
And argon, krypton, neon, radon, xenon, zinc and rhodium,
And chlorine, cobalt, carbon, copper, tungsten, tin and sodium.
These are the only ones of which the news has come to Harvard,
And there may be many others, but they haven't been discarvard.

To the tune of 'I am the very model of a modern major general'.

'The Elements' (1959), in John Emsley (ed.), *Nature's Building Blocks* (2001), 1–2.

Gottfried Wilhelm Leibniz

1646–1716

German philosopher and mathematician

1 One cannot explain words without making incursions into the sciences themselves, as is evident from dictionaries; and, conversely, one cannot present a science without at the same time defining its terms.

'Of the Division of the Sciences' (1765), Book 4, Chapter 21, in *New Essays on Human Understanding*, trans. and ed. Peter Remnant (1981), 522.

2 Now this supreme wisdom, united to a goodness that is no less infinite, cannot but have chosen the best. For as a lesser evil is a kind of good, even so a lesser good is a kind of evil if it stands in the way of a greater good; and there would be something to correct in the actions of God if it were possible to do better. As in mathematics, when there is no maximum nor minimum, in short nothing distinguished, everything is done equally, or when that is not possible nothing at all is done: so it may be said likewise in respect of perfect wisdom, which is no less orderly than mathematics, that if there were not the best (*optimum*) among all possible worlds, God would not have produced any.

Theodicy: Essays on the Goodness of God and Freedom of Man and the Origin of Evil (1710), 128.

3 It follows from the supreme perfection of God, that in creating the universe he has chosen the best possible plan, in which there is the greatest variety together with the greatest order; the best arranged ground, place, time; the most results produced in the most simple ways; the most of power, knowledge, happiness and goodness in the creatures that the universe could permit. For since all the possibles in the understanding of God laid claim to existence in proportion to their perfections, the actual world, as the resultant of all these claims, must be the most perfect possible. And without this it would not be possible to give a reason why things have turned out so rather than otherwise.

The Principles of Nature and Grace (1714), *The Philosophical Works of Leibnitz* (1890), ed. G. M. Duncan, 213–4.

4 Our reasonings are grounded upon *two great principles, that of contradiction*, in virtue of which we judge *false* that which involves a contradiction, and *true* that which is opposed or contradictory to the false.

The Monadology and Other Philosophical Writings (1714), trans. Robert Latta (1898), 235.

5 There are also two kinds of *truths*, those of *reasoning* and those of *fact*. Truths of reasoning are necessary and their opposite is impossible: truths of fact are contingent and their opposite is possible. When a truth is necessary, its reason can be found by analysis, resolving it into more simple ideas and truths, until we come to those which are primary.

The Monadology and Other Philosophical Writings (1714), trans. Robert Latta (1898), 235–6.

6 These principles have given me a way of explaining naturally the union or rather the mutual agreement [*conformité*] of the soul and the organic body. The soul follows its own laws, and the body likewise follows its own laws; and they agree with each other in virtue of the pre-established harmony between all substances, since they are all representations of one and the same universe.

The Monadology and Other Philosophical Writings (1714), trans. Robert Latta (1898), 262.

7 We should like Nature to go no further; we should like it to be finite, like our mind; but this is to ignore the greatness and majesty of the Author of things.

Letter to S. Clarke, 1715. Trans. M. Morris and G. H. R. Parkinson, *Leibniz: Philosophical Writings* (1973), 220.

8 According to their [Newton and his followers] doctrine, God Almighty wants to wind up his watch from time to time: otherwise it would cease to move. He had not, it seems, sufficient foresight to make it a perpetual motion. Nay, the machine of God's making, is so imperfect, according to these gentlemen; that he is obliged to clean it now and then by an extraordinary concourse, and even to mend it, as a clockmaker mends his work.

'Mr. Leibniz's First Paper' (1715). In H. G. Alexander (ed.), *The Leibniz-Clarke Correspondence* (1956), 11–2.

1 It is God who is the ultimate reason of things, and the Knowledge of God is no less the beginning of science than his essence and will are the beginning of things.

Letter on a General Principle Useful in Explaining the Laws of Nature (1687).

2 All the different classes of beings which taken together make up the universe are, in the ideas of God who knows distinctly their essential gradations, only so many ordinates of a single curve so closely united that it would be impossible to place others between any two of them, since that would imply disorder and imperfection. Thus men are linked with the animals, these with the plants and these with the fossils, which in turn merge with those bodies which our senses and our imagination represent to us as absolutely inanimate. And, since the law of continuity requires that when the essential attributes of one being approximate those of another all the properties of the one must likewise gradually approximate those of the other, it is necessary that all the orders of natural beings form but a single chain, in which the various classes, like so many rings, are so closely linked one to another that it is impossible for the senses or the imagination to determine precisely the point at which one ends and the next begins—all the species which, so to say, lie near to or upon the borderlands being equivocal, and endowed with characters which might equally well be assigned to either of the neighboring species. Thus there is nothing monstrous in the existence of zoophytes, or plant-animals, as Budaeus calls them; on the contrary, it is wholly in keeping with the order of nature that they should exist. And so great is the force of the principle of continuity, to my thinking, that not only should I not be surprised to hear that such beings had been discovered—creatures which in some of their properties, such as nutrition or reproduction, might pass equally well for animals or for plants, and which thus overturn the current laws based upon the supposition of a perfect and absolute separation of the different orders of coexistent beings which fill the universe;—not only, I say, should I not be surprised to hear that they had been discovered, but, in fact, I am convinced that there must be such creatures, and that natural history will perhaps some day become acquainted with them, when it has further studied that infinity of living things whose small size conceals them for ordinary observation and which are hidden in the bowels of the earth and the depths of the sea.

Lettre Prétendue de M. De Leibnitz, à M. Hermann dont M. Koenig a Cité le Fragment (1753), cxi-cxii, trans. in A. O. Lovejoy, *The Great Chain of Being: A Study of the History of an Idea* (1936), 144–5.

3 Imaginary numbers are a fine and wonderful refuge of the divine spirit, almost an amphibian between being and non-being.

Quoted in F. Klein, *Elementary Mathematics From an Advanced Standpoint* (1932), Vol. 1, 56.

4 *Nihil est sine ratione.*

There is nothing without a reason.

Attributed.

Georges Lemaître 1894–1966

Belgian astronomer

5 If the world has begun with a single quantum, the notions of space and time would altogether fail to have any meaning at the beginning; they would only begin to have a sensible meaning when the original quantum had been divided into a sufficient number of quanta. If this suggestion is correct, the beginning of the world happened a little before the beginning of space and time. I think that such a beginning of the world is far enough from the present order of Nature to be not at all repugnant. It may be difficult to follow up the idea in detail as we are not yet able to count the quantum packets in every case. For example, it may be that an atomic nucleus must be counted as a unique quantum, the atomic number acting as a kind of quantum number. If the future development of quantum theory happens to turn in that direction, we could conceive the beginning of the universe in the form of a unique atom, the atomic weight of which is the total mass of the universe. This highly unstable atom would divide in smaller and smaller atoms by a kind of super-radioactive process.

'The Beginning of the World from the Point of View of Quantum Theory', *Nature*, 1931, 127, 706.

1 The radius of space began at zero; the first stages of the expansion consisted of a rapid expansion determined by the mass of the initial atom, almost equal to the present mass of the universe. If this mass is sufficient, and the estimates which we can make indicate that this is indeed so, the initial expansion was able to permit the radius to exceed the value of the equilibrium radius. The expansion thus took place in three phases: a first period of rapid expansion in which the atom-universe was broken into atomic stars, a period of slowing-down, followed by a third period of accelerated expansion. It is doubtless in this third period that we find ourselves today, and the acceleration of space which followed the period of slow expansion could well be responsible for the separation of stars into extra-galactic nebulae.

'La formation des nébuleuses dans l'univers en expansion', *Comptes Rendus*, 1933, **196**, 903–4. Trans. Helge Kragh, *Cosmology and Controversy: The Historical Development of Two Theories of the Universe* (1996), 52.

2 As far as I see, such a theory [of the primeval atom] remains entirely outside any metaphysical or religious question. It leaves the materialist free to deny any transcendental Being. He may keep, for the bottom of space-time, the same attitude of mind he has been able to adopt for events occurring in non-singular places in space-time. For the believer, it removes any attempt to familiarity with God, as were Laplace's chiquenaude or Jeans' finger. It is consonant with the wording of Isaiah speaking of the 'Hidden God' hidden even in the beginning of the universe . . . Science has not to surrender in face of the Universe and when Pascal tries to infer the existence of God from the supposed infinitude of Nature, we may think that he is looking in the wrong direction.

'The Primeval atom Hypothesis and the Problem of Clusters of Galaxies', in R. Stoops (ed.), *La Structure et l'Évolution de l'Univers* (1958), 1–32. Trans. Helge Kragh, *Cosmology and Controversy: The Historical Development of Two Theories of the Universe* (1996), 60.

3 Scientific progress is the discovery of a more and more comprehensive simplicity . . . The previous successes give us confidence in the future of science: we become more and more conscious of the fact that the universe is cognizable.

In O. Godart and M. Heller (eds.), *Cosmology of Lemaitre* (1985), 162.

Nicolas Lemery 1645–1715

French chemist

4 A demonstrative and convincing proof that an *acid* does consist of pointed parts is, that not only all *acid salts* do *Crystallize* into edges, but all Dissolutions of different things, caused by *acid* liquors, do assume this figure in their *Crystallization*; these *Crystalls* consist of points differing both in length and bigness from one another, and this diversity must be attributed to the keener or blunter edges of the different sorts of *acids*.

A Course of Chymistry (1675), trans. W. Harris (1686), 24.

Sir John Edward Lennard-Jones 1894–1954

British scientist

1 Graduates engaged in post-graduate work are reminded that their Supervisor is a University Officer and when visiting him officially in that capacity they should dress as they would in visiting any other officers of the University or of their own College (e.g. a tutor). Gowns, however, need not be worn in the chemical laboratory.

Note from Lennard-Jones to his PhD student, Charles Coulson, 14 July 1933. Quoted in S. C. Altham and E. J. Bowen, 'Charles Alfred Coulson 1910–1974', *Biographical Memoirs of Fellows of the Royal Society*, 1974, 20, 78.

William J. le Noble 1928–

Dutch-born American chemist

2 It may sound like a lot of work to keep up with organic chemistry, and it is; however, those who haven't the time to do it become subject to decay in the ability to teach and to contribute to the Science—a sort of first-order process the half-life of which can't be much more than a year or two.

Highlights of Organic Chemistry: An Advanced Textbook (1974), 112.

Leonardo da Vinci 1452–1519

Italian scholar-artist

3 There is no result in nature without a cause; understand the cause and you will have no need of the experiment.

'Philosophy', in *The Notebooks of Leonardo da Vinci*, trans. E. MacCurdy, (1938) Vol. 1, 70.

4 Experience is never at fault; it is only your judgment that is in error in promising itself such results from experience as are not caused by our experiments. For having given a beginning, what follows from it must necessarily be a natural development of such a beginning, unless it has been subject to a contrary influence, while, if it is affected by any contrary influence, the result which ought to follow from the aforesaid beginning will be found to partake of this contrary influence in a greater or less degree in proportion as the said influence is more or less powerful than the aforesaid beginning.

'Philosophy', in *The Notebooks of Leonardo da Vinci*, trans. E. MacCurdy (1938), Vol. 1, 70.

5 The body of the earth is of the nature of a fish . . . because it draws water as its breath instead of air.

'Philosophy', in *The Notebooks of Leonardo da Vinci*, trans. E. MacCurdy (1938), Vol. 1, 70.

6 Nature being capricious and taking pleasure in creating and producing a continuous sucession of lives and forms because she knows that they serve to increase her terrestrial substance, is more ready and swift in her creating than time is in destroying, and therefore she has ordained that many animals shall serve as food one for the other; and as this does not satisfy her desire she sends forth frequently certain noisome and pestilential vapours and continual plagues upon the vast accumulations and herds of animals and especially upon human beings who increase very rapidly because other animals do not feed upon them.

'Philosophy', in *The Notebooks of Leonardo da Vinci*, trans. E. MacCurdy (1938), Vol. 1, 80.

7 Poor is the pupil who does not surpass his master.

'Aphorisms', in *The Notebooks of Leonardo da Vinci*, trans. E. MacCurdy (1938), Vol. 1, 98.

8 This work should commence with the conception of man, and should describe the nature of the womb, and how the child inhabits it, and in what stage it dwells there, and the manner of its quickening and feeding, and its growth, and what interval there is between one stage of growth and another, and what thing drives it forth from the body of the mother, and for what reason it sometimes emerges from the belly of its mother before the due time.

'Anatomy', in *The Notebooks of Leonardo da Vinci*, trans. E. MacCurdy (1938), Vol. 1, 139.

1 There are many occasions when the muscles that form the lips of the mouth move the lateral muscles that are joined to them, and there are an equal number of occasions when these lateral muscles move the lips of this mouth, replacing it where it cannot return of itself, because the function of muscle is to pull and not to push except in the case of the genitals and the tongue.

'Anatomy', in *The Notebooks of Leonardo da Vinci*, trans. E. MacCurdy (1938), Vol. 1, 152.

2 In the mountains of Parma and Piacenza, multitudes of shells and corals filled with worm-holes may be seen still adhering to the rocks, and when I was making the great horse at Milan a large sack of those which had been found in these parts was brought to my workshop by some peasants . . . The red stone of the mountains of Verona is found with shells all intermingled, which have become part of this stone . . . And if you should say that these shells have been and still constantly are being created in such places as these by the nature of the locality or by potency of the heavens in these spots, such an opinion cannot exist in brains possessed of any extensive powers of reasoning because the years of their growth are numbered upon the outer coverings of their shells; and both small and large ones may be seen; and these would not have grown without feeding, or fed without movement, and here [embedded in rock] they would not have been able to move . . . The peaks of the Apennines once stood up in a sea, in the form of islands surrounded by salt water . . . and above the plains of Italy where flocks of birds are flying today, fishes were once moving in large shoals.

'Physical Geography', in *The Notebooks of Leonardo da Vinci*, trans. E. MacCurdy (1938), Vol. 1, 355–6, 359.

3 Why are the bones of great fishes, and oysters and corals and various other shells and sea-snails, found on the high tops of mountains that border the sea, in the same way in which they are found in the depths of the sea?

'Physical Geography', in *The Notebooks of Leonardo da Vinci*, trans. E. MacCurdy (1938), Vol. 1, 361.

4 The [mechanical] bird I have described ought to be able by the help of the wind to rise to a great height, and this will prove to be its safety; since even if . . . revolutions [of the winds] were to befall it, it would still have time to regain a condition of equilibrium; provided that its various parts have a great power of resistance, so that they can safely withstand the fury and violence of the descent, by the aid of the defenses which I have mentioned; and its joints should be made of strong tanned hide, and sewn with cords of strong raw silk. And let no one encumber himself with iron bands, for these are very soon broken at the joints or else they become worn out, and consequently it is well not to encumber oneself with them.

'Flight', in *The Notebooks of Leonardo da Vinci*, trans. E. MacCurdy (1938), Vol. 1, 427.

5 A bird maintains itself in the air by imperceptible balancing, when near to the mountains or lofty ocean crags; it does this by means of the curves of the winds which as they strike against these projections, being forced to preserve their first impetus bend their straight course towards the sky with divers revolutions, at the beginning of which the birds come to a stop with their wings open, receiving underneath themselves the continual buffetings of the reflex courses of the winds.

'Flight', in *The Notebooks of Leonardo da Vinci*, trans. E. MacCurdy (1938), Vol. 1, 471.

6 A bird is an instrument working according to a mathematical law, which instrument it is within the capacity of man to reproduce, with all its movements.

'Flying Machine', from *The Notebooks of Leonardo da Vinci*, trans. E. MacCurdy (1938), Vol. 1, 511.

1 Although nature commences with reason and ends in experience it is necessary for us to do the opposite, that is to commence as I said before with experience and from this to proceed to investigate the reason.

'Movement and Weight', from *The Notebooks of Leonardo da Vinci*, trans. E. MacCurdy (1938), Vol. 1, 546.

2 Perspective is a most subtle discovery in mathematical studies, for by means of lines it causes to appear distant that which is near, and large that which is small.

Attributed.

Aldo Leopold 1886–1948

American conservationist

3 Twenty centuries of 'progress' have brought the average citizen a vote, a national anthem, a Ford, a bank account, and a high opinion of himself, but not the capacity to live in high density without befouling and denuding his environment, nor a conviction that such capacity, rather than such density, is the true test of whether he is civilized.

Game Management (1933), 423.

4 Conservation is getting nowhere because it is incompatible with our Abrahamic concept of land. We abuse land because we regard it as a commodity belonging to us. When we see land as a community to which we belong, we may begin to use it with love and respect. There is no other way for land to survive the impact of mechanized man, nor for us to reap from it the esthetic harvest it is capable, under science, of contributing to culture. That land is a community is the basic concept of ecology, but that land is to be loved and respected is an extension of ethics. That land yields a cultural harvest is a fact long known, but latterly often forgotten.

A Sand County Almanac, and Sketches Here and There (1949), viii-ix.

Max Lerner 1902–92

American author

5 Power politics existed before Machiavelli was ever heard of; it will exist long after his name is only a faint memory. What he did, like Harvey, was to recognize its existence and subject it to scientific study.

The Prince and the Discourses by Niccolò Machiavelli, with an Introduction by Max Lerner (1950), xliii.

Alaine René Le Sage 1668–1747

French novelist and dramatist

6 Facts are stubborn things.

Gil Blas (1748), trans. Tobias Smollett (1962 edn., with an introduction by Bergen Evans), 280.

John Leslie 1766–1832

British mathematician and natural philosopher

7 The true business of the philosopher, though not flattering to his vanity, is merely to ascertain, arrange and condense the facts.

'Dissertation Fifth: Exhibiting a General View of the Progress of Mathematical and Physical Science, chiefly during the Eighteenth Century', *Encyclopedia Britannica*, 7th edn. (1842), 743. In John Heilbron, *Weighing Imponderables and Other Quantitative Science around 1800* (1993), 17.

Leucippus fl. 5th century BC

Greek philosopher

8 Nothing occurs at random, but everything for a reason and by necessity.

Aetius 1.25.4. In G. S. Kirk, J. E. Raven and M. Schofield (eds.), *The Presocratic Philosophers* (1983), 420.

Primo Levi 1919–87

Italian chemist and writer

9 It is the destiny of wine to be drunk, and it is the destiny of glucose to be oxidized. But it was not oxidized immediately: its drinker kept it in his liver for more than a week, well curled up and tranquil, as a reserve aliment for a sudden effort; an effort that he was forced to make the following Sunday, pursuing a bolting horse. Farewell to the hexagonal structure: in the space of a few instants the skein was unwound and became glucose again, and this was dragged by the bloodstream all the way to a minute muscle fiber in the thigh, and here brutally split into two molecules of lactic acid, the grim harbinger of fatigue: only later, some minutes after, the panting of the lungs was able to supply the oxygen necessary to quietly oxidize the latter. So a new molecule of carbon dioxide returned to the atmosphere, and a parcel of the energy that the sun had handed to the vine-shoot passed from the state of chemical energy to that of mechanical energy, and thereafter settled down in the slothful condition of heat, warming up imperceptibly the air moved by the running and the blood of the runner. 'Such is life,' although rarely is it described in this manner: an inserting itself, a drawing off to its advantage, a parasitizing of the downward course of energy, from its noble solar form to the degraded one of low-temperature heat. In this downward course, which leads to equilibrium and thus death, life draws a bend and nests in it.

The Periodic Table (1975), trans. Raymond Rosenthal (1984), 192–3.

Richard Levins 1930– and **Richard Charles Lewontin** 1929–

American mathematician and biologist and American biologist

1 The organism cannot be regarded as simply the passive object of autonomous internal and external forces; it is also the subject of its own evolution.

The Dialectical Biologist (1985), 89.

Claude Lévi-Strauss 1908–90

French social anthropologist

2 Anthropology found its Galileo in Rivers, its Newton in Mauss.

Structural Anthropology (1958), 159.

George Henry Lewes 1817–78

British philosopher, naturalist and man of letters

3 Metaphysical ghosts cannot be killed, because they cannot be touched; but they may be dispelled by dispelling the twilight in which shadows and solidities are easily confounded. The Vital Principle is an entity of this ghostly kind; and although the daylight has dissipated it, and positive Biology is no longer vexed with its visitations, it nevertheless reappears in another shape in the shadowy region of mystery which surrounds biological and all other questions.

The History of Philosophy from Thales to Comte (1867), lxxxiv.

Peter Lewin 1935–

Canadian paediatrician

4 The possibility that the infective agent may not contain nucleic acid and consist only of a peptide or peptide-polysaccharide complex which has replication properties within susceptible cells is intriguing. If peptides, short-chain proteins, or peptide/fatty-acid/polysaccharide complexes activate nucleic-acid template activity in the host genes to produce identical infective particles, this would invalidate the accepted dogma of present-day molecular biology in which D.N.A. and R.N.A. templates control all biological activity.

'Scrapie: An Infective Peptide?', *The Lancet*, 1972, i, 748.

Clive Staples Lewis 1898–1963

British novelist and theologian

5 We may not be able to get certainty, but we can get probability, and half a loaf is better than no bread.

Christian Reflections (1967), 111.

Gilbert Newton Lewis 1875–1946

American chemist

1 There is always the danger in scientific work that some word or phrase will be used by different authors to express so many ideas and surmises that, unless redefined, it loses all real significance.

'Valence and Tautomerism', *Journal of the American Chemical Society*, 1913, 35, 1448.

2 I take it that a monograph of this sort belongs to the ephemera literature of science. The studied care which is warranted in the treatment of the more slowly moving branches of science would be out of place here. Rather with the pen of a journalist we must attempt to record a momentary phase of current thought, which may at any instant change with kaleidoscopic abruptness.

Valence and the Structure of Atoms and Molecules (1923), Preface.

3 In the year 1902 (while I was attempting to explain to an elementary class in chemistry some of the ideas involved in the periodic law) becoming interested in the new theory of the electron, and combining this idea with those which are implied in the periodic classification, I formed an idea of the inner structure of the atom which, although it contained certain crudities, I have ever since regarded as representing essentially the arrangement of electrons in the atom . . . In accordance with the idea of Mendeleéf, that hydrogen is the first member of a full period, I erroneously assumed helium to have a shell of eight electrons. Regarding the disposition in the positive charge which balanced the electrons in the neutral atom, my ideas were very vague; I believed I inclined at that time toward the idea that the positive charge was also made up of discrete particles, the localization of which determined the localization of the electrons.

Valence and the Structure of Atoms and Molecules (1923), 29–30.

4 I have no patience with attempts to identify science with measurement, which is but one of its tools, or with any definition of the scientist which would exclude a Darwin, a Pasteur or a Kekulé. The scientist is a practical man and his are practical aims. He does not seek the *ultimate* but the *proximate*. He does not speak of the last analysis but rather of the next approximation. His are not those beautiful structures so delicately designed that a single flaw may cause the collapse of the whole. The scientist builds slowly and with a gross but solid kind of masonry. If dissatisfied with any of his work, even if it be near the very foundations, he can replace that part without damage to the remainder. On the whole, he is satisfied with his work, for while science may never be wholly right it certainly is never wholly wrong; and it seems to be improving from decade to decade.

The Anatomy of Science (1926), 6–7.

5 Borel makes the amusing supposition of a million monkeys allowed to play upon the keys of a million typewriters. What is the chance that this wanton activity should reproduce exactly all of the volumes which are contained in the library of the British Museum? It certainly is not a large chance, but it may be roughly calculated, and proves in fact to be considerably larger than the chance that a mixture of oxygen and nitrogen will separate into the two pure constituents. After we have learned to estimate such minute chances, and after we have overcome our fear of numbers which are very much larger or very much smaller than those ordinarily employed, we might proceed to calculate the chance of still more extraordinary occurrences, and even have the boldness to regard the living cell as a result of random arrangement and rearrangement of its atoms. However, we cannot but feel that this would be carrying extrapolation too far. This feeling is due not merely to a recognition of the enormous complexity of living tissue but to the conviction that the whole trend of life, the whole process of building up more and more diverse and complex structures, which we call evolution, is the very opposite of that which we might expect from the laws of chance.

The Anatomy of Science (1926), 158–9.

1 It must be admitted that science has its castes. The man whose chief apparatus is the differential equation looks down upon one who uses a galvanometer, and he in turn upon those who putter about with sticky and smelly things in test tubes. But all of these, and most biologists too, join together in their contempt for the pariah who, not through a glass darkly, but with keen unaided vision, observes the massing of a thundercloud on the horizon, the petal as it unfolds, or the swarming of a hive of bees. And yet sometimes I think that our laboratories are but little earthworks which men build about themselves, and whose puny tops too often conceal from view the Olympian heights; that we who work in these laboratories are but skilled artisans compared with the man who is able to observe, and to draw accurate deductions from the world about him.

The Anatomy of Science (1926), 170–1.

2 It was not easy for a person brought up in the ways of classical thermodynamics to come around to the idea that gain of entropy eventually is nothing more nor less than loss of information.

Letter to Irving Langmuir, 5 August 1930. Quoted in Nathan Reingold, *Science in America: A Documentary History 1900–1939* (1981), 400.

3 ON LEWIS The members of the department became like the Athenians who, according to the Apostle Paul, 'spent their time in nothing else, but either to tell or to hear some new thing.' Any one who thought he had a bright idea rushed out to try it out on a colleague. Groups of two or more could be seen every day in offices, before blackboards or even in corridors, arguing vehemently about these 'brain storms.' It is doubtful whether any paper ever emerged for publication that had not run the gauntlet of such criticism. The whole department thus became far greater than the sum of its individual members.

J. Hildebrand, obituary of Gilbert Newton Lewis, *Biographical Memoirs of the National Academy of Science*, 1958, 31, 212.

Thomas Lewis 1881–1945

British physiologist and cardiologist

4 Inexact method of observation, as I believe, is one flaw in clinical pathology to-day. Prematurity of conclusion is another, and in part follows from the first; but in chief part an unusual craving and veneration for hypothesis, which besets the minds of most medical men, is responsible. Except in those sciences which deal with the intangible or with events of long past ages, no treatises are to be found in which hypothesis figures as it does in medical writings. The purity of a science is to be judged by the paucity of its recorded hypotheses. Hypothesis has its right place, it forms a working basis; but it is an acknowledged makeshift, and, at the best, of purpose unaccomplished. Hypothesis is the heart which no man with right purpose wears willingly upon his sleeve. He who vaunts his lady love, ere yet she is won, is apt to display himself as frivolous or his lady a wanton.

The Mechanism and Graphic Registration of the Heart Beat (1920), vii.

5 ON LEWIS The laboratory was an unattractive half basement and low ceilinged room with an inner dark room for the galvanometer and experimental animals. It was dark, crowded with equipment and uninviting. Into it came patients for electrocardiography, dogs for experiments, trays with coffee and buns for lunch. It was hot and dusty in summer and cold in winter. True a large fire burnt brightly in the winter but anyone who found time to warm his backside at it was not beloved by Lewis. It was no good to try and look out of the window for relaxation, for it was glazed with opaque glass. The scientific peaks were our only scenery, and it was our job to try and find the pathways to the top.

'Tribute to Sir Thomas Lewis', *University College Hospital Magazine* (1955), 40, 71.

Libau see **Libavius, Andreas**

Andreas Libavius 1560–1616

German chemist

1 Fermentation is the exhalation of a substance through the admixture of a ferment which, by virtue of its spirit, penetrates the mass and transforms it into its own nature.

Die Alchemie des Andreas Libavius, ein Lehrbuch der Chemie aus dem Jahre 1597, trans. E. Pietsch and A. Kotowski (1964), 103–4. Joseph S. Fruton, *Proteins, Enzymes, Genes: The Interplay of Chemistry and Biology* (1999), 119.

Georg Christoph Lichtenberg 1742–99

German natural philosopher

2 The most heated defenders of a science, who cannot endure the slightest sneer at it, are commonly those who have not made very much progress in it and are secretly aware of this defect.

Aphorisms, trans. R. J. Hollingdale (1990), 82.

3 The natural scientists of the previous age knew less than we do and believed they were very close to the goal: we have taken very great steps in its direction and now discover we are still very far away from it. With the most rational philosophers an increase in their knowledge is always attended by an increased conviction of their ignorance.

Aphorisms, trans. R. J. Hollingdale (1990), 89.

4 If we make a couple of discoveries here and there we need not believe things will go like this for ever. An acrobat can leap higher than a farm-hand, and one acrobat higher than another, yet the height no man can overleap is still very low. Just as we hit water when we dig in the earth, so we discover the incomprehensible sooner or later.

Aphorisms, trans. R. J. Hollingdale (1990), 92.

5 He marvelled at the fact that the cats had two holes cut in their fur at precisely the spot where their eyes were.

Aphorisms, trans. R. J. Hollingdale (1990), 108.

6 We have to believe that everything has a cause, as the spider spins its web in order to catch flies. But it does this before it knows there are such things as flies.

Aphorisms, trans. R. J. Hollingdale (1990), 112.

7 The construction of the universe is certainly very much easier to explain than is that of the plant.

Aphorisms, trans. R. J. Hollingdale (1990), 119.

8 If an angel were to tell us about his philosophy, I believe many of his statements might well sound like '2 × 2=13'.

Lichtenberg: Aphorisms & Letters (1969), 31.

9 The more experiences and experiments accumulate in the exploration of nature, the more precarious the theories become. But it is not always good to discard them immediately on this account. For every hypothesis which once was sound was useful for thinking of previous phenomena in the proper interrelations and for keeping them in context. We ought to set down contradictory experiences separately, until enough have accumulated to make building a new structure worthwhile.

Lichtenberg: Aphorisms & Letters (1969), 61.

Justus Von Liebig 1803–73

German chemist

1 *An diesen Apparate ist nichts neu als seine Einfachkeit und die vollkommene zu Verlaessigkeit, welche er gewaehst.*

In this apparatus is nothing new but its simplicity and thorough trustworthiness.

Comment on his revolutionary method of organic analysis.

Poggendorf's Annalen, 1831, **21**, 4. Trans. W. H. Brock.

2 The loveliest theories are being overthrown by these damned experiments; it's no fun being a chemist anymore.

Liebig to Berzelius, 22 July 1834. In J. Carrière (ed.), *Berzelius und Liebig: ihre Briefe* (1898), 94. Trans. W. H. Brock.

3 I have spent some months in England, have seen an awful lot and learned little. England is not a land of science, there is only a widely practised dilettantism, the chemists are ashamed to call themselves chemists because the pharmacists, who are despised, have assumed this name.

Liebig to Berzelius, 26 November 1837. In J. Carrière (ed.), *Berzelius und Liebig: ihre Briefe* (1898), 134. Trans. W. H. Brock.

4 I will now direct the attention of scientists to a previously unnoticed cause which brings about the metamorphosis and decomposition phenomena which are usually called decay, putrefaction, rotting, fermentation and moldering. This cause is the ability possessed by a body engaged in decomposition or combination, i.e. in chemical action, to give rise in a body in contact with it the same ability to undergo the same change which it experiences itself.

Annalen der Pharmacie, 1839, 30, 262. Trans. W. H. Brock.

5 There are various causes for the generation of force: a tensed spring, an air current, a falling mass of water, fire burning under a boiler, a metal that dissolves in an acid—one and the same effect can be produced by means of all these various causes. But in the animal body we recognise only *one* cause as the ultimate cause of all generation of force, and that is the reciprocal interaction exerted on one another by the constituents of the food and the oxygen of the air. The only known and ultimate cause of the vital activity in the animal as well as in the plant is a chemical process.

'Der Lebensprocess im Thiere und die Atmosphäre', *Annalen der Chemie und Pharmacie*, 1841, **41**, 215–7. Trans. Kenneth L. Caneva, *Robert Mayer and the Conservation of Energy* (1993), 78.

6 God has ordered all his Creation by Weight and Measure.

Written over the door of the first chemical laboratory in the world for students, Giessen, 1842.

Quoting, as did many others, from the *Wisdom of Solomon*, 11.20.

7 We may fairly judge of the commercial prosperity of a country from the amount of sulphuric acid it consumes.

Familiar Lectures on Chemistry (1843).

8 A time will come, when fields will be manured with a solution of glass (silicate of potash), with the ashes of burnt straw, and with the salts of phosphoric acid, prepared in chemical manufactories, exactly as at present medicines are given for fever and goitre.

Agricultural Chemistry (1847), 4th edn., 186.

9 When the state is shaken to its foundations by internal or external events, when commerce, industry and all trades shall be at a stand, and perhaps on the brink of ruin; when the property and fortune of all are shaken or changed, and the inhabitants of towns look forward with dread and apprehension to the future, then the agriculturalist holds in his hand the key to the money chest of the rich, and the savings-box of the poor; for political events have not the slightest influence on the natural law, which forces man to take into his system, daily, a certain number of ounces of carbon and nitrogen.

Reflecting on events of 1848.
Familiar Letters on Chemistry (1851), 3rd edn., 483.

1 Without an acquaintance with chemistry, the statesman must remain a stranger to the true vital interests of the state, to the means of its organic development and improvement; . . . The highest economic or material interests of a country, the increased and more profitable production of food for man and animals, . . . are most closely linked with the advancement and diffusion of the natural sciences, especially of chemistry.
Familiar Letters on Chemistry (1851), 3rd edn., 19.

2 I would . . . establish the conviction that Chemistry, as an independent science, offers one of the most powerful means towards the attainment of a higher mental cultivation; that the study of Chemistry is profitable, not only inasmuch as it promotes the material interests of mankind, but also because it furnishes us with insight into those wonders of creation which immediately surround us, and with which our existence, life, and development, are most closely connected.
Familiar Letters on Chemistry (1859), 4th edn., 1.

3 A manure containing several ingredients acts in this wise: The effect of all of them in the soil accommodates itself to that one among them which, in comparison to the wants of the plant, is present in the smallest quantity.
'Laws of Minimum', in *Natural Laws of Husbandry* (1863), 215.

4 The more fodder, the more flesh; the more flesh, the more manure; the more manure, the more grain.
Letters on the Utilization of London Sewage (1865)

5 The progress of mankind is due exclusively to the progress of natural sciences, not to morals, religion or philosophy.
Letter to Schoenbein, 1 August 1866. In *Liebig und Schoenbein: Briefwechsel* (1900), 221. Trans. W. H. Brock.

6 If you want to become a chemist, you will have to ruin your health. If you don't ruin your health studying, you won't accomplish anything these days in chemistry.

Liebig's advice to Kekulé.
Quoted in *Berichte der Deutschen Chemishen Gesellschaft*, 23, 1890. Trans. W. H. Brock.

7 I have learnt that all our theories are not Truth itself, but resting places or stages on the way to the conquest of Truth, and that we must be contented to have obtained for the strivers after Truth such a resting place which, if it is on a mountain, permits us to view the provinces already won and those still to be conquered.
Liebig to Gilbert, 25 December 1870. Rothamsted Archives. Quotation supplied by W. H. Brock.

8 If it is impossible to judge merit and guilt in the field of natural science, then it is not possible in any field, and historical research becomes an idle, empty activity.
Reden und Abhandlungen (1874). Trans. W. H. Brock.

9 ON LIEBIG Mr Justus Liebig is no doubt a very clever gentleman and a most profound chemist, but in our opinion he knows as much of agriculture as the horse that ploughs the ground, and there is not an old man that stands between the stilts of a plough in Virginia, that cannot tell him of facts totally at variance with his finest spun theories.
The Southern Planter, 1845, 3, 23.

10 ON LIEBIG Chemistry was gibberish in Latin and German, but in Leibig's mouth it became a powerful language.
Jakob Grimm and Wilhelm Grimm (eds.), *Deutsches Wörterbuch* (1854), Vol. 1, xxxi. Trans. W. H. Brock.

11 ON LIEBIG I suppose I should be run after for a professorship if I had studied at

Giessen, as it seems to be a settled point that no young man can be expected to know anything of chemistry unless he has studied with Liebig.

Said by Josiah Dwight Witney. In W. D. Miles, *American Chemists and Chemical Engineers* (1976), 508.

1 ON LIEBIG But for twenty years previous to 1847 a force had been at work in a little county town of Germany destined to effect the education of Christendom, and at the same time to enlarge the boundaries of human knowledge, first in chemistry and the allied branches, then in every other one of the natural sciences. The place was Giessen; the inventor Liebig; the method, a laboratory for instruction and research.

Quoted in Daniel Gilman, President of Johns Hopkins, in *University Problems in the United States* (1898), 120.

2 ON LIEBIG Liebig taught the world two great lessons. The first was that in order to teach chemistry it was necessary that students should be taken into a laboratory. The second lesson was that he who is to apply scientific thought and method to industrial problems must have a thorough knowledge of the sciences. The world learned the first lesson more readily than it learned the second.

Ira Remsen, Address to the Industrial Chemistry Society, Glasgow, 1910. Quoted in G. H. Getman, *The Life of Ira Remsen* (1980), 122.

3 ON LIEBIG A relation can be traced between the conditions in Europe and the early years of the last century that produced the great German investigator, Justus von Liebig, and the fact that present-day fourteen-year-old boys are frequently six feet tall, while 'petite' girls are now becoming rarer.

Hubert B. Vickery, 'Liebig and Proteins', *Journal of Chemical Education*, 1942, 19, 73.

4 ON LIEBIG Liebig himself seems to have occupied the role of a gate, or sorting-demon, such as his younger contemporary Clerk Maxwell once proposed, helping to concentrate energy into one favored room of the Creation at the expense of everything else.

Quoted in T. Pynchon, *Gravity's Rainbow* (1973), 411.

James Lind 1716–94

British physician

5 On the 20th of May 1747, I took twelve patients in the scurvy, on board the *Salisbury* at sea. Their cases were as similar as I could have them. They all in general had putrid gums, the spots and lassitude, with weakness of their knees. They lay together in one place, being a proper apartment for the sick in the fore-hold; and had one diet common to all, *viz*, water-gruel sweetened with sugar in the morning; fresh mutton-broth often times for dinner; at other times puddings, boiled biscuit with sugar, &c.; and for supper, barley and raisins, rice and currents, sago and wine, or the like. Two of these were ordered each a quart of cyder a-day. Two others took twenty-five gutta of *elixir vitriol* three times a-day, upon an empty stomach; using a gargle strongly acidulated with it for their mouths. Two others took two spoonfuls of vinegar three times a-day, upon an empty stomach; having their gruels and their other food well acidulated with it, as also the gargle for their mouth. Two of the worst patients, with the tendons in the ham rigid, (a symptom none of the rest had), were put under a course of sea-water. Of this they drank half a pint every day, and sometimes more or less as it operated, by way of gentle physic. Two others had each two oranges and one lemon given them every day. These they eat with greediness, at different times, upon an empty stomach. They continued but six days under this course, having consumed the quantity that could be spared. The two remaining patients, took the bigness of a nutmeg three times a-day, of an electuary recommended by an hospital-surgeon, made of garlic, mustard-seed, *rad. raphan.* balsam of *Peru*, and gum myrrh; using for common drink, barley-water well acidulated with tamarinds; by a decoction of which, with the addition of *cremor tartar*, they were gently purged three or four times during the course. The consequence was, that the most sudden and visible good effects were perceived from the use of the oranges and lemons; one of those who had taken them, being at the end of six days fit for duty.

A Treatise of the Scurvy (1753), 191–3.

Frederick Alexander Lindemann (Lord Cherwell) 1886–1957

German-born British mathematician and physicist

1 It is more important to know the properties of chlorine than the improprieties of Claudius!

Quoted in D. W. F. Hardie and J. D. Pratt, *A History of the Modern Chemical Industry* (1966), frontispiece.

Carl Linnaeus (von Linné) 1707–78

Swedish botanist and taxonomist

2 The first step in wisdom is to know the things themselves; this notion consists in having a true idea of the objects; objects are distinguished and known by classifying them methodically and giving them appropriate names. Therefore, classification and name-giving will be the foundation of our science.

Systema Naturae (1735), trans. M. S. J. Engel-Ledeboer and H. Engel (1964), 19.

3 There is no generation from an egg in the Mineral Kingdom. Hence no vascular circulation of the humours as in the remaining Natural Kingdoms.

Systema Naturae (1735), trans. M. S. J. Engel-Ledeboer and H. Engel (1964), 20.

4 Nature's economy shall be the base for our own, for it is immutable, but ours is secondary. An economist without knowledge of nature is therefore like a physicist without knowledge of mathematics.

'Tankar om grunden til oeconomien', 1740, 406. Trans. Lisbet Koerner, *Linnaeus: Nature and Nation* (1999), 103.

5 There are some viviparous *flies*, which bring forth 2,000 young. These in a little time would fill the air, and like clouds intercept the rays of the sun, unless they were devoured by birds, spiders, and many other animals.

Oeconomia Naturae, The Oeconomy of Nature. Trans. Benjamin Stillingfleet, *Miscellaneous Tracts Relating to Natural History* (1775), revised edition, 1777, 119.

6 We admit as many genera as there are different groups of natural species of which the fructification has the same structure.

Fundamenta Botanica (1736), 159. Trans. Gunnar Eriksson, 'Linnaeus the Botanist', in Tore Frängsmyr (ed.), *Linnaeus: The Man and his Work* (1983), 86.

7 Yet man does recognise himself [as an animal]. But I ask you and the whole world for a generic differentia between man and ape which conforms to the principles of natural history. I certainly know of none . . . If I were to call man ape or vice versa, I should bring down all the theologians on my head. But perhaps I should still do it according to the rules of science.

Linné to Gmelin, 14 January 1747, in G. H. T. Plieninger (ed.), *J. G. Gmelini Reliqua quae Supersunt Comercii Epistolici* (1861), 55. Trans. Gunnar Broberg, 'Linnaeus's Classification of Man', in Tore Frängsmyr (ed.), *Linnaeus: The Man and his Work* (1983), 172.

8 I well know what a splendidly great difference there is [between] a man and a *bestia* when I look at them from a point of view of morality. Man is the animal which the Creator has seen fit to honor with such a magnificent mind and has condescended to adopt as his favorite and for which he has prepared a nobler life; indeed, sent out for its salvation his only son; but all this belongs to another forum; it behooves me like a cobbler to stick to my last, in my own workshop, and as a naturalist to consider man and his body, for I know scarcely one feature by which man can be distinguished from apes, if it be not that all the apes have a gap between their fangs and their other teeth, which will be shown by the results of further investigation.

T. Fredbärj (ed.), *Menniskans Cousiner* (Valda Avhandlingar av Carl von Linné nr, 21) (1955), 4. Trans. Gunnar Broberg, 'Linnaeus's Classification of Man', in Tore Frängsmyr (ed.), *Linnaeus: The Man and his Work* (1983), 167.

1 Stones grow, plants grow, and live, animals grow live and feel.

Philosophia Botanica (1751), Introduction 1–4. Trans. Frans A. Stafleu, *Linnaeus and the Linnaeans: The Spreading of their Ideas in Systematic Botany, 1735–1789* (1971), 33.

2 A herbarium is better than any illustration; every botanist should make one.

Philosophia Botanica (1751), aphorism 11. Trans. Frans A. Stafleu, *Linnaeus and the Linnaeans: The Spreading of their Ideas in Systematic Botany, 1735–1789* (1971), 38.

3 Fragments of the natural method must be sought with the greatest care

This is the first and last desideratum among botanists.

Nature makes no jumps.

[*Natura non facit saltus*]

All taxa show relationships on all sides like the countries on a map of the world.

Philosophia Botanica (1751), aphorism 77. Trans. Frans A. Stafleu, *Linnaeus and the Linnaeans: The Spreading of their Ideas in Systematic Botany, 1735–1789* (1971), 45.

4 Of what use are the great number of petrifactions, of different species, shape and form which are dug up by naturalists? Perhaps the collection of such specimens is sheer vanity and inquisitiveness. I do not presume to say; but we find in our mountains the rarest animals, shells, mussels, and corals embalmed in stone, as it were, living specimens of which are now being sought in vain throughout Europe. These stones alone whisper in the midst of general silence.

Philosophia Botanica (1751), aphorism 132. Trans. Frans A. Stafleu, *Linnaeus and the Linnaeans: The Spreading of their Ideas in Systematic Botany, 1735–1789* (1971), 56.

5 The names of the plants ought to be stable [*certa*], consequently they should be given to stable genera.

Philosophia Botanica (1751), aphorism 151. Trans. Frans A. Stafleu, *Linnaeus and the Linnaeans: The Spreading of their Ideas in Systematic Botany, 1735–1789* (1971), 57.

6 There are as many species as the infinite being created diverse forms in the beginning, which, following the laws of generation, produced many others, but always similar to them: therefore there are as many species as we have different structures before us today.

Philosophia Botanica (1751), aphorism 157. Trans. Frans A. Stafleu, *Linnaeus and the Linneans: The Spreading of their Ideas in Systematic Botany, 1735–1789* (1971), 63.

7 The *species* and the *genus* are always the work of nature [i.e. specially created]; the *variety* mostly that of circumstance; the *class* and the *order* are the work of nature and art.

Philosophia Botanica (1751), aphorism 162. Trans. Frans A. Stafleu, *Linnaeus and the Linnaeans: The Spreading of their Ideas in Systematic Botany, 1735–1789* (1971), 67.

8 Botany is based on fixed genera.

Philosophia Botanica (1751), aphorism 209. Trans. Frans A. Stafleu, *Linnaeus and the Linnaeans: The Spreading of their Ideas in Systematic Botany, 1735–1789* (1971), 64.

9 All the species recognized by Botanists came forth from the Almighty Creator's hand, and the number of these is now and always will be exactly the same, while every day new and different florists' species arise from the true species so-called by Botanists, and when they have arisen they finally revert to the original forms. Accordingly to the former have been assigned by Nature fixed limits, beyond which they cannot go: while the latter display without end the infinite sport of Nature.

Philosophia Botanica (1751), aphorism 310. Trans. Frans A. Stafleu, *Linnaeus and the Linnaeans: The Spreading of their Ideas in Systematic Botany, 1735–1789* (1971), 90.

10 Nomenclature, the other foundation of botany, should provide the names as soon as the classification is made . . . If the names are unknown knowledge of the things also perishes . . . For a single genus, a single name.

Philosophia Botanica (1751), aphorism 210. Trans. Frans A. Stafleu, *Linnaeus and the Linnaeans: The Spreading of their Ideas in Systematic Botany, 1735–1789* (1971), 80.

1 In natural science the principles of truth ought to be confirmed by observation.

Philosophia Botanica (1751), final sentence. Trans. Frans A. Stafleu, *Linnaeus and the Linneans: The Spreading of their Ideas in Systematic Botany, 1735–1789* (1971), 31.

2 God created, Linnaeus ordered.

Quoted in Sten Lindroth, 'The Two Faces of Linnaeus', in Tore Frängsmyr (ed.), *Linnaeus: The Man and his Work* (1983), 22.

3 *Temporis fila.*

Child of time.

A favourite expression of Linnaeus. Tore Frängsmyr, 'Linnaeus as a Geologist', in Tore Frängsmyr (ed.), *Linnaeus: The Man and his Work* (1983), 143.

Gabriel Lippmann 1845–1921

French physicist

4 Everyone believes in the normal law, the experimenters because they imagine that it is a mathematical theorem, and the mathematicians because they think it is an experimental fact.

Conversation with Henri Poincaré. In Henri Poincaré, *Calcul des Probabilités* (1896), 171.

Joseph Lister 1827–1912

British surgeon

5 But when it has been shown by the researches of Pasteur that the septic property of the atmosphere depended not on the oxygen, or any gaseous constituent, but on minute organisms suspended in it, which owed their energy to their vitality, it occurred to me that decomposition in the injured part might be avoided without excluding the air, by applying as a dressing some material capable of destroying the life of the floating particles. Upon this principle I have based a practice.

'On the Antiseptic Principle in the Practice of Surgery', *The British Medical Journal*, 1867, ii, 246.

6 The frequency of disastrous consequences in compound fracture, contrasted with the complete immunity from danger to life or limb in simple fracture, is one of the most striking as well as melancholy facts in surgical practice.

'On a New Method of Treating Compound Fracture, Abscesses, etc: With Observations on the Conditions of Supperation', Part 1, *The Lancet*, 1867, 326.

7 Bearing in mind that it is from the vitality of the atmospheric particles that all the mischief arises, it appears that all that is requisite is to dress the wound with some material capable of killing these septic germs, provided that any substance can be found reliable for this purpose, yet not too potent as a caustic. In the course of the year 1864 I was much struck with an account of the remarkable effects produced by carbolic acid upon the sewage of the town of Carlisle, the admixture of a very small proportion not only preventing all odour from the lands irrigated with the refuse material, but, as it was stated, destroying the entozoa which usually infest cattle fed upon such pastures.

'On a New Method of Treating Compound Fracture, Abscesses, etc: With Observations on the Conditions of Supperation', Part 1, *The Lancet*, 1867, 327.

Clarence Cook Little 1888–1971

American geneticist

8 If we assume that there is only one enzyme present to act as an oxidizing agent, we must assume for it as many different degrees of activity as are required to explain the occurrence of the various colors known to mendelize (three in mice, yellow, brown, and black). If we assume that a different enzyme or group of enzymes is responsible for the production of each pigment we must suppose that in mice at least three such enzymes or groups of enzymes exist. To determine which of these conditions occurs in mice is not a problem for the biologist, but for the chemist. The biologist must confine his attention to determining the number of distinct agencies at work in pigment formation irrespective of their chemical nature. These agencies, because of their physiological behavior, the biologist chooses to call 'factors,' and attempts to learn what he can about their functions in the evolution of color varieties.

Experimental Studies of the Inheritance of Color in Mice (1913), 17–8.

John Locke 1632–1704

British philosopher

1 Truth scarce ever yet carried it by Vote any where at its first appearance: New Opinions are always suspected, and usually opposed, without any other Reason, but because they are not already common.

An Essay Concerning Human Understanding (1690). Edited by Peter Nidditch (1975), The Epistle Dedicatory, 4.

2 The Commonwealth of Learning is not at this time without Master-Builders, whose mighty Designs, in advancing the Sciences, will leave lasting Monuments to the Admiration of Posterity; But every one must not hope to be a Boyle, or a Sydenham; and in an Age that produces such Masters, as the Great—Huygenius, and the incomparable Mr. Newton, with some other of that Strain; 'tis Ambition enough to be employed as an Under-Labourer in clearing Ground a little, and removing some of the Rubbish, that lies in the way to Knowledge.

An Essay Concerning Human Understanding (1690). Edited by Peter Nidditch (1975), The Epistle to the Reader, 9–10.

3 The senses at first let in particular *Ideas*, and furnish the yet empty Cabinet: And the Mind by degrees growing familiar with some of them, they are lodged in the Memory, and Names got to them.

An Essay Concerning Human Understanding (1690). Edited by Peter Nidditch (1975), Book 1, Chapter 2, Section 15, 55.

4 Had you or I been born at the Bay of *Soldania*, possibly our Thoughts, and Notions, had not exceeded those brutish ones of the *Hotentots* that inhabit there: And had the *Virginia* King *Apochancana*, been educated in *England*, he had, perhaps been as knowing a Divine, and as good a Mathematician as any in it. The difference between him, and a more improved *English*-man, lying barely in this, That the exercise of his Facilities was bounded within the Ways, Modes, and Notions of his own Country, and never directed to any other or farther Enquiries.

An Essay Concerning Human Understanding (1690). Edited by Peter Nidditch (1975), Book 1, Chapter 4, Section 12, 92.

5 Every Man being conscious to himself, That he thinks, and that which his Mind is employ'd about whilst thinking, being the *Ideas*, that are there, 'tis past doubt, that Men have in their Minds several *Ideas*, such as are those expressed by the words, *Whiteness*, *Hardness*, *Sweetness*, *Thinking*, *Motion*, *Man*, *Elephant*, *Army*, *Drunkenness*, and others: It is in the first place then to be inquired, How he comes by them? I know it is a received Doctrine, That Men have native *Ideas*, and original Characters stamped upon their Minds, in their very first Being.

An Essay Concerning Human Understanding (1690). Edited by Peter Nidditch (1975), Book 2, Chapter 1, Section 1, 104.

6 Let us then suppose the Mind to be, as we say, white Paper, void of all Characters, without any *Ideas*; How comes it to be furnished? Whence comes it by that vast store, which the busy and boundless Fancy of Man has painted on it, with an almost endless variety? Whence has it all the materials of Reason and Knowledge? To this I answer, in one word, from *Experience:*

In that, all our Knowledge is founded; and from that it ultimately derives it self. Our Observation employ'd either about *external, sensible Objects; or about the internal Operations of our Minds, perceived and reflected on by our selves, is that, which supplies our Understandings with all the materials of thinking.*

An Essay Concerning Human Understanding (1690). Edited by Peter Nidditch (1975), Book 2, Chapter 1, Section 2, 104.

1 Nature never makes excellent things, for mean or no uses: and it is hardly to be conceived, that our infinitely wise Creator, should make so admirable a Faculty, as the power of Thinking, that Faculty which comes nearest the Excellency of his own incomprehensible Being, to be so idlely and uselesly employ'd, at least 1/4 part of its time here, as to think constantly, without remembering any of those Thoughts, without doing any good to it self or others, or being anyway useful to any other part of Creation.

An Essay Concerning Human Understanding (1690). Edited by Peter Nidditch (1975), Book 2, Chapter 1, Section 15, 113.

2 No Man's Knowledge here, can go beyond his Experience.

An Essay Concerning Human Understanding (1690). Edited by Peter Nidditch (1975), Book 2, Chapter 1, Section 19, 115.

3 The *Ideas of primary Qualities* of Bodies, are *Resemblances* of them, and their Patterns do really exist in the Bodies themselves; but the *Ideas, produced* in us *by* these *Secondary Qualities, have no resemblance* of them at all. There is nothing like our *Ideas*, existing in the Bodies themselves. They are in Bodies, we denominate from them, only a Power to produce those Sensations in us: And what is Sweet, Blue or Warm in *Idea*, is but the certain Bulk, Figure, and Motion of the insensible parts in the Bodies themselves, which we call so.

An Essay Concerning Human Understanding (1690). Edited by Peter Nidditch (1975), Book 2, Chapter 8, Section 15, 137.

4 The *Qualities* then that are in *Bodies* rightly considered, are of *Three sorts.*

First, the *Bulk, Figure, Number, Situation*, and *Motion, or Rest* of their solid Parts; those are in them, whether we perceive them or no; and when they are of that size, that we can discover them, we have by these an *Idea* of the thing, as it is in it self, as is plain in artificial things. These I call *primary Qualities.*

Secondly, The *Power* that is in any Body, *by* Reason of *its* insensible *primary Qualities*, to operate after a peculiar manner on any of our Senses, and thereby *produce in us* the *different Ideas* of several Colours, Sounds, Smells, Tastes, *etc.* These are usually called sensible Qualities.

Thirdly, The *Power* that is in any Body, *by* Reason of the particular Constitution of *its primary Qualities, to* make such a *change* in the *Bulk, Figure, Texture, and Motion of another Body*, as to make it operate on our Senses, differently from what it did before. Thus the Sun has a Power to make Wax white, and Fire to make Lead fluid. These are usually called Powers.

An Essay Concerning Human Understanding (1690). Edited by Peter Nidditch (1975), Book 2, Chapter 8, Section 23, 140–1.

5 We have hitherto considered those *Ideas*, in the reception whereof, the Mind is only passive, which are those simple ones received from *Sensation* and *Reflection* before-mentioned, whereof the Mind cannot make any one to it self, nor have any *Idea* which does not wholy consist of them. But as these simple *Ideas* are observed to exist in several Combinations united together; so the Mind has a power to consider several of them united together, as one *Idea*; and that not only as they are united in external Objects, but as it self has joined them. *Ideas* thus made up of several simple ones put together, I call *Complex*; such as are *Beauty, Gratitude, a Man, an Army, the Universe*; which tough complicated various simple *Ideas*, made up of simple ones, yet are, when the Mind pleases, considered each by if self, as one entire thing, and signified by one name.

An Essay Concerning Human Understanding (1690). Edited by Peter Nidditch (1975), Book 2, Chapter 12, Section 1, 163–4.

1 It is one thing, to shew a Man that he is in an Error; and another, to put him in possession of Truth.

An Essay Concerning Human Understanding (1690). Edited by Peter Nidditch (1975), Book 4, Chapter 7, Section 11, 602.

2 Such propositions are therefore called *Eternal Truths*, not because they are Eternal Propositions actually formed, and antecedent to the Understanding, that at any time makes them; nor because they are imprinted on the Mind from any patterns, that are any where out of the mind, and existed before: But because, being once made, about abstract *Ideas*, so as to be true, they will, whenever they can be supposed to be made again at any time, past or to come, by a Mind having those *Ideas*, always actually be true. For names being supposed to stand perpetually for the same ideas, and the same ideas having immutably the same habitudes one to another, Propositions concerning any abstract *Ideas* that are once true, must needs be *eternal Verities*.

An Essay Concerning Human Understanding (1690). Edited by Peter Nidditch (1975), Book 4, Chapter 11, Section 14, 638–9.

3 In our search after the Knowledge of *Substances*, our want of *Ideas*, that are suitable to such a way of proceeding, obliges us to a quite different method. We advance not here, as in the other (where our abstract *Ideas* are real as well as nominal Essences) by contemplating our *Ideas*, and considering their Relations and Correspondencies; that helps us very little, for the Reasons, and in another place we have at large set down. By which, I think it is evident, that Substances afford Matter of very little general Knowledge; and the bare Contemplation of their abstract *Ideas*, will carry us but a very little way in the search of Truth and Certainty. What then are we to do for the improvement of our *Knowledge in Substantial beings?* Here we are to take a quite contrary Course, the want of *Ideas* of their real essences sends us from our own Thoughts, to the Things themselves, as they exist.

An Essay Concerning Human Understanding (1690). Edited by Peter Nidditch (1975), Book 4, Chapter 12, Section 9, 644.

4 From whence it is obvious to conclude that, since our Faculties are not fitted to penetrate into the internal Fabrick and real Essences of Bodies; but yet plainly discover to us the Being of a GOD, and the Knowledge of our selves, enough to lead us into a full and clear discovery of our Duty, and great Concernment, it will become us, as rational Creatures, to imploy those Faculties we have about what they are most adapted to, and follow the direction of Nature, where it seems to point us out the way.

An Essay Concerning Human Understanding (1690). Edited by Peter Nidditch (1975), Book 4, Chapter 12, Section 11, 646.

5 Not that we may not, to explain any *Phenomena* of Nature, make use of any probable *Hypothesis* whatsoever: *Hypotheses*, if they are well made, are at least great helps to the Memory, and often direct us to new discoveries. But my Meaning is, that we should *not take up any one too hastily*, (which the Mind, that would always penetrate into the Causes of Things, and have Principles to rest on, is very apt to do,) till we have very well examined Particulars, and made several Experiments, in that thing which we would explain by our Hypothesis, and see whether it will agree to them all; whether our Principles will carry us quite through, and not be as inconsistent with one *Phenomenon* of Nature, as they seem to accommodate and explain another.

An Essay Concerning Human Understanding (1690). Edited by Peter Nidditch (1975), Book 4, Chapter 12, Section 13, 648.

1 Crooked things may be as stiff and unflexible as streight: and Men may be as positive and peremptory in Error as in Truth.

An Essay Concerning Human Understanding (1690). Edited by Peter Nidditch (1975), Book 4, Chapter 19, Section 11, 703.

2 All Men are liable to Errour, and most Men are in many Points, by Passion or Interest, under Temptation to it.

An Essay Concerning Human Understanding (1690). Edited by Peter Nidditch (1975), Book 4, Chapter 20, Section 17, 718.

Joseph Norman Lockyer

1836–1920

British astrophysicist and first editor of Nature

3 ON LOCKYER [Lockyer] . . . sometimes forgets he is only the editor and not the author of *Nature*.

J. W. L. Glaisher (ed.), *The Collected Mathematical Papers of Henry John Stephen Smith* (1894), Vol. 1, xliv.

Oliver Joseph Lodge 1851–1940

British physicist

4 Matter moves, but Ether is strained.

Continuity: The Presidential Address to the British Association (1913), 33.

5 If the 'Principle of Relativity' in an extreme sense establishes itself, it seems as if even Time would become discontinuous and be supplied in atoms, as money is doled out in pence or centimes instead of continuously;—in which case our customary existence will turn out to be no more really continuous than the events on a kinematograph screen;—while that great agent of continuity, the Ether of Space, will be relegated to the museum of historical curiosities.

Continuity: The Presidential Address to the British Association (1913), 40–1.

6 Genuine religion has its root deep down in the heart of humanity and in the reality of things. It is not surprising that by our methods we fail to grasp it: the actions of the Deity make no appeal to any special sense, only a universal appeal; and our methods are, as we know, incompetent to detect complete uniformity. There is a principle of Relativity here, and unless we encounter flaw or jar or change, nothing in us responds; we are deaf and blind therefore to the Immanent Grandeur, unless we have insight enough to recognise in the woven fabric of existence, flowing steadily from the loom in an infinite progress towards perfection, the ever-growing garment of a transcendent God.

Continuity: The Presidential Address to the British Association (1913), 92–3.

7 There is a conservation of matter and of energy, there may be a conservation of life; or if not of life, of something which transcends life.

Christopher: A Study in Human Personality (1918), 68.

8 There is no instrument for measuring the pressure of the Ether, which is probably millions of times greater: it is altogether too uniform for direct apprehension. A deep-sea fish has probably no means of apprehending the existence of water, it is too uniformly immersed in it: and that is our condition in regard to the Ether.

Ether and Reality: A Series of Discourses on the Many Functions of the Ether of Space (1925), 28.

9 The first thing to realise about the Ether is its absolute continuity.

Ether and Reality: A Series of Discourses on the Many Functions of the Ether of Space (1925), 35.

Jacques Loeb 1859–1924

German-born American biologist

10 CREATION OF LIFE.
The Startling Discovery of Prof. Loeb.
Lower Animals Produced by Chemical Means.
Process May Apply to the Human Species.
Immaculate Conception is Explained.

Wonderful Experiments Conducted at Woods Hole.

The Boston Herald, 26 November 1899, 17.

1 Through the discovery of Büchner, Biology was relieved of another fragment of mysticism. The splitting up of sugar into CO_2 and alcohol is no more the effect of a 'vital principle' than the splitting up of cane sugar by invertase. The history of this problem is instructive, as it warns us against considering problems as beyond our reach because they have not yet found their solution.

The Dynamics of Living Matter (1906), 22.

2 Will it be possible to solve these problems? It is certain that nobody has thus far observed the transformation of dead into living matter, and for this reason we cannot form a definite plan for the solution of this problem of transformation. But we see that plants and animals during their growth continually transform dead into living matter, and that the chemical processes in living matter do not differ in principle from those in dead matter. There is, therefore, no reason to predict that abiogenesis is impossible, and I believe that it can only help science if the younger investigators realize that experimental abiogenesis is the goal of biology.

The Dynamics of Living Matter (1906), 223.

3 For the metaphysical term 'will' we may in these instances safely substitute the chemical term 'photochemical action of light.'

The Mechanistic Conception of Life (1912), 30.

4 Since Pawlow [Pavlov] and his pupils have succeeded in causing the secretion of saliva in the dog by means of optic and acoustic signals, it no longer seems strange to us that what the philosopher terms an 'idea' is a process which can cause chemical changes in the body.

The Mechanistic Conception of Life (1912), 63.

5 When . . . the biologist is confronted with the fact that in the organism the parts are so adapted to each other as to give rise to a harmonious whole; and that the organisms are endowed with structures and instincts calculated to prolong their life and perpetuate their race, doubts as to the adequacy of a purely physiochemical viewpoint in biology may arise. The difficulties besetting the biologist in this problem have been rather increased than diminished by the discovery of Mendelian heredity, according to which each character is transmitted independently of any other character. Since the number of Mendelian characters in each organism is large, the possibility must be faced that the organism is merely a mosaic of independent hereditary characters. If this be the case the question arises: What moulds these independent characters into a harmonious whole? The vitalist settles this question by assuming the existence of a pre-established design for each organism and of a guiding 'force' or 'principle' which directs the working out of this design. Such assumptions remove the problem of accounting for the harmonious character of the organism from the field of physics or chemistry. The theory of natural selection invokes neither design nor purpose, but it is incomplete since it disregards the physiochemical constitution of living matter about which little was known until recently.

The Organism as a Whole: From a Physiochemical Viewpoint (1916), v-vi.

6 Man is a megalomaniac among animals—if he sees mountains he will try to imitate them by pyramids, and if he sees some grand process like evolution, and thinks it would be at all possible for him to be in on that game, he would irreverently have to have his whack at that too. That daring megalomania of his—has it not brought him to his present place?

'Application and Prospects', unpublished lecture, 1916. In Philip J. Pauly, *Controlling Life: Jacques Loeb and the Engineering idea in Biology* (1987), 179.

Friedrich August Johannes Loeffler 1852–1915

German microbiologist

1 If diphtheria is a disease caused by a microorganism, it is essential that three postulates be fulfilled. The fulfilment of these postulates is necessary in order to demonstrate strictly the parasitic nature of a disease:

1) The organism must be shown to be constantly present in characteristic form and arrangement in the diseased tissue.

2) The organism which, from its behaviour appears to be responsible for the disease, must be isolated and grown in pure culture.

3) The pure culture must be shown to induce the disease experimentally.

An early statement of Koch's postulates.

Mittheilungen aus den Kaiserliche Gesundheitsamt (1884) Vol. 2. Trans. T. D. Brock, *Robert Koch: A Life in Medicine and Bacteriology* (1988), 180.

Otto Loewi 1873–1961

German-born American physiologist and pharmacologist

See ***Dale, Henry Hallett*** *147:3.*

2 The night before Easter Sunday of that year (1920) I awoke, turned on the light, and jotted down a few notes on a tiny slip of thin paper. Then I fell asleep again. It occurred to me at six o'clock in the morning that during the night I had written down something most important, but I was unable to decipher the scrawl. The next night, at three o'clock, the idea returned. It was the design of an experiment to determine whether the hypothesis of chemical transmission that I had uttered seventeen years ago was correct. I got up immediately, went to the laboratory, and performed a simple experiment on a frog heart according to the nocturnal design. I have to describe this experiment briefly since its results became the foundation of the theory of chemical transmission of the nervous impulse. The hearts of two frogs were isolated, the first with its nerves, the second without. Both hearts were attached to Straub cannulas filled with a little Ringer solution. The vagus nerve of the first heart was stimulated for a few minutes. Then the Ringer solution that had been in the first heart during the stimulation of the vagus was transferred to the second heart. It slowed and its beats diminished just as if its vagus had been stimulated. Similarly, when the accelerator nerve was stimulated and the Ringer from this period transferred, the second heart speeded up and its beats increased. These results unequivocally proved that the nerves do not influence the heart directly but liberate from their terminals specific chemical substances which, in their turn, cause the well-known modifications of the function of the heart characteristic of the stimulation of its nerves.

'An Autobiographic Sketch', *Perspectives in Biology and Medicine*, 1960, 4, 17.

3 A drug is a substance which, if injected into a rabbit, produces a paper.

Quoted in Albert Szent-Gyorgyi, 'Some Reminiscences of My Life as a Scientist', *International Journal of Quantum Biology Symposium*, 1976, 3, 7.

Mikhail Vasilievich Lomonosov 1711–65

Russian poet, chemist and physicist

4 In general, the bigger a mountain the older it is. The biggest mountains were built before any others, because when they were built there was incomparably more flammable material within the Earth. Over the many thousands of years that have passed, the quantity of flammable material has doubtless decreased.

On the Strata of the Earth (1763), paragraph 119.

5 Nature uncovers the inner secrets of nature in two ways: one by the force of bodies operating outside it; the other by the very movements of its innards. The external actions are strong winds, rains, river currents, sea waves, ice, forest fires, floods; there is only one internal force—earthquake.

About the Layers of the Earth and other Works on Geology (1757), trans. A. P. Lapov (1949), 45.

1 And so many think incorrectly that everything was created by the Creator in the beginning as it is seen, that not only the mountains, valleys, and waters, but also various types of minerals occurred together with the rest of the world, and therefore it is said that it is unnecessary to investigate the reasons why they differ in their internal properties and their locations. Such considerations are very dangerous for the growth of all the sciences, and hence for natural knowledge of the Earth, particularly the art of mining, though it is very easy for those clever people to be philosophers, having learnt by heart the three words 'God so created' and to give them in reply in place of all reasons.

About the Layers of the Earth and other Works on Geology (1757), trans. A. P. Lapov (1949), 55.

Heinz London 1907–70

German-born British physicist

2 For the second law [of thermodynamics], I will burn at the stake.

Comment made to H. Montgomery during his time at Harwell. In D. Shoenberg's obituary of H. London, *Biographical Memoirs of Fellows of the Royal Society*, 1971, 17, 442.

Kathleen Yardley Lonsdale 1903–71

British crystallographer, chemist and physicist

3 It makes me feel both proud and rather humble that it shall be called Lonsdaleite. Certainly the name seems appropriate since the mineral only occurs in very small quantities (perhaps rare would be too flattering) and it is generally rather mixed up!

Letter to Clifford Frondel, who had named a meteoritic form of diamond after her. Lonsdale was a very tiny lady!

In Maureen M. Julian, 'Women in Crystallography', in G. Kass-Simon and P. Farnes (eds.), *Women of Science: Righting the Record* (1990), 356.

Edward N. Lorenz 1917–

American meteorologist

4 Predictability: Does the Flap of a Butterfly's Wings in Brazil Set off a Tornado in Texas?

Title of lecture given to the American Association for the Advancement of Science, Washington, 29 December 1972. In E. N. Lorenz, *The Essence of Chaos* (1993), 181.

Konrad Lorenz 1903–89

Austrian zoologist and ethnologist

5 It is a good morning exercise for a research scientist to discard a pet hypothesis every day before breakfast. It keeps him young.

On Aggression, trans. M. Latzke (1966), 8.

6 The neuro-physiological organization which we call instinct functions in a blindly mechanical way, particularly apparent when its function goes wrong.

On Aggression, trans. M. Latzke (1966), 73.

7 This is a classical example of the process which we call, with Tinbergen, a *redirected activity*. It is characterized by the fact that an activity is released by *one* object but discharged *at another*, because the first one, while presenting stimuli specifically eliciting the response, simultaneously emits others which inhibit its discharge. A human example is furnished by the man who is very angry with someone and hits the table instead of the other man's jaw, because inhibition prevents him from doing so, although his pent-up anger, like the pressure within a volcano, demands outlet.

On Aggression, trans. M. Latzke (1966), 145.

8 I believe—and human psychologists, particularly psycho-analysts should test this—that present-day civilized man suffers from insufficient discharge of his aggressive drive. It is more than probable that the evil effects of the human aggressive drives, explained by Sigmund Freud as the results of a special death wish, simply derive from the fact that in prehistoric times intra-specific selection bred into man a measure of aggression drive for which in the social order today he finds no adequate outlet.

On Aggression, trans. M. Latzke (1966), 209.

1 Historians will have to face the fact that natural selection determined the evolution of cultures in the same manner as it did that of species.

On Aggression, trans. M. Latzke (1966), 260.

2 In nature we find not only that which is expedient, but also everything which is not so inexpedient as to endanger the existence of the species.

On Aggression, trans. M. Latzke (1966), 260.

3 If you confine yourself to this Skinnerian technique, you study nothing but the learning apparatus and you leave out everything that is different in octopi, crustaceans, insects and vertebrates. In other words, you leave out everything that makes a pigeon a pigeon, a rat a rat, a man a man, and, above all, a healthy man healthy and a sick man sick.

'Some Psychological Concepts and Issues. A Discussion between Konrad Lorenz and Richard I Evans'. In Richard I. Evans, *Konrad Lorenz: The Man and his Ideas* (1975), 60.

4 Visualize yourself confronted with the task of killing, one after the other, a cabbage, a fly, a fish, a lizard, a guinea pig, a cat, a dog, a monkey and a baby chimpanzee. In the unlikely case that you should experience no greater inhibitions in killing the chimpanzee than in destroying the cabbage or the fly, my advice to you is to commit suicide at your earliest possible convenience, because you are a weird monstrosity and a public danger.

'The Enmity Between Generations and Its Probable Ethological Causes'. In Richard I. Evans, *Konrad Lorenz: The Man and his Ideas* (1975), 227.

Hermann Rudolph Lotze 1817–81

German biologist and philosopher

5 *Die durchschnittliche Lebensdauer einer physiologischen Wahrheit ist drei bis vier Jahre.*

The average lifespan of a physiological truth is three or four years.

Attributed.

Arthur Oncken Lovejoy 1873–1962

American philosopher and historian

6 Through the Middle Ages and down to the late eighteenth century, many philosophers, most men of science, and, indeed, most educated men, were to accept without question—the conception of the universe as a Great Chain of Being, composed of an immense, or—by the strict but seldom rigorously applied logic of the principle of continuity—of an infinite number of links ranging in hierarchical order from the meagerest kind of existents, which barely escape non-existence, through 'every possible' grade up to the ens perfectissimum—or, in a somewhat more orthodox version, to the highest possible kind of creature, between which and the Absolute Being the disparity was assumed to be infinite—every one of them differing from that immediately above and that immediately below it by the 'least possible' degree of difference.

The Great Chain of Being (1936), 59.

7 Next to the word 'Nature,' 'the Great Chain of Being' was the sacred phrase of the eighteenth century, playing a part somewhat analogous to that of the blessed word 'evolution' in the late nineteenth.

The Great Chain of Being (1936), 184.

Augusta Ada King, Countess of Lovelace 1815–52

British mathematician

1 The Analytical Engine *weaves algebraical patterns* just as the Jacquard loom weaves flowers and leaves.

Comment on Babbage's engines.
From 'Sketch of the Analytical Engine invented by Charles Babbage, Esq.' [by I. F. Menabrea with notes by Ada Lovelace], *Scientific Memoirs*, 1843, 3, 696.

2 The Analytical Engine has no pretensions whatever to *originate* anything. It can do whatever we *know how to order* it to perform. It can *follow* analysis; but it has no power of *anticipating* any analytical relations or truths. Its province is to assist us in making *available* what we are already acquainted with.

Comment on Babbage's engines.
From 'Sketch of the Analytical Engine invented by Charles Babbage, Esq.' [by I. F. Menabrea with notes by Ada Lovelace], *Scientific Memoirs*, 1843, 3, 722.

3 I want to put in something about Bernoulli's numbers, in one of my Notes, as an example of how the implicit function may be worked out by the engine, without having been worked out by human head & hands first. Give me the necessary data & formulae.

That brain of mine is something more than merely Mortal; as time will show; (if only my breathing & some other etceteras do not make too rapid a progress *towards* instead of *from* mortality). Before ten years are over, the Devil's in it if I haven't sucked out some of the life-blood from the mysteries of this universe.

I believe myself to possess a most singular combination of qualities exactly fitted to make me *pre-eminently* a discoverer of the *hidden realities* of nature . . . the belief has been *forced* upon me . . .

Firstly: Owing to some peculiarity in my nervous system, I have *perceptions* of some things, which no one else has . . . and *intuitive* perception of . . . things hidden from eyes, ears, & ordinary senses . . .

Secondly: my immense reasoning faculties;

Thirdly: my concentration faculty, by which I mean the power not only of throwing my whole energy & existence into whatever I choose, but also of bringing to bear on any one subject or idea, a vast apparatus from all sorts of apparently irrelevant & extraneous sources . . .

Well, here I have written what most people would call a remarkably *mad* letter; & yet certainly one of the most logical, sober-minded, cool, pieces of composition, (I believe), that I ever framed.

This First Child of Mine. In Dorothy Stein (ed.), *Ada: A Life and a Legacy* (1985), 87.

4 Imagination is the *Discovering* Faculty, pre-eminently . . . It is that which feels & discovers what *is*, the REAL which we see not, which exists not for our *senses* . . . Mathematical science shows what *is*. It is the language of unseen relations between things . . . Imagination too shows what *is* . . . Hence she is or should be especially cultivated by the truly Scientific, those who wish to enter into the worlds around us!

In Time I Will Do All, I Dare Say. In Dorothy Stein (ed.), *Ada: A Life and a Legacy* (1985), 129.

James Lovelock 1919–

British climatologist and inventor

5 No one who has experienced the intense involvement of computer modeling would deny that the temptation exists to use any data input that will enable one to continue playing what is perhaps the ultimate game of solitaire.

Gaia: A New Look at Life on Earth (1979), 137–8.

6 Any artist or novelist would understand—some of us do not produce their best when directed. We expect the artist, the novelist and the composer to lead solitary lives, often working at home. While a few of these creative individuals exist in institutions or universities, the idea of a majority of established novelists or painters working at the 'National Institute for Painting and Fine Art' or a university 'Department of Creative Composition' seems mildly amusing. By contrast, alarm greets the idea of a creative scientist working at home. A lone scientist is as unusual as a solitary termite and regarded as irresponsible or worse.

Homage to Gaia: The Life of an Independent Scholar (2000), 2.

1 Until that afternoon, my thoughts on planetary atmospheres had been wholly concerned with atmospheric analysis as a method of life detection and nothing more. Now that I knew the composition of the Martian atmosphere was so different from that of our own, my mind filled with wonderings about the nature of the Earth. If the air is burning, what sustains it at a constant composition? I also wondered about the supply of fuel and the removal of the products of combustion. It came to me suddenly, just like a flash of enlightenment, that to persist and keep stable, something must be regulating the atmosphere and so keeping it at its constant composition. Moreover, if most of the gases came from living organisms, then life at the surface must be doing the regulation.

Homage to Gaia: The Life of an Independent Scholar (2000), 253.

Joseph Lovering 1813–92

American mathematican

2 The great problem of the day is, how to subject all physical phenomena to dynamical laws. With all the experimental devices, and all the mathematical appliances of this generation, the human mind has been baffled in its attempts to construct a universal science of physics.

'President's Address', *Proceedings of the American Association for the Advancement of Science*, 1874, 23, 34–5.

Robert H. Lowie 1883–1957

Austrian-born American anthropologist

3 Since biological change occurs slowly and cultural changes occur in every generation, it is futile to try to explain the fleeting phenomena of culture by a racial constant. We can often explain them—in terms of contact with other peoples, of individual genius, of geography—but not by *racial* differences.

An Introduction to Cultural Anthropology (1934), 9.

John Lubbock (Lord Avebury) 1834–1913

British naturalist

4 Deprived, therefore, as regards this period, of any assistance from history, but relieved at the same time from the embarrassing interference of tradition, the archaeologist is free to follow the methods which have been so successfully pursued in geology—the rude bone and stone implements of bygone ages being to the one what the remains of extinct animals are to the other. The analogy may be pursued even further than this. Many mammalia which are extinct in Europe have representatives still living in other countries. Our fossil pachyderms, for instance, would be almost unintelligible but for the species which still inhabit some parts of Asia and Africa; the secondary marsupials are illustrated by their existing representatives in Australia and South America; and in the same manner, if we wish clearly to understand the antiquities of Europe, we must compare them with the rude implements and weapons still, or until lately, used by the savage races in other parts of the world. In fact, the Van Diemaner and South American are to the antiquary what the opossum and the sloth are to the geologist.

Pre-historic Times, as Illustrated by Ancient Remains, and the Manners and Customs of Modern Savages, 2nd Edition (1869), 416.

1 Savages have often been likened to children, and the comparison is not only correct but also highly instructive. Many naturalists consider that the early condition of the individual indicates that of the race,—that the best test of the affinities of a species are the stages through which it passes. So also it is in the case of man; the life of each individual is an epitome of the history of the race, and the gradual development of the child illustrates that of the species.

Pre-historic Times, as Illustrated by Ancient Remains, and the Manners and Customs of Modern Savages, 2nd Edition (1869), 558.

2 Where the untrained eye will see nothing but mire and dirt, Science will often reveal exquisite possibilities.

The Pleasure of Life (1887), 156.

Keith Lucas 1879–1916

British physiologist

3 The skeletal striated muscle cell of amphibia therefore resembles the cardiac striated muscle cell in the property of 'all or none' contraction.

'The 'All or None' Contraction of the Amphibian Skeletal Muscle Fibre', *Journal of Physiology*, 1909, 38, 133.

Lucian 115–80 BC

Greek rhetorician

4 *Deus ex machina.*

A God from the machine.

A phrase, originally written in Greek, used by Lucian on several occasions: *Philopseudes*, 29; *De Mercede Conductis*, 1; *Hermotimus*, 86.

Titus Lucretius c.100–c.55 BC

Graeco-Roman natural philosopher

5 Anything made out of destructible matter
Infinite time would have devoured before.
But if the atoms that make and replenish the world
Have endured through the immense span of the past
Their natures are immortal—that is clear.
Never can things revert to nothingness!

On the Nature of Things, trans. Anthony M. Esolen (1995), Book 1, lines 232–7, 31.

6 Things stand apart so far and differ, that
What's food for one is poison for another.

On the Nature of Things, trans. Anthony M. Esolen (1995), Book 4, lines 634–5, 31.

7 Here I most violently want you to
Avoid one fearful error, a vicious flaw.
Don't think that our bright eyes were made that we
Might look ahead; that hips and knees and ankles
So intricately bend that we might take
Big strides, and the arms are strapped to the sturdy shoulders
And hands are given for servants to each side
That we might use them to support our lives.
All other explanations of this sort
Are twisted, topsy-turvy logic, for
Nothing what is born produces its own use.
Sight was not born before the light of the eyes,
Nor were words and pleas created before the tongue
Rather the tongue's appearance long preceded
Speech, and the ears were formed far earlier than
The sound first heard. To sum up, all the members
Existed, I should think, *before* their use,
So use has not caused them to have grown.

On the Nature of Things, trans. Anthony M. Esolen (1995), Book 4, lines 820–8, 145.

8 Under what law each thing was created, and how necessary it is for it to continue under this, and how it cannot annul the strong rules that govern its lifetime.

On the Nature of Things, Book 5, line 56. Trans. R. W. Sharples.

1 And part of the soil is called to wash away
In storms and streams shave close and gnaw the rocks.
Besides, whatever the earth feeds and grows
Is restored to earth. And since she surely is
The womb of all things and their common grave,
Earth must dwindle, you see and take on growth again.

On the Nature of Things, trans. Anthony M. Esolen (1995), Book 5, lines 255–60, 166.

2 I return to the newborn world, and the soft-soil fields,
What their first birthing lifted to the shores
Of light, and trusted to the wayward winds.
First the Earth gave the shimmer of greenery
And grasses to deck the hills; then over the meadows
The flowering fields are bright with the color of springtime,
And for all the trees that shoot into the air.

On the Nature of Things, trans. Anthony M. Esolen (1995) Book 5, lines 777–84, 181.

3 Many animals even now spring out of the soil,
Coalescing from the rains and the heat of the sun.
Small wonder, then, if more and bigger creatures,
Full-formed, arose from the new young earth and sky.
The breed, for instance, of the dappled birds
Shucked off their eggshells in the springtime, as
Crickets in summer will slip their slight cocoons
All by themselves, and search for food and life.
Earth gave you, then, the first of mortal kinds,
For all the fields were soaked with warmth and moisture.

On the Nature of Things, trans. Anthony M. Esolen (1995), Book 5, lines 794–803, 181.

4 And many kinds of creatures must have died,
Unable to plant out new sprouts of life.
For whatever you see that lives and breathes and thrives
Has been, from the very beginning, guarded, saved
By its trickery or its swiftness or brute strength.
And many have been entrusted to our care,
Commended by their usefulness to us.
For instance, strength supports the savage lion;
Foxes rely on their cunning; deer their flight.

On the Nature of Things, trans. Anthony M. Esolen (1995), Book 5, lines 852–60, 183.

5 Huts they made then, and fire, and skins for clothing,
And a woman yielded to one man in wedlock . . .
. . . Common, to see the offspring they had made;
The human race began to mellow then.
Because of fire their shivering forms no longer
Could bear the cold beneath the covering sky.

On the Nature of Things, trans. Anthony M. Esolen (1995), Book 5, lines 1008–13, 187.

6 Moreover, within the hollows of the earth,
When from one quarter the wind builds up, lunges,
Muscles the deep caves with its headstrong power,
The earth leans hard where the force of wind has pressed it;
Then above ground, the higher the house is built,
The nearer it rises to the sky, the worse
Will it lean that way and jut out perilously,
The beams wrenched loose and hanging ready to fall.
And to think, men can't believe that for this world

Some time of death and ruin lies in wait,
Yet they see so great a mass of earth collapse!
And the winds pause for breath—that's lucky, for else
No force could rein things galloping to destruction.
But since they pause for breath, to rally their force,
Come building up and then fall driven back,
More often the earth will threaten ruin than
Perform it. The earth will lean and then sway back,
Its wavering mass restored to the right poise.
That explains why all houses reel, top floor
Most then the middle, and ground floor hardly at all.

On the Nature of Things, trans. Anthony M. Esolen (1995), Book 6, lines 558–77, 216.

Carl Freidrich Wilhelm Ludwig 1816–95

German physiologist

1 Scientific physiology has the task of determining the functions of the animal body and deriving them as a necessary consequence from its elementary conditions.

Lehrbuch der Physiologie des Menschens (1852), Vol 1, 1. Trans. Paul F. Cranefield, 'The Organic Physics of 1847 and the Biophysics of Today', *Journal of the History of Medicine and Allied Sciences*, 1957, 12, 410.

2 For the first time there was constructed with this machine [locomotive engine] a *self-acting* mechanism in which the interplay of forces took shape transparently enough to discern the connection between the heat generated and the motion produced. The great puzzle of the vital force was also immediately solved for the physiologist in that it became evident that it is more than a mere poetic comparison when one conceives of the coal as the food of the locomotive and the combustion as the basis for its life.

'Leid und Freude in der Naturforschung', *Die Gartenlaube* (1870), 359. Trans. Kenneth L. Caneva, *Robert Mayer and the Conservation of Energy* (1993), 145.

Martin Luther 1483–1546

German protestant theologian

3 Astrology is framed by the devil, to the end people may be scared from entering into the state of matrimony, and from every divine and human office and calling.

W. Hazlitt (trans. and ed.) *The Table Talk of Martin Luther*, (1857), 343.

André Lwoff 1902–94

French microbiologist

4 The mechanist is intimately convinced that a precise knowledge of the chemical constitution, structure, and properties of the various organelles of a cell will solve biological problems. This will come in a few centuries. For the time being, the biologist has to face such concepts as orienting forces or morphogenetic fields. Owing to the scarcity of chemical data and to the complexity of life, and despite the progresses of biochemistry, the biologist is still threatened with vertigo.

Problems of Morphogenesis in Ciliates: The Kinetosomes in Development, Reproduction and Evolution (1950), 92–3.

5 For the philosopher, order is the entirety of repetitions manifested, in the form of types or of laws, by perceived objects. Order is an intelligible relation. For the biologist, order is a sequence in space and time. However, according to Plato, all things arise out of their opposites. Order was born of the original disorder, and the long evolution responsible for the present biological order necessarily had to engender disorder.

An organism is a molecular society, and biological order is a kind of social order. Social order is opposed to revolution, which is an abrupt change of order, and to anarchy, which is the absence of order.

I am presenting here today both revolution and anarchy, for which I am fortunately not the only one responsible. However, anarchy cannot survive and prosper except in an ordered society, and revolution becomes sooner or later the new order. Viruses have not failed to follow the general law. They are strict parasites which, born of disorder, have created a very remarkable new order to ensure their own perpetuation.

'Interaction Among Virus, Cell, and Organism', Nobel Lecture, 11 December 1965. In *Nobel Lectures: Physiology or Medicine 1963–1970* (1972), 174.

1 *L'art du chercheur c'est d'abord de se trouver un bon patron.*

The researcher's art is first of all to find himself a good boss.

Cited in Review of *Advice to a Young Scientist* by P. B. Medawar. In Max Perutz, *Is Science Necessary?: Essays on Science and Scientists* (1991), 194.

Charles Lyell 1797–1875

British geologist

See ***Whewell, William*** *616:5*

2 Geology is the science which investigates the successive changes that have taken place in the organic and inorganic kingdoms of nature; it enquires into the causes of these changes, and the influence which they have exerted in modifying the surface and external structure of our planet.

Principles of Geology (1830–3), Vol. 1, 1.

3 Geology is intimately related to almost all the physical sciences, as is history to the moral. An historian should, if possible, be at once profoundly acquainted with ethics, politics, jurisprudence, the military art, theology; in a word, with all branches of knowledge, whereby any insight into human affairs, or into the moral and intellectual nature of man, can be obtained. It would be no less desirable that a geologist should be well versed in chemistry, natural philosophy, mineralogy, zoology, comparative anatomy, botany; in short, in every science relating to organic and inorganic nature. With these accomplishments the historian and geologist would rarely fail to draw correct and philosophical conclusions from the various monuments transmitted to them of former occurrences.

Principles of Geology (1830–3), Vol. 1, 2–3.

4 Geology differs as widely from cosmogony, as speculations concerning the creation of man differ from history.

Principles of Geology (1830–3), Vol. 1, 4.

5 We might expect . . . in the summer of the 'great year,' which we are now considering, that there would be a great predominance of tree-ferns and plants allied to the palms and arborescent grasses in the isles of the wide ocean, while the dicotyledenous plants and other forms now most common in temperate regions would almost disappear from the earth. Then might these genera of animals return, of which the memorials are preserved in the ancient rocks of our continents. The huge iguanodon might reappear in the woods, and the ichthyosaur in the sea, while the pterodactyle might flit again through umbrageous groves of tree-ferns. Coral reefs might be prolonged beyond the arctic circle, where the whale and narwal [sic] now abound. Turtles might deposit their eggs in the sand of the sea beach, where now the walrus sleeps, and where the seal is drifted on the ice-floe.

Principles of Geology (1830–3), Vol. 1, 123.

6 There is no foundation in geological facts, for the popular theory of the successive development of the animal and vegetable world, from the simplest to the most perfect forms.

Principles of Geology (1830–3), Vol. 1, 153.

7 The earth's becoming at a particular period the residence of human beings, was an era in the moral, not in the physical world—that our study and contemplation of the earth, and the laws which govern its animate productions, ought no more to be considered in the light of a disturbance or deviation from the system, than the discovery of the satellites of Jupiter should be regarded as a physical event in the history of those heavenly bodies, however influential they may have become from that time in advancing the progress of sound philosophy among men.

Principles of Geology (1830–3), Vol. 1, 163.

1 Each species may have had its origin in a single pair, or individual, where an individual was sufficient, and species may have been created in succession at such times and in such places as to enable them to multiply and endure for an appointed period, and occupy an appointed space on the globe.

Principles of Geology (1830–3), Vol. 2, 124.

2 Never was there a dogma more calculated to foster indolence, and to blunt the keen edge of curiosity, than . . . [the] assumption of the discordance between the former and the existing causes of change.

Principles of Geology (1830–3), Vol. 3, 2–3.

3 It must have appeared almost as improbable to the earlier geologists, that the laws of earthquakes should one day throw light on the origin of mountains, as it must to the first astronomers, that the fall of an apple should assist in explaining the motions of the moon.

Principles of Geology (1830–3), Vol. 3, 5.

4 As geologists, we learn that it is not only the present condition of the globe that has been suited to the accommodation of myriads of living creatures, but that many former states also have been equally adapted to the organization and habits of prior races of beings. The disposition of the seas, continents, and islands, and the climates have varied; so it appears that the species have been changed, and yet they have all been so modelled, on types analogous to those of existing plants and animals, as to indicate throughout a perfect harmony of design and unity of purpose. To assume that the evidence of the beginning or end of so vast a scheme lies within the reach of our philosophical inquiries, or even of our speculations, appears to us inconsistent with a just estimate of the relations which subsist between the finite powers of man and the attributes of an Infinite and Eternal Being.

Principles of Geology (1830–3), Vol. 3, 384–5.

5 In the course of this short tour, I became convinced that we must turn to the *New World* if we wish to see in perfection the oldest monuments of the earth's history, so far at least as relates to its earliest inhabitants.

Travels in North America (1845), Vol. 1, 19.

6 Etna presents us not merely with an image of the power of subterranean heat, but a record also of the vast period of time during which that power has been exerted. A majestic mountain has been produced by volcanic action, yet the time of which the volcanic forms the register, however vast, is found by the geologist to be of inconsiderable amount, even in the modern annals of the earth's history. In like manner, the Falls of Niagara teach us not merely to appreciate the power of moving water, but furnish us at the same time with data for estimating the enormous lapse of ages during which that force has operated. A deep and long ravine has been excavated, and the river has required ages to accomplish the task, yet the same region affords evidence that the sum of these ages is as nothing, and as the work of yesterday, when compared to the antecedent periods, of which there are monuments in the same district.

Travels in North America (1845), Vol. 1, 28–9.

7 However much we may enlarge our ideas of the time which has elapsed since the Niagara first began to drain the waters of the upper lakes, we have seen that this period was one only of a series, all belonging to the present zoological epoch; or that in which the living testaceous fauna, whether freshwater or marine, had already come into being. If such events can take place while the zoology of the earth remains almost stationary and unaltered, what ages may not be comprehended in those successive tertiary periods during which the Flora and Fauna of the globe have been almost entirely changed. Yet how subordinate a place in the long calendar of geological chronology do the successive tertiary periods themselves occupy! How much more enormous a duration must we assign to many antecedent revolutions of the earth and its inhabitants! No analogy can be found in the natural world to the immense scale of these divisions of past time, unless we contemplate the celestial spaces which have been measured by the astronomer.

Travels in North America (1845), Vol. 1, 51–2.

1 Man, whose organization is regarded as the highest, departs from the vertebrate archetype; and it is because the study of anatomy is usually commenced from, and often confined to, *his* structure, that a knowledge of the archetype has been so long hidden from anatomists.

'The Lexington Papers', *The Quarterly Review*, 1851, 89, 450–1.

2 I may conclude this chapter by quoting a saying of Professor Agassiz, that whenever a new and startling fact is brought to light in science, people first say, 'it is not true,' then that 'it is contrary to religion,' and lastly, 'that everybody knew it before.'

The Antiquity of Man (1863), 105.

3 Hitherto, no rival hypothesis has been proposed as a substitute for the doctrine of transmutation; for 'independent creation,' as it is often termed, or the direct intervention of the Supreme Cause, must simply be considered as an avowal that we deem the question to lie beyond the domain of science.

The Antiquity of Man (1863), 421.

4 It was a profound saying of Wilhelm Humboldt, that 'Man is man only by means of speech, but in order to invent speech he must be already man.'

The Antiquity of Man (1863), 468.

5 The question now at issue, whether the living species are connected with the extinct by a common bond of descent, will best be cleared up by devoting ourselves to the study of the actual state of the living world, and to those monuments of the past in which the relics of the animate creation of former ages are best preserved and least mutilated by the hand of time.

The Antiquity of Man (1863), 470.

6 So far from having a materialistic tendency, the supposed introduction into the earth at successive geological periods of life,—sensation,—instinct,—the intelligence of the higher mammalia bordering on reason,—and lastly the improvable reason of Man himself, presents us with a picture of the ever-increasing dominion of mind over matter.

The Antiquity of Man (1863), 506.

7 [My Book] will endeavour to establish the *principle[s] of reasoning* in . . . [geology]; and all my geology will come in as illustration of my views of those principles, and as evidence strengthening the system necessarily arising out of the admission of such principles, which . . . are neither more nor less than that *no causes whatever* have from the earliest time to which we can look back, to the present, ever acted, but those *now acting*; and that they never acted with different degrees of energy from that which they now exert.

Letter to Roderick Murchison Esq., 15 January 1829. In Mrs Lyell (ed.), *The Life, Letters and Journals of Sir Charles Lyell, Bart* (1881), Vol. 1, 234.

8 Probably there was a beginning—it is a metaphysical question, worthy a theologian—species have begun and ended—but the analogy is faint and distant.

Letter to [George] Poulett Scrope, 14 June 1830. Quoted in Mrs Lyell (ed.), *Life, Letters and Journals of Sir Charles Lyell, Bart* (1881), Vol. 1, 269.

1 In attempting to explain geological phenomena, the bias has always been on the wrong side; there has always been a disposition to reason *à priori* on the extraordinary violence and suddenness of changes, both in the inorganic crust of the earth, and in organic types, instead of attempting strenuously to frame theories in accordance with the ordinary operations of nature.

Letter to Rev. W. Whewell, 7 March 1837. Quoted in Mrs Lyell (ed.), *Life, Letters and Journals of Sir Charles Lyell, Bart* (1881), Vol. 2, 3.

2 I conceive that Lamarck was the first to bring it forward systematically & to 'go the whole orang' . . . Yet evolutionists 'cannot be pooh-poohed & ought not to be so.'

Letter to Huxley, 17 June 1859, Imperial College Archives, Huxley Papers, 6:20. Partly reprinted in Leonard G. Wilson (ed.), *Sir Charles Lyell's Scientific Journals on the Species Question* (1970), 314. Lyell used this expression again in 1863, this time to Darwin. See *The Correspondence of Charles Darwin* (1999), Vol. 11, 231.

3 'Time's noblest offspring is the last.' This line of Bp. Berkeley's expresses the real cause of the belief in progress in the animal creation.

Leonard G. Wilson (ed.), *Sir Charles Lyell's Scientific Journals on the Species Question* (1970), 162.

4 The ordinary naturalist is not sufficiently aware that when dogmatizing on what species are, he is grappling with the whole question of the organic world & its connection with the time past & with Man; that it involves the question of Man & his relation to the brutes, of instinct, intelligence & reason, of Creation, transmutation & progressive improvement or development. Each set of geological questions & of ethnological & zool. & botan. are parts of the great problem which is always assuming a new aspect.

Leonard G. Wilson (ed.), *Sir Charles Lyell's Scientific Journals on the Species Question* (1970), 164.

5 'In the economy of the world,' said the Scotch geologist [James Hutton], 'I can find no traces of a beginning, no prospect of an end.'

Charles Lyell, *Principles of Geology* (1830–3), Vol. 1, 63 (misquoted; see **Hutton, James** 309:6).

Robert Staughton Lynd

1892–1970

American sociologist

6 When the last Puritan has disappeared from the earth, the man of science will take his place as a killjoy, and we shall be given the same old advice but for different reasons.

Attributed.

Catharine Macaulay 1731–91

British historian

7 Pope has elegantly said a *perfect woman's but a softer man*. And if we take in the consideration, that there can be but one rule of moral excellence for beings made of the same materials, organized after the same manner, and subjected to similar laws of Nature, we must either agree with Mr. Pope, or we must reverse the proposition, and say, that *a perfect man is a woman formed after a coarser mold*.

Letter XXII. 'No Characteristic Difference in Sex'. In *Letters on Education with Observations on Religious and Metaphysical Subjects* (1790), 128.

Thomas Babington Macaulay

1800–59

British historian

8 [I can] scarcely write upon mathematics or mathematicians. Oh for words to express my abomination of the science.

Lamenting mathematics whilst an undergraduate at Cambridge, 1818.

Quoted in John Gascoigne, *Cambridge in the Age of Enlightenment* (1989), 272.

Ernest William MacBride 1866–1940

British zoologist

1 I have watched all the work going on there, and the more I see of it the more I am convinced that Mendelism has nothing to do with evolution.

Letter, 'Embryology and Evolution,' *Nature*, 1931, 127, 56.

Maclyn McCarty see Avery, Oswald T., et al.

Barbara McClintock 1902–92

American geneticist

2 When you know you're right, you don't care what others think. You know sooner or later it will come out in the wash.

Quoted in Claudia Wallis, 'Honoring a Modern Mendel', *Time*, 24 October 1983, 43–4.

3 They thought I was crazy, absolutely mad.

The National Academy of Sciences, 1944, in response to her theory that genes could 'jump' around on a chromosome; she later won the Nobel Prize for medicine or physiology.

Quoted in Claudia Wallis, 'Honoring a Modern Mendel', *Time*, 24 October 1983, 43.

4 With the tools and the knowledge, I could turn a developing snail's egg into an elephant. It is not so much a matter of chemicals because snails and elephants do not differ that much; it is a matter of timing the action of genes.

Quoted in Bruce Wallace, *The Search for the Gene* (1992), 176.

David Keith Chalmers MacDonald 1920–63

British physicist

5 It does appear that on the whole a physicist . . . tries to *reduce* his theory at all times to as few parameters as possible and is inclined to feel that a theory is a 'respectable' one, though by no means necessarily correct, if in principle it does offer reasonably specific means for its possible refutation. Moreover the physicist will generally arouse the irritation amongst fellow physicists if he is not prepared to abandon his theory when it clashes with subsequent experiments. On the other hand it would appear that the chemist regards theories—or perhaps better *his* theories (!)—as far less sacrosanct, and perhaps in extreme cases is prepared to modify them *continually* as each bit of new experimental evidence comes in.

'Discussion: Physics and Chemistry: Comments on Caldin's View of Chemistry', *British Journal of the Philosophy of Science*, 1960, 11, 222.

Ernst Mach 1838–1916

Czech physicist, physiologist and psychologist

6 Physics is experience, arranged in economical order.

'The Economical Nature of Physics' (1882), in *Popular Scientific Lectures*, trans. Thomas J. McCormack (1910), 197.

7 It would not become physical science to see in its self created, changeable, economical tools, molecules and atoms, realities behind phenomena . . . The atom must remain a tool for representing phenomena.

'The Economical Nature of Physics' (1882), in *Popular Scientific Lectures*, trans. Thomas J. McCormack (1910), 206–7.

8 To us investigators, the concept 'soul' is irrelevant and a matter for laughter. But matter is an abstraction of exactly the same kind, just as good and just as bad as it is. We know as much about the soul as we do of matter.

'Die Geschichte und die Wurzel des Satzes von der Erhaltung der Arbeit' (1872). Trans. Philip E. Jourdain, *History and Root of the Principle of the Conservation of Energy* (1911), 48.

1 Science itself, therefore, may be regarded as a minimal problem, consisting of the completest possible presentment of facts with the *least expenditure of thought*.

The Science of Mechanics: A Critical and Historical Account of Its Development (1893), trans. Thomas J. McCormack (1960), 490.

2 Thought experiment is in any case a necessary precondition for physical experiment. Every experimenter and inventor must have the planned arrangement in his head before translating it into fact.

'On Thought Experiments' (1897), in Erwin H. Hiebert (ed.), *Erkenntnis und Irrtum* (1905), trans. Thomas J. McCormack and Paul Foulkes (1976), 184.

3 The power of mathematics rests on its evasion of all unnecessary thought and on its wonderful saving of mental operations.

Quoted in Freeman Dyson, 'Mathematics in the Physical Sciences', *Scientific American*, September 1964, **211**, No. 3, 133.

Charles Macintosh 1766–1843

Scottish manufacturing chemist

4 ON **MACINTOSH** The number of travellers by gigs, the outside of coaches, and on horseback, have, since the introduction of railways, been prodigiously diminished; and as, in addition, the members of the medical faculty having lent their aid to run down the use of water-proof (apparently having found it decided enemy against their best friends colds and catarrhs), the use of the article [the Macintosh] in the form of cloaks, etc., has of late become comparatively extinct.

George Macintosh (ed.), *A Biographical Memoir of the late Charles Macintosh Esq FRS* (1847), 89.

Robert M. MacIver 1882–1970

British sociologist and political scientist

5 We are apt to think we know what time is because we can measure it, but no sooner do we reflect upon it than that illusion goes. So it appears that the range of the measureable is not the range of the knowable. There are things we can measure, like time, but yet our minds do not grasp their meaning. There are things we cannot measure, like happiness or pain, and yet their meaning is perfectly clear to us.

The Elements of Social Science (1921), 15–6.

Colin Maclaurin 1698–1746

British mathematician

6 It is not therefore the business of philosophy, in our present situation in the universe, to attempt to take in at once, in one view, the whole scheme of nature; but to extend, with great care and circumspection, our knowledge, by just steps, from sensible things, as far as our observations or reasonings from them will carry us, in our enquiries concerning either the greater motions and operations of nature, or her more subtile and hidden works. In this way Sir Isaac Newton proceeded in his discoveries.

An Account of Sir Isaac Newton's Philosophical Discoveries, in Four Books (1748), 19.

Colin MacLeod 1909–72

Canadian physician

See ***Avery, Oswald T., et al.***

Herbert Marshall McLuhan 1911–80

Canadian author and communications theorist

7 The medium is the message.

Understanding Media (1964), title of Chapter 1.

8 People who write obscurely are either unskilled in writing or up to mischief.

Science and Literature in Plato's Republic (1984) 52.

Herbert Marshall McLuhan 1911–80 and Quentin Fiore 1920–

Canadian author and media theorist and American social scientist

1 Ours is a brand-new world of allatonceness [all-at-once-ness]. 'Time' has ceased, 'space' has vanished. We now live in a global village . . . a simultaneous happening . . . The new electronic interdependence recreates the world in the image of a global village.

The Medium is the Massage (1967), 63–7.

William Maclure 1763–1840

British-born American geologist

2 In all speculations on the origin, or agents that have produced the changes on this globe, it is probable that we ought to keep within the boundaries of the probable effects resulting from the regular operations of the great laws of nature which our experience and observation have brought within the sphere of our knowledge. When we overleap those limits, and suppose a total change in nature's laws, we embark on the sea of uncertainty, where one conjecture is perhaps as probable as another; for none of them can have any support, or derive any authority from the practical facts wherewith our experience has brought us acquainted.

Observations on the Geology of the United States of America (1817), iv-v.

John Royden Maddox 1925–

British writer and broadcaster, editor of Nature, 1966–73 and 1980–95

3 The El Niño phenomenon, the geophysicists' equivalent of the universal solvent.

'Great Greenhouse in the Sky?', *Nature*, 1983, 306, 221.

François Magendie 1783–1855

French physician and physiologist

4 Facts, and facts alone, are the foundation of science . . . When one devotes oneself to experimental research it is in order to augment the sum of known facts, or to discover their mutual relations.

Précis élémentaire de Physiologie (1816), ii. Trans. J. M. D. Olmsted, *François Magendie: Pioneer in Experimental Physiology and Scientific Medicine in XIX Century France* (1944), 62.

5 I know that certain minds would regard as audacious the idea of relating the laws which preside over the play of our organs to those laws which govern inanimate bodies; but, although novel, this truth is none the less incontestable. To hold that the phenomena of life are entirely distinct from the general phenomena of nature is to commit a grave error, it is to oppose the continued progress of science.

Leçons sur les Phénomènes Physiques de la Vie (1836–38), Vol. 1, 6. Trans. J. M. D. Olmsted, *François Magendie* (1944), 203.

6 Every one is fond of comparing himself to something great and grandiose, as Louis XIV likened himself to the sun, and others have had like similes. I am more humble. I am a mere street scavenger (*chiffonier*) of science. With my hook in my hand and my basket on my back, I go about the streets of science, collecting what I find.

Quoted in Michael Foster, *Claude Bernard* (1899), 40.

Rabbi Moses Ben Maimon Maimonides c.1135–1204

Jewish philosopher and physician

7 Astrology is a sickness, not a science . . . It is a tree under the shade of which all sorts of superstitions thrive.

Attributed.

Henry James Sumner Maine 1822–88

British jurist and historian

8 Except the blind forces of Nature, nothing moves in this world which is not Greek in its origin.

Village Communities in the East and West (1871), 238.

Nicholas Malebranche 1638–1715

French philosopher

1 I do not believe there is anything useful which men can know with exactitude that they cannot know by arithmetic and algebra.

Oeuvres, Vol. 2, 292g. Trans. J. L. Heilbron, *Electricity in the 17th and 18th Centuries: A Study of Early Modern Physics* (1979), 42.

Bronislaw Kasper Malinowski 1884–1942

Polish-born British anthropologist

2 There are no peoples however primitive without religion and magic. Nor are there, it must be added at once, any savage races lacking either in the scientific attitude or in science, though this lack has been frequently attributed to them.

Magic, Science and Religion (1925), 17.

3 [Magic] enables man to carry out with confidence his important tasks, to maintain his poise and his mental integrity in fits of anger, in the throes of hate, of unrequited love, of despair and anxiety. The function of magic is to ritualize man's optimism, to enhance his faith in the victory of hope over fear. Magic expresses the greater value for man of confidence over doubt, of steadfastness over vacillation, of optimism over pessimism.

Magic, Science and Religion (1925), 90.

4 Coastal sailing as long as it is perfectly safe and easy commands no magic. Overseas expeditions are invariably bound up with ceremonies and ritual. Man resorts to magic only where chance and circumstances are not fully controlled by knowledge.

Culture (1931), 636.

Bo G. Malmström 1927–2000

Swedish biochemist

5 Life is order, death is disorder. A fundamental law of Nature states that spontaneous chemical changes in the universe tend toward chaos. But life has, during milliards (*American English* billions) of years of evolution, seemingly contradicted this law. With the aid of energy derived from the sun it has built up the most complicated systems to be found in the universe—living organisms. Living matter is characterized by a high degree of chemical organisation on all levels, from the organs of large organisms to the smallest constituents of the cell. The beauty we experience when we enjoy the exquisite form of a flower or a bird is a reflection of a microscopic beauty in the architecture of molecules.

'The Nobel Prize for Chemistry: Introductory Address'. *Nobel Lectures: Chemistry 1981–1990* (1992), 69.

Marcello Malpighi 1628–94

Italian physician, physiologist and microscopist

6 I have destroyed almost the whole race of frogs, which does not happen in that savage Batrachomyomachia of Homer. For in the anatomy of frogs, which, by favour of my very excellent colleague D. Carolo Fracassato, I had set on foot in order to become more certain about the membranous substance of the lungs, it happened to me to see such things that not undeservedly I can better make use of that (saying) of Homer for the present matter—

'I see with my eyes a work trusty and great.'

For in this (frog anatomy) owing to the simplicity of the structure, and the almost complete transparency of the vessels which admits the eye into the interior, things are more clearly shown so that they will bring the light to other more obscure matters.

De Pulmonibus (1661), trans. James Young, *Proceedings of the Royal Society of Medicine* (1929–30), Vol. 23, 7.

7 For Nature is accustomed to rehearse with certain large, perhaps baser, and all classes of wild (animals), and to place in the imperfect the rudiments of the perfect animals.

De Pulmonibus (1661), trans. James Young, *Proceedings of the Royal Society of Medicine* (1929–30), Vol. 23, 7.

1 The power of the eye could not be extended further in the opened living animal, hence I had believed that this body of the blood breaks into the empty space, and is collected again by a gaping vessel and by the structure of the walls. The tortuous and diffused motion of the blood in divers directions, and its union at a determinate place offered a handle to this. But the dried lung of the frog made my belief dubious. This lung had, by chance, preserved the redness of the blood in (what afterwards proved to be) the smallest vessels, where by means of a more perfect lens, no more there met the eye the points forming the skin called Sagrino, but vessels mingled annularly. And, so great is the divarication of these vessels as they go out, here from a vein, there from an artery, that order is no longer preserved, but a network appears made up of the prolongations of both vessels. This network occupies not only the whole floor, but extends also to the walls, and is attached to the outgoing vessel, as I could see with greater difficulty but more abundantly in the oblong lung of a tortoise, which is similarly membranous and transparent. Here it was clear to sense that the blood flows away through the tortuous vessels, that it is not poured into spaces but always works through tubules, and is dispersed by the multiplex winding of the vessels.

De Pulmonibus (1661), trans. James Young, *Proceedings of the Royal Society of Medicine* (1929–30), Vol. 23, 8.

2 Observation by means of the microscope will reveal more wonderful things than those viewed in regard to mere structure and connection: for while the heart is still beating the contrary (i.e., in opposite directions in the different vessels) movement of the blood is observed in the vessels,—though with difficulty,—so that the circulation of the blood is clearly exposed.

De Pulmonibus (1661), trans. James Young, *Proceedings of the Royal Society of Medicine* (1929–30), Vol. 23, 8.

3 Casting off the dark fog of verbal philosophy and vulgar medicine, which inculcate names alone . . . I tried a series of experiments to explain more clearly many phenomena, particularly those of physiology. In order that I might subject as far as possible the reasonings of the Galenists and Peripatetics to sensory criteria, I began, after trying experiments, to write dialogues in which a Galenist adduced the better-known and stronger reasons and arguments; these a mechanist surgeon refuted by citing to the contrary the experiments I had tried, and a third, neutral interlocutor weighed the reasons advanced by both and provided an opportunity for further progress.

'Malpighi at Pisa 1656–1659', in H. B. Adelmann (ed.), *Marcello Malpighi and the Evolution of Embryology* (1966), Vol. 1, 155–6.

4 I could clearly see that the blood is divided and flows through tortuous vessels and that it is not poured out into spaces, but is always driven through tubules and distributed by the manifold bendings of the vessels . . . [F]rom the simplicity Nature employs in all her works, we may conclude . . . that the network I once believed to be nervous [that is, sinewy] is really a vessel intermingled with the vesicles and sinuses and carrying the mass of blood to them or away from them . . . though these elude even the keenest sight because of their small size . . . From these considerations it is highly probable that the question about the mutual union and anastomosis of the vessels can be solved; for if Nature once circulates the blood within vessels and combines their ends in a network, it is probable that they are joined by anastomosis at other times too.

'The Return to Bologna 1659–1662', in H. B. Adelmann (ed.), *Marcello Malpighi and the Evolution of Embryology* (1966), Vol. 1, 194–5.

1 Nature . . . in order to carry out the marvelous operations [that occur] in animals and plants has been pleased to construct their organized bodies with a very large number of machines, which are of necessity made up of extremely minute parts so shaped and situated as to form a marvelous organ, the structure and composition of which are usually invisible to the naked eye without the aid of a microscope . . . Just as Nature deserves praise and admiration for making machines so small, so too the physician who observes them to the best of his ability is worthy of praise, not blame, for he must also correct and repair these machines as well as he can every time they get out of order.

'Reply to Doctor Sbaraglia in *Opera Posthuma*, 1697', in H. B. Adelmann (ed.), *Marcello Malpighi and the Evolution of Embryology* (1966), Vol. 1, 568.

2 Nature has but one plan of operation, invariably the same in the smallest things as well as in the largest, and so often do we see the smallest masses selected for use in Nature, that even enormous ones are built up solely by fitting these together. Indeed, all Nature's efforts are devoted to uniting the smallest parts of our bodies in such a way that all things whatsoever, however diverse they may be, which coalesce in the structure of living things construct the parts by means of a sort of compendium.

'On the Developmental Process', in H. B. Adelmann (ed.), *Marcello Malpighi and the Evolution of Embryology* (1966), Vol. 2, 843.

3 The seed is the fetus, in other words, a true plant with its parts (that is, its leaves, of which there are usually two, its stalk or stem, and its bud) completely fashioned.

'On the Developmental Process', in H. B. Adelmann (ed.), *Marcello Malpighi and the Evolution of Embryology* (1966), Vol. 2, 845.

4 The generation of seeds . . . is therefore marvelous and analogous to the other productions of living things. For first of all an umbilicus appears . . . Its extremity gradually expands and after gathering a colliquamentous ichor becomes analogous to an amnion . . . In the course of time the seed or fetus begins to become visible.

'On the Developmental Process', in H. B. Adelmann (ed.), *Marcello Malpighi and the Evolution of Embryology* (1966), Vol. 2, 850.

5 By its very nature the uterus is a field for growing the seeds, that is to say the ova, sown upon it. Here the eggs are fostered, and here the parts of the living [fetus], when they have further unfolded, become manifest and are made strong. Yet although it has been cast off by the mother and sown, the egg is weak and powerless and so requires the energy of the semen of the male to initiate growth. Hence in accordance with the laws of Nature, and like the other orders of living things, women produce eggs which, when received into the chamber of the uterus and fecundated by the semen of the male, unfold into a new life.

'On the Developmental Process', in H. B. Adelmann (ed.), *Marcello Malpighi and the Evolution of Embryology* (1966), Vol. 2, 861.

6 It is therefore proper to acknowledge that the first filaments of the chick pre-exist in the egg and have a deeper origin, exactly as [the embryo] in the eggs of plants.

'On the Formation of the Chick in the Egg' (1673), in H. B. Adelmann (ed.), *Marcello Malpighi and the Evolution of Embryology* (1966), Vol. 2, 945.

7 This, however, seems to be certain: the ichor, that is, the material I have mentioned that finally becomes red, exists before the heart begins to beat, but the heart exists and even beats before the blood reddens.

'On the Formation of the Chick in the Egg' (1673), in H. B. Adelmann (ed.), *Marcello Malpighi and the Evolution of Embryology* (1966), Vol. 2, 957.

8 Surgical knowledge depends on long practice, not from speculations.

'Letter to Borghese, July 27th 1689', in H. B. Adelmann (ed.), *The Correspondence of Marcello Malpighi* (1975), Vol. 4, 1486.

1 In such sad circumstances I but see myself exhalted by my own enemies, for in order to defeat some small works of mine they try to make the whole rational medicine and anatomy fall, as if I were myself these noble disciplines.

'Letter to Marescotti about the dispute with Sbaraglia and others, 1689(?)', in H. B. Adelmann (ed.), *The Correspondence of Marcello Malpighi* (1975), Vol. 4, 1561.

2 We are many small puppets moved by fate and fortune through strings unseen by us; therefore, if it is so as I think, one has to prepare oneself with a good heart and indifference to accept things coming towards us, because they cannot be avoided, and to oppose them requires a violence that tears our souls too deeply, and it seems that both fortune and men are always busy in affairs for our dislike because the former is blind and the latter only think of their interest.

'Letter to Bellini, Oct 17th 1689', in H. B. Adelmann (ed.), *The Correspondence of Marcello Malpighi* (1975), Vol. 4, 1534.

Thomas Robert Malthus

1766–1834

British political economist

3 I think I may fairly make two postulata.

First, that food is necessary to the existence of man.

Secondly, that the passion between the sexes is necessary, and will remain nearly in its present state.

These two laws ever since we have had any knowledge of mankind, appear to have been fixed laws of our nature; and, as we have not hitherto seen any alteration in them, we have no right to conclude that they will ever cease to be what they now are, without an immediate act of power in that Being who first arranged the system of the universe, and for the advantage of his creatures, still executes, according to fixed laws, all its various operations.

An Essay on the Principle of Population (1798). In E. A. Wrigley and David Souden (eds.), *The Works of Thomas Malthus* (1986), Vol. 1, 8.

4 Population, when unchecked, increases in a geometrical ratio. Subsistence increases only in an arithmetical ratio. A slight acquaintance with numbers will show the immensity of the first power in comparison of the second.

An Essay on the Principle of Population (1798). In E. A. Wrigley and David Souden (eds.), *The Works of Thomas Malthus* (1986), Vol. 1, 9.

5 The prodigious waste of human life occasioned by this perpetual struggle for room and food, was more than supplied by the mighty power of population, acting, in some degree, unshackled, from the constant habit of emigration.

An Essay on the Principle of Population (1798). In E. A. Wrigley and David Souden (eds.), *The Works of Thomas Malthus* (1986), Vol. 1, 21.

6 Famine seems to be the last, the most dreadful resource of nature. The power of population is so superior to the power in the earth to produce subsistence for man, that premature death must in some shape or other visit the human race. The vices of mankind are active and able ministers of depopulation. They are the precursors in the great army of destruction; and often finish the dreadful work themselves. But should they fail in this war of extermination, sickly seasons, epidemics, pestilence, and plague, advance in terrific array, and sweep off their thousands and ten thousands. Should success be still incomplete, gigantic inevitable famine stalks in the rear, and with one mighty blow, levels the population with the food of the world.

An Essay on the Principle of Population (1798). In E. A. Wrigley and David Souden (eds.), *The Works of Thomas Malthus* (1986), Vol. 1, 51–2.

Mao Zedong 1893–1976

Chinese Marxist leader

7 Where do correct ideas come from? Do they drop from the skies? No. Are they innate in the minds? No. They come from social practice, and from it alone; they come from three kinds of social practice, the struggle for production, the class struggle and scientific experiment.

Quotations from Chairman Mao Tse-Tung (1967), 116.

Herbert Marcuse 1898–1979

German-born American philosopher

1 By virtue of the way it has organized its technological base, contemporary industrial society tends to be totalitarian. For 'totalitarian' is not only a terroristic political coordination of society, but also a non-terroristic economic-technical coordination which operates through the manipulation of needs by vested interests. It thus precludes the emergence of an effective opposition against the whole. Not only a specific form of government or party rule makes for totalitarianism, but also a specific system of production and distribution which may well be compatible with a 'pluralism' of parties, newspapers, 'countervailing powers,' etc.

One Dimensional Man (1964), 3.

Robert Ranulph Marett 1866–1943

British social anthropologist

2 None of Darwin's particular doctrines will necessarily endure the test of time and trial. Into the melting-pot must they go as often as any man of science deems it fitting. But Darwinism as the touch of nature that makes the whole world kin can hardly pass away.

Anthropology (1912), 11.

Étienne Jules Marey 1830–1904

French physiologist

3 *La méthode graphique.*

The graphical method.

La méthode graphique dans les sciences expérimentales, et principalement en physiologie et en médecine (1878).

Maria the Jewess fl. 1st century

Hellenistic alchemist

4 One becomes two, two becomes three, and by means of the third and fourth achieves unity; thus two are but one.

Quoted in R. Patai, *The Jewish Alchemists* (1994), 66.

Jacques Maritain 1882–1973

French philosopher

5 Since science's competence extends to observable and measurable phenomena, not to the inner being of things, and to the means, not to the ends of human life, it would be nonsense to expect that the progress of science will provide men with a new type of metaphysics, ethics, or religion.

'Science and Ontology', *Bulletin of Atomic Scientists*, 1949, 5, 200.

Christopher Marlowe c.1564–93

British poet and playwright

6 FAUSTUS: How many heavens or spheres are there?
MEPHASTOPHILIS: Nine: the seven planets, the firmament, and the empyreal heaven.
FAUSTUS: But is there not *coelum igneum, et crystallinum?*
MEPH.: No Faustus, they be but fables.
FAUSTUS: Resolve me then in this one question: Why are not conjunctions, oppositions, aspects, eclipses all at one time, but in some years we have more, in some less?
MEPH.: Per inaequalem motum respectu totius.
FAUSTUS: Well, I am answered. Now tell me who made the world.
MEPH.: I will not.
FAUSTUS: Sweet Mephastophilis, tell me.
MEPH.: Move me not, Faustus.
FAUSTUS: Villain, have I not bound thee to tell me any thing?
MEPH.: Ay, that is not against our kingdom.
This is. Thou are damn'd, think thou of hell.
FAUSTUS: Think, Faustus, upon God that made the world!

MEPH.: Remember this.

Doctor Faustus: A 1604-Version Edition, edited by Michael Keefer (1991), Act II, Scene iii, lines 60–77, 43–4.

George Perkins Marsh 1801–82

American diplomat, scholar and conservationist

1 Sight is a faculty; seeing, an art.

Man and Nature (1864), 10.

2 Man is everywhere a disturbing agent. Wherever he plants his foot, the harmonies of nature are turned to discords.

Man and Nature (1864), 36.

3 The ravages committed by man subvert the relations and destroy the balance which nature had established between her organized and her inorganic creations; and she avenges herself upon the intruder, by letting loose upon her defaced provinces destructive energies hitherto kept in check by organic forces destined to be his best auxiliaries, but which he has unwisely dispersed and driven from the field of action. When the forest is gone, the great reservoir of moisture stored up in its vegetable mould is evaporated, and returns only in deluges of rain to wash away the parched dust into which that mould has been converted. The well-wooded and humid hills are turned to ridges of dry rock, which encumbers the low grounds and chokes the watercourses with its debris, and—except in countries favored with an equable distribution of rain through the seasons, and a moderate and regular inclination of surface—the whole earth, unless rescued by human art from the physical degradation to which it tends, becomes an assemblage of bald mountains, of barren, turfless hills, and of swampy and malarious plains. There are parts of Asia Minor, of Northern Africa, of Greece, and even of Alpine Europe, where the operation of causes set in action by man has brought the face of the earth to a desolation almost as complete as that of the moon; and though, within that brief space of time which we call 'the historical period,' they are known to have been covered with luxuriant woods, verdant pastures, and fertile meadows, they are now too far deteriorated to be reclaimable by man, nor can they become again fitted for human use, except through great geological changes, or other mysterious influences or agencies of which we have no present knowledge, and over which we have no prospective control. The earth is fast becoming an unfit home for its noblest inhabitant, and another era of equal human crime and human improvidence, and of like duration with that through which traces of that crime and that improvidence extend, would reduce it to such a condition of impoverished productiveness, of shattered surface, of climatic excess, as to threaten the depravation, barbarism, and perhaps even extinction of the species.

Man and Nature (1864), 42–3.

4 So long as the fur of the beaver was extensively employed as a material for fine hats, it bore a very high price, and the chase of this quadruped was so keen that naturalists feared its speedy consideration. When a Parisian manufacturer invented the silk hat, which soon came into almost universal use, the demand for beavers' fur fell off, and this animal—whose habits, as we have seen, are an important agency in the formation of bogs and other modifications of forest nature—immediately began to increase, reappeared in haunts which we had long abandoned, and can no longer be regarded as rare enough to be in immediate danger of extirpation. Thus the convenience or the caprice of Parisian fashion has unconsciously exercised an influence which may sensibly affect the physical geography of a distant continent.

Man and Nature (1864), 84.

5 The equation of animal and vegetable life is too complicated a problem for human intelligence to solve, and we can never know how wide a circle of disturbance we produce in the harmonies of nature when we throw the smallest pebble into the ocean of organic life.

Man and Nature (1864), 103.

1 The great question, whether man is of nature or above her.

Man and Nature (1864), 549.

2 Wherever modern Science has exploded a superstitious fable or even a picturesque error, she has replaced it with a grander and even more poetical truth.

'The Study of Nature', *The Christian Examiner*, 1860, 67, 40.

3 The *improvement* of forest trees is the work of centuries. So much more the reason for beginning *now*.

Letter to C. S. Sargent, 12 June 1879. In David Lowenthal, *George Perkins Marsh: Versatile Vermonter* (1958), 255.

Karl Heinrich Marx 1818–83

German political philosopher

4 Natural science will in time incorporate into itself the science of man, just as the science of man will incorporate into itself natural science: there will be one science.

Economic and Philosophic Manuscripts of 1844 (1844), 304.

5 *Die Religion ist der Seufzer der bedrängten Kreatur, das Gemüt einer herzlosen Welt, wie sie der Geist geistloser Zustände ist. Sie ist das* Opium *des Volks. Die Aufhebung der Religion als des* illusorischen *Glücks des Volks ist die Forderung seines* wirklichen *Glücks.*

Religion is the sigh of the oppressed creature, the heart of a heartless world, just as it is the spirit of spiritless conditions. It is the opium of the people. To abolish religion as the *illusory* happiness of the people is to demand their *real* happiness.

'Zur Kritik der Hegelschen Rechtsphilosophie. Einleitung' (1844), *Karl Marx Fredrich Engels* (1964), 378–9.

6 Men make their own history, but not just as they please. They do not choose the circumstances for themselves, but have to work upon circumstances as they find them, have to fashion the material handed down by the past. The legacy of the dead generations weighs like an alp upon the brains of the living.

The Eighteenth Brumaire of Louis Bonaparte (1852).

7 Darwin's book is very important and serves me as a basis in natural science for the class struggle in history. One has to put up with the crude English method of development, of course. Despite all deficiencies not only is the death-blow dealt here for the first time to 'teleology' in the natural sciences, but their rational meaning is empirically explained.

Marx to Lasalle, 16 January 1861. In *Marx-Engels Selected Correspondence, 1846–95*, trans. Donna Torr (1934), 125.

8 We know only a single science, the science of history. History can be contemplated from two sides, it can be divided into the history of nature and the history of mankind. However, the two sides are not to be divided off; as long as men exist the history of nature and the history of men are mutually conditioned.

Karl Marx and Frederick Engels, *The German Ideology* (1845–6), Vol. 1, 28. English translation 1965.

9 So far no chemist has ever discovered exchange-value either in a pearl or a diamond.

Capital: A Critique of Political Economy (1867), trans. Ben Fowkes (1976), Vol. 1, 177.

10 The philosophers have only interpreted the world in various ways, the point is to change it.

Epitaph on Marx's tombstone in Highgate Cemetery.

Theses on Feuerbach (1845), 5.

Karl Heinrich Marx 1818–83 and **Friedrich Engels** 1820–95

German political philosopher and German socialist

1 Life is not determined by consciousness, but consciousness by life.

The German Ideology [written 1845–1846]. Edited by R. Pascal (1938), 15.

Kirtley F. Mather 1888–1978

American geologist

2 The responsibility which rests upon man is proportional to the ability which he possesses and the opportunity which he faces. Perhaps that responsibility is no greater for him than was that of Notharctus or Eohippus or a trilobite, each in his own day, but because of man's unique abilities it is the greatest responsibility that has ever rested upon any of the earth's offspring.

Sons of the Earth: The Geologist's View of History (1930), 258.

Henry Maudsley 1835–1918

British psychiatrist

3 One most necessary function of the brain is to exert an inhibitory power over the nerve centres that lie below it, just as man exercises a beneficial control over his fellow animals of a lower order of dignity; and the increased irregular activity of the lower centres surely betokens a degeneration: it is like the turbulent, aimless action of a democracy without a head.

The Physiology and Pathology of Mind (1868), 94.

Pierre-Louis Moreau de Maupertuis 1698–1759

French mathematician and natural philosopher

4 In order to turn natural history into a true science, one would have to devote oneself to investigations capable of telling us not the particular shape of such and such an animal, but the general procedures of nature in the animal's production and preservation.

'Lettre sur le progress des sciences' in *Oeuvres de Mr. De Maupertuis* (1756), Vol. 2, 386. Quoted in Jacques Roger, *The Life Sciences in Eighteenth-Century French Thought*, ed. Keith R. Benson and trans. Robert Ellrich (1997), 392.

5 Might one not say that in the chance combination of nature's production, since only those endowed with certain relations of suitability could survive, it is no cause for wonder that this suitability is found in all species that exist today? Chance, one might say, produced an innumerable multitude of individuals; a small number turned out to be constructed in such fashion that the parts of the animal could satisfy its needs; in another, infinitely greater number, there was neither suitability nor order: all of the later have perished; animals without a mouth could not live, others lacking organs for reproduction could not perpetuate themselves: the only ones to have remained are those in which were found order and suitability; and these species, which we see today, are only the smallest part of what blind fate produced.

'Essai de Cosmologie' in *Oeuvres de Mr. De Maupertuis* (1756), Vol. 1, 11–12. Quoted in Jacques Roger, *The Life Sciences in Eighteenth-Century French Thought*, ed. Keith R. Benson and trans. Robert Ellrich (1997), 381.

James Clerk Maxwell 1831–79

British physicist

See ***Galton, Francis*** *240:3*

6 But when we face the great questions about gravitation Does it require time? Is it polar to the 'outside of the universe' or to anything? Has it any reference to electricity? or does it stand on the very foundation of matter—mass or inertia? then we feel the need of tests, whether they be comets or nebulae or laboratory experiments or bold questions as to the truth of received opinions.

Letter to Michael Faraday, 9 November 1857. In P. M. Harman (ed.), *The Scientific Letters and Papers of James Clerk Maxwell* (1990), Vol. 1, 1846–1862, 551–2.

1 Accordingly, we find Euler and D'Alembert devoting their talent and their patience to the establishment of the laws of rotation of the solid bodies. Lagrange has incorporated his own analysis of the problem with his general treatment of mechanics, and since his time M. Poinsôt has brought the subject under the power of a more searching analysis than that of the calculus, in which ideas take the place of symbols, and intelligent propositions supersede equations.

J. C. Maxwell on Louis Poinsôt (1777–1859) in 'On a Dynamical Top' (1857). In W. D. Niven (ed.), *The Scientific Papers of James Clerk Maxwell* (1890), Vol. 1, 248.

2 I have been battering away at Saturn, returning to the charge every now and then. I have effected several breaches in the solid ring, and now I am splash into the fluid one, amid a clash of symbols truly astounding. When I reappear it will be in the dusky ring, which is something like the state of the air supposing the siege of Sebastopol conducted from a forest of guns 100 miles one way, and 30,000 miles the other, and the shot never to stop, but go spinning away round a circle, radius 170,000 miles.

Letter to Lewis Campbell, 28 August 1857. In P. M. Harman (ed.), *The Scientific Letters and Papers of James Clerk Maxwell* (1990), Vol. 1, 1846–1862, 538.

3 My theory of electrical forces is that they are called into play in insulating media by *slight* electric displacements, which put certain small portions of the medium into a state of distortion which, being resisted by the *elasticity* of the medium, produces an electromotive force . . . I suppose the elasticity of the sphere to react on the electrical matter surrounding it, and press it downwards.

From the determination by Kohlrausch and Weber of the numerical relation between the statical and magnetic effects of electricity, I have determined the *elasticity* of the medium in air, and assuming that it is the same with the luminiferous ether I have determined the velocity of propagation of transverse vibrations.

The result is
193088 miles per second
(deduced from electrical & magnetic experiments).

Fizeau has determined the velocity of light
= 193118 miles per second
by direct experiment.

This coincidence is not merely numerical. I worked out the formulae in the country, before seeing Webers [sic] number, which is in millimetres, and I think we have now strong reason to believe, whether my theory is a fact or not, that the luminiferous and the electromagnetic medium are one.

Letter to Michael Faraday, 19 October 1861. In P. M. Harman (ed.), *The Scientific Letters and Papers of James Clerk Maxwell* (1990), Vol. 1, 1846–1862, 684–6.

4 We can scarcely avoid the inference that light consists in the transverse undulations of the same medium which is the cause of electric and magnetic phenomena.

'On Physical Lines of Force' (1862). In W. D. Niven (ed.), *The Scientific Papers of James Clerk Maxwell* (1890), Vol. 1, 500.

5 [Helmholtz] is not a philosopher in the exclusive sense, as Kant, Hegel, Mansel are philosophers, but one who prosecutes physics and physiology, and acquires therein not only skill in developing any desideratum, but wisdom to know what are the desiderata, e.g., he was one of the first, and is one of the most active, preachers of the doctrine that since all kinds of energy are convertible, the first aim of science at this time should be to ascertain in what way particular forms of energy can be converted into each other, and what are the equivalent quantities of the two forms of energy.

Letter to Lewis Campbell, 21 April 1862. In P. M. Harman (ed.), *The Scientific Letters and Papers of James Clerk Maxwell* (1990), Vol. 1, 711.

6 I hope you enjoy the absence of pupils . . . the total oblivion of them for definite intervals is a necessary condition for doing them justice at the proper time.

Letter to Lewis Campbell, 21st April 1862. In P. M. Harman (ed.), *The Scientific Letters and Papers of James Clerk Maxwell* (1990), Vol. 1, 712.

1 What, then, is light according to the electromagnetic theory? It consists of alternate and opposite rapidly recurring transverse magnetic disturbances, accompanied with electric displacements, the direction of the electric displacement being at the right angles to the magnetic disturbance, and both at right angles to the direction of the ray.

'A Dynamical Theory of the Electromagnetic Field' (1864). In W. D. Niven (ed.), *The Scientific Papers of James Clerk Maxwell* (1890), Vol. 2, 1862–1973, 195.

2 The theory I propose may therefore be called a theory of the *Electromagnetic Field* because it has to do with the space in the neighbourhood of the electric or magnetic bodies, and it may be called a *Dynamical* Theory, because it assumes that in the space there is matter in motion, by which the observed electromagnetic phenomena are produced.

'A Dynamical Theory of the Electromagnetic Field' (1865). In W. D. Niven (ed.), *The Scientific Papers of James Clerk Maxwell* (1890), Vol. 2, 527.

3 To pick a hole—say in the 2nd law of Θ^{cs}, that if two things are in contact the hotter cannot take heat from the colder without external agency.

Now let *A* & *B* be two vessels divided by a diaphragm and let them *contain* elastic molecules in a state of agitation which strike each other and the sides.

Let the number of particles be equal in *A* & *B* but let those in *A* have equal velocities, if oblique collisions occur between them their velocities will become unequal & I have shown that there will be velocities of all magnitudes in *A* and the same in *B* only the sum of the squares of the velocities is greater in *A* than in *B*.

When a molecule is reflected from the fixed diaphragm *CD* no work is lost or gained.

If the molecule instead of being reflected were allowed to go through a hole in *CD* no work would be lost or gained, only its energy would be transferred from the one vessel to the other.

Now conceive a finite being who knows the paths and velocities of all the molecules by simple inspection but who can do no work, except to open and close a hole in the diaphragm, by means of a slide without mass.

Let him first observe the molecules in *A* and when he sees one coming the square of whose velocity is less than the mean sq. vel. of the molecules in *B* let him open a hole & let it go into *B*. Next let him watch for a molecule in *B* the square of whose velocity is greater than the mean sq. vel. in *A* and when it comes to the hole let him draw and slide & let it go into *A*, keeping the slide shut for all other molecules.

Then the number of molecules in *A* & *B* are the same as at first but the energy in *A* is increased and that in *B* diminished that is the hot system has got hotter and the cold colder & yet no work has been done, only the intelligence of a very observant and neat fingered being has been employed.

Or in short if heat is the motion of finite portions of matter and if we can apply tools to such portions of matter so as to deal with them separately then we can take advantage of the different motion of different portions to restore a uniformly hot system to unequal temperatures or to motions of large masses.

Only we can't, not being clever enough.

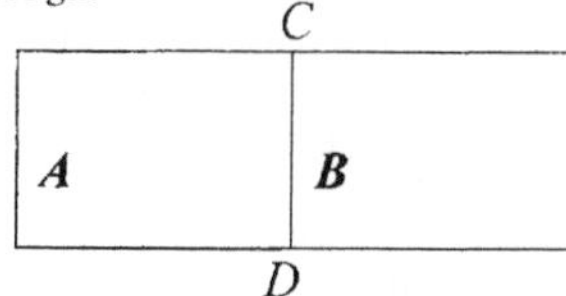

Letter to Peter Guthrie Tait, 11 December 1867. In P. M. Harman (ed.), *The Scientific*

Letters and Papers of James Clerk Maxwell (1995), Vol. 2, 331–2.

1 A strict materialist believes that everything depends on the motion of matter. He knows the form of the laws of motion though he does not know all their consequences when applied to systems of unknown complexity.

Now one thing in which the materialist (fortified with dynamical knowledge) believes is that if every motion great & small were accurately reversed, and the world left to itself again, everything would happen backwards the fresh water would collect out of the sea and run up the rivers and finally fly up to the clouds in drops which would extract heat from the air and evaporate and afterwards in condensing would shoot out rays of light to the sun and so on. Of course all living things would regrede from the grave to the cradle and we should have a memory of the future but not of the past.

The reason why we do not expect anything of this kind to take place at any time is our experience of irreversible processes, all of one kind, and this leads to the doctrine of a beginning & an end instead of cyclical progression for ever.

Letter to Mark Pattison, 7 April 1868. In P. M. Harman (ed.), *The Scientific Letters and Papers of James Clerk Maxwell* (1995), Vol. 2, 1862–1873, 360–1.

2 Any opinion as to the form in which the energy of gravitation exists in space is of great importance, and whoever can make his opinion probable will have made an enormous stride in physical speculation. The apparent universality of gravitation, and the equality of its effects on matter of all kinds are most remarkable facts, hitherto without exception; but they are purely experimental facts, liable to be corrected by a single observed exception. We cannot conceive of matter with negative inertia or mass; but we see no way of *accounting* for the proportionality of gravitation to mass by any legitimate method of demonstration. If we can see the tails of comets fly off in the direction opposed to the sun with an accelerated velocity, and if we believe these tails to be matter and not optical illusions or mere tracks of vibrating disturbance, then we must admit a force in that direction, and we may establish that it is caused by the sun if it always depends upon his position and distance.

Letter to William Huggins, 13 October 1868. In P. M. Harman (ed.), *The Scientific Letters and Papers of James Clerk Maxwell* (1995), Vol. 2, 1862–1873, 451–2.

3 It was a great step in science when men became convinced that, in order to understand the nature of things, they must begin by asking, not whether a thing is good or bad, noxious or beneficial, but of what kind it is? And how much is there of it? Quality and quantity were then first recognized as the primary features to be discovered in scientific inquiry.

'Address to the Mathematical and Physical Sections of the British Association', 1870. In W. D. Niven (ed.), *The Scientific Papers of James Clerk Maxwell* (1890), Vol. 2, 217.

4 The University of Cambridge, in accordance with that law of its evolution, by which, while maintaining the strictest continuity between the successive phases of its history, it adapts itself with more or less promptness to the requirements of the times, has lately instituted a course of Experimental Physics.

'Introductory Lecture on Experimental Physics', (1871). In W. D. Niven (ed.), *The Scientific Papers of James Clerk Maxwell* (1890), Vol. 2, 241.

5 For if there is any truth in the dynamical theory of gases the different molecules in a gas at uniform temperature are moving with very different velocities. Put such a gas into a vessel with two compartments [*A* and *B*] and make a small hole in the wall about the right size to let one molecule through. Provide a lid or stopper for this hole and appoint a doorkeeper, very intelligent and exceedingly quick, with microscopic eyes but still an essentially finite being.

Whenever he sees a molecule of great velocity coming against the door from *A* into *B* he is to let it through, but if the molecule happens to be going slow he is to keep the door shut. He is also to let slow molecules pass from *B* to *A* but not fast ones . . . In this way the temperature of *B* may be raised and that of *A* lowered without any expenditure of work, but only by the intelligent action of a mere guiding agent (like a pointsman on a railway with perfectly acting switches who should send the express along one line and the goods along another).

I do not see why even intelligence might not be dispensed with and the thing be made self-acting.

Moral The 2nd law of Thermodynamics has the same degree of truth as the statement that if you throw a tumblerful of water into the sea you cannot get the same tumblerful of water out again.

Letter to John William Strutt, 6 December 1870. In P. M. Harman (ed.), *The Scientific Letters and Papers of James Clerk Maxwell* (1995), Vol. 2, 582–3.

1 Very few of us can now place ourselves in the mental condition in which even such philosophers as the great Descartes were involved in the days before Newton had announced the true laws of the motion of bodies.

'Introductory Lecture on Experimental Physics', 1871. In W. D. Niven (ed.), *The Scientific Papers of James Clerk Maxwell* (1890), Vol. 2, 241.

2 The popularisation of scientific doctrines is producing as great an alteration in the mental state of society as the material applications of science are effecting in its outward life. Such indeed is the respect paid to science, that the most absurd opinions may become current, provided they are expressed in language, the sound of which recals [sic] some well-known scientific phrase.

'Introductory Lecture on Experimental Physics' (1871). In W. D. Niven (ed.), *The Scientific Papers of James Clerk Maxwell* (1890), Vol. 2, 242.

3 Science appears to us with a very different aspect after we have found out that it is not in lecture rooms only, and by means of the electric light projected on a screen, that we may witness physical phenomena, but that we may find illustrations of the highest doctrines of science in games and gymnastics, in travelling by land and by water, in storms of the air and of the sea, and wherever there is matter in motion.

'Introductory Lecture on Experimental Physics' (1871). In W. D. Niven (ed.), *The Scientific Papers of James Clerk Maxwell* (1890), Vol. 2, 243.

4 It is of great advantage to the student of any subject to read the original memoirs on that subject, for science is always most completely assimilated when it is in the nascent state.

A Treatise on Electricity and Magnetism (1873), Vol. 1, Preface, xiii-xiv.

5 This characteristic of modern experiments—that they consist principally of measurements,—is so prominent, that the opinion seems to have got abroad, that in a few years all the great physical constants will have been approximately estimated, and that the only occupation which will then be left to men of science will be to carry these measurements to another place of decimals . . . But we have no right to think thus of the unsearchable riches of creation, or of the untried fertility of those fresh minds into which these riches will continue to be poured.

'Introductory Lecture on Experimental Physics', 1871. In W. D. Niven (ed.), *The Scientific Papers of James Clerk Maxwell* (1890), Vol. 2, 244.

6 The experimental investigation by which Ampère established the law of the mechanical action between electric currents is one of the most brilliant achievements in science. The whole theory and experiment, seems as if it had leaped, full grown and full armed, from the brain of the 'Newton of

Electricity'. It is perfect in form, and unassailable in accuracy, and it is summed up in a formula from which all the phenomena may be deduced, and which must always remain the cardinal formula of electro-dynamics.

A Treatise on Electricity and Magnetism (1873), Vol. 2, 162.

1 In fact, whenever energy is transmitted from one body to another in time, there must be a medium or substance in which the energy exists after it leaves one body and before it reaches the other . . . and if we admit this medium as an hypothesis, I think it ought to occupy a prominent place in our investigations, and that we ought to endeavour to construct a mental representation of all the details of its action, and this has been my constant aim in this treatise.

A Treatise on Electricity and Magnetism (1873), Vol. 2, 438.

2 The equations of dynamics completely express the laws of the historical method as applied to matter, but the application of these equations implies a perfect knowledge of all the data. But the smallest portion of matter which we can subject to experiment consists of millions of molecules, not one of which ever becomes individually sensible to us. We cannot, therefore, ascertain the actual motion of any one of these molecules; so that we are obliged to abandon the strict historical method, and to adopt the statistical method of dealing with large groups of molecules . . . Thus molecular science teaches us that our experiments can never give us anything more than statistical information, and that no law derived from them can pretend to absolute precision. But when we pass from the contemplation of our experiments to that of the molecules themselves, we leave a world of chance and change, and enter a region where everything is certain and immutable.

'Molecules' (1873). In W. D. Niven (ed.), *The Scientific Papers of James Clerk Maxwell* (1890), Vol. 2, 374.

3 But in the heavens we discover by their light, and by their light alone, stars so distant from each other that no material thing can ever have passed from one to another; and yet this light, which is to us the sole evidence of the existence of these distant worlds, tells us also that each of them is built up of molecules of the same kinds as those which we find on earth. A molecule of hydrogen, for example, whether in Sirius or in Arcturus, executes its vibrations in precisely the same time. Each molecule, therefore, throughout the universe, bears impressed on it the stamp of a metric system as distinctly as does the metre of the Archives at Paris, or the double royal cubit of the Temple of Karnac . . . the exact quantity of each molecule to all others of same kind gives it, as Sir John Herschel has well said, the essential character of a manufactured article and precludes the idea of its being external and self-existent.

'Molecules', 1873. In W. D. Niven (ed.), *The Scientific Papers of James Clerk Maxwell* (1890), Vol. 2, 375–6.

4 For the evolution of science by societies the main requisite is the perfect freedom of communication between each member and any one of the others who may act as a reagent.

The gaseous condition is exemplified in the soiree, where the members rush about confusedly, and the only communication is during a collision, which in some instances may be prolonged by button-holing.

The opposite condition, the crystalline, is shown in the lecture, where the members sit in rows, while science flows in an uninterrupted stream from a source which we take as the origin. This is radiation of science.

Conduction takes place along the series of members seated round a dinner table, and fixed there for several hours, with flowers in the middle to prevent any cross currents.

The condition most favourable to life is an intermediate plastic or colloïdal condition, where the order of business is (1) Greetings and confused talk; (2) A short communication from one who has something to say and to show; (3) Remarks on the communication addressed to the Chair, introducing matters irrelevant to the communication but interesting to the members; (4) This lets each member see who is interested in his special hobby, and who is likely to help him; and leads to (5) Confused conversation and examination of objects on the table.

I have not indicated how this programme is to be combined with eating.

Letter to William Grylls Adams, 3 December 1873. In P. M. Harman (ed.), *The Scientific Letters and Papers of James Clerk Maxwell* (1995), Vol. 2, 1862–1873, 949–50.

1 If we betake ourselves to the statistical method, we do so confessing that we are unable to follow the details of each individual case, and expecting that the effects of widespread causes, though very different in each individual, will produce an average result on the whole nation, from a study of which we may estimate the character and propensities of an imaginary being called the Mean Man.

'Does the Progress of Physical Science tend to give any advantage to the opinion of necessity (or determinism) over that of the continuency of Events and the Freedom of the Will?' In P. M. Harman (ed.), *The Scientific Letters and Papers of James Clerk Maxwell* (1995), Vol. 2, 1862–1873, 818.

2 At quite uncertain times and places,
The atoms left their heavenly path,
And by fortuitous embraces,
Engendered all that being hath.
And though they seem to cling together,
And form 'associations' here,
Yet, soon or late, they burst their tether,
And through the depths of space career.

From 'Molecular Evolution', *Nature*, **8**, 1873. In Lewis Campbell and William Garnett, *The Life of James Clerk Maxwell* (1882), 637.

3 But I should be very sorry if an interpretation founded on a most conjectural scientific hypothesis were to get fastened to the text in Genesis . . . The rate of change of scientific hypothesis is naturally much more rapid than that of Biblical interpretations, so that if an interpretation is founded on such an hypothesis, it may help to keep the hypothesis above ground long after it ought to be buried and forgotten.

Letter to Rev. C. J. Ellicott, Bishop of Gloucester and Bristol, 22 November 1876. Quoted in Lewis Campbell and William Garnett, *The Life of James Clerk Maxwell* (1882), 394.

4 In your letter you apply the word imponderable to a molecule. Don't do that again. It may also be worth knowing that the aether cannot be molecular. If it were, it would be a gas, and a pint of it would have the same properties as regards heat, etc., as a pint of air, except that it would not be so heavy.

Letter to Lewis Campbell, September 1874. In Lewis Campbell and William Garnett, *The Life of James Clerk Maxwell* (1882), 391.

5 An Experiment, like every other event which takes place, is a natural phenomenon; but in a Scientific Experiment the circumstances are so arranged that the relations between a particular set of phenomena may be studied to the best advantage.

'General Considerations Concerning Scientific Apparatus', 1876. In W. D. Niven (ed.), *The Scientific Papers of James Clerk Maxwell* (1890), Vol. 2, 505.

6 Available energy is energy which we can direct into any desired channel. Dissipated energy is energy which we cannot lay hold of and direct at pleasure, such as the energy of the confused agitation of molecules which we call heat. Now, confusion, like the correlative term order, is not a property of material things in themselves, but only in relation to the mind which perceives them. A memorandum-book does not, provided it is neatly written, appear confused to an illiterate person, or to the owner who understands it thoroughly, but to any other person able to read it appears to be inextricably confused. Similarly the notion of dissipated energy could not occur to a being who could not turn any of the energies of nature to his own account, or to one who could trace the motion of every molecule and seize it at the right moment. It is only to a being in the intermediate stage, who can lay hold of some forms of energy while others elude his grasp, that energy appears to be passing inevitably from the available to the dissipated state.

'Diffusion', *Encyclopaedia Britannica* (1878). In W. D. Niven (ed.), *The Scientific Papers of James Clerk Maxwell* (1890), Vol. 2, 646.

1 'O tell me, when along the line
From my full heart the message flows,
What currents are induced in thine?
One click from thee will end my woes'.
Through many an Ohm the Weber flew,
And clicked the answer back to me,
'I am thy Farad, staunch and true,
Charged to a Volt with love for thee'.

From 'Valentine from A Telegraph Clerk ♂ to a Telegraph Clerk ♀'. In Lewis Campbell and William Garnett, *The Life of James Clerk Maxwell* (1882), 631.

2 In a University we are especially bound to recognise not only the unity of science itself, but the communion of the workers in science. We are too apt to suppose that we are congregated here merely to be within reach of certain appliances of study, such as museums and laboratories, libraries and lecturers, so that each of us may study what he prefers. I suppose that when the bees crowd round the flowers it is for the sake of the honey that they do so, never thinking that it is the dust which they are carrying from flower to flower which is to render possible a more splendid array of flowers, and a busier crowd of bees, in the years to come. We cannot, therefore, do better than improve the shining hour in helping forward the cross-fertilization of the sciences.

'The Telephone', *Nature*, **18**, 1878. In W. D. Niven (ed.), *The Scientific Papers of James Clerk Maxwell* (1890), Vol. 2, 743–4.

3 Thus science strips off, one after the other, the more or less gross materialisations by which we endeavour to form an objective image of the soul, till men of science, speculating, in their non-scientific intervals, like other men on what science may possibly lead to, have prophesied that we shall soon have to confess that the soul is nothing else than a function of certain complex material systems.

Review of B. Stewart and P. G. Tait's book on Paradoxical Philosophy, in *Nature*, **19**, 1878. In W. D. Niven (ed.), *The Scientific Papers of James Clerk Maxwell* (1890), Vol. 2, 760.

Joseph Mayer 1904–83

American chemist

4 What does one have to do to be called a scientist? I decided that anyone who spent on science more than 10% of his waking, thinking time for a period of more than a year would be called a scientist, at least for that year.

Quoted in 'The Way it Was', *Annual Review of Physical Chemistry*, 1982, 33, 1–2.

Julius Robert Mayer 1814–78

German physicist and physiologist

5 The fall of a given weight from a height of around 365 meters corresponds to the heating of an equal weight of water from 0° to 1°.

'Bemerkungen über die Kräfte der unbelebten Natur', *Annalen der Chemie und Pharmacie*, 1842, **42**:2, 29. Trans. Kenneth L. Caneva, *Robert Mayer and the Conservation of Energy* (1993), 25.

6 My position is perfectly definite. Gravitation, motion, heat, light, electricity and chemical action are one and the same object in various forms of manifestation.

Annalen der Chemie und der Pharmacie (1842). Trans. A. S. Eve and C. H. Creasey, *The Life and Work of John Tyndall* (1945), 94.

7 Truly I say to you, a single number has more genuine and permanent value than an expensive library full of hypotheses.

Letter to Griesinger, 20 July 1844. In Jacob J. Weyrauch (ed.), *Kleinere Schriften und Briefe von Robert Mayer, nebst Mittheilungen aus seinem Leben* (1893), 226. Trans. Kenneth L. Caneva, *Robert Mayer and the Conservation of Energy* (1993), 37.

1 The physiological combustion theory takes as its starting point the fundamental principle that the amount of heat that arises from the combustion of a given substance is an *invariable* quantity—i.e., one *independent* of the circumstances accompanying the combustion—from which it is more specifically concluded that the chemical effect of the combustible materials undergoes no quantitative change even as a result of the vital process, or that the living organism, with all its mysteries and marvels, is not capable of generating heat out of nothing.

Bemerkungen über das mechanische Aequivalent der Wärme [*Remarks on the Mechanical Equivalent of Heat*] (1851), 17–9. Trans. Kenneth L. Caneva, *Robert Mayer and the Conservation of Energy* (1993), 240.

John Maynard Smith 1920–2004

British geneticist and evolutionary biologist

2 It is an occupational risk of biologists to claim, towards the end of their careers, that the problems which they have not solved are insoluble.

'Popper's World', *The London Review of Books* (18–31 August 1983), 12.

John Mayow 1641–79

British physiologist and chemist

3 We do not doubt to assert, that air does not serve for the motion of the lungs, but rather to communicate something to the blood . . . It is very likely that it is the fine nitrous particles, with which the air abounds, that are communicated to the blood through the lungs.

Tractatus duo. Quorum prior agit de respiratione: alter de rachitude (1668), 43. Quoted in Robert G. Frank jr., *Harvey and the Oxford Physiologists* (1980), 228.

4 ON MAYOW Dr Mayow of All Souls College was a very ingenious Man, & an Excellent Scholar; but by resolving to marry a wife of a Great Fortune fatally miscarried to his unspeakable and insuperable Grief. For when he was one Year at ye Bath, and happening to lodge at ye same house with an Irish lady and her Daughter, as was pretended, who went for vast Fortunes, when ye old Lady feigning an occasion went to London, he courted the young one and quickly married her: but she proving nothing like ye Fortune yt he imagin'd he was so confounded with it yt he did not long enjoy his life.

Thomas Hearne, *Remarks and Collections of Thomas Hearne*, Vol. 1, ed. C. E. Doble (1885), 192.

Ernst Mayr 1904–2005

German-born American evolutionary biologist

5 A species consists of a group of populations which replace each other geographically or ecologically and of which the neighboring ones integrate or hybridise wherever they are in contact or which are potentially capable of doing so (with one or more of the populations) in those cases where contact is prevented by geographical or ecological barriers.

'Speciation Phenomena in Birds', *The American Naturalist*, 1940, 74, 256.

6 Two forms or species are *sympatric*, if they occur together, that is if their areas of distribution overlap or coincide. Two forms (or species) are *allopatric*, if they do not occur together, that is if they exclude each other geographically. The term allopatric is primarily useful in denoting geographic representatives.

Systematics and the Origin of Species: From the Viewpoint of a Zoologist (1942), 148–9.

7 A new species develops if a population which has become geographically isolated from its parental species acquires during this period of isolation characters which promote or guarantee reproductive isolation when the external barriers break down.

Systematics and the Origin of Species: From the Viewpoint of a Zoologist (1942), 155.

1 The reduced variability of small populations is not always due to accidental gene loss, but sometimes to the fact that the entire population was started by a single pair or by a single fertilized female. These 'founders' of the population carried with them only a very small proportion of the variability of the parent population. This 'founder' principle sometimes explains even the uniformity of rather large populations, particularly if they are well isolated and near the borders of the range of the species.

Systematics and the Origin of Species: From the Viewpoint of a Zoologist (1942), 237.

2 The assumptions of population thinking are diametrically opposed to those of the typologist. The populationist stresses the uniqueness of everything in the organic world. What is true for the human species,—that no two individuals are alike, is equally true for all other species of animals and plants . . . All organisms and organic phenomena are composed of unique features and can be described collectively only in statistical terms. Individuals, or any kind of organic entities, form populations of which we can determine the arithmetic mean and the statistics of variation. Averages are merely statistical abstractions, only the individuals of which the populations are composed have reality. The ultimate conclusions of the population thinker and of the typologist are precisely the opposite. For the typologist, the type (eidos) is real and the variation an illusion, while for the populationist the type (average) is an abstraction and only the variation is real. No two ways of looking at nature could be more different.

Darwin and the Evolutionary Theory in Biology (1959), 2.

3 Isolating mechanisms are biological properties of individuals that prevent the interbreeding of populations that are actually or potentially sympatric.

Animal Species and Evolution (1963), 91.

4 Biology can be divided into the study of proximate causes, the study of the physiological sciences (broadly conceived), and into the study of ultimate (evolutionary) causes, the subject of natural history.

The Growth of Biological Thought: Diversity, Evolution and Inheritance (1982), 67.

5 Most classifications, whether of inanimate objects or of organisms, are hierarchical. There are 'higher' and 'lower' categories, there are higher and lower ranks. What is usually overlooked is that the use of the term 'hierarchy' is ambiguous, and that two fundamentally different kinds of arrangements have been designated as hierarchical. A hierarchy can be either *exclusive* or *inclusive*. Military ranks from private, corporal, sergeant, lieutenant, captain, up to general are a typical example of an exclusive hierarchy. A lower rank is not a subdivision of a higher rank; thus, lieutenants are not a subdivision of captains. The *scala naturae*, which so strongly dominated thinking from the sixteenth to the eighteenth century, is another good illustration of an exclusive hierarchy. Each level of perfection was considered an advance (or degradation) from the next lower (or higher) level in the hierarchy, but did *not* include it.

The Growth of Biological Thought: Diversity, Evolution and Inheritance (1982), 205–6.

6 A species is a reproductive community of populations (reproductively isolated from others) that occupies a specific niche in nature.

The Growth of Biological Thought: Diversity, Evolution and Inheritance (1982), 273.

7 All interpretations made by a scientist are hypotheses, and all hypotheses are tentative. They must forever be tested and they must be revised if found to be unsatisfactory. Hence, a change of mind in a scientist, and particularly in a great scientist, is not only not a sign of weakness but rather evidence for continuing attention to the respective problem and an ability to test the hypothesis again and again.

The Growth of Biological Thought: Diversity, Evolution and Inheritance (1982), 831.

1 It is curious how often erroneous theories have had a beneficial effect for particular branches of science.

The Growth of Biological Thought: Diversity, Evolution and Inheritance (1982), 847.

2 There is more to biology than rats, *Drosophila, Caenorhabditis, and E. coli.*

Forword to Lynn Margulis and Dorion Sagan, *Acquiring Genomes: A Theory of the Origins of Species* (2003), xiv.

Bruce Mazlish 1923-

American historian

3 [T]he human desire to escape the flesh, which took one form in asceticism, might take another form in the creation of machines. Thus, the wish to rise above the bestial body manifested itself not only in angels but in mechanical creatures. Certainly, once machines existed, humans clearly attached to them feelings of escape from the flesh.

The Fourth Discontinuity: The Co-Evolution of Humans and Machines (1993), 218.

Margaret Mead 1901-78

American anthropologist

4 Samoa culture demonstrates how much the tragic or the easy solution of the Oedipus situation depends upon the inter-relationship between parents and children, and is not created out of whole cloth by the young child's biological impulses.

Male and Female: A Study of the Sexes in a Changing World (1949), 119.

5 And as I had my father's kind of mind—which was also his mother's—I learned that the mind is not sex-typed.

Blackberry Winter: My Earlier Years (1973), 54.

Peter Brian Medawar 1915-87

Brazilian-born British immunologist

6 Heredity proposes and development disposes.

'Postscript: D'Arcy Thompson and *Growth and Form*'. From Ruth D'Arcy Thompson, *D'Arcy Wentworth Thompson: The Scholar Naturalist 1860-1948* (1958), 225.

7 Is the Scientific Paper a Fraud?

'Is the Scientific Paper a Fraud?', *The Listener*, 12 September 1963, 377-8.

8 The lives of scientists, considered as Lives, almost always make dull reading. For one thing, the careers of the famous and the merely ordinary fall into much the same pattern, give or take an honorary degree or two, or (in European countries) an honorific order. It could be hardly otherwise. Academics can only seldom lead lives that are spacious or exciting in a worldly sense. They need laboratories or libraries and the company of other academics. Their work is in no way made deeper or more cogent by privation, distress or worldly buffetings. Their private lives may be unhappy, strangely mixed up or comic, but not in ways that tell us anything special about the nature or direction of their work. Academics lie outside the devastation area of the literary convention according to which the lives of artists and men of letters are intrinsically interesting, a source of cultural insight in themselves. If a scientist were to cut his ear off, no one would take it as evidence of a heightened sensibility; if a historian were to fail (as Ruskin did) to consummate his marriage, we should not suppose that our understanding of historical scholarship had somehow been enriched.

'J.B.S: A Johnsonian Scientist', *New York Review of Books*, 10 October 1968, reprinted in *Pluto's Republic* (1982), 263.

9 If the task of scientific methodology is to piece together an account of what scientists actually *do*, then the testimony of biologists should be heard with specially close attention. Biologists work very close to the frontier between bewilderment and understanding. Biology is complex, messy and richly various, like real life; it travels faster nowadays than physics or chemistry (which is just as well, since it has so much farther to go), and it travels nearer to the ground. It should therefore give us a specially direct and immediate insight into science in the making.

Induction and Intuition in Scientific Thought (1969), 1.

1 The fact that scientists do not consciously practice a formal methodology is very poor evidence that no such methodology exists. It could be said—has been said—that there is a distinctive methodology of science which scientists practice unwittingly, like the chap in Molière who found that all his life, unknowingly, he had been speaking prose.

Induction and Intuition in Scientific Thought (1969), 9.

2 Deductivism in mathematical literature and inductivism in scientific papers are simply the postures we choose to be seen in when the curtain goes up and the public sees us. The theatrical illusion is shattered if we ask what goes on behind the scenes. In real life discovery and justification are almost always different processes.

Induction and Intuition in Scientific Thought (1969), 26.

3 Scientific reasoning is a kind of dialogue between the possible and the actual, between what might be and what is in fact the case.

Induction and Intuition in Scientific Thought (1969), 48.

4 There is nothing distinctively scientific about the hypothetico-deductive process. It is not even distinctively intellectual. It is merely a scientific context for a much more general stratagem that underlies almost all regulative processes or processes of continuous control, namely *feedback*, the control of performance by the consequences of the act performed. In the hypothetico-deductive scheme the inferences we draw from a hypothesis are, in a sense, its logical output. If they are true, the hypothesis need not be altered, but correction is obligatory if they are false. The continuous feedback from inference to hypothesis is implicit in Whewell's account of scientific method; he would not have dissented from the view that scientific behaviour can be classified as appropriately under cybernetics as under logic.

Induction and Intuition in Scientific Thought (1969), 54–5.

5 The scientific method is a potentiation of common sense, exercised with a specially firm determination not to persist in error if any exertion of hand or mind can deliver us from it. Like other exploratory processes, it can be resolved into a dialogue between fact and fancy, the actual and the possible; between what could be true and what is in fact the case. The purpose of scientific enquiry is not to compile an inventory of factual information, nor to build up a totalitarian world picture of Natural Laws in which every event that is not compulsory is forbidden. We should think of it rather as a logically articulated structure of justifiable beliefs about nature. It begins as a story about a Possible World—a story which we invent and criticise and modify as we go along, so that it ends by being, as nearly as we can make it, a story about real life.

Induction and Intuition in Scientific Thought (1969), 59.

6 The intensity of a conviction that a hypothesis is true has no bearing on whether it is true or false.

Advice to a Young Scientist (1979), 39.

7 It is a common failing—and one that I have myself suffered from—to fall in love with a hypothesis and to be unwilling to take no for an answer. A love affair with a pet hypothesis can waste years of precious time. There is very often no finally decisive yes, though quite often there can be a decisive no.

Advice to a Young Scientist (1979), 73.

1 Scientists are people of very dissimilar temperaments doing different things in very different ways. Among scientists are collectors, classifiers and compulsive tidiers-up; many are detectives by temperament and many are explorers; some are artists and others artisans. There are poet-scientists and philosopher-scientists and even a few mystics.

'Hypothesis and Imagination', In *The Art of the Soluble* (1967), 132.

Lise Meitner 1878–1968

Austrian physicist

2 I will have nothing to do with a bomb!

On refusing an invitation to join a group of British scientists, including Otto Robert Frisch, to go to Los Alamos to work on the atomic bomb in 1943.

Ruth Sime, *Lise Meitner: A Life in Physics* (1996), 305.

3 No-one really thought of fission before its discovery.

Oral History Interview with Thomas S. Kuhn (Otto Robert Frisch present) 12 May 1963, Archive for the History of Quantum Physics, 18–20. Ruth Sime, *Lise Meitner: A Life in Physics* (1996), 371.

4 Science makes people reach selflessly for truth and objectivity; it teaches people to accept reality, with wonder and admiration, not to mention the deep awe and joy that the natural order of things brings to the true scientist.

Lecture, Austrian UNESCO Commision, 30 March 1953, in *Atomenergie und Frieden: Lise Meitner und Otto Hahn* (1953), 23–4. Trans. Ruth Sime, *Lise Meitner: A Life in Physics* (1996), 375.

5 O. Hahn and F. Strassmann have discovered a new type of nuclear reaction, the splitting into two smaller nuclei of the nuclei of uranium and thorium under neutron bombardment. Thus they demonstrated the production of nuclei of barium, lanthanum, strontium, yttrium, and, more recently, of xenon and caesium. It can be shown by simple considerations that this type of nuclear reaction may be described in an essentially classical way like the fission of a liquid drop, and that the fission products must fly apart with kinetic energies of the order of hundred million electron-volts each.

'Products of the Fission of the Urarium Nucleus', *Nature*, 1939, **143**, 471.

6 ON MEITNER The subject, cosmic physics, of her inaugural lecture was reported as 'cosmetic physics' in the press (more plausible with a female Dozent!).

Biographical Memoirs of Fellows of the Royal Society, 1970, **16**, 408.

7 ON MEITNER Lise Meitner: a physicist who never lost her humanity.

The inscription chosen on her tombstone in St. James' Church, Bramley, Hampshire.

Ruth Sime, *Lise Meitner: A Life in Physics* (1996), 380.

8 ON MEITNER *Einstein referred to Lise Meitner as:*

Our Marie Curie.

Ruth Sime, *Lise Meitner: A Life in Physics* (1996), 74–5.

9 ON MEITNER *Ihre Arbeit ist gekrönt worden mit dem Nobel Preis für Otto Hahn.*

Her work has been crowned by the Nobel Prize for Otto Hahn.

Said of her research.

Lise Meitner 1878–1968 and Otto Robert Frisch 1904–79

Austrian physicist and Austrian-born British nuclear physicist

10 In the discussion of the energies involved in the deformation of nuclei, the concept of surface tension of nuclear matter has been used and its value had been estimated from simple considerations regarding nuclear forces. It must be remembered, however, that the surface tension of a charged droplet is diminished by its charge, and a rough estimate shows that the surface tension of nuclei, decreasing with increasing nuclear charge, may become zero for atomic numbers of the order of 100. It seems therefore possible that the uranium nucleus has only small stability of form, and may, after neutron capture, divide itself into two nuclei of roughly equal size (the precise ratio of sizes depending on finer structural features and perhaps partly on chance). These two nuclei will repel each other and should gain a total kinetic energy of *c.* 200 Mev., as calculated from nuclear radius and charge. This amount of energy may actually be expected to be available from the difference in packing fraction between uranium and the elements in the middle of the periodic system. The whole 'fission' process can thus be described in an essentially classical way, without having to consider quantum-mechanical 'tunnel effects', which would actually be extremely small, on account of the large masses involved.

Lise Meitner and O. R. Frisch, 'Disintegration of Uranium by Neutrons: a New Type of Nuclear Reaction', *Nature*, 1939, **143**, 239.

Lord Melbourne see **Lamb, William**

Henry William Menard 1920–86

American earth scientist and marine geologist

1 It appears that the extremely important papers that trigger a revolution may not receive a proportionately large number of citations. The normal procedures of referencing are not used for folklore. A real scientific revolution, like any other revolution, is news. *The Origin of Species* sold out as fast as it could be printed and was denounced from the pulpit almost immediately. Sea-floor spreading has been explained, perhaps not well, in leading newspapers, magazines, books, and most recently in a color motion picture. When your elementary school children talk about something at dinner, you rarely continue to cite it.

'Citations in a Scientific Revolution', in R. Shagam et al., *Studies in Earth and Space Sciences: A Memoir in Honor of Harry Hammond Hess* (1972), 4.

Henry Louis Mencken 1880–1956

American writer and journalist

2 The value the world sets upon motives is often grossly unjust and inaccurate. Consider, for example, two of them: mere insatiable curiosity and the desire to do good. The latter is put high above the former, and yet it is the former that moves some of the greatest men the human race has yet produced: the scientific investigators. What animates a great pathologist? Is it the desire to cure disease, to save life? Surely not, save perhaps as an afterthought. He is too intelligent, deep down in his soul, to see anything praiseworthy in such a desire. He knows by life-long observation that his discoveries will do quite as much harm as good, that a thousand scoundrels will profit to every honest man, that the folks who most deserve to be saved will probably be the last to be saved. No man of self-respect could devote himself to pathology on such terms. What actually moves him is his unquenchable curiosity—his boundless, almost pathological thirst to penetrate the unknown, to uncover the secret, to find out what has not been found out before. His prototype is not the liberator releasing slaves, the good Samaritan lifting up the fallen, but the dog sniffing tremendously at an infinite series of rat-holes.

Prejudices (1923), 269–70.

3 Astronomers and physicists, dealing habitually with objects and quantities far beyond the reach of the senses, even with the aid of the most powerful aids that ingenuity has been able to devise, tend almost inevitably to fall into the ways of thinking of men dealing with objects and quantities that do not exist at all, e.g., theologians and metaphysicians. Thus their speculations tend almost inevitably to depart from the field of true science, which is that of precise observation, and to become mere soaring in the empyrean. The process works backward, too. That is to say, their reports of what they pretend actually to *see* are often very unreliable. It is thus no wonder that, of all men of science, they are the most given to flirting with theology. Nor is it remarkable that, in the popular belief, most astronomers end by losing their minds.

Minority Report: H. L. Mencken's Notebooks (1956), Sample 74, 60.

1 It is impossible to imagine the universe run by a wise, just and omnipotent God, but it is quite easy to imagine it run by a board of gods. If such a board actually exists it operates precisely like the board of a corporation that is losing money.

Minority Report: H. L. Mencken's Notebooks (1956), Sample 79, 63.

Johann Gregor Mendel 1822–84

Czech geneticist and monk

2 That no generally applicable law of the formulation and development of hybrids has yet been successfully formulated can hardly astonish anyone who is acquainted with the extent of the task and who can appreciate the difficulties with which experiments of this kind have to contend.

'Experiments on Plant Hybrids' (1865). In Curt Stern and Eva R. Sherwood (eds.), *The Origin of Genetics: A Mendel Source Book* (1966), 2.

3 When two plants, constantly different in one or several traits, are crossed, the traits they have in common are transmitted unchanged to the hybrids and their progeny, as numerous experiments have proven; a pair of differing traits, on the other hand, are united in the hybrid to form a new trait, which usually is subject to changes in the hybrids' progeny.

'Experiments on Plant Hybrids' (1865). In Curt Stern and Eva R. Sherwood (eds.), *The Origin of Genetics: A Mendel Source Book* (1966), 5.

4 Experiments on ornamental plants undertaken in previous years had proven that, as a rule, hybrids do not represent the form exactly intermediate between the parental strains. Although the intermediate form of some of the more striking traits, such as those relating to shape and size of leaves, pubescence of individual parts, and so forth, is indeed nearly always seen, in other cases one of the two parental traits is so preponderant that it is difficult or quite impossible, to detect the other in the hybrid. The same is true for *Pisum* hybrids. Each of the seven hybrid traits either resembles so closely one of the two parental traits that the other escapes detection, or is so similar to it that no certain distinction can be made. This is of great importance to the definition and classification of the forms in which the offspring of hybrids appear. In the following discussion those traits that pass into hybrid association entirely or almost entirely unchanged, thus themselves representing the traits of the hybrid, are termed *dominating* and those that become latent in the association, *recessive*. The word 'recessive' was chosen because the traits so designated recede or disappear entirely in the hybrids, but reappear unchanged in their progeny, as will be demonstrated later.

'Experiments on Plant Hybrids' (1865). In Curt Stern and Eva R. Sherwood (eds.), *The Origin of Genetics: A Mendel Source Book* (1966), 9.

5 In this generation, *along with the dominating* traits, the *recessive* ones also reappear, their individuality fully revealed, and they do so in the decisively expressed average proportion of 3:1, so that among each four plants of this generation three receive the dominating and one the recessive characteristic.

'Experiments on Plant Hybrids' (1865). In Curt Stern and Eva R. Sherwood (eds.), *The Origin of Genetics: A Mendel Source Book* (1966), 10.

6 If A denotes one of the two constant traits, for example, the dominating one, *a* the recessive, and the A*a* the hybrid form in which both are united, then the expression:

A + 2A*a* + *a*

Gives the series for the progeny of plants hybrid in a pair of differing traits.

'Experiments on Plant Hybrids' (1865). In Curt Stern and Eva R. Sherwood (eds.), *The Origin of Genetics: A Mendel Source Book* (1966), 16.

1 My experiments with single traits all lead to the same result: that from the seeds of hybrids, plants are obtained half of which in turn carry the hybrid trait (A*a*), the other half, however, receive the parental traits A and *a* in equal amounts. Thus, on the average, among four plants two have the hybrid trait A*a*, one the parental trait A, and the other the parental trait a. Therefore, 2A*a*+A+*a* or A + 2A*a* + *a* is the empirical simple series for two differing traits.

Letter to Carl Nägeli, 31st December 1866. In Curt Stern and Eva R. Sherwood (eds.), *The Origin of Genetics: A Mendel Source Book* (1966), 63.

Dmitry Ivanovich Mendeleev

1834–1907

Russian chemist

2 Why do they [Americans] quarrel, why do they hate Negroes, Indians, even Germans, why do they not have science and poetry commensurate with themselves, why are there so many frauds and so much nonsense? I cannot soon give a solution to these questions . . . It was clear that in the United States there was a development not of the best, but of the middle and worst sides of European civilization; the notorious general voting, the tendency to politics . . . all the same as in Europe. A new dawn is not to be seen on this side of the ocean.

The Oil Industry in the North American State of Pennsylvania and in the Caucasus (1877). In Russian. Trans. H. M. Leicester, 'Mendeleev's Visit to America', *Journal of Chemical Education*, 1957, **34**, 333.

3 When the elements are arranged in vertical columns according to increasing atomic weight, so that the horizontal lines contain analogous elements again according to increasing atomic weight, an arrangement results from which several general conclusions may be drawn.

'The Relations of the Properties to the Atomic Weights of the Elements', *Zeitschrift für Chemie*, 1869.

4 If all the elements are arranged in the order of their atomic weights, a periodic repetition of properties is obtained. This is expressed by the law of periodicity.

Principles of Chemistry (1905), Vol. 2, 17.

5 There must be some bond of union between mass and the chemical elements; and as the mass of a substance is ultimately expressed (although not absolutely, but only relatively) in the atom, a functional dependence should exist and be discoverable between the individual properties of the elements and their atomic weights. But nothing, from mushrooms to a scientific dependence can be discovered without looking and trying. So I began to look about and write down the elements with their atomic weights and typical properties, analogous elements and like atomic weights on separate cards, and soon this convinced me that the properties of the elements are in periodic dependence upon their atomic weights; and although I had my doubts about some obscure points, yet I have never doubted the universality of this law, because it could not possibly be the result of chance.

Principles of Chemistry (1905), Vol. 2, 18.

6 ON MENDELEEV I admit that Mendeleef has two wives, but I have only one Mendeleef.

Tsar of Russia to nobleman complaining of Mendeleef's technical bigamy.

In R. Oesper, *The Human Side of Scientists* (1975), 129.

Robert King Merton 1910–2003

American sociologist

1 The self-fulfilling prophecy is, in the beginning, a false definition of the situation evoking a new behavior which makes the originally false conception come true. The specious validity of the self-fulfilling prophecy perpetuates a reign of error. For the prophet will cite the actual course of events as proof that he was right from the very beginning . . . Such are the perversities of social logic.

The Self-Fulfilling Prophecy (1948), 477.

2 The institutional goal of science is the extension of certified knowledge. The technical methods employed toward this end provide the relevant definition of knowledge: empirically confirmed and logically consistent predictions. The institutional imperatives (mores) derive from the goal and the methods. The entire structure of technical and moral norms implements the final objective. The technical norm of empirical evidence, adequate, valid and reliable, is a prerequisite for sustained true prediction; the technical norm of logical consistency, a prerequisite for systematic and valid prediction. The mores of science possess a methodologic rationale but they are binding, not only because they are procedurally efficient, but because they are believed right and good. They are moral as well as technical prescriptions. Four sets of institutional imperatives—universalism, communism, disinterestedness, organized scepticism—comprise the ethos of modern science.

Social Theory and Social Structure (1957), 552–3.

3 Most institutions demand unqualified faith; but the institution of science makes skepticism a virtue.

Social Theory and Social Structure (1957), 547.

4 [The] complex pattern of the misallocation of credit for scientific work must quite evidently be described as 'the Matthew effect', for, as will be remembered, the Gospel According to St. Matthew puts it this way: For unto every one that hath shall be given, and he shall have abundance: but from him that hath not shall be taken away even that which he hath. Put in less stately language, the Matthew effect consists of the accruing of greater increments of recognition for particular scientific contributions to scientists of considerable repute and the withholding of such recognition from scientists who have not yet made their mark.

'The Matthew Effect in Science', *Science*, 1968, **159**, 58.

5 Science is public, not private, knowledge.

Science, Technology and Society in Seventeenth-century England (1988), 219.

6 We thus begin to see that the institutionalized practice of citations and references in the sphere of learning is not a trivial matter. While many a general reader—that is, the lay reader located outside the domain of science and scholarship—may regard the lowly footnote or the remote endnote or the bibliographic parenthesis as a dispensable nuisance, it can be argued that these are in truth central to the incentive system and an underlying sense of distributive justice that do much to energize the advancement of knowledge.

'The Matthew Effect in Science, II: Cumulative Advantage and the Symbolism of Intellectual Property', *Isis*, 1988, **79**, 621.

Elie Metchnikoff 1845–1916

Russian biologist and immunologist

7 One day when the whole family had gone to a circus to see some extraordinary performing apes, I remained alone with my microscope, observing the life in the mobile cells of a transparent star-fish larva, when a new thought suddenly flashed across my brain. It struck me that similar cells might serve in the defence of the organism against intruders. Feeling that there was in this something of surpassing interest, I felt so excited that I began striding up and down the room and even went to the seashore in order to collect my thoughts.

I said to myself that, if my supposition was true, a splinter introduced into the body of a star-fish larva, devoid of blood-vessels or of a nervous system, should soon be surrounded by mobile cells as is to be observed in a man who runs a splinter into his finger. This was no sooner said than done.

There was a small garden to our dwelling, in which we had a few days previously organised a 'Christmas tree' for the children on a little tangerine tree; I fetched from it a few rose thorns and introduced them at once under the skin of some beautiful star-fish larvae as transparent as water.

I was too excited to sleep that night in the expectation of the result of my experiment, and very early the next morning I ascertained that it had fully succeeded.

That experiment formed the basis of the phagocyte theory, to the development of which I devoted the next twenty-five years of my life.

In Olga Metchnikoff, *Life of Elie Metchnikoff 1845–1916* (1921), 116–7.

1 It is possible to state as a general principle that the mesodermic phagocytes, which originally (as in the sponges of our days) acted as digestive cells, retained their role to absorb the dead or weakened parts of the organism as much as different foreign intruders.

'Uber die Pathologische Bedeutung der Intracellularen Verduung', *Fortschritte der Medizin*, 1884, 17, 558–569. Trans. Alfred I. Tauber and Leon Chernyak, *Metchnikoff and the Origins of Immunology* (1991), 141.

Nicholas Metropolis 1915–99

American physicist and pioneer of scientific computing

2 But, contrary to the lady's prejudices about the engineering profession, the fact is that quite some time ago the tables were turned between theory and applications in the physical sciences. Since World War II the discoveries that have changed the world are not made so much in lofty halls of theoretical physics as in the less-noticed labs of engineering and experimental physics. The roles of pure and applied science have been reversed; they are no longer what they were in the golden age of physics, in the age of Einstein, Schrödinger, Fermi and Dirac.

'The Age of Computing: a Personal Memoir', *Dædalus*, 1992, 121, 120.

P. Michaelis

German geneticist

3 Not only do the various components of the cells form a living system, in which the capacity to live, react, and reproduce is dependent on the interactions of all the members of the system; but this living system is identical with the genetic system. The form of life is determined not only by the specific nature of the hereditary units but also by the structure and arrangement of the system. The whole system is more than the sum of its parts, and the effect of each of the components depends on and is influenced by all previous reactions, whose sequence is in turn determined by the whole idiotype.

'Cytoplasmic Inheritance in Epilobium and Its Theoretical Significance', *Advances in Genetics*, 1954, 6, 320.

John Michell 1724–93

British geophysicist and astronomer

4 The strata of the earth are frequently very much bent, being raised in some places, and depressed in others, and this sometimes with a very quick ascent or descent; but as these ascents and descents, in a great measure, compensate one another, if we take a large extent of country together, we may look upon the whole set of strata, as lying nearly horizontally. What is very remarkable, however, in their situation, is, that from most, if not all, large tracts of high and mountainous countries, the strata lie in a situation more inclined to the horizon, than the country itself, the mountainous countries being generally, if not always, formed out of the lower strata of earth. This situation of the strata may be not unaptly represented in the following manner. Let a number of leaves of paper, of several different sorts or colours, be pasted upon one another; then bending them up together into a ridge in the middle, conceive them to be reduced again to a level surface, by a plane so passing through them, as to cut off all the part that had been raised; let the middle now be again raised a little, and this will be a good general representation of most, if not of all, large tracts of mountainous countries, together with the parts adjacent, throughout the whole world.

'Conjectures Concerning the Cause, and Observations upon the Phenomena of Earthquakes', *Philosophical Transactions of the Royal Society of London*, 1760, 51, 584–5.

Albert Abraham Michelson

1852–1931

German-born American physicist

1 The more important fundamental laws and facts of physical science have all been discovered, and these are now so firmly established that the possibility of their ever being supplanted in consequence of new discoveries is exceedingly remote. Nevertheless, it has been found that there are apparent exceptions to most of these laws, and this is particularly true when the observations are pushed to a limit, *i.e.*, whenever the circumstances of experiment are such that extreme cases can be examined. Such examination almost surely leads, not to the overthrow of the law, but to the discovery of other facts and laws whose action produces the apparent exceptions. As instances of such discoveries, which are in most cases due to the increasing order of accuracy made possible by improvements in measuring instruments, may be mentioned: first, the departure of actual gases from the simple laws of the so-called perfect gas, one of the practical results being the liquefaction of air and all known gases; second, the discovery of the velocity of light by astronomical means, depending on the accuracy of telescopes and of astronomical clocks; third, the determination of distances of stars and the orbits of double stars, which depend on measurements of the order of accuracy of one-tenth of a second—an angle which may be represented as that which a pin's head subtends at a distance of a mile. But perhaps the most striking of such instances are the discovery of a new planet or observations of the small irregularities noticed by Leverrier in the motions of the planet Uranus, and the more recent brilliant discovery by Lord Rayleigh of a new element in the atmosphere through the minute but unexplained anomalies found in weighing a given volume of nitrogen. Many other instances might be cited, but these will suffice to justify the statement that 'our future discoveries must be looked for in the sixth place of decimals'.

Light Waves and Their Uses (1903), 23–4.

2 The velocity of light is one of the most important of the fundamental constants of Nature. Its measurement by Foucault and Fizeau gave as the result a speed greater in air than in water, thus deciding in favor of the undulatory and against the corpuscular theory. Again, the comparison of the electrostatic and the electromagnetic units gives as an experimental result a value remarkably close to the velocity of light—a result which justified Maxwell in concluding that light is the propagation of an electromagnetic disturbance. Finally, the principle of relativity gives the velocity of light a still greater importance, since one of its fundamental postulates is the constancy of this velocity under all possible conditions.

Studies in Optics (1927), 120.

1 The generalized theory of relativity has furnished still more remarkable results. This considers not only uniform but also accelerated motion. In particular, it is based on the impossibility of distinguishing an acceleration from the gravitation or other force which produces it. Three consequences of the theory may be mentioned of which two have been confirmed while the third is still on trial: (1) It gives a correct explanation of the residual motion of forty-three seconds of arc per century of the perihelion of Mercury. (2) It predicts the deviation which a ray of light from a star should experience on passing near a large gravitating body, the sun, namely, 1″.7. On Newton's corpuscular theory this should be only half as great. As a result of the measurements of the photographs of the eclipse of 1921 the number found was much nearer to the prediction of Einstein, and was inversely proportional to the distance from the center of the sun, in further confirmation of the theory. (3) The theory predicts a displacement of the solar spectral lines, and it seems that this prediction is also verified.

Studies in Optics (1927), 160–1.

Mary Midgley 1919–

British philosopher

2 When some portion of the biosphere is rather unpopular with the human race—a crocodile, a dandelion, a stony valley, a snowstorm, an odd-shaped flint—there are three sorts of human being who are particularly likely still to see point in it and befriend it. They are poets, scientists and children. Inside each of us, I suggest, representatives of all these groups can be found.

Animals and Why They Matter: A Journey Around the Species Barrier (1983), 145.

Ludwig Mies van der Rohe

1886–1969

German-born American architect

3 God is in the details.

'On Restraint in Design', *New York Herald Tribune*, 28 June 1959.

John Stuart Mill 1806–73

British philosopher and economist

4 To find fault with our ancestors for not having annual parliaments, universal suffrage, and vote by ballot, would be like quarrelling with the Greeks and Romans for not using steam navigation, when we know it is so safe and expeditious; which would be, in short, simply finding fault with the third century before Christ for not being the eighteenth century after. It was necessary that many other things should be thought and done, before, according to the laws of human affairs, it was possible that steam navigation should be thought of. Human nature must proceed step by step, in politics as well as in physics.

The Spirit of the Age (1831). Ed. Frederick A. von Hayek (1942), 48.

5 It is a law, that every event depends on the same law.

A System of Logic: Ratiocinative and Inductive (1843), Vol. 1, Book 2, Chapter 5, 396.

6 The validity of all the Inductive Methods depends on the assumption that every event, or the beginning of every phenomenon, must have some cause; some antecedent, upon the existence of which it is invariably and unconditionally consequent.

A System of Logic: Ratiocinative and Inductive (1843), Vol. 2, 107.

7 The ends of scientific classification are best answered, when the objects are formed into groups respecting which a greater number of general propositions can be made, and those propositions more important, than could be made respecting any other groups into which the same things could be distributed . . .

A classification thus formed is properly scientific or philosophical, and is commonly called a Natural, in contradistinction to a Technical or Artificial, classification or arrangement.

A System of Logic, Ratiocinative and Inductive (1843), Vol. 2, Book 4, Chapter 7, 302–3.

1 The laws and conditions of the production of wealth partake of the character of physical truths. There is nothing optional or arbitrary in them . . . It is not so with the Distribution of Wealth. That is a matter of human institution solely. The things once there, mankind, individually or collectively, can do with them as they like.

Principles of Political Economy (1848), Book 2, 199.

2 All silencing of discussion is an assumption of infallibility.

'On Liberty' (1859), in M. Warnock (ed.), *Utilitarianism* (1962), 143.

3 It is better to be a human being dissatisfied than a pig satisfied; better to be Socrates dissatisfied than a fool satisfied.

Utilitarianism (1861), 212.

4 There is a tolerably general agreement about what a university is not. It is not a place of professional education.

J. S. Mill in his inaugural address as Rector of St. Andrews in 1867. Quoted in M. Sanderson, *The Universities and British Industry 1850–1970* (1972), 5.

5 So true is it that unnatural generally means only uncustomary, and that everything which is usual appears natural. The subjection of women to men being a universal custom, any departure from it quite naturally appears unnatural.

The Subjection of Women (1869), 270.

George Armitage Miller 1920–

American neuropsychologist

6 What about the magical number seven? What about the seven wonders of the world, the seven seas, the seven deadly sins, the seven daughters of Atlas in the Pleiades, the seven ages of man, the seven levels of hell, the seven primary colors, the seven notes of the musical scale, and the seven days of the week? What about the seven-point rating scale, the seven categories for absolute judgment, the seven objects in the span of attention, and the seven digits in the span of immediate memory? For the present I propose to withhold judgment. Perhaps there is something deep and profound behind all these sevens, something just calling out for us to discover it. But I suspect that it is only a pernicious, Pythagorean coincidence.

The Magical Number Seven, Plus or Minus Two (1956), 42–3.

Hugh Miller 1802–56

British theologian and geologist

7 The six thousand years of human history form but a portion of the geologic day that is passing over us: they do not extend into the *yesterday* of the globe, far less touch the myriads of ages spread out beyond.

My Schools and Schoolmasters (1854), 3rd edition, 41.

8 Because science flourishes, must poesy decline? The compliant serves but to betray the weakness of the class who urge it.

Sketch Book of Popular Geology (1860), 80.

9 Poets need be in no degree jealous of the geologists. The stony science, with buried creations for its domains, and half an eternity charged with its annals, possesses its realms of dim and shadowy fields, in which troops of fancies already walk like disembodied ghosts in the old fields of Elysium, and which bid fair to be quite dark and uncertain enough for all the purposes of poesy for centuries to come.

Sketch Book of Popular Geology (1860), 82–3.

10 Nature is a vast tablet, inscribed with signs, each of which has its own significancy, and becomes poetry in the mind when read; and geology is simply the key by which myriads of these signs, hitherto indecipherable, can be unlocked and perused, and thus a new province added to the poetical domain.

Sketch Book of Popular Geology (1860), 87.

1 The development doctrines are doing much harm on both sides of the Atlantic, especially among intelligent mechanics, and a class of young men engaged in the subordinate departments of trade and the law. And the harm thus considerable in amount must be necessarily more than considerable in degree. For it invariably happens, that when persons in these walks become materialists, they become turbulent subjects and bad men.

The Footprints of the Creator (1861), lxvi.

2 That special substance according to whose mass and degree of development all the creatures of this world take rank in the scale of creation, is not *bone*, but *brain*.

The Footprints of the Creator (1861), 135.

3 No true geologist holds by the development hypothesis;—it has been resigned to sciolists and smatterers;—and there is but one other alternative. They began to be, *through the miracle of creation*. From the evidence furnished by these rocks we are shut down either to belief in *miracle*, or to the belief in something else infinitely harder of reception, and as thoroughly unsupported by testimony as it is contrary to experience. Hume is at length answered by the severe truths of the stony science.

The Footprints of the Creator (1861), 267.

4 The geologist, in those tables of stone which form his records, finds no examples of dynasties once passed away again returning. There has been no repetition of the dynasty of the fish, of the reptile, of the mammal. The dynasty of the future is to have glorified man for its inhabitant; but it is to be the dynasty—'*the kingdom*'—not of glorified man made in the image of God, but of God himself in the form of man.

The Testimony of the Rocks (1869), 142–3.

5 All geologic history is full of the beginning and the ends of species,—of their first and last days; but it exhibits no genealogies of development.

The Testimony of the Rocks (1869), 183, 1st Edition, 1857.

6 The science of the geologist seems destined to exert a marked influence on that of the natural theologian . . . Not only—to borrow from Paley's illustration—does it enable him to argue on the old grounds, from the contrivance exhibited in the *watch* found on the moor, that the watch could not have lain upon the moor for ever; but it establishes further, on different and more direct evidence, that there was a time when absolutely the watch was not there; nay, further, so to speak, that there was a previous time in which no watches existed at all, but only water-clocks; yet further, that there was at time in which there were not even water-clocks, but only sun-dials; and further, an earlier time still in which sun-dials were not, nor any measurers of time of any kind.

The Testimony of the Rocks (1869), 175–6.

7 The existing premises, wholly altered by geologic science, are no longer those of Hume. The foot-print in the sand—to refer to his happy illustration—does not now stand alone. Instead of one, we see many footprints, each in turn in advance of the print behind it, and on a higher level.

The Testimony of the Rocks (1869), 186–7, 1st Edition, 1857.

8 The primary rocks, . . . I regard as the deposits of a period in which the earth's crust had sufficiently cooled down to permit the existence of a sea, with the necessary denuding agencies,—waves and currents,—and, in consequence, of deposition also; but in which the internal heat acted so near the surface, that whatever was deposited came, as a matter of course, to be metamorphosed into semi-plutonic forms, that retained only the stratification. I dare not speak of the scenery of the period. We may imagine, however, a dark atmosphere of steam and vapour, which for age after age conceals the face of the sun, and through which the light of moon or star never penetrates; oceans of thermal water heated in a thousand centres to the boiling point; low, half-molten islands, dim through the fog, and scarce more fixed than the waves themselves, that heave and tremble under the impulsions of the igneous agencies; roaring geysers, that ever and anon throw up their intermittent jets of boiling fluid, vapour, and thick steam, from these tremulous lands; and, in the dim outskirts of the scene, the red gleam of fire, shot forth from yawning cracks and deep chasms, and that bears aloft fragments of molten rock and clouds of ashes. But should we continue to linger amid a scene so *featureless* and wild, or venture adown some yawning opening into the abyss beneath, where all is fiery and yet dark,—a solitary hell, without suffering or sin,—we would do well to commit ourselves to the guidance of a living poet of the true faculty,—Thomas Aird and see with his eyes.

Sketch Book of Popular Geology (1859), 238–9.

Robert Andrews Millikan 1868–1953

American physicist

1 The chemist in America has in general been content with what I have called a loafer electron theory. He has imagined the electrons sitting around on dry goods boxes at every corner [viz. the cubic atom], ready to shake hands with, or hold on to similar loafer electrons in other atoms.

'Atomism in Modern Physics', *Journal of the Chemical Society*, 1924, 1411.

2 The fact that Science walks forward on two feet, namely theory and experiment, is nowhere better illustrated than in the two fields for slight contributions to which you have done me the great honour of awarding me the Nobel Prize in Physics for the year 1923. Sometimes it is one foot that is put forward first, sometimes the other, but continuous progress is only made by the use of both—by theorizing and then testing, or by finding new relations in the process of experimenting and then bringing the theoretical foot up and pushing it on beyond, and so on in unending alterations.

'The Electron and the Light-quant from the Experimental Point of View', Nobel Lecture, 23 May 1924. In *Nobel Lectures: Physics 1922–1941* (1998), 54.

3 From that night on, the electron—up to that time largely the plaything of the scientist—had clearly entered the field as a potent agent in the supplying of man's commercial and industrial needs . . . The electronic amplifier tube now underlies the whole art of communications, and this in turn is at least in part what has made possible its application to a dozen other arts. It was a great day for both science and industry when they became wedded through the development of the electronic amplifier tube.

The Autobiography of Robert A. Millikan (1951), 136.

Charles Wright Mills 1916–62

American sociologist

4 The sociological imagination enables us to grasp history and biography and the relations between the two within society. That is its task and its promise. To recognize this task and this promise is the mark of the classic social analyst.

The Sociological Imagination (1959), 6.

John Milton 1608–74

English poet

5 He scarce had ceased when the superior fiend
Was moving toward the shore; his ponderous shield

Ethereal temper, massy, large and round,
Behind him cast; the broad circumference
Hung on his shoulders like the moon, whose orb
Through optic glass the Tuscan artist views
At evening from the top of Fésolè,
Or in Valdarno, to descry new lands,
Rivers or mountains in her spotty globe.

Paradise Lost, Books I and II (1667), edited by Anna Baldwin (1998), lines 283–91, p. 9.

1 Oh dark, dark, dark, amid the blaze of noon,
Irrecoverably dark, total Eclipse
Without all hope of day!

Samson Agonistes (1671), lines 80–2.

Hermann Minkowski 1864–1909

German mathematician

2 Henceforth space by itself, and time by itself, are doomed to fade away into mere shadows, and only a kind of union of the two will preserve independence.

In Peter Louis Galison, 'Minkowski's Space-Time: From Visual Thinking to the Absolute World', *Historical Studies in the Physical Sciences*, 1979, 10, 97.

3 The rigid electron is in my view a monster in relation to Maxwell's equations, whose innermost harmony is the principle of relativity . . . the rigid electron is no working hypothesis, but a working hindrance. Approaching Maxwell's equations with the concept of the rigid electron seems to me the same thing as going to a concert with your ears stopped up with cotton wool. We must admire the courage and the power of the school of the rigid electron which leaps across the widest mathematical hurdles with fabulous hypotheses, with the hope to land safely over there on experimental-physical ground.

In Arthur I. Miller, *Albert Einstein's Special Theory of Relativity* (1981), 350.

Marvin Lee Minsky 1927– and Seymour Papert 1928–

American computer scientist and South African mathematician

4 But just as astronomy succeeded astrology, following Kepler's discovery of planetary regularities, the discoveries of these many principles in empirical explorations of intellectual processes in machines should lead to a science, eventually.

Artificial Intelligence (1973), 25.

Alfred Ezra Mirsky 1900–74 and Linus Pauling 1901–94

American biochemist and American chemist

5 Our conception of a native protein molecule (showing specific properties) is the following. The molecule consists of one polypeptide chain which continues without interruption throughout the molecule (or, in certain cases, of two or more such chains); this chain is folded into a uniquely defined configuration, in which it is held by hydrogen bonds between the peptide nitrogen and oxygen atoms and also between the free amino and carboxyl groups of the diamino and dicarboxyl amino acid residues.

The characteristic specific properties of native proteins we attribute to their uniquely defined configurations.

The denatured protein molecule we consider to be characterized by the absence of a uniquely defined configuration.

'On the Structure of Native, Denatured, and Coagulated Proteins', *Proceedings of the National Academy of Sciences of the United States of America*, 1936, 22, 442–3.

Peter Dennis Mitchell 1920–92

British biochemist

6 Finally, to the theme of the respiratory chain, it is especially noteworthy that David Keilin's chemically simple view of the respiratory chain appears now to have been right all along—and he deserves great credit for having been so reluctant to become involved when the energy-rich chemical intermediates began to be so fashionable. This reminds me of the aphorism: 'The obscure we see eventually, the completely apparent takes longer'.

'David Keilin's Respiratory Chain Concept and its Chemiosmotic Consequences', Nobel Lecture, 8 December 1978. In *Nobel Lectures: Chemistry 1971–1980* (1993), 325.

Eilhard Mitscherlich 1794–1863

German chemist

1 An equal number of atoms, combined in the same way produce the same crystal forms, and the same crystal form does not depend on the nature of the atoms, but only on their number and mode of combination.

Kungliga Svenska Vetenskaps Akademiens Handlingar (1822). Trans. J. R. Partington, *A History of Chemistry*, Vol. 4 (1972), 210.

Alwin Mittasch 1869–1953

German Chemist

2 Chemistry without catalysis, would be a sword without a handle, a light without brilliance, a bell without sound.

R. E. Oesper, 'Alwin Mittasch', *Journal of Chemical Education*, 1948, **25**, 531–2.

Walter Hamilton Moberley 1881–1974

British philosopher

3 For God's sake, stop researching for a while and begin to think.

The Crisis in the University (1949), 183.

Paul Julius Möbius 1853–1907

German neurologist

4 Everybody is pathological to a certain degree . . . the more so the elevated his standing . . . only myth and cliche have that a person must be either sane or crazy.

Ausgewahlte Werke, Vol. 1 (1909), xi.

Hugo von Mohl 1805–72

German botanist

5 As I have already mentioned, wherever cells are formed, this tough fluid precedes the first solid structures that indicate the presence of future cells. Moreover, we must assume that this substance furnishes the material for the formation of the nucleus and of the primitive sac, not only because these structures are closely apposed to it, but also because they react to iodine in the same way. We must assume also that the organization of this substance is the process that inaugurates the formation of new cells. It therefore seems justifiable for me to propose a name that refers to its physiological function: I propose the word *protoplasma*.

H. Mohl, *Botanisch Zeitung*, 1846, **4**, col.73, trans. Henry Harris, *The Birth of the Cell* (1999), 75.

Abraham de Moivre 1667–1754

French mathematician and statistician

6 Further, the same Arguments which explode the Notion of Luck, may, on the other side, be useful in some Cases to establish a due comparison between Chance and Design: We may imagine Chance and Design to be, as it were, in Competition with each other, for the production of some sorts of Events, and many calculate what Probability there is, that those Events should be rather owing to the one than to the other.

Doctrine of Chances (1718), Preface, v.

Thomas Molyneux 1661–1733

Irish physician and zoologist

7 That no real Species of Living Creatures is so utterly extinct, as to be lost entirely out of the World, since it was first Created, is the Opinion of many Naturalists; and 'tis grounded on so good a Principle of Providence taking Care in general of all its Animal Productions, that it deserves our Assent. However great Vicissitudes may be observed to attend the Works of

Nature, as well as Humane Affairs; so that some entire Species of Animals, which have been formerly Common, nay even numerous in certain Countries; have, in Process of time, been so perfectly soft, as to become there utterly unknown; tho' at the same time it cannot be denyed, but the kind has been carefully preserved in some other part of the World.

'A Discourse concerning the Large Horns frequently found under Ground in Ireland, Concluding from them that the great American Deer, call'd a Moose, was formerly common in that Island: With Remarks on some other things Natural to that Country', *Philosophical Transactions of the Royal Society of London*, 1697, **19**, 489.

Jacques Lucien Monod 1910–76

French biologist

1 Chance and necessity.

Title of *Chance and Necessity: An Essay on the Natural Philosophy of Modern Biology*, trans. Austryn Wainhouse (1972).

2 Biology occupies a position among the sciences both marginal and central. Marginal because, the living world constituting only a tiny and very 'special' part of the universe, it does not seem likely that the study of living beings will ever uncover general laws applicable outside the biosphere. But if the ultimate aim of the whole of science is indeed, as I believe, to clarify man's relationship to the universe, then biology must be accorded a central position, since of all the disciplines it is the one that endeavours to go most directly to the heart of the problems that must be resolved before that of 'human nature' can even be framed in other than metaphysical terms.

Chance and Necessity: An Essay on the Natural Philosophy of Modern Biology, trans. Austryn Wainhouse (1972), 11.

3 A *totally* blind process can by definition lead to anything; it can even lead to vision itself.

Chance and Necessity: An Essay on the Natural Philosophy of Modern Biology, trans. Austryn Wainhouse (1972), 96.

4 There is and will remain a Platonic element in science which could not be taken away without ruining it. Among the infinite diversity of singular phenomena science can only look for invariants.

Chance and Necessity: An Essay on the Natural Philosophy of Modern Biology, trans. Austryn Wainhouse (1972), 100.

5 The fundamental biological variant is DNA. That is why Mendel's definition of the gene as the unvarying bearer of hereditary traits, its chemical identification by Avery (confirmed by Hershey), and the elucidation by Watson and Crick of the structural basis of its replicative invariance, are without any doubt the most important discoveries ever made in biology. To this must be added the theory of natural selection, whose certainty and full significance were established only by those later theories.

Chance and Necessity: An Essay on the Natural Philosophy of Modern Biology, trans. Austryn Wainhouse (1972), 102–3.

6 Even today a good many distinguished minds seem unable to accept or even to understand that from a source of noise natural selection could quite unaided have drawn all the music of the biosphere. Indeed natural selection operates *upon* the products of chance and knows no other nourishment; but it operates in a domain of very demanding conditions, from which chance is banned.

Chance and Necessity: An Essay on the Natural Philosophy of Modern Biology, trans. Austryn Wainhouse (1972), 114.

7 Evolution in the biosphere is therefore a necessarily irreversible process defining *a direction in time*; a direction which is the *same* as that enjoined by the law of increasing entropy, that is to say, the second law of thermodynamics. This is far more than a mere comparison: the second law is founded upon considerations *identical* to those which establish the irreversibility of evolution. Indeed, *it is legitimate to view the irreversibility of evolution as an expression of the second law in the biosphere.*

Chance and Necessity: An Essay on the Natural Philosophy of Modern Biology, trans. Austryn Wainhouse (1972), 118.

1 Every living being is *also* a fossil. Within it, all the way down to the microscopic structure of its proteins, it bears the traces if not the stigmata of its ancestry.

Chance and Necessity: An Essay on the Natural Philosophy of Modern Biology, trans. Austryn Wainhouse (1972), 150.

Lady Mary Wortley Montagu

1689–1762

British writer and traveller

2 A propos of Distempers, I am going to tell you a thing that I am sure will make you wish your selfe here. The Small Pox so fatal and so general amongst us is here entirely harmless by the invention of engrafting (which is the term they give it). There is a set of old Women who make it their business to perform the Operation.

Letter to Sarah Chiswell, 1 April 1717. In Robert Halsband (ed.), *The Complete Letters of the Lady Mary Wortley Montagu* (1965), Vol. 1, 338.

3 I am patriot enough to take pains to bring this usefull invention [smallpox inoculation] into fashion in England, and I should not fail to write to some of our Doctors very particularly about it, if I knew any one of 'em that I thought had Virtue enough to destroy such a considerable branch of Revenue for the good of Mankind, but that Distemper is too beneficial to them not to expose to all their Resentment the hardy wight that should undertake to put an end to it.

Letter to Sarah Chiswell, 1 April 1717. In Robert Halsband (ed.), *The Complete Letters of the Lady Mary Wortley Montagu* (1965), Vol. 1, 339.

Michel Eyquem de Montaigne

1533–92

French essayist and man of letters

4 Nature clasps all her creatures in a universal embrace; there is not one of them which she has not plainly furnished with all means necessary to the conservation of its being.

The Essays of Michel de Montaigne, Book 2, Chapter 12, 'Apology for Raymond Sebond', trans. M. A. Screech (1991), 509.

5 If atoms do, by chance, happen to combine themselves into so many shapes, why have they never combined together to form a house or a slipper? By the same token, why do we not believe that if innumerable letters of the Greek alphabet were poured all over the market-place they would eventually happen to form the text of the Iliad?

The Essays of Michel de Montaigne, Book 2, Chapter 12, 'Apology for Raymond Sebond', trans. M. A. Screech (1991), 612.

Charles Louis de Secondat, Baron de Montesquieu

1689–1755

French philosopher

6 Laws, in their most general signification, are the necessary relations arising from the nature of things. In this sense all beings have their laws: the Deity His laws, the material world its laws, the intelligences superior to man their laws, the beasts their laws, man his laws.

The Spirit of the Laws (1748), Vol. 1, book 1, 1.

Eliakim Hastings Moore

1862–1932

American mathematician

7 We lay down a fundamental principle of generalization by abstraction: *The existence of analogies between central features of various theories implies the existence of a general theory which underlies the particular theories and* *unifies them with respect to those central features.*

Introduction to a Form of General Analysis (1910), Preface, 1.

Conwy Lloyd Morgan 1852–1936

British psychologist

1 In no case may we interpret an action [of an animal] as the outcome of the exercise of a higher psychical faculty, if it can be interpreted as the outcome of the exercise of one which stands lower in the psychological scale.

Known as Morgan's canon, the principle of parsimony in animal research.

An Introduction to Comparative Psychology (1894), 53.

2 The primary aim, object, and purpose of consciousness is control. Consciousness in a mere automaton is a useless and unnecessary epiphenomenon.

An Introduction to Comparative Psychology (1894), 182.

Thomas Hunt Morgan 1866–1945

American biologist

3 The origin of an adaptive structure and the purposes it comes to fulfil are only chance combinations. Purposefulness is a very human conception for usefulness. It is usefulness looked at backwards. Hard as it is to imagine, inconceivably hard it may appear to many, that there is no direct relation between the origin of useful variations and the ends they come to serve, yet the modern zoologist takes his stand as a man of science on this ground. He may admit in secret to his father confessor, the metaphysician, that his poor intellect staggers under such a supposition, but he bravely carries forward his work of investigation along the only lines that he has found fruitful.

'For Darwin', *The Popular Science Monthly*, 1909, 74, 380.

4 In the modern interpretation of Mendelism, facts are being transformed into factors at a rapid rate. If one factor will not explain the facts, then two are involved; if two prove insufficient, three will sometimes work out. The superior jugglery sometimes necessary to account for the results may blind us, if taken too naively, to the common-place that the results are often so excellently 'explained' because the explanation was invented to explain them. We work backwards from the facts to the factors, and then, presto! explain the facts by the very factors that we invented to account for them. I am not unappreciative of the distinct advantages that this method has in handling the facts. I realize how valuable it has been to us to be able to marshal our results under a few simple assumptions, yet I cannot but fear that we are rapidly developing a sort of Mendelian ritual by which to explain the extraordinary facts of alternative inheritance. So long as we do not lose sight of the purely arbitrary and formal nature of our formulae, little harm will be done; and it is only fair to state that those who are doing the actual work of progress along Mendelian lines are aware of the hypothetical nature of the factor-assumption.

'What are 'Factors' in Mendelian Explanations?', *American Breeders Association*, 1909, 5, 365.

5 That the fundamental aspects of heredity should have turned out to be so extraordinarily simple supports us in the hope that nature may, after all, be entirely approachable. Her much-advertised inscrutability has once more been found to be an illusion due to our ignorance. This is encouraging, for, if the world in which we live were as complicated as some of our friends would have us believe we might well despair that biology could ever become an exact science.

The Physical Basis of Heredity (1919), 15.

6 Except for the rare cases of plastid inheritance, the *inheritance* of all known characters can be sufficiently accounted for by the presence of genes in the chromosomes. In a word the cytoplasm may be ignored genetically.

'Genetics and the Physiology of Development', *The American Naturalist*, 1926, 60, 491.

1 Realizing how often ingenious speculation in the complex biological world has led nowhere and how often the real advances in biology as well as in chemistry, physics and astronomy have kept within the bounds of mechanistic interpretation, we geneticists should rejoice, even with our noses on the grindstone (which means both eyes on the objectives), that we have at command an additional means of testing whatever original ideas pop into our heads.

'The Rise of Genetics', *Science*, 1932, 1969, 264.

2 Now that we locate them [genes] in the chromosomes are we justified in regarding them as material units; as chemical bodies of a higher order than molecules? Frankly, these are questions with which the working geneticist has not much concern himself, except now and then to speculate as to the nature of the postulated elements. There is no consensus of opinion amongst geneticists as to what the genes are—whether they are real or purely fictitious—because at the level at which the genetic experiments lie, it does not make the slightest difference whether the gene is a hypothetical unit, or whether the gene is a material particle. In either case the unit is associated with a specific chromosome, and can be localized there by purely genetic analysis. Hence, if the gene is a material unit, it is a piece of chromosome; if it is a fictitious unit, it must be referred to a definite location in a chromosome—the same place as on the other hypothesis. Therefore, it makes no difference in the actual work in genetics which point of view is taken. Between the characters that are used by the geneticist and the genes that his theory postulates lies the whole field of embryonic development.

'The Relation of Genetics to Physiology and Medicine', Nobel Lecture, 4 June 1934. In *Nobel Lectures, Physiology or Medicine 1922–1941* (1965), 315.

3 Certain students of genetics inferred that the Mendelian units responsible for the selected character were genes producing only a single effect. This was careless logic. It took a good deal of hammering to get rid of this erroneous idea. As facts accumulated it became evident that each gene produces not a single effect, but in some cases a multitude of effects on the characters of the individual. It is true that in most genetic work only one of these character-effects is selected for study—the one that is most sharply defined and separable from its contrasted character—but in most cases minor differences also are recognizable that are just as much the product of the same gene as is the major effect.

'The Relation of Genetics to Physiology and Medicine', Nobel Lecture, 4 June 1934. In *Nobel Lectures, Physiology or Medicine 1922–1941* (1965), 317.

John Morley (Lord Morley of Blackburn) 1838–1923

British man of letters

4 The next great task of science is to create a religion for humanity.

Quoted in Julian Huxley, *Essays of a Biologist* (1928), 235.

5 The Historic Method may be described as the comparison of the forms of an idea, or a usage, or a belief, at any given time, with the earlier forms from which they were evolved, or the later forms into which they were developed and the establishment from such a comparison, of an ascending and descending order among the facts. It consists in the explanation of existing parts in the frame of society by connecting them with corresponding parts in some earlier frame; in the identification of present forms in the past, and past forms in the present. Its main process is the detection of corresponding customs, opinions, laws, beliefs, among different communities, and a grouping of them into general classes with reference to some one common feature. It is a certain way of seeking answers to various questions of origin, resting on the same general doctrine of evolution, applied to moral and social forms, as that which is being applied with so much ingenuity to the series of organic matter.

On Compromise (1874), 22–3.

Anton Lazzaro Moro 1687–1764

Italian geologist

1 I want to note that, because there is the aforementioned difference between mountain and mountain, it will be appropriate, to avoid confusion, to distinguish one [type] from another by different terms; so I shall call the first *Primary* and the second *Secondary*.

De' Crostacei e degli altri Marini Corpi che si truovana su' monti (1740), 263, trans. Ezio Vaccari.

Bede Nairn Morris 1927–88

Australian immunologist and pathologist

2 We must avoid the impression that the qualification for election is that the prostate is larger than the brain.

Speaking on the age of election of fellows at the Australian Academy of Science, 1980.

Abraham Cressy Morrison 1884–1951

American chemist

3 The glimpses of chemical industry's services to man afforded by this book could be presented only by utilizing innumerable chemical products. The first outline of its plan began to take shape on chemically produced notepaper with the aid of a chemically-treated graphite held in a synthetic resin pencil. Early corrections were made with erasers of chemically compounded rubber. In its ultimate haven on the shelves of your bookcase, it will rest on a coating of chemical varnish behind a pane of chemically produced glass. Nowhere has it been separated from that industry's products.

Man in a Chemical World (1937), L'Envoi, 284.

Samuel Finley Breese Morse 1791–1872

American inventor and artist

See **Bible, The** 65:10

4 What hath God wrought.

First message sent by him over the electric telegraph, from Washington to Baltimore, 24 May 1844.

Henry Gwyn Jeffreys Moseley 1887–1915

English physicist

5 We have now got what seems to be definite proof that an X ray which spreads out in a spherical form from a source as a wave through the aether can when it meets an atom collect up all its energy from all round and concentrate it on the atom. It is as if when a circular wave on water met an obstacle, the wave were all suddenly to travel round the circle and disappear all round and concentrate its energy on attacking the obstacle. Mechanically of course this is absurd, but mechanics have in this direction been for some time a broken reed.

Letter to Margery Moseley, 2 February 1913. In J. L. Heilbron (ed.), *H. G. J. Moseley: The Life and Letters of an English Physicist 1887–1915* (1974), 201.

6 The whole subject of the X rays is opening out wonderfully, Bragg has of course got in ahead of us, and so the credit all belongs to him, but that does not make it less interesting. We find that an X ray bulb with a platinum target gives out a sharp line spectrum of five wavelengths which the crystal separates out as if it were a diffraction grating. In this way one can get pure monochromatic X rays. Tomorrow we search for the spectra of other elements. There is here a whole new branch of spectroscopy, which is sure to tell one much about the nature of an atom.

Letter to his mother, 18 May 1913. In J. L. Heilbron (ed.), *H. G. J. Moseley: The Life and Letters of an English Physicist 1887–1915* (1974), 205.

1 In the last four days I have got the spectrum given by Tantalum. Chromium. Manganese. Iron. Nickel. Cobalt. and Copper and part of the Silver spectrum. The chief result is that all the elements give the same kind of spectrum, the result for any metal being quite easy to guess from the results for the others. This shews that the insides of all the atoms are very much alike, and from these results it will be possible to find out something of what the insides are made up of.

Letter to his mother, 2 November 1913. In J. L. Heilbron (ed.), *H. G. J. Moseley: The Life and Letters of an English Physicist 1887–1915* (1974), 209.

Frederick Mosteller 1916–

American statistician

2 Although we often hear that data speak for themselves, their voices can be soft and sly.

Beginning Statistics with Data Analysis (1983), 234.

Frederick Mosteller 1916– and Robert R. Bush 1920–71

American statisticians

3 The main purpose of a significance test is to inhibit the natural enthusiasm of the investigator.

Selected Quantitative Techniques (1954), 331–2.

Hermann Joseph Muller

1890–1967

American geneticist

4 The central problem of biological evolution is the nature of *mutation*, but hitherto the occurrence of this has been wholly refractory and impossible to influence by artificial means, although a control of it might obviously place the process of evolution in our hands.

'The Recent Findings in Heredity' (unpublished lecture, 1916, Lilly Library), 3. Quoted in Elof Axel Carlson, *Genes, Radiation, and Society: The Life and Work of H. J. Muller* (1981), 104.

5 If these d'Hérelle bodies were really genes, fundamentally like our chromosome genes, they would give us an utterly new angle from which to attack the gene problem. They are filterable, to some extent isolable, can be handled in test-tubes, and their properties, as shown by their effects on the bacteria, can then be studied after treatment. It would be very rash to call these bodies genes, and yet at present we must confess that there is no distinction known between the genes and them. Hence we can not categorically deny that perhaps we may be able to grind genes in a mortar and cook them in a beaker after all. Must we geneticists become bacteriologists, physiological chemists and physicists, simultaneously with being zoologists and botanists? Let us hope so.

'Variation Due to Change in the Individual Gene', *The American Naturalist*, 1922, 56, 48–9.

6 The gene as the basis of life.

'The Gene as the Basis of Life', *Proceedings of the International Congress of Plant Sciences*, 1929, 1, 897–921.

7 As science is more and more subject to grave misuse as well as to use for human benefit it has also become the scientist's responsibility to become aware of the social relations and applications of his subject, and to exert his influence in such a direction as will result in the best applications of the findings in his own and related fields. Thus he must help in educating the public, in the broad sense, and this means first educating himself, not only in science but in regard to the great issues confronting mankind today.

'Message to University Students Studying Science', *Kagaku Asahi* 11, no. 6 (1951), 28–29. Quoted in Elof Axel Carlson, *Genes, Radiation, and Society: The Life and Work of H. J. Muller* (1981), 371.

1 We do not know of any enzymes or other chemical defined organic substances having specifically acting auto-catalytic properties such as to enable them to construct replicas of themselves. Neither was there a general principle known that would result in pattern-copying; if there were, the basis of life would be easier to come by. Moreover, there was no evidence to show that the enzymes were not products of hereditary determiners or genes, rather than these genes themselves, and they might even be products removed by several or many steps from the genes, just as many other known substances in the cell must be. However, the determiners or genes themselves must conduct, or at least guide, their own replication, so as to lead to the formation of genes just like themselves, in such wise that even their own mutations become incorporated in the replicas. And this would probably take place by some kind of copying of pattern similar to that postulated by Troland for the enzymes, but requiring some distinctive chemical structure to make it possible. By virtue of this ability of theirs to replicate, these genes—or, if you prefer, genetic material—contained in the nuclear chromosomes and in whatever other portion of the cell manifests this property, such as the chloroplastids of plants, must form the basis of all the complexities of living matter that have arisen subsequent to their own appearance on the scene, in the whole course of biological evolution. That is, this genetic material must underlie all evolution based on mutation and selective multiplication.

'Genetic Nucleic Acid', *Perspectives in Biology and Medicine*, 1961, **5**, 6–7.

2 Natural selection based on the differential multiplication of variant types cannot exist before there is material capable of replicating itself and its own variations, that is, before the origination of specifically genetic material or gene-material.

'Genetic Nucleic Acid', *Perspectives in Biology and Medicine*, 1961, **5**, 7.

Johannes Peter Müller 1801–58

German anatomist and physiologist

3 A good physiological experiment like a good physical one requires that it should present anywhere, at any time, under identical conditions, the same certain and unequivocal phenomena that can always be confirmed.

'Bestätigung des Bell'schen Lehrsatzes, dass die doppelten Wurzeln der Rückenmarksnerven verschiedene Functionen haben, durch neue und entscheidende Experimente' (1831). Trans. Edwin Clarke and C. D. O'Malley, *The Human Brain and Spinal Cord* (1968), 304.

4 Observation is simple, indefatigable, industrious, upright, without any preconceived opinion. Experiment is artificial, impatient, busy, digressive, passionate, unreliable. We see every day one experiment after another, the second outweighing the impression gained from the first, both, often enough, carried out by men who are neither much distinguished for their spirit, nor for carrying with them the truth of personality and self denial. Nothing is easier than to make a series of so-called interesting experiments. Nature can only in some way be forced, and in her distress, she will give her suffering answer. Nothing is more difficult than to explain it, nothing is more difficult than a valid physiological experiment. We consider as the first task of current physiology to point at it and comprehend it.

Inaugural lecture as docent of physiology at

the University of Bonn, 19 October 1824. Published in Johannes Müller, *Zur vergleichenden Physiologie des Gesichtssinnes des Menschen und der Thiere* (1826), 20.

Robert Sanderson Mulliken
1896–1986
American chemist and physicist

1 While it is never safe to affirm that the future of Physical Science has no marvels in store even more astonishing than those of the past, it seems probable that most of the grand underlying principles have been firmly established and that further advances are to be sought chiefly in the rigorous application of these principles to all the phenomena which come under our notice.

'Spectroscopy, Molecular Orbitals, and Chemical Bonding', Nobel Lecture, 12 December 1966. In *Nobel Lectures: Chemistry 1963–1970* (1972), 159.

2 I would like to emphasize strongly my belief that the era of computing chemists, when hundreds if not thousands of chemists will go to the computing machine instead of the laboratory for increasingly many facets of chemical information, is already at hand. There is only one obstacle, namely that someone must pay for the computing time.

'Spectroscopy, Molecular Orbitals, and Chemical Bonding', Nobel Lecture, 12 December 1966. In *Nobel Lectures: Chemistry 1963–1970* (1972), 159.

3 I decided that life rationally considered seemed pointless and futile, but it is still interesting in a variety of ways, including the study of science. So why not carry on, following the path of scientific hedonism? Besides, I did not have the courage for the more rational procedure of suicide.

Life of a Scientist (1989), 24.

Kary Banks Mullis 1944–
American biochemist

4 In 1944 Erwin Schroedinger, stimulated intellectually by Max Delbruck, published a little book called *What is life?* It was an inspiration to the first of the molecular biologists, and has been, along with Delbruck himself, credited for directing the research during the next decade that solved the mystery of how 'like begat like.' Max was awarded this Prize in 1969, and rejoicing in it, he also lamented that the work for which he was honored before all the peoples of the world was not something which he felt he could share with more than a handful. Samuel Beckett's contributions to literature, being honored at the same time, seemed to Max somehow universally accessible to anyone. But not his. In his lecture here Max imagined his imprisonment in an ivory tower of science.

'The Polymerase Chain Reaction', Nobel Lecture, 8 December 1993. In *Nobel Lectures: Chemistry 1991–1995* (1997), 103.

Lewis Mumford 1895–1990
American historian of technology

5 Looking back over the last thousand years, one can divide the development of the machine and the machine civilization into three successive but *over-lapping and interpenetrating phases*: eotechnic, paleotechnic, neotechnic . . . Speaking in terms of power and characteristic materials, the eotechnic phase is a water-and-wood complex: the paleotechnic phase is a coal-and-wood complex . . . The dawn-age of our modern technics stretches roughly from the year 1000 to 1750. It did not, of course, come suddenly to an end in the middle of the eighteenth century. A new movement appeared in industrial society which had been gathering headway almost unnoticed from the fifteenth century on: after 1750 industry passed into a new phase, with a different source of power, different materials, different objectives.

Technics and Civilisation (1934), 109.

6 Iron and coal dominated everywhere, from grey to black: the black boots, the black stove-pipe hat, the black coach or carriage, the black iron frame of the hearth, the black cooking pots and pans and stoves. Was it a mourning? Was it protective coloration? Was it mere depression of the senses? No matter what the original color of the paleotechnic milieu might be it was soon reduced by reason of the soot and cinders that accompanied its activities, to its characteristic tones, grey, dirty-brown, black.

Technics and Civilisation (1934), 163.

Roderick Impey Murchison

1792–1871

British geologist

1 Combining in our survey then, the whole range of deposits from the most recent to the most ancient group, how striking a succession do they present:—so various yet so uniform—so vast yet so connected. In thus tracing back to the most remote periods in the physical history of our continents, *one system of operations*, as the means by which many complex formations have been successively produced, the mind becomes impressed with the singleness of nature's laws; and in this respect, at least, geology is hardly inferior in simplicity to astronomy.

The Silurian System (1839), 574.

2 During cycles long anterior to the creation of the human race, and while the surface of the globe was passing from one condition to another, whole races of animals—each group adapted to the physical conditions in which they lived—were successively created and exterminated.

Siluria (1854), 5.

3 The earliest signs of living things, announcing as they do a high complexity of organization, entirely exclude the hypothesis of a *transmutation* from lower to higher grades of being. The first fiat of Creation which went forth, doubtlessly ensured the perfect adaptation of animals to the surrounding media; and thus, whilst the geologist recognizes a beginning, he can see in the innumerable facts of the eye of the earliest crustacean, the same evidences of Omniscience as in the completion of the vertebrate form.

Siluria (1854), 469.

4 We should greatly err, if we endeavoured to force all ancient nature into a close comparison with existing operations.

Siluria (1854), 481.

5 The order of . . . successive generations is indeed much more clearly proved than many a legend which has assumed the character of history in the hands of man; for the geological record is the work of God.

Siluria (1872), 476.

6 Placed as the fossils are in their several tiers of burial-places the one over the other; we have in them true witnesses of successive existences, whilst the historian of man is constantly at fault as to dates and even the sequence of events, to say nothing of the contradicting statements which he is forced to reconcile.

Siluria (1872), 476.

7 Physical geography and geology are inseparable scientific twins.

Address at the Anniversary Meeting of the Royal Geographical Society (1857), 55.

Gardner Murphy 1895–1979

American psychologist

8 It has become accepted doctrine that we must attempt to study the whole man. Actually we cannot study even a whole tree or a whole guinea pig. But it is a whole tree and a whole guinea pig that have survived and evolved, and we must make the attempt.

Personality (1947), 5.

Ernest Nagel 1901–85

Bohemian-born American philosopher of science

9 It is the desire for explanations that are at once systematic and controllable by factual evidence that generates science; and it is the organization and classification of knowledge on the basis of explanatory principles that is the distinctive goal of the sciences.

The Structure of Science (1961), 4.

Thomas Nagel 1937–

American philosopher

1 Without consciousness the mind-body problem would be much less interesting. With consciousness it seems hopeless.

Moral Questions (1979), 166.

David Ledbetter Nanney 1925–

American developmental geneticist

2 It appears unlikely that the role of the genes in development is to be understood so long as the genes are considered as dictatorial elements in the cellular economy. It is not enough to know what a gene does *when* it manifests itself. One must also know the mechanisms determining which of the many gene-controlled potentialities will be realized.

'The Role of the Cytoplasm in Heredity', in William D. McElroy and Bentley Glass (eds.), *A Symposium on the Chemical Basis of Heredity* (1957), 162.

Frederic Ogden Nash 1902–71

American poet

3 God in His wisdom made the fly
And then forgot to tell us why.

'The Fly' (1942), *Good Intentions* (1943), 220.

Joseph Needham 1900–95

British biochemist and historian of science

4 But Chinese civilization has the overpowering beauty of the wholly other, and only the wholly other can inspire the deepest love and the profoundest desire to learn.

The Grand Titration (1969), 176.

Jawaharlal Nehru (Pandit Nehru) 1889–1964

Indian prime minister

5 It is science alone that can solve the problems of hunger and poverty, of insanitation and illiteracy, of superstition and deadening custom and tradition, of vast resources running to waste, of a rich country inhabited by starving people . . . Who indeed could afford to ignore science today? At every turn we have to seek its aid . . . the future belongs to science and those who make friends with science.

Quoted in Atma Ram, 'The Making of Optical Glass in India: Its Lessons for Industrial Development', *Proceedings of the National Institute of Sciences of India*, 1961, 27, 564–5.

Hermann Walther Nernst 1864–1941

German chemist

6 One should avoid carrying out an experiment requiring more than 10 per cent accuracy.

Quoted in W. Jost, '45 Years of Physical Chemistry in Germany', *Annual Review of Physical Chemistry*, 1966, 17, 9.

7 *On examinations:Das Wissen ist der Tod der Forschung.*

Knowledge is the death of research.

Nernst's motto.

Erwin N. Hiebert, 'Hermann Walther Nernst', in C. C. Gillispie (ed.), *The Dictionary of Scientific Biography* (1981), Supplement, Vol. 15, 450.

8 ON NERNST Nernst was a great admirer of Shakespeare, and it is said that in a conference concerned with naming units after appropriate persons, he proposed that the unit of rate of liquid flow should be called the falstaff.

J. R. Partington, 'The Nernst Memorial Lecture', *Journal of the Chemical Society*, 1953 (Part 3), 2855.

John Alexander Reina Newlands 1837–98

British chemist

9 The eighth element, starting from a given one, is a kind of repetition of the first, like the eighth note of an octave in music.

'Letter to the Editor', *Chemical News*, 1864, 10, 94.

James Roy Newman 1907–66 and Byron S. Miller 1915–2003

American popular science writer and American lawyer

1 The self-regulating mechanism of the market place cannot always be depended upon to produce adequate results in scientific research.

The Control of Atomic Energy (1948), 17.

John Henry Newman 1801–90

British cardinal and theologian

2 To discover and to teach are distinct functions; they are also distinct gifts, and are not commonly found united in the same person.

Discourses on the Scope and Nature of University Education. Addressed to the Catholics of Dublin (1852), Preface, xii.

3 A science is not mere knowledge, it is knowledge which has undergone a process of intellectual digestion. It is the grasp of many things brought together in one, and hence is its power; for, properly speaking, it is Science that is power, not Knowledge.

Discourses on the Scope and Nature of University Education. Addressed to the Catholics of Dublin (1852), Discourse 5, 144.

4 Literature stands related to Man as Science stands to Nature; it is his history.

Discourses on the Scope and Nature of University Education. Addressed to the Catholics of Dublin (1852), Discourse 10, 353.

Isaac Newton 1642–1727

English natural philosopher, astronomer and mathematician

See ***Auden*** *27:5;* ***Hume, David*** *307:6;* ***Keynes, John Maynard*** *342:5;* ***Locke, John*** *392:2;* ***Smart, Christopher*** *553:8*

5 Ax:100 Every thing doth naturally persevere in yt state in wch it is unlesse it bee interrupted by some externall cause, hence . . . [a] body once moved will always keepe ye same celerity, quantity & determination of its motion.

Newton's 'Waste Book' (1665). Quoted in Richard Westfall, *Never at Rest: A Biography of Isaac Newton* (1980), 145.

6 This Excellent Mathematician having given us, in the Transactions of February last, an account of the cause, which induced him to think upon Reflecting Telescopes, instead of Refracting ones, hath thereupon presented the curious world with an Essay of what may be performed by such Telescopes; by which it is found, that Telescopical Tubes may be considerably shortened without prejudice to their magnifiying effect.

On his invention of the catadioptrical telescope, in a communication to the Royal Society.

'An Account of a New Catadioptrical Telescope Invented by Mr Newton', *Philosophical Transactions*, 1672, 7, 4004.

7 Our present work sets forth mathematical principles of philosophy. For the basic problem of philosophy seems to be to discover the forces of nature from the phenomena of motions and then to demonstrate the other phenomena from these forces. It is to these ends that the general propositions in books 1 and 2 are directed, while in book 3 our explanation of the system of the world illustrates these propositions.

The Principia: Mathematical Principles of Natural Philosophy (1687), 3rd edition (1726), trans. I. B. Cohen and Anne Whitman (1999), Preface to the first edition, 382.

8 Inherent force of matter is the power of resisting by which every body, so far as it is able, perseveres in its state either of resting or of moving uniformly straight forward.

The Principia: Mathematical Principles of Natural Philosophy (1687), 3rd edition (1726), trans. I. B. Cohen and Anne Whitman (1999), Definition 3, 404.

9 Impressed force is the action exerted on a body to change its state either of resting or of moving uniformly straight forward.

The Principia: Mathematical Principles of Natural Philosophy (1687), 3rd edition (1726), trans. I. B. Cohen and Anne Whitman (1999), Definition 4, 405.

1 Centripetal force is the force by which bodies are drawn from all sides, are impelled, or in any way tend, toward some point as to a center.

The Principia: Mathematical Principles of Natural Philosophy (1687), 3rd edition (1726), trans. I. B. Cohen and Anne Whitman (1999), Definition 5, 405.

2 Absolute, true, and mathematical time, in and of itself and of its own nature, without reference to anything external, flows uniformly and by another name is called duration. Relative, apparent, and common time is any sensible and external measure (precise or imprecise) of duration by means of motion; such as a measure—for example, an hour, a day, a month, a year—is commonly used instead of true time.

The Principia: Mathematical Principles of Natural Philosophy (1687), 3rd edition (1726), trans. I. B. Cohen and Anne Whitman (1999), Definitions, Scholium, 408.

3 Although time, space, place, and motion are very familiar to everyone, it must be noted that these quantities are popularly conceived solely with reference to the objects of sense perception.

The Principia: Mathematical Principles of Natural Philosophy (1687), 3rd edition (1726), trans. I. B. Cohen and Anne Whitman (1999), Definitions, Scholium, 408.

4 Absolute space, of its own nature without reference to anything external, always remains homogenous and immovable. Relative space is any movable measure or dimension of this absolute space; such a measure or dimension is determined by our senses from the situation of the space with respect to bodies and is popularly used for immovable space, as in the case of space under the earth or in the air or in the heavens, where the dimension is determined from the situation of the space with respect to the earth. Absolute and relative space are the same in species and in magnitude, but they do not always remain the same numerically. For example, if the earth moves, the space of our air, which in a relative sense and with respect to the earth always remains the same, will now be one part of the absolute space into which the air passes, now another part of it, and thus will be changing continually in an absolute sense.

The Principia: Mathematical Principles of Natural Philosophy (1687), 3rd edition (1726), trans. I. B. Cohen and Anne Whitman (1999), Definitions, Scholium, 408–9.

5 Every body perseveres in its state of being at rest or of moving uniformly straight forward, except insofar as it is compelled to change its state by forces impressed.

The Principia: Mathematical Principles of Natural Philosophy (1687), 3rd edition (1726), trans. I. B. Cohen and Anne Whitman (1999), Axioms, or Laws of Motion, Law 1, 416.

6 A change in motion is proportional to the motive force impressed and takes place along the straight line in which that force is impressed.

The Principia: Mathematical Principles of Natural Philosophy (1687), 3rd edition (1726), trans. I. B. Cohen and Anne Whitman (1999), Axioms, or Laws of Motion, Law 2, 416.

7 To any action there is always an opposite and equal reaction; in other words, the actions of two bodies upon each other are always equal and always opposite in direction.

The Principia: Mathematical Principles of Natural Philosophy (1687), 3rd edition (1726), trans. I. B. Cohen and Anne Whitman (1999), Axioms, or Laws of Motion, Law 3, 417.

8 I use the word ‘attraction’ here in a general sense for any endeavor whatever of bodies to approach one another, whether that endeavor occurs as a result of the action of the bodies either drawn toward one other or acting on one another by means of spirits emitted or whether it arises from the action of aether or of air or of any medium whatsoever—whether corporeal or incorporeal—in any way impelling toward one another the bodies floating therein. I use the word 'impulse' in the same general sense, considering in this treatise not the species of forces and their physical qualities but their quantities and mathematical proportions, as I have explained in the definitions.

The Principia: Mathematical Principles of Natural Philosophy (1687), 3rd edition (1726), trans. I. B. Cohen and Anne Whitman (1999), Book 1, Section 11, Scholium, 588.

1 I have presented principles of philosophy that are not, however, philosophical but strictly mathematical—that is, those on which the study of philosophy can be based. These principles are the laws and conditions of motions and of forces, which especially relate to philosophy . . . It still remains for us to exhibit system of the world from these same principles.

The Principia: Mathematical Principles of Natural Philosophy (1687), 3rd edition (1726), trans. I. B. Cohen and Anne Whitman (1999), Introduction to Book 3, 793.

2 No more causes of natural things should be admitted than are both true and sufficient to explain their phenomena.

The Principia: Mathematical Principles of Natural Philosophy (1687), 3rd edition (1726), trans. I. B. Cohen and Anne Whitman (1999), Book 3, Rules of Reasoning in Philosophy, Rule 1, 794.

3 Therefore, the causes assigned to natural effects of the same kind must be, so far as possible, the same.

The Principia: Mathematical Principles of Natural Philosophy (1687), 3rd edition (1726), trans. I. B. Cohen and Anne Whitman (1999), Book 3, Rules of Reasoning in Philosophy, Rule 2, 795.

4 Those qualities of bodies that cannot be intended and remitted [i.e., qualities that cannot be increased and diminished] and that belong to all bodies on which experiments can be made should be taken as qualities of all bodies universally.

The Principia: Mathematical Principles of Natural Philosophy (1687), 3rd edition (1726), trans. I. B. Cohen and Anne Whitman (1999), Book 3, Rules of Reasoning in Philosophy, Rule 3, 795.

5 In experimental philosophy, propositions gathered from phenomena by induction should be considered either exactly or very nearly true notwithstanding any contrary hypotheses, until yet other phenomena make such propositions either more exact or liable to exceptions.

The Principia: Mathematical Principles of Natural Philosophy (1687), 3rd edition (1726), trans. I. B. Cohen and Anne Whitman (1999), Book 3, Rules of Reasoning in Philosophy, Rule 4, 796.

6 He rules all things, not as the world soul but as the lord of all. And because of his dominion he is called Lord God *Pantokrator*. For '*god*' is a relative word and has reference to servants, and godhood is the lordship of God, not over his own body as is supposed by those for whom God is the world soul, but over servants. The supreme God is an eternal, infinite, and absolutely perfect being; but a being, however perfect, without dominion is not the Lord God.

The Principia: Mathematical Principles of Natural Philosophy (1687), 3rd edition (1726), trans. I. B. Cohen and Anne Whitman (1999), General Scholium, 940–1.

7 And from true lordship it follows that the true God is living, intelligent, and powerful; from the other perfections, that he is supreme, or supremely perfect. He is eternal and infinite, omnipotent and omniscient; that is, he endures from eternity to eternity; and he is present from infinity to infinity; he rules all things, and he knows all things that happen or can happen.

The Principia: Mathematical Principles of Natural Philosophy (1687), 3rd edition (1726), trans. I. B. Cohen and Anne Whitman (1999), General Scholium, 941.

1 Thus far I have explained the phenomena of the heavens and of our sea by the force of gravity, but I have not yet assigned a cause to gravity. Indeed, this force arises from some cause that penetrates as far as the centers of the sun and planets without any diminution of its power to act, and that acts not in proportion to the quantity of the *surfaces* of the particles on which it acts (as mechanical causes are wont to do) but in proportion to the quantity of *solid* matter, and whose action is extended everywhere to immense distances, always decreasing as the squares of the distances.

The Principia: Mathematical Principles of Natural Philosophy (1687), 3rd edition (1726), trans. I. B. Cohen and Anne Whitman (1999), General Scholium, 943.

2 I have not as yet been able to deduce from phenomena the reason for these properties of gravity, and I do not feign hypotheses [*Hypotheses non fingo*]. For whatever is not deduced from the phenomena must be called a hypothesis; and hypotheses, whether metaphysical or physical, or based on occult qualities, or mechanical, have no place in experimental philosophy. In this experimental philosophy, propositions are deduced from the phenomena, and are made general by induction. The impenetrability, mobility, and the impetus of bodies, and the laws of motion and the law of gravity have been found by this method. And it is enough that gravity really exists and acts according to the laws that we have set forth and is sufficient to explain all the motions of the heavenly bodies and of our sea.

The Principia: Mathematical Principles of Natural Philosophy (1687), 3rd edition (1726), trans. I. B. Cohen and Anne Whitman (1999), General Scholium, 943.

3 Thus you may multiply each stone 4 times & no more for they will then become oyles shining in ye dark and fit for magicall uses. You may ferment them with ☉ [gold] and ☽ [silver], by keeping the stone and metal in fusion together for a day, & then project upon metalls. This is the multiplication of ye stone in vertue. To multiply it in weight ad to it of ye first Gold whether philosophic or vulgar.

Praxis (c.1693), quoted in Betty Jo Teeter Dobbs, *The Janus Faces of Genius: The Role of Alchemy in Newton's Thought* (1991), 304.

4 My Design in this Book is not to explain the Properties of Light by Hypotheses, but to propose and prove them by Reason and Experiments: In order to which, I shall premise the following Definitions and Axioms.

Opticks (1704), Book 1, Part 1, Introduction, 1.

5 If the Humours of the Eye by old Age decay, so as by shrinking to make the *Cornea* and Coat of the *Crystalline Humour* grow flatter than before, the Light will not be refracted enough, and for want of a sufficient Refraction will not converge to the bottom of the Eye but to some place beyond it, and by consequence paint in the bottom of the Eye a confused Picture, and according to the Indistinctness of this Picture the Object will appear confused. This is the reason of the decay of sight in old Men, and shews why their Sight is mended by Spectacles. For those Convex glasses supply the defect of plumpness in the Eye, and by increasing the Refraction make the rays converge sooner, so as to convene distinctly at the bottom of the Eye if the Glass have a due degree of convexity. And the contrary happens in short-sighted Men whose Eyes are too plump. For the Refraction being now too great, the Rays converge and convene in the Eyes before they come at the bottom; and therefore the Picture made in the bottom and the Vision caused thereby will not be distinct, unless the Object be brought so near the Eye as that the place where the converging Rays convene may be removed to the bottom, or that the plumpness of the Eye be taken off and the Refractions diminished by a Concave-glass of a due degree of Concavity, or lastly that by Age the Eye grow flatter till it come to a due Figure:

For short-sighted Men see remote Objects best in Old Age, and therefore they are accounted to have the most lasting Eyes.

Opticks (1704), Book 1, Part 1, Axiom VII, 10–11.

1 From what has been said it is also evident, that the Whiteness of the Sun's Light is compounded all the Colours wherewith the several sorts of Rays whereof that Light consists, when by their several Refrangibilities they are separated from one another, do tinge Paper or any other white Body whereon they fall. For those Colours . . . are unchangeable, and whenever all those Rays with those their Colours are mix'd again, they reproduce the same white Light as before.

Opticks (1704), Book 1, Part 2, Exper. XV, 114.

2 And if one look through a Prism upon a white Object encompassed with blackness or darkness, the reason of the Colours arising on the edges is much the same, as will appear to one that shall a little consider it. If a black Object be encompassed with a white one, the Colours which appear through the Prism are to be derived from the Light of the white one, spreading into the Regions of the black, and therefore they appear in a contrary order to that, when a white Object is surrounded with black. And the same is to be understood when an Object is viewed, whose parts are some of them less luminous than others. For in the borders of the more and less luminous Parts, Colours ought always by the same Principles to arise from the Excess of the Light of the more luminous, and to be of the same kind as if the darker parts were black, but yet to be more faint and dilute.

Opticks (1704), Book 1, Part 2, Prop. VIII, Prob. III, 123.

3 Now if Light be reflected, not by impinging on the solid parts of Bodies, but by some other principle; it's probable that as many of its Rays as impinge on the solid parts of Bodies are not reflected but stifled and lost in the Bodies. For otherwise we must allow two sorts of Reflexions. Should all the Rays be reflected which impinge on the internal parts of clear Water or Crystal, those Substances would rather have a cloudy Colour than a clear Transparency. To make Bodies look black, it's necessary that many Rays be stopp'd, retained, and lost in them; and it seems not probable that any Rays can be stopp'd and stifled in them which do not impinge on their parts.

Opticks (1704), Book 2, Part 3, Prop. VIII, 69.

4 Do not the Rays which differ in Refrangibility differ also in Flexibity; and are they not by their different Inflexions separated from one another, so as after separation to make the Colours in the three Fringes above described? And after what manner are they inflected to make those Fringes?

Opticks (1704), Book 3, Query 2, 132–3.

5 Do not Bodies and Light act mutually upon one another; that is to say, Bodies upon Light in emitting, reflecting, refracting and inflecting it, and Light upon Bodies for heating them, and putting their parts into a vibrating motion wherein heat consists?

Opticks (1704), Book 3, Query 5, 133.

6 Is not Fire a Body heated so hot as to emit Light copiously? For what else is a red hot Iron than Fire? And what else is a burning Coal than red hot Wood?

Opticks (1704), Book 3, Query 9, 134.

7 Do not great Bodies conserve their heat the longest, their parts heating one another, and may not great dense and fix'd Bodies, when heated beyond a certain degree, emit Light so copiously, as by the Emission and Re-action of its Light, and the Reflexions and Refractions of its Rays within its Pores to grow still hotter, till it comes to a certain period of heat, such as is that of the Sun?

Opticks (1704), Book 3, Query 11, 135.

8 Are not all Hypotheses erroneous, in which Light is supposed to consist in

Pression or Motion, propagated through a fluid Medium? For in all these Hypotheses the Phaenomena of Light have been hitherto explain'd by supposing that they arise from new Modifications of the Rays; which is an erroneous Supposition.

Opticks, 2nd edition (1718), Book 3, Query 28, 337.

1 And for rejecting such a Medium, we have the Authority of those the oldest and most celebrated Philosophers of *Greece* and *Phoenicia*, who made a *Vacuum*, and Atoms, and the Gravity of Atoms, the first Principles of their Philosophy; tacitly attributing Gravity to some other Cause than dense Matter. Later Philosophers banish the Consideration of such a Cause out of natural Philosophy, feigning Hypotheses for explaining all things mechanically, and referring other Causes to Metaphysicks: Whereas the main Business of natural Philosophy is to argue from Phaenomena without feigning Hypotheses, and to deduce Causes from Effects, till we come to the very first Cause, which certainly is not mechanical; and not only to unfold the Mechanism of the World, but chiefly to resolve these and such like Questions. What is there in places almost empty of Matter, and whence is it that the Sun and Planets gravitate towards one another, without dense Matter between them? Whence is it that Nature doth nothing in vain; and whence arises all that Order and Beauty which we see in the World? . . . does it not appear from phaenomena that there is a Being incorporeal, living, intelligent, omnipresent, who in infinite space, as it were in his Sensory, sees the things themselves intimately, and thoroughly perceives them, and comprehends them wholly by their immediate presence to himself.

Opticks, 2nd edition (1718), Book 3, Query 28, 343–5.

2 The changing of Bodies into Light, and Light into Bodies, is very conformable to the Course of Nature, which seems delighted with Transmutations.

Opticks, 2nd edition (1718), Book 3, Query 30, 349.

3 And thus Nature will be very conformable to her self and very simple, performing all the great Motions of the heavenly Bodies by the Attraction of Gravity which intercedes those Bodies, and almost all the small ones of their Particles by some other attractive and repelling Powers which intercede the Particles. The *Vis inertiae* is a passive Principle by which Bodies persist in their Motion or Rest, receive Motion in proportion to the Force impressing it, and resist as much as they are resisted. By this Principle alone there never could have been any Motion in the World. Some other Principle was necessary for putting Bodies into Motion; and now they are in Motion, some other Principle is necessary for conserving the Motion.

Opticks, 2nd edition (1718), Book 3, Query 31, 372–3.

4 Seeing therefore the variety of Motion which we find in the World is always decreasing, there is a necessity of conserving and recruiting it by active Principles, such as are the cause of Gravity, by which Planets and Comets keep their Motions in their Orbs, and Bodies acquire great Motion in falling; and the cause of Fermentation, by which the Heart and Blood of Animals are kept in perpetual Motion and Heat; the inward Parts of the Earth are constantly warm'd, and in some places grow very hot; Bodies burn and shine, Mountains take fire, the Caverns of the Earth are blown up, and the Sun continues violently hot and lucid, and warms all things by his Light. For we meet with very little Motion in the World, besides what is owing to these active Principles.

Opticks, 2nd edition (1718), Book 3, Query 31, 375.

5 All these things being consider'd, it seems probable to me, that God in the Beginning form'd Matter in solid, massy, hard, impenetrable, moveable Particles, of such Sizes and Figures, and with such other Properties, and in such Proportion to Space, as most conduced to the End for which he form'd them; and that these primitive Particles being Solids, are incomparably harder than any porous Bodies compounded of them; even so very hard, as never to wear or break in pieces; no ordinary Power being able to divide what God himself made one in the first Creation. While the Particles continue entire, they may compose Bodies of one and the same Nature and Texture in all Ages: But should they wear away, or break in pieces, the Nature of Things depending on them, would be changed. Water and Earth, composed of old worn Particles and Fragments of Particles, would not be of the same Nature and Texture now, with Water and Earth composed of entire Particles in the Beginning.

Opticks, 2nd edition (1718), Book 3, Query 31, 375–6.

1 It seems to me farther, that these Particles have not only a *Vis inertiae*, accompanied with such passive Laws of Motion as naturally result from that Force, but also that they are moved by certain active Principles, such as that of Gravity, and that which causes Fermentation, and the Cohesion of Bodies. These Principles I consider, not as occult Qualities, supposed to result from the specifick Forms of Things, but as general Laws of Nature, by which the Things themselves are form'd; their Truth appearing to us by Phaenomena, though their Causes be not yet discover'd. For these are manifest Qualities, and their Causes only are occult.

Opticks, 2nd edition (1718), Book 3, Query 31, 376–7.

2 As in Mathematicks, so in Natural Philosophy, the Investigation of difficult Things by the Method of Analysis, ought ever to precede the Method of Composition. This Analysis consists in making Experiments and Observations, and in drawing general Conclusions from them by Induction, and admitting of no Objections against the Conclusions, but such as are taken from Experiments, or other certain Truths. For Hypotheses are not to be regarded in experimental Philosophy.

Opticks, 2nd edition (1718), Book 3, Query 31, 380.

3 For nature is a perpetuall circulatory worker, generating fluids out of solids, and solids out of fluids, fixed things out of volatile, & volatile out of fixed, subtile out of gross, & gross out of subtile, Some things to ascend & make the upper terrestriall juices, Rivers and the Atmosphere; & by consequence others to descend for a Requitall to the former. And as the Earth, so perhaps may the Sun imbibe this spirit copiously to conserve his Shineing, & keep the Planets from recedeing further from him. And they that will, may also suppose, that this Spirit affords or carryes with it thither the solary fewell & materiall Principle of Light; And that the vast aethereall Spaces between us, & the stars are for a sufficient repository for this food of the Sunn and Planets.

Letter to Oldenburg, 7 December 1675. In H. W. Turnbull (ed.), *The Correspondence of Isaac Newton*, 1661–1675 (1959), Vol. 1, 366.

4 If I have seen further it is by standing on ye sholders of Giants.

Letter to Hooke, 5 February 1675/6. In H. W.Turnbull (ed.), *The Correspondence of Isaac Newton*, 1, 1661–1675 (1959), Vol. 1, 416. Although this is the usual citation for this famous quotation, it is not the original. It appeared in Robert Burton's *The Anatomy of Melancholy* (1624), with which Newton was familiar, in a slightly different form: 'Pigmies placed on the shoulders of giants see more than the giants themselves'. It has even been traced to Bernard of Chartres in the early twelfth century and the Roman grammarian Priscian. Details of the earlier use of this quotation are given in Robert K. Merton, *On the Shoulders of Giants* (1965).

5 I see I have made my self a slave to Philosophy.

Letter to Henry Oldenberg, 18 November 1676. In H. W. Turnbull (ed.), *The*

Correspondence of Isaac Newton, 1676–1687 (1960), Vol. 2, 182.

1 [1] And first I suppose that there is diffused through all places an aethereal substance capable of contraction & dilatation, strongly elastick, & in a word, much like air in all respects, but far more subtile.

2. I suppose this aether pervades all gross bodies, but yet so as to stand rarer in their pores then in free spaces, & so much ye rarer as their pores are less . . .

3. I suppose ye rarer aether within bodies & ye denser without them, not to be terminated in a mathematical superficies, but to grow gradually into one another.

Letter to Robert Boyle, 28 February 1678/9. In H. W. Turnbull (ed.), *The Correspondence of Isaac Newton*, 1676–1687 (1960), Vol. 2, 289.

2 As I am writing, another illustration of ye generation of hills proposed above comes into my mind. Milk is as uniform a liquor as ye chaos was. If beer be poured into it & ye mixture let stand till it be dry, the surface of ye curdled substance will appear as rugged & mountanous as the Earth in any place.

Letter to Thomas Burnet, January 1680/1. In H. W. Turnbull (ed.), *The Correspondence of Isaac Newton*, 1676–1687 (1960), Vol. 2, 334.

3 Philosophy is such an impertinently litigious Lady that a man had as good be engaged in Law suits as have to do with her.

Letter to Edmond Halley, 20 June 1686. In H. W. Turnbull (ed.), *The Correspondence of Isaac Newton*, 1676–1687 (1960), Vol. 2, 437.

4 When I wrote my treatise about our System, I had an eye upon such principles as might work with considering men, for the belief of the Deity, & nothing can rejoice me more than to find it useful for that purpose.

Letter to Richard Bentley, 10 December 1692. In H. W. Turnbull (ed.), *The Correspondence of Isaac Newton*, 1676–1687 (1960), Vol. 3, 233.

5 I do not know what I may appear to the world, but to myself I seem to have been only like a boy playing on the seashore, and diverting myself in now and then finding a smoother pebble or a prettier shell than ordinary, whilst the great ocean of truth lay all undiscovered before me.

First reported in Joseph Spence, *Anecdotes, Observations and Characters, of Books and Men* (1820), Vol. 1 of 1966 edn, sect. 1259, p. 462. Said to have been addressed by Newton in 1727 (year of his death) to Chevalier Andrew Michael Ramsey (but DNB says he was in France at the time!). Quoted in David Brewster, *Memoirs of the Life, Writings, and Discoveries of Sir Isaac Newton* (1855), Vol. 2, 407.

6 No old Men (excepting Dr. Wallis) love Mathematicks.

Comment made by Newton to Whiston. Quoted in Richard Westfall, *Never at Rest: A Biography of Isaac Newton* (1980), 139.

7 In the beginning of the year 1665 I found the Method of approximating series & the Rule for reducing any dignity of any Bionomial into such a series. The same year in May I found the method of Tangents of Gregory & Slusius, & in November had the direct method of fluxions & the next year in January had the Theory of Colours & in May following I had entrance into ye inverse method of fluxions. And the same year I began to think of gravity extending to ye orb of the Moon & (having found out how to estimate the force with wch [a] globe revolving within a sphere presses the surface of the sphere) from Keplers rule of the periodic times of the Planets being in sesquialterate proportion of their distances from the center of their Orbs, I deduced that the forces wch keep the Planets in their Orbs must [be] reciprocally as the squares of their distances from the centers about wch they revolve: & thereby compared the force requisite to keep the Moon in her Orb with the force of gravity at the surface of the earth, & found them answer pretty nearly. All this was in the two plague years of 1665–1666. For in those days I was in the prime of my age for invention & minded

Mathematicks & Philosophy more then than at any time since.

Quoted in Richard Westfall, *Never at Rest: A Biography of Isaac Newton* (1980), 143.

1 By the help of the new *Analysis* Mr. Newton found out most of the Propositions in his *Principia Philosophiae*: but because the Ancients for making things certain admitted nothing in Geometry before it was demonstrated synthetically, he demonstrated the Propositions synthetically, that the Systeme of the Heavens might be founded upon good Geometry. And this makes it now difficult for unskilful men to see the Analysis by which those Propositions were found out.

'An Account of the Book entituled *Commercium Epistolicum*, *Philosophical Transactions* (1714–16), 29, 206.

2 Oh Diamond! Diamond! thou little knowest the mischief done!

To a dog, who knocked over a candle which set fire to some important papers and 'destroyed the almost finished labours of some years'.

First reported in Thomas Maude, *Wensley-Dale; or, Rural Contemplation* (1780). Probably apocryphal as there is no reference to any dog in any of the recollections of Newton's contemporaries. Quoted in D. Gjertsen, *The Newton Handbook* (1986), 177.

3 ON NEWTON His genius now began to mount upwards apace & shine out with more strength, & as he told me himself, he excelled particularly in making verses . . . In everything he undertook he discovered an application equal to the pregnancy of his parts & exceeded the most sanguine expectations his master had conceived of him.

John Conduitt on Newton as a schoolboy in Grantham.

Quoted in Richard Westfall, *Never at Rest: A Biography of Isaac Newton* (1980), 65.

4 ON NEWTON In the year 1666 he retired again from Cambridge . . . to his mother in Lincolnshire & whilst he was musing in a garden it came into his thought that the power of gravity (wch brought an apple from the tree to the ground) was not limited to a certain distance from the earth but that this power must extend much farther than was usually thought. Why not as high as the moon said he to himself & if so that must influence her motion & perhaps retain her in her orbit, whereupon he fell a calculating what would be the effect of that supposition but being absent from books & taking the common estimate in use among Geographers & our seamen before Norwood had measured the earth, that 60 English miles were contained in one degree of latitude on the surface of the Earth his computation did not agree with his theory & inclined him then to entertain a notion that together with the force of gravity there might be a mixture of that force wch the moon would have if it was carried along in a vortex.

John Conduitt, Memorandum of a conversation with Newton in August 1726. Quoted in Richard Westfall, *Never at Rest: A Biography of Isaac Newton* (1980), 154.

5 ON NEWTON In 1684 Dr Halley came to visit him at Cambridge, after they had been some time together, the Dr asked him what he thought the Curve would be that would be described by the Planets supposing the force of attraction towards the Sun to be reciprocal to the square of their distance from it. Sr Isaac replied immediately that it would be an Ellipsis, the Doctor struck with joy & amazement asked him how he knew it, why saith he I have calculated it, whereupon Dr Halley asked him for his calculation without any farther delay. Sr Isaac looked among his papers but could not find it, but he promised him to renew it, & then to send it him.

The origins of The Principia. *Without telling Newton, Halley was testing him about a hypothesis of planetary motion formulated ten years earlier by Robert Hooke.*

Newton's recollection as told to Abraham DeMoivre. Quoted in Richard Westfall, *Never at Rest: A Biography of Isaac Newton* (1980), 403.

6 ON NEWTON Here lies Sir Isaac Newton, Knight, who by a vigour of mind almost supernatural, first demonstrated, the motions and Figures of the Planets, the Paths of the comets, and the Tides of the Oceans . . . Let Mortals rejoice that there has existed such and so great an ornament of Nature.

Epitaph of Isaac Newton. On his tomb at Westminster Abbey.

1 ON **NEWTON** And from my pillow, looking forth by light
Of moon or favouring stars, I could behold
The antechapel where the statue stood
Of Newton with his prism and silent face,
The marble index of a mind for ever
Voyaging through strange seas of Thought, alone.

William Wordsworth, *The Prelude* (1850), Book 3, Lines 58–63.

Alex Nickon 1927–

German chemist

2 Chemists do have a good sense of humor but lose it when they serve as referees.

A. Nickon and E. F. Silversmith, *Organic Chemistry: The Name Game, Modern Coined Terms and Their Origins* (1987), 293, footnote.

Charles-Jules-Henri Nicolle 1866–1936

French bacteriologist

3 The native hospital in Tunis was the focal point of my research. Often, when going to the hospital, I had to step over the bodies of typhus patients who were awaiting admission to the hospital and had fallen exhausted at the door. We had observed a certain phenomenon at the hospital, of which no one recognized the significance, and which drew my attention. In those days typhus patients were accommodated in the open medical wards. Before reaching the door of the wards they spread contagion. They transmitted the disease to the families that sheltered them, and doctors visiting them were also infected. The administrative staff admitting the patients, the personnel responsible for taking their clothes and linen, and the laundry staff were also contaminated. In spite of this, once admitted to the general ward the typhus patient did not contaminate any of the other patients, the nurses or the doctors. I took this observation as my guide. I asked myself what happened between the entrance to the hospital and the wards. This is what happened: the typhus patient was stripped of his clothes and linen, shaved and washed. The contagious agent was therefore something attached to his skin and clothing, something which soap and water could remove. It could only be the louse. It was the louse.

'Investigations on Typhus', Nobel lecture, 1928. In *Nobel Lectures: Physiology or Medicine 1922–1941* (1965), 181.

4 And this is the ultimate lesson that our knowledge of the mode of transmission of typhus has taught us: Man carries on his skin a parasite, the louse. Civilization rids him of it. Should man regress, should he allow himself to resemble a primitive beast, the louse begins to multiply again and treats man as he deserves, as a brute beast. This conclusion would have endeared itself to the warm heart of Alfred Nobel. My contribution to it makes me feel less unworthy of the honour which you have conferred upon me in his name.

'Investigations on Typhus', Nobel Lecture, 1928. In *Nobel Lectures: Physiology or Medicine 1922–1941* (1965), 187.

Friedrich Nietzsche 1844–1900

German philosopher

5 Formerly one sought the feeling of the grandeur of man by pointing to his divine *origin*: this has now become a forbidden way, for at its portal stands the ape, together with other gruesome beasts, grinning knowingly as if to say: no further in this direction! One therefore now tries the opposite direction: the way mankind is *going* shall serve as proof of his grandeur and kinship with God. Alas this, too, is vain! At the end of this way stands the funeral urn of the *last* man and gravedigger (with the inscription '*nihil humani a me alienum puto*'). However high mankind may have evolved—and perhaps at the end it will stand even lower than at the beginning!—it cannot pass over into a higher order, as little as the ant and the earwig can at the end of its 'earthly course' rise up to kinship with God and eternal life. The becoming drags the has-been along behind it: why should an exception to this eternal spectacle be made on behalf of some little star or for any little species upon it! Away with such sentimentalities!

Daybreak: Thoughts on the Prejudices of Morality (1881), trans. R. J. Hollingdale (1982), 32.

1 God is dead. God remains dead. And we have killed him.

The Gay Science (1882), book 3, no. 125, trans. W. Kaufmann (1974), 181.

2 *Kein Sieger glaubt an den Zufall.*
No victor believes in chance.

The Gay Science (1882), book 3, no. 258, trans. W. Kaufmann (1974), 217.

3 One should not understand this compulsion to construct concepts, species, forms, purposes, laws ('a world of identical cases') as if they enabled us to fix the *real world*; but as a compulsion to arrange a world for ourselves in which our existence is made possible:—we thereby create a world which is calculable, simplified, comprehensible, etc., for us.

The Will to Power (Notes written 1883–1888), book 3, no. 521. Trans. W. Kaufmann and R. J. Hollingdale and ed. W. Kaufmann (1968), 282.

4 If the world may be thought of as a certain definite quantity of force and as a certain definite number of centers of force—and every other representation remains indefinite and therefore useless—it follows that, in the great dice game of existence, it must pass through calculable number of combinations. In infinite time, every possible combination would at some time or another be realized; more: it would be realized an infinite number of times. And since between every combination and its next recurrence all other possible combinations would have to take place, and each of these combination conditions of the entire sequence of combinations in the same series, a circular movement of absolutely identical series is thus demonstrated: the world as a circular movement that has already repeated itself infinitely often and plays its game *in infinitum*. This conception is not simply a mechanistic conception; for if it were that, it would not condition an infinite recurrence of identical cases, but a final state. *Because* the world has not reached this, mechanistic theory must be considered an imperfect and merely provisional hypothesis.

The Will to Power (Notes written 1883–1888), book 4, no. 1066. Trans. W. Kaufmann and R. J. Hollingdale and ed. W. Kaufmann (1968), 549.

5 And do you know what 'the world' is to me? Shall I show it to you in my mirror? This world: a monster of energy, without beginning, without end; a firm, iron magnitude of force that does not grow bigger or smaller, that does not expend itself but only transforms itself; as a whole, of unalterable size, a household without expenses or losses, but likewise without increase or income; enclosed by 'nothingness' as by a boundary; not by something blurry or wasted, not something endlessly extended, but set in a definite space as a definite force, and not a space that might be 'empty' here or there, but rather as force throughout, as a play of forces and waves of forces, at the same time one and many, increasing here and at the same time decreasing there; a sea of forces flowing and rushing together, eternally changing, eternally flooding back, with tremendous years of recurrence, with an ebb and a flood of its forms; out of the simplest forms striving toward the most complex, out of the stillest, most rigid, coldest forms toward the hottest, most turbulent, most self-contradictory, and then again returning home to the simple out of this abundance, out of the play of contradictions back to the joy of concord, still affirming itself in this uniformity of its courses and its years, blessing itself as that which must return eternally, as a becoming that knows no satiety, no disgust, no weariness: this, my *Dionysian* world of the eternally self-creating, the eternally self-destroying, this mystery world of the twofold voluptuous delight, my 'beyond good and evil,' without goal, unless the joy of the circle itself is a goal; without will, unless a ring feels good will toward itself—do you want a *name* for this world? A *solution* for all its riddles? A *light* for you, too, you best-concealed, strongest, most intrepid, most midnightly men?—*This world is the will to power—and nothing besides!* And you yourselves are also this will to power—and nothing besides!

The Will to Power (Notes written 1883–1888), book 4, no. 1067. Trans. W. Kaufmann and R. J. Hollingdale and ed. W. Kaufmann (1968), 549–50.

1 It is perhaps just dawning on five or six minds that physics, too, is only an interpretation and exegesis of the world (to suit us, if I may say so!) and *not* a world-explanation.

Beyond Good and Evil (1886). Trans. W. Kaufmann (ed.), *Basic Writings of Nietzsche* (1968), 211.

2 One should not wrongly reify 'cause' and 'effect,' as the natural scientists do (and whoever, like them, now 'naturalizes' in his thinking), according to the prevailing mechanical doltishness which makes the cause press and push until it 'effects' its end; one should use 'cause' and 'effect' only as pure concepts, that is to say, as conventional fictions for the purpose of designation and communication—*not* for explanation.

Beyond Good and Evil (1886). Trans. W. Kaufmann (ed.), *Basic Writings of Nietzsche* (1968), 219.

3 Since Copernicus, man seems to have got himself on an inclined plane—now he is slipping faster and faster away from the center into—what? into nothingness? into a '*penetrating* sense of his nothingness?' . . . all science, natural as well as *unnatural*—which is what I call the self-critique of knowledge—has at present the object of dissuading man from his former respect for himself, as if this had been but a piece of bizarre conceit.

On the Genealogy of Morals (1887), trans. W. Kaufmann and R. J. Hollingdale (1969), 155–6.

Florence Nightingale 1820–1910

British nurse and reformer

4 Let people who have to observe sickness and death look back and try to register in their observation the appearances which have preceded relapse, attack or death, and not assert that there were none, or that there were not the *right* ones. A want of the habit of observing conditions and an inveterate habit of taking averages are each of them often equally misleading.

Notes on Nursing: What it is, and What it is Not (1860), 67.

5 It may seem a strange principle to enunciate as the very first requirement in a Hospital that it should do the sick no harm.

Notes on Hospitals (1863), i.

6 ON NIGHTINGALE Her [Nightingale's] statistics were more than a study, they were indeed her religion. For her, Quetelet was the hero as scientist, and the presentation copy of his *Physique Sociale* is annotated by her on every page. Florence Nightingale believed—and in all the actions of her life acted upon that belief—that the administrator could only be successful if he were guided by statistical knowledge. The legislator—to say nothing of the politician—too often failed for want of this knowledge. Nay, she went further: she held that the universe—including human communities—was evolving in accordance with a divine plan; that it was man's business to endeavour to understand this plan and guide his actions in sympathy with it. But to understand God's thoughts, she held we must study statistics, for these are the measure of his purpose. Thus the study of statistics was for her a religious duty.

In Karl Pearson, *The Life, Letters and Labours of Francis Galton* (1924), Vol. 2, 414–5.

1 ON NIGHTINGALE A Lady with a Lamp shall stand
In the great history of the land,
A noble type of good,
Heroic womanhood.

Henry Wadsworth Longfellow, '*Santa Filomena*' (1857).

Henry E. Niles

2 'Causation' has been popularly used to express the condition of association, when applied to natural phenomena. There is no philosophical basis for giving it a wider meaning than partial or absolute association. In no case has it been proved that there is an inherent necessity in the laws of nature. Causation is correlation . . . [P]erfect correlation, *when based upon sufficient experience*, is causation in the scientific sense.

'Correlation, Causation and Wright's Theory of "Path Coefficients"', *Genetics*, 1922, 7, 259–61.

Alfred Bernhard Nobel 1833–96

Swedish chemist, engineer and industrialist

3 The whole of my remaining realizable estate shall be dealt with the following way: the capital, invested in safe securities by my executors, shall constitute a fund, the interest on which shall be annually distributed in the form of prizes to those who, during the preceding year, shall have conferred the greatest benefit on mankind. The said interest shall be divided into five equal parts, which shall be apportioned as follows: one part to the person who shall have made the most important discovery or invention within the field of physics; one part to the person who shall have made the most important chemical discovery or improvement; one part to the person who shall have made the most important discovery within the domain of physiology or medicine; one part to the person who shall have produced in the field of literature the most outstanding work in an ideal direction; and one part to the person who shall have done the most or the best work for fraternity between nations, for the abolition or reduction of standing armies and for the holding and promotion of peace congresses. The prizes for physics and chemistry shall be awarded by the Swedish Academy of Sciences; that for physiological or medical work by the Caroline Institute in Stockholm; that for literature by the Academy in Stockholm, and that for champions of peace by a committee of five persons to be elected by the Norwegian Storting. It is my express wish that in awarding the prizes no consideration whatever shall be given to the nationality of the candidates, but that the most worthy shall receive the prize, whether he be a Scandinavian or not.

Alfred Nobel's Will, The Nobel Foundation Website (www.nobel.se).

Novalis 1772–1801

German romantic poet

4 *Das Leben der Götter ist Mathematik.*
Mathematics is the Life of the Gods.

Attributed.

Cristiane Nüsselein-Volhard 1942– and Eric Wieshaus 1947–

German geneticist and American physiologist

5 In systemic searches for embryonic lethal mutants of *Drosophila melanogaster* we have identified 15 loci which when mutated alter the segmental patterns of the larva. These loci probably represent the majority of such genes in *Drosophila*. The phenotypes of the mutant embryos indicate that the process of segmentation involves at least three levels of spatial organization: the entire egg as developmental unit, a repeat unit with the length of two segments, and the individual segment.

'Mutations Affecting Segment Number and Polarity in *Drosophila*', *Nature*, 1980, **287**, 795.

Ronald Sydney Nyholm 1917–71

Australian chemist

1 Those of us who were familiar with the state of inorganic chemistry in universities twenty to thirty years ago will recall that at that time it was widely regarded as a dull and uninteresting part of the undergraduate course. Usually, it was taught almost entirely in the early years of the course and then chiefly as a collection of largely unconnected facts. On the whole, students concluded that, apart from some relationships dependent upon the Periodic table, there was no system in inorganic chemistry comparable with that to be found in organic chemistry, and none of the rigour and logic which characterised physical chemistry. It was widely believed that the opportunities for research in inorganic chemistry were few, and that in any case the problems were dull and uninspiring; as a result, relatively few people specialized in the subject . . . So long as inorganic chemistry is regarded as, in years gone by, as consisting simply of the preparations and analysis of elements and compounds, its lack of appeal is only to be expected. The stage is now past and for the purpose of our discussion we shall define inorganic chemistry today as the integrated study of the formation, composition, structure and reactions of the chemical elements and compounds, excepting most of those of carbon.

The Renaissance of Inorganic Chemistry (1956), 4–5. Inaugural Lecture delivered at University College, London, 1 March 1956.

2 If a little less time was devoted to the translation of letters by Julius Caesar describing Britain 2000 years ago and a little more time was spent on teaching children how to describe (in simple modern English) the method whereby ethylene was converted into polythene in 1933 in the ICI laboratories at Northwich, and to discussing the enormous social changes which have resulted from this discovery, then I believe that we should be training future leaders in this country to face the world of tomorrow far more effectively than we are at the present time.

Quoted in Nyholm's Obituary by D. P. Craig, *Biographical Memoirs of Fellows of the Royal Society*, 1972, **18**, 461.

Kenneth Page Oakley 1911–81

British anthropologist

3 The most satisfactory definition of man from the scientific point of view is probably Man the Tool-maker.

'The Earliest Tool-makers', *Antiquity*, 1956, 117, 4.

William of Ockham c.1285–1349

British philosopher

4 *From the village of Ockham in Surrey: Ockham's Razor:*

Frustra fit per plura, quod fieri potest per pauciora.

It is vain to do with more what can be done with less.

Summa logicae (*The Sum of All Logic*) [before 1324], Part I, chapter 12. William of Ockham borrowing from Petrus Aureolus, *The Eloquent Doctor*, 2 Sent. distinction 12, question 1.

5 *Pluralitas non est ponenda sine necessitate.*

A plurality (of reasons) should not be posited without necessity.

Quodlibeta (*Quodlibetal Questions*) [1324–1325], Quodlibet 6, question 10, trans. A. Freddoso (1991), Vol. 2, 521.

6 *Entia/Essentia non sunt multiplicanda praeter necessitatem.*

Entities should not be multiplied unnecessarily.

Attributed.

Eugene Pleasants Odum

1913–2002

American ecologist and ornithologist

1 Man has generally been preoccupied with obtaining as much 'production' from the landscape as possible, by developing and maintaining early successional types of ecosystems, usually monocultures. But, of course, man does not live by food and fiber alone; he also needs a balanced CO_2—O_2 atmosphere, the climactic buffer provided by oceans and masses of vegetation, and clean (that is, unproductive) water for cultural and industrial uses. Many essential life-cycle resources, not to mention recreational and esthetic needs, are best provided man by the less 'productive' landscapes. In other words, the landscape is not just a supply depot but is also the *oikos*—the home—in which we must live.

'The Strategy of Ecosystem Development. An Understanding of Ecological Succession Provides a Basis for Resolving Man's Conflict with Nature', *Science*, 1969, **164**, 266.

2 The Pacific coral reef, as a kind of oasis in a desert, can stand as an object lesson for man who must now learn that mutualism between autotrophic and heterotrophic components, and between producers and consumers in the societal realm, coupled with efficient recycling of materials and use of energy, are the keys to maintaining prosperity in a world of limited resources.

'The Emergence of Ecology as a New Integrative Discipline', *Science*, 1977, **195**, 1290.

Lorenz Oken 1779–1851

German natural philosopher

3 During its development the animal passes through all stages of the animal kingdom. The foetus is a representation of all animal classes in time.

Elements of Physiophilosophy (1833–45), 3rd German edition, trans. Alfred Tulk (1847), 491.

4 *Der ganze Mensch ist nur ein Wirbelbein.*

The whole of a human being is merely a vertebra.

Quoted in Henry Harris, *The Birth of the Cell* (1999), 62.

Henry Oldenburg c.1618–77

German-born British Secretary of the Royal Society of London

5 Whereas there is nothing more necessary for promoting the improvement of Philosophical Matters, than the communicating to such, as apply their Studies and Endeavours that way, such things as are discovered or put in practice by others; it is therefore thought fit to employ the Press, as the most proper way to gratifie those, whose engagement in such Studies, and delight in the advancement of Learning and profitable Discoveries, doth entitle them to the knowledge of what this Kingdom, or other parts of the World, do, from time to time, afford as well of the progress of the Studies, Labours, and attempts of the Curious and learned in things of this kind, as of their compleat Discoveries and performances: To the end, that such Productions being clearly and truly communicated, desires after solid and usefull knowledge may be further entertained, ingenious Endeavours and Undertakings cherished, and those, addicted to and conversant in such matters, may be invited and encouraged to search, try, and find out new things, impart their knowledge to one another, and contribute what they can to the Grand design of improving Natural knowledge, and perfecting all Philosophical Arts, and Sciences. All for the Glory of God, the Honour and

Advantage of these Kingdoms, and the Universal Good of Mankind.

'Introduction', *Philosophical Transactions* (1665), 1, 1–2.

1 The King saw them wth no common satisfaction, expressing his desire in no particular to have yt Stellar fish engraven and printed. We wish very much, Sir, yt you could procure for us a particular description of yesd Fish, viz. whether it be common there; what is observable in it when alive; what colour it then hath; what kind of motion in the water; what use it maketh of all that curious workmanship, wch Nature hath adorn'd it wth?

Letter to John Winthrop, Jr., 26 March 1670, about specimens sent by Winthrop to the Society.

In A. Rupert Hall & Marie Boas Hall (eds.), *The Correspondence of Henry Oldenburg* (1969), Vol. 6, 594.

Richard Dixon Oldham 1858–1936

British seismologist and geologist

2 Of all regions of the earth none invites speculation more than that which lies beneath our feet, and in none is speculation more dangerous; yet, apart from speculation, it is little that we can say regarding the constitution of the interior of the earth. We know, with sufficient accuracy for most purposes, its size and shape: we know that its mean density is about 5 1/2 times that of water, that the density must increase towards the centre, and that the temperature must be high, but beyond these facts little can be said to be known. Many theories of the earth have been propounded at different times: the central substance of the earth has been supposed to be fiery, fluid, solid, and gaseous in turn, till geologists have turned in despair from the subject, and become inclined to confine their attention to the outermost crust of the earth, leaving its centre as a playground for mathematicians.

'The Constitution of the Interior of the Earth, as Revealed by Earthquakes', *Quarterly Journal of the Geological Society*, 1906, **62**, 456.

3 Just as the spectroscope opened up a new astronomy by enabling the astronomer to determine some of the constituents of which distant stars are composed, so the seismograph, recording the unfelt motion of distant earthquakes, enables us to see into the earth and determine its nature with as great a certainty, up to a certain point, as if we could drive a tunnel through it and take samples of the matter passed through.

'The Constitution of the Interior of the Earth, as Revealed by Earthquakes', *Quarterly Journal of the Geological Society*, 1906, **62**, 456.

David Roger Oldroyd 1936–

Australian historian of science

4 For geologists . . . the word 'plate' is in somewhat the same position as 'skyscraper'. It is a dead (or almost dead) metaphor. Geologists know (or they think they know) what plates are, almost as well as the commuter knows what a skyscraper is.

Thinking about the Earth (1996), 307.

Everett C. Olson 1910–93

American vertebrate palaeontologist

5 Morphological information has provided the greatest single source of data in the formulation and development of the theory of evolution and that even now, when the preponderance of work is experimental, the basis for interpretation in many areas of study remains the form and relationships of structures.

'Morphology, Paleontology, and Evolution', in Sol Tax (ed.), *Evolution after Darwin*, Vol. 1, *The Evolution of Life* (1960), 524.

Walter J. Ong 1912–2003

American philosopher

6 Without writing, the literate mind would not and could not think as it does, not only when engaged in writing but normally even when it is composing its thoughts in oral form. More than any other single invention writing has transformed human consciousness.

Orality and Literacy: The Technologizing of the Word (1982), 78.

Heike Kamerlingh Onnes see Kamerlingh-Onnes, Heike

Julius Robert Oppenheimer

1904–67

American physicist

1 If atomic bombs are to be added as new weapons to the arsenals of a warring world, or to the arsenals of nations preparing for war, then the time will come when mankind will curse the names of Los Alamos and of Hiroshima.

Speech given at the acceptance of a Certificate of Appreciation from General Groves, Secretary of War, 16 October 1945. In Alice Kimball Smith and Charles Weiner (eds.), *Robert Oppenheimer: Letters and Recollections* (1980), 310–11.

2 If the radiance of a thousand suns
Were to burst at once into the sky
That would be like the splendour of the
Mighty One . . .
I am become Death,
The shatterer of worlds.

Oppenheimer quoted this after the bombing of Hiroshima and Nagasaki.

Sacred Hindu Epic, Bhagavad Gita. Quoted in A. Berry (ed.), *Harrap's Book of Scientific Anecdotes* (1989), 175.

3 The scientist is not responsible for the laws of nature, but it is a scientist's job to find out how these laws operate. It is the scientist's job to find ways in which these laws can serve the human will. However, it is not the scientist's job to determine whether a hydrogen bomb should be used. This responsibility rests with the American people and their chosen representatives.

Quoted in L. Wolpert and A. Richards (eds.), *A Passion for Science* (1988), 9.

4 A discovery in science, or a new theory, even when it appears most unitary and most all-embracing, deals with some immediate element of novelty or paradox within the framework of far vaster, unanalysed, unarticulated reserves of knowledge, experience, faith, and presupposition. Our progress is narrow; it takes a vast world unchallenged and for granted. This is one reason why, however great the novelty or scope of new discovery, we neither can, nor need, rebuild the house of the mind very rapidly. This is one reason why science, for all its revolutions, is conservative. This is why we will have to accept the fact that no one of us really will ever know very much. This is why we shall have to find comfort in the fact that, taken together, we know more and more.

Science and the Common Understanding (1954), 53–4.

5 Despite the vision and the far-seeing wisdom of our wartime heads of state, the physicists felt a peculiarly intimate responsibility for suggesting, for supporting, and in the end, in large measure, for achieving the realization of atomic weapons. Nor can we forget that these weapons, as they were in fact used, dramatized so mercilessly the inhumanity and evil of modern war. In some sort of crude sense which no vulgarity, no humor, no overstatement can quite extinguish, the physicists have known sin; and this is a knowledge which they cannot lose.

The Open Mind (1955), 88.

6 The great testimony of history shows how often in fact the development of science has emerged in response to technological and even economic needs, and how in the economy of social effort, science, even of the most abstract and recondite kind, pays for itself again and again in providing the basis for radically new technological developments. In fact, most people—when they think of science as a good thing, when they think of it as worthy of encouragement, when they are willing to see their governments spend substance upon it, when they greatly do honor to men who in science have attained some eminence—have in mind that the conditions of their life have been altered just by such technology, of which they may be reluctant to be deprived.

The Open Mind (1955), 89–90.

1 Today, it is not only that our kings do not know mathematics, but our philosophers do not know mathematics and—to go a step further—our mathematicians do not know mathematics.

'The Tree of Knowledge', *Harper's Magazine*, 1958, **217**, 55.

Nicole Oresme c.1320–82

French natural philosopher

2 But having considered everything which has been said, one could by this believe that the earth and not the heavens is so moved, and there is no evidence to the contrary. Nevertheless, this seems prima facie as much, or more, against natural reason as are all or several articles of our faith. Thus, that which I have said by way of diversion (*esbatement*) in this manner can be valuable to refute and check those who would impugn our faith by argument.

On the Book of the Heavens and the World of Aristotle [1377], bk. II, ch. 25, sect. 10, trans. A. D. Menut and A. J. Denomy, quoted in Marshall Clagett, *The Science of Mechanics in the Middle Ages* (1959), 606.

George Orwell (Eric Arthur Blair) 1903–50

British writer

See ***Salam, Abdus*** 532:5

3 For once Benjamin consented to break his rule, and he read out to her what was written on the wall. There was nothing there now except a single Commandment. It ran: ALL ANIMALS ARE EQUAL BUT SOME ANIMALS ARE MORE EQUAL THAN OTHERS.

Animal Farm (1945), 87.

4 'Who controls the past', ran the party slogan, 'controls the future: Who controls the present controls the past.'

Nineteen Eighty-Four (1949), 37.

Henry Fairfield Osborn 1857–1935

American palaeontologist

5 If the Weismann idea triumphs, it will be in a sense a triumph of fatalism; for, according to it, while we may indefinitely improve the forces of our education and surroundings, and this civilizing nurture will improve the individuals of each generation, its actual effects will not be cumulative as regards the race itself, but only as regards the environment of the race; each new generation must start *de novo*, receiving no increment of the moral and intellectual advance made during the lifetime of its predecessors. It would follow that one deep, almost instinctive motive for a higher life would be removed if the race were only superficially benefited by its nurture, and the only possible channel of actual improvement were in the selection of the fittest chains of race plasma.

'The Present Problem of Heredity', *The Atlantic Monthly*, 1891, **57**, 363.

6 Now it is a well-known principle of zoological evolution that an isolated region, if large and sufficiently varied in its topography, soil, climate and vegetation, will give rise to a diversified fauna according to the *law of adaptive radiation* from primitive and central types. Branches will spring off in all directions to take advantage of every possible opportunity of securing food. The modifications which animals undergo in this adaptive radiation are largely of mechanical nature, they are limited in number and kind by hereditary, stirp or germinal influences, and thus result in the independent evolution of similar types in widely-separated regions under the *law of parallelism or homoplasy*. This law causes the independent origin not only of similar genera but of similar families and even of our similar orders. Nature thus repeats herself upon a vast scale, but the similarity is never complete and exact.

'The Geological and Faunal Relations of Europe and America during the Tertiary Period and the Theory of the Successive Invasions of an African Fauna', *Science*, 1900, 11, 563–4.

Andreas Osiander see Copernicus, Nicholas

William Osler 1849–1919

Canadian physician

1 A desire to take medicine is, perhaps, the great feature which distinguishes man from other animals.

'Recent Advances in Medicine', *Science*, 1891, 17, 170.

2 There is no more potent antidote to the corroding influence of mammon than the presence in the community of a body of men devoted to science, living for investigation and caring nothing for the lust of the eyes and the pride of life.

'Teacher and Student' (1892). In *Aequanimitas with Other Addresses to Medical Students, Nurses and Practitioners of Medicine* (1904), 29.

3 Now of the difficulties bound up with the public in which we doctors work, I hesitate to speak in a mixed audience. Common sense in matters medical is rare, and is usually in inverse ratio to the degree of education.

'Teaching and Thinking' (1894). In *Aequanimitas with Other Addresses to Medical Students, Nurses and Practitioners of Medicine* (1904), 131.

4 In the Mortality Bills, pneumonia is an easy second, to tuberculosis; indeed in many cities the death-rate is now higher and it has become, to use the phrase of Bunyan 'the captain of the men of death.'

'Medicine in the Nineteenth Century' (1904). In *Aequanimitas with Other Addresses to Medical Students, Nurses and Practitioners of Medicine* (1904), 260.

5 Fed on the dry husks of facts, the human heart has a hidden want which science cannot supply.

Science and Immorality (1904), 76.

6 At the outset do not be worried about this big question—Truth. It is a very simple matter if each one of you starts with the desire to get as much as possible. No human being is constituted to know the truth, the whole truth, and nothing but the truth; and even the best of men must be content with fragments, with partial glimpses, never the full fruition. In this unsatisfied quest the attitude of mind, the desire, the thirst—a thirst that from the soul must arise!—the fervent longing, are the be-all and the end-all.

'The Student Life' (1905). In G. L. Keynes (ed.), *Selected Writings of Sir William Osler* (1951), 172.

7 The extraordinary development of modern science may be her undoing. Specialism, now a necessity, has fragmented the specialities themselves in a way that makes the outlook hazardous. The workers lose all sense of proportion in a maze of minutiae.

'The Old Humanities and the New Science' (1919). In G. L. Keynes (ed.), *Selected Writings of Sir William Osler* (1951), 27.

8 That man can interrogate as well as observe nature, was a lesson slowly learned in his evolution.

William Bennett Bean (ed.), *Sir William Osler: Aphorisms from his Bedside Teachings and Writings*, No. 86 (1950), 58.

9 We are all dietetic sinners; only a small percent of what we eat nourishes us, the balance goes to waste and loss of energy.

William Bennett Bean (ed.), *Sir William Osler: Aphorisms from his Bedside Teachings and Writings*, No. 191 (1950), 96.

10 Medicine is a science of uncertainty and an art of probability.

William Bennett Bean (ed.), *Sir William Osler: Aphorisms from his Bedside Teachings and Writings*, No. 265 (1950), 125.

11 Look wise, say nothing, and grunt. Speech was given to conceal thought.

William Bennett Bean (ed.), *Sir William Osler:*

Aphorisms from his Bedside Teachings and Writings, No. 267 (1950), 126.

1 Our bowels are outside of us—just a tucked-in portion.

William Bennett Bean (ed.), *Sir William Osler: Aphorisms from his Bedside Teachings and Writings*, No. 351 (1950), 145.

2 The future belongs to Science. More and more she will control the destinies of the nations. Already she has them in her crucible and on her balances.

Introduction to René Vallery-Radot, *The Life of Pasteur* (1919), xvi.

Carl Wilhelm Wolfgang Ostwald 1883–1943

Latvian chemist

See also **Curie, Marie** *143:6*

3 The description of some of the experiments, which are communicated here, was completely worked out at my writing-table, *before I had seen anything of the phenomena in question.* After making the experiments on the following day, it was found that nothing in the description required to be altered. I do not mention this from feelings of pride, but in order to make clear the extraordinary ease and security with which the relations in question can be considered on the principles of Arrhenius' theory of free ions. Such facts speak more forcibly then any polemics for the value of this theory.

Philosophical Magazine, 1891, 32, 156.

4 *Die nicht wässerigen Losungen leiten ja nicht.*

Non-aqueous solutions don't conduct.

Zeitschrift für Physikalische Chemie, 1901, **5**, 341. Trans. W. H. Brock.

5 A catalyst is a substance which alters the velocity of a chemical reaction without appearing in the final products.

'Über Katalyse', *Zeitschrift für Elektrochemie*, 1901, 7, 995–1004. Quoted in J. R. Partington, *A History of Chemistry*, Vol. 4 (1901), 599–600.

6 The only difference between elements and compounds consists in the supposed impossibility of proving the so-called elements to be compounds.

'Faraday Lecture: Elements and Compounds', *Journal of the Chemical Society*, 1904, **85**, 520.

7 What we call matter is only a complex of energies which we find together in the same place.

'Faraday Lecture: Elements and Compounds', *Journal of the Chemical Society*, 1904, **85**, 520.

8 I must consider the organizer as more important than the discoverer.

Lebenslinien, Part 3, 1927, 435.

9 The chemists work with inaccurate and poor measuring services, but they employ very good materials. The physicists, on the other hand, use excellent methods and accurate instruments, but they apply these to very inferior materials. The physical chemists combine both these characteristics in that they apply imprecise methods to impure materials.

Quoted in R. Oesper, *The Human Side of Scientists* (1975), 116.

10 $G=A-W$

Glück gleich Arbeit weniger Widerstand.

Happiness is equal to work minus resistance.

Quoted by E. P. Hillpern, a former assistant of Ostwald in E. P. Hillpern, 'Some Personal Qualities of Wilhelm Ostwald Recalled by a Former Student', *Chymia*, 1949, **2**, 59.

11 ON OSTWALD In 1912 I went to a book sale and bought ten books for fifty cents. One of the books was by Ostwald *The Scientific Foundations of Analytical Chemistry*. Ostwald wrote at the beginning of that book that analytical chemists are the maidservants of other chemists. This made quite an impression on me, because I didn't want to be a maidservant.

Told by Izaak Kolthoff, Dutch-American analytical chemist. Beckman Center interview 15 March 1984. Tape-recording deposited at The Chemical Heritage Foundation, Philadelphia. Quotation supplied by W. H. Brock.

Publius Ovidus Naso Ovid 43 BC–AD c.17

Roman poet

1 Ere land and sea and the all-covering sky
Were made, in the whole world the countenance
Of nature was the same, all one, well named
Chaos, a raw and undivided mass,
Naught but a lifeless bulk, with warring seeds
Of ill-joined elements compressed together.
No sun as yet poured light upon the world,
No waxing moon her crescent filled anew,
Nor in the ambient air yet hung the earth,
Self-balanced, equipoised, nor Ocean's arms
Embraced the long far margin of the land.
Though there were land and sea and air, the land
No foot could tread, no creature swim the sea,
The air was lightless; nothing kept its form,
All objects were at odds, since in one mass
Cold essence fought with hot, and moist with dry,
And hard with soft and light with things of weight.
This strife a god, with nature's blessing, solved;
Who severed land from sky and sea from land,
And from the denser vapours set apart
The ethereal sky; and, each from the blind heap
Resolved and freed, he fastened in its place
Appropriate in peace and harmony.

Metamorphoses, Book 1, The Creation, l. 5–25. In A. D. Melville, trans., *Ovid: Metamorphoses* (1986), 1.

2 The safest course lies in between.

Metamorphoses, Book 2, Phaethon, l. 137. In A. D. Melville, trans., *Ovid: Metamorphoses* (1986), 28.

Richard Owen 1804–92

British surgeon, comparative anatomist and palaeontologist

3 The laws of Coexistence;—the adaptation of structure to function; and to a certain extent the elucidation of natural affinities may be legitimately founded upon the examination of fully developed species;—But to obtain an insight into the laws of development,—the signification or *bedeutung*, of the parts of an animal body demands a patient examination of the successive stages of their development, in every group of Animals.

'Lecture Four, 9 May 1837', *The Hunterian Lectures in Comparative Anatomy, May–June 1837*, ed. Phillip Reid Sloan (1992), 191.

4 The combination of such characters, some, as the sacral ones, altogether peculiar among Reptiles, others borrowed, as it were, from groups now distinct from each other, and all manifested by creatures far surpassing in size the largest of existing reptiles, will, it is presumed, be deemed sufficient ground for establishing a distinct tribe or sub-order of Saurian Reptiles, for which I would propose the name of *Dinosauria*.

'Report on British Fossil Reptiles', *Report of the Eleventh Meeting of the British Association for the Advancement of Science* (1842), 103.

5 Analogue. A part or organ in one animal which has the same function as another part or organ in a different animal.

'Glossary', *Lectures on the Comparative Anatomy and Physiology of the Invertebrate Animals Delivered at the Royal College of Surgeons in 1843* (1843), 374.

6 Homologue. The same organ in different animals under every variety of form and function.

'Glossary', *Lectures on the Comparative*

Anatomy and Physiology of the Invertebrate Animals Delivered at the Royal College of Surgeons in 1843 (1843), 379.

1 The Archetypal idea was manifested in the flesh, under divers such modifications, upon this planet, long prior to the existence of those animal species that actually exemplify it. To what natural laws or secondary causes the orderly succession and progression of such organic phaenomena may have been committed we as yet are ignorant. But if, without derogation of the Divine power, we may conceive the existence of such ministers, and personify them by the term 'Nature,' we learn from the past history of our globe that she has advanced with slow and stately steps, guided by the archetypal light, amidst the wreck of worlds, from the first embodiment of the Vertebrate idea under its old Ichthyic vestment, until it became arrayed in the glorious garb of the Human form.

On the Nature of Limbs (1849), 86.

2 In Man the brain presents an ascensive step in development, higher and more strongly marked than that by which the preceding subclass was distinguished from the one below it. Not only do the cerebral hemispheres overlap the olfactory lobes and cerebellum, but they extend in advance of the one, and further back than the other. Their posterior development is so marked, that anatomists have assigned to that part the character of a third lobe; it is peculiar to the genus *Homo*, and equally peculiar is the 'posterior horn of the lateral ventricle,' and the 'hippocampus minor,' which characterize the hind lobe of each hemisphere. The superficial grey matter of the cerebrum, through the number and depth of the convolutions, attains its maximum of extent in Man. Peculiar mental powers are associated with this highest form of brain, and their consequences wonderfully illustrate the value of the cerebral character; according to my estimate of which, I am led to regard the genus *Homo*, as not merely a representative of a distinct order, but of a distinct subclass of the Mammalia, for which I propose a name of 'ARCHENCEPHALA.'

'On the Characters, Principles of Division, and Primary Groups of the Class MAMMALIA' (1857), *Journal of the Proceedings of the Linnean Society of London*, 1858, 2, 19–20.

Heinz R. Pagels 1939–88

American physicist and science writer

3 Physicists speak of the particle representation or the wave representation. Bohr's principle of complementarity asserts that there exist complementary properties of the same object of knowledge, one of which if known will exclude knowledge of the other. We may therefore describe an object like an electron in ways which are mutually exclusive—e.g., as wave or particle—without logical contradiction provided we also realize that the experimental arrangements that determine these descriptions are similarly mutually exclusive. Which experiment—and hence which description one chooses—is purely a matter of human choice.

The Cosmic Code: Quantum Physics as the Language of Nature (1982), 94.

4 Theoretical and experimental physicists are now studying nothing at all—the vacuum. But that nothingness contains all of being.

The Cosmic Code: Quantum Physics as the Language of Nature (1982), 279.

5 It is unlikely that we will ever see a star being born. Stars are like animals in the wild. We may see the very young, but never their actual birth, which is a veiled and secret event. Stars are born inside thick clouds of dust and gas in the spiral arms of the galaxy, so thick that visible light cannot penetrate them.

Perfect Symmetry: The Search for the Beginning of Time (1985), 44.

James Paget 1814–99

British surgeon and physiologist

1 Clinical science has as good a claim to the name and rights and self-subsistence of a science as any other department of biology.

Address by the President. *Transactions of the Clinical Society of London*, 1870, 3, xxxii.

Thomas Paine 1737–1809

English-born American writer

2 It is a fraud of the Christian system to call the sciences *human invention*; it is only the application of them that is human. Every science has for its basis a system of principles as fixed and unalterable as those by which the universe is regulated and governed. Man cannot make principles—he can only discover them.

The Age of Reason (1794), 27.

William Paley 1743–1805

British natural theologian

3 No anatomist ever discovered a system of organization, calculated to produce pain and disease; or, in explaining the parts of the human body, ever said, this is to irritate; this is to inflame; this duct is to convey the gravel to the kidneys; this gland to secrete the humour which forms the gout: if by chance he come at a part of which he knows not the use, the most he can say is, that it is useless; no one ever suspects that it is put there to incommode, to annoy, or torment.

The Principles of Moral and Political Philosophy (1785), Vol. 1, 79.

4 Who can refute a *sneer?*

The Principles of Moral and Political Philosophy (1785), Vol. 2, 114.

5 In crossing a heath, suppose I pitched my foot against a *stone*, and were asked how the stone came to be there, I might possibly answer, that, for any thing I knew to the contrary, it had lain there for ever: nor would it perhaps be very easy to shew the absurdity of this answer. But suppose I had found a *watch* upon the ground, and it should be enquired how the watch happened to be in that place, I should hardly think of the answer which I had before given, that, for any thing I knew, the watch might have always been there.

Natural Theology: or, Evidences of the Existence and Attributes of the Deity, Collected from the Appearances of Nature (1802), 1–2.

6 There cannot be design without a designer; contrivance without a contriver; order without choice; arrangement, without any thing capable of arranging; subserviency and relation to a purpose; means suitable to an end, and executing their office in accomplishing that end, without the end ever having been contemplated, or the means accommodated to it. Arrangement, disposition of parts, subserviency of means to an end, relation of instruments to use, imply the preference of intelligence and mind.

Natural Theology: or, Evidences of the Existence and Attributes of The Deity, Collected from the Appearances of Nature (1802), 12.

7 It is a happy world after all. The air, the earth, the water teem with delighted existence. In a spring noon, or a summer evening, on whichever side I turn my eyes, myriads of happy beings crowd upon my view. 'The insect youth are on the wing.' Swarms of new-born flies are trying their pinions in the air. Their sportive motions, their wanton mazes, their gratuitous activity testify their joy and the exultation they feel in their lately discovered faculties . . . The whole winged insect tribe, it is probable, are equally intent upon their proper employments, and under every variety of constitution, gratified, and perhaps equally gratified, by the offices which the author of their nature has assigned to them.

Natural Theology: or, Evidences of the Existence and Attributes of The Deity, Collected from the Appearances of Nature (1802), 490–1.

Bernard Palissy c.1510–90

French naturalist

1 I can assure you, reader, that in a very few hours, even during the first day, you will learn more natural philosophy about things contained in this book, than you could learn in fifty years by reading the theories and opinions of the ancient philosophers. Enemies of science will scoff at the astrologers: saying, where is the ladder on which they have climbed to heaven, to know the foundation of the stars? But in this respect I am exempt from such scoffing; for in proving my written reason, I satisfy sight, hearing, and touch: for this reason, defamers will have no power over me: as you will see when you come to see me in my little Academy.

The Admirable Discourses (1580), trans. Aurèle La Rocque (1957), 27.

Pyotr Simon Pallas 1741–1811

German naturalist

2 These two orders of mountains [Secondary and Tertiary] offer the most ancient chronicle of our globe, least liable to falsifications and at the same time more legible than the writing of the primitive ranges. They are Nature's archives, prior to even the most remote records and traditions that have been preserved for our observant century to investigate, comment on and bring to the light of day, and which will not be exhausted for several centuries after our own.

'Observations sur la Formation des Montagnes', *Acta Academiae Scientiarum Imperialis Petropolitanae* (1777) [1778], 46. Trans. Albert Carozzi.

Friedrich Adolf Paneth 1887–1958

Austrian chemist

3 [The unreactivity of the noble gas elements] belongs to the surest of experimental results.

Angewandte Chemie, 1924, 37, 421. Trans. in Hilde Hein and George E. Hein, 'The Chemistry of Noble Gases—A Modern Case Study in Experimental Science', *Journal of the History of Ideas*, 1966, 27, 420.

Paracelsus (Theophrastus Phillipus Aureolus Bombastus von Hohenheim) c.1493–1541

Swiss doctor and chemist

4 Every creature has its own food, and an appropriate alchemist with the task of dividing it . . . The alchemist takes the food and changes it into a tincture which he sends through the body to become blood and flesh. This alchemist dwells in the stomach where he cooks and works. The man eats a piece of meat, in which is both bad and good. When the meat reaches the stomach, there is the alchemist who divides it. What does not belong to health he casts away to a special place, and sends the good wherever it is needed. That is the Creator's decree . . . That is the virtue and power of the alchemist in man.

Volumen Medicinae Paramirum (c.1520), in *Paracelsus: Essential Readings*, edited by Nicholas Goodrick-Clarke (1990), 50–1.

5 Medicine rests upon four pillars—philosophy, astronomy, alchemy, and ethics. The first pillar is the philosophical knowledge of earth and water; the second, astronomy, supplies its full understanding of that which is of fiery and airy nature; the third is an adequate explanation of the properties of all the four elements—that is to say, of the whole cosmos—and an introduction into the art of their transformations; and finally, the fourth shows the physician those virtues which must stay with him up until his death, and it should support and complete the three other pillars.

Das Buch Paragranum (c.1529–30), in J. Jacobi (ed.), *Paracelsus: Selected Writings* (1951), 133–4.

6 And I do not take my medicines from the apothecaries; their shops are but foul sculleries, from which comes nothing but foul broths. As for you,

you defend your kingdom with belly-crawling and flattery. How long do you think this will last? . . . let me tell you this: every little hair on my neck knows more than you and all your scribes, and my shoebuckles are more learned than your Galen and Avicenna, and my beard has more experience than all your high colleges.

'Credo', in J. Jacobi (ed.), *Paracelsus: Selected Writings* (1951), 80.

1 Man is a seed and the world is his apple; and just as the seed fares in the apple, so does man fare in the world, which surrounds him.

'Man in the Cosmos', in J. Jacobi (ed.), *Paracelus: Selected Writings* (1951), 112.

2 For, as the element of water lies in the middle of the globe, so, the branches run out from the root in its circuit on all sides towards the plains and towards the light. From this root very many branches are born. One branch is the Rhine, another the Danube, another the Nile, etc.

'The Philosophy of the Generation of the Elements', Book the Fourth, Text II. In *The Hermetic and Alchemical Writings of Aureolus Philippus Theophrastus Bombast, of Hohenheim, called Paracelsus the Great*, trans. A. E. Waite (1894), Vol. 1, 232.

3 In matters eternal it is Belief that makes all works visible, in matters corporeal it is the light of Nature that reveals things invisible.

In Walter Pagel, *Paracelsus*, 2nd edition (1982), 54.

4 Alchemy is the art that separates what is useful from what is not by transforming it into its ultimate matter and essence.

In Walter Pagel, *Paracelsus*, 2nd edition (1982), 113.

5 *Buch der Nature.*

Book of Nature.

Phrase used by Paracelsus after Konrad of Megenberg's *Buch der Nature* (*c.*1350).

6 I prefer the spagyric chemical physicians, for they do not consort with loafers or go about gorgeous in satins, silks and velvets, gold rings on their fingers, silver daggers hanging at their sides and white gloves on their hands, but they tend their work at the fire patiently day and night. They do not go promenading, but seek their recreation in the laboratory, wear plain learthern dress and aprons of hide upon which to wipe their hands, thrust their fingers amongst the coals, into dirt and rubbish and not into golden rings. They are sooty and dirty like the smiths and charcoal burners, and hence make little show, make not many words and gossip with their patients, do not highly praise their own remedies, for they well know that the work must praise the master, not the master praise his work. They well know that words and chatter do not help the sick nor cure them . . . Therefore they let such things alone and busy themselves with working with their fires and learning the steps of alchemy. These are distillation, solution, putrefaction, extraction, calcination, reverberation, sublimination, fixation, separation, reduction, coagulation, tinction, etc.

Quoted in R. Oesper, *The Human Side of Scientists* (1975), 150.

Ambroise Paré 1510–90

French surgeon

7 *Je le pansay, Dieu le guarit.*

I treated him—God cured him.

The Apologie and Treatise, containing the Voyages made into Divers Places (first published in 4th edition of his Collected Works, 1585), ed. Geoffrey Keynes (1951), 88.

Vilfredo Pareto 1848–1923

Italian economist

8 Human behaviour reveals uniformities which constitute natural laws. If these uniformities did not exist, then there would be neither social science nor political economy, and even the study of history would largely be useless. In effect, if the future actions of men having nothing in common with their past actions, our knowledge of them, although possibly satisfying our curiosity by way of an interesting story, would be entirely useless to us as a guide in life.

Cours d'Economie Politique (1896–7), Vol. 2, 397.

Cyril Northcote Parkinson

1909–93

British historian

1 Work expands so as to fill the time available for its completion.

Parkinson's First Law.

Parkinson's Law or the Pursuit of Progress (1958), 4.

2 Parkinson's Law is a purely scientific discovery, inapplicable except in theory to the politics of the day. It is not the business of the botanist to eradicate the weeds. Enough for him if he can tell us just how fast they grow.

Parkinson's Law or the Pursuit of Progress (1958), 15.

3 Expenditure rises to meet income.

Parkinson's Second Law.

The Law and the Profits (1960), 3.

4 Expansion means complexity, and complexity decay.

Parkinson's Third Law.

In Laws and Outlaws (1962), 168.

Talcott Parsons 1902–79

American sociologist

5 A scientifically unimportant discovery is one which, however true and however interesting for other reasons, has no consequences for a system of theory with which scientists in that field are concerned.

The Structure of Social Action (1937), Vol. 1, 7.

6 Sociology should . . . be thought of as a science of action—of the ultimate common value element *in its relations* to the other elements of action.

The Structure of Social Action (1937), Vol. 1, 440.

Blaise Pascal 1623–62

French mathematician and philosopher

7 We are generally more effectually persuaded by reasons we have ourselves discovered than by those which have occurred to others.

Pensées (1670), Section 1, aphorism 18. In H. F. Stewart (ed.), *Pascal's Pensées* (1950), 11.

8 Let him look at that dazzling light hung aloft as an eternal lamp to lighten the universe; let him behold the earth, a mere dot compared with the vast circuit which that orb describes, and stand amazed to find that the vast circuit itself is but a very fine point compared with the orbit traced by the stars as they roll their course on high.

But if our vision halts there, let imagination pass beyond; it will fail to form a conception long before Nature fails to supply material. The whole visible world is but an imperceptible speck in the ample bosom of Nature. No notion comes near it. Though we may extend our thought beyond imaginable space, yet compared with reality we bring to birth mere atoms. Nature is an infinite sphere whereof the centre is everywhere, the circumference nowhere. In short, imagination is brought to silence at the thought, and that is the most perceptible sign of the all-power of God.

Let man reawake and consider what he is compared with the reality of things; regard himself lost in this remote corner of Nature; and from the tiny cell where he lodges, to wit the Universe, weigh at their true worth earth, kingdoms, towns, himself. What is a man face to face with infinity?

Pensées (1670), Section 1, aphorism 43. In H. F. Stewart (ed.), *Pascal's Pensées* (1950), 19.

9 Man is but a reed, the weakest thing in nature; but a thinking reed.

Pensées (1670), Section 1, aphorism 160. In H. F. Stewart (ed.), *Pascal's Pensées* (1950), 83.

10 We know that there is an infinite, and we know not its nature. As we know it to be false that numbers are finite, it is therefore true that there is a numerical infinity. But we know not of what kind; it is untrue that it is even, untrue that it is odd; for the addition of a unit does not change its nature; yet it is a number, and every number is odd or even (this certainly holds of every finite number). Thus we may quite well know that there is a God without knowing what He is.

Pensées (1670), Section 1, aphorism 223. In H. F. Stewart (ed.), *Pascal's Pensées* (1950), 117.

1 A game is on, at the other end of this infinite distance, and heads or tails will turn up. What will you wager? According to reason you cannot leave either; according to reason you cannot leave either undone . . . Yes, but wager you must; there is no option, you have embarked on it. So which will you have. Come. Since you must choose, let us see what concerns you least. You have two things to lose: truth and good, and two things to stake: your reason and your will, your knowledge and your happiness. And your nature has two things to shun: error and misery. Your reason does not suffer by your choosing one more than the other, for you must choose. That is one point cleared. But your happiness? Let us weigh gain and loss in calling heads that God is. Reckon these two chances: if you win, you win all; if you lose, you lose naught. Then do not hesitate, wager that He is.

Pensées (1670), Section 1, aphorism 223. In H. F. Stewart (ed.), *Pascal's Pensées* (1950), 117–9.

2 We know the truth not only through our reason but also through our heart. It is through the latter that we know first principles, and reason, which has nothing to do with it, tries in vain to refute them.

Pensées (1670), trans. A. J. Krailsheimer (1966), Section I, VI, aphorism 110, p. 58.

3 I have made this one [letter] longer than usual because I did not have the leisure to make it shorter.

The Provincial Letters, Letter XVI, to the Reverend Jesuit Fathers, 4 December 1656, ed. A. J. Krailsheimer (1967), 257.

Louis Pasteur 1822–95

French microbiologist

4 Descriptive anatomy is to physiology what geography is to history.

Quoted in Patrice Debré, *Louis Pasteur*, trans. Elborg Forster (1994), 283.

5 The universe is an asymmetrical entity. I am inclined to believe that life as it is manifested to us must be a function of the asymmetry of the universe or of the consequence of this fact. The universe is asymmetrical; for if one placed the entire set of bodies that compose the solar system, each moving in its own way, before a mirror, the image shown would not be superimposable on the reality.

René Vallery-Radot, *Vie de Pasteur* (1900), 79. Quoted in Patrice Debré, *Louis Pasteur*, trans. Elborg Forster (1994), 78.

6 I propose to provide proof . . . that just as always an alcoholic ferment, the yeast of beer, is found where sugar is converted into alcohol and carbonic acid, so always a special ferment, a lactic yeast, is found where sugar is transformed into lactic acid. And, furthermore, when any plastic nitrogenated substance is able to transform sugar into that acid, the reason is that it is a suitable nutrient for the growth of the [lactic] ferment.

Comptes Rendus, 1857, **45**, 913.

7 Are the atoms of the dextroacid (tartaric) grouped in the spirals of a right-hand helix or situated at the angles of an irregular tetrahedron, or arranged in such or such particular unsymmetrical fashion? We are unable to reply to these questions. But there can be no reason for doubting that the grouping of the atoms has an unsymmetrical arrangement with a non-superimposable image. It is not less certain that the atoms of the laevo-acid realize precisely an unsymmetrical arrangement of the inverse of the above.

Leçons de Chemie (1860), 25.

8 The artificial products do not have any molecular dissymmetry; and I could not indicate the existence of a more profound separation between the products born under the influence of life and all the others.

Quoted in Joseph S. Fruton, *Proteins, Enzymes, Genes: The Interplay of Chemistry and Biology* (1999), 135.

1 My present and most fixed opinion regarding the nature of alcoholic fermentation is this: The chemical act of fermentation is essentially a phenomenon correlative with a vital act, beginning and ending with the latter. I believe that there is never any alcoholic fermentation without their being simultaneously the organization, development, multiplication of the globules, or the pursued, continued life of globules which are already formed.

'Mémoire sur la fermentation alcoolique', *Annales de Chemie et de Physique*, 1860, **58**:3, 359–360. Translated in Joseph S. Fruton, *Proteins, Enzymes, Genes: The Interplay of Chemistry and Biology* (1999), 137.

2 Wine is the most healthful and most hygienic of beverages.

Études sur le Vin (1866), Vol.1, xx.

3 Every chemical substance, whether natural or artificial, falls into one of two major categories, according to the spatial characteristic of its form. The distinction is between those substances that have a plane of symmetry and those that do not. The former belong to the mineral, the latter to the living world.

Pasteur Vallery-Radot (ed.), *Oeuvres de Pasteur* (1922–1939), Vol. 1, 331. Quoted in Patrice Debré, *Louis Pasteur*, trans. Elborg Forster (1994), 261.

4 After death, life reappears in a different form and with different laws. It is inscribed in the laws of the permanance of life on the surface of the earth and everything that has been a plant and an animal will be destroyed and transformed into a gaseous, volatile and mineral substance.

Quoted in Patrice Debré, *Louis Pasteur*, trans. Elborg Forster (1994), 110.

5 As I show you this liquid, I too could tell you, 'I took my drop of water from the immensity of creation, and I took it filled with that fecund jelly, that is, to use the language of science, full of the elements needed for the development of lower creatures. And then I waited, and I observed, and I asked questions of it, and I asked it to repeat the original act of creation for me; what a sight it would be! But it is silent! It has been silent for several years, ever since I began these experiments. Yes! And it is because I have kept away from it, and am keeping away from it to this moment, the only thing that it has not been given to man to produce, I have kept away from it the germs that are floating in the air, I have kept away from it life, for life is the germ, and the germ is life.'

Quoted in Patrice Debré, *Louis Pasteur*, trans. Elborg Forster (1994), 169.

6 These microscopic organisms form an entire world composed of species, families and varieties whose history, which has barely begun to be written, is already fertile in prospects and findings of the highest importance. The names of these organisms are very numerous and will have to be defined and in part discarded. The word *microbe* which has the advantage of being shorter and carrying a more general meaning, and of having been approved by my illustrious friend, *M. Littré*, the most competent linguist in France, is one we will adopt.

Charles-Emile Sedillot, 'Influence de M. Pasteur sure les progres de la chirurgie' [Influence of Pasteur on the progress of surgery]. A paper read to the Académie de Medecine, March 1878

7 *Dans les champs de l'observation le hasard ne favorise que les esprits préparés.*

Where observation is concerned, chance favours only the prepared mind.

'Address given on the Inauguration of the Faculty of Science, University of Lille, 7 December 1854', in R. Vallery-Radot, *La Vie de Pasteur* (1900), XX.

8 As in the experimental sciences, truth cannot be distinguished from error as long as firm principles have not been established through the rigorous observation of facts.

Étude sur la maladie des vers à soie (1870), 39.

1 The only thing that can bring joy is work.

Pasteur Vallery-Radot (ed.), *Correspondance de Pasteur 1840–1895* (1940), Vol. 1, 293. Quoted in Patrice Debré, *Louis Pasteur*, trans. Elborg Forster (1994), 64.

2 I give them experiments and they respond with speeches.

Quoted in Patrice Debré, *Louis Pasteur*, trans. Elborg Forster (1994), 362.

3 Nothing is lost and nothing is created in the operations of art as those of nature.

Pasteur Vallery-Radot (ed.), *Correspondance de Pasteur 1840–1895* (1940), Vol. 1, 326. Quoted in Patrice Debré, *Louis Pasteur*, trans. Elborg Forster (1994), 90.

4 Science knows no country because knowledge belongs to humanity, and is the torch which illuminates the world. Science is the highest personification of the nation because that nation will remain the first which carries the furthest the works of thought and intelligence.

Toast at banquet of the International Congress of Sericulture, Milan, 1876. Quoted in Maurice B. Strauss, *Familiar Medical Quotations* (1968), 519.

5 Herrmann Pidoux and Armand Trousseau stated 'Disease exists within us, because of us, and through us', Pasteur did not entirely disagree, 'This is true for certain diseases', he wrote cautiously, only to add immedately: 'I do not think that it is true for all of them'.

Pasteur Vallery-Radot (ed.), *Oeuvres de Pasteur* (1922–1939), Vol. 6, 167. Quoted in Patrice Debré, *Louis Pasteur*, trans. Elborg Forster (1994), 261.

6 There is no such thing as a special category of science called applied science; there is science and its applications, which are related to one another as the fruit is related to the tree that has borne it.

Pasteur Vallery-Radot (ed.), *Correspondance de Pasteur 1840–1895* (1940), Vol. 1, 315. Quoted in Patrice Debré, *Louis Pasteur*, trans. Elborg Forster (1994), 84.

7 If science has no country, the scientist should have one, and ascribe to it the influence which his works may have in this world.

Speech at the Inauguration of the Institute Pasteur (1888). Quoted in René Vallery-Radot, *The Life of Pasteur* (1923), 444.

8 Analogy cannot serve as proof.

Quoted in Patrice Debré, *Louis Pasteur*, trans. Elborg Forster (1994), 260.

9 ON PASTEUR My son, all my life I have loved this science so deeply that I can now hear my heart beat for joy.

J. B. Biot to Pasteur after he had separated tartaric acid crystals.

Quoted in R. Oesper, *The Human Side of Scientists* (1975), 152.

Coventry Kersey Dighton Patmore 1823–96

British poet and essayist

10 Not greatly moved with awe am I
To learn that we may spy
Five thousand firmaments beyond our own.
The best that's known
Of the heavenly bodies does them credit small.
View'd close, the Moon's fair ball
Is of ill objects worst,
A corpse in Night's highway, naked, fire-scarr'd, accurst;
And now they tell
That the Sun is plainly seen to boil and burst
Too horribly for hell.
So, judging from these two,
As we must do,
The Universe, outside our living Earth,
Was all conceiv'd in the Creator's mirth,
Forecasting at the time Man's spirit deep,
To make dirt cheap.
Put by the Telescope!
Better without it man may see,
Stretch'd awful in the hush'd midnight,
The ghost of his eternity.

'The Two Deserts' (1880–85). *Poems*, intro. Basil Champneys (1906), 302.

Wolfgang Pauli 1900–58

Austrian-born Swiss physicist

1 The fact that XY thinks slowly is not serious, but that he publishes faster than he thinks is inexcusable.

Quoted in R. Oesper, *The Human Side of Scientists* (1975), 154.

2 Physics is very muddled again at the moment; it is much too hard for me anyway, and I wish I were a movie comedian or something like that and had never heard anything about physics.

Letter to R. Kronig, 21 May 1925, quoted in R. Kronig, 'The Turning Point', in M. Fierz and V. F. Weisskopf (eds.), *Theoretical Physics in the Twentieth Century. A Memorial Volume to Wolfgang Pauli* (1960). Trans. in M. Klein, *Letters on Wave Mechanics*, x.

3 There can never be two or more equivalent electrons in an atom, for which in a strong field the values of all the quantum numbers n, k_1, k_2 and m are the same. If an electron is present, for which these quantum numbers (in an external field) have definite values, then this state is 'occupied.'

Quoted in M. Fierz, 'Wolfgang Pauli', in C. C. Gillispie (ed.), *Dictionary of Scientific Biography* (1974), Vol. 10, 423.

4 *Shown a paper by a young theoretician, Pauli read it, shook his head sadly, and said:*

Das ist nicht einmal falsch.

That is not even wrong.

Attributed.

Linus Carl Pauling 1901–94

American chemist

5 With moth cytochrome C there are 30 differences and 74 identities. With bread yeast and humans, there are about 45 amino acids that are different and about 59 that are identical. Think how close together man and this other organism, bread yeast, are. What is the probability that in 59 positions the same choice out of 20 possibilities would have been made by accident? It is impossibly small. There is, there must be, a developmental explanation of this. The developmental explanation is that bread yeast and man have a common ancestor, perhaps two billion years ago. And so we see that not only are all men brothers, but men and yeast cells, too, are at least close cousins, to say nothing about men and gorillas or rhesus monkeys. It is the duty of scientists to dispel ignorance of such relationships.

'The Social Responsibilities of Scientists and Science', *The Science Teacher*, 1933, 33, 15.

6 The energy of a covalent bond is largely the energy of resonance of two electrons between two atoms. The examination of the form of the resonance integral shows that the resonance energy increases in magnitude with increase in the *overlapping* of the two atomic orbitals involved in the formation of the bond, the word 'overlapping' signifying the extent to which regions in space in which the two orbital wave functions have large values coincide . . . Consequently it is expected that *of two orbitals in an atom the one which can overlap more with an orbital of another atom will form the stronger bond with that atom, and, moreover, the bond formed by a given orbital will tend to lie in that direction in which the orbital is concentrated*

Nature of the Chemical Bond and the Structure of Molecules and Crystals (1939), 76.

7 It has been recognized that hydrogen bonds restrain protein molecules to their native configurations, and I believe that as the methods of structural chemistry are further applied to physiological problems it will be found that the significance of the hydrogen bond for physiology is greater than that of any other single structural feature.

Nature of the Chemical Bond and the Structure of Molecules and Crystals (1939), 265.

8 The scientist, if he is to be more than a plodding gatherer of bits of information, needs to exercise an active imagination. The scientists of the past whom we now recognize as great are those who were gifted with transcendental imaginative powers, and the part played by the imaginative faculty of his daily life is as least as important for the scientist as it is for the worker in any other field—much more important than for most. A good scientist thinks logically and accurately when conditions call for logical and accurate thinking—but so does any other good worker when he has a sufficient number of well-founded facts to serve as the basis for the accurate, logical induction of generalizations and the subsequent deduction of consequences.

'Imagination in Science', *Tomorrow*, December 1943, 38–9. Quoted in Barbara Marinacci (ed.), *Linus Pauling in His Own Words: Selected Writings, Speeches, and Interviews* (1995), 82.

1 If the structure that serves as a template (the gene or virus molecule) consists of, say, two parts, which are themselves complementary in structure, then each of these parts can serve as the mould for the production of a replica of the other part, and the complex of two complementary parts thus can serve as the mould for the production of duplicates of itself.

Molecular Architecture and the Processes of Life (1948), 10.

2 We may say that life has borrowed from inanimate processes the same mechanism used in producing these striking structures that are crystals.

'The Nature of Forces between Large Molecules of Biological Interest', *Nature*, 1948, 161, 708.

3 *Instead of collecting stamps, he collected dictionaries and encyclopaedias:*

Because you can learn more from them.

'Dr Linus Pauling, Atomic Architect', *Science Illustrated*, 1948, 3, 40.

4 Men will gather knowledge no matter what the consequences. Science will go on whether we are pessimistic or optimistic, as I am. More interesting discoveries than we can imagine will be made, and I am awaiting them, full of curiosity and enthusiasm.

'Dr Linus Pauling, Atomic Architect', *Science Illustrated*, 1948, 3, 40.

5 It is structure that we look for whenever we try to understand anything. All science is built upon this search; we investigate how the cell is built of reticular material, cytoplasm, chromosomes; how crystals aggregate; how atoms are fastened together; how electrons constitute a chemical bond between atoms. We like to understand, and to explain, observed facts in terms of structure. A chemist who understands why a diamond has certain properties, or why nylon or hemoglobin have other properties, because of the different ways their atoms are arranged, may ask questions that a geologist would not think of formulating, unless he had been similarly trained in this way of thinking about the world.

'The Place of Chemistry in the Integration of the Sciences', *Main Currents in Modern Thought*, 1950, 7, 110.

6 I think that the formation of [DNA's] structure by Watson and Crick may turn out to be the greatest development in the field of molecular genetics in recent years.

'Discussion des rapports de M Pauling', *Rep. Institut International de Chemie Solvay: Conference on Proteins*, 6–14 April 1953 (1953), 113.

7 During the time that [Karl] Landsteiner gave me an education in the field of immunology, I discovered that he and I were thinking about the serologic problem in very different ways. He would ask, *What do these experiments force us to believe about the nature of the world?* I would ask, *What is the most simple and general picture of the world that we can formulate that is not ruled out by these experiments?* I realized that medical and biological investigators were not attacking their problems in the same way that theoretical physicists do, the way I had been in the habit of doing.

'Molecular Disease', *Pfizer Spectrum*, 1958, 6:9, 234.

1 It will be possible, through the detailed determination of amino-acid sequences of hemoglobin molecules and of other molecules too, to obtain much information about the course of the evolutionary process, and to illuminate the question of the origin of species.

'Molecular Disease and Evolution'. Typescript of the Rudolph Virchow Lecture, dated 5 November 1962. Quoted in T. Hager, *Force of Nature: The Life of Linus Pauling* (1997), 541.

2 [Professor Pauling] confesses that he had harboured the feeling that sooner or later he would be the one to get the DNA structure; and although he was pleased with the double-helix, he 'rather wished the idea had been his'.

'The Need to Understand', *New Scientist*, 1971, 50, 755.

3 It was obvious—to me at any rate—what the answer was to why an enzyme is able to speed up a chemical reaction by as much as 10 million times. It had to do this by lowering the energy of activation—the energy of forming the activated complex. It could do this by forming strong bonds with the activated complex, but only weak bonds with the reactants or products.

Quoted in Thomas Hager, *Force of Nature: The Life of Linus Pauling* (1995), 284.

4 I try to identify myself with the atoms . . . I ask what I would do if I were a carbon atom or a sodium atom.

Comment made to George Gray, the Rockefeller's resident science writer and publicist. Quoted in Thomas Hager, *Force of Nature: The Life of Linus Pauling* (1995), 377.

5 Life . . . is a relationship between molecules.

Quoted in T. Hager, *Force of Nature: The Life of Linus Pauling* (1997), 542.

6 Well David, I have a lot of ideas and throw away the bad ones.

Said by Pauling to David Harker (a student of Pauling's in the 1930s) when he asked 'Dr Pauling, how do you have so many good ideas?'

Quoted in Thomas Hager, *Force of Nature: The Life of Linus Pauling* (1995), 529.

7 I have been especially fortunate for about 50 years in having two memory banks available—whenever I can't remember something I ask my wife, and thus I am able to draw on this auxiliary memory bank. Moreover, there is a second way in which I get ideas . . . I listen carefully to what my wife says, and in this way I often get a good idea. I recommend to . . . young people . . . that you make a permanent acquisition of an auxiliary memory bank that you can become familiar with and draw upon throughout your lives.

T. Goertzel and B. Goertzel, *Linus Pauling* (1995), 240.

8 ON PAULING If the double helix was so important, how come you didn't work on it?

Mrs Helen Pauling to her husband when Crick, Watson and Wilkins received the Nobel Prize.

Pauling at a History of Science conference in 1990. Quotation supplied by W. H. Brock.

Linus Pauling 1901–94 and **Robert B. Corey** 1897–1971

American chemists

9 The nucleic acids, as constituents of living organisms, are comparable in importance to proteins. There is evidence that they are involved in the processes of cell division and growth, that they participate in the transmission of hereditary characters, and that they are important constituents of viruses. An understanding of the molecular structure of the nucleic acids should be of value in the effort to understand the fundamental phenomena of life.

'A Proposed Structure for the Nucleic Acids', *Proceedings of the National Academy of Sciences*, 1953, 39, 84.

Linus Pauling, **Robert B. Corey**, and **H. R. Branson**

1 An amino acid residue (other than glycine) has no symmetry elements. The general operation of conversion of one residue of a single chain into a second residue equivalent to the first is accordingly a rotation about an axis accompanied by translation along the axis. Hence the only configurations for a chain compatible with our postulate of equivalence of the residues are helical configurations.

'The Structure of Proteins: Two Hydrogen-bonded Helical Configurations of the Polypeptide Chain', *Proceedings of the National Academy of Sciences of the United States of America*, 37, 1951, 206.

Ivan Petrovich Pavlov 1849–1936

Russian physiologist and psychologist

2 The digestive canal is in its task a complete chemical factory. The raw material passes through a long series of institutions in which it is subjected to certain mechanical and, mainly, chemical processing, and then, through innumerable side-streets, it is brought into the depot of the body. Aside from this basic series of institutions, along which the raw material moves, there is a series of lateral chemical manufactories, which prepare certain reagents for the appropriate processing of the raw material.

Speech to the Society of Russian Physicians in December 1874. Trans. in Daniel P. Todes, *Pavlov's Physiology Factory: Experiment, Interpretation, Laboratory Enterprise* (2002), 155.

3 We must painfully acknowledge that, precisely because of its great intellectual developments, the best of man's domesticated animals—the dog—most often becomes the victim of physiological experiments. Only dire necessity can lead one to experiment on cats—on such impatient, loud, malicious animals. During chronic experiments, when the animal, having recovered from its operation, is under lengthy observation, the dog is irreplaceable; moreover, it is extremely touching. It is almost a participant in the experiments conducted upon it, greatly facilitating the success of the research by its understanding and compliance.

'Vivisection' (1893), trans. in Daniel P. Todes, *Pavlov's Physiology Factory: Experiment, Interpretation, Laboratory Enterprise* (2002), 123.

4 Essentially only one thing in life interests us: our psychical constitution, the mechanism of which was and is wrapped in darkness. All human resources, art, religion, literature, philosophy and historical sciences, all of them join in bringing lights in this darkness. But man has still another powerful resource: natural science with its strictly objective methods. This science, as we all know, is making huge progress every day. The facts and considerations which I have placed before you at the end of my lecture are one out of numerous attempts to employ a *consistent*, purely scientific method of thinking in the study of the mechanism of the highest manifestations of life in the dog, the representative of the animal kingdom that is man's best friend.

'Physiology of Digestion', Nobel Lecture, 12 December 1904. In *Nobel Lectures: Physiology or Medicine 1901–1921* (1967), 134.

5 One can truly say that the irresistible progress of natural science since the time of Galileo has made its first halt before the study of the higher parts of the brain, the organ of the most complicated relations of the animal to the external world. And it seems, and not without reason, that now is the really critical moment for natural science; for the brain, in its highest complexity—the human brain—which created and creates natural science, itself becomes the object of this science.

Natural Science and Brain (1909), 120.

6 In the dog two conditions were found to produce pathological disturbances by functional interference, namely, an unusually acute clashing of the excitatory and inhibitory processes, and the influence of strong and extraordinary stimuli. In man precisely similar conditions constitute the usual causes of nervous and psychic disturbances. Different conditions productive of extreme excitation, such as intense grief or bitter insults, often lead, when the natural reactions are inhibited by the necessary restraint, to profound and prolonged loss of balance in nervous and psychic activity.

Conditioned Reflexes—An Investigation of the Physiological Activity of the Cerebral Cortex, trans. and ed. G. V. Anrep (1927), 397.

1 I am convinced that an important stage of human thought will have been reached when the physiological and the psychological, the objective and the subjective, are actually united, when the tormenting conflicts or contradictions between my consciousness and my body will have been factually resolved or discarded.

Physiology of the Higher Nervous Activity (1932), 93–4.

2 The nervous system is the most complex and delicate instrument on our planet, by means of which relations, connections are established between the numerous parts of the organism, as well as between the organism, as a highly complex system, and the innumerable, external influences. If the closing and opening of electric current is now regarded as an ordinary technical device, why should there be any objection to the idea that the same principle acts in this wonderful instrument? On this basis the constant connection between the external agent and the response of the organism, which it evokes, can be rightly called an unconditioned reflex, and the temporary connection—a conditioned reflex.

The Conditioned Reflex (1935), 249.

3 ON **PAVLOV** Pavlov's data on the two fundamental antagonistic nervous processes—stimulation and inhibition—and his profound generalizations regarding them, in particular, that these processes are parts of a united whole, that they are in a state of constant conflict and constant transition of the one to the other, and his views on the dominant role they play in the formation of the higher nervous activity—all those belong to the most established natural-scientific validation of the Marxist dialectal method. They are in complete accord with the Leninist concepts on the role of the struggle between opposites in the evolution, the motion of matter.

In E. A. Asratyan, *I. P. Pavlov: His Life and Work* (1953), 153.

Karl Pearson 1857–1936

British statistician

4 'Endow scientific research and we shall know the truth, when and where it is possible to ascertain it;' but the counterblast is at hand: 'To endow research is merely to encourage the research for endowment; the true man of science will not be held back by poverty, and if science is of use to us, it will pay for itself.' Such are but a few samples of the conflict of opinion which we find raging around us.

The Grammar of Science (1892), 5.

5 The unity of all science consists alone in its method, not in its material.

The Grammar of Science (1892), 15.

6 When every fact, every present or past phenomenon of that universe, every phase of present or past life therein, has been examined, classified, and co-ordinated with the rest, then the mission of science will be completed. What is this but saying that the task of science can never end till man ceases to be, till history is no longer made, and development itself ceases?

The Grammar of Science (1892), 15.

7 If I have put the case of science at all correctly, the reader will have recognised that modern science does much more than demand that it shall be left in undisturbed possession of what the theologian and metaphysician please to term its 'legitimate field'. It claims that the whole range of phenomena, mental as well as physical—the entire universe—is its field. It asserts that the scientific method is the sole gateway to the whole region of knowledge.

The Grammar of Science (1892), 29–30.

1 All great scientists have, in a certain sense, been great artists; the man with no imagination may collect facts, but he cannot make great discoveries.

The Grammar of Science (1892), 37.

2 The starting point of Darwin's theory of evolution is precisely the existence of those differences between individual members of a race or species which morphologists for the most part rightly neglect. The first condition necessary, in order that any process of Natural Selection may begin among a race, or species, is the existence of differences among its members; and the first step in an enquiry into the possible effect of a selective process upon any character of a race must be an estimate of the frequency with which individuals, exhibiting any given degree of abnormality with respect to that character, occur. The unit, with which such an enquiry must deal, is not an individual but a race, or a statistically representative sample of a race; and the result must take the form of a numerical statement, showing the relative frequency with which the various kinds of individuals composing the race occur.

Biometrika: A Journal for the Statistical Study of Biological Problems, 1901, 1, 1–2.

3 I look upon statistics as the handmaid of medicine, but on that very account I hold that it befits medicine to treat her handmaid with proper respect, and not to prostitute her services for controversial or personal purposes.

'On the Influence of the Sanatorium Treatment of Tuberculosis', *British Medical Journal*, 1910, i, 1517.

4 There is nothing opposed in Biometry and Mendelism. Your husband [W.F.R. Weldon] and I worked *that* out at Peppards [on the Chilterns] and you will see it referred in the *Biometrika* memoir. The Mendelian formula leads up to the 'ancestral law'. What we fought against was the slovenliness in applying Mendel's categories and asserting that such formulae apply in cases when they did not.

Letter to Mrs.Weldon, 12 April 1907. Quoted in M. E. Magnello, 'Karl Pearson's Mathematization of Inheritance: From Ancestral Heredity to Mendelian Genetics (1895–1909)', *Annals of Science*, 1998, 55, 89.

5 Medals are great encouragement to young men and lead them to feel their work is of value, I remember how keenly I felt this when in the 1890s I received the Darwin Medal and the Huxley Medal. When one is old, one wants no encouragement and one goes on with one's work to the extent of one's power, because it has become habitual.

Letter to Major Greenwood, 8 December 1933. Quoted in M. E. Magnello, 'Karl Pearson', in P. Armitage and T. Colton (eds.), *The Encyclopedia of Biostatistics* (1998), Vol. 4, 3314.

6 It is the old experience that a rude instrument in the hand of a master craftsman will achieve more than the finest tool wielded by the uninspired journeyman.

Quoted in *The Life, Letters and Labours of Francis Galton* (1930), Vol. 3A, 50.

Charles Sanders Peirce

1839–1914

American mathematician, logician and philospher

7 Every work of science great enough to be well remembered for a few generations affords some exemplification of the defective state of the art of reasoning of the time when it was written; and each chief step in science has been a lesson in logic.

'The Fixation of Belief' (1877). In Justus Buchler, *The Philosophy of Pierce* (1940), 6.

8 It is a common observation that a science first begins to be exact when it is quantitatively treated. What are called the exact sciences are no others than the mathematical ones.

On The Doctrine of Chances, with Later Reflections (1878), 61.

1 The rudest numerical scales, such as that by which the mineralogists distinguish different degrees of hardness, are found useful. The mere counting of pistils and stamens sufficed to bring botany out of total chaos into some kind of form. It is not, however, so much from counting as from measuring, not so much from the conception of number as from that of continuous quantity, that the advantage of mathematical treatment comes. Number, after all, only serves to pin us down to a precision in our thoughts which, however beneficial, can seldom lead to lofty conceptions, and frequently descend to pettiness.

On the Doctrine of Chances, with Later Reflections (1878), 61–2.

2 Science has hitherto been proceeding without the guidance of any rational theory of logic, and has certainly made good progress. It is like a computer who is pursuing some method of arithmetical approximation. Even if he occasionally makes mistakes in his ciphering, yet if the process is a good one they will rectify themselves. But then he would approximate much more rapidly if he did not commit these errors; and in my opinion, the time has come when science ought to be provided with a logic. My theory satisfies me; I can see no flaw in it. According to that theory universality, necessity, exactitude, in the *absolute* sense of these words, are unattainable by us, and do not exist in nature. There is an ideal law to which nature approximates; but to express it would require an endless series of modifications, like the decimals expressing surd. Only when you have asked a question in so crude a shape that continuity is not involved, is a perfectly true answer attainable.

Letter to G. F. Becker, 11 June 1893. Merrill Collection, Library of Congress. Quoted in Nathan Reingold, *Science in Nineteenth-Century America: A Documentary History* (1966), 231–2.

Edmund D. Pellegrino 1920–

American professor of medicine and medical ethics

3 Measurement has too often been the leitmotif of many investigations rather than the experimental examination of hypotheses. Mounds of data are collected, which are statistically decorous and methodologically unimpeachable, but conclusions are often trivial and rarely useful in decision making. This results from an overly rigorous control of an insignificant variable and a widespread deficiency in the framing of pertinent questions. Investigators seem to have settled for what is measurable instead of measuring what they would really like to know.

'Patient Care—Mystical Research or Researchable Mystique?', *Clinical Research*, 1964, 12, no. 4, 422.

Marzari Giuseppe Pencati

1779–1836

Italian geologist

4 One never finds fossil bones bearing no resemblance to human bones. Egyptian mummies, which are at least three thousand years old, show that men were the same then. The same applies to other mummified animals such as cats, dogs, crocodiles, falcons, vultures, oxen, ibises, etc. Species, therefore, do not change by degrees, but emerged after the new world was formed. Nor do we find intermediate species between those of the earlier world and those of today's. For example, there is no intermediate bear between our bear and the very different cave bear. To our knowledge, no spontaneous generation occurs in the present-day world. All organized beings owe their life to their fathers. Thus all records corroborate the globe's modernity. Negative proof: the barbarity of the human species four thousand years ago. Positive proof: the great revolutions and the floods preserved in the traditions of all peoples.

'Note prese al Corso di Cuvier. Corso di Geologia all'Ateneo nel 1805', quoted in Pietro Corsi, *The Age of Lamarck*, trans. J. Mandelbaum (1988), 183.

Granville Penn 1761–1844

British writer

1 Philosophers, if they have much imagination, are apt to let it loose as well as other people, and in such cases are sometimes led to mistake a fancy for a fact. Geologists, in particular, have very frequently amused themselves in this way, and it is not a little amusing to follow them in their fancies and their waking dreams. Geology, indeed, in this view, may be called a romantic science.

Conversations on Geology (1840), 5.

William Henry Perkin, Jr. 1860–1929

British chemist

2 Physical chemistry is all very well, but it does not apply to organic substances.

Quoted in L. E. Sutton's obituary of Nevil V. Sidgwick, *Proceedings of the Chemical Society*, 1958, 312.

Martin L. Perl 1927–

American physicist

3 My final remark to young women and men going into experimental science is that they should pay little attention to the speculative physics ideas of my generation. After all, if my generation has any really good speculative ideas, we will be carrying these ideas out ourselves.

'Reflections on the Discovery of the Tau Lepton', Nobel Lecture, 8 December 1995. In *Nobel Lectures: Physics 1991–1995* (1997), 193.

Claude Perrault 1613–88

French zoologist, physiologist and engineer

4 I would clarify that by 'animal' I understand a being that has feeling and that is capable of exercising life functions through a principle called soul; that the soul uses the body's organs, which are true machines, by virtue of its being the principal cause of the action of each of the machine's parts; and that although the placement that these parts have with respect to one another does scarcely anything else through the soul's mediation than what it does in pure machines, the entire machine nonetheless needs to be activated and guided by the soul in the same way as an organ, which, although capable of rendering different sounds through the placement of the parts of which it is composed, nonetheless never does so except through the guidance of the organist.

'La Méchanique des Animaux', in *Oeuvres Diverses de Physique et de Méchanique* (1721), Vol. 1, 329. Quoted in Jacques Roger, *The Life Sciences in Eighteenth-Century French Thought*, ed. Keith R. Benson and trans. Robert Ellrich (1997), 273–4.

Max Ferdinand Perutz 1914–2002

Austrian-born British molecular biologist

5 Women's liberation could have not succeeded if science had not provided them with contraception and household technology.

'The Impact of Science on Society: The Challenge for Education', in J. L. Lewis and P. J. Kelly (eds.), *Science and Technology and Future Human Needs* (1987), 18.

6 A discovery is like falling in love and reaching the top of a mountain after a hard climb all in one, an ecstasy not induced by drugs but by the revelation of a face of nature that no one has seen before and that often turns out to be more subtle and wonderful than anyone had imagined.

'True Science', review of Peter Medawar,

Advice to a Young Scientist (1980). In *The London Review of Books*, March 1981, 6.

1 For Christmas, 1939, a girl friend gave me a book token which I used to buy Linus Pauling's recently published *Nature of the Chemical Bond*. His book transformed the chemical flatland of my earlier textbooks into a world of three-dimensional structures.

'What Holds Molecules Together', in *I Wish I'd Made You Angry Earlier* (1998), 165.

2 The discovery of an interaction among the four hemes made it obvious that they must be touching, but in science what is obvious is not necessarily true. When the structure of hemoglobin was finally solved, the hemes were found to lie in isolated pockets on the surface of the subunits. Without contact between them how could one of them sense whether the others had combined with oxygen? And how could as heterogeneous a collection of chemical agents as protons, chloride ions, carbon dioxide, and diphosphoglycerate influence the oxygen equilibrium curve in a similar way? It did not seem plausible that any of them could bind directly to the hemes or that all of them could bind at any other common site, although there again it turned out we were wrong. To add to the mystery, none of these agents affected the oxygen equilibrium of myoglobin or of isolated subunits of hemoglobin. We now know that all the cooperative effects disappear if the hemoglobin molecule is merely split in half, but this vital clue was missed. Like Agatha Christie, Nature kept it to the last to make the story more exciting. There are two ways out of an impasse in science: to experiment or to think. By temperament, perhaps, I experimented, whereas Jacques Monod thought.

'The Second Secret of Life', in *I Wish I'd Made You Angry Earlier* (1998), 263–5.

3 What is known for certain is dull.

'My Commonplace Book', in *I Wish I'd Made You Angry Earlier* (1998), 314.

4 I rarely plan my research; it plans me.

'My Commonplace Book', in *I Wish I'd Made You Angry Earlier* (1998), 314.

Christoph Heinrich Pfaff

1773–1852

German natural philosopher

5 *I*. Animals have an electricity peculiar to themselves to which the name *animal electricity* is given. *II*. The organs in which animal electricity acts above all others, and by which it is distributed throughout the whole body, are the nerves, and the most important organ of secretion is the brain.

Über thierische Elektricität und Reizbarkeit. Ein Beytrag zu den neuesten Entdeckungen über diese Gegenstände (1795), 329. Quoted and trans. in Edwin Clarke and C. D. O'Malley, *The Human Brain and Spinal Cord* (1968), 180.

John Phillips 1800–74

British geologist

6 Life through many long periods has been manifested in a countless host of varying structures, all circumscribed by one general plan, each appointed to a definite place, and limited to an appointed duration. On the whole the earth has been thus more and more covered by the associated life of plants and animals, filling all habitable space with beings capable of enjoying their own existence or ministering to the enjoyment of others; till finally, after long preparation, a being was created capable of the wonderful power of measuring and weighing all the world of matter and space which surrounds him, of treasuring up the past history of all the forms of life, and considering his own relation to the whole. When he surveys this vast and co-ordinated system, and inquires into its history and origin, can he be at a loss to decide whether it be a work of Divine thought and wisdom, or the fortunate offspring of a few atoms of matter, warmed by the *anima mundi*, a spark of electricity, or an accidental ray of sunshine?

Life on the Earth: Its Origin and Succession (1860), 216–7.

Jean Piaget 1896–1980

Swiss psychologist

1 Chance . . . in the accommodation peculiar to sensorimotor intelligence, plays the same role as in scientific discovery. It is only useful to the genius and its revelations remain meaningless to the unskilled.

The Origin of Intelligence in the Child (1936), trans. Margaret Cook (1953), 303.

2 The fundamental hypothesis of genetic epistemology is that there is a parallelism between the progress made in the logical and rational organization of knowledge and the corresponding formative psychological processes. With that hypothesis, the most fruitful, most obvious field of study would be the reconstituting of human history—the history of human thinking in prehistoric man. Unfortunately, we are not very well informed in the psychology of primitive man, but there are children all around us, and it is in studying children that we have the best chance of studying the development of logical knowledge, physical knowledge, and so forth.

'Genetic Epistemology', *Columbia Forum*, 1969, 12, 4.

Jean Picard 1620–82

French astronomer

3 We must look to the heavens . . . for the measure of the earth.

Mesure de la terre (1671), 165. Trans. J. L. Heilbron, *Weighing Imponderables and Other Quantitative Science around 1800* (1993), 37.

Marc-Auguste Pictet 1752–1825

Swiss physicist

4 So-called extraordinary events always split into two extremes naturalists who have not witnessed them: those who believe blindly and those who do not believe at all. The latter have always in mind the story of the golden goose; if the facts lie slightly beyond the limits of their knowledge, they relegate them immediately to fables. The former have a secret taste for marvels because they seem to expand Nature; they use their imagination with pleasure to find explanations. To remain doubtful is given to naturalists who keep a middle path between the two extremes. They calmly examine facts; they refer to logic for help; they discuss probabilities; they do not scoff at anything, not even errors, because they serve at least the history of the human mind; finally, they report rather than judge; they rarely decide unless they have good evidence.

Quoted in Albert V. Carozzi, *Histoire des sciences de la terre entre 1790 et 1815 vue à travers les documents inédités de la Société de Physique et d'Histoire Naturelle de Genève* (1990), 175. Trans. Albert V. and Marguerite Carozzi.

Clements von Pirquet 1874–1929

Austrian immunologist

5 The conception that antibodies, which should protect against disease, are also responsible for the disease, sounds at first absurd. This has as its basis the fact that we are accustomed to see in disease only the harm done to the organism and to see in the antibodies solely antitoxic [protective] substances. One forgets too easily that the disease represents only a stage in the development of immunity, and that the organism often attains the advantage of immunity only by means of disease . . . Serum sickness represents, so to speak, an unnatural (artificial) form of disease.

C. von Pirquet and B. Schick, *Die Serumkrankheit* (1906), trans B. Schick, *Serum Sickness* (1951), 119–20.

Bartholomeo Pitiscus 1561–1613

German mathematician

6 And how admirable and rare an ornament, O good God, is mildenesse in a divine? And how much is it to be wished in this age, that all divines were mathematicians? that is men gentle and meeke.

Trigonometria (1595), trans. R. Handson (1614), Epistle Dedicatorie.

George Placzek 1905–55

Czech-born American physicist

1 I am an experimenter, or rather I used to be one. Then I stopped working, and since then people think I am a theoretician.

Quoted in Otto Frisch, *What Little I Remember* (1979), 105.

Max Karl Ernst Ludwig Planck 1858–1947

German physicist

See **Keynes, John Maynard** 342:4

2 Physical changes take place continuously, while chemical changes take place discontinuously. Physics deals chiefly with continuous varying quantities, while chemistry deals chiefly with whole numbers.

Treatise on Thermodynamics (1897), trans. Alexander Ogg (1903), 22, footnote.

3 If E is considered to be a continuously divisible quantity, this distribution is possible in infinitely many ways. We consider, however—this is the most essential point of the whole calculation—E to be composed of a well-defined number of equal parts and use thereto the constant of nature $h = 6.55 \times 10^{-27}$ erg sec. This constant multiplied by the common frequency ν of the resonators gives us the energy element ε in erg, and dividing E by ε we get the number P of energy elements which must be divided over the *N* resonators.

The original formulation of Planck's constant, published in 1900, often written $\varepsilon = h\nu$.

'On the theory of the energy distribution law of the normal spectrum', in D. ter Haar and Stephen G. Brush, trans., *Planck's Original Papers in Quantum Physics* (1972), 40.

4 The goal is nothing other than the coherence and completeness of the system not only in respect of all details, but also in respect of all physicists of all places, all times, all peoples, and all cultures.

Acht Vorlesungen (1910), 'Vorwort': 4. Translated in J. L. Heilbron, *The Dilemmas of an Upright Man* (1986), 51.

5 The assumption of an absolute determinism is the essential foundation of every scientific enquiry.

Physikalische Abhandlungen und Vorträge (1958), Vol. 3, 89. Translated in J. L. Heilbron, *The Dilemmas of an Upright Man* (1986) 66.

6 How do we discover the individual laws of Physics, and what is their nature? It should be remarked, to begin with, that we have no right to assume that any physical law exists, or if they have existed up to now, that they will continue to exist in a similar manner in the future. It is perfectly conceivable that one fine day Nature should cause an unexpected event to occur which would baffle us all; and if this were to happen we would be powerless to make any objection, even if the result would be that, in spite of our endeavors, we should fail to introduce order into the resulting confusion. In such an event, the only course open to science would be to declare itself bankrupt. For this reason, science is compelled to begin by the general assumption that a general rule of law dominates throughout Nature.

The Universe in the Light of Modern Physics, trans. W. H. Johnston (1931), 58–9.

7 Scientific discovery and scientific knowledge have been achieved only by those who have gone in pursuit of it without any practical purpose whatsoever in view.

Where is Science Going?, trans. James Murphy (1933), 137.

8 The quantum hypothesis will eventually find its exact expression in certain equations which will be a more exact formulation of the law of causality.

Where is Science Going?, trans. James Murphy (1933), 143.

9 Science cannot solve the ultimate mystery of nature. And that is because, in the last analysis, we ourselves are part of nature and therefore part of the mystery that we are trying to solve.

Where is Science Going?, trans. James Murphy (1933), Epilogue, 217.

1 My original decision to devote myself to science was a direct result of the discovery which has never ceased to fill me with enthusiasm since my early youth—the comprehension of the far from obvious fact that the laws of human reasoning coincide with the laws governing the sequences of the impressions we receive from the world about us; that, therefore, pure reasoning can enable man to gain an insight into the mechanism of the latter. In this connection, it is of paramount importance that the outside world is something independent from man, something absolute, and the quest for the laws which apply to this absolute appeared to me as the most sublime scientific pursuit in life.

'A Scientific Autobiography' (1948), in *Scientific Autobiography and Other Papers*, trans. Frank Gaynor (1950), 13.

2 A new scientific truth does not triumph by convincing its opponents and making them see the light, but rather because its opponents eventually die, and a new generation grows up that is familiar with it.

'A Scientific Autobiography' (1948), in *Scientific Autobiography and Other Papers*, trans. Frank Gaynor (1950), 33–4.

3 I had always looked upon the search for the absolute as the noblest and most worth while task of science.

'A Scientific Autobiography' (1948), in *Scientific Autobiography and Other Papers*, trans. Frank Gaynor (1950), 46.

4 The Theory of Relativity confers an absolute meaning on a magnitude which in classical theory has only a relative significance: the velocity of light. The velocity of light is to the Theory of Relativity as the elementary quantum of action is to the Quantum Theory: it is its absolute core.

'A Scientific Autobiography' (1948), in *Scientific Autobiography and Other Papers*, trans. Frank Gaynor (1950), 47.

5 An experiment is a question which science poses to Nature, and a measurement is the recording of Nature's answer.

'The Meaning and Limits of Exact Science' (1947), in *Scientific Autobiography and Other Papers*, trans. Frank Gaynor (1950), 110.

6 ON PLANCK There is a story that once, not long after he came to Berlin, Planck forgot which room had been assigned to him for a lecture and stopped at the entrance office of the university to find out. Please tell me, he asked the elderly man in charge, 'In which room does Professor Planck lecture today?' The old man patted him on the shoulder 'Don't go there, young fellow,' he said 'You are much too young to understand the lectures of our learned Professor Planck'.

Apocryphal.

Plato 427–347/8 BC

Greek philosopher

See ***Whitehead, Alfred North*** *621:7*

7 If someone separated the art of counting and measuring and weighing from all the other arts, what was left of each (of the others) would be, so to speak, insignificant.

Philebus 55e. Trans. R. W. Sharples.

8 They assembled together and dedicated these as the first-fruits of their love to Apollo in his Delphic temple, inscribing there those maxims which are on every tongue—'know thyself' and 'Nothing overmuch.'

Protagoras 343ab, trans. W. R. M. Lamb, in *Plato: Laches Protagoras Meno Euthydemus* (1924), 197.

9 'Unless,' said I [Socrates], either philosophers become kings in our states or those whom we now call our kings and rulers take to the pursuit of philosophy seriously and adequately, and there is a conjunction of these two things, political power and philosophic intelligence, while the motley horde of the natures who at present pursue either apart from the other are compulsorily excluded, there can be no cessation of troubles, dear Glaucon, for our states, nor, I fancy for the human race either. Nor, until this happens, will this constitution which we have been expounding in theory ever be put into practice within the limits of possibility and see the light of the sun.

The Republic 5 474ce, trans. P. Shorey (1930), Vol. 1, Book 5, 509.

1 'Yes,' he said. 'But these things (the solutions to problems in solid geometry such as the duplication of the cube) do not seem to have been discovered yet.' 'There are two reasons for this,' I said. 'Because no city holds these things in honour, they are investigated in a feeble way, since they are difficult; and the investigators need an overseer, since they will not find the solutions without one. First, it is hard to get such an overseer, and second, even if one did, as things are now those who investigate these things would not obey him, because of their arrogance. If however a whole city, which did hold these things in honour, were to oversee them communally, the investigators would be obedient, and when these problems were investigated continually and with eagerness, their solutions would become apparent.'

The Republic 7 528bc, trans. R. W. Sharples.

2 For it is obvious to everybody, I think, that this study [of astronomy] compels the soul to look upward and leads it away from things here to higher things.

The Republic 7 529a, trans. P. Shorey (1935), Vol. 2, Book 7, 179–81.

3 The qualities of number appear to lead to the apprehension of truth.

The Republic 7 525b, trans. P. Shorey (1935), Vol. 2, Book 7, 161.

4 Wherefore also these Kinds [elements] occupied different places even before the universe was organised and generated out of them. Before that time, in truth, all these were in a state devoid of reason or measure, but when the work of setting in order this Universe was being undertaken, fire and water and earth and air, although possessing some traces of their known nature, were yet disposed as everything is likely to be in the absence of God; and inasmuch as this was then their natural condition, God began by first marking them out into shapes by means of forms and numbers.

Timaeus 53ab, trans. R. G. Bury, in *Plato: Timaeus, Critias, Cleitophon, Menexenus, Epistles* (1929), 125–7.

5 Let no-one ignorant of geometry enter.

Said to have been inscribed above the door of Plato's Academy.

A. S. Riginos, *Platonica: the Anecdotes concerning the Life and Writings of Plato* (1976), 38–40.

6 ON PLATO It was Plato, according to Sosigenes, who set this as a problem for those concerned with these things, through what suppositions of uniform and ordered movements the appearances concerning the movements of the wandering heavenly bodies could be preserved.

Simplicius, *On Aristotle's On the Heavens*, 488.21. Trans. R. W. Sharples.

Robert Platt (Lord Platt of Grindleford) 1900–78

British physician

7 The unprecedented development of science and technology . . . so rapid that it is said that 90 per cent of the scientists which this country has ever produced are still living today.

Reflections on Medicine and Humanism: Linacre Lecture (1963), 328.

John Playfair 1748–1819

British mathematician and geologist

8 To trace the series of these revolutions, to explain their causes, and thus to connect together all the indications of change that are found in the mineral kingdom, is the proper object of a THEORY OF THE EARTH.

Illustrations of the Huttonian Theory of the Earth (1802), 2.

9 Thus we conclude, that the strata both primary and secondary, both those of ancient and those of more recent origin, have had their materials furnished from the ruins of former continents, from the dissolution of rocks, or the destruction of animal or vegetable bodies, similar, at least in some respects, to those that now occupy the surface of the earth.

Illustrations of the Huttonian Theory of the Earth (1802), 14–5.

1 Every river appears to consist of a main trunk, fed from a variety of branches, each running in a valley proportional to its size, and all of them together forming a system of vallies, communicating with one another, and having such a nice adjustment of their declivities that none of them join the principal valley on too high or too low a level; a circumstance which would be infinitely improbable if each of these vallies were not the work of the stream that flows in it.

Illustrations of the Huttonian Theory of the Earth (1802), 102.

2 The powers which tend to preserve, and those which tend to change the condition of the earth's surface, are never *in equilibrio*; the latter are, in all cases, the most powerful, and, in respect of the former, are like *living* in comparison of *dead* forces. Hence the law of decay is one which suffers no exception: The elements of all bodies were once loose and unconnected, and to the same state nature has appointed that they should all return . . . TIME performs the office of *integrating* the infinitesimal parts of which this progression is made up; it collects them into one sum, and produces from them an amount greater than any that can be assigned.

Illustrations of the Huttonian Theory of the Earth (1802), 116–7.

3 Amid all the revolutions of the globe, the economy of nature has been uniform, in this respect, as well as in so many others, and her laws are the only thing that have resisted the general movement. The rivers and the rocks, the seas and the continents, have been changed in all their parts; but the laws which direct those changes, and the rules to which they are subject, have remained invariably the same.

Illustrations of the Huttonian Theory of the Earth (1802), 421–2.

4 The Author of nature has not given laws to the universe, which, like the institutions of men, carry in themselves the elements of their own destruction; he has not permitted in his works any symptom of infancy or of old age, or any sign by which we may estimate either their future or their past duration. He may put an end, as he no doubt gave a beginning, to the present system at some determinate period of time; but we may rest assured, that this great catastrophe will not be brought about by the laws now existing, and that it is not indicated by any thing which we perceive.

'Biographical Account of the Late Dr James Hutton, F.R.S. Edin.' (read 1803), *Transactions of the Royal Society of Edinburgh*, 1805, **5**, 55.

5 The ridge of the Lammer-muir hills . . . consists of primary micaceous schistus, and extends from St Abb's head westward . . . The sea-coast affords a transverse section of this alpine tract at its eastern extremity, and exhibits the change from the primary to the secondary strata . . . Dr HUTTON wished particularly to examine the latter of these, and on this occasion Sir JAMES HALL and I had the pleasure to accompany him. We sailed in a boat from Dunglass . . . We made for a high rocky point or head-land, the SICCAR . . . On landing at this point, we found that we actually trode [sic] on the primeval rock . . . It is here a micaceous shistus, in beds nearly vertical, highly indurated, and stretching from S.E. to N.W. The surface of this rock . . . has a thin covering of red horizontal sandstone laid over it, . . . Here, therefore, the immediate contact of the two rocks is not only visible, but is curiously dissected and laid open by the action of the waves . . . On us who saw these phenomena for the first time, the impression will not easily be forgotten. The palpable evidence presented to us, of one of the most extraordinary and important facts in the natural history of the earth, gave a reality and substance to those theoretical speculations, which, however probable had never till now been directly authenticated by the testimony of the senses . . . What clearer evidence could we have had of the different formation of these rocks, and of the long interval which separated their formation, had we actually seen them emerging from the bosom of the deep? . . . The mind seemed to grow giddy by looking so far into the abyss of time; and while we listened with earnestness and admiration to the philosopher who was now unfolding to us the order and series of these wonderful events, we became sensible how much farther reason may sometimes go than imagination can venture to follow.

'Biographical Account of the Late Dr James Hutton, F.R.S. Edin.' (read 1803), *Transactions of the Royal Society of Edinburgh*, 1805, 5, 71–3.

Lyon Playfair (Baron Playfair of St Andrew) 1818–98

British chemist

1 I have now said enough to show you that it is indispensable for this country to have a scientific education in connexion with manufacturers, if we wish to outstrip the intellectual competition which now, happily for the world, prevails in all departments of industry. As surely as darkness follows the setting of the sun, so surely will England recede as a manufacturing nation, unless her industrial population become much more conversant with science than they are now.

'The Study of Abstract Science Essential to the Progress of Industry', *Records of the School of Mines*, 1852, 1, 48.

William Playfair 1759–1823

British publicist

2 No study is less alluring or more dry and tedious than statistics, unless the mind and imagination are set to work, or that the person studying is particularly interested in the subject; which last can seldom be the case with young men in any rank of life.

The Statistical Brewery (1801), 16.

Pliny the Elder (Gaius Plinius Secundus) c.23–79

Roman natural philosopher

3 It [the earth] alone remains immoveable, whilst all things revolve round it.

Natural History, 2, 11. Trans. H. Rackham, *Pliny: Natural History*, corrected edition (1949), Vol. 1, 177.

4 The largest land animal is the elephant, and it is the nearest to man in intelligence: it understands the language of its country and obeys orders, remembers duties that it has been taught, is pleased by affection and by marks of honour, nay more it possesses virtues rare even in man, honesty, wisdom, justice, also respect for the stars and reverence for the sun and moon.

Natural History, 8, 1. Trans. H. Rackham, *Pliny: Natural History* (1947), Vol. 3, 3.

5 The most disgraceful cause of the scarcity [of remedies] is that even those who know them do not want to point them out, as if they were going to lose what they pass on to others.

Natural History, 25, 16. Trans. R. W. Sharples.

Noël-Antoine Pluche 1688–1761

French populariser of science

6 It is not always the most brilliant speculations nor the choice of the most exotic materials that is most profitable. I prefer Monsieur de Réaumur busy exterminating moths by means of an oily fleece; or increasing fowl production by making them hatch without the help of their mothers, than Monsieur Bernouilli absorbed in algebra, or Monsieur Leibniz calculating the various advantages and disadvantages of the possible worlds.

Spectacle, 1, 475. Quoted in Camille Limoges, 'Noël-Antoine Pluche', in C. C. Gillispie (ed.), *Dictionary of Scientific Biography* (1974), Vol. 11, 43.

John Harold Plumb 1911–2001

British historian

1 Whether we like it or not, quantification in history is here to stay for reasons which the quantifiers themselves might not actively approve. We are becoming a numerate society: almost instinctively there seems now to be a greater degree of truth in evidence expressed numerically than in any literary evidence, no matter how shaky the statistical evidence, or acute the observing eye.

Is History Sick? (1973), 64.

Edgar Allan Poe 1809–49

American writer, poet and critic

2 Science! true daughter of Old Time thou art!
Who alterest all things with thy peering eyes.
Why preyest thou thus upon the poet's heart,
Vulture, whose wings are dull realities?
How should he love thee? or how deem thee wise,
Who wouldst not leave him in his wandering
To seek for treasure in the jewelled skies,
Albeit he soared with an undaunted wing?
Hast thou not dragged Diana from her car?
And driven the Hamadryad from the wood
To seek a shelter in some happier star?
Hast thou not torn the Naiad from her flood,
The Elfin from the green grass, and from me
The summer dread beneath the tamarind tree?

Sonnet—To Science (1829). In *The Complete Poems and Stories of Edgar Allan Poe* (1946), Vol. 1, 28.

3 Were the succession of stars endless, then the background of the sky would present us an uniform luminosity, like that displayed by the Galaxy—*since there could be absolutely no point, in all that background, at which would not exist a star*. The only mode, therefore, in which, under such a state of affairs, we could comprehend the *voids* which our telescopes find in innumerable directions, would be by supposing the distance of the invisible background so immense that no ray from it has yet been able to reach us at all.

'Eureka: An Essay on the Material and Spiritual Universe' (1848), in W. H. Auden (ed.), *Edgar Allan Poe: Selected Prose, Poetry, and Eureka* (1950), 557.

Jules Henri Poincaré 1854–1912

French philosopher and mathematician

See **Lippmann, Gabriel** 391:4

4 Experiment is the sole source of truth. It alone can teach us something new; it alone can give us certainty.

Science and Hypothesis (1902), trans. W. J. G. and preface by J. Larmor (1905), 140.

5 The man of science must work with method. Science is built up of facts, as a house is built of stones; but an accumulation of facts is no more a science than a heap of stones is a house.

Science and Hypothesis (1902), trans. W. J. G. and preface by J. Larmor (1905), 141.

6 It is often said that experiments should be made without preconceived ideas. That is impossible. Not only would it make every experiment fruitless, but even if we wished to do so, it could not be done. Every man has his own conception of the world, and this he cannot so easily lay aside. We must, for example, use language, and our language is necessarily steeped in preconceived ideas. Only they are unconscious preconceived ideas, which are a thousand times the most dangerous of all.

Science and Hypothesis (1902), trans. W. J. G. and preface by J. Larmor (1905), 143.

1 It is a misfortune for a science to be born too late when the means of observation have become too perfect. That is what is happening at this moment with respect to physical chemistry; the founders are hampered in their general grasp by third and fourth decimal places.

Science and Hypothesis (1902), trans. W. J. G. and preface by J. Larmor (1905), 181.

2 Geometry is not true, it is advantageous.

Science and Hypothesis (1902), in *The Foundations of Science: Science and Hypothesis, The Value of Science, Science and Method* (1946), trans. by George Bruce Halsted, 91.

3 The advance of science is not comparable to the changes of a city, where old edifices are pitilessly torn down to give place to new, but to the continuous evolution of zoologic types which develop ceaselessly and end by becoming unrecognisable to the common sight, but where an expert eye finds always traces of the prior work of the centuries past. One must not think then that the old-fashioned theories have been sterile and vain.

The Value of Science (1905), in *The Foundations of Science: Science and Hypothesis, The Value of Science, Science and Method* (1946), trans. by George Bruce Halsted, 208.

4 *All the scientist creates in a fact is the language in which he enunciates it*. If he predicts a fact, he will employ this language, and for all those who can speak and understand it, his prediction is free from ambiguity. Moreover, this prediction once made, it evidently does not depend upon him whether it is fulfilled or not.

The Value of Science (1905), in *The Foundations of Science: Science and Hypothesis, The Value of Science, Science and Method* (1946), trans. by George Bruce Halsted, 332.

5 I then began to study arithmetical questions without any great apparent result, and without suspecting that they could have the least connexion with my previous researches. Disgusted at my want of success, I went away to spend a few days at the seaside, and thought of entirely different things. One day, as I was walking on the cliff, the idea came to me, again with the same characteristics of conciseness, suddenness, and immediate certainty, that arithmetical transformations of indefinite ternary quadratic forms are identical with those of non-Euclidian geometry.

Science and Method (1908), trans. Francis Maitland (1914), 53–4.

6 Every phenomenon, however trifling it be, has a cause, and a mind infinitely powerful, and infinitely well-informed concerning the laws of nature could have foreseen it from the beginning of the ages. If a being with such a mind existed, we could play no game of chance with him; we should always lose.

Science and Method (1908), trans. Francis Maitland (1914), 65.

7 Why is it that showers and even storms seem to come by chance, so that many people think it quite natural to pray for rain or fine weather, though they would consider it ridiculous to ask for an eclipse by prayer.

Science and Method (1908), trans. Francis Maitland (1914), 68.

8 Consider now the Milky Way. Here also we see an innumerable dust, only the grains of this dust are no longer atoms but stars; these grains also move with great velocities, they act at a distance one upon another, but this action is so slight at great distances that their trajectories are rectilineal; nevertheless, from time to time, two of them may come near enough together to be deviated from their course, like a comet that passed too close to Jupiter. In a word, in the eyes of a giant, to whom our Suns were what our atoms are to us, the Milky Way would only look like a bubble of gas.

Science and Method (1908), trans. Francis Maitland (1914), 254–5.

1 It may be appropriate to quote a statement of Poincaré, who said (partly in jest no doubt) that there must be something mysterious about the normal law since mathematicians think it is a law of nature whereas physicists are convinced that it is a mathematical theorem.

Quoted in Mark Kac, *Statistical Independence in Probability, Analysis and Number Theory* (1959), 52.

2 *Les faits ne parlent pas.*

Facts do not speak.

Attributed.

Michael Polanyi 1891–1976

Hungarian-born British philosopher

3 Nobody knows more than a tiny fragment of science well enough to judge its validity and value at first hand. For the rest he has to rely on views accepted at second hand on the authority of a community of people accredited as scientists. But this accrediting depends in its turn on a complex organization. For each member of the community can judge at first hand only a small number of his fellow members, and yet eventually each is accredited by all. What happens is that each recognizes as scientists a number of others by whom he is recognized as such in return, and these relations form chains which transmit these mutual recognitions at second hand through the whole community. This is how each member becomes directly or indirectly accredited by all. The system extends into the past. Its members recognize the same set of persons as their masters and derive from this allegiance a common tradition, of which each carries on a particular strand.

Personal Knowledge (1958), 163.

4 Almost every major systematic error which has deluded men for thousands of years relied on practical experience. Horoscopes, incantations, oracles, magic, witchcraft, the cures of witch doctors and of medical practitioners before the advent of modern medicine, were all firmly established through the centuries in the eyes of the public by their supposed practical successes. The scientific method was devised precisely for the purpose of elucidating the nature of things under more carefully controlled conditions and by more rigorous criteria than are present in the situations created by practical problems.

Personal Knowledge (1958), 183.

5 The amount of knowledge which we can justify from evidence directly available to us can never be large. The overwhelming proportion of our factual beliefs continue therefore to be held at second hand through trusting others, and in the great majority of cases our trust is placed in the authority of comparatively few people of widely acknowledged standing.

Personal Knowledge (1958), 208.

John Charlton Polkinghorne 1930–

British clergyman and physicist

6 The test of a theory is its ability to cope with all the relevant phenomena, not its *a priori* 'reasonableness'. The latter would have proved a poor guide in the development of science, which often makes progress by its encounter with the totally unexpected and initially extremely puzzling.

'From DAMTP [Department of Applied Mathematics and Theoretical Physics] to Westcott House', *Cambridge Review*, 1981, 103, 61.

Jean Victor Poncelet 1788–1867

French mathematician

7 In fact, Gentlemen, no geometry without arithmetic, no mechanics without geometry . . . you cannot count upon success, if your mind is not sufficiently exercised on the forms and demonstrations of geometry, on the theories and calculations of arithmetic . . . In a word, the theory of proportions is for industrial teaching, what algebra is for the most elevated mathematical teaching.

Discours . . . à l'ouverture du cours de méchanique industrielle à Metz (1827), 2–3, trans. Ivor Grattan-Guinness.

Alexander Pope 1688–1744

English poet and satirist

1 One *Science* only will one *Genius* fit;
So *vast* is Art, so *narrow* Human Wit.

'An Essay on Criticism' (1711), lines 60–1. In John Butt (ed.), *The Poems of Alexander Pope* (1965), 146.

2 First follow NATURE, and your Judgement frame
By her just Standard, which is still the same:
Unerring Nature, still divinely bright.
One *clear*, *unchanged* and *universal* light.

'An Essay on Criticism' (1711), lines 68–71. In John Butt (ed.), *The Poems of Alexander Pope* (1965), 146.

3 A little *Learning* is a dang'rous Thing;
Drink deep, or taste not the *Pierian* Spring:
There *shallow Draughts* intoxicate the Brain,
And drinking *largely* sobers us again.

'An Essay on Criticism' (1711), lines 215–8. In John Butt (ed.), *The Poems of Alexander Pope* (1965), 151.

4 To Observations which ourselves we make,
We grow more partial for th' observer's sake.

'Moral Essays', Epistle I, to Richard Temple, Viscount Cobham (1734). In John Butt (ed.), *The Poems of Alexander Pope* (1965), 550.

5 Who shall decide, when Doctors disagree?

'Moral Essays', Epistle III, to Allen Lord Bathurst (1733). In John Butt (ed.), *The Poems of Alexander Pope* (1965), 570.

6 Who sees with equal eye, as God of all,
A hero perish or a sparrow fall,
Atoms or systems into ruin hurl'd,
And now a bubble burst, and now a world.

'An Essay on Man' (1733–4), Epistle I. In John Butt (ed.), *The Poems of Alexander Pope* (1965), 507.

7 Why has not Man a microscopic eye?
For this plain reason, Man is not a Fly.

'An Essay on Man' (1733–4), Epistle I. In John Butt (ed.), *The Poems of Alexander Pope* (1965), 511.

8 See, thro' this air, this ocean, and this earth,
All matter quick, and bursting into birth.
Above, how high progressive life may go!
Around, how wide! how deep extend below!
Vast chain of being, which from God began,
Natures æthereal, human, angel, man,
Beast, bird, fish, insect! what no eye can see,
No glass can reach! from Infinite to thee,
From thee to Nothing!—On superior pow'rs
Were we to press, inferior might on ours:
Or in the full creation leave a void,
Where, one step broken, the great scale's destroy'd:
From Nature's chain whatever link you strike,
Tenth or ten thousandth, breaks the chain alike.

'An Essay on Man' (1733–4), Epistle I. In John Butt (ed.), *The Poems of Alexander Pope* (1965), 513.

9 All Nature is but Art, unknown to thee;
All Chance, Direction, which thou canst not see;
All Discord, Harmony, not understood;
All partial Evil, universal Good:

And, spite of Pride, in erring Reason's spite,
One truth is clear, 'Whatever IS, is RIGHT.'

'An Essay on Man' (1733–4), Epistle I. In John Butt (ed.), *The Poems of Alexander Pope* (1965), 515.

1 Know then thyself, presume not God to scan;
The proper study of Mankind is Man.
Plac'd on this isthmus of a middle state,
A being darkly wise, and rudely great:
With too much knowledge for the Sceptic side,
With too much weakness for the Stoic's pride,
He hangs between; in doubt to act, or rest;
In doubt to deem himself a God, or Beast;
In doubt his Mind or Body to prefer,
Born but to die, and reas'ning but to err;
Alike in ignorance, his reason such,
Whether he thinks too little, or too much:
Chaos of Thought and Passion, all confus'd;
Still by himself abus'd, or disabus'd;
Created half to rise, and half to fall;
Great lord of all things, yet a prey to all;
Sole judge of Truth, in endless Error hurl'd:
The glory, jest, and riddle of the world!
. . . Superior beings, when of late they saw
A mortal Man unfold all Nature's law,
Admir'd such wisdom in an earthly shape,
And shew'd a NEWTON as we shew an Ape.

'An Essay on Man' (1733–4), Epistle II. In John Butt (ed.), *The Poems of Alexander Pope* (1965), 516–7.

2 This long disease, my life.

'Epistle to Dr Arbuthnot' (1735). In John Butt (ed.), *The Poems of Alexander Pope* (1965), 602.

3 Nature, and Nature's laws lay hid in Night.
God said, *Let Newton be*! and all was *Light*.
Epitaph intended for Sir Isaac Newton in Westminster Abbey.

In John Butt (ed.), *The Poems of Alexander Pope* (1965), 808.

Karl Raimund Popper 1902–94

Austrian-born British philosopher of science
See ***Kuhn, Thomas S.*** *352:5*

4 Psychologism is, I believe, correct only in so far as it insists upon what may be called 'methodological individualism' as opposed to 'methodological collectivism'; it rightly insists that the 'behaviour' and the 'actions' of collectives, such as states or social groups, must be reduced to the behaviour and to the actions of human individuals. But the belief that the choice of such an individualist method implies the choice of a psychological method is mistaken.

The Open Society and Its Enemies (1945), Vol. 2, 87.

5 The influence (for good or ill) of Plato's work is immeasurable. Western thought, one might say, has been either Platonic or anti-Platonic, but hardly ever non-Platonic.

The Open Society and its Enemies (1945).

6 But I shall certainly admit a system as empirical or scientific only if it is capable of being *tested* by experience. These considerations suggest that not the *verifiability* but the *falsifiability* of a system is to be taken as a criterion of demarcation. In other words: I shall not require of a scientific system that it shall be capable of being singled out, once and for all, in a positive sense; but I shall require that its logical form shall be such that it can be singled out, by means of empirical tests, in a negative sense: *it must be possible for an empirical scientific system to be refuted by experience*.

The Logic of Scientific Discovery (1959), 40–1.

7 There can be no ultimate statements in science: there can be no statements in

science which can not be tested, and therefore none which cannot in principle be refuted, by falsifying some of the conclusions which can be deduced from them.

The Logic of Scientific Discovery (1959), 47.

1 Science is not a system of certain, or well-established, statements; nor is it a system which steadily advances towards a state of finality . . . And our guesses are guided by the unscientific, the metaphysical (though biologically explicable) faith in laws, in regularities which we can uncover—discover. Like Bacon, we might describe our own contemporary science—'the method of reasoning which men now ordinarily apply to nature'—as consisting of 'anticipations, rash and premature' and as 'prejudices'.

The Logic of Scientific Discovery (1959), 278.

2 In so far as a scientific statement speaks about reality, it must be falsifiable: and in so far as it is not falsifiable, it does not speak about reality.

The Logic of Scientific Discovery (1959), 314.

3 I think that we shall have to get accustomed to the idea that we must not look upon science as a 'body of knowledge', but rather as a system of hypotheses; that is to say, as a system of guesses or anticipations which in principle cannot be justified, but with which we work as long as they stand up to tests, and of which we are never justified in saying that we know they are 'true' or 'more or less certain' or even 'probable'.

The Logic of Scientific Discovery (1959), 317.

4 It is easy to obtain confirmations, or verifications, for nearly every theory—if we look for confirmations. Confirmations should count only if they are the result *of risky predictions* . . . A theory which is not refutable by any conceivable event is non-scientific. Irrefutability is not a virtue of a theory (as people often think) but a vice. Every genuine *test* of a theory is an attempt to falsify it, or refute it.

Conjectures and Refutations: The Growth of Scientific Knowledge (1963), 36.

5 Thus science must begin with myths, and with the criticism of myths; neither with the collection of observations, nor with the invention of experiments, but with the critical discussion of myths, and of magical techniques and practices.

Conjectures and Refutations: The Growth of Scientific Knowledge (1963), 50.

6 The history of science, like the history of all human ideas, is a history of irresponsible dreams, of obstinacy, and of error. But science is one of the very few human activities—perhaps the only one—in which errors are systematically criticized and fairly often, in time, corrected. This is why we can say that, in science, we often learn from our mistakes, and why we can speak clearly and sensibly about making progress there.

Conjectures and Refutations: The Growth of Scientific Knowledge (1963), 216.

7 'Normal' science, in Kuhn's sense, exists. It is the activity of the non-revolutionary, or more precisely, the not-too-critical professional: of the science student who accepts the ruling dogma of the day . . . in my view the 'normal' scientist, as Kuhn describes him, is a person one ought to be sorry for . . . He has been taught in a dogmatic spirit: he is a victim of indoctrination . . . I can only say that I see a very great danger in it and in the possibility of its becoming normal . . . a danger to science and, indeed, to our civilization. And this shows why I regard Kuhn's emphasis on the existence of this kind of science as so important.

'Normal Science and its Dangers', in I. Lakatos and A. Musgrave (eds.), *Criticism and the Growth of Knowledge* (1970), 52–3.

8 Great Scientists . . . are men of bold ideas, but highly critical of their own ideas: they try to find whether their ideas are right by trying first to find whether they are not perhaps wrong. They work with bold conjectures and severe attempts at refuting their own conjectures.

'Replies to my Critics', in P. A. Schilpp (ed.), *The Philosophy of Karl Popper* (1974), Book 2, 977–8.

1 Propose theories which can be criticized. Think about possible decisive falsifying experiments—crucial experiments. But do not give up your theories too easily—not, at any rate, before you have critically examined your criticism.

'Replies to my Critics', in P. A. Schilpp (ed.), *The Philosophy of Karl Popper* (1974), Book 2, 984.

2 The difficulties connected with my criterion of demarcation (D) are important, but must not be exaggerated. It is vague, since it is a methodological rule, and since the demarcation between science and nonscience is vague. But it is more than sharp enough to make a distinction between many physical theories on the one hand, and metaphysical theories, such as psychoanalysis, or Marxism (in its present form), on the other. This is, of course, one of my main theses; and nobody who has not understood it can be said to have understood my theory.

The situation with Marxism is, incidentally, very different from that with psychoanalysis. Marxism was once a scientific theory: it predicted that capitalism would lead to increasing misery and, through a more or less mild revolution, to socialism; it predicted that this would happen first in the technically highest developed countries; and it predicted that the technical evolution of the 'means of production' would lead to social, political, and ideological developments, rather than the other way round.

But the (so-called) socialist revolution came first in one of the technically backward countries. And instead of the means of production producing a new ideology, it was Lenin's and Stalin's ideology that Russia must push forward with its industrialization ('Socialism is dictatorship of the proletariat plus electrification') which promoted the new development of the means of production.

Thus one might say that Marxism was once a science, but one which was refuted by some of the facts which happened to clash with its predictions (I have here mentioned just a few of these facts).

However, Marxism is no longer a science; for it broke the methodological rule that we must accept falsification, and it immunized itself against the most blatant refutations of its predictions. Ever since then, it can be described only as nonscience—as a metaphysical dream, if you like, married to a cruel reality.

Psychoanalysis is a very different case. It is an interesting psychological metaphysics (and no doubt there is some truth in it, as there is so often in metaphysical ideas), but it never was a science. There may be lots of people who are Freudian or Adlerian cases. But what prevents their theories from being scientific in the sense here described is, very simply, that they do not exclude any physically possible human behaviour. Whatever anybody may do is, in principle, explicable in Freudian or Adlerian terms. (Adler's break with Freud was more Adlerian than Freudian, but Freud never looked on it as a refutation of his theory.)

This point is very clear. Neither Freud nor Adler excludes any particular person's acting in any particular way, whatever the outward circumstances. Whether a man sacrificed his life to rescue a drowning child (a case of sublimation) or whether he murdered the child by drowning him (a case of repression) would not possibly be predicted or excluded by Freud's theory; *the theory was compatible with everything that could happen—even without any special immunization treatment.*

Thus while Marxism became non-scientific by its adoption of an immunizing strategy, psychoanalysis was immune to start with, and remained so.

'Replies to my Critics', in P. A. Schilpp (ed.), *The Philosophy of Karl Popper* (1974), Book 2, 984–5.

1 Some scientists find, or so it seems, that they get their best ideas when smoking; others by drinking coffee or whisky. Thus there is no reason why I should not admit that some may get their ideas by observing, or by repeating observations.

Realism and the Aim of Science (1983), 36.

2 Almost everyone . . . seems to be quite sure that the differences between the methodologies of history and of the natural sciences are vast. For, we are assured, it is well known that in the natural sciences we start from observation and proceed by induction to theory. And is it not obvious that in history we proceed very differently?

Yes, I agree that we proceed very differently. But we do so in the natural sciences as well.

In both we start from myths—from traditional prejudices, beset with error—and from these we proceed by criticism: by the critical elimination of errors. In both the role of evidence is, in the main, to correct our mistakes, our prejudices, our tentative theories—that is, to play a part in the critical discussion, in the elimination of error. By correcting our mistakes, we raise new problems. And in order to solve these problems, we invent conjectures, that is, tentative theories, which we submit to critical discussion, directed towards the elimination of error.

The Myth of the Framework: In Defence of Science and Rationality (1993), 140.

3 The philosopher of science is not much interested in the thought processes which lead to scientific discoveries; he looks for a logical analysis of the completed theory, including the relationships establishing its validity. That is, he is not interested in the context of discovery, but in the context of justification.

Attributed.

4 What really makes science grow is new ideas, including false ideas.

Attributed.

Cole Albert Porter 1891–1964

American composer and lyricist

5 You all have learned reliance
On the sacred teachings of Science,
So I hope, through life, you will never
decline
In spite of philistine
Defiance
To do what all good scientists do.
Experiment.
Make it your motto day and night.
Experiment
And it will lead you to the light.

From the song 'Experiment', in the 1933 musical *Nymph Errant*.

George Porter 1920–2002

British chemist

6 To solve a problem is to create new problems, new knowledge immediately reveals new areas of ignorance, and the need for new experiments. At least, in the field of fast reactions, the experiments do not take very long to perform.

'Flash Photolysis and Some of its Applications', Nobel Lecture, 11 December 1967. In *Nobel Lectures: Chemistry 1963–1970* (1972), 261.

7 To feed applied science by starving basic science is like economising on the foundations of a building so that it may be built higher. It is only a matter of time before the whole edifice crumbles.

'Lest the Edifice of Science Crumble', *New Scientist*, 1986, 111, 16.

Ezra Loomis Pound 1885–1972

American poet and critic

1 Quiet this metal!
Let the manes put off their terror, let them put off their aqueous bodies with fire.
Let them assume the milk-white bodies of agate.
Let them draw together the bones of the metal.

'The Alchemist: Chant for the Transmutation of Metal'. In T. S. Eliot (ed.), *Ezra Pound: Selected Poems* (1928), 61–2.

Frank Press 1924–

American geophysicist

2 Scientists have reaped rich rewards, they have sat high in government councils and have been blinded by the attractiveness of public life—all this because they happen to have been good killers.

'Science in the Age of Aquarius', *EOS—Transactions of the American Geophysical Union*, 1974, 55, 1026.

Joseph Prestwich 1812–96

British geologist

3 Time is in itself [not] a difficulty, but a time-rate, assumed on very insufficient grounds, is used as a master-key, whether or not it fits, to unravel all difficulties. What if it were suggested that the brick-built Pyramid of Hawâra had been laid brick by brick by a single workman? Given time, this would not be beyond the bounds of possibility. But Nature, like the Pharaohs, had greater forces at her command to do the work better and more expeditiously than is admitted by Uniformitarians.

'The Position of Geology', *The Nineteenth Century*, 1893, 34, 551.

Derek J. de Solla Price 1922–83

British-American historian of science

4 Using any reasonable definition of a scientist, we can say that 80 to 90 percent of all the scientists that have ever lived are alive now. Alternatively, any young scientist, starting now and looking back at the end of his career upon a normal life span, will find that 80 to 90 percent of all scientific work achieved by the end of the period will have taken place before his very eyes, and that only 10 to 20 percent will antedate his experience.

Little Science, Big Science (1963), 1–2.

5 It is clear that we cannot go up another two orders of magnitude as we have climbed the last five. If we did, we should have two scientists for every man, woman, child, and dog in the population, and we should spend on them twice as much money as we had. Scientific doomsday is therefore less than a century distant.

Little Science, Big Science (1963), 19.

Joseph Priestley 1733–1804

British chemist, theologian and natural philosopher

6 When air has been freshly and strongly tainted with putrefaction, so as to smell through the water, sprigs of mint have presently died, upon being put into it, their leaves turning black; but if they do not die presently, they thrive in a most surprizing manner. In no other circumstances have I ever seen vegetation so vigorous as in this kind of air, which is immediately fatal to animal life. Though these plants have been crouded in jars filled with this air, every leaf has been full of life; fresh shoots have branched out in various directions, and have grown much faster than other similiar plants, growing in the same exposure in common air.

This observation led me to conclude, that plants, instead of affecting the air in the same manner with animal respiration, reverse the effects of breathing, and tend to keep the atmosphere sweet and wholesome, when it is become noxious, in consequence on animals living and breathing, or dying and putrefying in it.

In order to ascertain this, I took a quantity of air, made thoroughly noxious, by mice breathing and dying in it, and divided it into two parts; one of which I put into a phial immersed in water; and to the other (which was contained in a glass jar, standing in water) I put a sprig of mint. This was about the beginning of August 1771, and after eight or nine days, I found that a mouse lived perfectly well in that part of the air, in which the sprig of mint had grown, but died the moment it was put into the other part of the same original quantity of air; and which I had kept in the very same exposure, but without any plant growing in it.

'Observations on Different Kinds of Air', *Philosophical Transactions*, 1772, 62, 193–4.

1 In completing one discovery we never fail to get an imperfect knowledge of others.

Experiments and Observations on Different Kinds of Air (1774), vii.

2 I have procured air [oxygen] . . . between five and six times as good as the best common air that I have ever met with.

Experiments and Observations on Different Kinds of Air (1775), Vol. 2, 48.

3 The feeling of it to my lungs was not sensibly different from that of common air; but I fancied that my breast felt peculiarly light and easy for some time afterwards. Who can tell but that, in time, this pure air may become a fashionable article in luxury. Hitherto only two mice and myself have had the privilege of breathing it.

Experiments and Observations on Different Kinds of Air (1775), Vol. 2, 102.

4 The contents of this section will furnish a very striking illustration of the truth of a remark, which I have more than once made in my philosophical writings, and which can hardly be too often repeated, as it tends greatly to encourage philosophical investigations viz. That more is owing to what we call *chance*, that is, philosophically speaking, to the observation of *events arising from unknown causes*, than to any proper *design*, or pre-conceived *theory* in this business. This does not appear in the works of those who write *synthetically* upon these subjects; but would, I doubt not, appear very strikingly in those who are the most celebrated for their philosophical acumen, did they write *analytically* and ingenuously.

'On Dephlogisticated Air, and the Constitution of the Atmosphere', in *The Discovery of Oxygen, Part 1, Experiments by Joseph Priestley 1775* (Alembic Club Reprint, 1894), 5.

5 There are, I believe, very few maxims in philosophy that have laid firmer hold upon the mind, than that air, meaning atmospherical air (free from various foreign matters, which were always supposed to be dissolved, and intermixed with it) is a *simple elementary substance*, indestructible, and unalterable, at least as much so as water is supposed to be. In the course of my enquiries, I was, however, soon satisfied that atmospherical air is not an unalterable thing; for that the phlogiston with which it becomes loaded from bodies burning in it, and animals breathing it, and various other chemical processes, so far alters and depraves it, as to render it altogether unfit for inflammation, respiration, and other purposes to which it is subservient; and I had discovered that agitation in water, the process of vegetation, and probably other natural processes, by taking out the superfluous phlogiston, restore it to its original purity.

'On Dephlogisticated Air, and the Constitution of the Atmosphere', in *The Discovery of Oxygen, Part 1, Experiments by Joseph Priestley 1775* (Alembic Club Reprint, 1894), 6.

1 But it is not given to every electrician to die in so glorious a manner as the justly envied Richmann.

G. W. Richmann had been killed by lightning while investigating its nature.

The History and Present State of Electricity, with Original Experiments (1767), 3rd edition (1775), Vol. 1, 108.

2 As every circumstance relating to so capital a discovery as this (the greatest, perhaps, that has been made in the whole compass of philosophy, since the time of Sir Isaac Newton) cannot but give pleasure to all my readers, I shall endeavour to gratify them with the communication of a few particulars which I have from the best authority.

The Doctor [Benjamin Franklin], after having published his method of verifying his hypothesis concerning the sameness of electricity with the matter lightning, was waiting for the erection of a spire in Philadelphia to carry his views into execution; not imagining that a pointed rod, of a moderate height, could answer the purpose; when it occurred to him, that, by means of a common kite, he could have a readier and better access to the regions of thunder than by any spire whatever. Preparing, therefore, a large silk handkerchief, and two cross sticks, of a proper length, on which to extend it, he took the opportunity of the first approaching thunder storm to take a walk into a field, in which there was a shed convenient for his purpose. But dreading the ridicule which too commonly attends unsuccessful attempts in science, he communicated his intended experiment to no body but his son, who assisted him in raising the kite.

The kite being raised, a considerable time elapsed before there was any appearance of its being electrified. One very promising cloud passed over it without any effect; when, at length, just as he was beginning to despair of his contrivance, he observed some loose threads of the hempen string to stand erect, and to avoid one another, just as if they had been suspended on a common conductor. Struck with this promising appearance, he immediately presented his knuckle to the key, and (let the reader judge of the exquisite pleasure he must have felt at that moment) the discovery was complete. He perceived a very evident electric spark. Others succeeded, even before the string was wet, so as to put the matter past all dispute, and when the rain had wetted the string, he collected electric fire very copiously. This happened in June 1752, a month after the electricians in France had verified the same theory, but before he had heard of any thing that they had done.

The History and Present State of Electricity, with Original Experiments (1767), 3rd edition (1775), Vol. 1, 216–7.

3 The greater is the circle of light, the greater is the boundary of the darkness by which it is confined. But, notwithstanding this, the more light we get, the more thankful we ought to be, for by this means we have the greater range for satisfactory contemplation. In time the bounds of light will be still farther extended; and from the infinity of the divine nature, and the divine works, we may promise ourselves an endless progress in our investigation of them: a prospect truly sublime and glorious.

Experiments and Observations with a Continuation of the Observations on Air (1781), Vol. 2, ix.

4 But nothing of a nature foreign to the duties of my profession [as a dissenting clergyman] engaged my attention while I was at Leeds so much as the prosecution of my experiments relating to *electricity*, and especially the doctrine of *air*. The last I was led into a consequence of inhabiting a house adjoining to a public brewery, where I first amused myself with making experiments on fixed air [CO_2] which I found ready made in the process of fermentation. When I removed from that house, I was under the necessity of making the fixed air for myself; and one experiment leading to another, as I have distinctly and faithfully noted in my various publications on the subject, I by degrees contrived a convenient apparatus for the purpose, but of the cheapest kind. When I began these experiments I knew very little of *chemistry*, and had in a manner no idea on the subject before I attended a course of chymical lectures delivered in the Academy at Warrington by Dr. Turner of Liverpool. But I have often thought that upon the whole, this circumstance was no disadvantage to me; as in this situation I was led to devise an apparatus and processes of my own, adapted to my peculiar views. Whereas, if I had been previously accustomed to the usual chemical processes, I should not have so easily thought of any other; and without new modes of operation I should hardly have discovered anything materially new.

Memoirs of Dr. Joseph Priestley, in the Year 1795 (1806), Vol. 1, 61–2.

1 On the whole, I cannot help saying that it appears to me not a little extraordinary, that a theory so new, and of such importance, overturning every thing that was thought to be the best established in chemistry, should rest on so very narrow and precarious a foundation, the experiments adduced in support of it being not only ambiguous or explicable on either hypothesis, but exceedingly few. I think I have recited them all, and that on which the greatest stress is laid, viz. That of the formation of water from the decomposition of the two kinds of air, has not been sufficiently repeated. Indeed it required so difficult and expensive an apparatus, and so many precautions in the use of it, that the frequent repetition of the experiment cannot be expected; and in these circumstances the practised experimenter cannot help suspecting the accuracy of the result and consequently the certainty of the conclusion.

Considerations on the Doctrine of Phlogiston (1796), 57–8.

George Prochaska 1749–1820

German anatomist and physiologist

2 The external impressions which are made on the sensorial nerves are very quickly transmitted along the whole length of the nerves, as far as their origin; and having arrived there, they are reflected by a certain law, and pass on to certain and corresponding motor nerves, through which, being again very quickly transmitted to muscles, they excite certain and definite motions. This part, in which, as in a centre, the sensorial nerves, as well as the motor nerves, meet and communicate, and in which the impressions made on the sensorial nerves are reflected on the motor nerves, is designated by a term, now adopted by most physiologists, the *sensorium commune*.

A Dissertation on the Functions of the Nervous System (1784), trans. and ed. Thomas Laycock (1851), 429.

3 But since the brain, as well as the cerebellum, is composed of many parts, variously figured, it is possible, that nature, which never works in vain, has destined those parts to various uses, so that the various faculties of the mind seem to require different portions of the cerebrum and cerebellum for their production.

A Dissertation on the Functions of the Nervous System (1784), trans. and ed. Thomas Laycock (1851), 446.

Protagoras c.481–c.411 BC

Greek philosopher and teacher

*See **Aristotle** 22:4*

4 Concerning the gods, I have no means of knowing either that they exist or that they do not exist, nor what sort of form they may have; there are many reasons why knowledge on this subject is not possible, owing to the lack of evidence and the shortness of human life.

Protagoras, fr. 1, quoted in E. Hussey, *The Pre-Socratics* (1972), 109.

1 Man is the measure of all things.

Protagoras, fr. 4, quoted in E. Hussey, *The Pre-Socratics* (1972), 109.

William Prout 1785–1850

British chemist and physician

2 If the views we have ventured to advance be correct, we may almost consider the πρώτη ὕλη of the ancients to be realised in hydrogen, an opinion, by the by, not altogether new. If we actually consider the specific gravities of bodies in their gaseous state to represent the number of volumes condensed into one; or in other words, the number of the absolute weight of a single volume of the first matter (πρώτη ὕλη) which they contain, which is extremely probable, multiples in weight must always indicate multiples in volume, and *vice versa*; and the specific gravities, or absolute weights of all bodies in a gaseous state, must be multiples of the specific gravity or absolute weight of the first matter (πρώτη ὕλη), because all bodies in the gaseous state which unite with one another unite with reference to their volume.

'Correction of a Mistake in the Essay on the Relation between the Specific Gravities of Bodies in their Gaseous State and the Weights of their Atoms', *Annals of Philosophy*, 1816, 7, 113.

3 That a free, or at least an unsaturated acid usually exists in the stomachs of animals, and is in some manner connected with the important process of digestion, seems to have been the general opinion of physiologists till the time of SPALLANZANI. This illustrious philosopher concluded, from his numerous experiments, that the gastric fluids, when in a perfectly natural state, are neither acid nor alkaline. Even SPALLANZANI, however, admitted that the contents of the stomach are very generally acid; and this accords not only with my own observation, but with that, I believe, of almost every individual who has made any experiments on the subject. . . . The object of the present communication is to show, that the acid in question is the *muriatic* [hydrochloric] *acid*, and that the salts usually met with in the stomach, are the alkaline muriates.

'On the Nature of the Acid and Saline Matters Usually Existing in the Stomachs of Animals', *Philosophical Transactions of the Royal Society of London*, 1824, **114**, 45–6.

4 I had come to the conclusion, that the principal alimentary matters might be reduced to the three great classes, namely the *saccharine*, the *oily* and the *albuminous*.

'On the Ultimate Composition of Simple Alimentary Substances; with some Preliminary Remarks on the Analysis of Organised Bodies in General', *Philosophical Transactions of the Royal Society*, 1827, 117, 357.

Ptolemy c.100–c.178

Egyptian astronomer and mathematician

5 All those who think it paradoxical that so great a weight as the earth should not waver or move anywhere seem to me to go astray by making their judgement with an eye to their own affects and not to the property of the whole. For it would not still appear so extraordinary to them, I believe, if they stopped to think that the earth's magnitude compared to the whole body surrounding it is in the ratio of a point to it. For thus it seems possible for that which is relatively least to be supported and pressed against from all sides equally and at the same angle by that which is absolutely greatest and homogeneous.

'The Almagest I', in *Ptolemy: the Almagest; Nicolaus Copernicus: On the Revolutions of the Heavenly Spheres; Johannes Kepler: Epitome of Copernican Astronomy: IV–V The Harmonies of the World: V*, trans. R. Catesby Taliaferra (1952), 11.

6 Therefore the solid body of the earth is reasonably considered as being the largest relative to those moving against it and as remaining unmoved in any direction by the force of the very small weights, and as it were absorbing their fall. And if it had some one common movement, the same as that of the other weights, it would clearly leave them all behind because of its much greater magnitude. And the animals and other weights would be left hanging in the air, and the earth would very quickly fall out of the heavens. Merely to conceive such things makes them appear ridiculous.

'The Almagest I', in *Ptolemy: the Almagest; Nicolaus Copernicus: On the Revolutions of the Heavenly Spheres; Johannes Kepler: Epitome of Copernican Astronomy: IV–V The Harmonies of the World: V*, trans. R. Catesby Taliaferra (1952), 11.

1 I know that I am mortal by nature, and ephemeral; but when I trace at my pleasure the windings to and fro of the heavenly bodies I no longer touch earth with my feet: I stand in the presence of Zeus himself and take my fill of ambrosia, food of the gods.

Epigram. *Almagest*. In Owen Gingerich, *The Eye of Heaven: Ptolemy, Copernicus, Kepler* (1993), 4.

Philip Pullman 1946–

British author

2 At Gabriel College there was a very holy object on the high altar of the Oratory, covered with a black velvet cloth . . . At the height of the invocation the Intercessor lifted the cloth to reveal in the dimness a glass dome inside which there was something too distant to see, until he pulled a string attached to a shutter above, letting a ray of sunlight through to strike the dome exactly. Then it became clear: a little thing like a weathervane, with four sails black on one side and white on the other, began to whirl around as the light struck it. It illustrated a moral lesson, the Intercessor explained, for the black of ignorance fled from the light, whereas the wisdom of white rushed to embrace it.

Parody of Crookes's radiometer.

Northern Lights (2001), 149.

Hilary Putnam 1926–

American philosopher

3 Science is wonderful at destroying metaphysical answers, but incapable of providing substitute ones. Science takes away foundations without providing a replacement. Whether we want to be there or not, science has put us in the position of having to live without foundations. It was shocking when Nietzsche said this, but today it is commonplace; *our* historical position—and no end to it is in sight—is that of having to philosophise without 'foundations'.

The Many Faces of Realism: The Paul Carus Lectures (1987), 29. See **Nietzsche**.

Lambert-Adolphe-Jacques Quetelet 1796–1874

Belgian statistician

4 It would appear . . . that moral phenomena, when observed on a great scale, are found to resemble physical phenomena; and we thus arrive, in inquiries of this kind, at the fundamental principle, *that the greater the number of individuals observed, the more do individual peculiarities, whether physical or moral, become effaced, and leave in a prominent point of view the general facts, by virtue of which society exists and is preserved.*

A Treatise on Man and the Development of his Faculties (1842). Reprinted with an introduction by Solomon Diamond (1969), 6.

5 *L'homme moyen.*

The average man.

The invention of the concept.

A Treatise on Man and the Development of his Faculties (1842). Reprinted with an introduction by Solomon Diamond (1969), 96.

6 The determination of the average man is not merely a matter of speculative curiosity; it may be of the most important service to the science of man and the social system. It ought necessarily to precede every other inquiry into social physics, since it is, as it were, the basis. The average man, indeed, is in a nation what the centre of gravity is in a body; it is by having that central point in view that we arrive at the apprehension of all the phenomena of equilibrium and motion.

A Treatise on Man and the Development of his Faculties (1842). Reprinted with an introduction by Solomon Diamond (1969), 96.

1 Whether statistics be an art or a science . . . or a scientific art, we concern ourselves little. It is the basis of social and political dynamics, and affords the only secure ground on which the truth or falsehood of the theories and hypotheses of that complicated science can be brought to the test.

Letters on the Theory of Probabilities (1846), trans. O. G. Downes (1849).

Willard van Orman Quine

1908–2000

American philosopher

2 The totality of our so-called knowledge and beliefs, from the most casual matters of geography and history to the profoundest laws of atomic physics or even of pure mathematics and logic, is a man-made fabric which impinges on experience only along the edges. Or, to change the figure, total science is like a field of force whose boundary conditions are experience. A conflict with experience at the periphery occasions readjustments in the interior of the field. Truth values have to be redistributed over some of our statements. Reevaluation of some statements entails reevaluation of others, because of their logical interconnections—the logical laws being in turn simply certain further statements of the system, certain further elements of the field.

From A Logical Point of View (1953), 42.

3 As an empiricist I continue to think of the conceptual scheme of science as a tool, ultimately, for predicting future experience in the light of past experience. Physical objects are conceptually imported into the situation as convenient intermediaries—not by definition in terms of experience, but simply as irreducible posits comparable, epistemologically, to the gods of Homer. For my part I do, qua lay physicist, believe in physical objects and not in Homer's gods; and I consider it a scientific error to believe otherwise. But in point of epistemological footing the physical objects and the gods differ only in degree and not in kind. Both sorts of entities enter our conception only as cultural posits. The myth of physical objects is epistemologically superior to most in that it has proved more efficacious than other myths as a device for working a manageable structure into the flux of experience.

From A Logical Point of View (1953), 44.

4 Science is not a substitute for common sense, but an extension of it.

'The Scope and Language of Science' (1954), reprinted in *The British Journal for the Philosophy of Science*, 1957, **8**, 2.

5 The scientist is indistinguishable from the common man in his sense of evidence, except that the scientist is more careful.

'The Scope and Language of Science' (1954), reprinted in *The British Journal for the Philosophy of Science*, 1957, **8**, 5.

6 Physics investigates the essential nature of the world, and biology describes a local bump. Psychology, human psychology, describes a bump on the bump.

Theories and Things (1981), 93.

7 Students of the heavens are separable into astronomers and astrologers as readily as the minor domestic ruminants into sheep and goats, but the separation of philosophers into sages and cranks seems to be more sensitive to frames of reference.

Theories and Things (1981), 192.

François Rabelais c.1494–1553

French priest and writer

1 *Natura vacuum abhorret.*

Nature abhors a vacuum.

Gargantua and Pantagruel (1532–64), book 1, chapter 5. Trans. Thomas Urquhart and Peter Le Motteux (1934), Vol. 1, 24.

2 *Science sans conscience n'est que le ruine de l'âme.*

Knowledge without conscience is but the ruine of the soule.

Gargantua and Pantagruel (1532–64), book 2, chapter 8. Trans. Thomas Urquhart and Peter Le Motteux (1934), Vol. 1, 204.

Efraim Racker 1913–91

Polish-born American biochemist

3 Rejoice when other scientists do not believe what you know to be true. It will give you extra time to work on it in peace. When they start claiming that they have discovered it before you, look for a new project.

'Resolution and Reconstitution of Biological Pathways from 1919 to 1984', *Federation Proceedings*, 1983, 12, 2902.

Kathleen Jessie Raine 1908–2003

British poet and critic

4 Chemistry dissolves the goddess in the alembic,
Venus, the white queen, the universal matrix,
Down to the molecular hexagons and carbon-chains.

'The Human Form Divine', in *The Collected Poems of Kathleen Raine* (1956), 86.

Walter Alexander Raleigh 1861–1922

British critic and essayist

5 In an examination those who do not wish to know ask questions of those who cannot tell.

'Some Thoughts on Examinations', in *Laughter from a Cloud* (1923), 120.

Srinivasa Ramanujan 1887–1920

Indian mathematican

6 *Replying to G. H. Hardy's suggestion that the number of a taxi (1729) was 'dull':*

No, it is a very interesting number; it is the smallest number expressible as a sum of two cubes in two different ways, the two ways being $1^3 + 12^3$ and $9^3 + 10^3$.

Proceedings of the London Mathematical Society, 26 May, 1921.

7 ON **RAMANUJAN** Plenty of mathematicians, Hardy knew, could follow a step-by-step discursus unflaggingly—yet counted for nothing beside Ramanujan. Years later, he would contrive an informal scale of natural mathematical ability on which he assigned himself a 25 and Littlewood a 30. To David Hilbert, the most eminent mathematician of the day, he assigned an 80. To Ramanujan he gave 100.

Robert Kanigel, *The Man who knew Infinity: A Life of the Genius Ramanujan* (1975), 226.

Andrew Crombie Ramsey 1814–91

British geologist

8 I cannot see of what use these slides can be to a field man. I don't believe in looking at a mountain through a microscope.

Archibald Geikie, *Memoir of Sir Andrew Crombie Ramsey* (1895), 343.

Frank Plumpton Ramsey 1903–30

British mathematician and logician

9 Logic issues in tautologies, mathematics in identities, philosophy in definitions; all trivial, but all part of the vital work of clarifying and organising our thought.

'Last Papers: Philosophy' (1929), in *The Foundations of Mathematics and Other Logical Essays* (1931), 264.

10 Science, history and politics are not suited for discussion except by experts.

Others are simply in the position of requiring more information; and, till they have acquired all available information, cannot do anything but accept on authority the opinions of those better qualified.

The Foundations of Mathematics and Other Logical Essays (1931), Epilogue, 287–8.

1 Where I seem to differ from some of my friends is in attaching little importance to physical size. I don't feel the least humble before the vastness of the heavens. The stars may be large, but they cannot think or love; and these are qualities which impress me far more than size does. I take no credit for weighing nearly seventeen stone.

The Foundations of Mathematics and Other Logical Essays (1931), Epilogue, 291.

John Herman Randall, Jr.

1889–1980

American historian and philosopher

2 An egg is a chemical process, but it is not a mere chemical process. It is one that is going places—even when, in our world of chance and contingency, it ends up in an omelet and not in a chicken. Though it surely be a chemical process, we cannot understand it adequately without knowing the kind of chicken it has the power to become.

'The Changing Impact of Darwin on Philosophy', *Journal of the History of Ideas*, 1961, 22, 457.

William John Macquorn Rankine 1820–72

British engineer and physicist

3 The law of the conservation of energy is already known, viz. that the sum of the actual and potential energies in the universe is unchangeable.

'On the General Law of the Transformation of Energy', *Philosophical Magazine*, 1853, 5, 106.

François-Vincent Raspail

1794–1878

French natural philosopher

4 *Omnis cellula e cellula.*

Every cell is derived from another cell.

Annales des Sciences naturelles, 1825, 6, 224 and 384. Trans. Henry Harris, *The Birth of the Cell* (1999), 33. See **Virchow, Rudolf** 598:4

John Ashworth Ratcliffe

1902–87

British physicist

5 There was, I think, a feeling that the best science was that done in the simplest way. In experimental work, as in mathematics, there was 'style' and a result obtained with simple equipment was more elegant than one obtained with complicated apparatus, just as a mathematical proof derived neatly was better than one involving laborious calculations. Rutherford's first disintegration experiment, and Chadwick's discovery of the neutron had a 'style' that is different from that of experiments made with giant accelerators.

'Physics in a University Laboratory Before and After World War II', *Proceedings of the Royal Society of London*, Series A, 1975, 342, 463.

David Malcolm Raup 1933–

American geophysicist and palaeontologist

6 A new theory is guilty until proven innocent, and the pre-existing theory is innocent until proven guilty . . . Continental drift was guilty until proven innocent.

The Nemesis Affair: A Story of the Death of the Dinosaurs and the Ways of Science (1986), 195–205.

Jerome R. Ravetz 1929–

American-born British historian and philosopher of science

7 As the world of science has grown in size and in power, its deepest problems have changed from the epistemological to the social.

Scientific Knowledge and its Social Problems (1971), 10.

John Ray 1627–1705

British naturalist

1 In order that an inventory of plants may be begun and a classification of them correctly established, we must try to discover criteria of some sort for distinguishing what are called 'species'. After a long and considerable investigation, no surer criterion for determining species had occurred to me than distinguishing features that perpetuate themselves in propagation from seed. Thus, no matter what variations occur in the individuals or the species, if they spring from the seed of one and the same plant, they are accidental variations and not such as to distinguish a species. For these variations do not perpetuate themselves in subsequent seeding. Thus, for example, we do not regard *caryophylli* with full or multiple blossoms as a species distinct from *caryophylli* with single blossoms, because the former owe their origin to the seed of the latter and if the former are sown from their own seed, they once more produce single-blossom *caryophylli*. But variations that never have as their source seed from one and the same species may finally be regarded as distinct species. Or, if you make a comparison between any two plants, plants which never spring from each other's seed and never, when their seed is sown, are transmuted one into the other, these plants finally are distinct species. For it is just as in animals: a difference in sex is not enough to prove a difference of species, because each sex is derived from the same seed as far as species is concerned and not infrequently from the same parents; no matter how many and how striking may be the accidental differences between them; no other proof that bull and cow, man and woman belong to the same species is required than the fact that both very frequently spring from the same parents or the same mother. Likewise in the case of plants, there is no surer index of identity of species than that of origin from the seed of one and the same plant, whether it is a matter of individuals or species. For animals that differ in species preserve their distinct species permanently; one species never springs from the seed of another nor vice versa.

Historia Plantarum (1686), Vol. 1, 40. Trans. Edmund Silk. Quoted in Barbara G. Beddall, 'Historical Notes on Avian Classification', *Systematic Zoology*, 1957, 6, 133–4.

2 A wonder then it must needs be, that there should be any Man found so stupid and forsaken of reason as to persuade himself, that this most beautiful and adorned world was or could be produced by the fortuitous concourse of atoms.

The Wisdom of God Manifested in the Works of the Creation (1691), 21–2.

3 And I believe there are many Species in Nature, which were never yet taken notice of by Man, and consequently of no use to him, which yet we are not to think were created in vain; but it's likely . . . to partake of the overflowing Goodness of the Creator, and enjoy their own Beings. But though in this sense it be not true, that all things were made for Man; yet thus far it is, that all the Creatures in the World may be some way or other useful to us, at least to exercise our Wits and Understandings, in considering and contemplating of them, and so afford us Subject of Admiring and Glorifying their and our Maker. Seeing them, we do believe and assert that all things were in some sense made for us, we are thereby obliged to make use of them for those purposes for which they serve us, else we frustrate this End of their Creation.

The Wisdom of God Manifested in the Works of the Creation (1691), 169–70.

4 A multitude of words doth rather obscure than illustrate, they being a burden to the memory, and the first apt to be forgotten, before we come to the last. So that he that uses many words for the explaining of any subject, doth, like the cuttle-fish, hide himself, for the most part, in his own ink.

The Wisdom of God Manifested in the Works of the Creation (1691).

1 But here it may be objected, that the present Earth looks like a heap of Rubbish and Ruines; And that there are no greater examples of confusion in Nature than Mountains singly or jointly considered; and that there appear not the least footsteps of any Art or Counsel either in the Figure and Shape, or Order and Disposition of Mountains and Rocks. Wherefore it is not likely they came so out of God's hands . . . To which I answer, That the present face of the Earth with all its Mountains and Hills, its Promontaries and Rocks, as rude and deformed as they appear, seems to me a very beautiful and pleasant object, and with all the variety of Hills, and Valleys, and Inequalities far more grateful to behold, than a perfectly level Countrey without any rising or protuberancy, to terminate the sight: As any one that hath but seen the Isle of Ely, or any the like Countrey must need acknowledge.

Miscellaneous Discourses Concerning the Dissolution and Changes of the World (1692), 165–6.

2 Many Species of Animals have been lost out of the World, which Philosophers and Divines are unwilling to admit, esteeming the Destruction of any one *Species* a Dismembring of the Universe, and rendring the World imperfect; whereas they think the Divine Providence is especially concerned, and solicitous to secure and preserve the Works of the Creation. And truly so it is, as appears, in that it was so careful to lodge all Land Animals in the Ark at the Time of the general Deluge; and in that, of all Animals recorded in Natural Histories, we cannot say that there hath been any one Species lost, no not of the most infirm, and most exposed to Injury and Ravine. Moreover, it is likely, that as there neither is nor can be any new Species of Animals produced, all proceeding from Seeds at first created; so Providence, without which one individual Sparrow falls not to the ground, doth in that manner watch over all that are created, that an entire Species shall not be lost or destroyed by any Accident. Now, I say, if these Bodies were sometimes the Shells and Bones of Fish, it will thence follow, that many *Species* have been lost out of the World . . . To which I have nothing to reply, but that there may be some of them remaining some where or other in the Seas, though as yet they have not come to my Knowledge. For though they may have perished, or by some Accident been destroyed out of our Seas, yet the Race of them may be preserved and continued still in others.

Three Physico-Theological Discourses (1713), Discourse II, 'Of the General Deluge, in the Days of Noah; its Causes and Effects', 172–3.

Lord Rayleigh see Strutt, John William

Herbert Harold Read 1889–1970

British geologist

3 I suggest that the best geologist is he who has seen most rocks.

The Granite Controversy: Geological Addresses Illustrating the Evolution of a Disputant (1957), 3.

4 It may be that it is only by the grace of granitization that we have continents to live on.

The Granite Controversy: Geological Addresses Illustrating the Evolution of a Disputant (1957), 373.

Ronald Wilson Reagan 1911–2004

American actor and president

5 *On evolution:*

Well, it is a theory, it is a scientific theory only, and it has in recent years been challenged in the world of science and is not yet believed in the scientific community to be as infallible as it was once believed. But if it was going to be taught in the schools, then I think that also the biblical theory of creation, which is not a theory but the biblical theory of creation, should also be taught.

Told to an audience during the 1980 Presidential Campaign. In M. Ruse, *Is Science Sexist and Other Problems in the Bio-Medical Sciences?* (1981), Opening vignette.

René-Antoine Ferchault de Réaumur 1683–1757

French natural philosopher

1 An infinity of these tiny animals defoliate our plants, our trees, our fruits . . . they attack our houses, our fabrics, our furniture, our clothing, our furs . . . He who in studying all the different species of insects that are injurious to us, would seek means of preventing them from harming us, would seek to cause them to perish, proposes for his goal important tasks indeed.

In J. B. Gough, 'René-Antoine Ferchault de Réaumur', in Charles Gillispie (ed.), *Dictionary of Scientific Biography* (1975), Vol. 11, 332.

2 Facts are certainly the solid and true foundation of all sectors of nature study . . . Reasoning must never find itself contradicting definite facts; but reasoning must allow us to distinguish, among facts that have been reported, those that we can fully believe, those that are questionable, and those that are false. It will not allow us to lend faith to those that are directly contrary to others whose certainty is known to us; it will not allow us to accept as true those that fly in the face of unquestionable principles.

Memoires pour Servir a l'Histoire des Insectes (1736), Vol. 2, xxxiv. Quoted in Jacques Roger, *The Life Sciences in Eighteenth-Century French Thought*, ed. Keith R. Benson and trans. Robert Ellrich (1997), 165.

Robert Recorde c.1510–58

British mathematician

3 I will sette as I doe often in woorke use, a paire of paralleles, or gemowe times of one lengthe, thus: =, bicause noe 2 thynges, can be moare equalle.

Introducing the sign for equality. (Gemowe means 'twin' as in Gemini.)

The Whetstone of Witte (1557). No pagination.

Martin John Rees 1942–

British astronomer

4 It's becoming clear that in a sense the cosmos provides the only laboratory where sufficiently extreme conditions are ever achieved to test new ideas on particle physics. The energies in the Big Bang were far higher than we can ever achieve on Earth. So by looking at evidence for the Big Bang, and by studying things like neutron stars, we are in effect learning something about fundamental physics.

In Lewis Wolpert and Alison Richards, *A Passion for Science* (1988), 33.

5 The physicist is like someone who's watching people playing chess and, after watching a few games, he may have worked out what the moves in the game are. But understanding the rules is just a trivial preliminary on the long route from being a novice to being a grand master. So even if we understand all the laws of physics, then exploring their consequences in the everyday world where complex structures can exist is a far more daunting task, and that's an inexhaustible one I'm sure.

In Lewis Wolpert and Alison Richards, *A Passion for Science* (1988), 37.

Hans Reichenbach 1891–1953

German-born American philosopher

6 Philosophy has proceeded from speculation to science.

The Rise of Scientific Philosophy (1951), vii.

7 The essence of knowledge is *generalization*. That fire can be produced by rubbing wood in a certain way is a knowledge derived by generalization from individual experiences; the statement means that rubbing wood in this way will *always* produce fire. The art of discovery is therefore the art of correct generalization.

The Rise of Scientific Philosophy (1951), 5.

1 The picture of scientific method drafted by modern philosophy is very different from traditional conceptions. Gone is the ideal of a universe whose course follows strict rules, a predetermined cosmos that unwinds itself like an unwinding clock. Gone is the ideal of the scientist who knows the absolute truth. The happenings of nature are like rolling dice rather than like revolving stars; they are controlled by probability laws, not by causality, and the scientist resembles a gambler more than a prophet. He can tell you only his best posits—he never knows beforehand whether they will come true. He is a better gambler, though, than the man at the green table, because his statistical methods are superior. And his goal is staked higher—the goal of foretelling the rolling dice of the cosmos. If he is asked why he follows his methods, with what title he makes his predictions, he cannot answer that he has an irrefutable knowledge of the future; he can only lay his best bets. But he can prove that they *are* best bets, that making them is the best he can do—and if a man does his best, what else can you ask of him?

The Rise of Scientific Philosophy (1951), 248–9.

2 It appears that the solution of the problem of time and space is reserved to philosophers who, like Leibniz, are mathematicians, or to mathematicians who, like Einstein, are philosophers.

Paul Arthur Schilpp (ed.), *Albert Einstein: Philosopher-Scientist* (1959), Vol. 1, 307.

3 The statement that although the past can be recorded, the future cannot, is translatable into the statistical statement: *Isolated states of order are always postinteraction states, never preinteraction states.*

'Cause and Effect: Producing and Recording', in Maria Reichenbach (ed.), *The Direction of Time* (1956), 155.

Thomas Reid 1710–96

British philosopher

4 The laws of nature are the rules according to which the effects are produced; but there must be a cause which operates according to these rules. The laws of navigation never navigated a ship. The rules of architecture never built a house.

'Essay 1 On the Phenomena of Nature', *Essays on the Active Powers of Man* (1785), 47.

Erasmus Reinhold 1511–53

German astronomer

5 Copernicus, the most learned man whom we are able to name other than Atlas and Ptolemy, even though he taught in a most learned manner the demonstrations and causes of motion based on observation, nevertheless fled from the job of constructing tables, so that if anyone computes from his tables, the computation is not even in agreement with his observations on which the foundation of the work rests. Therefore first I have compared the observations of Copernicus with those of Ptolemy and others as to which are the most accurate, but besides the bare observations, I have taken from Copernicus nothing other than traces of demonstrations. As for the tables of mean motion, and of prosthaphaereses and all the rest, I have constructed these anew, following absolutely no other reasoning than that which I have judged to be of maximum harmony.

Dedicatory Epistle to Albert, Marquis of Brandenburg, Duke of Prussia, *Prutenicae Tabulae* (1551), 1585 edition. Quoted in Owen Gingerich, *The Eye of Heaven: Ptolemy, Copernicus, Kepler* (1993), 227.

Johann Philipp Reis 1834–74

German physicist

1 Chemistry is the dirty part of physics.

Quoted in R. Oesper, *The Human Side of Scientists* (1975), 116.

Robert Remak 1815–65

German microscopist

2 The extracellular genesis of cells in animals seemed to me, ever since the publication of the cell theory [of Schwann], just as unlikely as the spontaneous generation of organisms. These doubts produced my observations on the multiplication of blood cells by division in bird and mammalian embryos and on the division of muscle bundles in frog larvae. Since then I have continued these observations in frog larvae, where it is possible to follow the history of tissues back to segmentation.

'Ueber extracellulare Entstehung thierischer Zellen und über Vermehrung derselben durch Theilung', *Archiv für Anatomie, Physiologie und Wissenschaftliche Medicin*, 1852, 1, 49–50. Quoted in Erwin H. Ackerknecht, *Rudolf Virchow: Doctor Statesman Anthropologist* (1953), 83–4.

Ira Remsen 1846–1927

American chemist

3 While reading in a textbook of chemistry, . . . I came across the statement, 'nitric acid acts upon copper.' I was getting tired of reading such absurd stuff and I determined to see what this meant. Copper was more or less familiar to me, for copper cents were then in use. I had seen a bottle marked 'nitric acid' on a table in the doctor's office where I was then 'doing time.' I did not know its peculiarities, but I was getting on and likely to learn. The spirit of adventure was upon me. Having nitric acid and copper, I had only to learn what the words 'act upon' meant . . . I put one of them [cent] on the table, opened the bottle marked 'nitric acid'; poured some of the liquid on the copper; and prepared to make an observation. But what was this wonderful thing which I beheld? The cent was already changed, and it was no small change either. A greenish blue liquid foamed and fumed over the cent and over the table. The air in the neighborhood of the performance became colored dark red. A great colored cloud arose. This was disagreeable and suffocating—how should I stop this? I tried to get rid of the objectionable mess by picking it up and throwing it out of the window, which I had meanwhile opened. I learned another fact—nitric acid not only acts upon copper but it acts upon fingers. The pain led to another unpremeditated experiment. I drew my fingers across my trousers and another fact was discovered. Nitric acid acts upon trousers. Taking everything into consideration, that was the most impressive experiment, and, relatively, probably the most costly experiment I have ever performed.

F. H. Getman, *The Life of Ira Remsen* (1940), 9.

4 Be a physical chemist, an organic chemist, an analytical chemist, if you will, but above all be a *Chemist*.

Cautionary warning to his students against overspecialisation.

F. H. Getman, *The Life of Ira Remsen* (1940), 71.

5 ON REMSEN Remsen never wore his hat inside the door for he had much the same respect for his laboratory that most of us have for a church.

F. H. Getman, *The Life of Ira Remsen* (1940), 68.

Nicholas Rescher 1928–

German-born American philosopher and educator

6 The inherent unpredictability of future scientific developments—the fact that no secure inference can be drawn from one state of science to another—has important implications for the issue of the limits of science. It means that *present-day science cannot speak for future science*: it is in principle impossible to make any secure inferences from the substance of science at one time about its substance at a significantly different time. The prospect of future scientific revolutions can never be precluded. We cannot say with unblinking confidence what sorts of resources and conceptions the science of the future will or will not use. Given that it is effectively impossible to predict the details of what future science will accomplish, it is no less impossible to predict in detail what future science will *not* accomplish. We can never confidently put this or that range of issues outside 'the limits of science', because we cannot discern the shape and substance of future science with sufficient clarity to be able to say with any assurance what it can and cannot do. Any attempt to set 'limits' to science—any advance specification of what science can and cannot do by way of handling problems and solving questions—is destined to come to grief.

The Limits of Science (1984), 102–3.

Roger Randall Dougan Revelle

1909–91

American oceanographer

1 We know less about the ocean's bottom than about the *moon's* back side.

Quoted in his obituary in *Physics Today*, February 1992, 120.

Jean Rey c.1582–1645

French chemist

2 The condensed air becomes attached to [the metallic calx], and adheres little by little to the smallest of its particles: thus its weight increases from the beginning to the end: but when all is saturated, it can take up no more.

The Increase in Weight of Tin and Lead on Calcination (1630), Alembic Club Reprint (1895), 52.

Georg Joachim Rheticus 1514–74

Austrian mathematician and astronomer

3 With regard to the apparent motions of the sun and moon, it is perhaps possible to deny what is said about the motion of the earth, although I do not see how the explanation of precession is to be transferred to the sphere of the stars. But if anyone desires to look either to the principal end of astronomy and the order and harmony of the system of the spheres or to ease and elegance and a complete explanation of the causes of the phenomena, by the assumption of no other hypotheses will he demonstrate the apparent motions of the remaining planets more neatly and correctly. For all these phenomena appear to be linked most nobly together, as by a golden chain; and each of the planets, by its position and order and every inequality of its motion, bears witness that the earth moves and that we who dwell upon the globe of the earth, instead of accepting its changes of position, believe that the planets wander in all sorts of motions of their own.

Narratio Prima (1540). Trans. Edward Rosen, *Three Copernican Treatises* (1939), 164–5.

Dickinson Woodruff Richards

1895–1973

American physician and physiologist

4 The problems are the ones that we have always known. The little gods are still with us, under different names. There is conformity: of technique, leading to repetition; of language, encouraging if not imposing conformity of thought. There is popularity: it is so easy to ride along on an already surging tide; to plant more seed in an already well-ploughed field; so hard to drive a new furrow into stony ground. There is laxness: the disregard of small errors [sic], of deviations, of the unexpected response; the easy worship of the smooth curve. There is also fear: the fear of speculation; the overprotective fear of being wrong. We are forgetful of the curious and wayward dialectic of science, whereby a well-constructed theory even if it is wrong, can bring a signal advance.

Presidential Address, 'Medical Priesthoods:

Past and Present', *Transactions of the Association of American Physicians*, 1962, 75, 6.

Ellen Swallow Richards see **Swallow, Ellen**

Ivor Armstrong Richards

1893–1979

British critic, poet and teacher

1 We believe a scientist because he can substantiate his remarks, not because he is eloquent and forcible in his enunciation. In fact, we distrust him when he seems to be influencing us by his manner.

Science and Poetry (1926), 24.

Joseph William Richards

1864–1921

British-born American chemist and metallurgist

2 Differentiation and specialization are the watchword, now, of all progress,—industrial, scientific, philosophical. The day is past, we all acknowledge, when one man, even he be Newton, can know all that is to be known; the day is also past when one scientific society can cover satisfactorily the whole field of scientific research . . . the analogue of the specialist in science is the *society which specializes.*

Editorial to the first edition of *The Transactions of the American Electrochemical Society*, 1902, 1, 1.

Theodore William Richards

1868–1928

American chemist

3 A large majority of the more active men in American universities, to whom one would look for important research, are so overburdened with hack and routine work and administrative detail that they have but little time and energy left over for investigation. It is usually impossible for the colleges to avoid this, for their funds have been provided for teaching, and not for research. The average professor in America could not use money for apparatus if he had it. His extra time must be spent on insignificant college tasks and in endeavoring to eke out his too limited income by outside work, not in research. He is, moreover, too much worried by pecuniary embarrassment to give his mind freely to abstract thought.

Letter to Alexander Agassiz, 9 June 1902. Quoted in Nathan Reingold and Ida H. Reingold, *Science in America: A Documentary History 1900–1939* (1981), 19.

4 On the other hand it seems to me that the making of a great original discovery in science is not unlike the writing of a great poem or the painting of a great picture. The thought and its execution must be hammered out by genius alone and without the multitude of administrative duties which cooperation would be likely to bring upon its professor . . . I agree with Mr. Carnegie entirely in his belief that from the individual exceptional man alone is any great addition to be expected to a sum of our original conceptions. Who can imagine Faraday as cooperating!?

Letter to Alexander Agassiz, 9 June 1902. Quoted in Nathan Reingold and Ida H. Reingold, *Science in America: A Documentary History 1900–1939* (1981), 28.

5 Our usual laboratory hours are 8.30 to 5.30. On rare occasions that time may be somewhat extended . . . As a rule I think it is not well to work too long in the laboratory. Some reading and thinking must be done. Our chemical laboratory [sic] is open in the evening for reading.

Letter to J. B. Dickson, 2 April 1915. Quoted in Sheldon J. Kopperl, 'T. W. Richards' Role in American Graduate Education in Chemistry', *Ambix*, 1976, 23, 171.

Charles F. Richter 1900–85

American physicist and seismologist

6 In the course of historical or statistical study of earthquakes in any given region it is frequently desirable to have a scale for rating these shocks in terms of their original energy, independently of the effects which may be produced at any particular point of observation. On the suggestion of Mr H. O. Wood, it is proposed to refer to such a scale as a 'magnitude' scale.

'An Instrumental Earthquake Magnitude Scale', *Bulletin of the Seismological Society of America*, 1935, 25, 1.

1 Here I turned to Dr. Gutenberg, who made the natural suggestion to plot amplitudes logarithmically. This improved the situation, at least in appearance. I was lucky, because logarithmic plots are a device of the devil.

'Response to Second Award of the Medal of the Seismological Society of America', *Bulletin of the Seismological Society of America*, 1977, 67, 1245.

Floyd Karker Richtmyer 1881–1939

American physicist

2 The whole history of physics proves that a new discovery is quite likely lurking in the next decimal place.

'The Romance of the Next Decimal Place', *Science*, 1932, 75, 3.

Paul Ricoeur 1913–

French philosopher

3 Psychoanalysis does not satisfy the standards of the sciences of observation, and the 'facts' it deals with are not verifiable by multiple independent observers.

'Technique and Nontechnique in Interpretation', trans. Willis Domingo. *The Conflict of Interpretations: Essays in Hermeneutics* (1974), edited by Don Ihde, 186.

Philip Rieff 1922–

American sociologist

4 It is exhilarating and yet terrifying to read Freud as a moralist, to see how compelling can be the judgment of a man who never preaches, leads us nowhere, assures us of nothing except perhaps that, having learned from him, the burden of misery we must find strength to carry will be somewhat lighter.

Freud: The Mind of the Moralist (1959), Preface, xi.

Georg Friedrich Bernhard Riemann 1826–66

German mathematician

5 The general notion of multiply extended magnitudes [. . .] remained entirely unworked. I have in the first place, therefore, set myself the task of constructing the notion of a multiply extended magnitude out of general notions of magnitude. It will follow from this that a multiply extended magnitude is capable of different metric relations, and consequently that space is only a particular case of a triply extended magnitude [. . .] Thus arises the problem, to discover the simplest matters of fact from which the metric relations of space may be determined.

Habilitation Thesis on Geometry (1854), opening. In *Werke* (1892), 272–3, trans. Ivor Grattan-Guinness.

Lionel Charles Robbins (Baron Robbins of Clare Market) 1898–1984

British economist

6 Economics is the science which studies human behaviour as a relationship between ends and scarce means which have alternative uses.

'The Subject Matter of Economics', *An Essay on the Nature and Significance of Economic Science* (1932), 15.

John D. Roberts 1918–

American chemist

7 We may imagine that a medieval traveler, in the course of his wanderings, saw a rhinoceros, and that, after his return to his home, he attempted to describe this strange beast to his friends. A convenient way for him to convey an approximately correct idea of the animal's appearance would be to say that the rhinoceros is intermediate between a dragon and a unicorn; for, as we may assume, the people to whom he was talking would have a fairly clear idea of what these two latter purely mythological creatures were supposed to look like. Similarly, a convenient way to give an approximate description of the true structure of benzene is to say that it is intermediate between the two purely imaginary Kekulé structures; for, just as the fictitious dragon and unicorn formerly called forth clear pictures to medieval men and women, so also the equally fictitious Kekulé structures now call forth clear pictures to modern organic chemists. In each case, therefore, the unfamiliar reality is explained by reference to familiar fiction.

Quoted in G. W. Wheland, *Resonance in Organic Chemistry* (1955), 4.

Jean-Baptiste René Robinet 1735–1820

French natural philosopher

1 There are only individuals, and no kingdoms or classes or genera or species.

De la Nature (1761–6), Vol. 4, 1–2. Trans. Ernst Mayr, *The Growth of Biological Thought: Diversity, Evolution & Inheritance* (1982), 264.

James Harvey Robinson 1863–1936 and Charles Austin Beard 1874–1948

American historians

2 It may well be that men of science, not kings, or warriors, or even statesmen are to be the heroes of the future.

The Development of Modern Europe (1908), Vol. 2, 421.

Robert Robinson 1886–1975

British chemist

3 *Robert Robinson commented, referring to Frederick Sanger:*

They are not chemists there, just a lot of paper hangers.

During the early 1950s, Todd and his co-workers at Cambridge introduced paper chromotography to identify the various degradation products of vitamin B12.

Obituary of 'Alexander Robertus Todd, O.M., Baron Todd of Trumpington', *Biographical Memoirs of Fellows of the Royal Society*, 2000, 46, 527.

Carl Ransome Rogers 1902–87

American psychologist

4 Unless man can make new and original adaptations to his environment as rapidly as his science can change the environment, our culture will perish.

On Becoming a Person: A Therapist's View of Psychotherapy (1961), 348.

Neil Rollinson 1960–

British poet

5 Your coffee grows cold on the kitchen table,
which means the universe is dying.
Your dress on the carpet is just a dress,
it has lost all sense of you now.
I open the window, the sky is dark
and the house is also cooling, the garden,
the summer lawn, all of it finding an equilibrium.
I watch an ice-cube melt in my wine,
the heat of the Chardonnay passing into the ice.
It means the universe is dying: the second law of thermodynamics.
Entropy rising.
Only the fridge struggles to turn things round
but even here there's hidden loss.
It hums in the corner, the only sound
on a quiet night. Outside, in the vast sky
stars are cooling. I think of the sun
consuming its fuel, the afternoon that is past
and your dress that only this morning was warm to my touch.

'Entropy' (*c.*1999). In *London Review of Books* 1999, 21, 28.

Jean-Baptiste Louis Romé de l'Isle 1736–90

French crystallographer

1 Crystallographic science does not consist in the scrupulous description of all the accidents of crystalline form, but in specifying, by the description of these forms, the more or less close relationship they have with each other.

Cristallographie, 1793, 1, 91.

Alfred Sherwood Romer 1894–1973

American palaeontologist

2 In vertebrate paleontology, increasing knowledge leads to triumphant loss of clarity.

Synapsid Evolution and Dentition. International Colloquium on the Evolution of Mammals, Brussels, 1962.

Wilhelm Conrad Röntgen 1845–1923

German physicist

3 If the hand be held between the discharge-tube and the screen, the darker shadow of the bones is seen within the slightly dark shadow-image of the hand itself . . . For brevity's sake I shall use the expression 'rays'; and to distinguish them from others of this name I shall call them 'X-rays'.

'On a New Kind of Rays', 1895. In Herbert S. Klickstein, *Wilhelm Conrad Röntgen: On a New Kind of Rays, A Bibliographic Study* (1966), 4.

Theodore Roosevelt 1858–1919

American President

4 There are no words that can tell the hidden spirit of the wilderness, that can reveal its mystery . . . The nation behaves well if it treats its natural resources as assets which it must turn over to the next generation increased and not impaired in value.

Inscribed on the wall of the American Museum of Memorial History in New York. Quoted in Peter J. Bowler, *The Fontana History of the Environmental Sciences* (1992), 437.

Wickliffe Rose 1862–1931

American scientific administrator

5 Make the peaks higher.

Unofficial Rockefeller motto.

In Raymond B. Fosdick, *Adventure in Giving. The Story of the General Education Board* (1962), 230.

Israel Rosenfield 1939–

American historian

6 Historians constantly rewrite history, reinterpreting (reorganizing) the records of the past. So, too, when the brain's coherent responses become part of a memory, they are organized anew as part of the structure of consciousness. What makes them memories is that they become part of that structure and thus form part of the sense of self; my sense of self derives from a certainty that my experiences refer back to me, the individual who is having them. Hence the sense of the past, of history, of memory, is in part the creation of the self.

The Strange, Familiar, and Forgotten: An Anatomy of Consciousness (1995), 87.

Ronald Ross 1857–1932

British physician and microbiologist

7 The Panama Canal was dug with a microscope.

Attributed.

8 This day relenting God
Hath placed within my hand
A wondrous thing; and God
Be praised. At His command,
Seeking His secret deeds
With tears and toiling breath,
I find thy cunning seeds,
O million-murdering Death.
I know this little thing
A myriad men will save.
O Death, where is thy sting?
Thy victory, O Grave?

Lines written after the discovery of the malaria parasite in an amopheline mosquito.

Philosophes (1910), 55.

Jean Rostand 1894–1977

French biologist and historian

1 *On tue un homme, on est un assassin. On tue des millions d'hommes, on est un conquérant. On les tue tous, on est un dieu.*

Kill a man, and you are an assassin. Kill millions of men and you are a conqueror. Kill everyone, and you are a god.

Pensées d'un biologiste (1939), 116.

Jean-Jacques Rousseau 1712–78

Swiss-born French philosopher

2 To study men, we must look close by; to study man, we must learn to look afar; if we are to discover essential characteristics, we must first observe differences.

Essai sur l'origine des langues (1781), 384.

Joseph Roux 1834–86

French priest

3 Science is for those who learn; poetry, for those who know.

Meditations of a Parish Priest (1886), trans. Isabel F. Hapgood, Part 1, no. 71, 43.

Wilhelm Roux 1850–1924

German experimental embryologist

4 *Lehre von den Ursachen der organischen Gestaltungen.*

Developmental mechanics . . . is the doctrine of the causes of organic forms.

Archiv für Entwickelungsmechanik der Organismen (1895), 1.

Henry Augustus Rowland 1848–1901

American physicist

5 There is no such thing as absolute truth and absolute falsehood. The scientific mind should never recognise the perfect truth or the perfect falsehood of any supposed theory or observation. It should carefully weigh the chances of truth and error and grade each in its proper position along the line joining absolute truth and absolute error.

'The Highest Aim of the Physicist: Presidential Address Delivered at the 2nd Meeting of the Society, October 28th, 1899', *Bulletin of the American Physical Society*, 1899, 1, 13.

Royal Society of London 1662–

6 *Nullius in Verba.*

On no man's word.

Motto of the Royal Society, also translated as 'nothing upon trust'. A contraction of the phrase from Horace, 'Nullius addictus iurare in verba magistri' (not bound to swear by the words of any master).

7 The business of their weekly Meetings shall be, To order, take account, consider, and discourse of Philosophical Experiments, and Observations: to read, hear, and discourse upon Letters, Reports, and other Papers containing Philosophical matters, as also to view, and discourse upon the productions and rarities of Nature, and Art: and to consider what to deduce from them, or how they may be improv'd for use, or discovery.

'An Abstract of the Statutes of the Royal Society', in Thomas Sprat, *History of the Royal Society* (1667), 145.

8 The experiment of transfusing the blood of one dog into another was made before the Society by Mr King and Mr Thomas Coxe, upon a little mastiff and a spaniel, with very good success, the former bleeding to death, and the latter receiving the blood of the other, and emitting so much of his own as to make him capable of receiving the other.

First Blood Transfusion Recorded in the Minutes of the Royal Society, 14 November 1666. Quoted in Marjorie Hope Nicolson, *Pepys's Diary and the New Science* (1965), 70.

Beverly Anne Rubik

American biophysicist

9 Mainstream biology may be suffering from what I call 'Physics envy' in aiming to reduce life to nothing but well known, typically Newtonian principles of physics and chemistry.

'From the Editor's Desk', *Frontier Perspectives*, 1991, 2, 3.

Rudy von Bitter Rucker 1946–

American mathematician and science fiction writer

1 What is the shape of space? Is it flat, or is it bent? Is it nicely laid out, or is it warped and shrunken? Is it finite, or is it infinite? Which of the following does space resemble more: (a) a sheet of paper, (b) an endless desert, (c) a soap bubble, (d) a doughnut, (e) an Escher drawing, (f) an ice cream cone, (g) the branches of a tree, or (h) a human body?

The Fourth Dimension: And How to Get There (1985), 91.

Eric Gustaf Rudberg

Swedish physicist

2 If the Easter pilgrims in Piazza San Pietro were to represent the carriers in a metal, then an insulator would resemble the Antarctic with one solitary traveller. In the abundance of carriers there is an enormous gap between conductors and insulators.

Nobel Presentation Speech for William Shockley, John Bardeen and Walter Houser Brattain for their researches on semi-conductors and their discovery of the transistor effect which earned them the Nobel Prize in Physics, 1956. *Nobel Lectures: Physics 1942–1962* (1964), 315–6.

William Albert Hugh Rushton 1901–80

British neurophysiologist

3 But he was also a man with great personal magnetism and considerable charm. ['The goal of this presentation', he confided to me before a 12 minute talk describing our work, 'is to impress, rather than inform.'] There were those who misjudged all of this as arrogance.

From an Obituary of W. A. H. Rushton, in *Vision Research*, 1982, 22, 614.

John Ruskin 1819–1900

British writer and critic

4 If only the Geologists would let me alone, I could do very well, but those dreadful Hammers! I hear the clink of them at the end of every cadence of the Bible verses.

Letter to Henry Acland, 24 May 1851.

5 The work of science is to *substitute* facts for appearances, and demonstrations for impressions.

The Stones of Venice (1853), Vol. 3, 36.

6 For a stone, when it is examined, will be found a mountain in miniature. The fineness of Nature's work is so great, that, into a single block, a foot or two in diameter, she can compress as many changes of form and structure, on a small scale, as she needs for her mountains on a large one; and, taking moss for forests, and grains of crystal for crags, the surface of a stone, in by far the plurality of instances, is more interesting than the surface of an ordinary hill; more fantastic in form and incomparably richer in colour,—the last quality being, in fact, so noble in most stones of good birth (that is to say, fallen from the crystalline mountain ranges).

Modern Painters, 4, Containing part 5 of Mountain Beauty (1860), 311.

Bertrand Arthur William Russell (Earl Russell of Kingston Russell) 1872–1970

British logician and philosopher

7 The world of mathematics, which you condemn, is really a beautiful world; it has nothing to do with life and death and human sordidness, but is eternal, cold and passionless. To me, pure mathematics is one of the highest forms of art; it has a sublimity quite special to itself, and an immense dignity derived from the fact that its world is exempt from change and time. I am quite serious in this. The only difficulty is that none but mathematicians can enter this enchanted region, and they hardly ever have a sense of beauty. And mathematics is the only thing we know of that is capable of perfection; in thinking about it we become Gods.

Letter to Helen Thomas, 30 December 1901. Quoted in Nicholas Griffin (ed.), *The Selected Letters of Bertrand Russell* (1992), Vol. 1, 224.

1 This method is, to define as the number of a class the class of all classes similar to the given class. Membership of this class of classes (considered as a predicate) is a common property of all the similar classes and of no others; moreover every class of the set of similar classes has to the set of a relation which it has to nothing else, and which every class has to its own set. Thus the conditions are completely fulfilled by this class of classes, and it has the merit of being determinate when a class is given, and of being different for two classes which are not similar. This, then, is an irreproachable definition of the number of a class in purely logical terms.

The Principles of Mathematics (1903), 115.

2 Arithmetic must be discovered in just the same sense in which Columbus discovered the West Indies, and we no more create numbers than he created the Indians.

The Principles of Mathematics (1903), 451.

3 Frege has the merit of . . . finding a third assertion by recognising the world of logic which is neither mental nor physical.

Our Knowledge of the External World (1914), 201.

4 Mathematics, rightly viewed, possesses not only truth, but supreme beauty—a beauty cold and austere, like that of sculpture.

'The Study of Mathematics' (1902). In *Mysticism and Logic and Other Essays* (1918), 60.

5 Mathematics may be defined as the subject in which we never know what we are talking about, nor whether what we are saying is true.

'Mathematics and Metaphysicians' (1901). In *Mysticism and Logic and Other Essays* (1918), 75.

6 We are . . . led to a somewhat vague distinction between what we may call 'hard' data and 'soft' data. This distinction is a matter of degree, and must not be pressed; but if not taken too seriously it may help to make the situation clear. I mean by 'hard' data those which resist the solvent influence of critical reflection, and by 'soft' data those which, under the operation of this process, become to our minds more or less doubtful.

Our Knowledge of the External World (1925), 75.

7 One of the main purposes of scientific inference is to justify beliefs which we entertain already; but as a rule they are justified with a difference. Our pre-scientific general beliefs are hardly ever without exceptions; in science, a law with exceptions can only be tolerated as a makeshift. Scientific laws, when we have reason to think them accurate, are different in form from the common-sense rules which have exceptions: they are always, at least in physics, either differential equations, or statistical averages. It might be thought that a statistical average is not very different from a rule with exceptions, but this would be a mistake. Statistics, ideally, are accurate laws about large groups; they differ from other laws only in being about groups, not about individuals. Statistical laws are inferred by induction from particular statistics, just as other laws are inferred from particular single occurrences.

The Analysis of Matter (1927), 191.

8 All the conditions of happiness are realized in the life of the man of science.

The Conquest of Happiness (1930), 146.

9 When it was first proposed to establish laboratories at Cambridge, Todhunter, the mathematician, objected that it was unnecessary for students to see experiments performed, since the results could be vouched for by their teachers, all of them of the highest character, and many of them clergymen of the Church of England.

The Scientific Outlook (1931), 74.

1 Sir Arthur Eddington deducts religion from the fact that atoms do not obey the laws of mathematics. Sir James Jeans deduces it from the fact that they do.

The Scientific Outlook (1931), 112.

2 I do not believe that science *per se* is an adequate source of happiness, nor do I think that my own scientific outlook has contributed very greatly to my own happiness, which I attribute to defecating twice a day with unfailing regularity. Science in itself appears to me neutral, that is to say, it increases men's power whether for good or for evil. An appreciation of the ends of life is something which must be superadded to science if it is to bring happiness, but only the kind of society to which science is apt to give rise. I am afraid you may be disappointed that I am not more of an apostle of science, but as I grow older, and no doubt as a result of the decay of my tissues, I begin to see the good life more and more as a matter of balance and to dread all over-emphasis upon any one ingredient.

Letter to W. W. Norton, Publisher, 27 January 1931. In *The Autobiography of Bertrand Russell, 1914–1944* (1968), Vol. 2, 200.

3 I conclude that, while it is true that science cannot decide questions of value, that is because they cannot be intellectually decided at all, and lie outside the realm of truth and falsehood. Whatever knowledge is attainable, must be attained by scientific methods; and what science cannot discover, mankind cannot know.

Religion and Science (1935), 243.

4 The fundamental concept in social science is Power, in the same sense in which Energy is the fundamental concept in physics.

Power: A New Social Analysis (1938), 10.

5 Science, by itself, cannot supply us with an ethic. It can show us how to achieve a given end, and it may show us that some ends cannot be achieved. But among ends that can be achieved our choice must be decided by other than purely scientific considerations. If a man were to say, 'I hate the human race, and I think it would be a good thing if it were exterminated,' we could say, 'Well, my dear sir, let us begin the process with you.' But this is hardly argument, and no amount of science could prove such a man mistaken.

'The Science to Save us from Science', *New York Times Magazine*, 19 March 1950. Reprinted in M. Gardner (ed.), *The Sacred Beetle: Great Essays in Science* (1984), 406–7.

6 I had at one time a very bad fever of which I almost died. In my fever I had a long consistent delirium. I dreamt that I was in Hell, and that Hell is a place full of all those happenings that are improbable but not impossible. The effects of this are curious. Some of the damned, when they first arrive below, imagine that they will beguile the tedium of eternity by games of cards. But they find this impossible, because, whenever a pack is shuffled, it comes out in perfect order, beginning with the Ace of Spades and ending with the King of Hearts. There is a special department of Hell for students of probability. In this department there are many typewriters and many monkeys. Every time that a monkey walks on a typewriter, it types by chance one of Shakespeare's sonnets. There is another place of torment for physicists. In this there are kettles and fires, but when the kettles are put on the fires, the water in them freezes. There are also stuffy rooms. But experience has taught the physicists never to open a window because, when they do, all the air rushes out and leaves the room a vacuum.

'The Metaphysician's Nightmare', *Nightmares*

of Eminent Persons and Other Stories (1954), 38–9.

1 Although this may seem a paradox, all exact science is dominated by the idea of approximation.

W. H. Auden and L. Kronenberger, *The Viking Book of Aphorisms* (1966).

2 ON **RUSSELL** Bertrand Russell had given a talk on the then new quantum mechanics, of whose wonders he was most appreciative. He spoke hard and earnestly in the New Lecture Hall. And when he was done, Professor Whitehead, who presided, thanked him for his efforts, and not least for 'leaving the vast darkness of the subject unobscured'.

Quoted in Robert Oppenheimer, *The Open Mind* (1955), 102.

Edward Stuart Russell

1887–1954

British biologist

3 From the point of view of the pure morphologist the recapitulation theory is an instrument of research enabling him to reconstruct probable lines of descent; from the standpoint of the student of development and heredity the fact of recapitulation is a difficult problem whose solution would perhaps give the key to a true understanding of the real nature of heredity.

Form and Function: A Contribution to the History of Animal Morphology (1916), 312–3.

Henry Norris Russell 1877–1957

American astrophysicist

4 The Sun is no lonelier than its neighbors; indeed, it is a very common-place star,—dwarfish, though not minute,—like hundreds, nay thousands, of others. By accident the brighter component of Alpha Centauri (which is double) is almost the Sun's twin in brightness, mass, and size. Could this Earth be transported to its vicinity by some supernatural power, and set revolving about it, at a little less than a hundred million miles' distance, the star would heat and light the world just as the Sun does, and life and civilization might go on with no radical change. The Milky Way would girdle the heavens as before; some of our familiar constellations, such as Orion, would be little changed, though others would be greatly altered by the shifting of the nearer stars. An unfamiliar brilliant star, between Cassiopeia and Perseus would be—the Sun. Looking back at it with our telescopes, we could photograph its spectrum, observe its motion among the stars, and convince ourselves that it was the same old Sun; but what had happened to the rest of our planetary system we would not know.

The Solar System and its Origin (1935), 2–3.

5 One of the most striking results of modern investigation has been the way in which several different and quite independent lines of evidence indicate that a very great event occurred about two thousand million years ago. The radio-active evidence for the age of meteorites; and the estimated time for the tidal evolution of the Moon's orbit (though this is much rougher), all agree in their testimony, and, what is far more important, the red-shift in the nebulae indicates that this date is fundamental, not merely in the history of our system, but in that of the material universe as a whole.

The Solar System and its Origin (1935), 137.

Lord John Russell 1792–1878

British Prime Minister

6 Astronomy is the science of the harmony of infinite expanse.

Attributed.

Ernest Rutherford (Baron Rutherford of Nelson) 1871–1937

New Zealand-born British physicist

7 I am a great believer in the simplicity of things and as you probably know I am inclined to hang on to broad & simple ideas like grim death until evidence is too strong for my tenacity.

Letter to Irving Langmuir, 10 June 1919. Quoted in Nathan Reingold and Ida H. Reingold, *Science in America: A Documentary History 1900–1939* (1981), 354.

1 We have seen that a proton of energy corresponding to 30,000 volts can effect the transformation of lithium into two fast α-particles, which together have an energy equivalent of more than 16 million volts. Considering the individual process, the output of energy in the transmutation is more than 500 times greater than the energy carried by the proton. There is thus a great gain of energy in the single transmutation, but we must not forget that on an average more than 1000 million protons of equal energy must be fired into the lithium before one happens to hit and enter the lithium nucleus. It is clear in this case that on the whole the energy derived from transmutation of the atom is small compared with the energy of the bombarding particles. There thus seems to be little prospect that we can hope to obtain a new source of power by these processes. It has sometimes been suggested, from analogy with ordinary explosives, that the transmutation of one atom might cause the transmutation of a neighbouring nucleus, so that the explosion would spread throughout all the material. If this were true, we should long ago have had a gigantic explosion in our laboratories with no one remaining to tell the tale. The absence of these accidents indicates, as we should expect, that the explosion is confined to the individual nucleus and does not spread to the neighbouring nuclei, which may be regarded as relatively far removed from the centre of the explosion.

The Transmutation of the Atom (1933), 23–4

2 The energy produced by the breaking down of the atom is a very poor kind of thing. Anyone who expects a source of power from the transformation of these atoms is talking moonshine.

New York Herald Tribune, 12 September 1933.

3 The year 1896 . . . marked the beginning of what has been aptly termed the heroic age of Physical Science. Never before in the history of physics has there been witnessed such a period of intense activity when discoveries of fundamental importance have followed one another with such bewildering rapidity.

'The Electrical Structure of Matter', *Reports of the British Association for the Advancement of Science* (1924), c2.

4 I came into the room, which was half-dark, and presently spotted Lord Kelvin in the audience and realised that I was in for trouble at the last part of my speech dealing with the age of the earth, where my views conflicted with his. To my relief, Kelvin fell fast asleep, but as I came to the important point, I saw the old bird sit up, open an eye and cock a baleful glance at me! Then a sudden inspiration came, and I said Lord Kelvin had limited the age of the earth, *provided no new source was discovered*. That prophetic utterance refers to what we are now considering tonight, radium! Behold! The old boy beamed upon me.

Quoted in A. S. Eve, *Rutherford: Being the Life and Letters of the Rt. Hon. Lord Rutherford, O.M.* (1939), 107.

5 We haven't the money, so we've got to think.

Quoted by R. V. Jones, *Bulletin of the Institute of Physics*, 1962, 13, No. 4, 102.

6 All science is either physics or stamp collecting.

Quoted in J. B. Birks, *Rutherford at Manchester* (1962), 108.

7 Should a young scientist working with me come to me after two years of such work and ask me what to do next, I would advise him to get out of science. After two years of work, if a man does not know what to do next, he will never make a real scientist.

Quoted in R. Oesper, *The Human Side of Scientists* (1975), 165.

1 Don't let me catch anyone talking about the Universe in my department.

Quoted in *Sage: A Life of J. D. Bernal* (1980), 160.

2 Gentlemen, now you will see that now you see nothing. And why you see nothing you will see presently.

Quoted in R. Oesper, *The Human Side of Scientists* (1975), 164.

Gilbert Ryle 1900–76

British philosopher

3 The dogma of the Ghost in the Machine . . . maintains that there exist both bodies and minds; that there occur physical processes and mental processes; that there are mechanical causes of corporeal movements and mental causes of corporeal movements.

The Concept of Mind (1949), 22.

Paul Sabatier 1854–1941

French chemist

4 Theories cannot claim to be indestructible. They are only the plough which the ploughman uses to draw his furrow and which he has every right to discard for another one, of improved design, after the harvest. To be this ploughman, to see my labours result in the furtherance of scientific progress, was the height of my ambition, and now the Swedish Academy of Sciences has come, at this harvest, to add the most brilliant of crowns.

'The Method of direct Hydrogenation by Catalysis', Nobel Lecture, 11 December 1912. *Noble Lectures in Chemistry 1901–1921* (1966), 230–1.

Julius von Sachs 1832–97

German botanist

5 I have repeatedly had cause to refer to certain resemblances between the phenomena of irritability in the vegetable kingdom and those of the animal body, thus touching a province of investigation which has hitherto been far too little cultivated. In the last instance, indeed, I might say animal and vegetable life must of necessity agree in all essential points, including the phenomena of irritability also, since it is established that the animal organism is constructed entirely and simply from the properties of these substances that all vital movements both of plants and animals are to be explained.

Lectures on the Physiology of Plants (1887), 600.

Carl Edward Sagan 1934–96

American astronomer and science fiction writer

See ***Turco, R. P. et al.***

6 The size and age of the Cosmos are beyond ordinary human understanding. Lost somewhere between immensity and eternity is our tiny planetary home.

Cosmos (1981), 4.

7 There is another approach to the extraterrestrial hypothesis of UFO origins. This assessment depends on a large number of factors about which we know little, and a few about which we know literally nothing. I want to make some crude numerical estimate of the probability that we are frequently visited by extraterrestrial beings.

Now, there is a range of hypotheses that can be examined in such a way. Let me give a simple example: Consider the Santa Claus hypothesis, which maintains that, in a period of eight hours or so on December 24–25 of each year, an outsized elf visits one hundred million homes in the United States. This is an interesting and widely discussed hypothesis. Some strong emotions ride on it, and it is argued that at least it does no harm.

We can do some calculations. Suppose that the elf in question spends one second per house. This isn't quite the usual picture—'Ho, Ho, Ho,' and so on—but imagine that he is terribly efficient and very speedy; that would explain why nobody ever sees him very much—only one second per house, after all. With a hundred million houses he has to spend three years just filling stockings. I have assumed he spends no time at all in going from house to house. Even with relativistic reindeer, the time spent in a hundred million houses is three years and not eight hours. This is an example of hypothesis-testing independent of reindeer propulsion mechanisms or debates on the origins of elves. We examine the hypothesis itself, making very straightforward assumptions, and derive a result inconsistent with the hypothesis by many orders of magnitude. We would then suggest that the hypothesis is untenable.

We can make a similar examination, but with greater uncertainty, of the extraterrestrial hypothesis that holds that a wide range of UFOs viewed on the planet Earth are space vehicles from planets of other stars.

The Cosmic Connection: An Extraterrestrial Perspective (1973), 200.

1 For myself, I like a universe that includes much that is unknown and, at the same time, much that is knowable. A universe in which everything is known would be static and dull, as boring as the heaven of some weak-minded theologians. A universe that is unknowable is no fit place for a thinking being. The ideal universe for us is one very much like the universe we inhabit. And I would guess that this is not really much of a coincidence.

'Can We know the Universe?' in M. Gardner (ed.), *The Sacred Beetle and Other Great Essays in Science* (1985), 109.

2 One of the great commandments of science is, 'Mistrust arguments from authority'. (Scientists, being primates, and thus given to dominance hierarchies, of course do not always follow this commandment.)

The Demon-Haunted World: Science as a Candle in the Dark (1996), 31.

Carl Sagan 1934–96 and **Richard P. Turco** 1943–

American astronomer and American atmospheric chemist

See ***Turco, R. P. et al.***

3 If the greenhouse effect is a blanket in which we wrap ourselves to keep warm, nuclear winter kicks the blanket off.

A Path Where No Man Thought: Nuclear Winter and the End of the Arms Race (1990), 24.

4 The prediction of nuclear winter is drawn not, of course, from any direct experience with the consequences of global nuclear war, but rather from an investigation of the governing physics. (The problem does not lend itself to full experimental verification—at least not more than once.)

A Path Where No Man Thought: Nuclear Winter and the End of the Arms Race (1990), 26.

Abdus Salam 1926–96

Pakistani nuclear physicist

5 *Between the frontiers of the three super-states Eurasia, Oceania, and Eastasia, and not permanently in possession of any of them, there lies a rough quadrilateral with its corners at Tangier, Brazzaville, Darwin, and Hongkong. These territories contain a bottomless reserve of cheap labour. Whichever power controls equatorial Africa, or the Middle East or Southern India or the Indonesian Archipelago, disposes also of the bodies of hundreds of millions of ill-paid and hardworking coolies, expended by their conquerors like so much coal or oil in the race to turn out more armaments, to capture more territory, to control more labour, to turn out more armaments, to capture more territory, to control . . .*

Thus George Orwell—in his only reference to the less-developed world.

I wish I could disagree with him. Orwell may have erred in not anticipating the withering of direct colonial controls within the 'quadrilateral' he speaks about; he may not quite have gauged the vehemence of urges to political self-assertion. Nor, dare I hope, was he right in the sombre picture of conscious and heartless exploitation he has painted. But he did not err in predicting persisting poverty and hunger and overcrowding in 1984 among the less privileged nations.

I would like to live to regret my words but twenty years from now, I am positive, the less-developed world will be as hungry, as relatively undeveloped, and as desperately poor, as today.

'The Less-Developed World: How Can We be Optimists?' (1964). Reprinted in *Ideals and Realities* (1984), xv-xvi. [George Orwell, *Nineteen Eighty Four* (1949), Ch. 9, but misquoted.]

1 It is good to recall that three centuries ago, around the year 1660, two of the greatest monuments of modern history were erected, one in the West and one in the East; St. Paul's Cathedral in London and the Taj Mahal in Agra. Between them, the two symbolize, perhaps better than words can describe, the comparative level of architectural technology, the comparative level of craftsmanship and the comparative level of affluence and sophistication the two cultures had attained at that epoch of history. But about the same time there was also created—and this time only in the West—a third monument, a monument still greater in its eventual import for humanity. This was Newton's Principia, published in 1687. Newton's work had no counterpart in the India of the Mughuls.

'Ideals and Realities' (1975). Reprinted in *Ideals and Realities* (1984), 48.

Jonas Edward Salk 1914–95

American microbiologist

2 When you inoculate children with a polio vaccine, you don't sleep well for two or three months.

Greer Williams, *Virus Hunters* (1960), 234.

3 It is safe, and you can't get safer than safe.

Comment when asked about the safety of his polio vaccine.

J. R. Wilson, *Margin of Safety* (1963), 98.

4 IT IS SAID TO AWAIT CERTAINTY IS TO AWAIT ETERNITY.

Telegram to Basil O'Connor, 8 November 1954. In J. S. Smith, *Patenting the Sun: Polio and the Salk Vaccine* (1990), 295.

5 I couldn't possibly have become a member of this Institute, you know, if I hadn't organized it myself.

Jonas Salk on the Salk Institute at La Jolla on the California coast. In R. Carter, *Breakthrough: The Saga of Jonas Salk* (1966), 413.

6 ON SALK No one had ever picked my brains about influenza so expertly as he did.

Sir Macfarlane Burnet on meeting the young Jonas Salk in Ann Arbor in 1943. In M. Burnet, *Changing Patterns: an Atypical Autobiography* (1968), 169.

Frederick Sanger 1918–et al.

British biochemist

7 A DNA sequence for the genome of bacteriophage ΦX174 of approximately 5,375 nucleotides has been determined using the rapid and simple 'plus and minus' method. The sequence identifies many of the features responsible for the production of the proteins of the nine known genes of the organism, including initiation and termination sites for the proteins and RNAs. Two pairs of genes are coded by the same region of DNA using different reading frames.

'Nucleotide Sequence of Bacteriophage ΦX174 DNA', *Nature*, 1977, **265**, 687. (With G. M. Air, B. G. Barrell, N. L. Brown, A. R. Coulson, J. C. Fiddes, C. A. Hutchison III, P. M. Slocombe & M. Smith.)

George Santayana 1863–1952

Spanish-American philosopher

8 The empiricist . . . thinks he believes only what he sees, but he is much better at believing than at seeing.

Scepticism and Animal Faith: An Introduction to a System of Philosophy (1923), 201.

9 Progress, far from consisting in change, depends on retentiveness. When change is absolute there remains no being to improve and no direction is set for possible improvement: and when experience is not retained, as among savages, infancy is perpetual. Those who cannot remember the past are condemned to repeat it.

The Life of Reason, or the Phases of Human Progress (1954), 82.

1 Science, then, is the attentive consideration of common experience; it is common knowledge extended and refined. Its validity is of the same order as that of ordinary perception, memory, and understanding. Its test is found, like theirs, in actual intuition, which sometimes consists in perception and sometimes in intent. The flight of science is merely longer from perception to perception, and its deduction more accurate of meaning from meaning and purpose from purpose. It generates in the mind, for each vulgar observation, a whole brood of suggestions, hypotheses, and inferences. The sciences bestow, as is right and fitting, infinite pains upon that experience which in their absence would drift by unchallenged or misunderstood. They take note, infer, and prophesy. They compare prophesy with event, and altogether they supply—so intent are they on reality—every imaginable background and extension for the present dream.

The Life of Reason, or the Phases of Human Progress (1954), 393.

Edward Sapir 1884–1939

German-born American linguist and anthropologist

2 Language is a guide to 'social reality.' Though language is not ordinarily thought of as essential interest to the students of social science, it powerfully conditions all our thinking about social problems and processes. Human beings do not live in the objective world alone, nor alone in the world of social activity as ordinarily understood, but are very much at the mercy of the particular language which has become the medium of expression for their society. It is quite an illusion to imagine that one adjusts to reality essentially without the use of language and that language is merely an incidental means of solving specific problems of communication or reflection. The fact of the matter is that the 'real world' is to a large extent unconsciously built up on the language habits of the group. No two languages are ever sufficiently similar to be considered as representing the same social reality. The worlds in which different societies live are distinct worlds, not merely the same world with different labels attached.

'The Status of Linguistics as a Science', *Language*, 1929, 5, 207–14. In David Mandelbaum (ed.), *Selected Writings of Edward Sapir in Language, Culture, and Personality* (1949), 162.

George Alfred Léon Sarton

1884–1956

Belgian-born American historian of science

3 It is true that most men of letters and I am sorry to add, not a few scientists, know science only by its material achievements, but ignore its spirit and see neither its internal beauty nor the beauty it extracts from the bosom of nature. Now I would say that to find in the works of science of the past, that which is not and cannot be superseded, is perhaps the most important part of our own quest. A true humanist must know the life of science as he knows the life of art and the life of religion.

'The Faith of a Humanist', *Isis*, 1920, 3, 5.

4 No history of civilization can be tolerably complete which does not give considerable space to the explanation of scientific progress. If we had any doubts about this, it would suffice to ask ourselves what constitutes the essential difference between our and earlier civilizations. Throughout the course of history, in every period, and in almost every country, we find a small number of saints, of great artists, of men of science. The saints of to-day are not necessarily more saintly than those of a thousand years ago; our artists are not necessarily greater than those of early Greece; they are more likely to be inferior; and of course, our men of science are not necessarily more intelligent than those of old; yet one thing is certain, their knowledge is at once more extensive and more accurate. *The acquisition and systematization of positive knowledge is the only human activity which is truly cumulative and progressive.* Our civilization is essentially different from earlier ones, because our knowledge of the world and of ourselves is deeper, more precise, and more certain, because we have gradually learned to disentangle the forces of nature, and because we have contrived, by strict obedience to their laws, to capture them and to divert them to the gratification of our own needs.

Introduction to the History of Science (1927), Vol. 1, 3-4.

1 The most ominous conflict of our time is the difference of opinion, of outlook, between men of letters, historians, philosophers, the so-called humanists, on the one side and scientists on the other. The gap cannot but increase because of the intolerance of both and the fact that science is growing by leaps and bounds.

The History of Science and the New Humanism (1931), 69.

Horace Bénédict de Saussure 1740-99

Swiss geologist

2 The theory of the earth is the science which describes and explains changes that the terrestrial globe has undergone from its beginning until today, and which allows the prediction of those it shall undergo in the future. The only way to understand these changes and their causes is to study the present-day state of the globe in order to gradually reconstruct its earlier stages, and to develop probable hypotheses on its future state. Therefore, the present state of the earth is the only solid base on which the theory can rely.

In Albert V. Carozzi, 'Forty Years of Thinking in Front of the Alps: Saussure's (1796) Unpublished Theory of the Earth', *Earth Sciences History*, **8**, 1989, 136.

George B. Schaller 1933-

German-born American zoologist

3 For what are the whales being killed? For a few hundred jobs and products that are not needed, since there are cheap substitutes. If this continues, it will be the end of living and the beginning of survival. The world is being totaled.

Attributed.

Carl Wilhelm Scheele 1742-86

Swedish chemist

4 I took a glass retort, capable of containing eight ounces of water, and distilled fuming spirit of nitre according to the usual method. In the beginning the acid passed over red, then it became colourless, and lastly again all red: no sooner did this happen, than I took away the receiver; and tied to the mouth of the retort a bladder emptied of air, which I had moistened in its inside with milk of lime *lac calcis*, (i.e. lime-water, containing more quicklime than water can dissolve) to prevent its being corroded by the acid. Then I continued the distillation, and the bladder gradually expanded. Here-upon I left every thing to cool, tied up the bladder, and took it off from the mouth of the retort.—I filled a ten-ounce glass with this air and put a small burning candle into it; when immediately the candle burnt with a large flame, of so vivid a light that it dazzled the eyes. I mixed one part of this air with three parts of air, wherein fire would not burn; and this mixture afforded air, in every respect familiar to the common sort. Since this air is absolutely necessary for the generation of fire, and makes about one-third of our common air, I shall henceforth, for shortness sake call it *empyreal* air, [literally *fire-air*:] the air which is unserviceable for the fiery phenomenon, and which makes abut two-thirds of common air, I shall for the future call *foul air* [literally *corrupted air*].

Chemische Abhandlung von der Luft und dem Feuer (1777), *Chemical Observations and Experiments on Air and Fire* (1780), trans. J. R. Forster, 34–5.

Friedrich Von Schlegel

1772–1829

German philosopher

1 If you want to penetrate into the heart of physics, then let yourself be initiated into the mysteries of poetry.

Friedrich Schlegel's Lucinde and the Fragments, trans. Peter Firchow (1971), 250.

Jacob Mathias Schleiden

1804–81

German botanist

2 Something is as little explained by means of a distinctive vital force as the attraction between iron and magnet is explained by means of the name magnetism. We must therefore firmly insist that in the organic natural sciences, and thus also in botany, absolutely nothing has yet been explained and the entire field is still open to investigation as long as we have not succeeded in reducing the phenomena to physical and chemical laws.

Grundzüge der Wissenschaftlichen Botanik nebst einer Methodologischen Einleitung als Anleitung zum Studium der Planze [Principles of Scientific Botany] (1842–3), Vol. 1, 49. Trans. Kenneth L. Caneva, *Robert Mayer and the Conservation of Energy* (1993), 108.

Maarten Schmidt 1929–

Dutch-born American astronomer

3 The unprecedented identification of the spectrum of an apparently stellar object in terms of a large red-shift suggests either of the two following explanations.

The stellar object is a star with a large gravitational red-shift. Its radius would then be of the order of 10km. Preliminary considerations show that it would be extremely difficult, if not impossible, to account for the occurrence of permitted lines and a forbidden line with the same red-shift, and with widths of only 1 or 2 per cent of the wavelength.

The stellar object is the nuclear region of a galaxy with a cosmological red-shift of 0.158, corresponding to an apparent velocity of 47,400 km/sec. The distance would be around 500 megaparsecs, and the diameter of the nuclear region would have to be less than 1 kiloparsec. This nuclear region would be about 100 times brighter optically than the luminous galaxies which have been identified with radio sources thus far. If the optical jet and component *A* of the radio source are associated with the galaxy, they would be at a distance of 50 kiloparsecs implying a time-scale in excess of 105 years. The total energy radiated in the optical range at constant luminosity would be of the order of 1059 ergs.

Only the detection of irrefutable proper motion or parallax would definitively establish 3*C* 273 as an object within our Galaxy. At the present time, however, the explanation in terms of an extragalactic origin seems more direct and less objectionable.

'3C 273: A Star-like Object with Large Red-Shift', *Nature*, 1963, **197**, 1040.

Robert Schoenfeld

German chemist

4 Each of us has read somewhere that in New Guinea pidgin the word for 'piano' is (I use English spelling) 'this fellow you hit teeth belonging to him he squeal all same pig'. I am inclined to doubt whether this expression is authentic; it looks just like the kind of thing a visitor to the Islands would facetiously invent. But I accept 'cut grass belong head belong me' for 'haircut' as genuine ... Such phrases seem very funny to us, and make us feel very superior to the ignorant foreigners who use long winded expressions for simple matters. And then it is our turn to name quite a simple thing, a small uncomplicated molecule consisting of nothing more than a measly 11 carbons, seven hydrogens, one nitrogen and six oxygens. We sharpen our pencils, consult our rule books and at last come up with 3-[(1, 3- dihydro-1, 3-dioxo-2H-isoindol-2-yl) oxy]-3-oxopropanoic acid. A name like that could drive any self-respecting Papuan to piano-playing.

The Chemist's English (1990), 3rd Edition, 57.

Rudolf Schoenheimer 1898–1941 and **David Rittenberg** 1906–70

German-born American biochemists

1 [The] weakness of biological balance studies has aptly been illustrated by comparison with the working of a slot machine. A penny brings forth one package of chewing gum; two pennies bring forth two. Interpreted according to the reasoning of balance physiology, the first observation is an indication of the conversion of copper into gum; the second constitutes proof.

'The Application of Isotopes to the Study of Intermediary Metabolism', *Science*, 1938, 87, 222.

Arthur Schopenhauer 1788–1860

German philosopher

2 Physics is unable to stand on its own feet, but needs a metaphysics on which to support itself, whatever fine airs it may assume towards the latter.

The World as Will and Representation, trans. E. F. J. Byrne (1958), Vol. 2, 172.

Erwin Schrödinger 1887–1961

Austrian theoretical physicist

3 It is so to speak part of the creed of the atomist that all the partial differential equations of mathematical physics . . . are incorrect in a strictly mathematical sense. For the mathematical symbol of the differential quotient describes the transition in the limit to arbitrarily small spatial variations, while we are convinced that in forming such 'physical' differential quotients we must stop at 'physically infinitely small' regions, i.e., at those that still always contain very many molecules; if we were to push the limiting process further, the quotients concerned, which up to then really were proceeding nearer and nearer toward a definite limit . . . would again begin to vary strongly.

'Zur Dynamik elastisch dekoppelter Punktsysteme', *Annalen der Physik*, 1914, 44. Trans. Walter Moore, *Schrödinger: Life and Thought* (1989), 76.

4 In this communication I wish first to show in the simplest case of the hydrogen atom (nonrelativistic and undistorted) that the usual rates for quantization can be replaced by another requirement, in which mention of 'whole numbers' no longer occurs. Instead the integers occur in the same natural way as the integers specifying the number of nodes in a vibrating string. The new conception can be generalized, and I believe it touches the deepest meaning of the quantum rules.

'Quantisierung als Eigenwertproblem', *Annalen der Physik*, 1926, 79, 361. Trans. Walter Moore, *Schrödinger: Life and Thought* (1989), 200–2.

5 Bohr's standpoint, that a space-time description is impossible, I reject *a limine*. Physics does not consist only of atomic research, science does not consist only of physics, and life does not consist only of science. The aim of atomic research is to fit our empirical knowledge concerning it into our other thinking. All of this other thinking, so far as it concerns the outer world, is active in space and time. If it cannot be fitted into space and time, then it fails in its whole aim and one does not know what purpose it really serves.

Letter to Willy Wien, 25 August 1926. Quoted in Walter Moore, *Schrödinger: Life and Thought* (1989), 226.

1 Every complete set of chromosomes contains the full code; so there are, as a rule, two copies of the latter in the fertilized egg cell, which forms the earliest stage of the future individual.

In calling the structure of the chromosome fibres a code-script we mean that the all-penetrating mind, once conceived by Laplace, to which every causal connection lay immediately open, could tell from their structure whether the egg would develop, under suitable conditions, into a black cock or into a speckled hen, into a fly or a maize plant, a rhododendron, a beetle, a mouse or a woman. To which we may add, that the appearances of the egg cells are very often remarkably similar; and even when they are not, as in the case of the comparatively gigantic eggs of birds and reptiles, the difference is not so much in the relevant structures as in the nutritive material which in these cases is added for obvious reasons.

But the term code-script is, of course, too narrow. The chromosome structures are at the same time instrumental in bringing about the development they foreshadow. They are law-code and executive power—or, to use another simile, they are architect's plan and builder's craft—in one.

What is Life? (1944), 21–2.

2 The great revelation of the quantum theory was that features of discreteness were discovered in the Book of Nature, in a context in which anything other than continuity seemed to be absurd according to the views held until then.

What is Life? (1944), 48.

3 [A living organism] . . . feeds upon negative entropy . . . Thus the device by which an organism maintains itself stationary at a fairly high level of orderliness (= fairly low level of entropy) really consists in continually sucking orderliness from its environment.

What is Life? (1944), 73–5.

4 I belong to those theoreticians who know by direct observation what it means to make a measurement. Methinks it were better if there were more of them.

Quoted in Walter Moore, *Schrödinger: Life and Thought* (1989), 58–9.

5 Science is a game—but a game with reality, a game with sharpened knives . . . If a man cuts a picture carefully into 1000 pieces, you solve the puzzle when you reassemble the pieces into a picture; in the success or failure, both your intelligences compete. In the presentation of a scientific problem, the other player is the good Lord. He has not only set the problem but also has devised the rules of the game—but they are not completely known, half of them are left for you to discover or to deduce. The experiment is the tempered blade which you wield with success against the spirits of darkness—or which defeats you shamefully. The uncertainty is how many of the rules God himself has permanently ordained, and how many apparently are caused by your own mental inertia, while the solution generally becomes possible only through freedom from its limitations.

Quoted in Walter Moore, *Schrödinger: Life and Thought* (1989), 348.

6 Thus, the task is, not so much to see what no one has yet seen; but to think what nobody has yet thought, about that which everybody sees.

Quoted in L. Bertalanffy, *Problems of Life* (1952)

7 If you cannot—in the long run—tell everyone what you have been doing, your doing has been worthless.

Attributed.

Theodor Ambrose Hubert Schwann 1810–82

German physiologist

8 The fibrous material and muscle were thus digested in the same way as the coagulated egg albumen, namely, by free acid in combination with another substance active in very small amounts. Since the latter really carries on the digestion of the most important animal nutrient materials, one might with justice apply to it the name *pepsin*.

'Ueber das Wesen des Verdauungsprocesses', *Archiv für Anatomie, Physiologie und Wissenschaftliche Medicin* (1836), 90–138. Trans. L. G. Wilson, 'The Discovery of Pepsin', in John F. Fulton and Leonard G. Wilson (eds.), *Selected Readings in the History of Physiology* (1966), 191.

1 The principal result of my investigation is that a uniform developmental principle controls the individual elementary units of all organisms, analogous to the finding that crystals are formed by the same laws in spite of the diversity of their forms.

Mikroskopische Untersuchungen über die Uebereinstimmung in der Struktur und dem Wachsthum der Thiere und Pflanzen (1839). *Microscopic Researches into the Accordance in the Structure and Growth of Animals and Plants*, trans. Henry Smith (1847), 1.

2 The elementary parts of all tissues are formed of cells in an analogous, though very diversified manner, so that it may be asserted, *that there is one universal principle of development for the elementary parts of organisms, however different, and that this principle is the formation of cells.*

Mikroskopische Untersuchungen über die Uebereinstimmung in der Struktur und dem Wachsthum der Thiere und Pflanzen (1839). *Microscopic Researches into the Accordance in the Structure and Growth of Animals and Plants*, trans. Henry Smith (1847), 165.

3 We set out, therefore, with the supposition that an organised body is not produced by a fundamental power which is guided in its operation by a definite idea, but is developed, according to blind laws of necessity, by powers which, like those of inorganic nature, are established by the very existence of matter. As the elementary materials of organic nature are not different from those of the inorganic kingdom, the source of the organic phenomena can only reside in another combination of these materials, whether it be in a peculiar mode of union of the elementary atoms to form atoms of the second order, or in the arrangement of these conglomerate molecules when forming either the separate morphological elementary parts of organisms, or an entire organism.

Mikroskopische Untersuchungen über die Uebereinstimmung in der Struktur und dem Wachsthum der Thiere und Pflanzen (1839). *Microscopic Researches into the Accordance in the Structure and Growth of Animals and Plants*, trans. Henry Smith (1847), 190–1.

4 The cause of nutrition and growth resides not in the organism as a whole but in the separate elementary parts—the cells.

Mikroskopische Untersuchungen über die Uebereinstimmung in der Struktur und dem Wachsthum der Thiere und Pflanzen (1839). *Microscopic Researches into the Accordance in the Structure and Growth of Animals and Plants*, trans. Henry Smith (1847), 192.

Agostino Scilla 1629–1700

Italian geologist

5 It would indeed be a great delusion, if we stated that those sports of Nature [we find] enclosed in rocks are there by chance or by some vague creative power. Ah, that would be superficial indeed! In reality, those shells, which once were alive in water and are now dead and decomposed, were made thus by time not Nature; and what we now find as very hard, figured stone, was once soft mud and which received the impression of the shape of a shell, as I have frequently demonstrated.

La vana speculazione disingannata del senso (1670), trans. Ezio Vaccari, 83–4.

George Julius Poulett Scrope 1797–1876

British geologist and politician

6 Geologists have usually had recourse for the explanation of these changes to the supposition of sundry violent and extraordinary catastrophes, cataclysms, or general revolutions having occurred in the physical state of the earth's surface.

As the idea imparted by the term Cataclysm, Catastrophe, or Revolution, is extremely vague, and may comprehend any thing you choose to imagine, it answers for the time very well as an explanation; that is, it stops further inquiry. But it also has had the disadvantage of effectually stopping the advance of science, by involving it in obscurity and confusion.

Considerations on Volcanoes (1825), iv.

1 It has hitherto been a serious impediment to the progress of knowledge, that is in investigating the origin or causes of natural productions, recourse has generally been had to the examination, both by experiment and reasoning, of what *might be* rather than *what is*. The laws or processes of nature we have every reason to believe invariable. Their *results* from time to time vary, according to the combinations of influential circumstances; but the process remains the same. Like the poet or the painter, the chemist may, and no doubt often does, create combinations which nature never produced; and the possibility of such and such processes giving rise to such and such results, is no proof whatever that they were ever in natural operation.

Considerations on Volcanoes (1825), 243.

2 It is very remarkable that while the words *Eternal*, *Eternity*, *Forever*, are constantly in our mouths, and applied without hesitation, we yet experience considerable difficulty in contemplating any definite term which bears a very large proportion to the brief cycles of our petty chronicles. There are many minds that would not for an instant doubt the God of Nature to have *existed from all Eternity*, and would yet reject as preposterous the idea of going back a million of years in the History of *His Works*. Yet what is a million, or a million million, of solar revolutions to an Eternity?

Memoir on the Geology of Central France (1827), 165.

3 The frontispiece of Mr. Lyell's book is enough to throw a Wernerian into fits.

Review of Murchison's *Silurian System*, *Quarterly Review*, 1839, 64, 112.

4 We cannot see how the evidence afforded by the unquestioned progressive development of organised existence—crowned as it has been by the recent creation of the earth's greatest wonder, MAN, can be set aside, or its seemingly necessary result withheld for a moment. When Mr. Lyell finds, as a witty friend lately reported that there had been found, a *silver-spoon in grauwacke*, or a locomotive engine in mica-schist, then, but not sooner, shall we enrol ourselves disciples of the Cyclical Theory of Geological formations.

Review of Murchison's *Silurian System*, *Quarterly Review*, 1839, 64, 112–3.

5 The leading idea which is present in all our [geological] researches, and which accompanies every fresh observation, the sound of which to the ear of the student of Nature seems echoed from every part of her works, is—Time!—Time!—Time!

The Geology and Extinct Volcanoes of Central France (1858), 2nd edition, 208–9.

Paul B. Sears 1891–1990

American ecologist and conservationist

6 The rise of the ecologist almost exactly parallels the decline of the naturalist.

'Some Notes on the Ecology of Ecologists', *The Scientific Monthly*, 1956, 3, 23.

Ivan Mikhaylovich Sechenov 1829–1905

Russian physiologist and psychologist

7 My task was to show the psychologists that it is possible to apply physiological knowledge to the phenomena of psychical life.

'Reflexes of the Brain', *Selected Works* (1935), 335–6.

Adam Sedgwick 1785–1873

British geologist

1 But I think that in the repeated and almost entire changes of organic types in the successive formations of the earth—in the absence of mammalia in the older, and their very rare appearance (and then in forms entirely unknown to us) in the newer secondary groups—in the diffusion of warm-blooded quadrupeds (frequently of unknown genera) through the older tertiary systems—in their great abundance (and frequently of known genera) in the upper portions of the same series—and, lastly, in the recent appearance of man on the surface of the earth (now universally admitted)—in one word, from all these facts combined, we have a series of proofs the most emphatic and convincing,—that the existing order of nature is not the last of an uninterrupted succession of mere physical events derived from laws now in daily operation: but on the contrary, that the approach to the present system of things has been gradual, and that there has been a progressive development of organic structure subservient to the purposes of life.

'Address to the Geological Society, delivered on the Evening of the 18th of February 1831', *Proceedings of the Geological Society*, 1834, 1, 305–6.

2 Considered as a mere question of physics, (and keeping all moral considerations entirely out of sight,) the appearance of man is a geological phenomenon of vast importance, indirectly modifying the whole surface of the earth, breaking in upon any supposition of zoological continuity, and utterly unaccounted for by what we have any right to call the laws of nature.

'Address to the Geological Society, delivered on the Evening of the 18th of February 1831', *Proceedings of the Geological Society*, 1834, 1, 306.

3 Volcanic action is essentially paroxysmal; yet Mr. Lyell will admit no greater paroxysms than we ourselves have witnessed—no periods of feverish spasmodic energy, during which the very framework of nature has been convulsed and torn asunder. The utmost movements that he allows are a slight quivering of her muscular integuments.

'Address to the Geological Society, delivered on the Evening of the 18th of February 1831', *Proceedings of the Geological Society of London*, 1834, 1, 307.

4 The world is not as it was when it came from its Maker's hands. It has been modified by many great revolutions, brought about by an inner mechanism of which we very imperfectly comprehend the movements; but of which we gain a glimpse by studying their effects: and their many causes still acting on the surface of our globe with undiminished power, which are changing, and will continue to change it, as long as it shall last.

Letter 1 to William Wordsworth. Quoted in the appendix to W. Wordsworth, *A Complete Guide to the Lakes, Comprising Minute Direction for the Tourist, with Mr Wordsworth's Description of the Scenery of the County and Three Letters upon the Geology of the Lake District* (1842), 6.

5 Among the older records, we find chapter after chapter of which we can read the characters, and make out their meaning: and as we approach the period of man's creation, our book becomes more clear, and nature seems to speak to us in language so like our own, that we easily comprehend it. But just as we begin to enter on the history of physical changes going on before our eyes, and in which we ourselves bear a part, our chronicle seems to fail us—a leaf has been torn out from nature's record, and the succession of events is almost hidden from our eyes.

Letter 1 to William Wordsworth. Quoted in the appendix to W. Wordsworth, *A Complete Guide to the Lakes, Comprising Minute Direction for the Tourist, with Mr Wordsworth's Description of the Scenery of the County and Three Letters upon the Geology of the Lake District* (1842), 14.

1 We must in imagination sweep off the drifted matter that clogs the surface of the ground; we must suppose all the covering of moss and heath and wood to be torn away from the sides of the mountains, and the green mantle that lies near their feet to be lifted up; we may then see the muscular integuments, and sinews, and bones of our mother Earth, and so judge of the part played by each of them during those old convulsive movements whereby her limbs were contorted and drawn up into their present posture.

Letter 2 to William Wordsworth. Quoted in the appendix to W. Wordsworth, *A Complete Guide to the Lakes, Comprising Minute Direction for the Tourist, with Mr Wordsworth's Description of the Scenery of the County and Three Letters upon the Geology of the Lake District* (1842), 15.

2 We cannot take one step in geology without drawing upon the fathomless stores of by-gone time.

Letter 2 to William Wordsworth. Quoted in the appendix to W. Wordsworth, *A Complete Guide to the Lakes, Comprising Minute Direction for the Tourist, with Mr Wordsworth's Description of the Scenery of the County and Three Letters upon the Geology of the Lake District* (1842), 18–9.

3 The powers of nature are never in repose; her work never stands still.

Letter 3 to William Wordsworth. Quoted in the appendix to W. Wordsworth, *A Complete Guide to the Lakes, Comprising Minute Direction for the Tourist, with Mr Wordsworth's Description of the Scenery of the County and Three Letters upon the Geology of the Lake District* (1842), 49.

4 [*Vestiges* begins] from principles which are at variance with all sober inductive truth. The sober facts of geology shuffled, so as to play a rogue's game; phrenology (that sinkhole of human folly and prating coxcombry); spontaneous generation; transmutation of species; and I know not what; all to be swallowed, without tasting and trying, like so much horse-physic!! Gross credulity and rank infidelity joined in unlawful marriage, and breeding a deformed progeny of unnatural conclusions!

Letter to Charles Lyell, 9 April 1845. In John Willis Clark and Thomas McKenny Hughes (eds.), *The Life and Letters of the Reverend Adam Sedgwick* (1890), Vol. 2, 83.

5 If the [*Vestiges*] be true, the labours of sober induction are in vain; religion is a lie; human law is a mass of folly, and a base injustice; morality is moonshine; our labours for the black people of Africa were works of madmen; and man and woman are only better beasts!

Letter to Charles Lyell, 9 April 1845. In John Willis Clark and Thomas McKenny Hughes (eds.), *The Life and Letters of the Reverend Adam Sedgwick* (1890), Vol. 2, 84.

Emilio Gino Segrè 1905–89

Italian-born American physicist

6 The most striking impression was that of an overwhelming bright light. I had seen under similar conditions the explosion of a large amount—100 tons—of normal explosives in the April test, and I was flabbergasted by the new spectacle. We saw the whole sky flash with unbelievable brightness in spite of the very dark glasses we wore. Our eyes were accommodated to darkness, and thus even if the sudden light had been only normal daylight it would have appeared to us much brighter than usual, but we know from measurements that the flash of the bomb was many times brighter than the sun. In a fraction of a second, at our distance, one received enough light to produce a sunburn. I was near Fermi at the time of the explosion, but I do not remember what we said, if anything. I believe that for a moment I thought the explosion might set fire to the atmosphere and thus finish the earth, even though I knew that this was not possible.

Enrico Fermi: Physicist (1970), 147.

Hans Selye 1907–82

Austrian-born Canadian physician and endocrinologist

1 The fairest thing we can experience is the mysterious. It is the fundamental emotion which stands at the cradle of true science. He who knows it not, and can no longer wonder, no longer feel amazement, is as good as dead. We all had this priceless talent when we were young. But as time goes by, many of us lose it. The true scientist never loses the faculty of amazement. It is the essence of his being.

Newsweek, 31 March 1958.

Lucius Annaeus Seneca (the elder) *c.*60 BC–*c.*37 AD

Roman rhetorician and writer

2 *Quaedam remedia graviora ipsis periculis sunt.*

Some cures are worse than the dangers they combat.

Controversiae, 6.7. In M. Winterbottom (ed.), *The Elder Seneca* (1974), Vol. 1, 520.

Antoine Étienne Reynaud Augustin Serres 1786–1868

French comparative anatomist and embryologist

3 As our researches have made clear, an animal high in the organic scale only reaches this rank by passing through all the intermediate states which separate it from the animals placed below it. Man only becomes man after traversing transitional organisatory states which assimilate him first to fish, then to reptiles, then to birds and mammals.

Annales des Sciences Naturelles, 1834, **2 (ii)**, 248. Trans. in E. S. Russell, *Form and Function* (1916), 82.

François-Joseph Servois

1767–1847

French mathematician

4 [The infinitely small] neither have nor can have theory; it is a dangerous instrument in the hands of beginners [. . .] *anticipating*, for my part, *the judgement of posterity*, I would dare predict that this method will be accused one day, and rightly, of having retarded the progress of the mathematical sciences.

Annales des Mathématiques Pures et Appliquées, 1814–5, 5, 148.

William Shakespeare 1564–1616

English playwright and poet

5 Plutus himself,
That knows the tinct and multiplying med'cine,
Hath not in nature's mystery more science
Than I have in this ring.

All's Well that Ends Well (1603–4), V, iii.

6 In nature's infinite book of secrecy
A little I can read.

Antony and Cleopatra (1606–7), I, ii.

7 LEPIDUS: What manner o' thing is your crocodile?
ANTONY: It is shaped, sir, like itself, and it is as broad as it hath breadth. It is just so high as it is, and moves with it own organs. It lives by that which nourisheth it, and the elements once out of it, it transmigrates.
LEPIDUS: What colour is it of?
ANTONY: Of its own colour, too.
LEPIDUS: 'Tis a strange serpent.
ANTONY: 'Tis so, and the tears of it are wet.

Antony and Cleopatra (1606–7), II, vii.

8 The soul and body rive not more in parting
Than greatness going off.

Antony and Cleopatra (1606–7), IV, xiii.

9 It is as easy to count atomies as to resolve the propositions of a lover.

As You Like it (1599), III, ii.

1 There's nothing situate under heaven's eye
But hath his bond in earth, in sea, in sky.
The beasts, the fishes, and the wingèd fowls
Are their males' subjects and at their controls.
Man, more divine, the master of all these,
Lord of the wide world and wild wat'ry seas,
Indu'd with intellectual sense and souls,
Of more pre-eminence than fish and fowls,
Are masters to their females, and their lords;
Then let your will attend on their accords.
The Comedy of Errors (1594), II, i.

2 QUEEN: Thou know'st 'tis common—all that lives must die,
Passing through nature to eternity.
HAMLET: Ay, madam, it is common.
Hamlet (1601), I, ii.

3 There are more things in heaven and earth, Horatio,
Than are dreamt of in your philosophy.
Hamlet (1601), I, v.

4 For to define true madness,
What is't but to be nothing else but mad?
Hamlet (1601), II, ii.

5 There is nothing either good or bad, but thinking makes it so.
Hamlet (1601), II, ii.

6 O God, I could be bounded in a nutshell and count myself a king of infinite space, were it not that I have bad dreams.
Hamlet (1601), II, ii.

7 It goes so heavily with my disposition that this goodly frame, the earth, seems to me a sterile promontory. This most excellent canopy the air, look you, this brave o'erhanging, this majestic roof fretted with golden fire—why, it appears no other thing to me than a foul and pestilent congregation of vapours. What a piece of work is a man! How noble in reason, how infinite in faculty, in form and moving, how express and admirable, in action, how like an angel! in apprehension, how like a god—the beauty of the world, the paragon of animals! And yet to me, what is this quintessence of dust? Man delights not me—no, nor woman neither, though by your smiling you seem to say so.
Hamlet (1601), II, ii.

8 GLENDOWER: I can call spirits from the vasty deep.
HOTSPUR: Why, so can I, or so can any man;
But will they come when you do call for them?
Henry IV, Part 1 (1597), III, i.

9 Oh God! that one might read the book of fate,
And see the revolution of the times
Make mountains level, and the continent,
Weary of solid firmness, melt itself
Into the sea.
Henry V (1599), I, ii.

10 Creatures that by a rule in nature teach
The act of order to a peopled kingdom.
They have a king and officers of sorts;
Where some, like magistrates, correct at home,
Others, like merchants, venture trade abroad,
Others, like soldiers, armèd in their stings,
Make boot upon the summer's velvet buds;
Which pillage they with merry march bring home
To the tent-royal of their emperor.
Who, busied in his majesty, surveys
The singing masons building roofs of gold;
The civil citizens kneading up the honey;
The poor mechanic porters crowding in
Their heavy burdens at his narrow gate;

The sad-eyed justice, with his surly hum,
Delivering o'er to executors pale
The lazy yawning drone.
Henry V (1599), I, ii.

1 CALPURNIA: When beggars die there are no comets seen;
The heavens themselves blaze forth the death of princes.
CAESAR: Cowards die many times before their deaths;
The valiant never taste of death but once.
Of all the wonders that I have yet heard,
It seems to me most strange that men should fear,
Seeing that death, a necessary end,
Will come when it will come.
Julius Caesar (1599), II, ii.

2 And nature must obey necessity.
Julius Caesar (1599), IV, iii.

3 This is the excellent foppery of the world: that when we are sick in fortune—often the surfeits of our own behaviour—we make guilty of our disasters the sun, the moon, and stars, as if we were villains on necessity, fools by heavenly compulsion, knaves, thieves, and treachers by spherical predominance, drunkards, liars, and adulterers, by an enforced obedience of planetary influence, and all that we are evil in, by a divine thrusting on. An admirable evasion of whoremaster man, to lay his goatish disposition on the charge of a star! My father compounded with my mother under the Dragon's tail and my nativity was under Ursa Major, so that it follows that I am rough and lecherous. Fut! I should have been that I am had the maidenliest star in the firmament twinkled on my bastardizing.
King Lear (1605–6), I, ii.

4 I'll teach you differences.
King Lear (1605–6), 1, iv.

5 Blow, winds, and crack your cheeks! Rage, blow,
You cataracts and hurricanoes, spout
Till you have drench'd our steeples, drowned the cocks!
You sulph'rous and thought-executing fires,
Vaunt-couriers of oak-cleaving thunderbolts,
Singe my white head; and thou all-shaking thunder,
Strike flat the thick rotundity o'th' world,
Crack nature's moulds, all germens spill at once
That makes ingrateful man.
King Lear (1605–6), III, ii.

6 These are begot in the ventricle of memory, nourished in the womb of *pia mater*.
Love's Labour's Lost (1595), IV, ii.

7 There's no art
To find the mind's construction in the face.
Macbeth (1606), I, iv.

8 MACBETH: Canst thou not minister to a mind diseased,
Pluck from the memory a rooted sorrow,
Raze out the written troubles of the brain,
And with some sweet oblivious antidote
Cleanse the stuffed bosom of that perilous stuff
Which weighs upon the heart?
DOCTOR: Therein the patient
Must minister to himself.
MACBETH: Throw physic to the dogs; I'll none of it.
Macbeth (1606), V, iii.

9 Thou hast nor youth nor age,
But as it were an after-dinner's sleep
Dreaming on both; for all thy blessèd youth
Becomes as agèd, and doth beg the alms
Of palsied eld.
Measure for Measure (1604), III, i.

10 CLAUDIO: Death is a fearful thing.
ISABELLA: And shamed life a hateful.
CLAUDIO: Ay, but to die, and go we know not where;
To lie in cold obstruction, and to rot;

This sensible warm motion to become
A kneaded clod; and the delighted spirit
To bathe in fiery floods, or to reside
In thrilling region of thick-ribbèd ice;
To be imprisioned in the viewless winds,
And blown with restless violence round about
The pendant world; or to be worst than worst
Of those lawless and incertain thought
Imagine howling—'tis too horrible!
The weariest and most loathèd worldly life
That age, ache, penury, and imprisionment
Can lay on nature is a paradise
To what we fear of death.

Measure for Measure (1604), III, i.

1 Therefore the moon, the governess of floods,
Pale in her anger washes all the air,
That rheumatic diseases do abound;
And through this distemperature we see
The seasons alter: hoary-headed frosts
Fall in the fresh lap of the crimson rose.

A Midsummer Night's Dream (1595–6), II, i.

2 For there was never yet philosopher
That could endure the toothache patiently,
However they have writ the style of gods,
And made a push at chance and sufferance.

Much Ado about Nothing (1598–9), V, i.

3 Be not afeard. The isle is full of noises,
Sounds, and sweet airs, that give delight and hurt not.
Sometimes a thousand twangling instruments
Will hum about mine ears; and sometime voices
That if I then had waked after long sleep
Will make me sleep again; and then, in dreaming
The clouds methought would open and show riches
Ready to drop upon me, that, when I waked,
I cried to dream again.

The Tempest (1611), III, ii.

4 Our revels are now ended. These our actors,
As I foretold you, were all spirits, and
Are melted into air, into thin air;
And like the baseless fabric of this vision,
The cloud-capp'd towers, the gorgeous palaces,
The solemn temples, the great globe itself,
Yea, all which it inherit, shall dissolve.
And like this insubstantial pageant faded,
Leave not a rack behind. We are such stuff
As dreams are made on; and our little life
Is rounded with a sleep.

The Tempest (1611), IV, i.

5 One touch of nature makes the whole world kin.

Troilus and Cressida (1602), III, iii.

6 SIR TOBY: Does not our lives consist of the four elements?
SIR ANDREW: Faith, so they say; but I think it rather consists of eating and drinking.
SIR TOBY: Thou'rt a scholar; let us therefore eat and drink.

Twelfth Night (1601), II, iii.

7 Roses have thorns, and silver fountains mud.
Clouds and eclipses stain both moon and sun,
And loathsome canker lives in sweetest bud.
All men make faults, and even I in this.

Sonnet 35.

8 Since brass, nor stone, nor earth, nor boundless sea,
But bad mortality o'ersways their power,
How with this rage shall beauty hold a plea,
Whose action is no stronger than a flower?

Sonnet 65.

George Bernard Shaw

1856–1950

British playwright, poet and critic

1 Very nice sort of place, Oxford, I should think, for people that like that sort of place. They teach you to be a gentleman there. In the polytechnic they teach you to be an engineer or such like. See?

Man and Superman: A Comedy and a Philosophy (1903), Act 2, 50.

2 There is at bottom only one genuinely scientific treatment for all diseases, and that is to stimulate the phagocytes.

The Doctor's Dilemma (1911), Act 1, 24.

3 You can be a thorough-going Neo-Darwinian without imagination, metaphysics, poetry, conscience, or decency. For 'Natural Selection' has no moral significance: it deals with that part of evolution which has no purpose, no intelligence, and might more appropriately be called accidental selection, or better still, Unnatural Selection, since nothing is more unnatural than an accident. If it could be proved that the whole universe had been produced by such Selection, only fools and rascals could bear to live.

Back to Methuselah: A Metabiological Pentateuch (1921), Penguin edition (1939), xlii.

4 Science is always wrong. It never solves a problem without creating ten more.

Attributed.

Mary Wollstonecraft Shelley

1797–1851

British novelist

5 It was the secrets of heaven and earth that I desired to learn.

Frankenstein (1818), Ch. 2, ed. M. K. Joseph (1971), 37.

6 I beheld the wretch—the miserable monster whom I had created.

Frankenstein (1818), Ch. 5, ed. M. K. Joseph (1971), 58.

7 All men hate the wretched; how, then, must I be hated, who am miserable beyond all living things! Yet you, my creator, detest and spurn me, thy creature, to whom thou are bound by ties only dissoluble by the annihilation of one of us.

Frankenstein (1818), Ch. 10, ed. M. K. Joseph (1971), 99.

8 Every where I see bliss, from which I alone am irrevocably excluded.

Frankenstein (1818), Ch. 10, ed. M. K. Joseph (1971), 100.

9 Teach him to think for himself? Oh, my God, teach him rather to think like other people!

On her son's education.

In Matthew Arnold, *Essays in Criticism*, Second Series (1888).

Percy Bysshe Shelley 1792–1822

British romantic poet

10 I am the daughter of earth and water,
And the nursling of the sky;
I pass through the pores of the ocean and shores;
I change, but I cannot die.
For after the rain when with never a stain,
The pavilion of Heaven is bare,
And the winds and sunbeams with their convex gleams,
Build up the blue dome of air,
I silently laugh at my own cenotaph,
And out of the caverns of rain,
Like a child from the womb, like a ghost from the tomb,
I arise and unbuild it again.

The Cloud (1820). In K. Raine (ed.), *Shelley* (1974), 289.

11 The One remains, the many change and pass;
Heaven's light forever shines, Earth's shadows fly;
Life, like a dome of many-coloured glass,
Stains the white radiance of Eternity,
Until Death tramples it to fragments.

Adonais (1821), St. 52. In K. Raine (ed.), *Shelley* (1974), 209.

Charles Scott Sherrington

1857–1952
British physiologist

1 The terminal path may, to distinguish it from internuncial common paths, be called *the final common path*. The motor nerve to a muscle is a collection of such final common paths.

'Correlation of Reflexes and the Principle of the Common Path', *Report of the Seventy-Fourth Meeting of the British Association for the Advancement of Science* (1904), 730.

2 This integrative action in virtue of which the nervous system unifies from separate organs an animal possessing solidarity, an individual, is the problem before us.

The Integrative Action of the Nervous System (1906), 2.

3 If we denote excitation as an end-effect by the sign *plus* (+), and inhibition as end-effect by the sign *minus* (−), such a reflex as the scratch-reflex can be termed a reflex of double-sign, for it develops excitatory end-effect and then inhibitory end-effect even during the duration of the exciting stimulus.

The Integrative Action of the Nervous System (1906), 83.

4 With the nervous system intact the reactions of the various parts of that system, the 'simple reflexes', are ever combined into great unitary harmonies, actions which in their sequence one upon another constitute in their continuity what may be termed the 'behaviour'.

The Integrative Action of the Nervous System (1906), 237.

5 The role of inhibition in the working of the central nervous system has proved to be more and more extensive and more and more fundamental as experiment has advanced in examining it. Reflex inhibition can no longer be regarded merely as a factor specially developed for dealing with the antagonism of opponent muscles acting at various hinge-joints. Its role as a coordinative factor comprises that, and goes beyond that. In the working of the central nervous machinery inhibition seems as ubiquitous and as frequent as is excitation itself. The whole quantitative grading of the operations of the spinal cord and brain appears to rest upon mutual interaction between the two central processes 'excitation'and 'inhibition', the one no less important than the other. For example, no operation can be more important as a basis of coordination for a motor act than adjustment of the quantity of contraction, e.g. of the number of motor units employed and the intensity of their individual tetanic activity. This now appears as the outcome of nice co-adjustment of excitation and inhibition upon each of all the individual units which cooperate in the act.

'Inhibition as a Coordinative Factor', Nobel Lecture, 12 December 1932. *Nobel Lectures: Physiology or Medicine 1922–1941* (1965), 288.

6 The brain seems a thoroughfare for nerve-action passing its way to the motor animal. It has been remarked that Life's aim is an act not a thought. To-day the dictum must be modified to admit that, often, to refrain from an act is no less an act than to commit one, because inhibition is coequally with excitation a nervous activity.

The Brain and its Mechanism (1933), 10.

7 The brain is waking and with it the mind is returning. It is as if the Milky Way entered upon some cosmic dance. Swiftly the head-mass becomes an enchanted loom where millions of flashing shuttles weave a dissolving pattern, always a meaningful pattern though never an abiding one.

Man on His Nature (1940), 225.

8 Natural knowledge has not forgone emotion. It has simply taken for itself new ground of emotion, under impulsion from and in sacrifice to that one of its 'values', Truth.

Man on His Nature (1940), 404.

1 That our being should consist of *two* fundamental elements [physical and psychical] offers I suppose no greater inherent improbability than that it should rest on one only.

The Integrative Action of the Nervous System (1947), Foreword to 1947 Edition, xx.

Nevil Vincent Sidgwick

1873–1952

British chemist

2 [The] structural theory is of extreme simplicity. It assumes that the molecule is held together by links between one atom and the next: that every kind of atom can form a definite small number of such links: that these can be single, double or triple: that the groups may take up any position possible by rotation round the line of a single but not round that of a double link: finally that with all the elements of the first short period [of the periodic table], and with many others as well, the angles between the valencies are approximately those formed by joining the centre of a regular tetrahedron to its angular points. No assumption whatever is made as to the mechanism of the linkage. Through the whole development of organic chemistry this theory has always proved capable of providing a different structure for every different compound that can be isolated. Among the hundreds of thousands of known substances, there are never more isomeric forms than the theory permits.

Presidential Address to the Chemical Society, 16 April 1936, *Journal of the Chemical Society*, 1936, 533.

3 There was once an Editor of the Chemical Society, given to dogmatic expressions of opinion, who once duly said firmly that 'isomer' was wrong usage and 'isomeride' was correct, because the ending 'er' always meant a 'do-er'. 'As in water?' snapped Sidgwick.

Obituary of Nevil Vincent Sidgwick by L. E Sutton, *Proceedings of the Chemical Society*, 1958, 318.

Benjamin Silliman 1779–1864

American chemist and geologist

4 People have now-a-days got a strange opinion that everything should be taught by lectures. Now, I cannot see that lectures can do so much good as reading the books from which the lectures are taken. I know nothing that can best be taught by lectures, except where experiments are to be shewn. You may teach chemistry by lectures.

Elements of Chemistry (1830)

5 Science has thus, most unexpectedly, placed in our hands a new power of great but unknown energy. It does not wake the winds from their caverns; nor give wings to water by the urgency of heat; nor drive to exhaustion the muscular power of animals; nor operate by complicated mechanism; nor summon any other form of gravitating force, but, by the simplest means—the mere contact of metallic surfaces of small extent, with feeble chemical agents, a power everywhere diffused through nature, but generally concealed from our senses, is mysteriously evolved, and by circulation in insulated wires, it is still more mysteriously augmented, a thousand and a thousand fold, until it breaks forth with incredible energy.

Commenting on 'The Notice of the Electro-Magnetic Machine of Mr. Thomas Davenport, of Brandon, near Rutland, Vermont, U.S.', *The Annals of Electricity, Magnetism, & Chemistry; and Guardian of Experimental Science*, 1838, 2, 263.

George Gaylord Simpson

1902–84

American palaeontologist

6 The attempted synthesis of paleontology and genetics, an essential part of the present study, may be particularly surprising and possibly hazardous. Not long ago, paleontologists felt that a geneticist was a person who shut himself in a room, pulled down the shades, watched small flies disporting themselves in milk bottles, and thought that he was studying nature. A pursuit so removed from the realities of life, they said, had no significance for the true biologist. On the other hand, the geneticists said that paleontology had no further contributions to make to biology, that its only point had been the completed demonstration of the truth of evolution, and that it was a subject too purely descriptive to merit the name 'science'. The paleontologist, they believed, is like a man who undertakes to study the principles of the internal combustion engine by standing on a street corner and watching the motor cars whiz by.

Tempo and Mode in Evolution (1944), 1.

1 In summary, very large populations may differentiate rapidly, but their sustained evolution will be at moderate or slow rates and will be mainly adaptive. Populations of intermediate size provide the best conditions for sustained progressive and branching evolution, adaptive in its main lines, but accompanied by inadaptive fluctuations, especially in characters of little selective importance. Small populations will be virtually incapable of differentiation or branching and will often be dominated by random inadaptive trends and peculiarly liable to extinction, but will be capable of the most rapid evolution as long as this is not cut short by extinction.

Tempo and Mode in Evolution (1944), 70–1.

2 The theory here developed is that mega-evolution normally occurs among small populations that become preadaptive and evolve continuously (without saltation, but at exceptionally rapid rates) to radically different ecological positions. The typical pattern involved is probably this: A large population is fragmented into numerous small isolated lines of descent. Within these, inadaptive differentiation and random fixation of mutations occur. Among many such inadaptive lines one or a few are preadaptive, i.e., some of their characters tend to fit them for available ecological stations quite different from those occupied by their immediate ancestors. Such groups are subjected to strong selection pressure and evolve rapidly in the further direction of adaptation to the new status. The very few lines that successfully achieve this perfected adaptation then become abundant and expand widely, at the same time becoming differentiated and specialized on lower levels within the broad new ecological zone.

Tempo and Mode in Evolution (1944), 123.

3 The meaning of human life and the destiny of man cannot be separable from the meaning and destiny of life in general. 'What is man?' is a special case of 'What is life?' Probably the human species is not intelligent enough to answer either question fully, but even such glimmerings as are within our powers must be precious to us. The extent to which we can hope to understand ourselves and to plan our future depends in some measure on our ability to read the riddles of the past. The present, for all its awesome importance to us who chance to dwell in it, is only a random point in the long flow of time. Terrestrial life is one and continuous in space and time. Any true comprehension of it requires the attempt to view it whole and not in the artificial limits of any one place or epoch. The processes of life can be adequately displayed only in the course of life throughout the long ages of its existence.

The Meaning of Evolution: A Study of the History of Life and of its Significance for Man (1949), 9.

4 Life arose as a living molecule or protogene, the progression from this stage to that of the ameba is at least as great as from ameba to man. All the essential problems of living organisms are already solved in the one-celled (or, as many now prefer to say, noncellular) protozoan and these are only elaborated in man or the other multicellular animals. The step from nonlife to life may not have been so complex, after all, and that from cell to multicellular organism is readily comprehensible. The change from protogene to protozoan was probably the most complex that has occurred in evolution, and it may well have taken as long as the change from protozoan to man.

The Meaning of Evolution: A Study of the History of Life and of its Significance for Man (1949), 16.

1 Because intelligence is our own most distinctive feature, we may incline to ascribe superior intelligence to the basic primate plan, or to the basic plan of the mammals in general, but this point requires some careful consideration. There is no question at all that most mammals of today are more intelligent than most reptiles of today. I am not going to try to define intelligence or to argue with those who deny thought or consciousness to any animal except man. It seems both common and scientific sense to admit that ability to learn, modification of action according to the situation, and other observable elements of behavior in animals reflect their degrees of intelligence and permit us, if only roughly, to compare these degrees. In spite of all difficulties and all the qualifications with which the expert (quite properly) hedges his conclusions, it also seems sensible to conclude that by and large an animal is likely to be more intelligent if it has a larger brain at a given body size and especially if its brain shows greater development of those areas and structures best developed in our own brains. After all, we *know* we are intelligent, even though we wish we were more so.

The Meaning of Evolution: A Study of the History of Life and of its Significance for Man (1949), 78.

2 Scientists and particularly the professional students of evolution are often accused of a bias toward mechanism or materialism, even though believers in vitalism and in finalism are not lacking among them. Such bias as may exist is inherent in the method of science. The most successful scientific investigation has generally involved treating phenomena *as if* they were purely materialistic, rejecting any metaphysical hypothesis as long as a physical hypothesis seems possible. The method works. The restriction is necessary because science is confined to physical means of investigation and so it would stultify its own efforts to postulate that its subject is not physical and so not susceptible to its methods.

The Meaning of Evolution: A Study of the History of Life and of its Significance for Man (1949), 127.

3 It is still false to conclude that man is *nothing but* the highest animal, or the most progressive product of organic evolution. He is also a fundamentally new sort of animal and one in which, although organic evolution continues on its way, a fundamentally new sort of evolution has also appeared. The basis of this new sort of evolution is a new sort of heredity, the inheritance of learning. This sort of heredity appears modestly in other mammals and even lower in the animal kingdom, but in man it has incomparably fuller development and it combines with man's other characteristics unique in degree with a result that cannot be considered unique only in degree but must also be considered unique in kind.

The Meaning of Evolution: A Study of the History of Life and of its Significance for Man (1949), 286.

4 The meaning that we are seeking in evolution is its meaning to us, to man. The ethics of evolution must be human ethics. It is one of the many unique qualities of man, the new sort of animal, that he is the only ethical animal. The ethical need and its fulfillment are also products of evolution, but they have been produced in man alone.

The Meaning of Evolution: A Study of the History of Life and of its Significance for Man (1949), 309.

1 Man has risen, not fallen. He can choose to develop his capacities as the highest animal and to try to rise still farther, or he can choose otherwise. The choice is his responsibility, and his alone. There is no automatism that will carry him upward without choice or effort and there is no trend solely in the right direction. Evolution has no purpose; man must supply this for himself. The means to gaining right ends involve both organic evolution and human evolution, but human choice as to what *are* the right ends must be based on human evolution.

The Meaning of Evolution: A Study of the History of Life and of its Significance for Man (1949), 310.

2 Human judgment is notoriously fallible and perhaps seldom more so than in facile decisions that a character has no adaptive significance because we do not know the use of it.

The Major Features of Evolution (1953), 166.

3 The science of systematics has long been affected by profound philosophical preconceptions, which have been all the more influential for being usually covert, even subconscious.

The Major Features of Evolution (1953), 340.

4 To put it crudely but graphically, the monkey who did not have a realistic perception of the tree branch he jumped for was soon a dead monkey—and therefore did not become one of our ancestors.

This View of Life: The World of an Evolutionist (1963), 98.

5 The search for historical laws is, I maintain, mistaken in principle.

'Historical Science', C. C. Albritton (ed.), *The Fabric of Geology* (1963), 29.

6 A rill in a barnyard and the Grand Canyon represent, in the main, stages of valley erosion that began some millions of years apart.

'Uniformitarianism. An Inquiry into Principle, Theory, and Method in Geohistory and Biohistory', M. K. Hecht and W. C. Steere (eds.), *Essays in Evolution and Genetics in Honor of Theodosius Dobzhansky* (1970), 83.

Upton Beall Sinclair 1878–1968

American novelist and polemicist

7 And as for other men, who worked in tank-rooms full of steam, and in some of which there were open vats near the level of the floor, their peculiar trouble was that they fell into the vats; and when they were fished out, there was never enough of them left to be worth exhibiting,—sometimes they would be overlooked for days, till all but the bones of them had gone out into the world as Durham's Pure Leaf Lard!

This contributed to the passing of the Pure Food Act of 1906.

The Jungle (1906), 117.

Giovanni Benedetto Sinibaldi 1594–1658

Italian physician

8 When experience has proved a physical fact, one must give up reasoning.

Geneanthropeiae siue de Hominis Generatione Decateuchon (1642), Column 604. Quoted in Jacques Roger, *The Life Sciences in Eighteenth-Century French Thought*, ed. Keith R. Benson and trans. Robert Ellrich (1997), 22.

Willem de Sitter 1872–1934

Dutch astronomer

9 I am afraid all we can do is to accept the paradox and try to accommodate ourselves to it, as we have done to so many paradoxes lately in modern physical theories. We shall have to get accustomed to the idea that the change of the quantity R, commonly called the 'radius of the universe', and the evolutionary changes of stars and stellar systems are two different processes, going on side by side without any apparent connection between them. After all the 'universe' is an hypothesis, like the atom, and must be allowed the freedom to have properties and to do things which would be contradictory and impossible for a finite material structure.

Kosmos (1932), 133.

Bhurrhus Frederic Skinner

1904–90

American psychologist

1 The Law of Inhibition. The strength of a reflex may be decreased through presentation of a second stimulus which has no other relation to the effector involved.

The Behavior of Organisms: An Experimental Analysis (1938), 17.

2 Instead of saying that a man behaves because of the consequences which *are* to follow his behavior, we simply say that he behaves because of the consequences which *have* followed similar behavior in the past. This is, of course, the Law of Effect or operant conditioning.

Science and Human Behavior (1953), 87.

3 The hypothesis that man is not free is essential to the application of scientific method to the study of human behavior. The free inner man who is held responsible for the behavior of the external biological organism is only a prescientific substitute for the kinds of causes which are discovered in the course of a scientific analysis.

Science and Human Behavior (1953), 447.

4 The real question is not whether machines think but whether men do.

Contingencies of Reinforcement (1969), 288.

Herman Skolnik 1914–94

American chemist

5 Science and engineering students presumably are left to learn about their literature in the same way they learn about sex.

'Learning for Life', *Journal of Chemical Information and Computer Sciences*, 1981, 21 (4), 2A.

Jonathan Michael Wyndham Slack 1949–

British developmental biologist

6 If it is good to teach students about the chemical industry then why is it not good to assign ethical qualities to substances along with their physical and chemical ones? We might for instance say that CS [gas] is a *bad* chemical because it can only ever be used by a few people with something to protect against many people with nothing to lose. Terylene or indigotin are *neutral* chemicals. Under capitalism their production is an exploitive process, under socialism they are used for the common good. Penicillin is a *good* chemical.

Quoted in T. Pateman (ed.), *Countercourse* (1972), 215.

Hans Sloane 1660–1753

British physician, natural historian and collector

7 The knowledge of Natural-History, being Observation of Matters of Fact, is more certain than most others, and in my slender Opinion, less subject to Mistakes than Reasonings, Hypotheses, and Deductions are; . . . These are things we are sure of, so far as our Senses are not fallible; and which, in probability, have been ever since the Creation, and will remain to the End of the World, in the same Condition we now find them.

A Voyage to the Islands Madera, Barbados, Nieves, S. Christophers and Jamaica: With the Natural History of the Herbs and Trees, Four-footed Beasts, Fishes, Birds, Insects, Reptiles, &c. of the Last of those Islands (1707), Vol. 1, 1.

Christopher Smart 1722–71

British poet

8 For FRICTION is inevitable because the Universe is FULL of God's works.
For the PERPETUAL MOTION is in all works of Almighty GOD.
For it is not so in the engines of man, which are made of dead materials, neither indeed can be.
For the Moment of bodies, as it is used, is a false term—bless God ye Speakers on the Fifth of November.
For Time and Weight are by their several estimates.

For I bless GOD in the discovery of the LONGITUDE direct by the means of GLADWICK.
For the motion of the PENDULUM is the longest in that it parries resistance.
For the WEDDING GARMENTS of all men are prepared in the SUN against the day of acceptation.
For the wedding Garments of all women are prepared in the MOON against the day of their purification.
For CHASTITY is the key of knowledge as in Esdras, Sir Isaac Newton & now, God be praised, in me.
For Newton nevertheless is more of error than of the truth, but I am of the WORD of GOD.

From 'Jubilate Agno' (*c.*1758–1763), in N. Callan (ed.), *The Collected Poems of Christopher Smart* (1949), Vol. 1, 276.

Samuel Smiles 1812–1904

British author and social reformer

1 This extraordinary metal, the soul of every manufacture, and the mainspring perhaps of civilised society.

Of iron.

Men of Invention and Industry (1884), Ch. 4.

Adam Smith 1723–90

British political economist

2 People of the same trade seldom meet together, even for merriment and diversion, but the conversation ends in a conspiracy against the public, or in some contrivance to raise prices. It is impossible indeed to prevent such meetings, by any law which either could be executed, or would be consistent with liberty and justice.

An Enquiry into the Nature and Causes of the Wealth of Nations (1776). In R. H. Campbell and A. S. Skinner (eds.), *An Enquiry into the Nature and Causes of the Wealth of Nations* (1976), Vol. 1, Book 1, Chapter 10, Part 2, 145.

3 Science is the great antidote to the poison of enthusiasm and superstition.

An Enquiry into the Nature and Causes of the Wealth of Nations (1776). In R. H. Campbell and A. S. Skinner (eds.), *An Enquiry into the Nature and Causes of the Wealth of Nations* (1976), Vol. 2, Book 5, Article 3, 769.

Homer William Smith 1895–1962

American physiologist

4 The responsibility for maintaining the composition of the blood in respect to other constituents devolves largely upon the kidneys. It is no exaggeration to say that the composition of the blood is determined not by what the mouth ingests but by what the kidneys keep; they are the master chemists of our internal environment, which, so to speak, they synthesize in reverse. When, among other duties, they excrete the ashes of our body fires, or remove from the blood the infinite variety of foreign substances which are constantly being absorbed from our indiscriminate gastrointestinal tracts, these excretory operations are incidental to the major task of keeping our internal environment in an ideal, balanced state. Our glands, our muscles, our bones, our tendons, even our brains, are called upon to do only one kind of physiological work, while our kidneys are called upon to perform an innumerable variety of operations. Bones can break, muscles can atrophy, glands can loaf, even the brain can go to sleep, without immediately endangering our survival, but when the kidneys fail to manufacture the proper kind of blood neither bone, muscle, gland nor brain can carry on.

'The Evolution of the Kidney', *Lectures on the Kidney* (1943), 3.

5 Superficially, it might be said that the function of the kidneys is to make urine; but in a more considered view one can say that the kidneys make the stuff of philosophy itself.

'The Evolution of the Kidney', *Lectures on the Kidney* (1943), 4.

6 There are those who say that the human kidney was created to keep the blood pure, or more precisely, to keep our internal environment in an ideal balanced state. This I must deny. I grant that the human kidney is a marvelous organ, but I cannot grant that it was purposefully designed to excrete urine or to regulate the composition of the blood or to subserve the physiological welfare of *Homo sapiens* in any sense. Rather I contend that the human kidney manufactures the kind of urine that it does, and it maintains the blood in the composition which that fluid has, because this kidney has a certain functional architecture; and it owes that architecture not to design or foresight or to any plan, but to the fact that the earth is an unstable sphere with a fragile crust, to the geologic revolutions that for six hundred million years have raised and lowered continents and seas, to the predacious enemies, and heat and cold, and storms and droughts; to the unending succession of vicissitudes that have driven the mutant vertebrates from sea into fresh water, into desiccated swamps, out upon the dry land, from one habitation to another, perpetually in search of the free and independent life, perpetually failing, for one reason or another, to find it.

From Fish to Philosopher (1953), 210–1.

Sydney Smith 1771–1845

British clergyman and wit

1 Oh, don't tell me of facts, I never believe facts; you know, Canning said nothing was so fallacious as facts, except figures.

Lady Holland, *A Memoir of The Reverend Sydney Smith* (1854), 253.

2 Science is his forte and omniscience is his foible.

With reference to William Whewell.
Quoted in I. Todhunter (ed.), *William Whewell: An Account of His Writings With Selections from his Literary and Scientific Correspondence* (1876), Vol. 1, 410.

Theobald Smith 1859–1934

American bacteriologist

3 While the nature of Texas fever is by no means made clear as yet, we are able to affirm that ticks can produce it. Whether the disease can be transmitted by any other agency must be decided by future investigations. Meanwhile the evidence accumulated thus far seems to favor very strongly the dictum: No ticks, no Texas fever.

'Investigations of the Infectious Diseases of Animals', in *Report of the Bureau of Animal Industry* (1889–90), 98.

4 The universality of parasitism as an offshoot of the predatory habit negatives the position taken by man that it is a pathological phenomenon or a deviation from the normal processes of nature. The pathological manifestations are only incidents in a developing parasitism. As human beings intent on maintaining man's domination over nature we may regard parasitism as pathological insofar as it becomes a drain upon human resources. In our efforts to protect ourselves we may make every kind of sacrifice to limit, reduce, and even eliminate parasitism as a factor in human life. Science attempts to define the terms on which this policy of elimination may or may not succeed. We must first of all thoroughly understand the problem, put ourselves in possession of all the facts in order to estimate the cost. Too often it has been assumed that parasitism was abnormal and that it needed only a slight force to reestablish what was believed to be a normal equilibrium without parasitism. On the contrary, biology teaches us that parasitism is a normal phenomenon and if we accept this view we shall be more ready to pay the price of freedom as a permanent and ever recurring levy of nature for immunity from a condition to which all life is subject. The greatest victory of man over nature in the physical realm would undoubtedly be his own delivery from the heavy encumbrance of parasitism with which all life is burdened.

Parasitism and Disease (1934), 4.

1 Research cannot be forced very much. There is always danger of too much foliage and too little fruit.

Letter to Professor Simon H. Gage. Quoted in Paul Franklin Clark, 'Theobald Smith, Student of Disease (1859–1934)', *Journal of the History of Medicine and Allied Sciences*, 1959, 14, 492.

William Smith 1769–1839

British geologist

2 [My] numberless observations . . . made on the Strata . . . [have] made me confident of their uniformity throughout this Country & [have] led me to conclude that the same regularity . . . will be found to extend to every part of the Globe for Nature has done nothing by piecemeal. [T]here is no inconsistency in her productions. [T]he Horse never becomes an Ass nor the Crab an Apple by any intermixture or artificial combination whatever[. N]or will the Oak ever degenerate into an Ash or an Ash into an Elm. [H]owever varied by Soil or Climate the species will still be distinct on this ground. [T]hen I argue that what is found here may be found elsewhere[.] When proper allowances are made for such irregularities as often occur and the proper situation and natural agreement is well understood I am satisfied there will be no more difficulty in ascertaining the true quality of the Strata and the place of its possition [*sic*] than there is now in finding the true Class and Character of Plants by the Linean [*sic*] System.

Natural Order of the Strata in England and Wales Accurately Delineated and Described, unpublished manuscript, Department of Geology, University of Oxford, 1801, f. 7v.

3 The principles of Geology like those of geometry must begin at a point, through two or more of which the Geometrician draws a line and by thus proceeding from point to point, and from line to line, he constructs a map, and so proceeding from local to general maps, and finally to a map of the world. Geometricians founded the science of Geography, on which is based that of Geology.

Abstract View of Geology, page proofs of unpublished work, Department of Geology, University of Oxford, i.

4 Organized Fossils are to the naturalist as coins to the antiquary; they are the antiquities of the earth; and very distinctly show its gradual regular formation, with the various changes of inhabitants in the watery element.

Stratigraphical System of Organized Fossils (1817), ix-x.

Charles Percy Snow (Lord Snow of the City of Leicester) 1905–80

British writer and government administrator

See ***Davy, Humphry*** *168:4*

5 The future of chemistry rests and must rest, with physics.

'Chemistry', in H. Wright, *University Studies* (1933), 125.

6 I felt I was moving among two groups [literary intellectuals and scientists]—comparable in intelligence, identical in race, not grossly different in social origin, earning about the same incomes, who had almost ceased to communicate at all, who in intellectual, moral and psychological climate had so little in common that instead of going from Burlington House or South Kensington to Chelsea, one might have crossed an ocean.

The Two Cultures: The Rede Lecture (1959), 2.

7 Literary intellectuals at one pole—at the other scientists, and as the most representative, the physical scientists. Between the two a gulf of mutual incomprehension—sometimes (particularly among the young) hostility and dislike, but most of all lack of understanding.

The Two Cultures: The Rede Lecture (1959), 4.

8 A good many times I have been present at gatherings of people who, by the standards of the traditional culture, are thought highly educated and who have with considerable gusto been expressing their incredulity at the illiteracy of scientists. Once or twice I have been provoked and have asked the company how many of them could describe the Second Law of Thermodynamics. The response was cold: it was also negative. Yet I was asking something which is about the scientific equivalent of: *Have you read a work of Shakespeare's?*

The Two Cultures: The Rede Lecture (1959), 14–5.

1 One day at Fenner's (the university cricket ground at Cambridge), just before the last war, G. H. Hardy and I were talking about Einstein. Hardy had met him several times, and I had recently returned from visiting him. Hardy was saying that in his lifetime there had only been two men in the world, in all the fields of human achievement, science, literature, politics, anything you like, who qualified for the Bradman class. For those not familiar with cricket, or with Hardy's personal idiom, I ought to mention that 'the Bradman class' denoted the highest kind of excellence: it would include Shakespeare, Tolstoi, Newton, Archimedes, and maybe a dozen others. Well, said Hardy, there had only been two additions in his lifetime. One was Lenin and the other Einstein.

Variety of Men (1966), 87.

2 Einstein, twenty-six years old, only three years away from crude privation, still a patent examiner, published in the *Annalen der Physik* in 1905 five papers on entirely different subjects. Three of them were among the greatest in the history of physics. One, very simple, gave the quantum explanation of the photoelectric effect—it was this work for which, sixteen years later, he was awarded the Nobel prize. Another dealt with the phenomenon of Brownian motion, the apparently erratic movement of tiny particles suspended in a liquid: Einstein showed that these movements satisfied a clear statistical law. This was like a conjuring trick, easy when explained: before it, decent scientists could still doubt the concrete existence of atoms and molecules: this paper was as near to a direct proof of their concreteness as a theoretician could give. The third paper was the special theory of relativity, which quietly amalgamated space, time, and matter into one fundamental unity. This last paper contains no references and quotes no authority. All of them are written in a style unlike any other theoretical physicist's. They contain very little mathematics. There is a good deal of verbal commentary. The conclusions, the bizarre conclusions, emerge as though with the greatest of ease: the reasoning is unbreakable. It looks as though he had reached the conclusions by pure thought, unaided, without listening to the opinions of others. To a surprisingly large extent, that is precisely what he had done.

Variety of Men (1966), 100–1.

3 I think, on the whole that scientists make slightly better husbands and fathers than most of us, and I admire them for it.

Quoted in I. Langmuir, *Langmuir: The Man and the Scientist* (1962), 97.

4 I should never have made a *good* scientist, but I should have made a perfectly adequate one.

Interview with John Halperin. C. P. Snow, *An Oral Biography*, (1983), 11.

Frederick Soddy 1877–1956

British chemist

5 The fact remains that, if the supply of energy failed, modern civilization would come to an end as abruptly as does the music of an organ deprived of wind. [But] . . . the still unrecognized 'energy problem' . . . awaits the future.

Matter and Energy (1912), 251.

6 The same algebraic sum of positive and negative charges in the nucleus, when the arithmetical sum is different, gives what I call 'isotopes' or 'isotopic elements', because they occupy the same place in the periodic table. They are chemically identical, and save only as regards the relatively few physical properties which depend upon atomic mass directly, physically identical also.

'Intra-atomic Charge', *Nature*, 1913, **92**, 400.

1 The energy available for each individual man is his income, and the philosophy which can teach him to be content with penury should be capable of teaching him also the uses of wealth.

Science and Life: Aberdeen Addresses (1920), 6.

2 [The human control of atomic energy could] virtually provide anyone who wanted it with a private sun of his own.

'Advances in the Study of Radio-active Bodies', Two lectures delivered at the Royal Institution on 15 and 18 May 1915. Quoted in Thaddeus Trenn, 'The Central Role of Energy in Soddy's Holistic and Critical Approach to Nuclear Science, Economics, and Social Responsibility', *British Journal for the History of Science*, 1979, **42**, 261.

3 [The blame for the future plight of civilization] must rest on scientific men, equally with others, for being incapable of accepting the responsibility for the profound social upheavals which their own work primarily has brought about in human relationships.

Quoted in Thaddeus Trenn, 'The Central Role of Energy in Soddy's Holistic and Critical Approach to Nuclear Science, Economics, and Social Responsibility', *British Journal for the History of Science*, 1979, **42**, 261.

4 The dropping of the Atomic Bomb is a very deep problem . . . Instead of commemorating Hiroshima we should celebrate . . . man's triumph over the problem [of transmutation], and not its first misuse by politicians and military authorities.

Address to New Europe Group meeting on the third anniversary of the Hiroshima bomb. Quoted in New Europe Group, *In Commemoration of Professor Frederick Soddy* (1956), 6–7.

5 There has been no discovery like it in the history of man. It puts into man's hands the key to using the fundamental energy of the universe.

Address to New Europe Group meeting on the third anniversary of the Hiroshima bomb. Quoted in New Europe Group, *In Commemoration of Professor Frederick Soddy* (1956), 7.

6 Mankind has always drawn from outside sources of energy. This island was the first to harness coal and steam. But our present sources stand in the ratio of a million to one, compared with any previous sources. The release of atomic energy will change the whole structure of society.

Address to New Europe Group meeting on the third anniversary of the Hiroshima bomb. Quoted in New Europe Group, *In Commemoration of Professor Frederick Soddy* (1956), 7.

William Johnson Sollas

1849–1936

British geologist

7 If catastropic geology had at times pushed Nature to almost indecent extremes of haste, uniformitarian geology, on the other hand, had erred in the opposite direction, and pictured Nature when she was 'young and wantoned [sic] in her prime', as moving with the lame sedateness of advanced middle age. It became necessary, therefore, as Dr. Haughton expresses it, 'to hurry up the phenomena'.

The Age of the Earth and other Geological Studies, n.d. [1905?], 303.

8 The age of the earth was thus increased from a mere score of millions [of years] to a thousand millions and more, and the geologist who had before been bankrupt in time now found himself suddenly transformed into a capitalist with more millions in the bank than he knew how to dispose of . . . More cautious people, like myself, too cautious, perhaps, are anxious first of all to make sure that the new [radioactive] clock is not as much too fast as Lord Kelvin's was too slow.

1921 British Association for the

Advancement of Science symposium on 'The Age of the Earth', reported in *Nature*, 1921, 108, 282.

Mary Fairfax Greig Somerville
1780–1872
British mathematician

1 Astronomy affords the most extensive example of the connection of the physical sciences. In it are combined the sciences of number and quantity, or rest and motion. In it we perceive the operation of a force which is mixed up with everything that exists in the heavens or on earth; which pervades every atom, rules the motions of animate and inanimate beings, and is as sensible in the descent of a rain-drop as in the falls of Niagara; in the weight of the air, as in the periods of the moon.

On the Connexion of the Physical Sciences (1858), 1.

2 Science, regarded as the pursuit of truth, which can only be attained by patient and unprejudiced investigation, wherein nothing is to be attempted, nothing so minute as to be justly disregarded, must ever afford occupation of consummate interest, and subject of elevated meditation.

On the Connexion of the Physical Sciences (1858), 2–3.

3 So numerous are the objects which meet our view in the heavens, that we cannot imagine a point of space where some light would not strike the eye;—innumerable stars, thousands of double and multiple systems, clusters in one blaze with their tens of thousands of stars, and the nebulae amazing us by the strangeness of their forms and the incomprehensibility of their nature, till at last, from the limit of our senses, even these thin and airy phantoms vanish in the distance.

On the Connexion of the Physical Sciences (1858), 420.

4 Who shall declare the time allotted to the human race, when the generations of the most insignificant insect also existed for unnumbered ages? Yet man is also to vanish in the ever-changing course of events. The earth is to be burnt up, and the elements are to melt with fervent heat—to be again reduced to chaos—possibly to be renovated and adorned for other races of beings. These stupendous changes may be but cycles in those great laws of the universe, where all is variable but the laws themselves and He who has ordained them.

Physical Geography (1848), Vol. 1, 2–3.

Arnold Sommerfeld 1868–1951
German physicist

5 After the discovery of spectral analysis no one trained in physics could doubt the problem of the atom would be solved when physicists had learned to understand the language of spectra. So manifold was the enormous amount of material that has been accumulated in sixty years of spectroscopic research that it seemed at first beyond the possibility of disentanglement. An almost greater enlightenment has resulted from the seven years of Röntgen spectroscopy, inasmuch as it has attacked the problem of the atom at its very root, and illuminates the interior. What we are nowadays hearing of the language of spectra is a true 'music of the spheres' in order and harmony that becomes ever more perfect in spite of the manifold variety. The theory of spectral lines will bear the name of Bohr for all time. But yet another name will be permanently associated with it, that of Planck. All integral laws of spectral lines and of atomic theory spring originally from the quantum theory. It is the mysterious *organon* on which Nature plays her music of the spectra, and according to the rhythm of which she regulates the structure of the atoms and nuclei.

Atombau und Spektrallinien (1919), viii, *Atomic Structure and Spectral Lines*, trans. Henry L. Brose (1923), viii.

6 If you want to be a physicist, you must do three things—first, study mathematics, second, study more mathematics, and third, do the same.

Interview with Paul H. Kirkpatrick. In Daniel J. Kevles, *The Physicists* (1978), 200.

Lazzaro Spallanzani 1729–99

Italian natural philosopher

1 Let us now recapitulate all that has been said, and let us conclude that by hermetically sealing the vials, one is not always sure to prevent the birth of the animals in the infusions, boiled or done at room temperature, if the air inside has not felt the ravages of fire. If, on the contrary, this air has been powerfully heated, it will never allow the animals to be born, unless new air penetrates from outside into the vials. This means that it is indispensable for the production of the animals that they be provided with air which has not felt the action of fire. And as it would not be easy to prove that there were no tiny eggs disseminated and floating in the volume of air that the vials contain, it seems to me that suspicion regarding these eggs continues, and that trial by fire has not entirely done away with fears of their existence in the infusions. The partisans of the theory of ovaries will always have these fears and will not easily suffer anyone's undertaking to demolish them.

Nouvelles Recherches sur les Découvertes Microscopiques, et la Génération des Corps Organisés (1769), 134–5. Quoted in Jacques Roger, *The Life Sciences in Eighteenth-Century French Thought*, ed. Keith R. Benson and trans. Robert Ellrich (1997), 510–1.

Herbert Spencer 1820–1903

British philosopher and psychologist

See ***James, William*** 321:2

2 Organs, faculties, powers, capacities, or whatever else we call them; grow by use and diminish from disuse, it is inferred that they will continue to do so. And if this inference is unquestionable, then is the one above deduced from it—that humanity must in the end become completely adapted to its conditions—unquestionable also. Progress, therefore, is not an accident, but a necessity.

Social Statics: Or, The Conditions Essential to Human Happiness Specified, and the First of them Developed (1851), 65.

3 A function to each organ, and each organ to its own function, is the law of all organization.

Social Statics: Or, The Conditions Essential to Human Happiness Specified, and the First of them Developed (1851), 274.

4 No physiologist who calmly considers the question in connection with the general truths of his science, can long resist the conviction that different parts of the cerebrum subserve different kinds of mental action. Localization of function is the law of all organization whatever: separateness of duty is universally accompanied with separateness of structure: and it would be marvellous were an exception to exist in the cerebral hemispheres.

The Principles of Psychology (1855), 607.

5 The advance from the simple to the complex, through a process of successive differentiations, is seen alike in the earliest changes of the Universe to which we can reason our way back, and in the earliest changes which we can inductively establish; it is seen in the geologic and climatic evolution of the Earth; it is seen in the unfolding of every single organism on its surface, and in the multiplication of kinds of organisms; it is seen in the evolution of Humanity, whether contemplated in the civilized individual, or in the aggregate of races; it is seen in the evolution of Society in respect alike of its political, its religious, and its economical organization; and it is seen in the evolution of all those endless concrete and abstract products of human activity which constitute the environment of our daily life. From the remotest past which Science can fathom, up to the novelties of yesterday, that in which Progress essentially consists, is the transformation of the homogeneous into the heterogeneous.

Progress: Its Law and Cause (1857), 35.

1 Now, we propose in the first place to show, that this law of organic progress is the law of all progress. Whether it be in the development of the Earth, in the development in Life upon its surface, in the development of Society, of Government, of Manufactures, of Commerce, of Language, Literature, Science, Art, this same evolution of the simple into the complex, through a process of continuous differentiation, holds throughout. From the earliest traceable cosmical changes down to the latest results of civilization, we shall find that the transformation of the homogeneous into the heterogeneous is that in which Progress essentially consists.

'Progress: Its Law and Cause', *Westminster Review*, 1857, 67, 446–7.

2 Though, probably, no competent geologist would contend that the European classification of strata is applicable to all other parts of the globe, yet most, if not all geologists, write as though it were so.

'Illogical Geology', *The Universal Review*, 1859, 2, 54.

3 If there be an order in which the human race has mastered its various kinds of knowledge, there will arise in every child an aptitude to acquire these kinds of knowledge in the same order. So that even were the order intrinsically indifferent, it would facilitate education to lead the individual mind through the steps traversed by the general mind. But the order is *not* intrinsically indifferent; and hence the fundamental reason why education should be a repetition of civilization in little.

Education: Intellectual, Moral and Physical (1861), 76.

4 During human progress, every science is evolved out of its corresponding art.

Education: Intellectual, Moral and Physical (1861), 77.

5 Science is organized knowledge.

Education: Intellectual, Moral and Physical (1861), 77.

6 Evolution is a change from an indefinite, incoherent homogeneity, to a definite, coherent heterogeneity; through continuous differentiations and integrations.

Parodied by T. P. Kirkman as: 'Evolution is a change from nohowish, untalkaboutable, all-alikeness, to a somehowish and in-general-talkaboutable not-all-alikeness, by continuous somethingelseifications, and sticktogetherations'. Quoted by Spencer himself, First Principles *(1887), 5th edn., 565.*

First Principles (1862), 216.

7 The broadest and most complete definition of Life will be—*The continuous adjustment of internal relations to external relations.*

Principles of Biology (1864–7), Vol. 1, Part 1, Section 30, 80.

8 Now if the individuals of a species are thus necessarily made unlike, in countless ways and degrees—if the complicated sets of rhythms which we call their functions, though similar in their general characters, are dissimilar in their details—if in one individual the amount of action in a particular direction is greater than in any other individual, or if here a peculiar combination gives a resulting force which is not found elsewhere; then, among all the individuals, some will be less liable than others to have their equilibria overthrown by a particular incident force, previously unexperienced. Unless the change in the environment is of so violent a kind as to be universally fatal to the species, it must affect more or less differently the slightly different moving equilibria which the members of the species present.

It cannot but happen that some will be more stable than others, when exposed to this new or altered factor. That is to say, it cannot but happen that those individuals whose functions are most out of equilibrium with the modified aggregate of external forces, will be those to die; and that those will survive whose functions happen to be most nearly in equilibrium with the modified aggregate of external forces.

But this survival of the fittest, implies multiplication of the fittest.

Principles of Biology (1864–7), Vol. 1, Part 3, Section 164, 444.

1 This survival of the fittest, which I have here sought to express in mechanical terms, is that which Mr. Darwin has called 'natural selection, or the preservation of favoured races in the struggle for life'.

Principles of Biology (1864–7), Vol. 1, Part 3, Section 165, 444–5.

2 Throughout the whole animal kingdom, from the *Coelenterata* upwards, the first stage of evolution is the same. Equally in the germ of a polype and in the human ovum, the aggregated mass of cells out of which the creature is to arise, gives origin to a peripheral layer of cells, slightly differing from the rest which they include; and this layer subsequently divides into two—the inner, lying in contact with the included yolk, being called the mucous layer, and the outer, exposed to surrounding agencies, being called the serous layer: or, in the terms used by Prof. Huxley, in describing the development of the *Hydrozoa*—the endoderm and ectoderm.

Illustrations of Universal Progress; A Series of Discussions (1865), 404.

3 Intellectual progress is by no one trait so adequately characterized, as by development of the idea of causation.

The Data of Ethics (1879), 47.

4 Every science begins by accumulating observations, and presently generalizes these empirically; but only when it reaches the stage at which its empirical generalizations are included in a rational generalization does it become developed science.

The Data of Ethics (1879), 61.

5 I can only congratulate myself on leaving no descendants.

Said to Lord Rayleigh, who added the comment: 'Not bad for the prophet of evolution'.

R. J. Strutt, *Life of John William Strutt, Third Baron Lord Rayleigh* (1968), 396.

Spinoza 1632–77

Dutch philosopher

6 It is of the nature of Reason to perceive things under a certain species of eternity [sub specie aeternitatis].

Ethics, Part 2, prop 44. In Edwin Curley (ed.), *The Collected Works of Spinoza* (1985), Vol. 1, 481.

7 Nothing happens in Nature which can be attributed to any defect in it, for Nature is always the same, and its virtue and power of acting are everywhere one and the same, that is, the laws and rules of Nature, according to which all things happen, and change from one form to another, are always and everywhere the same. So the way of understanding the nature of anything, of whatever kind, must also be the same, namely, through the universal laws and rules of Nature.

Ethics, Part 3, Preface. In Edwin Curley (ed.), *The Collected Works of Spinoza* (1985), Vol. 1, 492.

8 That eternal and infinite being we call God, *or* Nature [*Deus sive Natura*], acts from the same necessity from which he exists.

Ethics, Part 4, Preface. In Edwin Curley (ed.), *The Collected Works of Spinoza* (1985), Vol. 1, 544.

Thomas Sprat 1635–1713

British historian of the Royal Society of London

9 As for what belongs to the *Members* themselves, that are to constitute the *Society*: It is to be noted, that they have freely admitted Men of different Religions, Countries, and Professions of Life. This they were oblig'd to do, or else they would come far short of the largeness of their own Declarations. For they openly profess, not to lay the Foundation of an *English*, *Scotch*, *Irish*, *Popish*, or *Protestant* Philosophy; but a Philosophy of *Mankind*.

The History of the Royal Society (1667), 62–3.

1 [In the *Royal* Society, there] has been, a constant Resolution, to reject all the amplifications, digressions, and swellings of style: to return back to the primitive purity, and shortness, when men deliver'd so many *things*, almost in an equal number of *words*. They have exacted from all their members, a close, naked, natural way of speaking; positive expressions; clear senses; a native easiness: bringing all things as near the Mathematical plainness, as they can: and preferring the language of Artizans, Countrymen, and Merchants, before that, of Wits, or Scholars.

The History of the Royal Society (1667), 113.

2 If to be the *Author* of *new things*, be a crime; how will the first Civilizers of *Men*, and makers of *Laws*, and Founders of *Governments* escape? Whatever now delights us in the Works of *Nature*, that excells the rudeness of the first Creation, is *New*. Whatever we see in Cities, or Houses, above the first wildness of Fields, and meaness of Cottages, and nakedness of Men, had its time, when this imputation of *Novelty*, might as well have bin laid to its charge. It is not therefore an offence, to profess the introduction of *New things*, unless that which is introduc'd prove pernicious in itself; or cannot be brought in, without the extirpation of others, that are better.

The History of the Royal Society (1667), 322.

3 *Invention* is an *Heroic* thing, and plac'd above the reach of a low, and vulgar *Genius*. It requires an *active*, a bold, a nimble, a restless *mind*: a thousand difficulties must be contemn'd with which a mean heart would be broken: many *attempts* must be made to no purpose: much *Treasure* must sometimes be scatter'd without any return: much violence, and vigour of thoughts must attend it: some irregularities, and excesses must be granted it, that would hardly be pardon'd by the severe *Rules of Prudence*.

The History of the Royal Society (1667), 392.

Christian Konrad Sprengel
1750–1816
German biologist

4 Nature appears not to have intended that any flower should be fertilized by its own pollen.

Das entdeckte Geheimniss der Natur im Bau und in der Befructung der Blumen (1793), 43. Quoted in Lawrence J. King, 'Christian Konrad Sprengel', in C. C. Gillispie (ed.), *Dictionary of Scientific Biography* (1975), Vol. 12, 588.

John Collings Squire 1884–1958
British poet and man of letters

5 It did not last : the Devil howling 'Ho, Let Einstein be,' restored the status quo.

'In Continuation of Pope on Newton', in J. C. Squire, *Poems in One Volume* (1926), 218. See **Pope, Alexander** 502:3.

Georg Ernst Stahl 1660–1734
German physician and chemist

6 Briefly, in the act of composition, as an instrument there intervenes and is most potent, fire, flaming, fervid, hot; but in the very substance of the compound there intervenes, as an ingredient, as it is commonly called, as a material principle and as a constituent of the whole compound the material and principle of fire, not fire itself. This I was the first to call phlogiston.

Specimen Beccherianum (1703). Trans. J. R. Partington, *A History of Chemistry* (1961), Vol. 2, 668.

7 Chymia, or Alchemy and Spagyrism, is the art of resolving compound bodies into their principles and of combining these again.

Fundamenta Chymiae (1720). Trans. J. R. Partington, *A History of Chemistry* (1961), Vol. 2, 664.

1 In my youth I often asked what could be the use and necessity of smelting by putting powdered charcoal at the bottom of the furnace. Nobody could give me any other reason except that the metal and especially lead, could bury itself in the charcoal and so be protected against the action of the bellows which would calcine or dissipate it. Nevertheless it is evident that this does not answer the question. I accordingly examined the operation of a metallurgical furnace and how it was used. In assaying some litharge [lead oxide], I noticed each time a little charcoal fell into the crucible, I always obtained a bit of lead . . . I do not think up to the present time foundry-men ever surmised that in the operation of founding with charcoal there was something [phlogiston] which became corporeally united with the metal.

Traité de Soufre (1766), 64. French translation published 1766, first published in German in 1718.

Anthony Standen

American science writer

2 The velocity of light occupies an extraordinary place in modern physics. It is *lèse-majesté* to make any criticism of the velocity of light. It is a sacred cow within a sacred cow, and it is just about the Absolutest Absolute in the history of human thought.

Science is a Sacred Cow (1950), 73.

Ernest Henry Starling 1866–1927

British physiologist

3 The specific character of the greater part of the toxins which are known to us (I need only instance such toxins as those of tetanus and diphtheria) would suggest that the substances produced for effecting the correlation of organs within the body, through the intermediation of the blood stream, might also belong to this class, since here also specificity of action must be a distinguishing characteristic. These chemical messengers, however, or 'hormones' (from ὁρμάω, I excite or arouse), as we might call them, have to be carried from the organ where they are produced to the organ which they affect by means of the blood stream and the continually recurring physiological needs of the organism must determine their repeated production and circulation through the body.

'The Chemical Correlation of the Functions of the Body', *The Lancet*, 1905, **ii**, 340.

4 The law of the heart is thus the same as the law of muscular tissue generally, that *the energy of contraction, however measured, is a function of the length of the muscle fibre.*

The Linacre Lecture on the Law of the Heart (1918), 142.

5 In physiology, as in all other sciences, no discovery is useless, no curiosity misplaced or too ambitious, and we may be certain that every advance achieved in the quest of pure knowledge will sooner or later play its part in the service of man.

The Linacre Lecture on the Law of the Heart (1918), 147.

6 Only by following out the injunction of our great predecessor [William Harvey] to search out and study the secrets of Nature by way of experiment, can we hope to attain to a comprehension of 'the wisdom of the body and the understanding of the heart,' and thereby to the mastery of disease and pain, which will enable us to relieve the burden of mankind.

'The Wisdom of the Body', *The Lancet*, 1923, 205, 870.

Hermann Staudinger 1881–1965

German chemist

7 The most fundamental difference between compounds of low molecular weight and macromolecular compounds resides in the fact that the latter may exhibit properties that cannot be deduced from a close examination of the low molecular weight materials. Not very different structures can be obtained from a few building blocks; but if 10,000 or 100,000 blocks are at hand, the most varied structures become possible, such as houses or halls, whose special structure cannot be predicted from the constructions that are possible with only a few building blocks . . . Thus, a chromosome can be viewed as a material whose macromolecules possess a well defined arrangement, like a living room in which each piece of furniture has its place.

Quoted in R. Oesper, *The Human Side of Scientists* (1975), 175.

Stendhal (Marie-Henri Beyle) 1783–1842

French writer

1 It is terrifying to think how much research is needed to determine the truth of even the most unimportant fact.

Attributed.

Nicolaus Steno 1638–86

Danish naturalist and priest

2 I have said that the investigation for which the teeth of the shark had furnished an opportunity, was very near an end . . . But thereafter, while I was examining more carefully these details of both places and bodies [sedimentary deposits and shells], these day by day presented points of doubt to me as they followed one another in indissoluble connection, so that I saw myself again and again brought back to the starting-place, as it were, when I thought I was nearest the goal. I might compare those doubts to the heads of the Lernean Hydra, since when one of them had been got rid of, numberless others were born; at any rate, I saw that I was wandering about in a sort of labyrinth, where the nearer one approaches the exit, the wider circuits does one tread.

The Prodromus of Nicolaus Steno's Dissertation Concerning a Solid Body enclosed by Process of Nature within a Solid (1669), trans. J. G. Winter (1916), 206.

3 In the case of those solids, whether of earth, or rock, which enclose on all sides and contain crystals, selenites, marcasites, plants and their parts, bones and the shells of animals, and other bodies of this kind which are possessed of a smooth surface, these same bodies had already become hard at the time when the matter of the earth and rock containing them was still fluid. And not only did the earth and rock not produce the bodies contained in them, but they did not even exist as such when those bodies were produced in them.

The Prodromus of Nicolaus Steno's Dissertation Concerning a Solid Body enclosed by Process of Nature within a Solid (1669), trans. J. G. Winter (1916), 218.

4 It . . . [can] be easily shown:

1. That all present mountains did not exist from the beginning of things.

2. That there is no growing of mountains.

3. That the rocks or mountains have nothing in common with the bones of animals except a certain resemblance in hardness, since they agree in neither matter nor manner of production, nor in composition, nor in function, if one may be permitted to affirm aught about a subject otherwise so little known as are the functions of things.

4. That the extension of crests of mountains, or chains, as some prefer to call them, along the lines of certain definite zones of the earth, accords with neither reason nor experience.

5. That mountains can be overthrown, and fields carried over from one side of a high road across to the other; that peaks of mountains can be raised and lowered, that the earth can be opened and closed again, and that other things of this kind occur which those who in their reading of history wish to escape the name of credulous, consider myths.

The Prodromus of Nicolaus Steno's Dissertation Concerning a Solid Body enclosed by Process of

Nature within a Solid (1669), trans. J. G. Winter (1916), 232–4.

1 Scripture and Nature agree in this, that all things were covered with water; how and when this aspect began, and how long it lasted, Nature says not, Scripture relates. That there was a watery fluid, however, at a time when animals and plants were not yet to be found, and that the fluid covered all things, is proved by the strata of the higher mountains, free from all heterogeneous material. And the form of these strata bears witness to the presence of a fluid, while the substance bears witness to the absence of heterogeneous bodies. But the similarity of matter and form in the strata of mountains which are different and distant from each other, proves that the fluid was universal.

The Prodromus of Nicolaus Steno's Dissertation Concerning a Solid Body enclosed by Process of Nature within a Solid (1669), trans. J. G. Winter (1916), 263–4.

Gunther Siegmund Stent 1924–

American molecular biologist

2 Thus a eukaryotic cell may be thought of as an empire directed by a republic of sovereign chromosomes in the nucleus. The chromosomes preside over the outlying cytoplasm in which formerly independent but now subject and degenerate prokaryotes carry out a variety of specialized service functions.

Molecular Genetics: An Introductory Narrative (1971), 622.

George Miller Sternberg 1838–1915

American physician and bacteriologist

See **Metchnikoff, Elie**

3 It has occurred to me that possibly the white corpuscles may have the office of picking up and digesting bacterial organisms when by any means they find their way into the blood. The propensity exhibited by the leukocytes for picking up inorganic granules is well known, and that they may be able not only to pick up but to assimilate, and so dispose of, the bacteria which come in their way does not seem to me very improbable in view of the fact that amoebae, which resemble them so closely, feed upon bacteria and similar organisms.

'A Contribution to the Study of the Bacterial Organisms Commonly Found Upon Exposed Mucous Surfaces and in the Alimentary Canal of Healthy Individuals', *Studies from the Biological Laboratory*, 1883, 2, 175.

Laurence Sterne 1713–68

British novelist

4 It is the nature of an hypothesis, when once a man has conceived it, that it assimilates every thing to itself, as proper nourishment; and, from the first moment of your begetting it, it generally grows the stronger by every thing you see, hear, read, or understand.

The Life and Opinions of Tristram Shandy Gentleman (1759–67), Penguin edition (1997), 121–2.

5 Sciences may be learned by rote, but Wisdom not.

The Life and Opinions of Tristram Shandy Gentleman (1759–67), Penguin edition (1997), 324.

Adlai Ewing Stevenson 1900–65

American politician

6 Nature is neutral. Man has wrested from nature the power to make the world a desert or make the deserts bloom. There is no evil in the atom; only in men's souls.

Speech: Hartford Connecticut, 18 September 1952. Quoted in A. Stevenson, *Speeches* (1953), 129.

Robert Louis Balfour Stevenson 1850–94

Scottish poet, novelist and essayist

7 The world is so full of a number of things,
I'm sure we should all be as happy as kings.

'Happy Thought', in *A Child's Garden of Verses* (1885), 28.

Stephen M. Stigler 1941–

American historian and statistician

1 Beware of the problem of testing too many hypotheses; the more you torture the data, the more likely they are to confess, but confessions obtained under duress may not be admissible in the court of scientific opinion.

'Testing Hypotheses or fitting Models? Another Look at Mass Extinctions'. In Matthew H. Nitecki and Antoni Hoffman (eds.), *Neutral Models in Biology* (1987), 148.

Alfred Stock 1876–1946

German chemist

2 The vacuum-apparatus requires that its manipulators constantly handle considerable amounts of mercury. Mercury is a strong poison, particularly dangerous because of its liquid form and noticeable volatility even at room temperature. Its poisonous character has been rather lost sight of during the present generation. My co-workers and myself found from personal experience—confirmed on many sides when published—that protracted stay in an atmosphere charged with only 1/100 of the amount of mercury required for its saturation, sufficed to induce chronic mercury poisoning. This first reveals itself as an affection of the nerves, causing headaches, numbness, mental lassitude, depression, and loss of memory; such are very disturbing to one engaged in intellectual occupations.

Hydrides of Boron and Silicon (1933), 203.

3 All statements about the hydrides of boron earlier than 1912, when Stock began to work upon them, are untrue.

Quoted in N.V. Sidgwick, *The Chemical Elements and their Compounds* (1950), Vol. 1, 338.

David Ross Stoddart 1937–

British geographer

4 Much of the geographical work of the past hundred years . . . has either explicitly or implicitly taken its inspiration from biology, and in particular Darwin. Many of the original Darwinians, such as Hooker, Wallace, Huxley, Bates, and Darwin himself, were actively concerned with geographical exploration, and it was largely facts of geographical distribution in a spatial setting which provided Darwin with the germ of his theory.

'Darwin's Impact on Geography', *Annals of the Association of American Geographers*, 1966, 56, 683.

George Gabriel Stokes 1819–1903

Anglo-Irish physicist and mathematician

5 But we have reason to think that the annihilation of work is no less a physical impossibility than its creation, that is, than perpetual motion.

'On the Change of Refrangibility of Light' (1852). In *Mathematical and Physical Papers* (1901), Vol. 3, 397.

Tom Stoppard 1937–

Czech-born British playwright

6 ROS: Eternity is a terrible thought. I mean, where's it going to end?

Rosencrantz and Guildenstern are Dead (1967), Act 2, 51.

7 THOMASINA: If you could stop every atom in its position and direction, and if your mind could comprehend all the actions thus suspended, then if you were really, *really* good at algebra you could write the formula for all the future; and although nobody can be so clever as to do it, the formula must exist just as if one could.

Arcadia (1993), Act 1, Scene 1, 13.

8 BERNARD: Oh, you're going to zap me with penicillin and pesticides. Spare me that and I'll spare you the bomb and aerosols. But don't confuse progress with perfectibility. A great poet is always timely. A great philosopher is an urgent need. There's no rush for Isaac Newton. We were quite happy with Aristotle's cosmos. Personally, I preferred it. Fifty-five crystal spheres

geared to God's crankshaft is my idea of a satisfying universe. I can't think of anything more trivial than the speed of light. Quarks, quasars—big bangs, black holes—who gives a shit? How did you people con us out of all that status? All that money? And why are you so pleased with yourselves?
CHLOE: Are you against penicillin, Bernard?
BERNARD: Don't feed the animals.

Arcadia (1993), Act 2, Scene 5, 87.

John William Strutt (3rd Baron Rayleigh) 1842–1919

British physicist

1 The history of science teaches only too plainly the lesson that no single method is absolutely to be relied upon, that sources of error lurk where they are least expected, and that they may escape the notice of the most experienced and conscientious worker.

'Transactions of the Sections', *Reports of the British Association for the Advancement of Science* (1883), 438.

2 The history of this paper suggests that highly speculative investigations, especially by an unknown author, are best brought before the world through some other channel than a scientific society, which naturally hesitates to admit into its printed records matters of uncertain value. Perhaps one may go further and say that a young author who believes himself capable of great things would usually do well to secure the favourable recognition of the scientific world by work whose scope is limited and whose value is easily judged, before embarking upon higher flights.

'On the Physics of Media that are Composed of Free and Perfectly Elastic Molecules in a State of Motion', *Philosophical Transactions*, 1892, 183, 560.

3 One's instinct is at first to try and get rid of a discrepancy, but I believe that experience shows such an endeavour to be a mistake. What one ought to do is to magnify a small discrepancy with a view to finding out the explanation.

General Monthly Meeting, on Argon, 1 April 1895, *Proceedings of the Royal Institution*, 1895, 14, 525.

4 I want to get back again from Chemistry to Physics as soon as I can. The second-rate men seem to know their place so much better.

R. J. Strutt, *John William Strutt, Third Baron Rayleigh* (1924), 222.

Robert John Strutt (4th Baron Rayleigh) 1875–1947

British physicist

5 The trouble is that all the investigators proceeded in exactly the same spirit, the spirit that is of scientific curiosity, and with no possibility of telling whether the issue of their work would prove them to be fiends, or dreamers, or angels.

'The Presidential Address: Part II Science and Warfare', *Reports of the British Association for the Advancement of Science* (1938), 18–9.

Eduard Suess 1831–1914

Austrian geologist

6 If we imagine an observer to approach our planet from outer space, and, pushing aside the belts of red-brown clouds which obscure our atmosphere, to gaze for a whole day on the surface of the earth as it rotates beneath him, the feature, beyond all others most likely to arrest his attention would be the wedge-like outlines of the continents as they narrow away to the South.

The Face of the Earth (1904), Vol. 1, 1.

7 [W]e are prone to forget that the planet may be measured *by* man, but not *according* to man.

The Face of the Earth (1904), Vol. 1, 17.

8 The breaking up of the terrestrial globe, this it is we witness. It doubtless began a long time ago, and the brevity of human life enables us to contemplate it without dismay. It is not only in the great mountain ranges that the traces of this process are found. Great segments of the earth's crust have sunk hundreds, in some cases, even thousands, of feet deep, and not the slightest inequality of the surface remains to indicate the fracture; the different nature of the rocks and the discoveries made in mining alone reveal its presence. Time has levelled all.

The Face of the Earth (1904), Vol. 1, 604.

William Graham Sumner

1840–1910

American sociologist, economist and social Darwinist

1 I suppose that the first chemists seemed to be very hard-hearted and unpoetical persons when they scouted the glorious dream of the alchemists that there must be some process for turning base metals into gold. I suppose that the men who first said, in plain, cold assertion, there is no fountain of eternal youth, seemed to be the most cruel and cold-hearted adversaries of human happiness. I know that the economists who say that if we could transmute lead into gold, it would certainly do us no good and might do great harm, are still regarded as unworthy of belief. Do not the money articles of the newspapers yet ring with the doctrine that we are getting rich when we give cotton and wheat for gold rather than when we give cotton and wheat for iron?

'The Forgotten Man' (1883). In *The Forgotten Man and Other Essays* (1918), 468.

2 If we put together all that we have learned from anthropology and ethnography about primitive men and primitive society, we perceive that the first task of life is to live. Men begin with acts, not with thoughts.

Folkways: A Study of the Sociological Importance of Usages, Manners, Customs, Mores and Morals (1907), 2.

3 Darwin was as much of an emancipator as was Lincoln.

Attributed.

Walter Stanborough Sutton

1877–1916

American geneticist

4 I may finally call attention to the probability that the association of paternal and maternal chromosomes in pairs and their subsequent separation during the reducing division as indicated above may constitute the physical basis of the Mendelian law of heredity.

'On the Morphology of the Chromosome Group in Brachystola Magna', *Biological Bulletin*, 1902, **4**, 39.

Ellen Henrietta Swallow

1842–1911

American chemist and ecologist

5 For this knowledge of right living, we have sought a new name. . . . As theology is the science of religious life, and biology the science of [physical] life . . . so let *Oekology* be henceforth the science of [our] normal lives . . . the worthiest of all the applied sciences which *teaches the principles on which to found . . . healthy . . . and happy life.*

Quoted in Robert Clarke (ed.), *Ellen Swallow: The Woman Who Founded Ecology* (1973), 120.

Jan Swammerdam 1637–80

Dutch naturalist

6 Herewith I offer you the Omnipotent Finger of God in the anatomy of a louse: wherein you will find miracles heaped on miracles and will see the wisdom of God clearly manifested in a minute point.

Letter to Melchisedec Thévenot, April 1678. In G. A. Lindeboom (ed.), *The Letters of Jan Swammerdam to Melchisedec Thévenot* (1975), 104–5.

Jonathan Swift 1667–1745

Anglo-Irish writer and Anglican clergyman

1 The Vermin only teaze and pinch
Their foes superior by an Inch.
So, Naturalists observe, a Flea
Hath smaller Fleas that on him prey,
And these have smaller Fleas to bite
'em.
And so proceed *ad infinitum.*

On Poetry: A Rhapsody (1735), lines 339–44.

2 *Of the Laputans:*

They have likewise discovered two lesser stars, or satellites, which revolve about Mars, whereof the innermost is distant from the centre of the primary planet exactly three of his diameters, and the outermost five; the former revolves in the space of ten hours, and the latter in twenty one and a half.

Gulliver's Travels (1726), Penguin edition (1967), Part III, Chapter 3, 213.

3 This Academy [at Lagado] is not an entire single Building, but a Continuation of several Houses on both Sides of a Street; which growing waste, was purchased and applied to that Use.

I was received very kindly by the Warden, and went for many Days to the Academy. Every Room hath in it one or more Projectors; and I believe I could not be in fewer than five Hundred Rooms.

The first Man I saw was of a meagre Aspect, with sooty Hands and Face, his Hair and Beard long, ragged and singed in several Places. His Clothes, Shirt, and Skin were all of the same Colour. He had been Eight Years upon a Project for extracting Sun-Beams out of Cucumbers, which were to be put into Vials hermetically sealed, and let out to warm the Air in raw inclement Summers. He told me, he did not doubt in Eight Years more, that he should be able to supply the Governor's Gardens with Sunshine at a reasonable Rate; but he complained that his Stock was low, and interested me to give him something as an Encouragement to Ingenuity, especially since this had been a very dear Season for Cucumbers. I made him a small Present, for my Lord had furnished me with Money on purpose, because he knew their Practice of begging from all who go to see them.

I saw another at work to calcine Ice into Gunpowder; who likewise shewed me a Treatise he had written concerning the Malleability of Fire, which he intended to publish.

There was a most ingenious Architect who had contrived a new Method for building Houses, by beginning at the Roof, and working downwards to the Foundation; which he justified to me by the life Practice of those two prudent Insects the Bee and the Spider.

In another Apartment I was highly pleased with a Projector, who had found a device of plowing the Ground with Hogs, to save the Charges of Plows, Cattle, and Labour. The Method is this: In an Acre of Ground you bury at six Inches Distance, and eight deep, a quantity of Acorns, Dates, Chestnuts, and other Masts or Vegetables whereof these Animals are fondest; then you drive six Hundred or more of them into the Field, where in a few Days they will root up the whole Ground in search of their Food, and make it fit for sowing, at the same time manuring it with their Dung. It is true, upon Experiment they found the Charge and Trouble very great, and they had little or no Crop. However, it is not doubted that this Invention may be capable of great Improvement.

I had hitherto seen only one Side of the Academy, the other being appropriated to the Advancers of speculative Learning.

Some were condensing Air into a dry tangible Substance, by extracting the Nitre, and letting the acqueous or fluid Particles percolate: Others softening Marble for Pillows and Pin-cushions. Another was, by a certain Composition of Gums, Minerals, and Vegetables outwardly applied, to prevent the Growth of Wool upon two young lambs; and he hoped in a reasonable Time to propagate the Breed of naked Sheep all over the Kingdom.

Gulliver's Travels (1726), Penguin edition (1967), Part III, Chapter 5, 223.

1 In the school of political projectors, I was but ill entertained, the professors appearing, in my judgment, wholly out of their senses; which is a scene that never fails to make me melancholy. These unhappy people were proposing schemes for persuading monarchs to choose favourites upon the score of their wisdom, capacity, and virtue; of teaching ministers to consult the public good; of rewarding merit, great abilities, and eminent services; of instructing princes to know their true interest, by placing it on the same foundation with that of their people; of choosing for employment persons qualified to exercise them; with many other wild impossible chimeras, that never entered before into the heart of man to conceive, and confirmed in me the old observation, that there is nothing so extravagant and irrational which some philosophers have not maintained for truth.

Gulliver's Travels (1726), Penguin edition (1967), Part III, Chapter 6, 232.

2 The greatest Inventions were produced in Times of Ignorance; as the use of the Compass, Gunpowder, and printing; and by the dullest Nation, as the Germans.

'Thoughts on Various Subjects', in William Alfred Eddy (ed.), *Satires and Personal Writings* (1932), 407.

3 That was excellently observ'd, say I, when I read a Passage in an Author, where his Opinion agrees with mine. When we differ, there I pronounce him to be mistaken.

'Thoughts on Various Subjects', in William Alfred Eddy (ed.), *Satires and Personal Writings* (1932), 416.

4 That the universe was formed by a fortuitous concourse of atoms, I will no more believe than that the accidental jumbling of the alphabet would fall into a most ingenious treatise of philosophy.

Attributed.

Thomas Sydenham 1624–89

English physician

See ***Locke, John*** *392:2*

5 All that Anatomie can doe is only to shew us the gross and sensible parts of the body, or the vapid and dead juices all which, after the most diligent search, will be noe more able to direct a physician how to cure a disease than how to make a man; for to remedy the defects of a part whose organicall constitution and that texture whereby it operates, he cannot possibly know, is alike hard, as to make a part which he knows not how is made. Now it is certaine and beyond controversy that nature performs all her operations on the body by parts so minute and insensible that I thinke noe body will ever hope or pretend, even by the assistance of glasses or any other intervention, to come to a sight of them, and to tell us what organicall texture or what kinde of ferment (for whether it be done by one or both of these ways is yet a question and like to be soe always notwithstanding all the endeavours of the most accurate dissections) separate any part of the juices in any of the viscera, or tell us of what liquors the particles of these juices are, or if this could be donne (which it is never like to be) would it at all contribute to the cure of the diseases of those very parts which we so perfectly knew.

'Anatomie' (1668). Quoted in Kenneth Dewhurst (ed.), *Dr. Thomas Sydenham (1624–1689): His Life and Original Writings* (1966), 85–6.

6 *Acute* [diseases] meaning those of which God is the author, chronic meaning those that originate in ourselves.

'Epistolary Dissertation to Dr. Cole', in *The Works of Thomas Sydenham, M.D.* (1850), trans. by R. G. Latham, Vol. 2, 68.

1 This is all very fine, but it won't do—Anatomy—botany—Nonsense! Sir, I know an old woman in Covent Garden, who understands botany better, and as for anatomy, my butcher can dissect a joint full as well; no, young man, all that is stuff; you must go to the bedside, it is there alone you can learn disease.!

Comment to Hans Sloane on Robert Boyle's letter of introduction describing Sloane as a 'ripe scholar, a good botanist, a skilful anatomist'.

Quoted in John D. Comrie, 'Life of Thomas Sydenham, M. D.', in Comrie (ed.), *Selected Works of Thomas Sydenham* (1922), 2.

2 Physick, says Sydenham, is not to bee learned by going to Universities, but hee is for taking apprentices; and says one had as good send a man to Oxford to learn shoemaking as practising physick.

Diary of the Rev. John Ward, M. A. (1648–1769), ed. Charles Severn (1839), 242.

James Joseph Sylvester 1814–97

British mathematician

3 Chemistry has the same quickening and suggestive influence upon the algebraist as a visit to the Royal Academy, or the old masters may be supposed to have on a Browning or a Tennyson. Indeed it seems to me that an exact homology exists between painting and poetry on the one hand and modern chemistry and modern algebra on the other. In poetry and algebra we have the pure idea elaborated and expressed through the vehicle of language, in painting and chemistry the idea enveloped in matter, depending in part on manual processes and the resources of art for its due manifestation.

Attributed.

Thomas Szasz 1920–

Hungarian-born American psychiatrist

4 Formerly, when religion was strong and science weak, men mistook magic for medicine; now, when science is strong and religion weak, men mistake medicine for magic.

The Second Sin (1973), 115.

Albert Szent-Györgyi 1893–1986

Hungarian-born American biochemist

5 I called it ignose, not knowing which carbohydrate it was. This name was turned down by my editor. 'God-nose' was not more successful, so in the end 'hexuronic acid' was agreed upon. To-day the substance is called 'ascorbic acid' and I will use this name.

Studies on Biological Oxidation and Some of its Catalysts (C4 Dicarboxylic Acids, Vitamin C and P Etc.) (1937), 73.

6 All living organisms are but leaves on the same tree of life. The various functions of plants and animals and their specialized organs are manifestations of the same living matter. This adapts itself to different jobs and circumstances, but operates on the same basic principles. Muscle contraction is only one of these adaptations. In principle it would not matter whether we studied nerve, kidney or muscle to understand the basic principles of life. In practice, however, it matters a great deal.

'Muscle Research', *Scientific American*, 1949, 180 (6), 22.

7 Research is to see what everybody has seen and think what nobody has thought.

Bioenergetics (1957), 57.

8 Why does man behave like perfect idiot? This is the problem I wish to deal with.

The Crazy Ape (1970), 11.

9 Through the ages, man's main concern was life after death. Today, for the first time, we find we must ask questions about whether there will be life before death.

The Crazy Ape (1970), 18.

10 One death is a tragedy, 100,000 deaths are statistics.

The Crazy Ape (1970), 29.

1 If any student comes to me and says he wants to be useful to mankind and go into research to alleviate human suffering, I advise him to go into charity instead. Research wants real egotists who seek their own pleasure and satisfaction, but find it in solving the puzzles of nature.

Vignette, 'Why do Research?', *American Journal of Physics*, 1975, **43** (5), 427.

2 [A vitamin is] a substance you get sick from if you don't eat it.

In Ralph W. Moss, *Free Radical* (1988), 78.

3 To regulate something always requires two opposing factors. You cannot regulate by a single factor. To give an example, the traffic in the streets could not be controlled by a green light or a red light alone. It needs a green light and a red light as well. The ratio between retine and promine determines whether there is any motion, any growth, or not. Two different inclinations have to be there in readiness to make the cells proliferate.

In Ralph W. Moss, *Free Radical* (1988), 186.

4 I always tried to live up to Leo Szilard's commandment, 'don't lie if you don't have to.' I had to. I filled up pages with words and plans I knew I would not follow. When I go home from my laboratory in the late afternoon, I often do not know what I am going to do the next day. I expect to think that up during the night. How could I tell them what I would do a year hence?

In Ralph W. Moss, *Free Radical* (1988), 217.

Leo Szilard 1898–1964

Hungarian-born American physicist and biologist

5 I have been asked whether I would agree that the tragedy of the scientist is that he is able to bring about great advances in our knowledge, which mankind may then proceed to use for purposes of destruction. My answer is that this is not the tragedy of the scientist; it is the tragedy of mankind.

S. R. Weart and G. W. Sallard (eds.), *Leo Szilard: His Version of the Facts* (1978), 229.

Hippolyte-Adolphe Taine
1828–93

French philosopher and historian

6 The more I study the things of the mind the more mathematical I find them. In them as in mathematics it is a question of quantities; they must be treated with precision. I have never had more satisfaction than in proving this in the realms of art, politics and history.

Notes made after the completion of the third chapter of Vol. 3 of *La Révolution*, 22 April 1883. In E. Sparvel-Bayly (trans.), *Life and Letters of H. Taine* (1902–1908), Vol. 3, 239.

Peter Guthrie Tait 1831–1901

British mathematician and physicist

7 It is very desirable to have a word to express the *Availability* for work of the heat in a given magazine; a term for that possession, the waste of which is called *Dissipation*. Unfortunately the excellent word *Entropy*, which Clausius has introduced in this connexion, is applied by him to the negative of the idea we most naturally wish to express. It would only confuse the student if we were to endeavour to invent another term for our purpose. But the necessity for some such term will be obvious from the beautiful examples which follow. And we take the liberty of using the term Entropy in this altered sense . . . The entropy of the universe tends continually to zero.

Sketch of Thermodynamics (1868), 100–2.

8 If it were possible for a metaphysician to be a golfer, he might perhaps occasionally notice that his ball, instead of moving forward in a vertical plane (like the generality of projectiles, such as brickbats and cricket balls), skewed away gradually to the right. If he did notice it, his methods would naturally lead him to content himself with his caddies's remark—'ye heeled that yin,' or 'Ye jist sliced it.' . . . But a scientific man is not to be put off with such flimsy verbiage as that. He *must* know more. What is 'Heeling', what is 'slicing', and why would either operation (if it could be thoroughly carried out) send a ball as if to cover point, thence to long slip, and finally behind back-stop? These, as Falstaff said, are 'questions to be asked.'

'The Unwritten Chapter on Golf', *Nature*, 1887, 36, 502.

1 Your printers have made but one blunder,
Correct it instanter, and then for the thunder!
We'll see in a jiffy if this Mr S[pencer]
Has the ghost of a claim to be thought a good fencer.
To my vision his merits have still seemed to dwindle,
Since I have found him allied with the great Dr T[yndall]
While I have, for my part, grown cockier and cockier,
Since I found an ally in yourself, Mr L[ockyer]
And am always, in consequence, thoroughly willin',
To perform in the pages of *Nature*'s M[acmillan].

Postcard from Tait to Lockyer, editor of *Nature*, cited by H. Dingle, *Nature*, 1969, 224, 829.

Bert Leston Taylor 1866–1921

American novelist, humorist and newspaper columist

2 Behold the mighty dinosaur,
Famous in prehistoric lore,
Not only for his power and strength
But for his intellectual length.
You will observe by these remains
The creature had two sets of brains—
One in his head (the usual place),
The other at his spinal base.
Thus he could reason 'A priori'
As well as 'A posteriori'.
No problem bothered him a bit
He made both head and tail of it.
So wise was he, so wise and solemn,
Each thought filled just a spinal column.
If one brain found the pressure strong
It passed a few ideas along.
If something slipped his forward mind
'Twas rescued by the one behind.
And if in error he was caught
He had a saving afterthought.
As he thought twice before he spoke
He had no judgment to revoke.
Thus he could think without congestion
Upon both sides of every question.
Oh, gaze upon this model beast
Defunct ten million years at least.

'The Dinosaur: A Poem' (1912). In E. H. Colbert (ed.), *The Dinosaur Book* (1951), 78.

Edward Teller 1908–2003

Hungarian-born American physicist

3 Science attempts to find logic and simplicity in nature. Mathematics attempts to establish order and simplicity in human thought.

The Pursuit of Simplicity (1980), 17.

4 One may say that predictions are dangerous particularly for the future.

Edward Teller, Wendy Teller and Wilson Talley, *Conversations from the Dark Side of Physics* (1991), 235.

Howard Martin Temin 1934–94 and Satoshi Mizutani

American virologists

5 These results demonstrate that there is a new polymerase inside the virions of RNA tumour viruses. It is not present in supernatants of normal cells but is present in virions of avian sarcoma and leukaemia RNA tumour viruses. The polymerase seems to catalyse the incorporation of deoxyribonucleotide triphosphates into DNA from an RNA template. Work is being performed to characterize further the reaction and the product. If the present results and Baltimore's results with Rauscher leukaemia virus are upheld, they will constitute strong evidence that the DNA provirus hypothesis is correct and that RNA tumour viruses have a DNA genome when they are in virions. This result would have strong implications for theories of viral carcinogenesis and, possibly, for theories of information transfer in other biological systems.

Howard Temin and Satoshi Mizutani, 'RNA-dependent DNA Polymerase in Virions of Rous Sarcoma Virus', *Nature*, 1970, **226**, 1213.

William Temple 1881–1944

British theologian and Archbishop of Canterbury

1 Science has its being in a perpetual mental restlessness.

'Poetry and Science', in W. H. Hadow, *Essays and Studies by Members of the English Association* (1932), Vol. 17, 12.

2 I prefer a God who once and for all impressed his will upon creation, to one who continually busied about modifying what he had already done.

Attributed.

Alfred, Lord Tennyson 1809–92

British poet

3 Here about the beach I wandered,
nourishing a youth sublime
With the fairy tales of science, and the
long result of Time.

Locksley Hall (1842). In Christopher Ricks (ed.), *The Poems of Tennyson* (1987), Vol. 2, 121.

4 When I dipt into the future far as
human eye could see;
Saw the Vision of the world, and all the
wonders that would be.

Locksley Hall (1842). In Christopher Ricks (ed.), *The Poems of Tennyson* (1987), Vol. 2, 121.

5 Saw the heavens fill with commerce,
argosies of magic sails,
Pilots of the purple twilight, dropping
down with costly bales;
Heard the heavens fill with shouting,
and there rained a ghastly dew
From the nations' airy navies grappling
in the central blue.

Locksley Hall (1842). In Christopher Ricks (ed.), *The Poems of Tennyson* (1987), Vol. 2, 126.

6 Till the war-drum throbbed no longer,
and the battle-flags were furled
In the Parliament of man, the
Federation of the world.

Locksley Hall (1842). In Christopher Ricks (ed.), *The Poems of Tennyson* (1987), Vol. 2, 126.

7 Science moves, but slowly slowly,
creeping on from point to point.

Locksley Hall (1842). In Christopher Ricks (ed.), *The Poems of Tennyson* (1987), Vol. 2, 127.

8 Knowledge comes, but wisdom lingers.

Locksley Hall (1842). In Christopher Ricks (ed.), *The Poems of Tennyson* (1987), Vol. 2, 127.

9 Forward, forward let us range,
Let the great world spin for ever down
the ringing grooves of change.

Locksley Hall (1842). In Christopher Ricks (ed.), *The Poems of Tennyson* (1987), Vol. 2, 130.

10 Every moment dies a man,
Every moment one is born.

The Vision of Sin (1842), verse IV. In Christopher Ricks (ed.), *The Poems of Tennyson* (1987), Vol. 2, 160. See **Babbage, Charles** 34:7.

11 This world was once a fluid haze of
light,
Till toward the centre set the starry
tides,
And eddied into suns, that wheeling
cast
The planets: then the monster, then
the man.

The Princess (1847), canto II. In Christopher Ricks (ed.), *The Poems of Tennyson* (1987), Vol. 2, 209.

12 Are God and Nature then at strife,
That Nature lends such evil dreams?
So careful of the type she seems,
So careless of the single life; . . .
'So careful of the type', but no.
From scarpèd cliff and quarried stone
She cries, 'A thousand types are gone:
I care for nothing, all shall go' . . .
Man, her last work, who seemed so
fair,
Such splendid purpose in his eyes,
Who rolled the psalm to wintry skies,
Who built him fanes of fruitless prayer,
Who trusted God was love indeed

And love Creation's final law—
Though Nature red in tooth and claw
With ravine, shrieked against his creed.

In Memoriam A. H. H. (1850), from cantos LV-LVI. In Christopher Ricks (ed.), *The Poems of Tennyson* (1987), Vol. 2, 371–3.

1 There rolls the deep where grew the tree.
O earth, what changes hast thou seen!
There where the long street roars, hath been
The stillness of the central sea.
The hills are shadows, and they flow
From form to form, and nothing stands;
They melt like mist, the solid lands,
Like clouds they shape themselves and go.

In Memoriam A. H. H. (1850), canto CXXIII. In Christopher Ricks (ed.), *The Poems of Tennyson* (1987), Vol. 2, 442–3.

2 Evolution ever climbing after some ideal good,
And Reversion ever dragging Evolution in the mud.

Locksley Hall Sixty Years After (1886). In Christopher Ricks (ed.), *The Poems of Tennyson* (1987), Vol. 3, 157.

3 Science grows and Beauty dwindles.

Locksley Hall Sixty Years After (1886). In Christopher Ricks (ed.), *The Poems of Tennyson* (1987), Vol. 3, 158.

Thales c.625–c.547 BC

Greek natural philosopher

4 All things are from water and all things are resolved into water.

Aëtius 1.3.1., in H. Diels, *Doxographi Graeci* (1879), 276. Trans. R. W. Sharples.

Theophrastus c.372–c.287 BC

Greek botanist and philosopher

5 Spontaneous generation, to put the matter simply, takes place in smaller plants, especially in those that are annuals and herbaceous. But still it occasionally occurs too in larger plants whenever there is rainy weather or some peculiar condition of air or soil; for thus it is said that the silphium sprang up in Libya when a murky and heavy sort of wet weather condition occurred, and that the timber growth which is now there has come from some similar reason or other; for it was not there in former times.

De Causis Plantarum 1.5.1, in *Theophrastus: De Causis Plantarum Book One: Text, Critical Apparatus, Translation, and Commentary*, trans. Robert Ewing Dengler (1927), 31.

F. K. Johannes Thiele 1865–1918

German chemist

6 In the beginning, there was benzene!

Motto above the door of Thiele's office—Wieland often spoke of his teacher Thiele's dislike of the chemistry of natural products.

Quoted in R. Huisgen, 'The Wieland Memorial Lecture: Heinrich Wieland', *Proceedings of the Chemical Society*, 1958, 214.

Lewis Thomas 1913–93

American physician

7 The uniformity of the earth's life, more astonishing than its diversity, is accountable by the high probability that we derived, originally, from some single cell, fertilized in a bolt of lightning as the earth cooled. It is from the progeny of this parent cell that we take our looks; we still share genes around, and the resemblance of the enzymes of grasses to those of whales is a family resemblance.

The Lives of a Cell: Notes of a Biology Watcher (1980), 5.

8 Music is the effort we make to explain to ourselves how our brains work. We listen to Bach transfixed because this is listening to a human mind.

The Medusa and the Snail: More Notes of a Biology Watcher (1980), 154.

D'Arcy Wentworth Thompson 1860–1948

British zoologist and classical scholar

9 Cell and tissue, shell and bone, leaf and flower, are so many portions of matter, and it is in obedience to the laws of physics that their particles have been moved, moulded and confirmed. They are no exception to the rule that God always geometrizes. Their problems of form are in the first instance mathematical problems, their problems of growth are essentially physical problems, and the morphologist is, *ipso facto*, a student of physical science.

On Growth and Form (1917), 7–8.

1 For the harmony of the world is made manifest in Form and Number, and the heart and soul and all the poetry of Natural Philosophy are embodied in the concept of mathematical beauty.

On Growth and Form (1917), Epilogue, 778–9.

2 When I was young Science walked hand-in-hand with Art; now she walks arm-in-arm with Trade.

Clifford Dobell, 'D'Arcy Wentworth Thompson 1860–1948', *Obituary Notices of Fellows of the Royal Society 1948–1949* (1949), 613.

3 Sooner or later nature does everything that is physically possible.

On Growth and Form (1917).

George Paget Thomson 1892–1975

British physicist

4 The goddess of learning is fabled to have sprung full-grown from the brain of Zeus, but it is seldom that a scientific conception is born in its final form, or owns a single parent. More often it is the product of a series of minds, each in turn modifying the ideas of those that came before, and providing material for those that came after. The electron is no exception.

'Electronic Waves', Nobel Lecture, 7 June 1938. *Nobel Lectures: Physics 1922–1941* (1998), 397.

5 Probably every physicist would believe in a creation [of the universe] if the Bible had not unfortunately said something about it many years ago and made it seem old fashioned.

'Continuous Creation and the Edge of Space', *New Republic*, 1951, **124**, 21–2.

Henry Thomson 1820–1904

British surgeon

6 Why should we limit by dogma or otherwise man's liberty to select his food and drink? . . . [T]he great practical rule of life in regard of human diet will not be found in enforcing limitation of the sources of food which Nature has abundantly provided.

Diet in Relation to Age and Activity (1886), 20.

James Thomson 1700–48

British poet

7 Even *Light itself*, which every thing displays,
Shone undiscover'd, till his brighter Mind
Untwisted all the shining Robe of Day.

On Newton's Opticks.

A Poem Sacred to the Memory of Sir Isaac Newton (1727), 10.

8 Let NEWTON, *pure Intelligence*, whom GOD
To Mortals lent, to trace his boundless Works
From Laws sublimely simple, speak thy Fame
In all Philosophy.

The Seasons (1746), 130.

John Thomson 1765–1846

British physician and surgeon

9 In physical science the discovery of new facts is open to every blockhead with patience, manual dexterity, and acute senses; it is less effectually promoted by genius than by co-operation, and more frequently the result of accident than of design.

Review of 'An Account of the Life, Lectures, and Writings of William Cullen, M.D. Professor of the Practice of Physic in the University of Edinburgh', *The Edinburgh Review*, 1832, **55**, 461.

John Arthur Thomson 1861–1933

Scottish naturalist and writer

10 When science makes minor mysteries disappear, greater mysteries stand confessed. For one object of delight whose emotional value science has inevitably lessened—as Newton damaged the rainbow for Keats—science gives back double. To the grand primary impressions of the world-power, the immensities, the pervading order, and the universal flux, with which the man of feeling has been nurtured from of old, modern science has added thrilling impressions of manifoldness, intricacy, uniformity, inter-relatedness, and evolution. Science widens and clears the emotional window. There are great vistas to which science alone can lead, and they make for elevation of mind. The opposition between science and feeling is largely a misunderstanding. As one of our philosophers has remarked, science is in a true sense 'one of the humanities.'

J. Arthur Thomson (ed.), *The Outline of Science: A Plain Story Simply Told* (1921/2), Vol. 2, Science and Modern Thought, 787.

Joseph John Thomson 1856–1940

British physicist

1 The study of . . . simple cases would, I think, often be of advantage even to students whose mathematical attainments are sufficient to enable them to follow the solution of the more general cases. For in these simple cases the absence of analytical difficulties allows attention to be more easily concentrated on the physical aspects of the question, and thus gives the student a more vivid idea and a more manageable grasp of the subject than he would be likely to attain if he merely regarded electrical phenomena through a cloud of analytical symbols.

Elements of the Mathematical Theory of Electricity and Magnetism (1895), v-vi.

2 With the discovery and study of Cathode rays, Röntgen rays and Radio-activity a new era has begun in Physics.

Conduction of Electricity through Gases (1903), vi.

3 From the point of view of the physicist, a theory of matter is a policy rather than a creed; its object is to connect or co-ordinate apparently diverse phenomena, and above all to suggest, stimulate and direct experiment. It ought to furnish a compass which, if followed, will lead the observer further and further into previously unexplored regions.

The Corpuscular Theory of Matter (1907), 1.

4 The ether is not a fantastic creation of the speculative philosopher; it is as essential to us as the air we breathe.

'Address of the President of the British Association for the Advancement of Science', *Science*, 27 August 1909, 30, 267.

5 By research in pure science I mean research made without any idea of application to industrial matters but solely with the view of extending our knowledge of the Laws of Nature. I will give just one example of the 'utility' of this kind of research, one that has been brought into great prominence by the War—I mean the use of X-rays in surgery. Now, not to speak of what is beyond money value, the saving of pain, or, it may be, the life of the wounded, and of bitter grief to those who loved them, the benefit which the state has derived from the restoration of so many to life and limb, able to render services which would otherwise have been lost, is almost incalculable. Now, how was this method discovered? It was not the result of a research in applied science starting to find an improved method of locating bullet wounds. This might have led to improved probes, but we cannot imagine it leading to the discovery of X-rays. No, this method is due to an investigation in pure science, made with the object of discovering what is the nature of Electricity. The experiments which led to this discovery seemed to be as remote from 'humanistic interest'—to use a much misappropriated word—as anything that could well be imagined. The apparatus consisted of glass vessels from which the last drops of air had been sucked, and which emitted a weird greenish light when stimulated by formidable looking instruments called induction coils. Near by, perhaps, were great coils of wire and iron built up into electro-magnets. I know well the impression it made on the average spectator, for I have been occupied in experiments of this kind nearly all my life, nothwithstanding the advice, given in perfect good faith, by non-scientific visitors to the laboratory, to put that aside and spend my time on something useful.

Speech made on behalf of a delegation from the Conjoint Board of Scientific Studies in 1916 to Lord Crewe, then Lord President of the Council. In George Paget Thomson, *J J Thomson and the Cavendish Laboratory in His Day* (1965), 167–8.

1 I have paid special attention to those properties of the Positive Rays which seem to throw light on the problems of the structure of molecules and atoms and the question of chemical combination . . . I am convinced that as yet we are only at the beginning of the harvest of results which will elucidate the process of chemical combination, and thus bridge over the most serious gap which now exists between Physics and Chemistry.

Rays of Positive Electricity and their Application to Chemical Analyses (1921), v.

2 If the modern conception of the atom is correct the barrier which separated physics from chemistry has been removed.

The Electron in Chemistry (1923), Preface.

Thomas Thomson 1773–1852

Scottish chemist

3 ON **THOMSON** This work belongs to those few productions from which science will derive no advantage whatever. Much of the experimental part, even of the fundamental experiments, appears to have been made at the writing-desk; and the greatest civility which his contemporaries can show its author, is to forget it was ever published.

Attack of Berzelius on Dr. Thomson's Attempt to Establish the First Principles of Chemistry by Experiment.

In *Jahresbericht*, 1827, 6, 77. Quoted in English in *Philosophical Magazine*, 1828, 4, 451.

William Thomas Thomson

(Baron Kelvin of Largs) 1824–1907

Scottish mathematician and physicist

4 When 'thermal agency' is thus spent in conducting heat through a solid, what becomes of the mechanical effect which it might produce? Nothing can be lost in the operations of nature—no energy can be destroyed.

'An Account of Carnot's Theory of the Motive Power of Heat; with Numerical Results Deduced from Regnault's Experiments on Steam' (1849). In *Mathematical and Physical Papers* (1882–1911), Vol. 1, 118.

5 A perfect thermo-dynamic engine is such that, whatever amount of mechanical effect it can derive from a certain thermal agency; if an equal amount be spent in working it backwards, an equal reverse thermal effect will be produced.

'An Account of Carnot's Theory of the Motive Power of Heat; with Numerical Results Deduced from Regnault's Experiments on Steam' (1849). In *Mathematical and Physical Papers* (1882–1911), Vol. 1, 119.

6 Considering it as thus established, that heat is not a substance, but a dynamical form of mechanical effect, we perceive that there must be an equivalence between mechanical work and heat, as between cause and effect.

'On the Dynamical Theory of Heat, with Numerical Results Deduced from Mr. Joule's Equivalent of a Thermal Unit, and M. Regnault's Observations on Steam' (1851). In *Mathematical and Physical Papers* (1882–1911), Vol. 1, 175.

7 The whole theory of the motive power of heat is founded on the two following propositions, due respectively to Joule, and to Carnot and Clausius.

PROP. I. (Joule).—When equal quantities of mechanical effect are produced by any means whatever from purely thermal sources, or lost in purely thermal effects, equal quantities of heat are put out of existence or are generated.

PROP. II. (Carnot and Clausius).—If an engine be such that, when it is worked backwards, the physical and mechanical agencies in every part of its motions are all reversed, it produces as much mechanical effect as can be produced by any thermo-dynamic engine, with the same temperatures of source and refrigerator, from a given quantity of heat.

'On the Dynamical Theory of Heat, with Numerical Results Deduced from Mr Joule's Equivalent of a Thermal Unit, and M. Regnault's Observations on Steam' (1851). In *Mathematical and Physical Papers* (1882), Vol. 1, 178.

1 It is impossible for a self-acting machine, unaided by any external agency, to convey heat from one body to another at a higher temperature.

'On the Dynamical Theory of Heat, with Numerical Results Deduced from Mr Joule's Equivalent of a Thermal Unit, and M. Regnault's Observations on Steam' (1851). In *Mathematical and Physical Papers* (1882), Vol. 1, 181.

2 Mechanical action may be derived from heat, and heat may be generated by mechanical action, by means of forces either acting between contiguous parts of bodies, or due to electric excitation; but in no other way known, or even conceivable, in the present state of science. Hence thermo-dynamics falls naturally into two divisions, of which the subjects are respectively, *the relation of heat to the forces acting between contiguous parts of bodies*, and *the relation of heat to electrical agency.*

'On the Dynamical Theory of Heat, with Numerical Results Deduced from Mr Joule's Equivalent of a Thermal Unit, and M. Regnault's Observations on Steam' (1851). Part VI, 'Thermo Electric Currents' (1854). In *Mathematical and Physical Papers* (1882), Vol. 1, 232.

3 The following general conclusions are drawn from the propositions stated above, and known facts with reference to the mechanics of animal and vegetable bodies:—

There is at present in the material world a universal tendency to the dissipation of mechanical energy.

Any *restoration* of mechanical energy, without more than an equivalent of dissipation, is impossible in inanimate material processes, and is probably never effected by means of organized matter, either endowed with vegetable life, or subjected to the will of an animated creature.

Within a finite period of time past the earth must have been, and within a finite period of time to come the earth must again be, unfit for the habitation of man as at present constituted, unless operations have been, or are to be performed, which are impossible under the laws to which the known operations going on at present in the material world are subject.

'On a Universal Tendency in Nature to the Dissipation of Mechanical Energy', *Proceedings of the Royal Society of Edinburgh*, 1852, 3, 141–2. In *Mathematical and Physical Papers* (1882–1911), Vol. 1, 513–4.

4 The result would inevitably be a state of universal rest and death, if the universe were finite and left to obey existing laws. But it is impossible to conceive a limit to the extent of matter in the universe; and therefore science points rather to an endless progress, through an endless space, of action involving the transformation of potential energy into palpable motion and thence into heat, than to a single finite mechanism, running down like a clock, and stopping for ever.

'On the Age of the Sun's Heat' (1862), *Popular Lectures and Addresses* (1891), Vol. 1, 349–50.

5 It would be a very wonderful, but not an absolutely incredible result, that volcanic action has never been more violent on the whole than during the last two or three centuries; but it is as certain that there is now less volcanic energy in the whole earth than there was a thousand years ago, as it is that there is less gunpowder in a 'Monitor' after she has been seen to discharge shot and shell, whether at a nearly equable rate or not, for five hours without receiving fresh supplies, than there was at the beginning of the action.

'On the Secular Cooling of the Earth', *Transactions of the Royal Society of Edinburgh*, 1864, 23, 159.

1 The 'Doctrine of Uniformity' in Geology, as held by many of the most eminent of British Geologists, assumes that the earth's surface and upper crust have been nearly as they are at present in temperature, and other physical qualities, during millions of millions of years. But the heat which we know, by observation, to be now conducted out of the earth yearly is so great, that if *this* action has been going on with any approach to uniformity for 20,000 million years, the amount of heat lost out of the earth would have been about as much as would heat, by 100 Cent., a quantity of ordinary surface rock of 100 times the earth's bulk. This would be more than enough to melt a mass of surface rock equal in bulk to the *whole earth.* No hypothesis as to chemical action, internal fluidity, effects of pressure at great depth, or possible character of substances in the interior of the earth, possessing the smallest vestige of probability, can justify the supposition that the earth's upper crust has remained nearly as it is, while from the whole, or from any part, of the earth, so great a quantity of heat has been lost.

'The "Doctrine of Uniformity" in Geology Briefly Refuted' (1866), *Popular Lectures and Addresses* (1891), Vol. 2, 6–7.

2 To take one of the simplest cases of the dissipation of energy, the conduction of heat through a solid—consider a bar of metal warmer at one end than the other and left to itself. To avoid all needless complication, of taking loss or gain of heat into account, imagine the bar to be varnished with a substance impermeable to heat. For the sake of definiteness, imagine the bar to be first given with one half of it at one uniform temperature, and the other half of it at another uniform temperature. Instantly a diffusing of heat commences, and the distribution of temperature becomes continuously less and less unequal, tending to perfect uniformity, but never in any finite time attaining perfectly to this ultimate condition. This process of diffusion could be perfectly prevented by an army of Maxwell's 'intelligent demons'* stationed at the surface, or interface as we may call it with Prof. James Thomson, separating the hot from the cold part of the bar.

* The definition of a 'demon', according to the use of this word by Maxwell, is an intelligent being endowed with free will, and fine enough tactile and perceptive organisation to give him the faculty of observing and influencing individual molecules of matter.

'The Kinetic Theory of the Dissipation of Energy', *Nature*, 1874, 9, 442.

3 If, then, the motion of every particle of matter in the universe were precisely reversed at any instant, the course of nature would be simply reversed for ever after. The bursting bubble of foam at the foot of a waterfall would reunite and descend into the water; the thermal motions would reconcentrate their energy, and throw the mass up the fall in drops re-forming into a close column of ascending water. Heat which had been generated by the friction of solids and dissipated by conduction, and radiation, and radiation with absorption, would come again to the place of contact, and throw the moving body back against the force to which it had previously yielded. Boulders would recover from the mud materials required to rebuild them into their previous jagged forms, and would become reunited to the mountain peak from which they had formerly broken away. And if also the materialistic hypothesis of life were true, living creatures would grow backwards, with conscious knowledge of the future but no memory of the past, and would become again unborn.

'The Kinetic Theory of the Dissipation of Energy', *Nature*, 1874, 9, 442.

1 In physical science a first essential step in the direction of learning any subject is to find principles of numerical reckoning and practicable methods for measuring some quality connected with it. I often say that when you can measure what you are speaking about and express it in numbers you know something about it; but when you cannot measure it, when you cannot express it in numbers, your knowledge is of a meagre and unsatisfactory kind: it may be the beginning of knowledge, but you have scarcely, in your thoughts, advanced to the stage of *science*, whatever the matter may be.

'Electrical Units of Measurement' (1883), *Popular Lectures and Addresses* (1891), Vol. 1, 80–1.

2 The idea of an atom has been so constantly associated with incredible assumptions of infinite strength, absolute rigidity, mystical actions at a distance, and individuality, that chemists and many other reasonable naturalists of modern times, losing all patience with it, have dismissed it to the realms of metaphysics, and made it smaller than 'anything we can conceive.' But if atoms are inconceivably small, why are not all chemical actions infinitely swift? Chemistry is powerless to deal with this question, and many others of paramount importance, if barred by the hardness of its fundamental assumptions, from contemplating the atom as a real portion of matter occupying a finite space, and forming not an immeasurably small constituent of any palpable body.

Sir William Thomson and Peter Guthrie Tait, *A Treatise on Natural Philosophy* (1883), Vol. 1, Part 2, 495.

3 Some people say they cannot understand a million million. Those people cannot understand that twice two makes four. That is the way I put it to people who talk to me about the incomprehensibility of such large numbers. I say *finitude* is incomprehensible, the infinite in the universe *is* comprehensible. Now apply a little logic to this. Is the negation of infinitude incomprehensible? What would you think of a universe in which you could travel one, ten, or a thousand miles, or even to California, and then find it comes to an end? Can you suppose an end of matter or an end of space? The idea is incomprehensible. Even if you were to go millions and millions of miles the idea of coming to an end is incomprehensible. You can understand one thousand per second as easily as you can understand one per second. You can go from one to ten, and then times ten and then to a thousand without taxing your understanding, and then you can go on to a thousand million and a million million. You can all understand it.

'The Wave Theory of Light' (1884), *Popular Lectures and Addresses* (1891), Vol. 1, 322.

4 We may consequently regard it as certain that, neither by natural agencies of inanimate matter, nor by the operations arbitrarily effected by animated Creatures, can there be any change produced in the amount of mechanical energy in the Universe.

Draft of 'On a Universal Tendency . . . ', PA 137, Kelvin Collection, Cambridge University Library. In Crosbie Smith, *The Science of Energy* (1998), 139.

Henry David Thoreau 1817–62

American essayist, poet and philosopher

5 Our life is frittered away by detail . . . Simplify, simplify.

Walden; or Life in the Woods (1854), 99.

6 It appears to be law that you cannot have a deep sympathy with both man and nature.

Walden; Or Life in the Woods (1854)

Kip Stephen Thorne 1940–

American physicist and astrophysicist

1 Of all the conceptions of the human mind from unicorns to gargoyles to the hydrogen bomb perhaps the most fantastic is the black hole: a hole in space with a definite edge over which anything can fall and nothing can escape; a hole with a gravitational field so strong that even light is caught and held in its grip; a hole that curves space and warps time.

Cosmology + 1: Readings from Scientific American (1977), 63.

Friedrich Tiedemann 1781–1861

German anatomist and physiologist

2 Considered from the standpoint of chemistry, living bodies appear to us as laboratories of chemical processes, for they undergo perpetual changes in their material substrate. They draw materials from the outside world and combine them with the mass of their liquid and solid parts.

Physiologie des Menschen (1830), Vol. 1, 'Allgemeine Betrachtungen der organischen Körper', 34. Trans. in Kenneth L. Caneva, *Robert Mayer and the Conservation of Energy* (1993), 71.

Edward Bradford Titchener 1867–1927

British-born American psychologist

3 An experiment is an observation that can be repeated, isolated and varied. The more frequently you can *repeat* an observation, the more likely are you to see clearly what is there and to describe accurately what you have seen. The more strictly you can *isolate* an observation, the easier does your task of observation become, and the less danger is there of your being led astray by irrelevant circumstances, or of placing emphasis on the wrong point. The more widely you can *vary* an observation, the more clearly will the uniformity of experience stand out, and the better is your chance of discovering laws.

A Text-Book of Psychology (1909), 20.

4 The great difference between science and technology is a difference of initial attitude. The scientific man follows his method whithersoever it may take him. He seeks acquaintance with his subject-matter, and he does not at all care about what he shall find, what shall be the content of his knowledge when acquaintance-with is transformed into knowledge-about. The technologist moves in another universe; he seeks the attainment of some determinate end, which is his sole and obsessing care; and he therefore takes no heed of anything that he cannot put to use as means toward that end.

Systematic Psychology: Prolegomena (1929), 66.

Alexander Robertus Todd (Lord Todd of Trumpington) 1907–97

British chemist and biochemist

5 I am an organic chemist, albeit one who adheres to the definition of organic chemistry given by the great Swedish chemist Berzelius, namely, the chemistry of substances found in living matter, and my science is one of the more abstruse insofar as it rests on concepts and employs a jargon neither of which is a part of everyday experience. Nevertheless, organic chemistry deals with matters of truly vital importance and in some of its aspects with which I myself have been particularly concerned it may prove to hold the keys to Life itself.

'Synthesis in the Study of Nucleotides', Nobel Lecture, 11 December 1957. In *Nobel Lectures: Chemistry 1942–1962* (1964), 522.

6 Of the nucleosides from deoxyribonucleic acids, all that was known with any certainty [in the 1940s] was that they were 2-deoxy-D-ribosides of the bases adenine, guanine, thymine and cytosine and it was assumed that they were structurally analogous to the ribonucleosides. The chemistry of the nucleotides—the phosphates of the nucleosides—was in a correspondingly primitive state. It may well be asked why the chemistry of these groups of compounds was not further advanced, particularly since we recognize today that they occupy a central place in the history of the living cell. True, their full significance was for a long time unrecognized and emerged only slowly as biochemical research got into its stride but I think a more important reason is to be found in the physical properties of compounds of the nucleotide group. As water-soluble polar compounds with no proper melting points they were extremely difficult to handle by the classic techniques of organic chemistry, and were accordingly very discouraging substances to early workers. It is surely no accident that the major advances in the field have coincided with the appearance of new experimental techniques such as paper and ion-exchange chromatography, paper electrophoresis, and countercurrent distribution, peculiarly appropriate to the compounds of this group.

'Synthesis in the Study of Nucleotides', Nobel Lecture, 11 December 1957. In *Nobel Lectures: Chemistry 1942–1962* (1964), 524.

1 After having a wash I proceeded to the bar where—believe it or not—there was a white-coated barman who was not only serving drinks but also cigarettes! I hastened forward and rather timidly said 'Can I have some cigarettes?'

'What's your rank?' was the slightly unexpected reply.

'I am afraid I haven't got one,' I answered.

'Nonsense—everyone who comes here has a rank.'

'I'm sorry but I just don't have one.'

'Now that puts me in a spot,' said the barman, 'for orders about cigarettes in this camp are clear—twenty for officers and ten for other ranks. Tell me what exactly are you?'

Now I really wanted those cigarettes so I drew myself up and said 'I am the Professor of Chemistry at Manchester University.'

The barman contemplated me for about thirty seconds and then said 'I'll give you five.'

Since that day I have had few illusions about the importance of professors!

Following a 1941 visit to the Defence Research Establishment at Porton to watch a demonstration of a new chemical weapon for use against tanks.

A Time to Remember: The Autobiography of a Chemist (1983), 59.

2 ON TODD Doesn't it strike you as odd
That a commonplace fellow like Todd
Should spell if you please,
His name with two Ds.
When one is sufficient for God.

Quoted by M. G. De St. V. Atkins, *The Times*, 22 January 1997, from memory of a conversation with 'an American tribiologist' who recalled it as current when he was an undergraduate at Christ's College.

3 ON TODD A. R. Todd
Thinks he's God.
N. F. Mott
Says he's not.

Quoted by William Lord in *The Times*, 22 January 1997, as illustrating the rivalry between chemistry and physics departments at Cambridge.

Robert Bentley Todd 1809–60 and **William Bowman** 1816–92

British physician and British surgeon

4 The nature of the connexion between the mind and nervous matter has ever been, and must continue to be, the deepest mystery in physiology; and they who study the laws of Nature, as ordinances of God, will regard it as one of those secrets of his counsels 'which Angels desire to look into.'

The Physiological Anatomy and Physiology of Man (1845), Vol. 1, 262.

Ernest R. Toon

American chemist

1 Chemistry is like a majestic skyscraper. The concrete secure foundation of chemistry consists of countless experimentally observed facts. The theories, principles and laws developed from these observations are like an elevator which runs from the bottom to the top of the edifice.

Ernest R. Toon and George L. Ellis (eds.), *Foundations of Chemistry* (1968), 1.

Rodolphe Töpffer 1799–1846

Swiss writer

2 I always love geology. In winter, particularly, it is pleasant to listen to theories about the great mountains one visited in the summer; or about the Flood or volcanoes; about great catastrophes or about blisters; above all about fossils . . . Everywhere there are hypotheses, but nowhere truths; many workmen, but no experts; priests, but no God. In these circumstances each man can bring his hypothesis like a candle to a burning altar, and on seeing his candle lit declare 'Smoke for smoke, sir, mine is better than yours'. It is precisely for this reason that I love geology.

Nouvelles Genévoises (1910), 306. First edition, 1841.

William Whiteman Carlton Topley 1886–1944

British bacteriologist

3 Committees are dangerous things that need most careful watching. I believe that a research committee can do one useful thing and one only. It can find the workers best fitted to attack a particular problem, bring them together, give them the facilities they need, and leave them to get on with the work. It can review progress from time to time, and make adjustments; but if it tries to do more, it will do harm.

Attributed.

Stephen Edelston Toulmin 1922– and June Goodfield 1927–

British philosopher and British historian

4 The picture of the natural world we all take for granted today, has one remarkable feature, which cannot be ignored in any study of the ancestry of science: it is a *historical* picture.

The Discovery of Time (1965), Penguin edition (1967), 17.

Joseph Pitton de Tournefort 1656–1708

French botanist

5 We have several stones whose generation is incomprehensible unless it is supposed that they come from some kind of seed, if I may be permitted to use this term; that is to say, from a germ in which the organic particles of these stones are enclosed '*en petit*', just as those of the largest plants are enclosed in the germs of their grains.

Histoire de l'Académie Royale des Sciences Année: Avec les Mémoires de Mathématique et de Physique (1702), 230.

Thomas Traherne c.1637–1674

English mystical poet and clergyman

6 He that knows the secrets of nature with Albertus Magnus, or the motions of the heavens with Galileo, or the cosmography of the moon with Hevelius, or the body of man with Galen, or the nature of diseases with Hippocrates, or the harmonies in melody with Orpheus, or of poesy with Homer, or of grammar with Lilly, or of whatever else with the greatest artist; he is nothing if he knows them merely for talk or idle speculation, or transient and external use. But he that knows them for value, and knows them his own, shall profit infinitely.

Bertram Dobell (ed.), *Centuries of Meditations* (1908), The third century, No. 41, 189–90.

Abraham Trembley 1710–84

Swiss zoologist

1 It was on the 25th November 1740 that I cut the first polyp. I put the two parts in a flat glass, which only contained water to the height of four to five lignes. It was thus easy for me to observe these portions of the polyp with a fairly powerful lens.

I shall indicate farther on the precautions I took in making my experiments on these cut polyps and the technique I adopted to cut them. It will suffice to say here that I cut the polyp concerned transversely, a little nearer the anterior than the posterior end. The first part was thus a little shorter than the second.

The instant that I cut the polyp, the two parts contracted so that at first they only appeared like two little grains of green matter at the bottom of the glass in which I put them—for green, as I have already said, is the colour of the first polyps that I possessed. The two parts expanded on the same day on which I separated them. They were very easy to distinguish from one another. The first had its anterior end adorned with the fine threads that serve the polyp as legs and arms, which the second had none.

The extensions of the first part was not the only sign of life that it gave on the same day that it was separated from the other. I saw it move its arms; and the next day, the first time I came to observe it, I found that it had changed its position; and shortly afterwards I saw it take a step. The second part was extended as on the previous day and in the same place. I shook the glass a little to see if it were still alive. This movement made it contract, from which I judged that it was alive. Shortly afterwards it extended again. On the following days I saw the same thing.

Mémoires, pour servir à l'histoire d'un genre de polyps d'eau douce à bras en forme de cornes (1744), 7–16. Trans. John R. Baker, *Abraham Trembley of Geneva: Scientist and Philosopher 1710–1784* (1952), 31.

2 Several times every day I observed the portions of the polyp with a magnifying glass. On the 4th December, that is to say on the ninth day after having cut the polyp, I seemed in the morning to be able to perceive, on the edges of the anterior end of the second part (the part that had neither head nor arms), three little points arising from those edges. They immediately made me think of the horns that serve as the legs and arms of the polyp. Nevertheless I did not want to decide at once that these were actually arms that were beginning to grow. Throughout the next day I continually observed these points: this excited me extremely, and I awaited with impatience the moment when I should know with certainty what they were. At last, on the following day, they were so big that there was no longer any room for doubt that they were actually arms growing at the anterior extremity of this second part. The next day two more arms started to grow out, and a few days later three more. The second part thus had eight of them, and these were all in a short time as long as those of the first part, that is to say as long as those the polyp possessed before it was cut. I then no longer found any difference between the second part and a polyp that had never been cut. I had remarked the same thing about the first part since the day after the operation. When I observed them with the magnifying glass with all the attention of which I was capable, each of the two appeared perceptibly to be a complete polyp, and they performed all the functions that were known to me: they extended, contracted, and walked.

Mémoires, pour servir à l'histoire d'un genre de polyps d'eau douce à bras en forme de cornes (1744), 7–16. Trans. John R. Baker, *Abraham*

Trembley of Geneva: Scientist and Philosopher 1710–1784 (1952), 32.

Lionel Trilling 1905–75

American literary critic

1 Now Freud may be right or he may be wrong in the place he gives to biology in human fate, but I think we must stop to consider whether this emphasis on biology, whether correct or incorrect, is not so far from being a reactionary idea that it is actually a liberating idea. It proposes to us that culture is not all-powerful. It suggests that there is a residue of human quality beyond the reach of cultural control, and that this residue of human quality, elemental as it may be, serves to bring culture itself under criticism and keeps it from being absolute.

Freud and the Crisis of our Culture (1955), 48.

Joshua Trimmer 1795–1857

British geologist

2 The personal adventures of a geologist would form an amusing narrative. He is trudging along, dusty and weather-beaten, with his wallet at his back, and his hammer on his shoulder, and he is taken for a stone-mason travelling in search of work. In mining-countries, he is supposed to be in quest of mines, and receives many tempting offers of shares in the 'Wheel Dream', or the 'Golden Venture';—he has been watched as a smuggler; it is well if he has not been committed as a vagrant, or apprehended as a spy, for he has been refused admittance to an inn, or has been ushered into the room appropriated to ostlers and postilions. When his fame has spread among the more enlightened part of the community of a district which he has been exploring, and inquiries are made of the peasantry as to the habits and pursuits of the great philosopher who has been among them, and with whom they have become familiar, it is found that the importance attached by him to shells and stones, and such like trumpery, is looked upon as a species of derangement, but they speak with delight of his affability, sprightliness, and good-humour. They respect the strength of his arm, and the weight of his hammer, as they point to marks which he inflicted on the rocks, and they recount with wonder his pedestrian performances, and the voracious appetite with which, at the close of a long day's work he would devour the coarsest food that was set before him.

Practical Geology and Mineralogy: With Instructions for the Qualitative Analysis of Minerals (1841), 31–2.

Wilfrid Batten Lewis Trotter 1872–1939

British surgeon and sociologist

3 It is the function of notions in science to be useful, to be interesting, to be verifiable and to acquire value from any one of these qualities. Scientific notions have little to gain as science from being forced into relation with that formidable abstraction, 'general truth'.

'The Commemoration of Great Men', *British Medical Journal* 1932, 1, 32.

4 The mind likes a strange idea as little as the body likes a strange protein, and resists it with a similar energy. It would not perhaps be too fanciful to say that a new idea is the most quickly acting antigen known to science. If we watch ourselves honestly, we shall often find that we have begun to argue against a new idea even before it has been completely stated. I have no doubt that that last sentence has already met with repudiation—and shown how quickly the defence mechanism gets to work.

The Collected Papers of Wilfred Trotter F.R.S. (1941), 186.

John Trowbridge 1843–1923

British physicist

5 We have one great guiding principle which, like the pillar of cloud by day, and the pillar of fire by night, will conduct us, as Moses and the Israelites were once conducted, to an eminence from which we can survey the promised scientific future. That principle is the conservation of energy.

'What is Electricity?', *Popular Science Monthly*, November 1884, 26, 77.

Richard P. Turco 1943- et al.

American atmospheric chemist

1 Global nuclear war could have a major impact on climate—manifested by significant surface darkening over many weeks, subfreezing land temperatures persisting for up to several months, large perturbations in global circulation patterns, and dramatic changes in local weather and precipitation rates—a harsh 'nuclear winter' in any season.

'Nuclear Winter: Global Consequences of Multiple Nuclear Explosions', *Science*, 1983, 222, 1290.

Alan Mathison Turing 1912–54

British mathematician

2 The automatic computing engine now being designed at N. P. L. [National Physics Laboratory] is a typical large scale electronic digital computing machine. In a single lecture it will not be possible to give much technical detail of this machine, and most of what I shall say will apply equally to any other machine of this type now being planned. From the point of view of the mathematician the property of being digital should be of greater interest than that of being electronic. That it is electronic is certainly important because these machines owe their high speed to this, and without the speed it is doubtful if financial support for their construction would be forthcoming. But this is virtually all that there is to be said on that subject. That the machine is digital however has more subtle significance. It means firstly that numbers are represented by sequences of digits which can be as long as one wishes. One can therefore work to any desired degree of accuracy. This accuracy is not obtained by more careful machining of parts, control of temperature variations, and such means, but by a slight increase in the amount of equipment in the machine.

Lecture to the London Mathematical Society, 20 February 1947. Quoted in B. E. Carpenter and R. W. Doran (eds.), *A. M. Turing's Ace Report of 1946 and Other Papers* (1986), 106.

3 It has been said that computing machines can only carry out the processes that they are instructed to do. This is certainly true in the sense that if they do something other than what they were instructed then they have just made some mistake. It is also true that the intention in constructing these machines in the first instance is to treat them as slaves, giving them only jobs which have been thought out in detail, jobs such that the user of the machine fully understands what in principle is going on all the time. Up till the present machines have only been used in this way. But is it necessary that they should always be used in such a manner? Let us suppose we have set up a machine with certain initial instruction tables, so constructed that these tables might on occasion, if good reason arose, modify those tables. One can imagine that after the machine had been operating for some time, the instructions would have altered out of all recognition, but nevertheless still be such that one would have to admit that the machine was still doing very worthwhile calculations. Possibly it might still be getting results of the type desired when the machine was first set up, but in a much more efficient manner. In such a case one would have to admit that the progress of the machine had not been foreseen when its original instructions were put in. It would be like a pupil who had learnt much from his master, but had added much more by his own work. When this happens I feel that one is obliged to regard the machine as showing intelligence.

Lecture to the London Mathematical Society,

20 February 1947. Quoted in B. E. Carpenter and R. W. Doran (eds.), *A. M. Turing's Ace Report of 1946 and Other Papers* (1986), 122–3.

1 In other words then, if a machine is expected to be infallible, it cannot also be intelligent.

Lecture to the London Mathematical Society, 20 February 1947. Quoted in B. E. Carpenter and R. W. Doran (eds.), *A. M. Turing's Ace Report of 1946 and Other Papers* (1986), 124.

2 I propose to consider the question, 'Can machines think?'

'Computing Machinery and Intelligence', *Mind*, 1950, **59**, 433.

3 Science is a Differential Equation. Religion is a Boundary Condition.

Comment made on a postcard sent to Robin O. Gandy. Reproduced in Andrew Hodges, *Alan Turing: The Enigma* (1983), 513.

Sherry Turkle 1948–

American sociologist

4 Under pressure from the computer, the question of mind in relation to machine is becoming a central cultural preoccupation. It is becoming for us what sex was to the Victorians—threat and obsession, taboo and fascination.

The Second Self: Computers and the Human Spirit (1984), 313.

Alfred Edwin Howard Tutton 1864–1938

British crystallographer

5 The beauty of crystals lies in the planeness of their faces.

The Natural History of Crystals (1924), 5.

Mark Twain (Samuel Langhorne Clemens) 1835–1910

American writer

6 There is something fascinating about science. One gets such wholesome returns of conjectures out of such trifling investment of fact.

Life on the Mississippi (1883), 208.

7 'I was reading an article about "Mathematics". Perfectly pure mathematics. My own knowledge of mathematics stops at "twelve times twelve," but I enjoyed that article immensely. I didn't understand a word of it; but facts, or what a man believes to be facts, are always delightful. That mathematical fellow believed in his facts. So do I. Get your facts first, and'—the voice dies away to an almost inaudible drone—'then you can distort 'em as much as you please.'

'An Interview with Mark Twain', in Rudyard Kipling, *From Sea to Sea* (1899), Vol. 2, 180.

8 Scientists have odious manners, except when you prop up their theory; then you can borrow money of them.

'The Bee', *c.*1902. In *What is Man? And Other Essays* (1917), 283.

9 Man is the Reasoning Animal. Such is the claim. I think it is open to dispute. Indeed, my experiments have proven to me that he is the Unreasoning Animal. Note his history, as sketched above. It seems plain to me that whatever he is he is *not* a reasoning animal. His record is the fantastic record of a maniac. I consider that the strongest count against his intelligence is the fact that with that record back of him he blandly sets himself up as the head animal of the lot: whereas by his own standards he is the bottom one.

In truth, man is incurably foolish. Simple things which the other animals easily learn, he is incapable of learning. Among my experiments was this. In an hour I taught a cat and a dog to be friends. I put them in a cage. In another hour I taught them to be friends with a rabbit. In the course of two days I was able to add a fox, a goose, a squirrel and some doves. Finally a monkey. They lived together in peace; even affectionately.

Next, in another cage I confined an Irish Catholic from Tipperary, and as soon as he seemed tame I added a Scotch Presbyterian from Aberdeen. Next a Turk from Constantinople; a Greek Christian from Crete; an Armenian; a Methodist from the wilds of Arkansas; a Buddhist from China; a

Brahman from Benares. Finally, a Salvation Army Colonel from Wapping. Then I stayed away two whole days. When I came back to note results, the cage of Higher Animals was all right, but in the other there was but a chaos of gory odds and ends of turbans and fezzes and plaids and bones and flesh—not a specimen left alive. These Reasoning Animals had disagreed on a theological detail and carried the matter to a Higher Court.

Letters From the Earth, ed. Bernard Devoto (1938), 227–8.

1 I stand almost with the others. They believe the world was made for man, I believe it likely that it was made for man; they think there is proof, astronomical mainly, that it was made for man, I think there is evidence only, not proof, that it was made for him. It is too early, yet, to arrange the verdict, the returns are not all in. When they are all in, I think that they will show that the world was made for man; but we must not hurry, we must patiently wait till they are all in.

Attributed.

Edward Burnett Tylor 1832–1917

British anthropologist

2 A first step in the study of civilization is to dissect it into details, and to classify these in their proper groups. Thus, in examining weapons, they are to be classed under spear, club, sling, bow and arrow, and so forth; among textile arts are to be ranged matting, netting, and several grades of making and weaving threads; myths are divided under such headings as myths of sunrise and sunset, eclipse-myths, earthquake-myths, local myths which account for the names of places by some fanciful tale, eponymic myths which account for the parentage of a tribe by turning its name into the name of an imaginary ancestor; under rites and ceremonies occur such practices as the various kinds of sacrifice to the ghosts of the dead and to other spiritual beings, the turning to the east in worship, the purification of ceremonial or moral uncleanness by means of water or fire. Such are a few miscellaneous examples from a list of hundreds . . . To the ethnographer, the bow and arrow is the species, the habit of flattening children's skulls is a species, the practice of reckoning numbers by tens is a species. The geographical distribution of these things, and their transmission from region to region, have to be studied as the naturalist studies the geography of his botanical and zoological species.

Primitive Culture (1871), Vol. 1, 7.

3 The thesis which I venture to sustain, within limits, is simply this, that the savage state in some measure represents an early condition of mankind, out of which the higher culture has gradually been developed or evolved, by processes still in regular operation as of old, the result showing that, on the whole, progress has far prevailed over relapse.

Primitive Culture (1871), Vol. 1, 28.

John Tyndall 1820–93

British physicist

4 The first experiment a child makes is a physical experiment: the suction-pump is but an imitation of the first act of every new-born infant.

'On the Study of Physics', From a Lecture delivered in the Royal Institution of Great Britain in the Spring of 1854. *Fragments of Science for Unscientific People: A Series of Detached Essays, Lectures, and Reviews* (1892), Vol. 1, 283.

5 A few days ago, a Master of Arts, who is still a young man, and therefore the recipient of a modern education, stated to me that until he had reached the age of twenty he had never been taught anything whatever regarding natural phenomena, or natural law. Twelve years of his life previously had been spent exclusively amongst the ancients. The case, I regret to say, is typical. Now we cannot, without prejudice to humanity, separate the present from the past.

'On the Study of Physics', From a Lecture delivered in the Royal Institution of Great Britain in the Spring of 1854. *Fragments of Science for Unscientific People: A Series of Detached Essays, Lectures, and Reviews* (1892), Vol. 1, 284–5.

1 The wintry clouds drop spangles on the mountains. If the thing occurred once in a century historians would chronicle and poets would sing of the event; but Nature, prodigal of beauty, rains down her hexagonal ice-stars year by year, forming layers yards in thickness. The summer sun thaws and partially consolidates the mass. Each winter's fall is covered by that of the ensuing one, and thus the snow layer of each year has to sustain an annually augmented weight. It is more and more compacted by the pressure, and ends by being converted into the ice of a true glacier, which stretches its frozen tongue far down beyond the limits of perpetual snow. The glaciers move, and through valleys they move like rivers.

The Glaciers of the Alps & Mountaineering in 1861 (1911), 247.

2 The law of conservation rigidly excludes both creation and annihilation. Waves may change to ripples, and ripples to waves,—magnitude may be substituted for number, and number for magnitude,—asteroids may aggregate to suns, suns may resolve themselves into florae and faunae, and florae and faunae melt in air,—the flux of power is eternally the same. It rolls in music through the ages, and all terrestrial energy,—the manifestations of life, as well as the display of phenomena, are but the modulations of its rhythm.

Heat Considered as a Mode of Motion (1863), 434.

3 We live *in* the sky, not *under* it.

'Climbing in Search of the Sky', *Fortnightly Review*, no. XXXVII (1 January 1870), 13.

4 Their business [those who believe in evolution] is not with the possible, but the actual—not with a world which *might* be, but with a world that *is*. This they explore with a courage not unmixed with reverence, and according to methods which, like the quality of a tree, are tested by their fruits. They have but one desire—to know the truth. They have but one fear—to believe a lie.

'Scientific Use of the Imagination', Discourse Delivered Before the British Association at Liverpool, 16 September 1870. *Fragments of Science for Unscientific People: A Series of Detached Essays, Lectures, and Reviews* (1892), Vol. 2, 134.

5 In our day grand generalizations have been reached. The theory of the origin of species is but one of them. Another, of still wider grasp and more radical significance, is the doctrine of the Conservation of Energy, the ultimate philosophical issues of which are as yet but dimly seem—that doctrine which 'binds nature fast in fate' to an extent not hitherto recognized, exacting from every antecedent its equivalent consequent, and bringing vital as well as physical phenomena under the dominion of that law of causal connexion which, so far as the human understanding has yet pierced, asserts itself everywhere in nature.

'Address Delivered Before The British Association Assembled at Belfast', 19 August 1874. *Fragments of Science for Unscientific People: A Series of Detached Essays, Lectures, and Reviews* (1892), Vol. 2, 180–1.

6 Believing, as I do, in the continuity of nature, I cannot stop abruptly where our microscopes cease to be of use. Here the vision of the mind authoritatively supplements the vision of the eye. By a necessity engendered and justified by science I cross the boundary of the experimental evidence, and discern in that Matter which we, in our ignorance of its latent powers, and notwithstanding our professed reverence for its Creator, have hitherto covered with opprobrium, the promise and potency of all terrestrial Life.

'Address Delivered Before The British Association Assembled at Belfast', 19 August 1874. *Fragments of Science for Unscientific*

People: A Series of Detached Essays, Lectures, and Reviews (1892), Vol. 2, 191.

1 We can trace the development of a nervous system, and correlate with it the parallel phenomena of sensation and thought. We see with undoubting certainty that they go hand in hand. But we try to soar in a vacuum the moment we seek to comprehend the connexion between them . . . Man the *object* is separated by an impassable gulf from man the *subject*.

'Address Delivered Before The British Association Assembled at Belfast', 19 August 1874. *Fragments of Science for Unscientific People: A Series of Detached Essays, Lectures, and Reviews* (1892), Vol. 2, 194–5.

2 It is not possible for me to purchase intellectual peace at the price of intellectual death.

'Address Delivered Before The British Association Assembled at Belfast', 19 August 1874. *Fragments of Science for Unscientific People: A Series of Detached Essays, Lectures, and Reviews* (1892), Vol. 2, 200.

3 The mutton in the study gathered over it a thick blanket of *Penicillium*. On the 13th [December 1875] it had assumed a light brown colour as if by a faint admixture of clay; but the infusion became transparent. The 'clay' here was the slime of dead or dormant *Bacteria*, the cause of their quiescence being the blanket of *Penicillium*. I found no active life in this tube, while all the others swarmed with *Bacteria*. In every case where the mould was thick and coherent the *Bacteria* died, or became dormant, and fell to the bottom of the sediment . . . The *Bacteria* which manufacture a green pigment appear to be uniformly victorious in their fight with the *Penicillium*.

Letter to T. H. Huxley, December 1875. Royal Institution Tyndall Manuscripts.

4 I allude to Mr. Charles Darwin, the Abrabam of scientific men—a searcher as obedient to the command of truth as was the patriarch to the command of God.

'Science and Man', Presidential Address Delivered before the Birmingham and Midland Institute, 1877. *Fragments of Science for Unscientific People: A Series of Detached Essays, Addresses, and Reviews* (1879), Vol. 2, 368.

5 By teaching us how to cultivate each ferment in its purity—in other words, by teaching us how to rear the individual organism apart from all others,—Pasteur has enabled us to avoid all these errors. And where this isolation of a particular organism has been duly effected it grows and multiplies indefinitely, but no change of it into another organism is ever observed. In Pasteur's researches the Bacterium remained a Bacterium, the Vibrio a Vibrio, the Penicillium a Penicillium, and the Torula a Torula. Sow any of these in a state of purity in an appropriate liquid; you get it, and it alone, in the subsequent crop. In like manner, sow smallpox in the human body, your crop is smallpox. Sow there scarlatina, and your crop is scarlatina. Sow typhoid virus, your crop is typhoid—cholera, your crop is cholera. The disease bears as constant a relation to its contagium as the microscopic organisms just enumerated do to their germs, or indeed as a thistle does to its seed.

'Fermentation, and its Bearings on Surgery and Medicine', in *Essays on the Floating-Matter of the Air in Relation to Putrefaction and Infection* (1881), 264.

6 ON TYNDALL In fact a favourite problem of his [Tyndall] is—Given the molecular forces in a mutton chop, deduce Hamlet or Faust therefrom. He is confident that the Physics of the Future will solve this easily.

Letter from T. H. Huxley to Herbert Spencer, 3 August 1861. In *Life and Letters of Thomas Henry Huxley* (1903), Vol. 1, 333.

Edward Tyson 1650–1708

British physician and anatomist

7 Nature when more shy in one, hath more freely confest and shewn herself in another; and a Fly sometimes hath given greater light towards the true knowledge of the structure and the uses of the Parts in Humane Bodies, than an often repeated dissection of the same might have done . . . We must not therefore think the meanest of the Creation vile or useless, since that in them in lively Characters (if we can but read) we may find the knowledge of a Deity and ourselves . . . In every Animal there is a world of wonders; each is a Microcosme or a world in it self.

Phocæna, or the Anatomy of a Porpess, dissected at Gresham College: With a Præliminary Discourse Concerning Anatomy, and a Natural History of Animals (1680), 2–3.

John Hoyer Updike 1932–

American novelist and man of letters

1 Neutrinos, they are very small
They have no charge and have no mass
And do not interact at all.

'Cosmic Gall', in *Telephone Poles and Other Poems* (1964), 5.

2 'But in the binary system,' Dale points out, handing back the squeezable glass, 'the alternative to one isn't minus one, it's zero. That's the beauty of it, mechanically.' 'O.K. Gotcha. You're asking me, What's this minus one? I'll tell you. It's a *plus one moving backward in time*. This is all in the space-time foam, inside the Planck duration, don't forget. The dust of points gives birth to time, and time gives birth to the dust of points. Elegant, huh? It *has* to be. It's blind chance, plus pure math. They're proving it, every day. Astronomy, particle physics, it's all coming together. Relax into it, young fella. It feels great. Space-time foam.'

Roger's Version (1986), Part 5, Chapter 2, 293–4.

James Ussher (Archbishop of Armagh) 1581–1656

Anglo-Irish prelate

3 But if any one, well seen in the knowledge, not onely of Sacred and exotick History, but of Astronomical Calculation, and the old Hebrew Kalendar, shall apply himself to these studies, I judge it indeed difficult, but not impossible for such a one to attain, not onely the number of years, but even, of dayes from the Creation of the World.

The Annals of the World (1658), Epistle to the reader.

4 In the beginning God created Heaven and Earth . . . Which beginning of time, according to our Chronologie, fell upon the entrance of the night preceding the twenty third day of *Octob.* in the year of the Julian Calendar, 710 [or 4004 B.C.]. Upon the first day therefore of the world, or *Octob.* 23. being our Sunday, God, together with the highest Heaven, created the Angels. Then having finished, as it were, the roofe of this building, he fell in hand with the foundation of this wonderfull Fabrick of the World, he fashioned this lowermost Globe, consisting of the Deep, and of the Earth; all the Quire of Angels singing together and magnifying his name therefore . . . And when the Earth was void and without forme, and darknesse covered the face of the Deepe, on the very middle of the first day, the light was created; which God severing from the darknesses, called the one day, and the other night.

The Annals of the World (1658), 1.

James William Valentine 1926–

American paleontologist

5 It is the intertwined and interacting mechanisms of evolution and ecology, each of which is at the same time a product and a process, that are responsible for life as we see it, and as it has been.

Evolutionary Paleoecology of the Marine Biosphere (1973), 58.

Antonio Vallisneri 1661–1730

Italian physician and naturalist

6 One will see a layer of smooth stones, popularly called *fluitati* [diluvium], and over these another layer of smaller pebbles, thirdly sand, and finally earth, and you will see this repeatedly . . . up to the summit of the Mountain. This clearly shows that the order has been caused by many floods, not just one.

De' Corpi Marini che su Monti si Trovano (1721), 57. Trans. Ezio Vaccari.

Henry Vaughan 1622–95

British poet

1 I saw Eternity the other night,
Like a great Ring of pure and endless light,
All calm, as it was bright;
And round beneath it, Time, in hours, days, years,
Driv'n by the spheres
Like a vast shadow mov'd; in which the world
And all her train were hurl'd.

'The World', in *Silex Scintillans* (1650), 91.

Thorstein Bunde Veblen 1857–1929

American economist and sociologist

2 Invention is the mother of necessity.

The Instinct of Workmanship: And the State of the Industrial Arts (1914), 316.

Vladímir Ivanovich Vernadsky 1863–1945

Russian mineralogist and geochemist

3 An organism is involved with the environment to which it is not only adapted but which is adapted to it as well.

'Problems of Biogeochemistry. Works of the Biogeochemical Laboratory', *Nauka*, 1980, 16, 22. Trans. A. V. Lapo.

4 The crust of erosion is always linked to life.

'Problems of Biogeochemistry. Works of the Biogeochemical Laboratory', *Nauka*, 1980, 16, 123. Trans. A. V. Lapo.

Andreas Vesalius 1514–64

Flemish physician and anatomist

5 It was above all in the period after the devastating incursions of the Goths that all branches of knowledge which previously had flourished gloriously and been practiced in the proper manner, began to deteriorate. This happened first of all in Italy where the most fashionable physicians, spurning surgery as did the Romans of old, assigned to their servants such surgical work as their patients seemed to require and merely exercised a supervision over them in the manner of architects.

De Humani Corporis Fabrica Libri Septem (1543), Book I, i, Preface to Charles V. Trans. William Frank Richardson, *On the Fabric of the Human Body* (1998), Book I, xlviii.

6 At this point, however, I have no intention whatever of criticizing the false teachings of Galen, who is easily first among the professors of dissection, for I certainly do not wish to start off by gaining a reputation for impiety toward him, the author of all good things, or by seeming insubordinate to his authority. For I am well aware how upset the practitioners (unlike the followers of Aristotle) invariably become nowadays, when they discover in the course of a single dissection that Galen has departed on two hundred or more occasions from the true description of the harmony, function, and action of the human parts, and how grimly they examine the dissected portions as they strive with all the zeal at their command to defend him. Yet even they, drawn by their love of truth, are gradually calming down and placing more faith in their own not ineffective eyes and reason than in Galen's writings.

De Humani Corporis Fabrica Libri Septem (1543), Book I, iv, Preface to Charles V. Trans. William Frank Richardson, *On the Fabric of the Human Body* (1998), Book I, liv.

7 Of all the constituents of the human body, bone is the hardest, the driest, the earthiest, and the coldest; and, excepting only the teeth, it is devoid of sensation. God, the great Creator of all things, formed its substance to this specification with good reason, intending it to be like a foundation for the whole body; for in the fabric of the human body bones perform the same function as do walls and beams in houses, poles in tents, and keels and ribs in boats.

Bones Differentiated by Function

Some bones, by reason of their strength, form as it were props for the body; these include the tibia, the femur, the spinal vertebrae, and most of the bony framework. Others are like bastions, defense walls, and ramparts, affording natural protection to other parts; examples are the skull, the spines and transverse processes of the vertebrae, the breast bone, the ribs. Others stand in front of the joints between certain bones, to ensure that the joint does not move too loosely or bend to too acute an angle. This is the function of the tiny bones, likened by the professors of anatomy to the size of a sesame seed, which are attached to the second internode of the thumb, the first internode of the other four fingers and the first internodes of the five toes. The teeth, on the other hand, serve specifically to cut, crush, pound and grind our food, and similarly the two ossicles in the organ of hearing perform a specifically auditory function.

De Humani Corporis Fabrica Libri Septem (1543), Book I, Ch. 1, 1. Trans. William Frank Richardson, *On the Fabric of the Human Body* (1998), Book I, 1.

1 Nature, the parent of all things, designed the human backbone to be like a keel or foundation. It is because we have a backbone that we can walk upright and stand erect. But this was not the only purpose for which Nature provided it; here, as elsewhere, she displayed great skill in turning the construction of a single member to a variety of different uses.

It Provides a Path for the Spinal Marrow, Yet is Flexible

Firstly, she bored a hole through the posterior region of the bodies of all the vertebrae, thus fashioning a suitable pathway for the spinal marrow which would descend through them.

Secondly, she did not make the backbone out of one single bone with no joints. Such a unified construction would have afforded greater stability and a safer seat for the spinal marrow since, not having joints, the column could not have suffered dislocations, displacements, or distortions. If the Creator of the world had paid such attention to resistance to injury and had subordinated the value and importance of all other aims in the fabric of parts of the body to this one, he would certainly have made a single backbone with no joints, as when someone constructing an animal of wood or stone forms the backbone of one single and continuous component. Even if man were destined only to bend and straighten his back, it would not have been appropriate to construct the whole from one single bone. And in fact, since it was necessary that man, by virtue of his backbone, be able to perform a great variety of movements, it was better that it be constructed from many bones, even though as a result of this it was rendered more liable to injury.

De Humani Corporis Fabrica Libri Septem (1543), Book I, Ch. 14, 57–8. Trans. William Frank Richardson, *On the Fabric of the Human Body* (1998), Book I, 138.

2 When I undertake the dissection of a human cadaver I pass a stout rope tied like a noose beneath the lower jaw and through the two zygomas up to the top of the head, either more toward the forehead or more toward the occiput according as I want the cadaver to hang with its head up or down. The longer end of the noose I run through a pulley fixed to a beam in the room so that I may raise or lower the cadaver as it hangs there or may turn it round in any direction to suit my purpose; and should I so wish I can allow it to recline at an angle upon a table, since a table can easily be placed underneath the pulley. This is how the cadaver was suspended for drawing all the muscle tables . . . though while that one was being drawn the rope was passed around the occiput so as to show the muscles in the neck. If the lower jaw has been removed in the course of dissection, or the zygomas have been broken, the hollows for the temporal muscles will nonetheless hold the noose sufficiently firmly. You must take care not to put the noose around the neck, unless some of the muscles connected to the occipital bone have already been cut away. It is best to suspend the cadaver like this because a human body lying on a table is very difficult to turn over on to its chest or its back.

De Humani Corporis Fabrica Libri Septem (1543), Book II, Ch. 24, 268. Trans. William Frank Richardson, *On the Fabric of the Human Body* (1999), Book II, 234.

1 For, however much we may clench our teeth in anger, we cannot but confess, in opposition to Galen's teaching but in conformity with the might of Aristotle's opinion, that the size of the orifice of the hollow vein at the right chamber of the heart is greater than that of the body of the hollow vein, no matter where you measure the latter. Then the following chapter will show the falsity of Galen's view that the hollow vein is largest at the point where it joins the hump of the liver.

De Humani Corporis Fabrica Libri Septem (1543), Book III, Ch. 6, 275. Trans. William Frank Richardson, *On the Fabric of the Human Body* (2002), Book III, 45.

2 As for Galen's netlike plexus, I do not need to pass on a lot of misinformation about it here, as I am quite sure that I have examined the whole system of the cerebral vessels. There is no occasion for making things up, since we are certain that Galen was deluded by his dissection of ox brains and described the cerebral vessels, not of a human but of oxen.

De Humani Corporis Fabrica Libri Septem (1543), Book III, Ch. 14, 310. Trans. William Frank Richardson, *On the Fabric of the Human Body* (2002), Book III, 140.

3 The source and origin of the nerves is the brain and spinal marrow, and hence some nerves originate from the brain and some from the spinal marrow. Some . . . experts set down the heart as the origin of the nerves and some the hard membrane that envelops the brain; none of them, however, thought it was the liver or any other viscus of that kind . . . Aristotle in particular, and quite a few others, thought that the nerves took origin from the heart.

De Humani Corporis Fabrica Libri Septem (1543), Book IV, Ch. 1, 315. Trans. William Frank Richardson, *On the Fabric of the Human Body* (2002), Book IV, 160.

4 Galen never inspected a human uterus.

De Humani Corporis Fabrica Libri Septem (1543), 532. Quoted and trans. in C. D. O'Malley, *Andreas Vesalius of Brussels* (1964), 142.

5 I strive that in public dissection the students do as much as possible so that if even the least trained of them must dissect a cadaver before a group of spectators, he will be able to perform it accurately with his own hands; and by comparing their studies one with another they will properly understand this part of medicine.

De Humani Corporis Fabrica Libri Septem (1543), 547. Quoted and trans. in C. D. O'Malley, *Andreas Vesalius of Brussels* (1964), 144.

6 How many things have been accepted on the word of Galen.

De Humani Corporis Fabrica Libri Septem (1543), 642. Quoted and trans. in C. D. O'Malley, *Andreas Vesalius of Brussels* (1964), 179.

7 However much the pits may be apparent, yet none, as far as can be comprehended by the senses, passes through the septum of the heart from the right ventricle into the left. I have not seen even the most obscure passages by which the septum of the ventricles is pervious, although they are mentioned by professors of anatomy since they are convinced that blood is carried from the right ventricle into the left. As a result—as I shall declare more openly elsewhere—I am in no little doubt regarding the function of the heart in this part.

De Humani Corporis Fabrica Libri Septem, revised edition (1555), 734. Quoted and trans. in C. D. O'Malley, *Andreas Vesalius of Brussels* (1964), 281.

1 I am not accustomed to say anything with certainty after only one or two observations.

Epistola, Rationem, Modumque Propinandi Radicis Chynae Decocti (1546), 141. Quoted and trans in C. D. O'Malley, *Andreas Vesalius of Brussels* (1964), 116.

Giambattista Vico 1668–1744

Italian philosopher

2 Our Science comes to be at once a history of the ideas, the customs, and the deeds of mankind. From these three we shall derive the principles of the history of human nature, which we shall show to be the principles of universal history, which principles it seems hitherto to have lacked.

The New Science (1744), Chapter 2, para 368. Trans. in Thomas Goddard Bergin and Max Harold Fisch (eds.), *The New Science of Giambattista Vico* (1970), 73.

Rudolf Carl Virchow 1821–1902

German pathologist

3 The physicians surely are the natural advocates of the poor and the social problem largely falls within their scope.

'The Aims of the Journal "Medical Reform"' (1848), in L. J. Rather (ed.), *Collected Essays on Public Health and Epidemiology* (1985), Vol. 1, 4.

4 Medicine is a social science, and politics is nothing more than medicine on a large scale.

'The Charity Physician' (1848), in L. J. Rather (ed.), *Collected Essays on Public Health and Epidemiology* (1985), Vol. 1, 33.

5 I definitely deny that any pathological process, i.e. any life-process taking place under unfavourable circumstances, is able to call forth qualitatively new formations lying beyond the customary range of forms characteristic of the species. *All pathological formations are either degenerations, transformations, or repetitions of typical physiological structures.*

'Cellular-Pathologie', *Archiv für pathologische Anatomie und Physiologie und für klinische Medizin*, 1855, **8**, 13–4. Trans. Lelland J. Rather, 'Cellular Pathology', in *Disease, Life, and Man: Selected Essays by Rudolf Virchow* (1958), 81.

6 No matter how we twist and turn we shall always come back to the cell. The eternal merit of Schwann does not lie in his cell theory that has occupied the foreground for so long, and perhaps will soon be given up, but in his description of the development of the various tissues, and in his demonstration that this development (hence all physiological activity) is in the end traceable back to the cell. Now if pathology is nothing but physiology with obstacles, and diseased life nothing but healthy life interfered with by all manner of external and internal influences then pathology too must be referred back to the cell.

'Cellular-Pathologie', *Archiv für pathologische Anatomie und Physiologie und für klinische Medizin*, 1855, **8**, 15. Trans. Lelland J. Rather, 'Cellular Pathology', in *Disease, Life, and Man: Selected Essays by Rudolf Virchow* (1958), 81.

7 All of our experience indicates that life can manifest itself only in a concrete form, and that it is bound to certain substantial loci. These loci are cells and cell formations. But we are far from seeking the last and highest level of understanding in the morphology of these loci of life. Anatomy does not exclude physiology, but physiology certainly presupposes anatomy. The phenomena that the physiologist investigates occur in special organs with quite characteristic anatomical arrangements; the various morphological parts disclosed by the anatomist are the bearers of properties or, if you will, of forces probed by the physiologist; when the physiologist has established a law, whether through physical or chemical investigation, the anatomist can still proudly state: This is the structure in which the law becomes manifest.

'Cellular-Pathologie', *Archiv für pathologische Anatomie und Physiologie und für klinische Medizin*, 1855, **8**, 19. Trans. Lelland J. Rather, 'Cellular Pathology', in *Disease, Life, and Man: Selected Essays by Rudolf Virchow* (1958), 84.

1 Life thus forms a long, unbroken chain of generations, in which the child becomes the mother, and the effect becomes the cause.

'On the Mechanistic Interpretation of Life' (1858), in *Disease, Life, and Man: Selected Essays by Rudolf Virchow*, trans. Lelland J. Rather (1958), 116.

2 I uphold my own rights, and therefore I also recognize the rights of others. This is the principle I act upon in life, in politics and in science. We owe it to ourselves to defend our rights, for it is the only guarantee for our individual development, and for our influence upon the community at large. Such a defence is no act of vain ambition, and it involves no renunciation of purely scientific aims. For, if we would serve science, we must extend her limits, not only as far as our own knowledge is concerned, but in the estimation of others.

'Preface to the First edition' (1858), *Cellular Pathology*, trans. Frank Chance (1860), x.

3 Just as a tree constitutes a mass arranged in a definite manner, in which, in every single part, in the leaves as in the root, in the trunk as in the blossom, cells are discovered to be the ultimate elements, so is it also with the forms of animal life. *Every animal presents itself as a sum of vital unities*, every one of which manifests all the characteristics of life. The characteristics and unity of life cannot be limited to any one particular spot in a highly developed organism (for example, to the brain of man), but are to be found only in the definite, constantly recurring structure, which every individual element displays. Hence it follows that the structural composition of a body of considerable size, a so-called individual, always represents a kind of social arrangement of parts, an arrangement of a social kind, in which a number of individual existences are mutually dependent, but in such a way, that every element has its own special action, and, even though it derive its stimulus to activity from other parts, yet alone effects the actual performance of its duties.

Lecture I 'Cells and the Cellular Theory' (1858), *Cellular Pathology*, trans. Frank Chance (1860), 13–14.

4 Where a cell arises, there a cell must have previously existed (*omnis cellula e cellula*), just as an animal can spring only from an animal, a plant only from a plant. In this manner, although there are still a few spots in the body where absolute demonstration has not yet been afforded, the principle is nevertheless established, that in the whole series of living things, whether they be entire plants or animal organisms, or essential constituents of the same, an eternal law of *continuous development* prevails.

Lecture II 'Physiological Tissues' (1858), *Cellular Pathology*, trans. Frank Chance (1860), 27–8. See **Raspail, François-Vincent** 514:4

5 *Mikroskopisch sehen lernen.*

Learn to see microscopically.

'Festnummer zu Ehren Rudolf Virchow', *Deutsche Medicinische Wochenschrift*, 1891, **42**, 1166. Quoted in Erwin H. Ackerknecht, *Rudolf Virchow: Doctor Statesman Anthropologist* (1953), 21.

6 The body is a cell state in which every cell is a citizen. Disease is merely the conflict of the citizens of the state brought about by the action of external forces.

Attributed.

7 ON **VIRCHOW** Like teeth husbands are hard to get and while we have them they give us a great deal of pain and trouble. But once we lose them they leave a wide gap.

Frau Virchow after the death of her husband.

Quotation attributed to Frau Virchow and supplied by W. H. Brock.

Virgil (Publius Vergilius Maro) 70–19 BC

Roman poet

1 *Felix, qui potuit rerum cognoscere causas.*

Blessed is he who has been able to win knowledge of the causes of things.

The Georgics, Book 2, l. 490. In *Virgil*, Vol. 1, *Eclogues, Georgics Aeneid I-VI*, trans. H. Rushton Fairclough (1916), 150.

John Hasbrouck van Vleck 1899–1980

American quantum physicist

2 It's pretty hard for me to lecture in French. I had to go to the Riviera afterwards to recuperate; I don't know what the audience had to do.

Said of his eight lectures to l'Institute Henri Poincaré. Quoted in his obituary, *Biographical Memoirs of Fellows of the Royal Society*, 1982, 28, 643.

Carl Vogt 1817–95

German physician and naturalist

3 Every natural scientist who thinks with any degree of consistency at all will, I think, come to the view that all those capacities that we understand by the phrase psychic activities (*Seelenthätigkeiten*) are but functions of the brain substance; or, to express myself a bit crudely here, that thoughts stand in the same relation to the brain as gall does to the liver or urine to the kidneys. To assume a soul that makes use of the brain as an instrument with which it can work as it pleases is pure nonsense; we would then be forced to assume a special soul for every function of the body as well.

Physiologische Briefe für Gelbildete alle Stände (1845–1847), 3 parts, 206. Trans. in Frederick Gregory, *Scientific Materialism in Nineteenth Century Germany* (1977), 64.

Jacob Volhard 1834–1910

German chemist

4 Before delivering your lectures, the manuscript should be in such a perfect form that, if need be, it could be set in type. Whether you follow the manuscript during the delivery of the lecture is purely incidental. The essential point is that you are thus master of the subject matter.

Advice to his son. Quoted in R. Oesper, *The Human Side of Scientists* (1975), 185.

Alessandro Giuseppe Antonio Anastasio Volta 1745–1827

Italian physicist

5 ON VOLTA In the beginning of the year 1800 the illustrious professor conceived the idea of forming a long column by piling up, in succession, a disc of copper, a disc of zinc, and a disc of wet cloth, with scrupulous attention to not changing this order. What could be expected beforehand from such a combination? Well, I do not hesitate to say, this apparently inert mass, this bizarre assembly, this pile of so many couples of unequal metals separated by a little liquid is, in the singularity of effect, the most marvellous instrument which men have yet invented, the telescope and the steam engine not excepted.

François Arago, 'Eloge for Volta' (1831), in *Oeuvres Complètes de François Arago* (1854), Vol. 1, 219–20.

6 ON VOLTA An alarm-bell to experimenters in every part of Europe.

On Volta's battery.

H. Davy, 'Historical Sketch of Electrical Discovery' (1810), in J. Davy (ed.), *The Collected Works of Sir Humphry Davy* (1840), Vol. 8, 271.

Voltaire (François Marie Arouet) 1694–1778

French philosopher and writer

7 A Frenchman who arrives in *London*, will find Philosophy, like every Thing else, very much chang'd there. He had left the World a *plenum*, and he now finds it a *vacuum*. At *Paris* the Universe is seen, compos'd of Vortices of subtile Matter; but nothing like it is seen in *London*. In *France*, 'tis the Pressure of

the Moon that causes the Tides; but in *England* 'tis the Sea that gravitates towards the Moon; so what when you think that the Moon should make it flood with us, those Gentlemen fancy it should be Ebb, which, very unluckily, cannot be prov'd. For to be able to do this, 'tis necessary the Moon and the Tides should have been enquir'd into, at the very instant of the Creation.

'Letter XIV. On Descartes and Sir Isaac Newton', in *Letters Concerning the English Nation* (1733), 109–10.

1 Heroes of physics, Argonauts of our time
Who leaped the mountains, who crossed the seas . . .
You have confirmed in uncomfortable places
What Newton knew without leaving his study.

Discours en Vers sur l'Homme (1734), Quatrième discours: de la Modération (1738). Quoted in and trans. J. L. Heilbron, *Weighing Imponderables and Other Quantitative Science around 1800* (1993), 224.

2 If you have seen with your own eyes a mountain move forward in a plain; that is, a huge rock of this mountain breaking off and covering entire fields; a whole castle sunken into the earth; a disappeared river later coming out from its abyss; clear marks of a vast amount of water which have once flooded countries today inhabited, and a hundred vestiges of other revolutions, then one is much more prone to believe that great changes altered the face of the earth than would be a lady from Paris who knows only that the place where her house was built was once plowable land. However, a lady from Naples who has seen the buried ruins of Herculaneum, is even less vulnerable to the prejudice which makes us believe that everything has always been the way it is today.

Dictionnaire philosophique (1764), 'Changements arrivées dans le globe', in *Oeuvres Complètes de Voltaire* (1846–1853), Vol. 7, 13. Trans. Albert V. and Marguerite Carozzi.

3 When I read some forty years ago that shells from Syria were found in the Alpes, I said, I admit, in a rather joking way, that these shells had apparently been carried by pilgrims on their return from Jerusalem. Mr. Buffon reprimanded me rather sharply in his *Theory of the Earth*, p. 281. I did not want to lose his friendship for peanuts; however, I am still of the same opinion because the impossibility of the formation of mountains by the sea is demonstrated to me.

Les Singularités de la Nature (1768), in *Oeuvres Complètes de Voltaire* (1877–1885), Vol. 26, 408. Trans. Albert V. and Marguerite Carozzi.
See **Buffon, Comte de** 101:2

4 In fact, no opinion should be with fervour. No one holds with fervour that 7 × 8 = 56 because it can be shown to be the case. Fervour is only necessary in commending an opinion which is doubtful or demonstrably false.

Quoted in Max Perutz, *Is Science Necessary?* (1991), 196.

Hugo de Vries 1848–1935

Dutch botanist and plant geneticist

5 [A plant] does not change itself gradually, but remains unaffected during all succeeding generations. It only throws off new forms, which are sharply contrasted with the parent, and which are from the very beginning as perfect and as constant, as narrowly defined, and as pure of type as might be expected of any species.

Species and Varieties: Their Origin and Mutation (1905), 28–9.

Conrad Hal Waddington 1905–75

British embryologist, geneticist and philosopher of science

6 *Science is the organised attempt of mankind to discover how things work as causal systems.* The scientific attitude of mind is an interest in such questions. It can be contrasted with other attitudes, which have different interests; for instance the magical, which attempts to make things work not as material systems but as immaterial forces which can be controlled by spells; or the religious, which is interested in the world as revealing the nature of God.

The Scientific Attitude (1941), Foreword, 9.

George Wald 1906–97

American biochemist

1 Evolution advances, not by a priori design, but by the selection of what works best out of whatever choices offer. We are the products of editing, rather than of authorship.

'The Origin of Optical Activity', *Annals of the New York Academy of Sciences*, 1957, **69**, 367.

2 Years ago I used to worry about the degree to which I specialized. Vision is limited enough, yet I was not really working on vision, for I hardly made contact with visual sensations, except as signals, nor with the nervous pathways, nor the structure of the eye, except the retina. Actually my studies involved only the rods and cones of the retina, and in them only the visual pigments. A sadly limited peripheral business, fit for escapists. But it is as though this were a very narrow window through which at a distance, one can only see a crack of light. As one comes closer the view grows wider and wider, until finally looking through the same narrow window one is looking at the universe. It is like the pupil of the eye, an opening only two to three millimetres across in daylight, but yielding a wide angle of view, and manoeuvrable enough to be turned in all directions. I think this is always the way it goes in science, because science is all one. It hardly matters where one enters, provided one can come closer, and then one does not see less and less, but more and more, because one is not dealing with an opaque object, but with a window.

Scientific American, 1960s, attributed.

3 Judging from our experience upon this planet, such a history, that begins with elementary particles, leads perhaps inevitably toward a strange and moving end: a creature that knows, a science-making animal, that turns back upon the process that generated him and attempts to understand it. Without his like, the universe could be, but not be known, and that is a poor thing. Surely this is a great part of our dignity as men, that we can know, and that through us matter can know itself; that beginning with protons and electrons, out of the womb of time and the vastnesses of space, we can begin to understand; that organized as in us, the hydrogen, the carbon, the nitrogen, the oxygen, those 16–21 elements, the water, the sunlight—all having become us, can begin to understand what they are, and how they came to be.

'The Origins of Life', *Proceedings of the National Academy of Sciences of the United States of America*, 1964, **52**, 609–10.

4 I have often had cause to feel that my hands are cleverer than my head. That is a crude way of characterizing the dialectics of experimentation. When it is going well, it is like a quiet conversation with Nature. One asks a question and gets an answer; then one asks the next question, and gets the next answer. An experiment is a device to make Nature speak intelligibly. After that one has only to listen.

'The Molecular Basis of Visual Excitation', Nobel Lecture, 12 December 1967. *Nobel Lectures: Physiology or Medicine 1963–1970* (1972), 292.

Wilhelm von Waldeyer-Hartz 1836–1921

German anatomist

5 The axis cylinders of all nerve fibers (motor, secretory, sensitive and sensory, conducting centrifugally or centripetally) have been proved to proceed *directly* from the *cells*. A connection with a fiber network, or an origin from such a network, does not take place.

'Über einige neuere Forschungen im Gebiete der Anatomie des Centralnervensystems', *Deutsche Medizinische Wochenschrift*, 1891,

17, 1352. Trans. Edwin Clarke and L. S. Jacyna, *Nineteenth Century Origins of Neuroscientific Concepts* (1987), 99.

Alfred Russel Wallace 1823–1913

British naturalist

1 In all works on Natural History, we constantly find details of the marvellous adaptation of animals to their food, their habits, and the localities in which they are found. But naturalists are now beginning to look beyond this, and to see that there must be some other principle regulating the infinitely varied forms of animal life. It must strike every one, that the numbers of birds and insects of different groups having scarcely any resemblance to each other, which yet feed on the same food and inhabit the same localities, cannot have been so differently constructed and adorned for that purpose alone. Thus the goat-suckers, the swallows, the tyrant fly-catchers, and the jacamars, all use the same kind of food, and procure it in the same manner: they all capture insects on the wing, yet how entirely different is the structure and the whole appearance of these birds!

A Narrative of Travels on the Amazon and Rio Negro (1853), 83–4.

2 Every species has come into existence coincident both in time and space with a pre-existing closely allied species.

'On the Law Which has Regulated the Introduction of New Species', *The Annals and Magazine of Natural History, Including Zoology, Botany and Geology*, 1855, 16, 186.

3 I was suffering from a sharp attack of intermittent fever, and every day during the cold and succeeding hot fits had to lie down for several hours, during which time I had nothing to do but to think over any subjects then particularly interesting me. One day something brought to my recollection Malthus's 'Principles of Population', which I had read about twelve years before. I thought of his clear exposition of 'the positive checks to increase'—disease, accidents, war, and famine—which keep down the population of savage races to so much lower an average than that of more civilized peoples. It then occurred to me that these causes or their equivalents are continually acting in the case of animals also; and as animals usually breed much more rapidly than does mankind, the destruction every year from these causes must be enormous in order to keep down the numbers of each species, since they evidently do not increase regularly from year to year, as otherwise the world would long ago have been densely crowded with those that breed most quickly. Vaguely thinking over the enormous and constant destruction which this implied, it occurred to me to ask the question, Why do some die and some live? And the answer was clearly, that on the whole the best fitted live. From the effects of disease the most healthy escaped; from enemies, the strongest, the swiftest, or the most cunning; from famine, the best hunters or those with the best digestion; and so on. Then it suddenly flashed upon me that this self-acting process would necessarily *improve the race*, because in every generation the inferior would inevitably be killed off and the superior would remain—that is, *the fittest would survive* . . . The more I thought over it the more I became convinced that I had at length found the long-sought-for law of nature that solved the problem of the origin of species.

My Life (1905), Vol. 1, 361–2.

4 We have also here an acting cause to account for that balance so often observed in nature,—a deficiency in one set of organs always being compensated by an increased development of some others—powerful wings accompanying weak feet, or great velocity making up for the absence of defensive weapons; for it has been shown that all varieties in which an unbalanced deficiency occurred could not long continue their existence. The action of this principle is exactly like that of the centrifugal governor of the steam engine, which checks and corrects any irregularities almost before they become evident; and in like manner no unbalanced deficiency in the animal kingdom can ever reach any conspicuous magnitude, because it would make itself felt at the very first step, by rendering existence difficult and extinction almost sure soon to follow.

'On the Tendency of Varieties to Depart Indefinitely from the Original Type', *Journal of the Proceedings of the Linnean Society, Zoology*, 1858, 3, 61–2.

1 The other book you may have heard of and perhaps read, but it is not one perusal which will enable any man to appreciate it. I have read it through five or six times, each time with increasing admiration. It will live as long as the 'Principia' of Newton. It shows that nature is, as I before remarked to you, a study that yields to none in grandeur and immensity. The cycles of astronomy or even the periods of geology will alone enable us to appreciate the vast depths of time we have to contemplate in the endeavour to understand the slow growth of life upon the earth. The most intricate effects of the law of gravitation, the mutual disturbances of all the bodies of the solar system, are simplicity itself compared with the intricate relations and complicated struggle which have determined what forms of life shall exist and in what proportions. Mr. Darwin has given the world a *new science*, and his name should, in my opinion, stand above that of every philosopher of ancient or modern times. The force of admiration can no further go!!!

Letter to George Silk, 1 September 1860. In *My Life* (1905), Vol. 1, 372–3.

2 It is for such inquiries the modern naturalist collects his materials; it is for this that he still wants to add to the apparently boundless treasures of our national museums, and will never rest satisfied as long as the native country, the geographical distribution, and the amount of variation of any living thing remains imperfectly known. He looks upon every species of animal and plant now living as the individual letters which go to make up one of the volumes of our earth's history; and, as a few lost letters may make a sentence unintelligible, so the extinction of the numerous forms of life which the progress of cultivation invariably entails will necessarily render obscure this invaluable record of the past. It is, therefore, an important object, which governments and scientific institutions should immediately take steps to secure, that in all tropical countries colonised by Europeans the most perfect collections possible in every branch of natural history should be made and deposited in national museums, where they may be available for study and interpretation. If this is not done, future ages will certainly look back upon us as a people so immersed in the pursuit of wealth as to be blind to higher considerations. They will charge us with having culpably allowed the destruction of some of those records of Creation which we had it in our power to preserve; and while professing to regard every living thing as the direct handiwork and best evidence of a Creator, yet, with a strange inconsistency, seeing many of them perish irrecoverably from the face of the earth, uncared for and unknown.

'On the Physical Geography of the Malay Archipelago', *Journal of the Royal Geographical Society*, 1863, 33, 234.

3 To say that mind is a product or function of protoplasm, or of its molecular changes, is to use words to which we can attach no clear conception. You cannot have, in the whole, what does not exist in any of the parts; and those who argue thus should put forth a definite conception of matter, with clearly enunciated properties, and show, that the necessary result of a certain complex arrangement of the elements or atoms of that matter, will be the production of self-consciousness. There is no escape from this dilemma—either all matter is conscious, or consciousness is something distinct from matter, and in the latter case, its presence in material forms is a proof of the existence of conscious beings, outside of, and independent of, what we term matter. The foregoing considerations lead us to the very important conclusion, that matter is essentially force, and nothing but force; that matter, as popularly understood, does not exist, and is, in fact, philosophically inconceivable. When we touch matter, we only really experience sensations of resistance, implying repulsive force; and no other sense can give us such apparently solid proofs of the reality of matter, as touch does. This conclusion, if kept constantly present in the mind, will be found to have a most important bearing on almost every high scientific and philosophical problem, and especially on such as relate to our own conscious existence.

'The Limits of Natural Selection as Applied to Man', the last chapter of *Contributions to the Theory of Natural Selection* (1870), 365–6.

1 We may, I think, draw a yet higher and deeper teaching from the phenomena of degeneration. We seem to learn from it the absolute necessity of labour and effort, of struggle and difficulty, of discomfort and pain, as the condition of all progress, whether physical or mental, and that the lower the organism the more need there is of these ever-present stimuli, not only to effect progress, but to avoid retrogression. And if so, does not this afford us the nearest attainable solution of the great problem of the origin of evil? What we call evil is the *essential* condition of progress in the lower stages of the development of conscious organisms, and will only cease when the mind has become so thoroughly healthy, so well balanced, and so highly organised, that the happiness derived from mental activity, moral harmony, and the social affections, will itself be a sufficient stimulus to higher progress and to the attainment of a more perfect life.

'Two Darwinian Essays', *Nature*, 1880, **22**, 142.

2 There is, I conceive, no contradiction in believing that mind is at once the cause of matter and of the development of individualised human minds through the agency of matter. And when, further on, [Mr Frederick F. Cook] asks, 'Does mortality give consciousness to spirit, or does spirit give consciousness for a limited period to mortality?' I would reply, 'Neither the one nor the other; but, mortality is the means by which a permanent individuality is given to spirit.'

'Harmony of Spiritualism and Science', *Light*, 1885, **5**, 352.

3 None but a naturalist can understand the intense excitement I experienced when I at length captured it [a hitherto unknown species of butterfly]. On taking it out of my net and opening the glorious wings, my heart began to beat violently, the blood rushed to my head, and I felt much more like fainting than I have done when in apprehension of immediate death. I had a headache the rest of the day, so great was the excitement produced by what will appear to most people a very inadequate cause.

The Malay Archipelago (1890), 257–8.

4 Can any thoughtful person admit for a moment that, in a society so constituted that these overwhelming contrasts of luxury and privation are looked upon as necessities, and are treated by the Legislature as matters with which it has practically nothing to do, there is the smallest probability that we can deal successfully with such tremendous social problems as those which involve the marriage tie and the family relation as a means of promoting the physical and moral advancement of the race? What a mockery to still further whiten the sepulchre of modern society, in which is hidden 'all manner of corruption,' with schemes for the moral and physical advancement of the race!

'Human Selection', *Fortnightly Review* 1890, 48, 330.

1 No! What we need are not prohibitory marriage laws, but a reformed society, an educated public opinion which will teach individual duty in these matters. And it is to the women of the future that I look for the needed reformation. Educate and train women so that they are rendered independent of marriage as a means of gaining a home and a living, and you will bring about natural selection in marriage, which will operate most beneficially upon humanity. When all women are placed in a position that they are independent of marriage, I am inclined to think that large numbers will elect to remain unmarried—in some cases, for life, in others, until they encounter the man of their ideal. I want to see women the selective agents in marriage; as things are, they have practically little choice. The only basis for marriage should be a disinterested love. I believe that the unfit will be gradually eliminated from the race, and human progress secured, by giving to the pure instincts of women the selective power in marriage. You can never have that so long as women are driven to marry for a livelihood.

'Heredity and Pre-Natal Influences. An Interview With Dr. Alfred Russel Wallace', *Humanitarian*, 1894, 4, 87.

2 The essential character of a species in biology is, that it is a group of living organisms, separated from all other such groups by a set of distinctive characters, having relations to the environment not identical with those of any other group of organisms, and having the power of continuously reproducing its like. Genera are merely assemblages of a number of these species which have a closer resemblance to each other in certain important and often prominent characters than they have to any other species.

'The Method of Organic Evolution', *Fortnightly Review*, 1895, 57, 441.

3 We claim to be more moral than other nations, and to conquer and govern and tax and plunder weaker peoples for *their* good! While robbing them we actually claim to be benefactors! And then we wonder, or profess to wonder, why other Governments hate us! Are they not fully justified in hating us? Is it surprising that they seek every means to annoy us, that they struggle to get navies to compete with us, and look forward to a time when some two or three of them may combine together and thoroughly humble and cripple us? And who can deny that any just Being, looking at all the nations of the earth with impartiality and thorough knowledge, would decide that we deserve to be humbled, and that it might do us good?

'Practical Politics', *The Clarion*, 30 September 1904, 1.

4 For nearly twelve years I travelled and lived mostly among uncivilised or completely savage races, and I became convinced that they all possessed good qualities, some of them in a very remarkable degree, and that in all the great characteristics of humanity they are wonderfully like ourselves. Some, indeed, among the brown Polynesians especially, are declared by numerous independent and unprejudiced observers, to be physically, mentally, and intellectually our equals, if not our superiors; and it has always seemed to me one of the disgraces of our civilisation that these fine people have not in a single case been protected from contamination by the vices and follies of our more degraded classes, and allowed to develope their own social and political organism under the advice of some of our best and wisest men and the protection of our world-wide power. That would have been indeed a worthy trophy of our civilisation. What we have actually done, and left undone, resulting in the degradation and lingering extermination of so fine a people, is one of the most pathetic of its tragedies.

'The Native Problem in South Africa and Elsewhere', *Independent Review*, 1906, 11, 182.

1 I have long since come to see that no one deserves either praise or blame for the ideas that come to him, but only for the actions resulting therefrom. Ideas and beliefs are certainly not voluntary acts. They come to us—we hardly know *how* or *whence*, and once they have got possession of us we can not reject or change them at will. It is for the common good that the promulgation of ideas should be free—uninfluenced by either praise or blame, reward or punishment. But the *actions* which result from our ideas may properly be so treated, because it is only by patient thought and work, that new ideas, if good and true, become adopted and utilized; while, if untrue or if not adequately presented to the world, they are rejected or forgotten.

'The Origin of the Theory of Natural Selection', *Popular Science Monthly*, 1909, 74, 400.

2 Truth is born into this world only with pangs and tribulations, and every fresh truth is received unwillingly. To expect the world to receive a new truth, or even an old truth, without challenging it, is to look for one of those miracles which do not occur.

'Alfred Russel Wallace: An interview by W. B. Northrop', *The Outlook*, 1913, 105, 622.

Bruce Wallace 1920– and **Theodosius Dobzhansky** 1900–75

American geneticist and Russian-born American geneticist

3 The assumption we have made . . . is that marriages and the union of gametes occur at random. The validity of this assumption may now be examined. 'Random mating' obviously does not mean promiscuity; it simply means, as already explained above, that in the choice of mates for marriage there is neither preference for nor aversion to the union of persons similar or dissimilar *with respect to a given trait or gene*. Not all gentlemen prefer blondes or brunettes. Since so few people know what their blood type is, it is even safer to say that the chances of mates being similar or dissimilar in blood type are determined simply by the incidence of these blood types in a given Mendelian population.

Radiation, Genes and Man (1960), 107.

Graham Wallas 1858–1932

British political scientist and psychologist

4 The little girl had the making of a poet in her who, being told to be sure of her meaning before she spoke, said 'How can I know what I think till I see what I say?'

The Art of Thought (1926), 106.

Horace Walpole 1717–97

British writer and collector

5 This discovery indeed is almost of that kind which I call *serendipity*, a very expressive word, which as I have nothing better to tell you, I shall endeavour to explain to you: you will understand it better by the derivation than by the definition. I once read a silly fairy tale, called *The Three Princes of Serendip*: as their highnesses travelled, they were always making discoveries, by accidents and sagacity, of things which they were not in quest of: for instance, one of them discovered that a mule blind of the right eye had travelled the same road lately, because the grass was eaten only on the left side, where it was worse than on the right—now do you understand *serendipity*?

Letter to Sir Horace Mann, 28 January 1754. In W. S. Lewis, Warren Hunting Smith and George L. Lam (eds.), *Horace Walpole's Correspondence with Sir Horace Mann* (1960), Vol. 20, 407–8.

1 Oh! But I have better news for you, Madam, if you have any patriotism as citizen of this world and wish its longevity. Mr. Herschel has found out that our globe is a comely middle-aged personage, and has not so many wrinkles as seven stars, who are evidently our seniors. Nay, he has discovered that the Milky Way is not only a mob of stars, but that there is another dairy of them still farther off, whence, I conclude, comets are nothing but pails returning from milking, instead of balloons filled with inflammable air.

Letter to the Countess of Upper Ossory, 4 July 1785. W. S. Lewis (ed.), *Horace Walpole's Correspondence with the Countess of Upper Ossory* (1965), Vol. 33, 474.

Wang Ch'ung AD 27–C.100

Chinese philosopher

2 As for the formation of matter, it is never the product of sudden events, but always the outcome of gradual change.

On Equilibrium (1929), trans. Yang Jing-Yi, 71.

Ian Warden

Australian journalist

3 To connect the dinosaurs, creatures of interest to everyone but the veriest dullard, with a spectacular extra-terrestrial event like the deluge of meteors . . . seems a little like one of those plots that a clever publisher might concoct to guarantee enormous sales. All the Alvarez-Raup theories lack is some sex and the involvement of the Royal family and the whole world would be paying attention to them.

The Canberra Times, 20 May 1984.

Edward Waring 1736–98

British mathematician

4 *Is mihi semper dicendus est inventor, qui primus evulgaverit, vel saltem cum amicis communicaverit.*

I should always call inventor him who first publishes, or at least communicates [the idea] to his friends.

Meditationes Analyticae (1785), ii-iii.

Johannes Eugenius Bülow Warming 1841–1924

Danish botanist

5 The term 'community' implies a diversity but at the same time a certain organized uniformity in the units. The units are the many individual plants that occur in every community, whether this be a beech-forest, a meadow, or a heath. Uniformity is established when certain atmospheric, terrestrial, and any of the other factors discussed in Section I are co-operating, and appears either because a certain, defined economy makes its impress on the community as a whole, or because a number of different growth-forms are combined to form a single aggregate which has a definite and constant guise.

Oecology of Plants: An Introduction to the Study of Plant Communities (1909), 91–2.

William Alexander Waters 1903–85

British chemist

6 The struggle between the unitary and dualistic theories of chemical affinity, which raged for nearly a century, was a form of civil war between inorganic and organic chemists.

Physical Aspects of Inorganic Chemistry (1935), 7.

James Dewey Watson 1928–

American molecular biologist

7 Imagination comes first in both artistic and scientific creations, but in science there is only one answer and that has to be correct.

'Discoverers of the Double Helix', *The Daily Telegraph*, 27 April 1987. In Max Perutz (ed.), *Is Science Necessary: Essays on Science and Scientists* (1991), 182.

8 We should first look at the evidence that DNA itself is not the direct template that orders amino acid sequences. Instead, the genetic information of DNA is transferred to another class of molecules which then serve as the protein templates. These intermediate templates are molecules of ribonucleic acid (RNA), large polymeric molecules chemically very similar to DNA. Their relation to DNA and protein is usually summarized by the *central dogma*, a flow scheme for genetic information first proposed some twenty years ago.

Molecular Biology of the Gene (1965), 281–2.

1 One could not be a successful scientist without realizing that, in contrast to the popular conception supported by newspapers and mothers of scientists, a goodly number of scientists are not only narrow-minded and dull, but also just stupid.

Attributed.

James Dewey Watson 1928– and Francis Crick 1916–2004

American molecular biologist and British molecular biologist

2 We wish to suggest a structure for the salt of deoxyribose nucleic acid (D.N.A.). This structure has novel features which are of considerable biological interest.

'Molecular Structure of Nucleic Acids', *Nature*, 1953, 171, 737.

3 We wish to put forward a radically different structure for the salt of deoxyribose nucleic acid. This structure has two helical chains each coiled round the same axis (see diagram).

'Molecular Structure of Nucleic Acids', *Nature*, 1953, 171, 737.

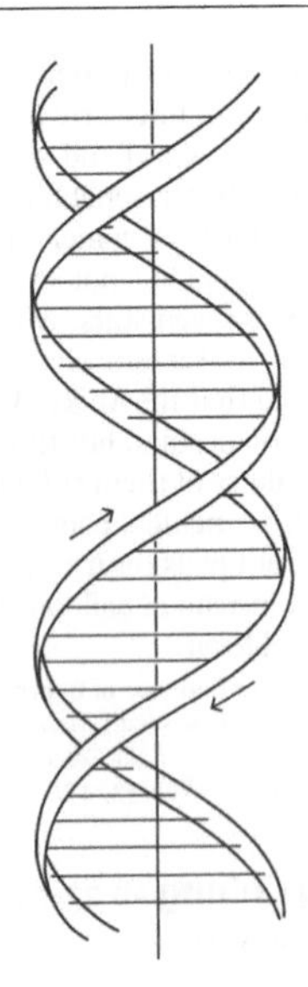

4 The novel feature of the structure is the manner in which the two chains are held together by the purine and pyrimidine bases. The planes of the bases are perpendicular to the fibre axis. They are joined together in pairs, a single base from one chain being hydrogen-bonded to a single base from the other chain, so that the two lie side by side with identical z-co-ordinates. One of the pair must be a purine and the other a pyrimidine for bonding to occur. The hydrogen bonds are made as follows: purine position 1 to pyrimidine position 1; purine position 6 to pyrimidine position 6.

'Molecular Structure of Nucleic Acids', *Nature*, 1953, 171, 737.

5 It has been found experimentally that the ratio of the amounts of adenine to thymine, and the ratio of guanine to cytosine, are always very close to unity for deoxyribose nucleic acid.

'Molecular Structure of Nucleic Acids', *Nature*, 1953, 171, 737.

6 It has not escaped our notice that the specific pairing we have postulated immediately suggests a possible copying mechanism for the genetic material.

'Molecular Structure of Nucleic Acids', *Nature*, 1953, 171, 737.

7 We should like to propose instead that the specificity of DNA self replication is accomplished without recourse to specific protein synthesis and that each of our complementary DNA chains serves as a template or mould for the formation onto itself of a new companion chain.

J. D. Watson and F. H. C. Crick, 'The Structure of DNA', *Cold Spring Harbor Symposium on Quantitative Biology*, 1953, 18, 128.

1 The information reported in this section [about the two different forms, A and B, of DNA] was very kindly reported to us prior to its publication by Drs Wilkins and Franklin. We are most heavily indebted in this respect to the Kings College Group, and we wish to point out that without this data the formation of the picture would have been most unlikely, if not impossible.

'The Complementary Structure of Deoxyribonucleic Acid', *Proceedings of the Royal Society of London, Series A*, 1954, 223, 82, footnote.

2 If the actual order of the bases on one of the pair of chains were given, one could write down the exact order of the bases on the other one, because of the specific pairing. Thus one chain is, as it were, the complement of the other, and it is this feature which suggests how the deoxyribonucleic acid molecule might duplicate itself.

'Genetic Implications of the Structure of Deoxyribonucleic Acid', *Nature*, 1958, 171, 965–6.

John Broadus Watson 1878–1958

American psychologist

3 Psychology, as the behaviorist views it, is a purely objective, experimental branch of natural science which needs introspection as little as do the sciences of chemistry and physics. It is granted that the behavior of animals can be investigated without appeal to consciousness. Heretofore the viewpoint has been that such data have value only in so far as they can be interpreted by analogy in terms of consciousness. The position is taken here that the behavior of man and the behavior of animals must be considered in the same plane.

Psychology as the Behaviorist Views It (1913), 176.

4 I would rather see the behavior of one white rat observed carefully from the moment of birth until death than to see a large volume of accurate statistical data on how 2,000 rats learned to open a puzzle box.

Introduction to G. V. Hamilton and Kenneth Macgowan, *What Is Wrong with Marriage?* (1929), xx.

Thomas John Watson, Sr. 1874–1956

American industrialist

5 I think that there is a world market for about five computers.

Quoted in C. Cerf and V. Navasky (eds.), *The Experts Speak* (1984), 208.

James Watt 1736–1819

Scottish inventor and engineer

6 About 6 or 8 years ago My Ingenious friend Mr John Robinson having [contrived] conceived that a fire engine might be made without a Lever—by Inverting the Cylinder & placing it above the mouth of the pit proposed to me to make a model of it which was set about by having never Compleated & I [being] having at that time Ignorant little knoledge of the machine however I always thought the Machine Might be applied to [more] other as valuable purposes [than] as drawing Water.

The words in square brackets were crossed out by Watt.

Entry in notebook (1765), in Eric Robinson and Douglas McKie (eds.), *Partners in Science: Letters of James Watt and Joseph Black* (1970), 434.

7 ON WATT Those who consider James Watt only as a great practical mechanic form a very erroneous idea of his character: he was equally distinguished as a natural philosopher and a chemist, and his inventions demonstrate his profound knowledge of those sciences, and that peculiar characteristic of genius, the union of them for practical application.

Said by Humphry Davy, *Proceedings of the Public Meeting held at Freemasons' Hall, on the 18th June, 1824, for Erecting a Monument to the Late James Watt* (1824), 8.

Evelyn Arthur St. John Waugh
1903–66
British novelist

1 March 15th
Imperial Banquet for Welcoming the English
Cruelty to Animals
MENU OF FOODS

VITAMIN A	VITAMIN E
Tin Sardines	*Spiced Turkey*
VITAMIN B	VITAMIN F
Roasted Beef	*Sweet Puddings*
VITAMIN C	VITAMIN G
Small Roasted Suckling Porks	*Coffee*
VITAMIN D	VITAMIN H
Hot Sheep and Onions	*Jam*

Black Mischief (1932), 1962 edn., 170.

2 Why, only last term we sent a man who had never been in a laboratory in his life as a senior Science Master to one of our leading public schools. He came [to our agency] wanting to do private coaching in music. He's doing very well, I believe.

Decline and Fall (1928), 1962 edn., 25.

Warren Weaver 1894–1978
American mathematician and administrator

3 Among the studies to which the [Rockefeller] Foundation is giving support is a series in a relatively new field, which may be called molecular biology, in which delicate modern techniques are being used to investigate ever more minute details of certain life processes.

'Molecular Biology', *Annual Report of the Rockefeller Foundation* (1938), 203–4. Reprinted in a letter to *Science*, 6 November 1970, 170, 582.

4 The century of biology upon which we are now well embarked is no matter of trivialities. It is a movement of really heroic dimensions, one of the great episodes in man's intellectual history. The scientists who are carrying the movement forward talk in terms of nucleo-proteins, of ultracentrifuges, of biochemical genetics, of electrophoresis, of the electron microscope, of molecular morphology, of radioactive isotopes. But do not be misled by these horrendous terms, and above all do not be fooled into thinking this is mere gadgetry. This is the dependable way to seek a solution of the cancer and polio problems, the problems of rheumatism and of the heart. This is the knowledge on which we must base our solution of the population and food problems. This is the understanding of life.

Letter to H. M. H. Carsan, 17 June 1949. Quoted in Raymond B. Fosdick, *The Story of the Rockefeller Foundation* (1952), 166.

Beatrice Martha Webb
1858–1943
British socialist and economist

5 [The religion of science was] an implicit faith that by the methods of physical science, and by these methods alone, could be solved all the problems arising out of the relation of man to man and of man towards the universe.

My Apprenticeship (1926), 89.

6 The one who stays in my mind as the ideal man of science is, not Huxley or Tyndall, Hooker or Lubbock, still less my friend, philosopher and guide Herbert Spencer, but Francis Galton, whom I used to observe and listen to—I regret to add, without the least reciprocity—with rapt attention. Even to-day I can conjure up, from memory's misty deep, that tall figure with its attitude of perfect physical and mental poise; the clean-shaven face, the thin, compressed mouth with its enigmatical smile; the long upper lip and firm chin, and, as if presiding over the whole personality of the man, the prominent dark eyebrows from beneath which gleamed, with penetrating humour, contemplative grey eyes. Fascinating to me was Francis Galton's all-embracing but apparently impersonal beneficence. But, to a recent and enthusiastic convert to the scientific method, the most relevant of Galton's many gifts was the unique contribution of three separate and distinct processes of the intellect; a continuous curiosity about, and rapid apprehension of individual facts, whether common or uncommon; the faculty for ingenious trains of reasoning; and, more admirable than either of these, because the talent was wholly beyond my reach, the capacity for correcting and verifying his own hypotheses, by the statistical handling of masses of data, whether collected by himself or supplied by other students of the problem.

My Apprenticeship (1926), 134–5.

Max Weber 1864–1920

German economist and sociologist

1 Inspiration in the field of science by no means plays any greater role, as academic conceit fancies, than it does in the field of mastering problems of practical life by a modern entrepreneur. On the other hand, and this also is often misconstrued, inspiration plays no less a role in science than it does in the realm of art.

'Wissenschaft als Beruf', *Gessammelte Aufsätze zur Wissenschaftslehre* (1922), 524–5. Originally a Speech at Munich University. Translated as 'Science as a Vocation', reprinted in H. H. Gerth and C. Wright-Mills (eds.), *Max Weber* (1974), 136.

2 An inner devotion to the task, and that alone, should lift the scientist to the height of dignity of the subject he pretends to serve.

'Wissenschaft als Beruf', *Gessammelte Aufsätze zur Wissenschaftslehre* (1922), 524–5. Originally a Speech at Munich University. Translated as 'Science as a Vocation', reprinted in H. H. Gerth and C. Wright-Mills (eds.), *Max Weber* (1974), 137.

3 In science, each of us knows that what he has accomplished will be antiquated in ten, twenty, fifty years. That is the fate to which science is subjected; it is the very *meaning* of scientific work, to which it is devoted in a quite specific sense, as compared with other spheres of culture for which in general the same holds. Every scientific 'fulfilment' raises new 'questions'; it *asks* to be 'surpassed' and outdated. Whoever wishes to serve science has to resign himself to this fact. Scientific works certainly can last as 'gratifications' because of their artistic quality, or they may remain important as a means of training. Yet they will be surpassed scientifically—let that be repeated—for it is our common fate and, more our common goal. We cannot work without hoping that others will advance further than we have. In principle, this progress goes on *ad infinitum*.

'Wissenschaft als Beruf', *Gessammelte Aufsätze zur Wissenschaftslehre* (1922), 524–5. Originally a Speech at Munich University. Translated as 'Science as a Vocation', reprinted in H. H. Gerth and C. Wright-Mills (eds.), *Max Weber* (1974), 138.

4 Who—aside from certain big children who are indeed found in the natural sciences—still believes that the findings of astronomy, biology, physics, or chemistry could teach us anything about the *meaning* of the world?

'Wissenschaft als Beruf', *Gessammelte Aufsätze zur Wissenschaftslehre* (1922), 524–5. Originally a Speech at Munich University. Translated as 'Science as a Vocation', reprinted in H. H. Gerth and C. Wright-Mills (eds.), *Max Weber* (1974), 142

Carl Wedl 1815–91

German histologist

5 The Frog—that arch-martyr to science—affords the most convenient subject.

Rudiments of Pathological Histology (1855), trans. and ed. by George Busk, 15.

Alfred Lothar Wegener 1880–1930

German geophysicist

6 The first concept of continental drift first came to me as far back as 1910, when considering the map of the world, under the direct impression produced by the congruence of the coast lines on either side of the Atlantic. At first I did not pay attention to the ideas because I regarded it as improbable. In the fall of 1911, I came quite accidentally upon a synoptic report in which I learned for the first time of palaeontological evidence for a former land bridge between Brazil and Africa. As a result I undertook a cursory examination of relevant research in the fields of geology and palaeontology, and this provided immediately such weighty corroboration that a conviction of the fundamental soundness of the idea took root in my mind.

The Origins of Continents and Oceans, 4th edition (1929), trans. John Biram (1966), 1.

1 It is a strange fact, characteristic of the incomplete state of our present knowledge, that totally opposing conclusions are drawn about prehistoric conditions on our planet, depending on whether the problem is approached from the biological or the geophysical viewpoint.

The Origins of Continents and Oceans, 4th edition (1929), trans. John Biram (1966), 5.

2 South America must have lain alongside Africa and formed a unified block which was split in two in the Cretaceous; the two parts must then have become increasingly separated over a period of millions of years like pieces of a cracked ice floe in water.

The Origins of Continents and Oceans, 4th edition (1929), trans. John Biram (1966), 17.

3 In the whole of geophysics there is probably hardly another law of such clarity and reliability as this—that there are two preferential levels for the world's surface which occur in alternation side by side and are represented by the continents and the ocean floors, respectively. It is therefore very surprising that scarcely anyone has tried to explain this law.

The Origins of Continents and Oceans, 4th edition (1929), trans. John Biram (1966), 37.

4 The Newton of drift theory has not yet appeared. His absence need cause no anxiety; the theory is still young and still often treated with suspicion. In the long run, one cannot blame a theoretician for hesitating to spend time and trouble on explaining a law about whose validity no unanimity prevails.

The Origins of Continents and Oceans, 4th edition (1929), trans. John Biram (1966), 167.

5 *The forces which displace continents are the same as those which produce great fold-mountain ranges.* Continental drift, faults and compressions, earthquakes, volcanicity, transgression cycles and polar wandering are undoubtedly connected causally on a grand scale. Their common intensification in certain periods of the earth's history shows this to be true. However, what is cause and what effect, only the future will unveil.

The Origins of Continents and Oceans, 4th edition (1929), trans. John Biram (1966), 179.

Charles Eugène Wegmann

1896–1982

Swiss geologist

6 Publicity floodlights are not always favorable for the development of sciences.

'L'analyse structurale en géologie', *Actualité scientifique, Sciences de la Terre* (1951), 55–64, trans. Albert V. and Marguerite Carozzi.

7 Therefore, these [geotectonic] models cannot be expected to assume that the deeper parts of the earth's crust were put together and built in a simpler way. The myth about the increasing simplicity with depth results from a general pre-scientific trend according to which the unknown or little known has to be considered simpler than the known. Many examples of this myth occur in the history of geology as, for instance, the development of views on the nature of the seafloor from the past to the present.

'Stockwerktektonik und Modelle von Esteinsdifferentiation', in *Geotektonisches Symposium zu Ehren von Hans Stille, als Festschrift zur Vollendung seines 80, Lebensjahres* (1956), 17, trans. Albert V. and Marguerite Carozzi.

1 Some authors seem to believe that hypotheses are the natural product of observations as is the case for a pear tree which produces pears; therefore only one exists which is 'the real and the good one'.

'Anatomie comparée des hypothèses sur les plissements de couverture (le Jura plissée)', *The Bulletin of the Geological Institutions of the University of Uppsala* (1961), Vol. 40, 180–1, trans. Albert V. and Marguerite Carozzi.

Alvin Weinberg 1915–

American physicist

2 We nuclear people have made a Faustian bargain with society. On the one hand, we offer . . . an inexhaustible source of energy . . . But the price that we demand of society for this magical energy source is both a vigilance and a longevity of our social institutions that we are quite unaccustomed to.

Social Institutions and Nuclear Energy (1972), 33.

Steven Weinberg 1933–

American physicist

3 In the beginning there was an explosion. Not an explosion like those familiar on earth, starting from a definite center and spreading out to engulf more and more of the circumambient air, but an explosion which occurred simultaneously everywhere, filling all space from the beginning, with every particle of matter rushing apart from every other particle. 'All space' in this context may mean either all of an infinite universe, or all of a finite universe which curves back on itself like the surface of a sphere. Neither possibility is easy to comprehend, but this will not get in our way; it matters hardly at all in the early universe whether space is finite or infinite.

At about one-hundredth of a second, the earliest time about which we can speak with any confidence, the temperature of the universe was about a hundred thousand million (10^{11}) degrees Centigrade. This is much hotter than in the center of even the hottest star, so hot, in fact, that none of the components of ordinary matter, molecules, or atoms, or even the nuclei of atoms, could have held together. Instead, the matter rushing apart in this explosion consisted of various types of the so-called elementary particles, which are the subject of modern high-energy nuclear physics.

The First Three Minutes: A Modern View of the Origin of the Universe (1977), 5.

4 It is almost irresistible for humans to believe that we have some special relation to the universe, that human life is not just a more-or-less farcical outcome of a chain of accidents reaching back to the first three minutes, but that we were somehow built in from the beginning.

The First Three Minutes: A Modern View of the Origin of the Universe (1977), 154.

5 The effort to understand the universe is one of the very few things that lifts human life a little above the level of farce, and gives it some of the grace of tragedy.

The First Three Minutes: A Modern View of the Origin of the Universe (1977), 155.

6 Our job in physics is to see things simply, to understand a great many complicated phenomena in a unified way, in terms of a few simple principles.

'Conceptual Foundations of the Unified Theory of Weak and Electromagnetic Interactions', Nobel Lecture, 8 December 1979. *Nobel Lectures: Physics 1971–1980* (1992), 543.

7 I suppose that I tend to be optimistic about the future of physics. And nothing makes me more optimistic than the discovery of broken symmetries. In the seventh book of the *Republic*, Plato describes prisoners who are chained in a cave and can see only shadows that things outside cast on the cave wall. When released from the cave at first their eyes hurt, and for a while they think that the shadows they saw in the cave are more real than the objects they now see. But eventually their vision clears, and they can understand how beautiful the real world is. We are in such a cave, imprisoned by the limitations on the sorts of experiments we can do. In particular, we can study matter only at relatively low temperatures, where symmetries are likely to be spontaneously broken, so that nature does not appear very simple or unified. We have not been able to get out of this cave, but by looking long and hard at the shadows on the cave wall, we can at least make out the shapes of symmetries, which though broken, are exact principles governing all phenomena, expressions of the beauty of the world outside.

'Conceptual Foundations of the Unified Theory of Weak and Electromagnetic Interactions', Nobel Lecture, 8 December 1979. *Nobel Lectures: Physics 1971–1980* (1992), 556.

1 Philosophy of Science is about as much use to scientists as ornithology is to birds.

Attributed.

August Friedrich Leopold Weismann 1834–1914

German biologist

2 The Continuity of the Germ-plasm.

The Continuity of the Germ-plasm as the Foundations of a Theory of Heredity (1885).

3 The development of the nucleoplasm during ontogeny may be to some extent compared to an army composed of corps, which are made up of divisions, and these of brigades, and so on. The whole army may be taken to represent the nucleoplasm of the germ-cell: the earliest cell-division . . . may be represented by the separation of the two corps, similarly formed but with different duties: and the following cell-divisions by the successive detachment of divisions, brigades, regiments, battalions, companies, etc.; and as the groups become simpler so does their sphere of action become limited.

'The Continuity of the Germ-plasm as the Foundation of a Theory of Heredity' (1885), *Essays upon Heredity and Kindred Biological Problems* (1891), Vol. 1, 195.

4 The nature of heredity is based upon the transmission of nuclear substance with a specific molecular constitution. This substance is the specific nucleoplasm of the germ-cell, to which I have given the name of germ-plasm.

Trans. Joseph S. Fruton, *Proteins, Enzymes, Genes: The Interplay of Chemistry and Biology* (1999), 391.

Victor Weisskopf 1908–2002

American physicist

5 To understand hydrogen is to understand all of physics.

Attributed by John S. Rigden, *The Essential Element* (2002), 253.

Arthur Wellesley, Duke of Wellington 1769–1852

British statesman and commander

6 But of this I can assure you that there is not a movement of any body of Men however small whether on Horse-back or on foot, nor an operation or March of any description nor any Service in the field that is not formed upon some mathematical principle, and in the performance of which the knowledge and practical application of the mathematicks will be found not only useful but necessary. The application of the Mathematicks to Gunnery, Fortification, Tactics, the survey and knowledge of formal Castrenantion etc. cannot be acquired without study.

Duke of Wellington to his son Douro, 1826. Quoted in *A Selection of the Private Correspondence of the First Duke of Wellington* (1952), 44.

Abraham Gottlob Werner 1749–1817

German geologist

7 I would rather have a mineral ill-classified and well-described, than well-classified and ill-described.

On the External Characters of Minerals (1774), xxix, trans. Albert V. and Marguerite Carozzi.

1 It is obvious that we know with certainty, that the *Floëtz* [layered] and primitive mountains have been produced by a series of precipitations and depositions formed in succession; that they took place from water which covered the globe, existing always more or less generally, and containing the different substances which have been produced from them.

New Theory of the Formation of Veins (1809), 110–1.

Alfred Werner 1866–1919

Swiss chemist

2 I should like to call the number of atom groups, with which an elementary atom coordinates . . . to form a complex radical, the *coordination number* of the atom in question . . . We must differentiate between valence number and coordination number. The valence number indicates the maximum number of monovalent atoms which can be bound directly to the atom in question *without the participation of other elementary atoms* . . . Perhaps this concept [of coordination number] is destined to serve as a basis for the theory of the constitution of inorganic compounds, just as valence theory formed the basis for the constitutional theory of carbon compounds.

'Beitrag zur Konstitution anorganischer Verbindungen', *Zeitschrift für anorganische Chemie*, 1893, 3, 267–330. Translated in George G. Kauffman (ed.), *Classics in Coordination Chemistry: Part 1: The Selected Papers of Alfred Werner* (1968), 84–7.

3 *Nun wie gehts?*

How goes it?

Werner's perennial salutation to research students, hence his nickname, Professor Nunwiegehts.

Quoted in R. Oesper, *The Human Side of Scientists* (1975), 188.

4 Gentlemen and ladies, this is ordinary alcohol, sometimes called ethanol; it is found in all fermented beverages. As you well know, it is considered by many to be poisonous, a belief in which I do not concur. If we subtract from it one CH_2-group we arrive at this colorless liquid, which you see in this bottle. It is sometimes called methanol or wood alcohol. It is certainly more toxic than the ethanol we have just seen. Its formula is CH_3OH. If, from this, we subtract the CH_2-group, we arrive at a third colorless liquid, the final member of this homologous series. This compound is hydrogen hydroxide, best known as water. It is the most poisonous of all.

R. Oesper, *The Human Side of Scientists* (1975), 189.

5 Your aim is no better than your knowledge of chemistry.

On being shot at by a Polish student whom Werner had failed in an examination.

G. B. Kauffman, *Alfred Werner* (1966), 59.

6 Chemistry must become the astronomy of the molecular world.

Autograph Quotation for a Charity, 1905. In G. B. Kaufman, *Alfred Werner* (1966).

7 There is no such thing as chemistry for medical students! Chemistry is chemistry!

G. B. Kauffman, *Alfred Werner* (1966), 60.

Frank Henry Westheimer 1912–

American chemist

8 Surprisingly, history is much more difficult than chemistry.

Louis Hammett Symposium, 31 August 1983, *Advances in Physical Organic Chemistry*, 1985, 21, 2.

John Archibald Wheeler 1911–

American physicist

9 The only thing harder to understand than a law of statistical origin would be a law that is not of statistical origin, for then there would be no way for it—or its progenitor principles—to come into being. On the other hand, when we view each of the laws of physics—and no laws are more magnificent in scope or better tested—as at bottom statistical in character, then we are at last able to forego the idea of a law that endures from everlasting to everlasting.

'Law without Law' (1979), in John Archibald Wheeler and Wojciech Hubert Zurek (eds.), *Quantum Theory and Measurement* (1983), 203.

1 Time ends. That is the lesson of the 'big bang'. It is also the lesson of the black hole, closer at hand and more immediate object of study.

The black hole is a completely collapsed object. It is mass without matter.

The Cheshire cat in *Alice in Wonderland* faded away leaving behind only its grin. A star that falls into an already existing black hole, or that collapses to make a new black hole, fades away. Of the star, of its matter and of its sunspots and solar prominences, all trace disappears. There remains behind only gravitational attraction, the attraction of disembodied mass.

'The Lesson of the Black Hole', *Proceedings of the American Philosophical Society*, 1981, 125, 25.

2 The universe came into being in a big bang, before which, Einstein's theory instructs us, there was no before. Not only particles and fields of force had to come into being at the big bang, but the laws of physics themselves, and this by a process as higgledy-piggledy as genetic mutation or the second law of thermodynamics.

'The Computer and the Universe', *International Journal of Theoretical Physics*, 1982, 21, 565.

William Whewell 1794–1866

British natural scientist, historian and philosopher of science

See **Faraday, Michael** 208:4, 209:1, 209:2.

3 I am persuaded that there is not in the nature of science anything unfavourable to religious feelings, and if I were not so persuaded I should be much puzzled to account for our being invested, as we so amply are, with the facilities that lead us to the discovery of scientific truth. It would be strange if our Creator should be found to be urging us on in a career which tended to be a forgetfulness of him.

Letter to H. J. Rose, 19 November 1826. Quoted in I. Todhunter (ed.), *William Whewell: An Account of His Writings with Selections From His Literary and Scientific Correspondence* (1876), Vol. 2, 76.

4 While the unique crystal stands on its shelf unmeasured by the goniometer, unslit by the optical lapidary, unanalysed by the chemist,—it is merely a piece of furniture, and has no more right to be considered as anything pertaining to science, than a curious china tea-cup on a chimney-piece.

'Report on the Progress and Present State of Mineralogy', *Report of the British Association for the Advancement of Science* (1831–32), 364–5.

5 Have the changes which lead us from one geologic state to another been, on a long average uniform in their intensity, or have they consisted of epochs of paroxysmal and catastrophic action, interposed between periods of comparative tranquillity? These two opinions will probably for some time divide the geological world into two sects, which may perhaps be designated as the *Uniformitarians* and the *Catastrophists.*

'Review of Charles Lyell's Principles of Geology', *Quarterly Review*, 1832, 47, 126.

6 It is a peculiar feature in the fortune of principles of such high elementary generality and simplicity as characterise the laws of motion, that when they are once firmly established, or supposed to be so, men turn with weariness and impatience from all questionings of the grounds and nature of their authority. We often feel disposed to believe that truths so clear and comprehensive are necessary conditions, rather than empirical attributes of their subjects: that they are legible by their own axiomatic light, like the first truths of geometry, rather than discovered by the blind gropings of experience.

An Introduction to Dynamics (1832), x.

1 By science, then, I understand the consideration of all subjects, whether of a pure or mixed nature, capable of being reduced to measurement and calculation. All things comprehended under the categories of space, time and number properly belong to our investigations; and all phenomena capable of being brought under the semblance of a law are legitimate objects of our inquiries.

Report of the British Association for the Advancement of Science (1833), xxviii.

2 I have considered the two terms you want to substitute for *eisode* and *exode*, and upon the whole I am disposed to recommend instead of them *anode* and *cathode*. These words may signify *eastern* and *western* way, just as well as the longer compounds which you mention . . . I may mention too that *anodos* and *cathodos* are good genuine Greek words, and not compounds coined for the purpose.

Letter to Michael Faraday, 25 April 1834. Quoted in I. Todhunter (ed.), *William Whewell: An Account of His Writings with Selections From His Literary and Scientific Correspondence* (1876), Vol. 2, 179.

3 The tendency of the sciences has long been an increasing proclivity of separation and dismemberment . . . The mathematician turns away from the chemist; the chemist from the naturalist; the mathematician, left to himself divides himself into a pure mathematician and a mixed mathematician, who soon part company . . . And thus science, even mere physical science, loses all traces of unity. A curious illustration of this result may be observed in the want of any name by which we can designate the students of the knowledge of the material world collectively. We are informed that this difficulty was felt very oppressively by the members of the British Association for the Advancement of Science, at their meetings at York, Oxford and Cambridge, in the last three summers. There was no general term by which these gentlemen could describe themselves with reference to their pursuits . . . some ingenious gentleman [William Whewell] proposed that, by analogy with *artist*, they might form *Scientist*, and added that there could be no scruple . . . when we have words such as *sciolist*, *economist*, and *atheist*—but this was not generally palatable.

Review of Mrs Somerville, 'On the Connexion of the Physical Sciences', *The Quarterly Review*, 51, 1834, 58–61.

4 The principles which constituted the triumph of the preceding stages of the science, may appear to be subverted and ejected by the later discoveries, but in fact they are, (so far as they were true), taken up into the subsequent doctrines and included in them. They continue to be an essential part of the science. The earlier truths are not expelled but absorbed, not contradicted but extended; and the history of each science, which may thus appear like a succession of revolutions, is, in reality, a series of developments.

History of the Inductive Sciences (1837) Vol. 1, 10.

5 All palaetiological sciences, all speculations which attempt to ascend from the present to the remote past, by the chain of causation, do also, by an inevitable consequence, urge us to look for the beginning of the state of things which we thus contemplate; but in none of these cases have men been able, by the aid of science, to arrive at a beginning which is homogeneous with the known course of events. The first origin of language, of civilization, of law and government, cannot be clearly made out by reasoning and research; and just as little, we may expect, will a knowledge of the origin of the existing and extinct species of plants and animals, be the result of physiological and geological investigation.

History of the Inductive Sciences (1837), Vol. 3, 581.

6 The mystery of creation is not within the range of [Nature's] legitimate territory; [Nature] says nothing, but she points upwards.

History of the Inductive Sciences (1837), Vol. 3, 588.

1 As we cannot use physician for a cultivator of physics, I have called him a *physicist*. We need very much a name to describe a cultivator of science in general. I should incline to call him a *Scientist*. Thus we might say, that as an Artist is a Musician, Painter or Poet, a Scientist is a Mathematician, Physicist, or Naturalist.

The Philosophy of the Inductive Sciences (1840), Vol. 1, cxiii.

2 We may best hope to understand the nature and conditions of real knowledge, by studying the nature and conditions of the most certain and stable portions of knowledge which we already possess: and we are most likely to learn the best methods of discovering truth, by examining how truths, now universally recognised, have really been discovered.

The Philosophy of the Inductive Sciences (1840), Vol. 1, 3–4.

3 We have here spoken of the prediction of facts *of the same kind* as those from which our rule was collected. But the evidence in favour of our induction is of a much higher and more forcible character when it enables us to explain and determine cases of a *kind different* from those which were contemplated in the formation of our hypothesis. The instances in which this has occurred, indeed, impress us with a conviction that the truth of our hypothesis is certain. No accident could give rise to such an extraordinary coincidence. No false supposition could, after being adjusted to one class of phenomena, so exactly represent a different class, when the agreement was unforeseen and contemplated. That rules springing from remote and unconnected quarters should thus leap to the same point, can only arise from *that* being where truth resides.

The Philosophy of the Inductive Sciences (1840), Vol. 2, 230.

4 [A] theory is a Fact; a Fact is a familiar Theory.

The Philosophy of the Inductive Sciences (1847), Vol. 1, 40.

5 The ponderous instrument of synthesis, so effective in his [Newton's] hands, has never since been grasped by one who could use it for such purposes; and we gaze at it with admiring curiosity, as on some gigantic implement of war, which stands idle among the memorials of ancient days, and makes us wonder what manner of man he was who could wield as a weapon what we can hardly lift as a burden.

History of the Inductive Sciences (1857), Vol. 2, 128.

6 The use of every organ has been discovered by starting from the assumption that it must have been *some* use.

History of the Inductive Sciences (1857), Vol. 3, 385.

7 *Time*, inexhaustible and ever accumulating his efficacy, can undoubtedly do much for the theorist in geology; but *Force*, whose limits we cannot measure, and whose nature we cannot fathom, is also a power never to be slighted: and to call in the one to protect us from the other, is equally presumptuous, to whichever of the two our superstition leans. To invoke Time, with ten thousand earthquakes, to overturn and set on edge a mountain-chain, should the phenomena indicate the change to have been sudden and not successive, would be ill excused by pleading the obligation of first appealing to known causes.

History of the Inductive Sciences (1857), Vol. 3, 513–4.

8 In truth, we know causes only by their effects; and in order to learn the nature of the causes which modify the earth, we must study them through all ages of their action, and not select arbitrarily the period in which we live as the standard for all other epochs.

History of the Inductive Sciences (1857), Vol. 3, 514.

1 Astronomy, as the science of cyclical motions, has nothing in common with Geology. But look at Astronomy where she has an analogy with Geology; consider our knowledge of the heavens as a palaetiological science;—as the study of a past condition, from which the present is derived by causes acting in time. Is there no evidence of a beginning, or of a progress?

History of the Inductive Sciences (1857), Vol. 3, 516.

2 It is a wrong business when the younger cultivators of science put out of sight and deprecate what their predecessors have done; but obviously that is the tendency of Huxley and his friends . . . It is very true that Huxley was bitter against the Bishop of Oxford, but I was not present at the debate. Perhaps the Bishop was not prudent to venture into a field where no eloquence can supersede the need for precise knowledge. The young naturalists declared themselves in favour of Darwin's views which tendency I saw already at Leeds two years ago. I am sorry for it, for I reckon Darwin's book to be an utterly *unphilosophical* one.

Letter to James D. Forbes, 24 July 1860. Trinity College Cambridge, Whewell Manuscripts.

Gilbert White 1720–93

British naturalist and clergyman

3 It has been my misfortune never to have had any neighbours whose studies have led them towards the pursuit of natural knowledge; so that, for want of a companion to quicken my industry and sharpen my attention, I have made but slender progress in a kind of information to which I have been attached from my childhood.

Letter to Thomas Pennant, 4 August 1767. In *The Natural History and Antiquities of Selborne* (1789), 27.

4 All the summer long is the swallow a most instructive pattern of unwearied industry and affection; for, from morning to night, while there is a family to be supported, she spends the whole day in skimming close to the ground, and exerting the most sudden turns and quick evolutions. Avenues, and long walks under hedges, and pasture-fields, and mown meadows where cattle graze, are her delight, especially if there are trees interspersed; because in such spots insects most abound. When a fly is taken a smart snap from her bill is heard, resembling the noise at the shutting of a watch case; but the motion of the mandibles are too quick for the eye.

Letter to Daines Barrington, 29 January 1774. In *The Natural History and Antiquities of Selborne* (1789), 169–70.

5 It is curious to observe with what different degrees of architectonic skill Providence has endowed birds of the same genus, and so nearly correspondent in their general mode of life! for while the swallow and the house-martin discover the greatest address in raising and securely fixing crusts or shells of loam as *cunabula* for their young, the bank-martin terebrates a round and regular hole in the sand or earth, which is serpentine, horizontal, and about two feet deep. At the inner end of this burrow does this bird deposit, in a good degree of safety, her rude nest, consisting of fine grasses and feathers, usually goose-feathers, very inartificially laid together.

Letter to Daines Barrington, 26 February 1774. In *The Natural History and Antiquities of Selborne* (1789), 176.

6 But in nothing are swifts more singular than in their early retreat. They retire, as to the main body of them, by the tenth of August, and sometimes a few days sooner: and every straggler invariably withdraws by the twentieth, while their congeners, all of them, stay till the beginning of October; many of them all through that month, and some occasionally to the beginning of November. This early retreat is mysterious and wonderful, since that time is often the sweetest season in the year. But, what is more extraordinary, they begin to retire still earlier in the most southerly parts of Andalusia, where they can be no ways influenced by any defect of heat; or, as one might suppose, defect of food. Are they regulated in their motions with us by failure of food, or by a propensity to moulting, or by a disposition to rest after so rapid a life, or by what? This is one of those incidents in natural history that not only baffles our searches, but almost eludes our guesses!

Letter to Daines Barrington, 28 September 1774. In *The Natural History and Antiquities of Selborne* (1789), 184.

1 The most insignificant insects and reptiles are of much more consequence, and have much more influence in the oeconomy of Nature, than the incurious are aware of; and are mighty in their effect, from their minuteness, which renders them less an object of attention; and from their numbers and fecundity. Earth-worms, though in appearance a small and despicable link in the chain of Nature, yet, if lost, would make a lamentable chasm.

Letter to Daines Barrington, 20 May 1777. In *The Natural History and Antiquities of Selborne* (1789), 216.

2 The old Sussex tortoise, that I have mentioned to you so often, is become my property. I dug it out of its winter dormitory in March last, when it was enough awakened to express its resentments by hissing; and, packing it in a box with earth, carried it eighty miles in post-chaises. The rattle and hurry of the journey so perfectly roused it that, when I turned it out on a border, it walked twice down to the bottom of my garden; however, in the evening, the weather being cold, it buried it-self in the loose mound, and continues still concealed . . . When one reflects on the state of this strange being, it is a matter of wonder to find that Providence should bestow such a profusion of days, such a seeming waste of longevity, on a reptile that appears to relish it so little as to squander more than two-thirds of its existence in joyless stupor, and be lost to all sensation for months together in the profoundest of slumbers.

Letter to Daines Barrington, 21 April 1780. In *The Natural History and Antiquities of Selborne* (1789), 261–2.

3 Nature, who is a great economist, converts the recreation of one animal to the support of another.

Attributed.

Lynn T. White Jr. 1907–87

American historian of science

4 The greatest spiritual revolutionary in Western history, Saint Francis, proposed what he thought was an alternative Christian view of nature and man's relation to it: he tried to substitute the idea of the equality of all creatures, including man, for the idea of man's limitless rule of creation. He failed. Both our present science and our present technology are so tinctured with orthodox Christian arrogance toward nature that no solution for our ecologic crisis can be expected from them alone. Since the roots of our trouble are so largely religious, the remedy must also be essentially religious, whether we call it that or not. We must rethink and refeel our nature and destiny. The profoundly religious, but heretical, sense of the primitive Franciscans for the spiritual autonomy of all parts of nature may point a direction. I propose Francis as a patron saint for ecologists.

The Historical Roots of our Ecologic Crisis (1967), 1207.

Alfred North Whitehead 1861–1947

British mathematician and philosopher

5 It is a profoundly erroneous truism, repeated by all copy-books and by eminent people when they are making speeches, that we should cultivate the habit of thinking of what we are doing. The precise opposite is the case.

Civilization advances by extending the number of important operations which we can perform without thinking about them. Operations of thought are like cavalry charges in a battle—they are strictly limited in number, they require fresh horses, and must only be made at decisive moments.

An Introduction to Mathematics (1911), 61.

1 It is a safe rule to apply that, when a mathematical or philosophical author writes with a misty profoundity, he is talking nonsense.

An Introduction to Mathematics (1911), 227.

2 But to come very near to a true theory, and to grasp its precise application, are two very different things, as the history of science teaches us. Everything of importance has been said before by somebody who did not discover it.

The Organisation of Thought (1917), 127.

3 Our problem is, in fact, to fit the world to our perceptions, and not our perceptions to the world.

The Organisation of Thought: Educational and Scientific (1917), 228.

4 A science which hesitates to forget its founders is lost.

'The Organisation of Thought', in *The Aims of Education and Other Essays* (1929), 162.

5 The aim of science is to seek the simplest explanations of complex facts. We are apt to fall into the error of thinking that the facts are simple because simplicity is the goal of our quest. The guiding motto in the life of every natural philosopher should be, Seek simplicity and distrust it.

The Concept of Nature (1920), 163.

6 A few generations ago the clergy, or to speak more accurately, large sections of the clergy were the standing examples of obscurantism. Today their place has been taken by scientists.

The Function of Reason (1929), 34–5.

7 The safest general characterization of the European philosophical tradition is that it consists of a series of footnotes to Plato.

Process and Reality (1929), 39.

8 The laws of physics are the decrees of fate.

Science and the Modern World (1926), 13.

9 Familiar things happen, and mankind does not bother about them. It requires a very unusual mind to undertake the analysis of the obvious.

Science and the Modern World (1938), 15.

10 The Science of Pure Mathematics in its modern developments may claim to be the most original creation of the human spirit.

Science and the Modern World (1938), 32.

11 I will not go so far as to say that to construct a history of thought without profound study of the mathematical ideas of successive epochs is like omitting Hamlet from the play which is named after him. That would be claiming too much. But it is certainly analogous to cutting out the part of Ophelia. This simile is singularly exact. For Ophelia is quite essential to the play, she is very charming—and a little mad. Let us grant that the pursuit of mathematics is a divine madness of the human spirit, a refuge from the goading urgency of contingent happenings.

Science and the Modern World (1938), 33.

12 We forget how strained and paradoxical is the view of nature which modern science imposes on our thoughts.

Science and the Modern World (1925), 104.

13 A single tree by itself is dependent upon all the adverse chances of shifting circumstances. The wind stunts it: the variations in temperature check its foliage: the rains denude its soil: its leaves are blown away and are lost for the purpose of fertilisation. You may obtain individual specimens of fine trees either in exceptional circumstances, or where human cultivation had intervened. But in nature the normal way in which trees flourish is by their association in a forest. Each tree may lose something of its individual perfection of growth, but they mutually assist each other in preserving the conditions of survival. The soil is preserved and shaded; and the microbes necessary for its fertility are neither scorched, nor frozen, nor washed away. A forest is the triumph of the organisation of mutually dependent species.

Science and the Modern World (1926), 296–7.

1 The originality of mathematics consists in the fact that in mathematical science connections between things are exhibited which, apart from the agency of human reason, are extremely unobvious.

Science and the Modern World (1938), 32.

2 The greatest invention of the nineteenth century was the invention of the method of invention.

Science and the Modern World (1938), 141.

3 All science as it grows toward perfection becomes mathematical in its ideas.

Attributed.

4 Mathematics as a science commenced when first someone, probably a Greek, proved propositions about any things or about some things, without specification of definite particular things. These propositions were first enunciated by the Greeks for geometry; and, accordingly, geometry was the great Greek mathematical science.

Attributed.

5 People make the mistake of talking about 'natural laws'. There are no natural laws. There are only temporary habits of nature.

Attributed.

6 Fools act on imagination without knowledge, pedants act on knowledge without imagination.

Attributed.

7 Aristotle invented science, but destroyed philosophy.

Attributed.

Charles Otis Whitman 1842–1910

American zoologist

8 Other things being equal, the investigator is always the best instructor. The highest grade of instruction in any science can only be furnished by one who is thoroughly imbued with the scientific spirit, and who is actually engaged in original work.

Quoted in Frank R. Lillie, *The Woods Hole Marine Biological Laboratory* (1944), 37–8.

Walt Whitman 1819–92

American poet, journalist and essayist

9 When I heard the learn'd astronomer,
When the proofs, the figures, were ranged in columns before me,
When I was shown the charts and diagrams, to add, divide, and measure them,
When I sitting heard the astronomer where he lectured with much applause in the lecture-room,
How soon unaccountable I became tired and sick,
Till rising and gliding out I wander'd off by myself,
In the mystical moist night-air, and from time to time,
Look'd up in perfect silence at the stars.

Leaves of Grass, edited by Jerome Loving (1990), 214.

Lancelot Law Whyte 1896–1972

British philosopher and science writer

10 Both science and art have to do with ordered complexity.

The Griffin, 1957, 6, No 10.

Robert Whytt 1714–66

British physician and neurophysiologist

11 That many very remarkable changes and involuntary motions are suddenly produced in the body by various affections of the mind, is undeniably evinced from a number of facts. Thus fear often causes a sudden and uncommon flow of pale urine. Looking much at one troubled with sore eyes, has sometimes affected the spectator with the same disease.—Certain sounds cause a shivering over the whole body.—The noise of a bagpipe has raised in some persons an inclination to make urine.—The sudden appearance of any frightful object, will, in delicate people, cause an uncommon palpitation of the heart.—The sight of an epileptic person agitated with convulsions, has brought on an epilepsy; and yawning is so very catching, as frequently to be propagated through whole companies.

An Essay on the Vital and Other Involuntary Motions of Animals (1751), 253–4.

1 If feeling be not a property of matter, but owing to a superior principle, it must follow, that the motions of the heart, and other muscles of animals, after being separated from their bodies, are to be ascribed to this principle; and that any difficulties which may appear in this matter are owing to our ignorance of the nature of the soul, of the manner of its existence, and of its wonderful union with, and action upon the body.

An Essay on the Vital and Other Involuntary Motions of Animals (1751), 389–90.

Egon Wiberg 1901–76

German chemist

2 Chlorine is a poisonous gas. In case I should fall over unconscious in the following demonstration involving chlorine, please pick me up and carry me into the open air. Should this happen, the lecture for the day will be concluded.

Quoted in R. Oesper, *The Human Side of Scientists* (1975), 192.

Norbert Wiener 1894–1964

American mathematician

3 The way of pure research is opposed to all the copy-book maxims concerning the virtues of industry and a fixed purpose, and the evils of guessing, but it is damned useful when it comes off. It is the diametrical opposite of Edison's reputed method of trying every conceivable expedient until he hit the right one. It requires, not diligence, but experience, information, and a good nose for the essence of a problem.

Letter to Paul de Kruif, 3 August 1933. Quoted in Nathan Reingold, *Science in America: A Documentary History 1900–1939* (1981), 409.

4 It is clear that the degradation of the position of the scientist as an independent worker and thinker to that of a morally irresponsible stooge in a science-factory has proceeded even more rapidly and devastatingly than I had expected. This subordination of those who ought to think to those who have the administrative power is ruinous for the morale of the scientist, and quite to the same extent it is ruinous to the quality of the subjective scientific output of the country.

'A Rebellious Scientist after Two Years', *Bulletin of the Atomic Scientists*, 1948, 4, 338.

5 We have decided to call the entire field of control and communication theory, whether in the machine or in the animal, by the name *Cybernetics*, which we form from the Greek κυβερνήτης or *steersman*. In choosing this term, we wish to recognize that the first significant paper on feedback mechanisms is an article on governors, which was published by Clerk Maxwell in 1868, and that *governor* is derived from a Latin corruption of κυβερνήτης. We also wish to refer to the fact that the steering engines of a ship are indeed one of the earliest and best-developed forms of feedback mechanisms.

Cybernetics (1948), 19.

6 Besides electrical engineering theory of the transmission of messages, there is a larger field [cybernetics] which includes not only the study of language but the study of messages as a means of controlling machinery and society, the development of computing machines and other such automata, certain reflections upon psychology and the nervous system, and a tentative new theory of scientific method.

Cybernetics (1948).

1 It is my thesis that the physical functioning of the living individual and the operation of some of the newer communication machines are precisely parallel in their analogous attempts to control entropy through feedback. Both of them have sensory receptors as one stage in their cycle of operation: that is, in both of them there exists a special apparatus for collecting information from the outer world at low energy levels, and for making it available in the operation of the individual or of the machine. In both cases these external messages are not taken *neat*, but through the internal transforming powers of the apparatus, whether it be alive or dead. The information is then turned into a new form available for the further stages of performance. In both the animal and the machine this performance is made to be effective on the outer world. In both of them, their *performed* action on the outer world, and not merely their *intended* action, is reported back to the central regulatory apparatus.

The Human Use of Human Beings: Cybernetics and Society (1954), 26–7.

2 It is popular to believe that the age of the individual and, above all, of the free individual, is past in science. There are many administrators of science and a large component of the general population who believe that mass attacks can do anything, and even that ideas are obsolete. Behind this drive to the mass attack there are a number of strong psychological motives. Neither the public or the big adminstrator has too good an understanding of the inner continuity of science, but they both have seen its world-shaking consequences, and they are afraid of it. Both of them wish to decerebrate the scientist, even as the Byzantine State emasculated its civil servants. Moreover, the great administrator who is not sure of his own intellectual level can aggrandize himself only by cutting his scientific employees down to size.

I am a Mathematician (1956), Epilogue, 363–4.

3 The future offers very little hope for those who expect that our new mechanical slaves will offer us a world in which we may rest from thinking. Help us they may, but at the cost of supreme demands upon our honesty and our intelligence. The world of the future will be an ever more demanding struggle against the limitations of our intelligence, not a comfortable hammock in which we can lie down to be waited upon by our robot slaves.

God & Golem, Inc (1964), 73–4.

Eugene Paul Wigner 1902–95

Hungarian-born American physicist

4 Physics does not endeavour to explain nature. In fact, the great success of physics is due to a restriction of its objectives: it only endeavours to explain the *regularities* in the behavior of objects.

'Events, Laws of Nature, and Invariance Principles', Nobel Lecture, 12 December 1963. *Nobel Lectures: Physics 1963–1970* (1972), 6.

5 With thermodynamics, one can calculate almost everything crudely; with kinetic theory, one can calculate fewer things, but more accurately; and with statistical mechanics, one can calculate almost nothing exactly.

Edward B. Stuart, Alan J. Brainard and Benjamin Gal-Or (eds.), *A Critical Review of Thermodynamics* (1970), 205.

6 I believe that the present laws of physics are at least incomplete without a translation into terms of mental phenomena.

'Physics and the Explanation of Life', *Foundations of Physics* 1970, 1, 35–45.

7 Physics is becoming so unbelievably complex that it is taking longer and longer to train a physicist. It is taking so long, in fact, to train a physicist to the place where he understands the nature of physical problems that he is already too old to solve them.

Attributed.

Oscar Wilde 1854–1900

Irish poet, dramatist and wit

1 *On the Niagara Falls:*

It would be more impressive if it flowed the other way.

Quoted in 'Professors, Politics, and Palaver', *Science*, 19 August 1977, 197, 742.

John Wilkins 1614–72

British bishop and natural philosopher

2 Whatever is *Natural* doth by that appear, adorned with all imaginable *Elegance* and *Beauty*. There are such inimitable gildings and embroideries in the smallest seeds of Plants, but especially in the parts of Animals, in the head or eye of a small Fly: such accurate order and symmetry in the frame of the most minute creatures, a *Lowse* or a *Mite*, as no man were able to conceive without seeing of them. Whereas the most curious works of Art, the sharpest finest Needle, doth appear as a blunt rough bar of iron, coming from the furnace or the forge. The most accurate engravings or embossments, seem such rude bungling deformed works, as if they had been done with a Mattock or a Trowel.

Of the Principles and Duties of Natural Religion (1675), 80.

Samuel Wilks 1824–1911

British physician

3 Being also in accord with Goethe that discoveries are made by the age and not by the individual, I should consider the instances to be exceedingly rare of men who can be said to be living before their age, and to be the repository of knowledge quite foreign to the thought of the time. The rule is that a number of persons are employed at a particular piece of work, but one being a few steps in advance of the others is able to crown the edifice with his name, or, having the ability to generalise already known facts, may become in time to be regarded as their originator. Therefore it is that one name is remembered whilst those of coequals have long been buried in obscurity.

'Historical Notes on Bright's Disease, Addison's Disease, and Hodgkin's Disease', *Guy's Hospital Reports*, 1877, **22**, 259–260.

Thomas Willis 1621–75

British physician

4 But the office of the Cerebral seems to be for the animal Spirits to supply some Nerves; by which involuntary actions (such as are the beating of the Heart, easie respiration, the Concoction of the Aliment, the protrusion of the Chyle, and many others) which are made after a constant manner unknown to us, or whether we will or no, are performed.

Anatomy of the Brain and Nerves (1664), trans. Samuel Pordage (1681), reprinted in William Feindel (ed.), *Thomas Willis: Anatomy of the Brain and Nerves* (1965), Vol. 2, 111.

5 That the Anatomy of the Nerves yields more pleasant and profitable Speculations, than the Theory of any parts besides in the animated Body: for from hence the true and genuine Reasons are drawn of very many Actions and Passions that are wont to happen in our Body, which otherwise seem most difficult and unexplicable; and no less from this Fountain the hidden Causes of Diseases and their Symptoms, which commonly are ascribed to the Incantations of Witches, may be found out and clearly laid open. But as to our observations about the Nerves, from our following Discourse it will plainly appear, that I have not trod the paths or footsteps of others, nor repeated what hath been before told.

Anatomy of the Brain and Nerves (1664), trans. Samuel Pordage (1681), reprinted in William Feindel (ed.), *Thomas Willis: Anatomy of the Brain and Nerves* (1965), Vol. 2, 125.

1 To describe all the several pairs of the spinal Nerves, and to rehearse all their branchings, and to unfold the uses and actions of them, would be a work of an immense labour and trouble: and as this *Neurologie* cannot be learned nor understood without an exact knowledge of the Muscles, we may justly here forbear entring upon its particular institution.

Anatomy of the Brain and Nerves (1664), trans. Samuel Pordage (1681), reprinted in William Feindel (ed.), *Thomas Willis: Anatomy of the Brain and Nerves* (1965), Vol. 2, 178.

Ian Wilmut 1944– et al.

British embryologist

2 Fertilization of mammalian eggs is followed by successive cell divisions and progressive differentiation, first into the early embryo and subsequently into all of the cell types that make up the adult animal. Transfer of a single nucleus at a specific stage of development, to an enucleated unfertilized egg, provided an opportunity to investigate whether cellular differentiation to that stage involved irreversible genetic modification. The first offspring to develop from a differentiated cell were born after nuclear transfer from an embryo-derived cell line that had been induced to become quiescent. Using the same procedure, we now report the birth of live lambs from three new cell populations established from adult mammary gland, fetus and embryo. The fact that a lamb was derived from an adult cell confirms that differentiation of that cell did not involve the irreversible modification of genetic material required for development to term. The birth of lambs from differentiated fetal and adult cells also reinforces previous speculation that by inducing donor cells to become quiescent it will be possible to obtain normal development from a wide variety of differentiated cells.

Announcement of 'Dolly'.

I. Wilmut, A. E. Schnieke, J. McWhir, et al., 'Viable Offspring Derived from Fetal and Adult Mammalian Cells', *Nature*, 1997, 385, 810.

Edmund Beecher Wilson

1856–1939

American biologist

3 The precise equivalence of the chromosomes contributed by the two sexes is a physical correlative of the fact that the two sexes play, on the whole, equal parts in hereditary transmission, and it seems to show that the chromosomal substance, the chromatin, is to be regarded as the physical basis of inheritance. Now, chromatin is known to be closely similar to, if not identical with, a substance known as nuclein ($C_{29}H_{49}N_9O_{22}$, according to Miescher), which analysis shows to be a tolerably definite chemical composed of nucleic acid (a complex organic acid rich in phosphorus) and albumin. And thus we reach the remarkable conclusion that inheritance may, perhaps, be effected by the physical transmission of a particular chemical compound from parent to offspring.

An Atlas of the Fertilization and Karyokinesis of the Ovum (1895), 4.

4 During the half-century that has elapsed since the enunciation of the cell-theory by Schleiden and Schwann, in 1838–39, it has become ever more clearly apparent that the key to all ultimate biological problems must, in the last analysis, be sought in the cell. It was the cell-theory that first brought the structure of plants and animals under one point of view by revealing their common plan of organization. It was through the cell-theory that Kölliker and Remak opened the way to an understanding of the nature of embryological development, and the law of genetic continuity lying at the basis of inheritance. It was the cell-theory again which, in the hands of Virchow and Max Schultze, inaugurated a new era in the history of physiology and pathology, by showing that all the various functions of the body, in health and in disease, are but the outward expression of cell-activities. And at a still later day it was through the cell-theory that Hertwig, Fol, Van Beneden, and Strasburger solved the long-standing riddle of the fertilization of the egg, and the mechanism of hereditary transmission. No other biological generalization, save only the theory of organic evolution, has brought so many apparently diverse phenomena under a common point of view or has accomplished more for the unification of knowledge. The cell-theory must therefore be placed beside the evolution-theory as one of the foundation stones of modern biology.

The Cell in Development and Inheritance (1896), 1.

1 These facts show that mitosis is due to the co-ordinate play of an extremely complex system of forces which are as yet scarcely comprehended. Its purpose is, however, as obvious as its physiological explanation is difficult. *It is the end of mitosis to divide every part of the chromatin of the mother-cell equally between the daughter-nuclei.* All the other operations are tributary to this. We may therefore regard the mitotic figure as essentially an apparatus for the distribution of the hereditary substance, and in this sense as the especial instrument of inheritance.

The Cell in Development and Inheritance (1896), 86.

2 Not a single visible phenomenon of cell-division gives even a remote suggestion of qualitative division. All the facts, on the contrary, indicate that the division of the chromatin is carried out with the most exact equality.

The Cell in Development and Inheritance (1896), 306.

3 The nucleus cannot operate without a cytoplasmic field in which its peculiar powers may come into play; but this field is created and moulded by itself. Both are necessary to *development*; the nucleus alone suffices for the *inheritance* of specific possibilities of development.

The Cell in Development and Inheritance (1896), 327.

Edward Osborne Wilson 1929-

American naturalist and sociobiologist

4 Why do we study insects? Because, together with man, hummingbirds and the bristlecone pine, they are among the great achievements of organic evolution.

The Insect Societies (1971), 1.

5 How can altruism, which by definition reduces personal fitness, possibly evolve by natural selection? The answer is kinship: if the genes causing the altruism are shared by two organisms because of common descent, and if the altruistic act by one organism increases the joint contribution of these genes to the next generation, the propensity to altruism will spread through the gene pool. This occurs even though the altruist makes less of a solitary contribution to the gene pool as the price of its altruistic act.

Sociobiology (1975), 3-4.

6 True spite is a commonplace in human societies, undoubtedly because human beings are keenly aware of their own blood lines and have the intelligence to plot intrigue. Human beings are unique in the degree of their capacity to lie to other members of their own species.

Sociobiology (1975), 119.

7 The key to the sociobiology of mammals is milk.

Sociobiology (1975), 456.

8 Can the cultural evolution of higher ethical values gain a direction and momentum of its own and completely replace genetic evolution? I think not. The genes hold culture on a leash. The leash is very long, but inevitably values will be constrained in accordance with their effects in the human gene pool. The brain is a product of evolution. Human behaviour—like the deepest capacities for emotional response which drive and guide it—is the circuitous technique by which human genetic material has been and will be kept intact. Morality has no other demonstrable ultimate function.

On Human Nature (1978), 167.

1 Marxism is sociobiology without biology . . . Although Marxism was formulated as the enemy of ignorance and superstition, to the extent that it has become dogmatic it has faltered in that commitment and is now mortally threatened by the discoveries of human sociobiology.

On Human Nature (1978), 191.

2 The elements of human nature are the learning rules, emotional reinforcers, and hormonal feedback loops that guide the development of social behaviour into certain channels as opposed to others. Human nature is not just the array of outcomes attained in existing societies. It is also the potential array that might be achieved through conscious design by future societies. By looking over the realized social systems of hundreds of animal species and deriving the principles by which these systems have evolved, we can be certain that all human choices represent only a tiny subset of those theoretically possible. Human nature is, moreover, a hodgepodge of special genetic adaptations to an environment largely vanished, the world of the Ice-Age hunter-gatherer.

On Human Nature (1978),196.

3 The naturalist is a civilized hunter.

Biophilia (1984), 1.

4 Hands-on experience at the critical time, not systematic knowledge, is what counts in the making of a naturalist. Better to be an untutored savage for a while, not to know the names or anatomical detail. Better to spend long stretches of time just searching and dreaming.

Naturalist (1994), 11–12.

5 Preferring a search for objective reality over revelation is another way of satisfying religious hunger. It is an endeavor almost as old as civilization and intertwined with traditional religion, but it follows a very different course—a stoic's creed, an acquired taste, a guidebook to adventure plotted across rough terrain. It aims to save the spirit, not by surrender but by liberation of the human mind. Its central tenet, as Einstein knew, is the unification of knowledge. When we have unified enough certain knowledge, we will understand who we are and why we are here. If those committed to the quest fail, they will be forgiven. When lost, they will find another way.

Consilience: The Unity of Knowledge (1998), 5.

6 The totality of life, known as the biosphere to scientists and creation to theologians, is a membrane of organisms wrapped around Earth so thin it cannot be seen edgewise from a space shuttle, yet so internally complex that most species composing it remain undiscovered. The membrane is seamless. From Everest's peak to the floor of the Mariana Trench, creatures of one kind or another inhabit virtually every square inch of the planetary surface.

The Future of Life (2002), 3.

James Harold Wilson (Baron Wilson of Rievaulx) 1916–95

British Prime Minister

7 The Britain that is going to be forged in the white heat of this revolution will be no place for restrictive practices or out of date methods, on either side of industry.

Speech at the Labour Party Conference, Scarborough, 1 October 1963.

8 If there was one word I could use to identify modern socialism it was 'science'.

The Relevance of British Socialism (1964). See A. Cottrell, *Physics Bulletin*, March 1976, 102.

John Tuzo Wilson 1908–93

Canadian geologist and geophysicist

1 But no other theory can explain so much. Continental drift is without a cause or a physical theory. It has never been applied to any but the last part of geological time.

'Geophysics and Continental Growth', *American Scientist*, 1959, **47**, 23.

David L. Wingate 1935–

British gastroenterologist

2 Even if a scientific model, like a car, has only a few years to run before it is discarded, it serves its purpose for getting from one place to another.

'Complex Clocks', *Digestive Diseases and Sciences*, 1983, **28**, 1139.

Johannes Wislicenus 1835–1902

German chemist

3 If molecules can be structurally identical and yet possess dissimilar properties, this can be explained only on the ground that the difference is due to a different arrangement of the atoms in space.

Annalen der Chemie, 1873, **166**, 47, translated in A. Ihde, *The Development of Modern Chemistry* (1964), 326.

4 It is one of the signs of the times that modern chemists hold themselves bound and consider themselves in a position to give an explanation for everything, and when their knowledge fails them to make sure of supernatural explanations. Such a treatment of scientific subjects, not many degrees removed from a belief in witches and spirit-rapping, even Wislicenus considers permissible.

H. Kolbe, 'Sign of the Times', *Journal für Praktische Chemie*, 1877, **15**, 473. Trans. W. H. Brock.

William Dwight Witney 1827–94

American linguist

5 A noteworthy and often-remarked similarity exists between the facts and methods of geology and those of linguistic study. The science of language is, as it were, the geology of the most modern period, the Age of the Man, having for its task to construct the history of development of the earth and its inhabitants from the time when the proper geological record remains silent . . . The remains of ancient speech are like strata deposited in bygone ages, telling of the forms of life then existing, and of the circumstances which determined or affected them; while words are as rolled pebbles, relics of yet more ancient formations, or as fossils, whose grade indicates the progress of organic life, and whose resemblances and relations show the correspondence or sequence of the different strata; while, everywhere, extensive denudation has marred the completeness of the record, and rendered impossible a detailed exhibition of the whole course of development.

Language and the Study of Language (1867), 47.

Ludwig Wittgenstein 1889–1951

Austrian-born British philosopher

6 *Die Welt ist alles, was der Fall ist.*

The world is everything that is the case.

Tractatus logico-philosophicus (1921) (1955 edn.), sect. 1, p. 31.

7 *Die Grenzen meiner Sprache bedeuten die Grenzen meiner Welt.*

The limits of my language mean the limits of my world.

Tractatus logico-philosophicus (1921) (1955 edn.), sect. 5.6, p. 149.

8 *Wovon man nicht sprechen kann, darüber muss man schweigen.*

Whereof one cannot speak, thereof one must be silent.

Tractatus logico-philosophicus (1921) (1955 edn.), sect. 7, p. 189.

9 If a lion could talk, we could not understand him.

Philosophical Investigations (1953), trans. G. E. M. Anscombe, 223.

1 What a curious attitude scientists have: 'We still don't know that; but it is knowable and it is only a matter of time before we get to know it!' As if that went without saying.

Quoted in S. Hilmy, *The Later Wittgenstein: The Emergence of a New Philosophical Method* (1987), 220.

Friedrich Wöhler 1800–82

German chemist

2 In a manner of speaking, I can no longer hold my chemical water. I must tell you that I can make urea without the use of kidneys of any animal, be it man or dog. Ammonium cyanate is urea.

Wöhler to Berzelius, 22 February 1828. In O. Wallach (ed.), *Briefwechsel zwischen J. Berzelius und F. Wöhler* (1901), Vol. 1, 206. Trans. W. H. Brock.

3 Research gave the unexpected result that, by combination of cyanic acid with ammonia, urea is formed. A noteworthy fact since it furnishes an example of the artificial production of an organic—indeed, a so-called animal—substance from inorganic materials!

'Über Künstliche Bildung des Harnstoffs', *Annalen der Physik und Chemie*, 1828, 12, 253. Trans. *Quarterly Journal of Science*, 1828, 25, 491.

4 Organic chemistry nowadays almost drives one mad. To me it appears like a primeval tropical forest full of the most remarkable things; a dreadful, endless jungle into which one dare not enter, for there seems no way out.

Wöhler to Berzelius, 1832. Quoted in J. R. Partington, *A History of Chemistry* (1901), Vol. 4, 327.

5 1839—The fermentation satire
THE MYSTERY OF ALCOHOLIC FERMENTATION RESOLVED
(Preliminary Report by Letter)
Schwindler

I am about to develop a new theory of wine fermentation . . . Depending on the weight, these seeds carry fermentation to completion somewhat less than as in the beginning, which is understandable . . . I shall develop a new theory of wine fermentation [showing] what simple means Nature employs in creating the most amazing phenomena. I owe it to the use of an excellent microscope designed by Pistorius.

When brewer's yeast is mixed with water the microscope reveals that the yeast dissolves into endless small balls, which are scarcely 1/800th of a line in diameter . . . If these small balls are placed in sugar water, it can be seen that they consist of the eggs of animals. As they expand, they burst, and from them develop small creatures that multiply with unbelievable rapidity in a most unheard of way. The form of these animals differs from all of the 600 types described up until now. They possess the shape of a Beinsdorff still (without the cooling apparatus). The head of the tube is a sort of proboscis, the inside of which is filled with fine bristles 1/2000th of a line long. Teeth and eyes are not discernible; however, a stomach, intestinal canal, anus (a rose red dot), and organs for secretion of urine are plainly discernible. From the moment they are released from the egg one can see these animals swallow the sugar from the solution and pass it to the stomach. It is digested immediately, a process recognized easily by the resultant evacuation of excrements. In a word, these infusors eat sugar, evacuate ethyl alcohol from the intestinal canal, and carbon dioxide from the urinary organs. The bladder, in the filled state, has the form of a champagne bottle; when empty, it is a small button . . . As soon as the animals find no more sugar present, they eat each other up, which occurs through a peculiar manipulation; everything is digested down to the eggs which pass unchanged through the intestinal canal. Finally, one again has fermentable yeast, namely the seed of the animals, which remain over.

'Das enträthselte Geheimiss der geistigen Gährung', *Annalen der Pharmacie und*

Chemie, 1839, **29**, 100–104; adapted from English translation by Ralph E. Oesper, *The Human Side of Scientists* (1975), 203–5.

1 If in the citation of work that we have both done together only one of us is named, and especially in a journal [*Annalen der Chemie*] in which both are named on the title page, about which everyone knows that you are the actual editor, and this editor allows that to happen and does not show the slightest consideration to report it, then everyone will conclude that this represents an agreement between us, that the work is yours alone, and that I am a jackass.

Wöhler to Liebig, 15 November 1840. In A. W. Hofmann (ed.), *Aus Justus Liebigs und Friedrich Wöhlers Briefwechsel* (1888), Vol. 1, 166. Trans. W. H. Brock.

2 To wage war with Marchand or anyone else again will benefit nobody and bring little profit to science. You consume yourself in this way, you ruin your liver and eventually your nerves with Morrison pills. Imagine the year 1900 when we have disintegrated into carbonic acid, ammonia and water and our bone substance is perhaps once more a constituent of the bones of the dog who defiles our graves. Who will then worry his head as to whether we have lived in peace or anger, who then will know about your scientific disputes and of your sacrifice of health and peace of mind for science? Nobody. But your good ideas and the discoveries you have made, cleansed of all that is extraneous to the subject, will still be known and appreciated for many years to come. But why am I trying to advise the lion to eat sugar.

Wöhler to Liebig, 9 March 1843. In A. W. Hofmann (ed.), *Aus Justus Liebigs und Friedrich Wöhlers Briefwechsel* (1888), Vol. 1, 224. Trans. R. Oesper, *The Human Side of Scientists* (1975), 205.

3 I cannot let the year run out without sending you a sign of my continued existence and to extend my sincere wishes for the well-being of you and your dear ones in the New Year. We will not be able to send New Year greetings much longer; but even when we have passed away and have long since decomposed, the bonds that united us in life will remain and we shall be remembered as a not too common example of two men, who truly without envy and jealousy, contended and struggled in the same field, yet nevertheless remained always closely bound in friendship.

Liebig to Wöhler, 31 December 1871. Quoted in R. Oesper, *The Human Side of Scientists* (1975), 206.

William Hyde Wollaston

1766–1828

British chemist and physicist

4 I am further inclined to think, that when our views are sufficiently extended, to enable us to reason with precision concerning the proportions of elementary atoms, we shall find the arithmetical relation alone will not be sufficient to explain their mutual action, and that we shall be obliged to acquire a geometric conception of their relative arrangement in all three dimensions of solid extension.

'On Super-acid and Sub-acid salts', *Philosophical Transactions of the Royal Society of London*, 1808, 101.

5 ON **WOLLASTON** Wollaston may be compared to Dalton for originality of view & was far his superior in accuracy. He was an admirable manipulator, steady, cautious & sure. His judgement was cool.—His views sagacious.—His inductions made with care, slowly formed & seldom renounced. He had much of the same spirit of Philosophy as Cavendish, he applied science to purposes of profit & for many years sold manufactured platinum. He died very rich.

J. Z. Fullmer, 'Davy's Sketches of his Contemporaries', *Chymia*, 1967, **12**, 134.

John Woodward 1665–1728

British geologist

6 In the *Choice* of . . . Things, *neglect not any*, tho' the most *ordinary* and *trivial*;

the *Commonest* Peble or Flint, Cockle or Oyster-shell, Grass, Moss, Fern or Thistle, will be as *useful*, and as *proper to be gathered* and sent, as any the rarest production of the Country. Only take care to choose of each the *fairest of its kind*, and such as are *perfect* or *whole*.

Brief Instructions for Making Observations in all Parts of the World (1696), 10.

1 For the *time* of making *Observations* none can ever be amiss; there being no *season*, nor indeed hardly any *place* where in some Natural Thing or other does not present it self worthy of Remark: yea there are some things that require Observation *all the Year round*, as Springs, Rivers, *&c*. Nor is there any *Season* amiss for the *gathering* Natural Things. Bodies of one kind or other presenting themselves at *all times*, and in Winter as well as Summer.

Brief Instructions for Making Observations in all Parts of the World (1696), 10–1.

2 During the time of the Deluge, whilst the Water was out upon, and covered the Terrestrial Globe, . . . all Fossils whatever that had before obtained any Solidity, were totally dissolved, and their constituent Corpuscles all disjoyned, their Cohesion perfectly ceasing . . . [A]nd, to be short, all Bodies whatsoever that were either upon the Earth, or that consituted the Mass of it, if not quite down to the Abyss, yet at least to the greatest depth we ever dig: I say all these were assumed up promiscuously into the Water, and sustained in it, in such a manner that the Water, and Bodies in it, together made up one common confused Mass. That at length all the Mass that was thus borne up in the Water, was again precipitated and subsided towards the bottom. That this subsidence happened generally, and as near as possibly could be expected in so great a Confusion, according to the laws of Gravity.

An Essay Toward A Natural History of the Earth (1695), 74–5.

3 But the Presidence of that mighty Power . . . its particular Agency and Concern therein: and its Purpose and Design . . . will more evidently appear, when I shall have proved . . . That the said Earth, though not indifferently and alike fertil in all parts of it, was yet generally much more fertil than ours is . . . That its Soil was more luxuriant, and teemed forth its Productions in far greater plenty and abundance than the present Earth does . . . That when Man was *fallen*, and had abandoned his primitive Innocence, the Case was much altered: and a far different Scene of Things presented; that generous Vertue, masculine Bravery, and prudent Circumspection which he was before Master of, now deserting him . . . and a strange imbecility immediately seized and laid hold of him: he became pusillanimous, and was easily ruffled with every little Passion within: supine, and as openly exposed to any Temptation or Assault from without. And now these exuberant Productions of the Earth became a continued Decoy and Snare unto him.

An Essay Toward a Natural History of the Earth (1695), 84–6.

4 I know well there are those who would have the Study of Nature restrained wholly to Observations; without ever proceeding further. But due Consideration, and a deeper Insight into Things, would soon have undeceived and made them sensible of their error. Assuredly, that man who should spend his whole life in amassing together stone, timber, and other materials for building, without ever at the making any use, or raising any fabrick out of them, might well be reputed very fantastic and extravagant. And a like censure would be his due, who should be perpetually heaping up of natural collections without design of building a structure of philosophy out of them, or advancing some propositions that might turn to the benefit and advantage of the world. This is in reality the true and only proper end of collections, of observations, and natural history: and they are of no manner of use or value without it.

An Attempt Toward a Natural History of the Fossils of England (1729), xiii-xiv.

Robert Burns Woodward

1917–79

American chemist

1 It is well to remember that most arguments in favor of not trying an experiment are too flimsily based.

Quoted in a lecture published in *Experientia Supplementum II* (1955), 226.

2 The synthesis of substances occurring in Nature, perhaps in greater measure than activities in any other area of organic chemistry, provides a measure of the conditions and powers of science.

'Synthesis', in A. Todd (ed.), *Perspectives in Organic Chemistry* (1956), 155.

3 That Brobdingnagian molecule, tobacco mosaic virus.

'Synthesis', in A. Todd (ed.), *Perspectives in Organic Chemistry* (1956), 175.

4 Organic chemistry has literally placed a new nature beside the old. And not only for the delectation and information of its devotees; the whole face and manner of society has been altered by its products. We are clothed, ornamented and protected by forms of matter foreign to Nature; we travel and are propelled, in, on and by them. Their conquest of our powerful insect enemies, their capacity to modify the soil and control its microscopic flora, their ability to purify and protect our water, have increased the habitable surface of the earth and multiplied our food supply; and the dramatic advances in synthetic medicinal chemistry comfort and maintain us, and create unparalleled social opportunities (and problems).

'Synthesis', in A. Todd (ed.), *Perspectives in Organic Chemistry* (1956), 180.

5 Why can the chemist not take the requisite numbers of atoms and simply put them together? The answer is that the chemist never has atoms at his disposal, and if he had, the direct combination of the appropriate numbers of atoms would lead only to a Brobdingnagian potpourri of different kinds of molecules, having a vast array of different structures. What the chemist has at hand always consists of substances, themselves made up of molecules, containing defined numbers of atoms in ordered arrangements. Consequently, in order to synthesize any one substance, his task is that of combining, modifying, transforming, and tailoring known substances, until the total effect of his manipulations is the conversion of one or more forms of matter into another.

'Art and Science in the Synthesis of Organic Compounds: Retrospect and Prospect', in Maeve O'Connor (ed.), *Pointers and Pathways in Research* (1963), 28.

6 The structure known, but not yet accessible by synthesis, is to the chemist what the unclimbed mountain, the uncharted sea, the untilled field, the unreached planet, are to other men . . . The unique challenge which chemical synthesis provides for the creative imagination and the skilled hand ensures that it will endure as long as men write books, paint pictures, and fashion things which are beautiful, or practical, or both.

'Art and Science in the Synthesis of Organic Compounds: Retrospect and Prospect', in Maeve O'Connor (ed.), *Pointers and Pathways in Research* (1963), 41.

7 We all know that enforced propinquity often leads on to greater intimacy.

'Recent Advances in the Chemistry of Natural Products', *Pure and Applied Chemistry*, 1968, 17, 545.

8 I have always been very fond of mathematics—for one short period, I even toyed with the possibility of abandoning chemistry in its favour. I enjoyed immensely both its conceptual and formal beauties, and the precision and elegance of its relationships and transformations. Why then did I not succomb to its charms? . . . because by and large, mathematics lacks the *sensuous* elements which play so large a role in my attraction to chemistry. I love crystals, the beauty of their forms and formation; liquids, dormant, distilling, sloshing! The fumes, the odors—good or bad, the rainbow of colours; the gleaming vessels of every size, shape and purpose. Much as I might *think* about chemistry, it would not exist for me without these physical, visual, tangible sensuous things.

Arthur Clay Cope Address, Chicago, 28 August 1973. O. T. Benfey and P. J. T. Morris (eds.), *Robert Burns Woodward. Architect and Artist in the World of Molecules* (2001), 427.

1 ON WOODWARD He never got drunk, he never got tired, and he never perspired.

Harvard chemistry students' axioms.

John D. Roberts, *The Right Place at the Right Time* (1990), 52.

William Wordsworth 1770–1850

British romantic poet

See ***Newton, Isaac*** *463:1;* ***Sedgwick, Adam*** *541:4*

2 Those to whom the harmonious doors
Of Science have unbarred celestial stores,
To whom a burning energy has given
That other eye which darts thro' earth and heaven,
Roams through all space and unconfined,
Explores the illimitable tracts of mind,
And piercing the profound of time can see
Whatever man has been and man can be.

An Evening Walk (1793). In E. de Selincourt (ed.), *The Poetical Works of William Wordsworth* (1940), Vol. 1, 13.

3 Sweet is the lore which Nature brings;
Our meddling intellect
Mis-shapes the beauteous forms of Things:—
We murder to dissect.

'The Tables Turned' (1798). In *The Works of William Wordsworth* (1994), 481.

4 And 'tis my faith that every flower
Enjoys the air it breathes.

Lines Written in Early Spring (1798). In *The Works of William Wordsworth* (1994), Book 4, 482.

5 Come forth into the light of things,
Let Nature be your Teacher.

The Tables Turned (1798). In *The Works of William Wordsworth* (1994), Book 4, 381.

6 The remotest discoveries of the Chemist, the Botanist, or the Mineralogist, will be as proper objects of the Poet's art as any upon which it can be employed, if the time should ever come when these things shall be familiar to us, and the relations under which they are contemplated by the followers of these respective sciences shall be manifestly and palpably material to us as enjoying and suffering beings.

W. J. B. Owen (ed.), *Preface to the Lyrical Ballads* (1957), 124–5. First published in 1800.

7 Lost in a gloom of uninspired research.

The Excursion (1814). In *The Works of William Wordsworth* (1994), Book 4, 810.

8 An inventive age
Has wrought, if not with speed of magic, yet
To most strange issues. I have lived to mark
A new and unforeseen creation rise
From out the labours of a peaceful Land
Wielding her potent enginery to frame
And to produce, with appetite as keen
As that of war, which rests not night or day.

The Excursion (1814). In *The Works of William Wordsworth* (1994), Book 8, 875.

9 'To every Form of being is assigned'
Thus calmly spoke the venerable Sage,
'An *active* Principle:—howe'er removed
From sense and observation, it subsists
In all things, in all natures; in the stars
Of azure heaven, the unenduring clouds,
In flower and tree, in every pebbly stone
That paves the brooks, the stationary rocks,

The moving waters, and the invisible
air.'

The Excursion (1814). In *The Works of William Wordsworth* (1994), Book 9, 884.

1 To the solid ground of nature trusts the
Mind that builds for aye.

'A Volant Tribe of Bards on Earth are Found' (1823). In *The Works of William Wordsworth* (1994), 259. This appeared as the masthead on *Nature* until 30 March 1963.

2 Man now presides
In power, where once he trembled in his weakness;
Science advances with gigantic strides;
But are we aught enriched in love and meekness?

To the Planet Venus (1838). In *The Works of William Wordsworth* (1994), Book 4, 281.

Christopher Wren 1632–1723

British mathematician and architect

3 For, Mathematical Demonstrations being built upon the impregnable Foundations of Geometry and Arithmetick, are the only Truths, that can sink into the Mind of Man, void of all Uncertainty; and all other Discourses participate more or less of Truth, according as their Subjects are more or less capable of Mathematical Demonstration.

Inaugural lecture of Christopher Wren in his chair of astronomy at Gresham College in 1657. From *Parentelia* (1741), reprinted 1951, 200–1.

4 ON WREN Sir Christopher Wren
Said, 'I am going to dine with some men.
'If anybody calls
'Say I am designing St. Paul's'

E. C. Bentley, *Biography for Beginners* (1905).

Milton Wright 1828–1917

American churchman

5 Men will never fly, because flying is reserved for angels.

Attributed.

Sewall Wright 1889–1988

American population geneticist and evolutionary theorist

6 Finally in a large population, divided and subdivided into partially isolated local races of small size, there is a continually shifting differentiation among the latter (intensified by local differences in selection but occurring under uniform and static conditions) which inevitably brings about an indefinitely continuing, irreversible, adaptive, and much more rapid evolution of the species. Complete isolation in this case, and more slowly in the preceding, originates new species differing for the most part in nonadaptive parallel orthogenetic lines, in accordance with the conditions. It is suggested, in conclusion, that the differing statistical situations to be expected among natural species are adequate to account for the different sorts of evolutionary processes which have been described, and that, in particular, conditions in nature are often such as to bring about the state of poise among opposing tendencies on which an indefinitely continuing evolutionary process depends.

'Evolution in Mendelian Populations', *Genetics*, 1931, 16, 158.

7 It need scarcely be pointed out that with such a mechanism complete isolation of portion of a species should result relatively rapidly in specific differentiation, and one that is not necessarily adaptive. The effective inter-group competition leading to adaptive advance may be between species rather than races. Such isolation is doubtless usually geographic in character at the outset but may be clinched by the development of hybrid sterility. The usual difference of the chromosome complements of related species puts the importance of chromosome aberration as an evolutionary process beyond question, but, as I see it, this importance is not in the character differences which they bring (slight in balanced types), but rather in leading to the sterility of hybrids and thus making permanent the isolation of two groups.

How far do the observations of actual species and their subdivisions conform to this picture? This is naturally too large a subject for more than a few suggestions.

That evolution involves non-adaptive differentiation to a large extent at the subspecies and even the species level is indicated by the kinds of differences by which such groups are actually distinguished by systematics. It is only at the subfamily and family levels that clear-cut adaptive differences become the rule. The principal evolutionary mechanism in the origin of species must thus be an essentially nonadaptive one.

Proceedings of the Sixth International Congress of Genetics: Ithaca, New York, 1932, Vol. 1 (1932), 363–4.

1 I have attempted to form a judgment as to the conditions for evolution based on the statistical consequences of Mendelian heredity. The most general conclusion is that evolution depends on a certain balance among its factors. There must be a gene mutation, but an excessive rate gives an array of freaks, not evolution; there must be selection, but too severe a process destroys the field of variability, and thus the basis for further advance; prevalence of local inbreeding within a species has extremely important evolutionary consequences, but too close inbreeding leads merely to extinction. A certain amount of crossbreeding is favorable but not too much. In this dependence on balance the species is like a living organism. At all levels of organization life depends on the maintenance of a certain balance among its factors.

Proceedings of the Sixth International Congress of Genetics: Ithaca, New York, 1932, Vol. 1 (1932), 365.

2 It seems to me that the view toward which we are tending is that the specificity in gene action is always a chemical specificity, probably the production of enzymes which guide metabolic processes along particular channels. A given array of genes thus determines the production of a particular kind of protoplasm with particular properties—such, for example, as that of responding to surface forces by the formation of a special sort of semipermeable membrane, and that of responding to trivial asymmetries in the play of external stimuli by polarization, with consequent orderly quantitative gradients in all physiologic processes. Different genes may now be called into play at different points in this simple pattern, either through the local formation of their specific substrates for action, or by activation of a mutational nature. In either case the pattern becomes more complex and qualitatively differentiated. Successive interactions of differentiated regions and the calling into play of additional genes may lead to any degree of complexity of pattern in the organism as a largely self-contained system. The array of genes, assembled in the course of evolution, must of course be one which determines a highly self-regulatory system of reactions. On this view the genes are highly specific chemically, and thus called into play only under very specific conditions; but their morphological effects, if any, rest on quantitative influences of immediate or remote products on growth gradients, which are resultants of all that has gone on before in the organism.

'Genetics of Abnormal Growth in the Guinea Pig', *Cold Spring Harbor Symposia on Quantitative Biology*, 1934, 2, 142.

3 It is the task of science, as a collective human undertaking, to describe from the *external* side, (on which alone agreement is possible), such statistical regularity as there is in a world in which every event has a unique aspect, and to indicate where possible the limits of such description. It is not part of its task to make imaginative interpretation of the internal aspect of reality—what it is like, for example, to be a lion, an ant or an ant hill, a liver cell, or a hydrogen ion. The only qualification is in the field of introspective psychology in which each human being is both observer and observed, and regularities may be established by comparing notes. Science is thus a limited venture. It must act as if all phenomena were deterministic at least in the sense of determinable probabilities. It cannot properly explain the behaviour of an amoeba as due partly to surface and other physical forces and partly to what the amoeba wants to do, with out danger of something like 100 per cent duplication. It must stick to the former. It cannot introduce such principles as creative activity into its interpretation of evolution for similar reasons. The point of view indicated by a consideration of the hierarchy of physical and biological organisms, now being bridged by the concept of the gene, is one in which science deliberately accepts a rigorous limitation of its activities to the description of the external aspects of events. In carrying out this program, the scientist should not, however, deceive himself or others into thinking that he is giving an account of all of reality. The unique inner creative aspect of every event necessarily escapes him.

'Gene and Organism', *American Naturalist*, 1953, **87**, 17.

1 The Darwinian process of continued interplay of a random and a selective process is not intermediate between pure chance and pure determinism, but qualitatively utterly different from either in its consequences.

'Comments on the Preliminary Working Papers of Eden and Waddington'. In P. Moorhead and M. Kaplan (eds.), *Mathematical Challenges to the Neo-Darwinian Interpretation of Evolution* (1967), 117.

Dorothy M. Wrinch 1894–1976

British mathematician, biologist and chemist

2 First they said my [cyclol] structure [of proteins] couldn't exist. Then when it was found in Nature they said it couldn't be synthesized in a laboratory. Then when it was synthesized they said it wasn't important anyway.

Quoted in Maureen M. Julian in G. Kass-Simon and Patricia Farnes (eds.), *Women of Science* (1990), 368.

Wilhelm Wundt 1832–1920

German psychologist

3 The endeavour to observe oneself must inevitably introduce changes into the course of mental events,—changes which could not have occurred without it, and whose usual consequence is that the very process which was to have been observed disappears from consciousness.

Principles of Physiological Psychology (1873), 1904 ed., Vol. 1, 5.

4 Psychological introspection goes hand in hand with the methods of experimental physiology. If one wants to put the main emphasis on the characteristic of the method, our science, experimental psychology, is to be distinguished from the ordinary mental philosophy [*Seelenlehre*], based purely on introspection.

Grundzüge der physiologischen Psychologie [Principles of Physiological Psychology] (1874), 2–3. Trans. K. Danziger, *Constructing the Subject: Historical Origins of Psychological Research* (1990), 206.

Charles Adolphe Wurtz 1817–84

French chemist

5 *La chimie est une science française; elle fut constitutée par Lavoisier d'immortelle mémoire.*

Chemistry is a French Science. It was founded by Lavoisier of immortal memory.

'Histoire des doctrines chimiques', *Dictionnaire de Chimie Pure et Appliquée* (1868), Vol. 1, i.

Hans Wynberg

Dutch chemist

1 The recent ruling by the Supreme Court restricting obscenity in books, magazines and movies, requires that we re-examine our own journals for lewd contents. The recent chemical literature provides many examples of words and concepts whose double meaning and thinly veiled overtones are an affront to all clean chemists. What must a layman think of 'coupling constants', 'tickling techniques', or indeed 'increased overlap'? The bounds of propriety are surely exceeded when heterocyclic chemists discuss homoenolization.

Chemical Engineering News, 8 October 1973, 68.

Ernst Ludwig Wynder 1922–99 and Evarts Ambrose Graham 1883–1957

German-born American physician and American surgeon

2 Excessive and prolonged use of tobacco, especially cigarettes, seems to be an important factor in the induction of bronchiogenic carcinoma. Among 605 men with bronchiogenic carcinoma, other than adenocarcinoma, 96.5 per cent were moderately heavy to chain smokers for many years, compared with 73.7 per cent among the general male hospital population without cancer. Among the cancer group 51.2 per cent were excessive or chain smokers compared to 19.1 per cent in the general hospital group without cancer.

'Tobacco Smoking as a Possible Etiologic Factor in Bronchiogenic Carcinoma', *The Journal of the American Medical Association*, 1950, 143, 336,

Rosalyn Sussman Yalow 1921–

American medical physicist

3 Radioimmunoassay (RIA) is simple in principle.

'Radioimmunoassay: A Probe for the Fine Structure of Biologic Systems', Nobel Lecture, 1977. In *Nobel Lectures: Physiology or Medicine 1971–1980* (1992), 450.

4 The first telescope opened the heavens; the first microscope opened the world of the microbes; radioisotopic methodology, as exemplified by RIA, has shown the potential for opening new vistas in science and medicine.

'Radioimmunoassay: A Probe for the Fine Structure of Biologic Systems', Nobel Lecture, 1977. In *Nobel Lectures: Physiology or Medicine 1971–1980* (1992), 465.

5 The Nobel Prize gives you an opportunity to make a fool of yourself in public.

Attributed.

Ye Zi-qi late Yuan or early Ming Dynasty

Chinese scholar

6 By the agitation of water and silt, and their gradual accumulation and consolidation . . . the rocks were formed gradually by the evolution of sediments in water.

Cao Mu Zi (1959), trans. Yang, Jing-Yi, 1.

Ellis Leon Yochelson 1928–

American palaeontologist

7 The geologist strides across the landscape to get the big picture, but the paleontologist stays at one spot or shuffles along looking at the ground for his pet objects.

'Fossils—The How and Why of Collecting and Storing', *Proceedings of the Biological Society of Washington*, 1969, 82, 590.

8 Scientists are supposed to make predictions, probably to prove they are human and can be as mistaken as anyone else. Long-range predictions are better to make because the audience to whom the prediction was made is no longer around to ask questions. The alternative . . . is to make conflicting predictions, so that one prediction may prove right.

'Fossils—The How and Why of Collecting and Storing', *Proceedings of the Biological Society of Washington*, 1969, 82, 597.

William Jay Youmans 1838–1901

American editor and writer

1 In order that the relations between science and the age may be what they ought to be, the world at large must be made to feel that science is, in the fullest sense, a ministry of good to all, not the private possession and luxury of a few, that it is the best expression of human intelligence and not the abracadabra of a school, that it is a guiding light and not a dazzling fog.

'Hindrances to Scientific Progress', *The Popular Science Monthly*, November 1890, 38, 121.

Edward Young 1683–1765

British poet, dramatist and literary critic

2 Distinguisht *Link* in Being's endless Chain!
Midway from *Nothing* to the *Deity*!

The Complaint: or, Night-Thoughts on Life, Death, & Immortality (1742), Night I, l. 73–4, ed. Stephen Cornford (1989), 39.

3 An undevout astronomer is mad!

The Complaint: or, Night-Thoughts on Life, Death, & Immortality (1742), Night IX, l. 773, ed. Stephen Cornford (1989), 277.

John Zachary Young 1907–97

British zoologist

4 What would be the use of a neuroscience that cannot tell us anything about love?

Programs of the Brain (1978), 143.

5 A marine protozoan is an aqueous salty system in an aqueous salty medium, but a man is an aqueous salty system in a medium in which there is but little water and most of that poor in salts.

Quoted in Larry R. Squire (ed.), *The History of Neuroscience in Autobiography* (1996), Vol. 1, 558.

Thomas Young 1773–1829

British physician and physicist

6 Proposition VIII. When two Undulations, from different Origins, coincide either perfectly or very nearly in Direction, their joint effect is a Combination of the Motions belonging to each.

'On the Theory of Light and Colours' (read 1801), *Philosophical Transactions*, 1802, 92, 34.

7 Proposition IX. *Radiant light consists in Undulations of the Luminiferous Ether.*

'On the Theory of Light and Colours' (read 1801), *Philosophical Transactions*, 1802, 92, 44.

8 But it will be found . . . that one universal law prevails in all these phenomena. Where two portions of the same light arrive in the eye by different routes, either exactly or very nearly in the same direction, the appearance or disappearance of various colours is determined by the greater or less difference in the lengths of the paths.

Lecture XIV. 'Of Physical Optics'. In *A Syllabus of a Course of Lectures on Natural and Experimental Philosophy* (1802), 112–4.

9 Suppose a number of equal waves of water to move upon the surface of a stagnant lake, with a certain constant velocity, and to enter a narrow channel leading out of the lake. Suppose then another similar cause to have excited another equal series of waves, which arrive at the same time, with the first. Neither series of waves will destroy the other, but their effects will be combined: if they enter the channel in such a manner that the elevations of one series coincide with those of the other, they must together produce a series of greater joint elevations; but if the elevations of one series are so situated as to correspond to the depressions of the other, they must exactly fill up those depressions. And the surface of the water must remain smooth; at least I can discover no alternative, either from theory or from experiment.

A Reply to the Animadversions of the Edinburgh Reviewers on Some Papers Published in the Philosophical Transactions (1804), 17–8.

10 Exper. I. I made a small hole in a window-shutter, and covered it with a piece of thick paper, which I perforated with a fine needle. For greater convenience of observation I placed a small looking-glass without the window-shutter, in such a position as to reflect the sun's light, in a direction nearly horizontal, upon the opposite wall, and to cause the cone of diverging light to pass over a table on which were several little screens of card-paper. I brought into the sunbeam a slip of card, about one-thirtieth of an inch in breadth, and observed its shadow, either on the wall or on other cards held at different distances. Besides the fringes of colour on each side of the shadow, the shadow itself was divided by similar parallel fringes, of smaller dimensions, differing in number, according to the distance at which the shadow was observed, but leaving the middle of the shadow always white. Now these fringes were the joint effects of the portions of light passing on each side of the slip of card and inflected, or rather diffracted, into the shadow. For, a little screen being placed a few inches from the card, so as to receive either edge of the shadow on its margin, all the fringes which had before been observed in the shadow on the wall, immediately disappeared, although the light inflected on the other side was allowed to retain its course, and although this light must have undergone any modification that the proximity of the other edge of the slip of card might have been capable of occasioning . . . Nor was it for want of a sufficient intensity of light that one of the two portions was incapable of producing the fringes alone; for when they were both uninterrupted, the lines appeared, even if the intensity was reduced to one-tenth or one-twentieth.

'Experiments and Calculations Relative to Physical Optics' (read 1803), *Philosophical Transactions*, 1804, 94, 2–3.

1 If we seek for the simplest arrangement, which would enable it [the eye] to receive and discriminate the impressions of the different parts of the spectrum, we may suppose three distinct sensations only to be excited by the rays of the three principal pure colours, falling on any given point of the retina, the red, the green, and the violet; while the rays occupying the intermediate spaces are capable of producing mixed sensations, the yellow those which belong to the red and green, and the blue those which belong to the green and violet.

'Chromatics', in *Supplement to the Fourth, Fifth, and Sixth Editions of the Encyclopædia Britannica* (1824), Vol. 3, 142.

2 ON YOUNG I must not pass by Dr. Young called Phaenomenon Young at Cambridge. A man of universal erudition, & almost universal accomplishments. Had he limited himself to any one department of knowledge, he must have been first in that department. But as a mathematician, a scholar, a hieroglyphist, he was eminent; & he knew so much that it is difficult to say what he did not know. He was a most amiable & good-tempered man; too fond, perhaps, of the society of persons of rank for a true philosopher.

J. Z. Fullmer, 'Davy's Sketches of his Contemporaries', *Chymia*, 1967, 12, 135.

3 ON YOUNG [Young] was afterwards accustomed to say, that at no period of his life was he particularly fond of repeating experiments, or even of very frequently attempting to originate new ones; considering that, however necessary to the advancement of science, they demanded a great sacrifice of time, and that when the fact was once established, that time was better employed in considering the purposes to which it might be applied, or the principles which it might tend to elucidate.

Hudson Gurney, *Memoir of the Life of Thomas Young, M.D. F.R.S.* (1831), 12–3.

Adam Zaluziansky c.1558–1613

Bohemian physicist and botanist

4 It is customary to connect Medicine with Botany, yet scientific treatment demands that we should consider each separately. For the fact is that in every art, theory must be disconnected and separated from practice, and the two must be dealt with singly and individually in their proper order before they are united. And for that reason, in order that Botany, which is, as it were, a special branch of Natural Philosophy [*Physica*], may form a unit by itself before it can be brought into connection with other sciences, it must be divided and unyoked from Medicine.

Methodi herbariae libri tres (1592), translated in Agnes Arber, *Herbals: Their Origin and Evolution*, 2nd edition (1938), 144.

David Zeaman 1921–

American psychologist

1 One of the differences between the natural and the social sciences is that in the natural sciences, each succeeding generation stands on the shoulders of those that have gone before, while in the social sciences, each generation steps in the faces of its predecessors.

Skinner's Theory of Teaching Machines (1959), 167.

Erick Christopher Zeeman 1925–

British mathematician

2 The scientist has to take 95 per cent of his subject on trust. He has to because he can't possibly do all the experiments, therefore he has to take on trust the experiments all his colleagues and predecessors have done. Whereas a mathematician doesn't have to take *anything* on trust. Any theorem that's proved, he doesn't believe it, really, until he goes through the proof himself, and therefore he knows his whole subject from scratch. He's absolutely 100 per cent certain of it. And that gives him an extraordinary conviction of certainty, and an arrogance that scientists don't have.

In Lewis Wolpert and Alison Richards, *A Passion for Science* (1988), 61.

Pieter Zeeman 1865–1943

Dutch physicist

3 In August, 1896, I exposed the sodium flame to large magnetic forces by placing it between the poles of a strong electromagnet. Again I studied the radiation of the flame by means of Rowland's mirror, the observations being made in the direction perpendicular to the lines of force. Each line, which in the absence of the effect of the magnetic forces was very sharply defined, was now broadened. This indicated that not only the original oscillations, but also others with greater and again others with smaller periods of oscillation were being radiated by the flame. The change was however very small. In an easily produced magnetic field it corresponded to a thirtieth of the distance between the two sodium lines, say two tenths of an Angstrom, a unit of measure whose name will always recall to physicists the meritorious work done by the father of my esteemed colleague.

'Light Radiation in a Magnetic Field', Nobel Lecture, 2 May 1903. In *Nobel Lectures: Physics 1901–1921* (1967), 34–5.

4 Firm support has been found for the assertion that electricity occurs at thousands of points where we at most conjectured that it was present. Innumerable electrical particles oscillate in every flame and light source. We can in fact assume that every heat source is filled with electrons which will continue to oscillate ceaselessly and indefinitely. All these electrons leave their impression on the emitted rays. We can hope that experimental study of the radiation phenomena, which are exposed to various influences, but in particular to the effect of magnetism, will provide us with useful data concerning a new field, that of atomistic astronomy, as

Lodge called it, populated with atoms and electrons instead of planets and worlds.

'Light Radiation in a Magnetic Field', Nobel Lecture, 2 May 1903. In *Nobel Lectures: Physics 1901–1921* (1967), 40.

Zeno of Elea c.490–c.425 BC

Greek philosopher and mathematician

1 The second [argument about motion] is the so-called Achilles, and it amounts to this, that in a race the quickest runner can never overtake the slowest, since the pursuer must first reach the point whence the pursued started, so that the slower must always hold a lead.

Aristotle, *Physics*, 239b, 14–6. In Jonathan Barnes (ed.), *The Complete Works of Aristotle* (1984), Vol. 1, 404.

2 The third [argument of motion is] to the effect that the flying arrow is at rest, which result follows from the assumption that time is composed of moments: if this assumption is not granted, the conclusion will not follow.

Aristotle, *Physics*, 239b, 30–1. In Jonathan Barnes (ed.), *The Complete Works of Aristotle* (1984), Vol. 1, 405.

Zhuangzi see Chuang Tzu

John M. Ziman 1925–2005

New Zealand-born British physicist

3 The physics of undergraduate text-books is 90% true.

Attributed.

Hans Zinsser 1878–1940

American bacteriologist

4 Infectious disease is one of the few genuine adventures left in the world.

Rats, Lice and History (1934), 13.

5 But however secure and well-regulated civilized life may become, bacteria, Protozoa, viruses, infected fleas, lice, ticks, mosquitoes, and bedbugs will always lurk in the shadows ready to pounce when neglect, poverty, famine, or war lets down the defenses.

Rats, Lice and History (1934), 13–4.

Conway Zirkle 1895–1972

American biologist

6 The Johns Hopkins University certifies that John Wentworth does *not know* anything but Biochemistry. Please pay no attention to any pronouncements he may make on any other subject, particularly when he joins with others of his kind to save the world from something or other. However, he worked hard for this degree and is potentially a most valuable citizen. Please treat him kindly.

On re-evaluating the training of scientists to make it more realistic and applicable to the outside world.

'Our Splintered Learning and the Nature of Scientists', *Science*, 1955, **121**, 516.

Keyword Index

Printed in the USA/Agawam, MA
September 14, 2021

781182.004